For centuries, number theorists have refined their intuition by computing examples. The advent of computers and (especially) sophisticated algorithms has gradually led to the emergence of algorithmic number theory as a distinct field. This young discipline has been shaped by strong connections to computer science, cryptography, and other parts of mathematics. One of its charms is that mathematical ideas often lead to better algorithms. Another striking feature is that the algorithmic worldview has led to fascinating new mathematical ideas and questions.

This volume contains twenty survey articles on topics in algorithmic number theory, written by leading experts in the field. The first two are introductory, aiming to entice the reader into pursuing the subject more deeply. The next eight cover core areas of the field: factoring, primality, smooth numbers, lattices, elliptic curves, algebraic number theory, and fast arithmetic algorithms. The remaining ten articles survey specific topics, often with a distinctive perspective, including cryptography, Arakelov class groups, computational class field theory, zeta functions over finite fields, arithmetic geometry, and modular forms.

Mathematical Sciences Research Institute
Publications

# 44

Algorithmic Number Theory:
Lattices, Number Fields, Curves and Cryptography

# Mathematical Sciences Research Institute Publications

1. Freed/Uhlenbeck: *Instantons and Four-Manifolds*, second edition
2. Chern (ed.): *Seminar on Nonlinear Partial Differential Equations*
3. Lepowsky/Mandelstam/Singer (eds.): *Vertex Operators in Mathematics and Physics*
4. Kac (ed.): *Infinite Dimensional Groups with Applications*
5. Blackadar: *K-Theory for Operator Algebras*, second edition
6. Moore (ed.): *Group Representations, Ergodic Theory, Operator Algebras, and Mathematical Physics*
7. Chorin/Majda (eds.): *Wave Motion: Theory, Modelling, and Computation*
8. Gersten (ed.): *Essays in Group Theory*
9. Moore/Schochet: *Global Analysis on Foliated Spaces*, second edition
10–11. Drasin/Earle/Gehring/Kra/Marden (eds.): *Holomorphic Functions and Moduli*
12–13. Ni/Peletier/Serrin (eds.): *Nonlinear Diffusion Equations and Their Equilibrium States*
14. Goodman/de la Harpe/Jones: *Coxeter Graphs and Towers of Algebras*
15. Hochster/Huneke/Sally (eds.): *Commutative Algebra*
16. Ihara/Ribet/Serre (eds.): *Galois Groups over* $\mathbb{Q}$
17. Concus/Finn/Hoffman (eds.): *Geometric Analysis and Computer Graphics*
18. Bryant/Chern/Gardner/Goldschmidt/Griffiths: *Exterior Differential Systems*
19. Alperin (ed.): *Arboreal Group Theory*
20. Dazord/Weinstein (eds.): *Symplectic Geometry, Groupoids, and Integrable Systems*
21. Moschovakis (ed.): *Logic from Computer Science*
22. Ratiu (ed.): *The Geometry of Hamiltonian Systems*
23. Baumslag/Miller (eds.): *Algorithms and Classification in Combinatorial Group Theory*
24. Montgomery/Small (eds.): *Noncommutative Rings*
25. Akbulut/King: *Topology of Real Algebraic Sets*
26. Judah/Just/Woodin (eds.): *Set Theory of the Continuum*
27. Carlsson/Cohen/Hsiang/Jones (eds.): *Algebraic Topology and Its Applications*
28. Clemens/Kollár (eds.): *Current Topics in Complex Algebraic Geometry*
29. Nowakowski (ed.): *Games of No Chance*
30. Grove/Petersen (eds.): *Comparison Geometry*
31. Levy (ed.): *Flavors of Geometry*
32. Cecil/Chern (eds.): *Tight and Taut Submanifolds*
33. Axler/McCarthy/Sarason (eds.): *Holomorphic Spaces*
34. Ball/Milman (eds.): *Convex Geometric Analysis*
35. Levy (ed.): *The Eightfold Way*
36. Gavosto/Krantz/McCallum (eds.): *Contemporary Issues in Mathematics Education*
37. Schneider/Siu (eds.): *Several Complex Variables*
38. Billera/Björner/Green/Simion/Stanley (eds.): *New Perspectives in Geometric Combinatorics*
39. Haskell/Pillay/Steinhorn (eds.): *Model Theory, Algebra, and Geometry*
40. Bleher/Its (eds.): *Random Matrix Models and Their Applications*
41. Schneps (ed.): *Galois Groups and Fundamental Groups*
42. Nowakowski (ed.): *More Games of No Chance*
43. Montgomery/Schneider (eds.): *New Directions in Hopf Algebras*
44. Buhler/Stevenhagen (eds.): *Algorithmic Number Theory: Lattices, Number Fields, Curves and Cryptography*
45. Jensen/Ledet/Yui: *Generic Polynomials: Constructive Aspects of the Inverse Galois Problem*
46. Rockmore/Healy (eds.): *Modern Signal Processing*
47. Uhlmann (ed.): *Inside Out: Inverse Problems and Applications*
48. Gross/Kotiuga: *Electromagnetic Theory and Computation: A Topological Approach*
49. Darmon/Zhang (eds.): *Heegner Points and Rankin L-Series*
50. Bao/Bryant/Chern/Shen (eds.): *A Sampler of Riemann–Finsler Geometry*
51. Avramov/Green/Huneke/Smith/Sturmfels (eds.): *Trends in Commutative Algebra*
52. Goodman/Pach/Welzl (eds.): *Combinatorial and Computational Geometry*
53. Schoenfeld (ed.): *Assessing Mathematical Proficiency*
54. Hasselblatt (ed.): *Dynamics, Ergodic Theory, and Geometry*
55. Pinsky/Birnir (eds.): *Probability, Geometry and Integrable Systems*

Volumes 1–4, 6–8, and 10–27 are published by Springer-Verlag

# Algorithmic Number Theory:
## Lattices, Number Fields, Curves and Cryptography

*Edited by*

J. P. Buhler
P. Stevenhagen

J. P. Buhler
CCR and Reed College
4320 Westerra Ct., San Diego, CA 92121
jpb@reed.edu

P. Stevenhagen
Mathematisch Instituut, Universiteit Leiden
Postbus 9512, 2300 RA Leiden, The Netherlands
psh@math.leidenuniv.nl

Silvio Levy (*Series Editor*)
Mathematical Sciences Research Institute
17 Gauss Way, Berkeley, CA 94720
levy@msri.org

The Mathematical Sciences Research Institute wishes to acknowledge support by the National Science Foundation and the *Pacific Journal of Mathematics* for the publication of this series.

---

CAMBRIDGE UNIVERSITY PRESS
Cambridge, New York, Melbourne, Madrid, Cape Town, Singapore, São Paulo, Delhi

Cambridge University Press
32 Avenue of the Americas, New York, NY 10013-2473, USA

www.cambridge.org
Information on this title: www.cambridge.org/9780521808545

© Mathematical Sciences Research Institute 2008

This publication is in copyright. Subject to statutory exception and to the provisions of relevant collective licensing agreements, no reproduction of any part may take place without the written permission of Cambridge University Press.

First published 2008

Printed in the United States of America

*A catalog record for this publication is available from the British Library.*

*Library of Congress Cataloging in Publication data*

Algorithmic number theory / edited by J. P. Buhler and P. Stevenhagen.
  p. cm. – (Mathematical Sciences Research Institute publications ; 44)
Includes bibliographical references and index.
ISBN 978-0-521-80854-5 (hardback)
 1. Number theory. 2. Algorithms. 3. Algebraic fields–Data processing. 4. Number theory–Data processing. 5. Factorization (Mathematics) 6. Lattice theory. 7. Curves, Elliptic. 8. Class field theory. I. Buhler, Joe P., 1950– II. Stevenhagen, P., 1963– III. Title. IV. Series.

QA241.S8295 2008
512.7–dc22                                                                                               2008031327

ISBN    978-0-521-80854-5 hardback

Cambridge University Press has no responsibility for the persistence or accuracy of URLs for external or third-party Internet Web sites referred to in this publication and does not guarantee that any content on such Web sites is, or will remain, accurate or appropriate. Information regarding prices, travel timetables, and other factual information given in this work are correct at the time of first printing, but Cambridge University Press does not guarantee the accuracy of such information thereafter.

# Contents

| | |
|---|---|
| Preface | *page* ix |
| Solving the Pell equation<br>HENDRIK W. LENSTRA, JR. | 1 |
| Basic algorithms in number theory<br>JOE BUHLER AND STAN WAGON | 25 |
| Smooth numbers and the quadratic sieve<br>CARL POMERANCE | 69 |
| The number field sieve<br>PETER STEVENHAGEN | 83 |
| Four primality testing algorithms<br>RENÉ SCHOOF | 101 |
| Lattices<br>HENDRIK W. LENSTRA, JR. | 127 |
| Elliptic curves<br>BJORN POONEN | 183 |
| The arithmetic of number rings<br>PETER STEVENHAGEN | 209 |
| Smooth numbers: computational number theory and beyond<br>ANDREW GRANVILLE | 267 |
| Fast multiplication and its applications<br>DANIEL J. BERNSTEIN | 325 |
| Elementary thoughts on discrete logarithms<br>CARL POMERANCE | 385 |
| The impact of the number field sieve on the discrete logarithm problem in finite fields<br>OLIVER SCHIROKAUER | 397 |
| Reducing lattice bases to find small-height values of univariate polynomials<br>DANIEL J. BERNSTEIN | 421 |
| Computing Arakelov class groups<br>RENÉ SCHOOF | 447 |

| | |
|---|---:|
| Computational class field theory<br>HENRI COHEN AND PETER STEVENHAGEN | 497 |
| Protecting communications against forgery<br>DANIEL J. BERNSTEIN | 535 |
| Algorithmic theory of zeta functions over finite fields<br>DAQING WAN | 551 |
| Counting points on varieties over finite fields of small characteristic<br>ALAN G. B. LAUDER AND DAQING WAN | 579 |
| Congruent number problems and their variants<br>JAAP TOP AND NORIKO YUI | 613 |
| An introduction to computing modular forms using modular symbols<br>WILLIAM A. STEIN | 641 |

# Preface

Our subject arises out of two roots of mathematical thought: fascination with properties of whole numbers and the urge to compute. Number theory and computer science flowered vividly during the last quarter of the twentieth century, and the synergy at their intersection was striking. Algorithmic number theory emerged as an exciting field in its own right, containing deep insights and having surprising applications.

In the fall of 2000 the Mathematical Sciences Research Institute, Berkeley, hosted a one-semester program on algorithmic number theory. Its opening workshop, cosponsored by the Clay Mathematics Institute, featured many foundational and survey talks. During the meeting, it was noted that there was a dearth of sources for newcomers to the field. After the conference, some of the speakers agreed to write articles based on their talks, and we were drafted to edit the volume.

A few authors turned in drafts promptly, some retaining the tutorial focus and tone of the original talks, while others were full-blown tutorials or surveys. Many authors (including the editors) dallied. Additional articles were solicited, to provide more coherence and to incorporate newer results that couldn't be ignored (most notably, the polynomial-time primality algorithm due to Manindra Agrawal, Neeraj Kayal, and Nitin Saxena). This led to complications that might have been expected for a volume with 20 substantial articles, 15 authors, and 650 pages. These have finally run their course, and we are delighted that the volume is ready to see the light of day.

We do apologize to the authors who responded promptly, and can only hope that they will be compensated by the greater breadth and interest of the volume in which their contributions appear.

The articles in the volume can be loosely categorized as follows. The first two articles are introductory, and are more elementary than their successors — they attempt to entice the reader into pursuing the ideas more deeply. The next eight articles provide surveys of central topics, including smooth numbers, factoring, primality testing, lattices, elliptic curves, algebraic number theory, and fast arithmetic algorithms. The remaining ten articles study specific topics more deeply, including cryptography, computational algebraic number theory, modular forms, and arithmetic geometry.

Although the articles in this volume are surveys in the broadest sense, the word should not be taken to mean an encyclopedic treatment that captures current conventional wisdom. We prefer the term overviews, and the articles have a distinctive and in some cases even nonstandard perspective.

It remains our pleasant duty to thank a number of institutions and people. Most obviously, the authors have produced many fascinating pages, sure to inspire others to pursue the subject. We thank the Clay Institute and MSRI for their generous funding for the workshop that provided the initial spark for this volume. We thank Cambridge University Press and MSRI for their support and patience during the production of this volume, and we especially thank Silvio Levy for his extensive efforts on this volume. John Voight took notes (by typing nearly real-time TEX into his laptop) at most of the talks at the workshop, and these were valuable to some of the authors.

Finally, Hendrik Lenstra has long been a source of pervasive and brilliant inspiration to the entire field of algorithmic number theory, and this volume is no exception: in addition to two distinctive articles, he has provided much-appreciated advice over the years to the editors and to virtually all of the other authors.

Joe Buhler<br>
Peter Stevenhagen<br>
San Diego, May 2008

# Solving the Pell equation

HENDRIK W. LENSTRA, JR.

ABSTRACT. We illustrate recent developments in computational number theory by studying their implications for solving the Pell equation. We shall see that, if the solutions to the Pell equation are properly represented, the traditional continued fraction method for solving the equation can be significantly accelerated. The most promising method depends on the use of smooth numbers. As with many algorithms depending on smooth numbers, its run time can presently only conjecturally be established; giving a rigorous analysis is one of the many open problems surrounding the Pell equation.

## 1. Pell's equation

The *Pell equation* is the equation

$$x^2 = dy^2 + 1,$$

to be solved in positive integers $x$, $y$ for a given nonzero integer $d$. For example, for $d = 5$ one can take $x = 9$, $y = 4$. We shall always assume that $d$ is positive but not a square, since otherwise there are clearly no solutions.

The English mathematician John Pell (1611–1685) has nothing to do with the equation. Euler (1707–1783) mistakenly attributed to Pell a solution method that had in fact been found by another English mathematician, William Brouncker (1620–1684), in response to a challenge by Fermat (1601–1665); but attempts to change the terminology introduced by Euler have always proved futile.

Pell's equation has an extraordinarily rich history, to which Weil [1984] is the best guide; see also [Dickson 1920, Chapter XII; Konen 1901; Whitford 1912]. Brouncker's method is in substance identical to a method that was known to Indian mathematicians at least six centuries earlier. As we shall see, the equation

---

This paper appeared in slightly different form in *Notices Amer. Math. Soc.* **49** (2002), 182–192, with the permission of MSRI and the editors of the present volume.

also occurred in Greek mathematics, but no convincing evidence that the Greeks could solve the equation has ever emerged.

A particularly lucid exposition of the "Indian" or "English" method of solving the Pell equation is found in Euler's *Algebra* [Euler 1770, Abschnitt 2, Capitel 7]. Modern textbooks usually give a formulation in terms of continued fractions, which is also due to Euler (see for example [Niven et al. 1991, Chapter 7]). Euler, as well as his Indian and English predecessors, appears to take it for granted that the method always produces a solution. That is true, but it is not obvious — all that is obvious is that *if* there is a solution, the method will find one. Fermat was probably in possession of a proof that there is a solution for every $d$ (see [Weil 1984, Chapter II, § XIII]), and the first to publish such a proof was Lagrange (1736–1813) [1773].

One may rewrite Pell's equation as

$$(x + y\sqrt{d}) \cdot (x - y\sqrt{d}) = 1,$$

so that finding a solution comes down to finding a nontrivial unit of the ring $\mathbb{Z}[\sqrt{d}]$ of norm 1; here the norm $\mathbb{Z}[\sqrt{d}]^* \to \mathbb{Z}^* = \{\pm 1\}$ between unit groups multiplies each unit by its conjugate, and the units $\pm 1$ of $\mathbb{Z}[\sqrt{d}]$ are considered trivial. This reformulation implies that once one knows a solution to Pell's equation, one can find infinitely many. More precisely, if the solutions are ordered by magnitude, then the $n$-th solution $x_n$, $y_n$ can be expressed in terms of the first one, $x_1$, $y_1$, by

$$x_n + y_n\sqrt{d} = (x_1 + y_1\sqrt{d})^n.$$

Accordingly, the first solution $x_1$, $y_1$ is called the *fundamental solution* to the Pell equation, and *solving* the Pell equation means finding $x_1$, $y_1$ for given $d$. By abuse of language, we shall also refer to $x + y\sqrt{d}$ instead of the pair $x$, $y$ as a solution to Pell's equation and call $x_1 + y_1\sqrt{d}$ the fundamental solution.

One may view the solvability of Pell's equation as a special case of *Dirichlet's unit theorem* from algebraic number theory, which describes the structure of the group of units of a general ring of algebraic integers [Stevenhagen 2008a]; for the ring $\mathbb{Z}[\sqrt{d}]$, it is the product of $\{\pm 1\}$ and an infinite cyclic group.

As an example, consider $d = 14$. One has

$$\sqrt{14} = 3 + \cfrac{1}{1 + \cfrac{1}{2 + \cfrac{1}{1 + \cfrac{1}{3 + \sqrt{14}}}}},$$

so the continued fraction expansion of $3+\sqrt{14}$ is purely periodic with period length 4. Truncating the expansion at the end of the first period, one finds that the fraction

$$3 + \cfrac{1}{1+\cfrac{1}{2+\cfrac{1}{\frac{1}{1}}}} = \frac{15}{4}$$

is a fair approximation to $\sqrt{14}$. The numerator and denominator of this fraction yield the fundamental solution $x_1 = 15$, $y_1 = 4$; indeed one has $15^2 = 14 \cdot 4^2 + 1$. Furthermore, one computes $(15 + 4\sqrt{14})^2 = 449 + 120\sqrt{14}$, so $x_2 = 449$, $y_2 = 120$; and so on. One finds:

| $n$ | $x_n$ | $y_n$ |
|---|---|---|
| 1 | 15 | 4 |
| 2 | 449 | 120 |
| 3 | 13455 | 3596 |
| 4 | 403201 | 107760 |
| 5 | 12082575 | 3229204 |
| 6 | 362074049 | 96768360 |

The shape of the table reflects the exponential growth of $x_n$ and $y_n$ with $n$.

For general $d$, the continued fraction expansion of $[\sqrt{d}] + \sqrt{d}$ is again purely periodic, and the period displays a symmetry similar to the one visible for $d = 14$. If the period length is even, one proceeds as above; if the period length is odd, one truncates at the end of the *second* period [Buhler and Wagon 2008].

## 2. The cattle problem

An interesting example of the Pell equation, both from a computational and from a historical perspective, is furnished by the *cattle problem* of Archimedes (287–212 B.C.). A manuscript containing this problem was discovered by Lessing (1729–1781) in the Wolffenbüttel library, and published by him in 1773 (see [Lessing 1773; Heiberg 1913, pp. 528–534]). It is now generally credited to Archimedes [Fraser 1972; Weil 1984]. In twenty-two Greek elegiac distichs, the problem asks for the number of white, black, dappled, and brown bulls and cows belonging to the Sun god, subject to several arithmetical restrictions. A version in English heroic couplets, published in [Archimedes 1999], is shown on page 4. In modern mathematical notation the problem is no less elegant. Writing $x$, $y$, $z$, $t$ for the numbers of white, black, dappled, and brown bulls,

# PROBLEM

*that Archimedes conceived in verse
and posed to the specialists at Alexandria
in a letter to Eratosthenes of Cyrene.*

The Sun god's cattle, friend, apply thy care
to count their number, hast thou wisdom's share.
They grazed of old on the Thrinacian floor
of Sic'ly's island, herded into four,
colour by colour: one herd white as cream,
the next in coats glowing with ebon gleam,
brown-skinned the third, and stained with spots the last.
Each herd saw bulls in power unsurpassed,
in ratios these: count half the ebon-hued,
add one third more, then all the brown include;
thus, friend, canst thou the white bulls' number tell.
The ebon did the brown exceed as well,
now by a fourth and fifth part of the stained.
To know the spotted — all bulls that remained —
reckon again the brown bulls, and unite
these with a sixth and seventh of the white.
Among the cows, the tale of silver-haired
was, when with bulls and cows of black compared,
exactly one in three plus one in four.
The black cows counted one in four once more,
plus now a fifth, of the bespeckled breed
when, bulls withal, they wandered out to feed.
The speckled cows tallied a fifth and sixth
of all the brown-haired, males and females mixed.
Lastly, the brown cows numbered half a third
and one in seven of the silver herd.
Tell'st thou unfailingly how many head
the Sun possessed, o friend, both bulls well-fed
and cows of ev'ry colour — no-one will
deny that thou hast numbers' art and skill,
though not yet dost thou rank among the wise.
But come! also the foll'wing recognise.
Whene'er the Sun god's white bulls joined the black,
their multitude would gather in a pack
of equal length and breadth, and squarely throng
Thrinacia's territory broad and long.
But when the brown bulls mingled with the flecked,
in rows growing from one would they collect,
forming a perfect triangle, with ne'er
a diff'rent-coloured bull, and none to spare.
Friend, canst thou analyse this in thy mind,
and of these masses all the measures find,
go forth in glory! be assured all deem
thy wisdom in this discipline supreme!

respectively, one reads in lines 8–16 the restrictions
$$x = (\tfrac{1}{2} + \tfrac{1}{3})y + t,$$
$$y = (\tfrac{1}{4} + \tfrac{1}{5})z + t,$$
$$z = (\tfrac{1}{6} + \tfrac{1}{7})x + t.$$
Next, for the numbers $x'$, $y'$, $z'$, $t'$ of cows of the same respective colors, the poet requires in lines 17–26
$$x' = (\tfrac{1}{3} + \tfrac{1}{4})(y + y'), \qquad z' = (\tfrac{1}{5} + \tfrac{1}{6})(t + t'),$$
$$y' = (\tfrac{1}{4} + \tfrac{1}{5})(z + z'), \qquad t' = (\tfrac{1}{6} + \tfrac{1}{7})(x + x').$$
Whoever can solve the problem thus far is called merely competent by Archimedes; to win the prize for supreme wisdom, one should also meet the conditions formulated in lines 33–40 that $x + y$ be a *square* and that $z + t$ be a *triangular number*.

The first part of the problem is just linear algebra, and there is indeed a solution in *positive* integers. The general solution to the first three equations is given by $(x, y, z, t) = m \cdot (2226, 1602, 1580, 891)$, where $m$ is a positive integer. The next four equations turn out to be solvable if and only if $m$ is divisible by 4657; with $m = 4657 \cdot k$ one has
$$(x', y', z', t') = k \cdot (7206360, 4893246, 3515820, 5439213).$$

The true challenge is now to choose $k$ such that $x + y = 4657 \cdot 3828 \cdot k$ is a square and $z + t = 4657 \cdot 2471 \cdot k$ is a triangular number. From the prime factorization $4657 \cdot 3828 = 2^2 \cdot 3 \cdot 11 \cdot 29 \cdot 4657$ one sees that the first condition is equivalent to $k = al^2$, where $a = 3 \cdot 11 \cdot 29 \cdot 4657$ and $l$ is an integer. Since $z + t$ is a triangular number if and only if $8(z + t) + 1$ is a square, we are led to the equation $h^2 = 8(z + t) + 1 = 8 \cdot 4657 \cdot 2471 \cdot al^2 + 1$, which is the Pell equation $h^2 = dl^2 + 1$ for
$$d = 2 \cdot 3 \cdot 7 \cdot 11 \cdot 29 \cdot 353 \cdot (2 \cdot 4657)^2 = 410\,286\,423\,278\,424.$$

Thus, by Lagrange's theorem, the cattle problem admits infinitely many solutions.

In 1867 the otherwise unknown German mathematician C. F. Meyer set out to solve the equation by the continued fraction method [Dickson 1920, p. 344]. After 240 steps in the continued fraction expansion for $\sqrt{d}$ he had still not detected the period, and he gave up. He may have been a little impatient; it was later discovered that the period length equals 203254; see [Grosjean and De Meyer 1991]. The first to solve the cattle problem in a satisfactory way was A. Amthor in 1880 (see [Krumbiegel and Amthor 1880]). Amthor did *not* directly apply the continued fraction method; what he did do we shall discuss below. Nor did he spell out the decimal digits of the fundamental solution to the Pell equation or the corresponding solution of the cattle problem. He did show that, in the smallest solution to the cattle problem, the total number of cattle is given by a number of 206545 digits; of the four leading digits 7766 that he gave, the fourth was wrong, due to the use of insufficiently precise logarithms. The full number occupies forty-seven pages of computer printout, reproduced in reduced size on twelve pages of the *Journal of Recreational Mathematics* [Nelson 1980/81]. In abbreviated form, it reads

$$77602714\ldots237983357\ldots55081800,$$

each of the six dots representing 34420 omitted digits.

Several nineteenth century German scholars were worried that so many bulls and cows might not fit on the island of Sicily, contradicting lines 3 and 4 of the poem; but, as Lessing remarked, the Sun god, to whom the cattle belonged, will have coped with it.

The story of the cattle problem shows that the continued fraction method is not the last word on the Pell equation.

## 3. Efficiency

We are interested in the *efficiency* of solution methods for the Pell equation. Thus, how much time does a given algorithm for solving the Pell equation take? Here *time* is to be measured in a realistic way, which reflects, for example, that large positive integers are more time-consuming to operate with than small ones; technically, one counts *bit operations*. The input to the algorithm is $d$, and the running time estimates are accordingly expressed as functions of $d$. If one supposes that $d$ is specified in binary or in decimal, then the *length of the input* is approximately proportional to $\log d$. An algorithm is said to run in *polynomial time* if there is a positive real number $c_0$ such that for all $d$ the running time is at most $(1 + \log d)^{c_0}$, in other words, if the time that it takes the algorithm to *solve* the Pell equation is not much greater than the time required to *write down* the equation.

How fast is the continued fraction method? Can the Pell equation be solved in polynomial time? The central quantity that one needs to consider in order to answer such questions is the *regulator* $R_d$, which is defined by

$$R_d = \log(x_1 + y_1\sqrt{d}),$$

where $x_1 + y_1\sqrt{d}$ denotes, as before, the fundamental solution to Pell's equation. The regulator coincides with what in algebraic number theory would be called the regulator of the kernel of the norm map $\mathbb{Z}[\sqrt{d}]^* \to \mathbb{Z}^*$. From

$x_1 - y_1\sqrt{d} = 1/(x_1 + y_1\sqrt{d})$ one deduces that $0 < x_1 - y_1\sqrt{d} < 1/(2\sqrt{d})$, and combining this with $x_1 + y_1\sqrt{d} = e^{R_d}$, one finds that

$$\frac{e^{R_d}}{2} < x_1 < \frac{e^{R_d}}{2} + \frac{1}{4\sqrt{d}}, \qquad \frac{e^{R_d}}{2\sqrt{d}} - \frac{1}{4d} < y_1 < \frac{e^{R_d}}{2\sqrt{d}}.$$

This shows that $R_d$ is very close to $\log(2x_1)$ and to $\log(2y_1\sqrt{d})$. That is, if $x_1$ and $y_1$ are to be represented in binary or in decimal, then $R_d$ is approximately proportional to the *length of the output* of any algorithm solving the Pell equation. Since the time required for spelling out the output is a lower bound for the total running time, we may conclude: *there exists $c_1$ such that any algorithm for solving the Pell equation takes time at least $c_1 R_d$.* Here $c_1$ denotes, just as do $c_2, c_3, \ldots$ below, a positive real number that does not depend on $d$.

The continued fraction method almost meets this lower bound. Let $l$ be the period length of the continued fraction expansion of $[\sqrt{d}] + \sqrt{d}$ if that length is even and twice that length if it is odd. Then one has

$$\frac{\log 2}{2} \cdot l < R_d < \frac{\log(4d)}{2} \cdot l;$$

see [Lenstra 1982, (11.4)]. Thus $R_d$ and $l$ are approximately proportional. Using this, one estimates easily that the time taken by a straightforward implementation of the continued fraction method is at most $R_d^2 \cdot (1 + \log d)^{c_2}$ for suitable $c_2$; and a more refined implementation, which depends on the fast Fourier transform, reduces this to $R_d \cdot (1 + \log d)^{c_3}$ for suitable $c_3$; see [Schönhage 1971]. We conclude that the latter version of the continued fraction method is optimal, apart from a logarithmic factor.

In view of these results it is natural to ask how the regulator grows as a function of $d$. It turns out that it fluctuates wildly. One has

$$\log(2\sqrt{d}) < R_d < \sqrt{d} \cdot (\log(4d) + 2),$$

the lower bound because of the inequality $y_1 < e^{R_d}/(2\sqrt{d})$ above and the upper bound by [Hua 1942]. The gap between the two bounds is very large, but it cannot be helped: if $d$ ranges over numbers of the form $k^2 - 1$, for which one has $x_1 = k$ and $y_1 = 1$, then $R_d - \log(2\sqrt{d})$ tends to 0; and one can show that there exist an infinite set $D$ of $d$'s and a constant $c_4$ such that all $d \in D$ have $R_d = c_4\sqrt{d}$. In fact, if $d_0, d_1$ are integers greater than 1 and $d_0$ is not a square, then there exists a positive integer $m = m(d_0, d_1)$ such that $D = \{d_0 d_1^{2n} : n \in \mathbb{Z}, n \geq m\}$ has this property for some $c_4 = c_4(d_0, d_1)$.

It is believed that for most $d$ the upper bound is closer to the truth. More precisely, a folklore conjecture asserts that there is a set $D$ of nonsquare positive

integers that has density 1 in the sense that $\lim_{x\to\infty} \#\{d \in D : d \leq x\}/x = 1$, and that satisfies
$$\lim_{d \in D} \frac{\log R_d}{\log \sqrt{d}} = 1.$$
This conjecture, however, is wide open. The same is true for the much weaker conjecture that $\limsup_d (\log R_d)/\log \sqrt{d}$, with $d$ ranging over the *squarefree* integers $> 1$, is *positive*.

If the folklore conjecture is true, then for most $d$ the factor $R_d$ entering the running time is about $\sqrt{d}$, which is an exponential function of the length $\log d$ of the input.

Combining the preceding results, one concludes that the continued fraction method takes time at most $\sqrt{d} \cdot (1 + \log d)^{c_5}$; that conjecturally it is exponentially slow for *most* values of $d$; and that *any* method for solving the Pell equation that spells out $x_1$ and $y_1$ in full is exponentially slow for *infinitely many* $d$ and will therefore fail to run in polynomial time.

If we want to improve upon the continued fraction method, then we need a way of representing $x_1$ and $y_1$ that is more compact than the decimal or binary notation.

## 4. Amthor's solution

Amthor's solution to the cattle problem depended on the observation that the number $d = 410\,286423\,278424$ can be written as $(2 \cdot 4657)^2 \cdot d'$, where $d' = 4\,729494$ is squarefree. Hence, if $x$, $y$ solves the Pell equation for $d$, then $x$, $2 \cdot 4657 \cdot y$ solves the Pell equation for $d'$ and will therefore for some $n$ be equal to the $n$-th solution $x'_n$, $y'_n$ (say) of that equation:
$$x + 2 \cdot 4657 \cdot y \cdot \sqrt{d'} = (x'_1 + y'_1 \sqrt{d'})^n.$$
This reduces the cattle problem to two easier problems: first, solving the Pell equation for $d'$; and second, finding the least value of $n$ for which $y'_n$ is divisible by $2 \cdot 4657$.

Since $d'$ is much smaller than $d$, Amthor could use the continued fraction algorithm for $d'$. In a computation that could be summarized in three pages, as in [Krumbiegel and Amthor 1880], he found the period length to be 92 and $x'_1 + y'_1 \sqrt{d'}$ to be given by

$u = 109\,931986\,732829\,734979\,866232\,821433\,543901\,088049$

$\quad + \; 50549\,485234\,315033\,074477\,819735\,540408\,986340 \cdot \sqrt{4\,729494}.$

In order to save space, one can write

$u = \left(300\,426607\,914281\,713365 \cdot \sqrt{609} + 84\,129507\,677858\,393258 \cdot \sqrt{7766}\right)^2.$

$$w = 300\,426607\,914281\,713365 \cdot \sqrt{609} + 84\,129507\,677858\,393258 \cdot \sqrt{7766}$$

$$k_j = (w^{4658 \cdot j} - w^{-4658 \cdot j})^2 / 368\,238304 \qquad (j = 1, 2, 3, \ldots)$$

| $j$-th solution | bulls | cows | all cattle |
|---|---|---|---|
| white   | $10\,366482 \cdot k_j$ | $7\,206360 \cdot k_j$ | $17\,572842 \cdot k_j$ |
| black   | $7\,460514 \cdot k_j$  | $4\,893246 \cdot k_j$ | $12\,353760 \cdot k_j$ |
| dappled | $7\,358060 \cdot k_j$  | $3\,515820 \cdot k_j$ | $10\,873880 \cdot k_j$ |
| brown   | $4\,149387 \cdot k_j$  | $5\,439213 \cdot k_j$ | $9\,588600 \cdot k_j$ |
| all colors | $29\,334443 \cdot k_j$ | $21\,054639 \cdot k_j$ | $50\,389082 \cdot k_j$ |

All solutions to the cattle problem of Archimedes.

This is derived from the identity $x + y\sqrt{d} = (\sqrt{(x-1)/2} + \sqrt{(x+1)/2})^2$, which holds whenever $x^2 = dy^2 + 1$. The regulator is found to be $R_{d'} \doteq 102.101583$.

In order to determine the least feasible value for $n$, Amthor developed a little theory, which one would nowadays cast in the language of finite fields and rings. Using that $p = 4657$ is a prime number for which the Legendre symbol $\left(\frac{d'}{p}\right)$ equals $-1$, he deduced from his theory that the least value for $n$ divides $p+1 = 4658$; had he been a little more careful, he would have found that it must divide $(p+1)/2 = 2329 = 17 \cdot 137$ (see [Vardi 1998]). In any case, trying a few divisors, one discovers that the least value for $n$ is actually *equal* to 2329. One has $R_d = 2329 \cdot R_{d'} \doteq 237794.586710$.

The conclusion is that the fundamental solution to the Pell equation for $d$ itself is given by $x_1 + y_1 \sqrt{d} = u^{2329}$, with $u$ as just defined. Amthor failed to put everything together, but I did this for the convenience of the reader: for the first time in history, *all* infinitely many solutions to the cattle problem displayed in a handy little table! It does, naturally, not contain the full decimal expansion of any of the numbers asked for, but what it does contain should be considered more enlightening. For example, it enables the reader not only to verify easily that the total number of cattle in the smallest solution has 206545 decimal digits and equals 77602714...55081800, but also to discover that the number of dappled bulls in the 1494 195300th solution equals 111111...000000, a number of 308 619694 367813 digits. (Finding the middle digits is probably much harder.) There is no doubt that Archimedes, who wrote a lengthy epistle about the representation of large numbers to King Gelon (see [Dijksterhuis 1956] or [Heiberg 1913, pp. 215–259]), would have been pleased and satisfied by the solution as expressed in the table.

## 5. Power products

Suppose one wishes to solve the Pell equation $x^2 = dy^2 + 1$ for a given value of $d$. From Amthor's approach to the cattle problem we learn that for two reasons it may be wise to find the smallest divisor $d'$ of $d$ for which $d/d'$ is a square: it saves time when performing the continued fraction algorithm, and it saves both time and space when expressing the final answer. There is no known algorithm for finding $d'$ from $d$ that is essentially faster than factoring $d$. In addition, if we want to determine *which* power of the fundamental solution for $d'$ yields the fundamental solution for $d$ — that is, the number $n$ from the previous section — we also need to know the prime factorization of $\sqrt{d/d'}$, as well as the prime factorization of $p - \left(\frac{d'}{p}\right)$ for each prime $p$ dividing $\sqrt{d/d'}$. Thus, if one wants to solve the Pell equation, one may as well start by factoring $d$. Known factoring algorithms may not be very fast for large $d$, but for most values of $d$ they are still expected to be orders of magnitudes faster than any known method for solving the Pell equation [Stevenhagen 2008b].

Let it now be assumed that $d$ is *squarefree*, and write $x_1 + y_1\sqrt{d}$ for the fundamental solution of the Pell equation, which is a unit of $\mathbb{Z}[\sqrt{d}]$. Then $x_1 + y_1\sqrt{d}$ may still be a proper power in the field $\mathbb{Q}(\sqrt{d})$ of fractions of $\mathbb{Z}[\sqrt{d}]$. For example, the least $d$ with $y_1 > 6$ is $d = 13$, for which one has $x_1 = 649$, $y_1 = 180$, and

$$649 + 180\sqrt{13} = \left(\frac{3 + \sqrt{13}}{2}\right)^6.$$

Also in the case $d = 109$, which Fermat posed as a challenge problem in 1657, the fundamental solution is a sixth power:

$$158\,070671\,986249 + 15\,140424\,455100\sqrt{109} = \left(\frac{261 + 25\sqrt{109}}{2}\right)^6.$$

However, this is as far as it goes: it is an elementary exercise in algebraic number theory to show that if $n$ is a positive integer for which $x_1 + y_1\sqrt{d}$ has an $n$-th root in $\mathbb{Q}(\sqrt{d})$, then $n = 1, 2, 3$, or $6$, the case $n = 2$ being possible only for $d \equiv 1, 2$, or $5 \bmod 8$, and the cases $n = 3$ and $6$ only for $d \equiv 5 \bmod 8$. Thus, for large squarefree $d$ one cannot expect to save much space by writing $x_1 + y_1\sqrt{d}$ as a power. This is also true when one allows the root to lie in a composite of quadratic fields, as we did for the cattle problem.

Let $d$ again be an arbitrary positive integer that is not a square. Instead of powers, we consider *power products* in $\mathbb{Q}(\sqrt{d})$, that is, expressions of the form

$$\prod_{i=1}^{t} (a_i + b_i\sqrt{d})^{n_i}$$

where $t$ is a nonnegative integer, $a_i$, $b_i$, $n_i$ are integers, $n_i \neq 0$, and for each $i$ at least one of $a_i$ and $b_i$ is nonzero. We define the *length* of such an expression to be

$$\sum_{i=1}^{t} \left(\log |n_i| + \log(|a_i| + |b_i|\sqrt{d})\right).$$

This is roughly proportional to the amount of bits needed to specify the numbers $a_i$, $b_i$, and $n_i$. Each power product represents a nonzero element of $\mathbb{Q}(\sqrt{d})$, and that element can be expressed uniquely as $(a + b\sqrt{d})/c$, with $a, b, c \in \mathbb{Z}$, $\gcd(a, b, c) = 1$, $c > 0$. However, the number of bits of $a, b, c$ will typically grow linearly with the exponents $|n_i|$ themselves rather than with their logarithms. So one avoids using the latter representation and works directly with the power products instead.

Several fundamental issues are raised by the representation of elements as power products. For example, can we recognize whether two power products represent the same element of $\mathbb{Q}(\sqrt{d})$, by means of a polynomial time algorithm? Here "polynomial time" means, as before, that the run time is bounded by a polynomial function of the length of the input, which in this case equals the sum of the lengths of the two given power products. Similarly, can we decide in polynomial time whether a given power product represents an element of the form $a + b\sqrt{d}$ with $a, b \in \mathbb{Z}$, that is, an element of $\mathbb{Z}[\sqrt{d}]$? If it does, can we decide whether one has $a^2 - db^2 = 1$ and $a, b > 0$, so that we have a solution to Pell's equation, and can we compute the residue classes of $a$ and $b$ modulo a given positive integer $m$, all in polynomial time?

All questions just raised have affirmative answers, even in the context of general algebraic number fields. Algorithms proving this were exhibited by Guoqiang Ge [1993; 1994]. In particular, one can efficiently decide whether a given power product represents a solution to Pell's equation, and if it does, one can efficiently compute any desired number of "least significant" decimal digits of that solution; taking the logarithm of the power product, one can do the same for the *leading* digits, and for the *number* of decimal digits, except possibly in the probably very rare cases that $a$ or $b$ is excessively close to a power of 10. There is *no* known polynomial time algorithm for deciding whether a given power product represents the *fundamental* solution to Pell's equation.

## 6. Infrastructure

Suppose now that, given $d$, we are not asking for the fundamental solution $x_1 + y_1\sqrt{d}$ to Pell's equation, but for a power product in $\mathbb{Q}(\sqrt{d})$ that represents it. The following theorem summarizes essentially all that is rigorously known about the smallest length of such a power product and about algorithms for finding one.

THEOREM. *There are positive real numbers $c_6$ and $c_7$ with the following properties.*

(a) *For each positive integer $d$ that is not a square there exists a power product that represents the fundamental solution to Pell's equation and that has length at most $c_6 \cdot (\log d)^2$.*

(b) *The problem of computing a power product representing the fundamental solution to Pell's equation is "polynomial time equivalent" to the problem of computing an integer $\tilde{R}_d$ with $|R_d - \tilde{R}_d| < 1$.*

(c) *There is an algorithm that given $d$ computes a power product representing the fundamental solution to Pell's equation in time at most*

$$(R_d^{1/2} + \log d)(1 + \log d)^{c_7}.$$

Part (a) of the theorem, which is taken from [Buchmann et al. 1995], implies that the question we are asking does admit a brief answer, so that there is no obvious obstruction to the existence of a polynomial time algorithm for *finding* such an answer.

Part (b), which is not formulated too rigorously, asserts the existence of two polynomial time algorithms. The first takes as input a power product

$$\prod_i (a_i + b_i \sqrt{d})^{n_i}$$

representing the fundamental solution to the Pell equation and gives as output an integer approximation to the regulator. There is no surprise here, one just uses the formula $R_d = \sum_i n_i \log |a_i + b_i \sqrt{d}|$ and applies a polynomial time algorithm for approximating logarithms; [Brent 1976]. The second algorithm takes as input the number $d$ as well as an integer approximation $\tilde{R}_d$ to $R_d$, and it computes a power product representing the fundamental solution to Pell's equation. Since the algorithm runs in polynomial time, the length of the output is polynomially bounded, and this is in fact the way part (a) of the theorem is proved.

The key notion underlying the second algorithm is that of "infrastructure", a word coined by Shanks [1972] to describe a certain multiplicative structure that he detected within the period of the continued fraction expansion of $\sqrt{d}$. It was subsequently shown in [Lenstra 1982] that this period can be "embedded" in a circle group of "circumference" $R_d$, the embedding preserving the cyclical structure. In the modern terminology of Arakelov theory, one may describe that circle group as the kernel of the natural map $\text{Pic}^0 \overline{\mathbb{Z}[\sqrt{d}]} \to \text{Pic}\,\mathbb{Z}[\sqrt{d}]$ from the group of "metrized line bundles of degree 0" on the "arithmetic curve" corresponding to $\mathbb{Z}[\sqrt{d}]$ to the usual class group of invertible ideals. By means of Gauss's reduced binary quadratic forms one can do explicit computations in $\mathrm{Pic}^0\,\mathbb{Z}[\sqrt{d}]$ and in its circle subgroup. For a fuller explanation of these notions and their algorithmic use we refer to the literature [Buchmann et al. 1995; Lenstra 1982; Schoof 1982; 2008; Shanks 1972; Williams 2002].

The equivalence stated in part (b) of the theorem has an interesting feature that is not commonly encountered in the context of equivalences. Namely, one may achieve an improvement by going "back and forth". Thus, starting from a power product representing the fundamental solution, one can first use it to compute $\tilde{R}_d$, and next use $\tilde{R}_d$ to find a *second* power product, possibly of smaller length than the initial one. And conversely, starting from any rough approximation to $R_d$ one can compute a power product and use it to compute $R_d$ to any desired accuracy.

The algorithm referred to in part (c) is the fastest rigorously proven algorithm for computing a power product as desired. Its run time is roughly the square root of the run time of the continued fraction algorithm. It again makes use of the infrastructure just discussed, combining it with a search technique that is known as the "baby step-giant step" method. The power product coming out of the algorithm may not have a very small length, but one can easily do something about this by using the algorithms of part (b). Our estimates for $R_d$ show that the run time is at most $d^{1/4} \cdot (1 + \log d)^{c_8}$ for some $c_8$; here the exponent $1/4$ can be improved to $1/5$ if one is willing to assume certain generalized Riemann hypotheses [Schoof 1982]. According to an unpublished result of Ulrich Vollmer, part (c) is valid with $c_7 = 1 + \varepsilon$ for all $\varepsilon > 0$ and all $d$ exceeding a bound depending on $\varepsilon$; this represents a slight improvement over the result obtained in [Buchmann and Vollmer 2006].

Mathematically the infrastructure methods have great interest. Algorithmically one conjectures that something faster is available. But as we shall see, the final victory may belong to the infrastructure.

## 7. Smooth numbers

The algorithms for solving Pell's equation that we saw so far have an exponential run time as a function of $\log d$. One prefers to have an algorithm whose run time is polynomial in $\log d$. The method that we shall now discuss is believed to have a run time that is halfway between exponential and polynomial. Like many subexponential algorithms in number theory, it makes use of *smooth numbers*, that is, nonzero integers that up to sign are built up from small prime factors. Smooth numbers have been used with great success in the design of algorithms for factoring integers and for computing discrete logarithms in multiplicative groups of rings [Pomerance 2008a; 2008b] Here we shall see how they can be used for the solution of Pell's equation as well.

Instead of giving a formal description, we illustrate the algorithm on the case $d = 4\,729494 = 2 \cdot 3 \cdot 7 \cdot 11 \cdot 29 \cdot 353$ derived from the cattle problem. The computation is less laborious and more entertaining than the expansion of $\sqrt{d}$ in a continued fraction performed by Amthor. We shall explain the method on an intuitive level only; readers desirous to see its formal justification should acquaint themselves with the basic theorems of algebraic number theory [Stevenhagen 2008a; 2008b].

The smooth numbers that the algorithm operates with are not ordinary integers, but elements of the ring $\mathbb{Z}[\sqrt{d}]$, with $d$ as just chosen. There is a natural way of extending the notion of smoothness to such numbers. Namely, for $\alpha = a + b\sqrt{d} \in \mathbb{Q}(\sqrt{d})$, with $a, b \in \mathbb{Q}$, write $\alpha' = a - b\sqrt{d}$. Then $\alpha \mapsto \alpha'$ yields an automorphism of the field $\mathbb{Q}(\alpha)$ and the ring $\mathbb{Z}[\alpha]$, and the *norm* map $N \colon \mathbb{Q}(\sqrt{d}) \to \mathbb{Q}$ defined by $N(\alpha) = \alpha\alpha' = a^2 - db^2$ respects multiplication. It is now natural to expect that an element $\alpha$ of $\mathbb{Z}[\sqrt{d}]$ is smooth if and only if $\alpha'$ is smooth; so one may as well pass to their product $N(\alpha)$, which is an ordinary integer, and *define* $\alpha$ to be smooth if $|N(\alpha)|$ is built up from prime numbers that lie below a certain bound. The size of this bound depends on the circumstances; in the present computation we choose it empirically.

The first step in the algorithm is to find a good supply of smooth numbers $a + b\sqrt{d}$ in $\mathbb{Z}[\sqrt{d}]$, or, equivalently, pairs of integers $a, b$ for which $a^2 - db^2$ is smooth. One does this by trying $b = 1, 2, 3, \ldots$ in succession, and trying integers $a$ in the neighborhood of $b\sqrt{d}$; then $|a^2 - db^2|$ is fairly small, which increases its chance to be smooth. For example, with $b = 1$ one finds for $a$ near $b\sqrt{d} \doteq 2174.74$ the following smooth values of $a^2 - d$:

$$2156^2 - d = -2 \cdot 7 \cdot 11 \cdot 17 \cdot 31, \qquad 2178^2 - d = 2 \cdot 3 \cdot 5 \cdot 11 \cdot 43,$$
$$2162^2 - d = -2 \cdot 5^3 \cdot 13 \cdot 17, \qquad 2184^2 - d = 2 \cdot 3 \cdot 7 \cdot 31^2,$$
$$2175^2 - d = 3 \cdot 13 \cdot 29, \qquad 2187^2 - d = 3 \cdot 5^2 \cdot 23 \cdot 31.$$

For $b = 2, 3, 4$, one finds, restricting to values of $a$ that are coprime to $b$:

$$4329^2 - 2^2 d = -3 \cdot 5 \cdot 17^2 \cdot 41, \qquad 4399^2 - 2^2 d = 5^2 \cdot 13 \cdot 31 \cdot 43,$$
$$4341^2 - 2^2 d = -3 \cdot 5 \cdot 17^3, \qquad 6514^2 - 3^2 d = -2 \cdot 5^3 \cdot 13 \cdot 41,$$
$$4351^2 - 2^2 d = 5^2 \cdot 23^2, \qquad 6524^2 - 3^2 d = -2 \cdot 5 \cdot 7 \cdot 41,$$
$$4363^2 - 2^2 d = 13^2 \cdot 17 \cdot 41, \qquad 6538^2 - 3^2 d = 2 \cdot 7 \cdot 13 \cdot 23 \cdot 43,$$
$$4389^2 - 2^2 d = 3 \cdot 5 \cdot 7 \cdot 11 \cdot 13 \cdot 23, \qquad 8699^2 - 4^2 d = 17 \cdot 41.$$

The prime numbers occurring in these sixteen factorizations are the small prime factors 2, 3, 7, 11, 29 of $d$, as well as the prime numbers $p \leq 43$ with $\left(\frac{d}{p}\right) = 1$. It is only the latter primes that matter, and there are seven of them: 5, 13, 17, 23, 31, 41, and 43. It is important that the number of smooth expressions $a^2 - db^2$ exceeds the number of those primes, which is indeed the case: $16 > 7$. If one uses only the prime numbers up to 31 and the eight factorizations that do not contain 41 or 43, there is still a good margin: $8 > 5$. Thus, one decides to work with the "smoothness bound" 31.

The next step is to write down the prime *ideal* factorizations of the eight numbers $(a + b\sqrt{d})/(a - b\sqrt{d})$. Consider, for example, the case $a = 2162$, $b = 1$. Since $2162^2 - d$ contains a factor 13, the element $2162 + \sqrt{d}$ has a prime ideal factor of norm 13, and from $2162 \equiv 4 \mod 13$ one sees that this is the prime ideal $\mathfrak{p}_{13} = (13, 4 + \sqrt{d})$; it is the kernel of the ring homomorphism $\mathbb{Z}[\sqrt{d}] \to \mathbb{Z}/13\mathbb{Z}$ sending $\sqrt{d}$ to $-4 \pmod{13}$. The conjugate prime ideal $\mathfrak{q}_{13} = (13, 4 - \sqrt{d})$ then occurs in $2162 - \sqrt{d}$. Likewise, $2162 + \sqrt{d}$ is divisible by the cube of the prime ideal $\mathfrak{p}_5 = (5, 2 + \sqrt{d})$ and by $\mathfrak{p}_{17} = (17, 3 + \sqrt{d})$, and $2162 - \sqrt{d}$ by $\mathfrak{q}_5^3 \mathfrak{q}_{17}$, where $\mathfrak{q}_5 = (5, 2 - \sqrt{d})$ and $\mathfrak{q}_{17} = (17, 3 - \sqrt{d})$. Finally, $2162 + \sqrt{d}$ has the prime ideal factor $(2, \sqrt{d})$, but since 2 divides $d$, this prime ideal equals its own conjugate, so it cancels when one divides $2162 + \sqrt{d}$ by its conjugate. Altogether one finds the prime ideal factorization

$$((2162 + \sqrt{d})/(2162 - \sqrt{d})) = (\mathfrak{p}_5/\mathfrak{q}_5)^3 \cdot (\mathfrak{p}_{13}/\mathfrak{q}_{13}) \cdot (\mathfrak{p}_{17}/\mathfrak{q}_{17}).$$

As a second example, consider $a = 4351$, $b = 2$. We have $4351^2 - 2^2 d = 5^2 \cdot 23^2$, and from $4351/2 \equiv -2 \mod 5$ one sees that $4351 + 2\sqrt{d}$ belongs to $\mathfrak{q}_5$ rather than $\mathfrak{p}_5$. Similarly, $4351/2 \equiv 2 \mod 23$ implies that it belongs to $\mathfrak{p}_{23} = (23, 2 + \sqrt{d})$. Writing $\mathfrak{q}_{23} = (23, 2 - \sqrt{d})$, one obtains

$$((4351 + 2\sqrt{d})/(4351 - 2\sqrt{d})) = (\mathfrak{p}_5/\mathfrak{q}_5)^{-2} \cdot (\mathfrak{p}_{23}/\mathfrak{q}_{23})^2.$$

Doing this for all eight pairs $a$, $b$, one arrives at this table:

|  | 5 |  | 13 |  | 17 |  | 23 | 31 |
|---|---|---|---|---|---|---|---|---|
| $2156 + \sqrt{d}$ | 0 | 0 | −1 | 0 | −1 | 1 | 0 | 0 |
| $2162 + \sqrt{d}$ | 3 | 1 | 1 | 0 | 0 | 1 | 0 | −9 |
| $2175 + \sqrt{d}$ | 0 | 1 | 0 | 0 | 0 | 0 | −2 | 9 |
| $2184 + \sqrt{d}$ | 0 | 0 | 0 | 0 | 2 | 0 | 0 | 5 |
| $2187 + \sqrt{d}$ | 2 | 0 | 0 | 1 | −1 | −1 | 0 | 10 |
| $4341 + 2\sqrt{d}$ | −1 | 0 | 3 | 0 | 0 | 0 | 0 | 3 |
| $4351 + 2\sqrt{d}$ | −2 | 0 | 0 | 2 | 0 | 0 | 1 | −5 |
| $4389 + 2\sqrt{d}$ | 1 | 1 | 0 | −1 | 0 | −1 | 2 | 0 |

The first row lists the prime numbers $p$ we are using. The first column lists the eight expressions $\alpha = a + b\sqrt{d}$. In the $\alpha$-th row and the $p$-th column, one finds the exponent of $\mathfrak{p}_p/\mathfrak{q}_p$ in the prime ideal factorization of $\alpha/\alpha'$; here $\mathfrak{p}_p$, $\mathfrak{q}_p$ are as above, with $\mathfrak{p}_{31} = (31, 14 + \sqrt{d})$ and $\mathfrak{q}_{31} = (31, 14 - \sqrt{d})$. Thus, each $\alpha$ gives rise to an "exponent vector" that belongs to $\mathbb{Z}^5$.

The third step in the algorithm is finding linear relations with integer coefficients between the eight exponent vectors. The set of such relations forms a free abelian group of rank 3, which is 8 minus the rank of the $8 \times 5$ matrix formed by the eight vectors. A set of three independent generators for the relation group is given in the last three columns of the preceding table; in general, one can find such a set by applying techniques of linear algebra over $\mathbb{Z}$; see [Lenstra 2008, Section 14].

In the final step of the algorithm one inspects the relations one by one. Consider for example the first relation. It expresses that the sum of the exponent vectors corresponding to $2156 + \sqrt{d}$ and $2162 + \sqrt{d}$ equals the sum of the exponent vectors for $2187 + \sqrt{d}$ and $4389 + 2\sqrt{d}$. In other words, if we put

$$\alpha = \frac{(2156 + \sqrt{d}) \cdot (2162 + \sqrt{d})}{(2187 + \sqrt{d}) \cdot (4389 + 2\sqrt{d})},$$

then the element $\varepsilon = \alpha/\alpha'$ has all exponents in its prime ideal factorization equal to 0. This is the same as saying that $\varepsilon$ is a unit $x + y\sqrt{d}$ of the ring $\mathbb{Z}[\sqrt{d}]$; also, the norm $\varepsilon\varepsilon' = x^2 - dy^2$ of this unit equals $N(\alpha)/N(\alpha') = 1$, so we obtain an integral solution to Pell's equation $x^2 - dy^2 = 1$, except that it is uncertain whether $x$ and $y$ are positive. We can write $\varepsilon = \alpha/\alpha' = \alpha^2/N(\alpha)$, where the prime factorization of $N(\alpha)$ is available from the factorizations of $a^2 - db^2$ that we started with; one finds in this manner the following two power product representations of $\varepsilon$:

$$\varepsilon = \frac{(2156 + \sqrt{d}) \cdot (2162 + \sqrt{d}) \cdot (2187 - \sqrt{d}) \cdot (4389 - 2\sqrt{d})}{(2156 - \sqrt{d}) \cdot (2162 - \sqrt{d}) \cdot (2187 + \sqrt{d}) \cdot (4389 + 2\sqrt{d})}$$

$$= \frac{3^2 \cdot 23^2 \cdot (2156 + \sqrt{d})^2 \cdot (2162 + \sqrt{d})^2}{2^2 \cdot 17^2 \cdot (2187 + \sqrt{d})^2 \cdot (4389 + 2\sqrt{d})^2}.$$

In the second representation, $\varepsilon$ is visibly a square, or, equivalently, $N(\alpha)$ is a square; this is a bad sign, since it is certain to happen when $\varepsilon = 1$, in which case one has $\alpha \in \mathbb{Q}$, $N(\alpha) = \alpha^2$, $x = 1$, and $y = 0$. That is indeed what occurs here. (Likewise, it would have been a bad sign if $\varepsilon$ were visibly $-d$ times a square; this is certain to happen if $\varepsilon = -1$.) In the present case, the numbers are small enough that one can directly verify that $\varepsilon = 1$. For larger power products, one can decide whether $\varepsilon$ equals $\pm 1$ by computing $\log |\varepsilon|$ to a suitable precision and proving that the logarithm of a positive unit of $\mathbb{Z}[\sqrt{d}]$ cannot be close to 0 without being equal to 0.

Thus, the first relation disappointingly gives rise to a trivial solution to the Pell equation. The reader may check that the unit

$$\frac{29^2 \cdot (4351 + 2\sqrt{d})^2 \cdot (4389 + 2\sqrt{d})^4}{5^4 \cdot 7^2 \cdot 11^2 \cdot 23^4 \cdot (2175 + \sqrt{d})^4}$$

obtained from the second relation is also equal to 1. The third relation yields the unit

$$\eta = \frac{2^4 \cdot 5^{14} \cdot (2175 + \sqrt{d})^{18} \cdot (2184 + \sqrt{d})^{10} \cdot (2187 + \sqrt{d})^{20} \cdot (4341 + 2\sqrt{d})^6}{3^{27} \cdot 7^5 \cdot 29^9 \cdot 31^{20} \cdot (2162 + \sqrt{d})^{18} \cdot (4351 + 2\sqrt{d})^{10}}.$$

Since this is not visibly a square, we can be certain that it is not 1. Since it is positive, it is not $-1$ either. So $\eta$ is of the form $x + y\sqrt{d}$, where $x, y \in \mathbb{Z}$ satisfy $x^2 - dy^2 = 1$ and $y \neq 0$; thus, $|x|, |y|$ solve Pell's equation. From the power product, one computes the logarithm of the unit to be about 102.101583. This implies that $\eta > 1$, so that $\eta$ is the largest of the four numbers $\eta, \eta' = 1/\eta, -\eta$, and $-\eta'$; in other words, $x + y\sqrt{d}$ is the largest of the four numbers $\pm x \pm y\sqrt{d}$, which is equivalent to $x$ and $y$ being *positive*. In general one can achieve this by first replacing $\eta$ by $-\eta$ if $\eta$ is negative, and next by $\eta'$ if $\eta < 1$.

We conclude that the power product defining $\eta$ does represent a solution to Pell's equation. The next question is whether it is the *fundamental* solution. In the present case we can easily confirm this, since from Amthor's computation we know that $R_d \doteq 102.101583$, and the logarithm of any *non*fundamental solution would be at least $2 \cdot R_d$. Therefore, $\eta$ is equal to the solution $u$ found by Amthor, and it is indeed fundamental. In particular, the numbers $\log \eta \doteq 102.101583$ and $\log u \doteq 102.101583$ are exactly equal, not just to a precision of six decimals.

The power product representation we found for $\eta$ is a little more compact than the standard representation we gave for $u$. Indeed, its length, as defined earlier, is about 93.099810, as compared to $R_d \doteq 102.101583$ for $u$. The power product

$$\frac{(2175 + \sqrt{d})^{18}}{(2175 - \sqrt{d})^{18}} \cdot \frac{(2184 + \sqrt{d})^{10}}{(2184 - \sqrt{d})^{10}} \cdot \frac{(2187 + \sqrt{d})^{20}}{(2187 - \sqrt{d})^{20}}$$

$$\cdot \frac{(4341 + 2\sqrt{d})^6}{(4341 - 2\sqrt{d})^6} \cdot \frac{(2162 - \sqrt{d})^{18}}{(2162 + \sqrt{d})^{18}} \cdot \frac{(4351 - 2\sqrt{d})^{10}}{(4351 + 2\sqrt{d})^{10}},$$

which also represents $u$, has length about 125.337907.

## 8. Performance

The smooth numbers method for solving Pell's equation exemplified in the previous section can be extended to any value of $d$. There is unfortunately not much one can currently prove either about the run time or about the correctness of the method. Regarding the run time, however, one can make a reasonable conjecture.

For $x > e$, write
$$L(x) = \exp \sqrt{(\log x) \log \log x}.$$

The conjecture is that, for some positive real number $c_9$ and all $d > 2$, the smooth numbers method runs in time at most $L(d)^{c_9}$. This is, at a doubly logarithmic level, the exact average of $x^{c_9} = \exp(c_9 \log x)$ and $(\log x)^{c_9} = \exp(c_9 \log \log x)$; so conjecturally, the run time of the smooth numbers method is in a sense halfway between exponential time and polynomial time.

The main ingredient in the heuristic reasoning leading to the conjecture is the following theorem: for fixed positive real numbers $c$, $c'$, and $x \to \infty$, the probability for a random positive integer $\leq x^{c'}$ (drawn from a uniform distribution) to have all its prime factors $\leq L(x)^c$ equals $1/L(x)^{c'/(2c)+o(1)}$. This theorem [Pomerance 2008b; Granville 2008] explains the importance of the function $L$ in the analysis of algorithms depending on smooth numbers. Other ingredients of the heuristic run time analysis are the belief that the expressions $a^2 - db^2$ that one hopes to be smooth are so with the same probability as if they were random numbers, and the belief that the units produced by the algorithm have a substantial probability of being different from $\pm 1$. These beliefs appear to be borne out in practice.

Probably one can take $c_9 = 3/\sqrt{8} + \varepsilon$ in the conjecture just formulated, for any $\varepsilon > 0$ and all $d$ exceeding a bound depending on $\varepsilon$; one has $3/\sqrt{8} \doteq 1.06066$. One of the bottlenecks is the time spent on solving a large sparse linear system over $\mathbb{Z}$. If one is very optimistic about developing a better algorithm for doing this, it may be possible to achieve 1 instead of $3/\sqrt{8}$.

The smooth numbers method needs to be supplemented with an additional technique if one wishes to be reasonably confident that the unit it produces is the fundamental solution to Pell's equation. We forgo a discussion of this technique, since there is no satisfactory method for testing whether it achieves its purpose. More precisely, there is currently no known way of verifying in subexponential time that a solution to the Pell equation that is given by means of a power product is the fundamental one. The most promising technique for doing this employs the *analytic class number formula*, but its effectiveness depends on the truth of the *generalized Riemann hypothesis*. The latter hypothesis, abbreviated "GRH", asserts that there does not exist an algebraic number field whose associated zeta function has a complex zero with real part greater than $\frac{1}{2}$. The GRH can also be used to corroborate the heuristic run time analysis, albeit in a probabilistic setting. This leads to the following theorem.

THEOREM. *There is a probabilistic algorithm that for some positive real number $c_{10}$ has the following properties.*

(a) *Given any positive integer $d$ that is not a square, the algorithm computes a positive integer $R$ that differs by less than 1 from some positive integer multiple $m \cdot R_d$ of $R_d$.*
(b) *If the GRH is true, then (a) is valid with $m = 1$.*
(c) *If the GRH is true, then for each $d > 2$ the expected run time of the algorithm is at most $L(d)^{c_{10}}$.*

The algorithm referred to in the theorem is *probabilistic* in the sense that it employs a random number generator; every time the random number generator is called, it draws, in unit time, a random bit from the uniform distribution, independently of previously drawn bits. The run time and the output of a probabilistic algorithm depend not only on the input, but also on the random bits that are drawn; so given the input, they may be viewed as random variables. In the current case, the expectation of the run time for fixed $d$ is considered in part (c) of the theorem, and (a) and (b) describe what we know about the output. In particular, the algorithm always terminates, and if GRH is true, then it is guaranteed to compute an integer approximation to the regulator.

The theorem just stated represents the efforts of several people; up-to-date lists of references being provided in [Vollmer 2002; 2003]. According to the latter work, one may take $c_{10} = 3/\sqrt{8} + \varepsilon$ for any $\varepsilon > 0$ and all $d$ exceeding a bound depending on $\varepsilon$.

The last word on algorithms for solving Pell's equation has not been spoken yet. Very recently, a *quantum algorithm* was exhibited [Hallgren 2002; Schmidt and Vollmer 2005] that computes, in polynomial time, a power product representing the fundamental solution. This algorithm depends on infrastructure, but not on smooth numbers. For practical purposes, the smooth numbers method will remain preferable until quantum computers become available.

## Acknowledgments

This article was written while I held the 2000–2001 HP-MSRI Visiting Research Professorship. I thank Sean Hallgren, Mike Jacobson, Jr., and Ulrich Vollmer for answering my questions, and Bart de Smit for providing numerical assistance. A special word of thanks is due to Hugh Williams, whose version [2002] of the same story contains many details omitted in mine.

# References

[Archimedes 1999] Archimedes, *The cattle problem*, in English verse by S. J. P. Hillion and H. W. Lenstra jr., Mercator, Santpoort, 1999.

[Brent 1976] R. P. Brent, "Fast multiple-precision evaluation of elementary functions", *J. Assoc. Comput. Mach.* **23**:2 (1976), 242–251.

[Buchmann and Vollmer 2006] J. Buchmann and U. Vollmer, "A Terr algorithm for computations in the infrastructure of real-quadratic number fields", *J. Théor. Nombres Bordeaux* **18**:3 (2006), 559–572.

[Buchmann et al. 1995] J. Buchmann, C. Thiel, and H. Williams, "Short representation of quadratic integers", pp. 159–185 in *Computational algebra and number theory* (Sydney, 1992), edited by W. Bosma and A. van der Poorten, Math. Appl. **325**, Kluwer Acad. Publ., Dordrecht, 1995.

[Buhler and Wagon 2008] J. P. Buhler and S. Wagon, "Basic algorithms in number theory", pp. 25–68 in *Surveys in algorithmic number theory*, edited by J. P. Buhler and P. Stevenhagen, Math. Sci. Res. Inst. Publ. **44**, Cambridge University Press, New York, 2008.

[Dickson 1920] L. E. Dickson, *History of the theory of numbers*, vol. II, Diophantine analysis, Carnegie Institution, Washington, DC, 1920.

[Dijksterhuis 1956] E. J. Dijksterhuis (editor), *The Arenarius of Archimedes with glossary*, Textus minores **21**, Brill, Leiden, 1956.

[Euler 1770] L. Euler, *Vollständige Anleitung zur Algebra*, Zweyter Theil, Kays. Acad. der Wissenschaften, St. Petersburg, 1770. Reprinted in *Opera mathematica*, ser. I, vol. 1, Teubner, Leipzig, 1911; translated as *Elements of algebra*, Springer, New York, 1984.

[Fraser 1972] P. M. Fraser, *Ptolemaic Alexandria*, Oxford University Press, Oxford, 1972.

[Ge 1993] G. Ge, *Algorithms related to multiplicative representations*, Ph.D. thesis, University of California, Berkeley, 1993.

[Ge 1994] G. Ge, "Recognizing units in number fields", *Math. Comp.* **63**:207 (1994), 377–387.

[Granville 2008] A. Granville, "Smooth numbers: computational number theory and beyond", pp. 267–323 in *Surveys in algorithmic number theory*, edited by J. P. Buhler and P. Stevenhagen, Math. Sci. Res. Inst. Publ. **44**, Cambridge University Press, New York, 2008.

[Grosjean and De Meyer 1991] C. C. Grosjean and H. E. De Meyer, "A new contribution to the mathematical study of the cattle-problem of Archimedes", pp. 404–453 in *Constantin Carathéodory: an international tribute*, vol. I, edited by T. M. Rassias, World Sci. Publishing, Teaneck, NJ, 1991.

[Hallgren 2002] S. Hallgren, "Polynomial-time quantum algorithms for Pell's equation and the principal ideal problem", pp. 653–658 in *Proceedings of the Thirty-Fourth Annual ACM Symposium on Theory of Computing*, ACM, New York, 2002.

[Heiberg 1913] J. L. Heiberg (editor), *Archimedis opera omnia cum commentariis Eutocii*, vol. II, Teubner, Leipzig, 1913. Reprinted Stuttgart, 1972.

[Hua 1942] L.-K. Hua, "On the least solution of Pell's equation", *Bull. Amer. Math. Soc.* **48** (1942), 731–735. Reprinted as pp. 119–123 in *Selected papers*, Springer, New York, 1983.

[Konen 1901] H. Konen, *Geschichte der Gleichung $t^2 - Du^2 = 1$*, S. Hirzel, Leipzig, 1901.

[Krumbiegel and Amthor 1880] B. Krumbiegel and A. Amthor, "Das Problema Bovinum des Archimedes", *Historisch-literarische Abteilung der Zeitschrift für Mathematik und Physik* **25** (1880), 121–136, 153–171.

[Lagrange 1773] J.-L. de la Grange, "Solution d'un problème d'arithmétique", *Mélanges de philosophie et de math. de la Société Royale de Turin* **4** (1766–1769) (1773), 44–97. This paper was written and submitted for publication in 1768, and it appeared in 1773; see [Weil 1984, pp. 314–315]. Reprinted in Lagrange's *Œuvres*, vol. I, Gauthier-Villars, Paris, 1867, 669–731.

[Lenstra 1982] H. W. Lenstra, Jr., "On the calculation of regulators and class numbers of quadratic fields", pp. 123–150 in *Journées arithmétiques* (Exeter, 1980), edited by J. V. Armitage, London Math. Soc. Lecture Note Ser. **56**, Cambridge Univ. Press, Cambridge, 1982.

[Lenstra 2008] H. W. Lenstra, Jr., "Lattices", pp. 127–181 in *Surveys in algorithmic number theory*, edited by J. P. Buhler and P. Stevenhagen, Math. Sci. Res. Inst. Publ. **44**, Cambridge University Press, New York, 2008.

[Lessing 1773] G. E. Lessing, "Zur Griechischen Anthologie", pp. 419–446 in *Zur Geschichte und Litteratur: Aus den Schätzen der Herzoglichen Bibliothek zu Wolfenbüttel, Zweyter Beytrag*, Fürstl. Waysenhaus-Buchhandlung, Braunschweig, 1773. Appears on pp. 99–115 of his *Sämtliche Schriften*, edited by K. Lachmann, 3rd ed., v. 12, G. J. Göschen, Leipzig, 1897, reprinted by de Gruyter, Berlin 1968.

[Nelson 1980/81] H. L. Nelson, "A solution to Archimedes' cattle problem", *J. Recreational Math.* **13**:3 (1980/81), 162–176.

[Niven et al. 1991] I. Niven, H. S. Zuckerman, and H. L. Montgomery, *An introduction to the theory of numbers*, Wiley, New York, 1991.

[Pomerance 2008a] C. Pomerance, "Elementary thoughts on discrete logarithms", pp. 385–396 in *Surveys in algorithmic number theory*, edited by J. P. Buhler and P. Stevenhagen, Math. Sci. Res. Inst. Publ. **44**, Cambridge University Press, New York, 2008.

[Pomerance 2008b] C. Pomerance, "Smooth numbers and the quadratic sieve", pp. 69–81 in *Surveys in algorithmic number theory*, edited by J. P. Buhler and P.

Stevenhagen, Math. Sci. Res. Inst. Publ. **44**, Cambridge University Press, New York, 2008.

[Schmidt and Vollmer 2005] A. Schmidt and U. Vollmer, "Polynomial time quantum algorithm for the computation of the unit group of a number field", pp. 475–480 in *STOC'05: Proceedings of the 37th Annual ACM Symposium on Theory of Computing*, ACM, New York, 2005. Extended abstract; full version, Technische Univ. Darmstadt preprint TI-04-01; available at www.cdc.informatik.tu-darmstadt.de /reports/TR/TI-04-01.qalg_unit_group.pdf.

[Schönhage 1971] A. Schönhage, "Schnelle Berechnung von Kettenbruchentwicklungen", *Acta Inform.* **1** (1971), 139–144.

[Schoof 1982] R. J. Schoof, "Quadratic fields and factorization", pp. 235–286 in *Computational methods in number theory*, vol. II, edited by H. W. Lenstra, Jr. and R. Tijdeman, Math. Centre Tracts **155**, Math. Centrum, Amsterdam, 1982.

[Schoof 2008] R. J. Schoof, "Computing Arakelov class groups", pp. 447–495 in *Surveys in algorithmic number theory*, edited by J. P. Buhler and P. Stevenhagen, Math. Sci. Res. Inst. Publ. **44**, Cambridge University Press, New York, 2008.

[Shanks 1972] D. Shanks, "The infrastructure of a real quadratic field and its applications", pp. 217–224 in *Proceedings of the Number Theory Conference* (Boulder, CO, 1972), Univ. Colorado, Boulder, Colo., 1972.

[Stevenhagen 2008a] P. Stevenhagen, "The arithmetic of number rings", pp. 209–266 in *Surveys in algorithmic number theory*, edited by J. P. Buhler and P. Stevenhagen, Math. Sci. Res. Inst. Publ. **44**, Cambridge University Press, New York, 2008.

[Stevenhagen 2008b] P. Stevenhagen, "The number field sieve", pp. 83–100 in *Surveys in algorithmic number theory*, edited by J. P. Buhler and P. Stevenhagen, Math. Sci. Res. Inst. Publ. **44**, Cambridge University Press, New York, 2008.

[Vardi 1998] I. Vardi, "Archimedes' cattle problem", *Amer. Math. Monthly* **105**:4 (1998), 305–319.

[Vollmer 2002] U. Vollmer, "An accelerated Buchmann algorithm for regulator computation in real quadratic fields", pp. 148–162 in *Algorithmic Number Theory, ANTS-V*, edited by C. Fieker and D. R. Kohel, Lecture Notes in Computer Science **2369**, Springer, New York, 2002.

[Vollmer 2003] U. Vollmer, *Rigorously analyzed algorithms for the discrete logarithm problem in quadratic number fields*, Ph.D. thesis, Technische Univ. Darmstadt, Fachbereich Informatik, 2003. Available at http://elib.tu-darmstadt.de/diss/000494/.

[Weil 1984] A. Weil, *Number theory: an approach through history*, Birkhäuser, Boston, 1984.

[Whitford 1912] E. E. Whitford, *The Pell equation*, self-published, New York, 1912.

[Williams 2002] H. C. Williams, "Solving the Pell equation", pp. 397–435 in *Number theory for the millennium* (Urbana, IL, 2000), vol. 3, edited by M. A. Bennett et al., A K Peters, Natick, MA, 2002.

HENDRIK W. LENSTRA, JR.
MATHEMATISCH INSTITUUT
UNIVERSITEIT LEIDEN
POSTBUS 9512
2300 RA LEIDEN
THE NETHERLANDS
　hwl@math.leidenuniv.nl

# Basic algorithms in number theory

JOE BUHLER AND STAN WAGON

| | | | |
|---|---|---|---|
| **Algorithmic complexity** | 26 | Continued fractions | 45 |
| Multiplication | 26 | Rational approximation | 48 |
| Exponentiation | 28 | **Modular polynomial equations** | 51 |
| Euclid's algorithm | 30 | Cantor–Zassenhaus | 52 |
| Primality | 31 | Equations modulo $p^n$ | 53 |
| Quadratic nonresidues | 36 | Chinese remainder theorem | 57 |
| Factoring, Pollard $\rho$ | 36 | **Quadratic extensions** | 57 |
| Discrete logarithms | 38 | Cipolla | 58 |
| Modular square roots | 40 | Lucas–Lehmer | 59 |
| Diophantine equations | 42 | Units in quadratic fields | 61 |
| **Euclid's algorithm** | 42 | Smith–Cornacchia | 64 |
| Extended Euclid | 43 | **Bibliography** | 66 |

## 1. Introduction

Our subject combines the ancient charms of number theory with the modern fascination with algorithmic thinking. Newcomers to the field can appreciate this conjunction by studying the many elementary pearls in the subject. The aim here is to describe a few of these gems with the combined goals of providing background for subsequent articles in this volume, and luring the reader into pursuing full-length treatments of the subject, such as [Bach and Shallit 1996; Bressoud and Wagon 2000; Cohen 1993; Crandall and Pomerance 2005; Knuth 1981; von zur Gathen and Gerhard 2003; Shoup 2005].

Many details will be left to the reader, and we will assume that he or she knows (or can look up) basic facts from number theory, algebra, and elementary programming.

We tend to focus more on the mathematics and less on the sometimes fascinating algorithmic details. However, the subject is grounded in, and motivated by, examples; one can learn interesting and surprising things by actually implementing algorithms in number theory. Implementing almost any of the algorithms here in a modern programming language isn't too hard; we encourage budding number theorists to follow the venerable tradition of their predecessors: write programs and think carefully about the output.

## 2. Algorithmic complexity

Algorithms take input and produce output. The *complexity* of an algorithm $A$ is a function $C_A(n)$, defined to be the maximum, over all input $I$ of size at most $n$, of the cost of running $A$ on input $I$. Cost is often measured in terms of the number of "elemental operations" that the algorithm performs and is intended, in suitable contexts, to approximate the running time of actual computer programs implementing these algorithms.

A formalization of these ideas requires precise definitions for "algorithm," "input," "output," "cost," "elemental operation," and so on. We will give none.

Instead, we consider a series of number-theoretic algorithms and discuss their complexity from a fairly naive point of view. Fortunately, this informal and intuitive approach is usually sufficient for purposes of algorithmic number theory. More precise foundations can be found in many texts on theoretical computer science or algorithmic complexity such as [Garey and Johnson 1979; Hopcroft and Ullman 1979; Kozen 2006].

The first problem arises in elementary school.

PROBLEM 1. MULTIPLICATION: Given integers $x$ and $y$, find their product $xy$.

From the algorithmic perspective, the problem is woefully underspecified. We interpret it in the following natural (but by no means only possible) way. An algorithm that solves MULTIPLICATION takes two strings of symbols as input and writes a string of symbols as its output. The input strings are base $b$ representation of integers $x$ and $y$, where $b > 1$ is fixed, and in practice one might expect $b = 2, 10, 2^{32}$, or $2^{64}$. The algorithm follows a well-defined procedure in which the next step is determined by the current state of the computation; one might imagine a program written in an idealized form of your favorite computer language that has access to unlimited memory. Its output string represents the base-$b$ representation of the product $xy$.

The natural notion of the size of an integer $x$ is the total number of symbols (base-$b$ digits) in the input, perhaps augmented by a small constant to allow for delimiting the integer and specifying its sign. For definiteness, we define the base-$b$ size of $x$ to be

$$\operatorname{size}_b(x) := 1 + \lceil \log_b(1 + |x|) \rceil,$$

where $\log_b$ is the logarithm to the base $b$, and $\lceil u \rceil$ is the ceiling of $u$ — the smallest integer greater than or equal to $u$.

The size of an integer $x$ is $O(\log|x|)$, where $g(x) = O(f(x))$ is a shorthand statement saying that $g$ is in the class of functions such that there is a constant $C$ with $|g(x)| \leq C|f(x)|$ for sufficiently large $x$. Note that $O(\log_a x) = O(\log_b x)$ for $a, b > 1$. In particular, if we are interested in complexity only up to a constant factor the choice of $b > 1$ is irrelevant.

The usual elementary school multiplication algorithm uses $O(n^2)$ digit operations to multiply two input integers of size $n$. More precisely, if $x$ and $y$ have size $n$, then approximately $n^2$ digit-sized multiplications and $n$ additions of $n$-digit intermediate products are required. Since adding $n$-digit integers takes time $O(n)$, the overall complexity of multiplying $n$-digit integers using this algorithm is $O(n^2) + n \cdot O(n) = O(n^2)$. Notice that the $O$-notation gracefully summarizes an upper bound on the running time of the algorithm — the complexity $O(n^2)$ is independent of the base $b$, the precise details of measuring the size of an integer, the definition of size of two inputs (as the maximum of the two integer inputs, or the total of their sizes), and so on.

An algorithm $A$ is said to take *polynomial time* if its complexity $C_A(n)$ is $O(n^k)$ for some integer $k$. Although this is a flexible definition, with unclear relevance to computational practice, it has proved to be remarkably robust. In fact, it is sometimes reasonable to take "polynomial time" as synonymous with "efficient," and in any case the notion has proved useful in both theory and practice.

Once it is known that a problem can be solved in polynomial time, it is interesting to find the smallest possible exponent $k$. Several improvements to the $O(n^2)$ multiplication algorithm are known, and the current state of the art is a striking algorithm of Schönhage [1971] that takes time $O(n \log n \log \log n)$ to multiply two $n$-digit integers. This is sometimes written inexactly as $O(n^{1+\varepsilon})$, where $\varepsilon$ denotes an arbitrarily small positive number. Note that $O(n)$ is an obvious lower bound since the input has size $O(n)$ and in order to multiply integers it is necessary to read them. Algorithms that use the Schönhage algorithm, or related ones, are said to use *fast arithmetic*, and algorithms that are close to obvious lower bounds are sometimes said to be *asymptotically fast*. Many such algebraic and arithmetic algorithms are known (see [Bernstein 2008] for examples), and they are becoming increasingly important in computational practice.

The elemental operations above act on single digits (i.e., bits if $b = 2$), and the resulting notion of complexity is sometimes called *bit complexity*. In other contexts it might be more useful to assume that any arithmetic operation takes constant time on integers of arbitrary size; this might be appropriate, for example, if all integers are known to fit into a single computer word. When complexity of an algorithm is defined by counting arithmetic operations, the results is said to be the *arithmetic complexity* of the algorithm. In this model the cost of a single multiplication is $O(1)$, reminding us that complexity estimates depends dramatically on the underlying assumptions.

PROBLEM 2. EXPONENTIATION: Given $x$ and a nonnegative integer $n$, compute $x^n$.

Again, the problem is underspecified as it stands. We will assume that $x$ is an element of a set that has a well-defined operation (associative with an identity element) that is written multiplicatively; moreover, we will measure cost as the number of such operations required to compute $x^n$ on input $x$ and $n$. The size of the input will be taken to be the size of the integer $n$.

Although $x^{16}$ can be computed with 15 multiplications in an obvious way, it is faster to compute it by 4 squarings. More generally, the binary expansion $n = \sum a_i 2^i$, with $a_i \in \{0, 1\}$, implies that

$$x^n = x^{a_0} (x^2)^{a_1} (x^4)^{a_2} \cdots \qquad (2\text{-}1)$$

which suggests a clever way to interleave multiplications and squarings:

RIGHT-TO-LEFT EXPONENTIATION

    **Input:** $x$ as above, and a nonnegative integer $n$
    **Output:** $x^n$
    1. $y := 1$
    2. While $n > 0$
        if $n$ is odd, $y := xy$     // $a_i$ is 1
        $x := x^2$, $n := \lfloor n/2 \rfloor$
    3. Return $y$

Here ":=" denotes assignment of values to variables, "//" indicates a comment, "1" denotes the identity for the operation, and the floor $\lfloor u \rfloor$ is the largest integer less than or equal to $u$. The correctness of the algorithm is reasonably clear from equation (2-1) since $x^{2^k}$ is multiplied into $y$ if and only if the $k$th bit $a_k$ of the binary expansion of $n$ is nonzero. This can be proved more formally by showing by induction that at the beginning of Step 2, $X^N = x^n y$ holds, where $X$ and $N$ denote the initial values of the variables $x$ and $n$. When $n$ is 0 equation says that $X^N = y$, so that $y$ is the desired power.

The usual inductive definition of $\text{Exp}(x, n) := x^n$ gives an obvious recursive algorithm:

$$\text{Exp}(x, n) = \begin{cases} 1 & \text{if } n = 0, \\ \text{Exp}(x^2, n/2) & \text{if } n > 0 \text{ is even}, \\ x \cdot \text{Exp}(x^2, (n-1)/2) & \text{if } n \text{ is odd.} \end{cases} \qquad (2\text{-}2)$$

Experienced programmers often implement recursive versions of algorithms because of their elegance and obvious correctness, and when necessary convert them to equivalent, and perhaps faster, iterative (nonrecursive) algorithms. If this is done to the recursive program the result is to Right-to-Left algorithm above.

Curiously, if the inductive definition is replaced by the mathematically equivalent algorithm in which squaring follows the recursive calls,

$$\text{Exp}(x, n) = \begin{cases} 1 & \text{if } n = 0, \\ \text{Exp}(x, n/2)^2 & \text{if } n > 0 \text{ is even,} \\ x \cdot \text{Exp}(x, (n-1)/2)^2 & \text{if } n \text{ is odd,} \end{cases} \quad (2\text{-}3)$$

then the corresponding iterative algorithm is genuinely different.

LEFT-TO-RIGHT EXPONENTIATION
**Input:** $x$, a nonnegative integer $n$, a power of two $m = 2^a$ such that $m/2 \leq n < m$
**Output:** $x^n$
1. $y := 1$
2. While $m > 1$
    $m := \lfloor m/2 \rfloor$, $y := y^2$
    If $n \geq m$ then $y := xy$, $n := n - m$
3. Return $y$

Correctness follows inductively by proving that at the beginning of Step 2, $n < m$ and $y^m x^n = x^N$. In contrast to the earlier algorithm, this version consumes the bits $a_i$ in the binary expansion of $n$ starting with the leftmost (most significant) bit.

The complexity of any of the versions of this algorithm (collectively called EXP in the sequel) is $O(\log n)$ since the number of operations is bounded by $2 \cdot \text{size}_2(n)$. As will be seen, this remarkable efficiency has numerous applications in algorithmic number theory. Note that the naive idea of computing $x^n$ by repeatedly multiplying by $x$ takes time $O(n)$, which is exponential in the input size.

REMARK 1. In a specific but important practical case the left-to-right version of EXP is better than the right-to-left version. Suppose that our operation is "multiplying modulo $N$" and that $x$ is small relative to $N$. Then multiplying by the original $x$ is likely to take less time than modular multiplication by an arbitrary integer $X$ in the range $0 \leq X < N$. The left-to-right version preserves the original $x$ (though the squarings involve arbitrary integers), whereas the right-to-left version modifies $x$ and hence performs almost all operations on arbitrary elements. In other words, with a different computational model (bit complexity, with the specific underlying operation "multiply modulo $N$", and $x$ small) the left-to-right algorithm, either recursive or iterative, is significantly better than right-to-left exponentiation.

REMARK 2. If the underlying operation is multiplication of integers, the bit complexity of computing $x^n$ is exponential, since the output has size that is exponential in the input size $\log n$. Any algorithm will be inefficient, illustrating yet again the dependence on the underlying computational model.

This discussion of calculating powers barely scratches the surface of a great deal of theoretical and practical work. The overwhelming importance of exponentiation has led to many significant practical improvements; perhaps the most basic is to replace the base-2 expansion of $n$ with a base-$b$ expansion for $b$ a small power of 2. On the theoretical side, there are interesting results on finding the absolutely smallest possible number of operations required to compute $x^n$ [Knuth 1981].

PROBLEM 3. GCD: Given positive integers $a$ and $b$, find the largest integer that is a divisor of both $a$ and $b$.

The greatest common divisor (GCD) is denoted $\gcd(a,b)$. Perhaps the most famous number-theoretic algorithm of all is due to Euclid.

EUCLID'S ALGORITHM
   **Input:** Positive integers $a$ and $b$
   **Output:** $\gcd(a,b)$
   While $b > 0$
        $\{a,b\} := \{b, a \bmod b\}$
   Return $a$

Here $r = a \bmod b$ is the remainder when $a$ is divided by $b$, i.e., the unique integer $r$, $0 \leq r < b$, such that there is a $q$ with $a = qb + r$. The simultaneous assignment statement $\{a,b\} = \{b, a \bmod b\}$ could be implemented in a more prosaic programming language by something along the lines of the three statements *temp* $:= b$, $b = a \bmod b$, $a = $ *temp*. The correctness of the algorithm can be verified by showing that the GCD of $a$ and $b$ doesn't change at each step of the algorithm, and that when $a$ becomes divisible by $b$, then $b$ is the GCD.

Just as with multiplication, the remainder $a \bmod b$ can be found in time $\log^2(\max(a,b))$ with straightforward algorithms, and time $\log^{1+\varepsilon} \max(a,b)$ for any positive $\varepsilon$ if asymptotically fast algorithms are used. (Here $\log^r x$ is shorthand for $(\log x)^r$.) It isn't too hard to work out that $a > b$ after one step of the algorithm, and then the smallest number is halved in (at most) two steps of the algorithm. This means that the number of times that the loop is executed (the number of remainder operations) is bounded by $O(\log \max(a,b))$. Thus the algorithm has complexity $O(k^3)$ on $k$-bit input, or $O(k^{2+\varepsilon})$ if fast arithmetic is used. We will have much much more to say about Euclid's algorithm later.

If $\gcd(a, b) = 1$ then $a$ and $b$ are said to be *relatively prime*, or *coprime*.

PROBLEM 4. PRIMALITY: Given a positive integer $n > 1$, is $n$ is a prime?

The is an example of a *decision problem*, for which the output is either "yes" or "no."

Perhaps the most straightforward algorithm for PRIMALITY is the trial division method: for $d = 2, 3, \ldots$, test whether $n$ is divisible by $d$. If $n$ is divisible by some $d \le \sqrt{n}$ then $n$ is composite; if $n$ is not divisible by any $d \le \sqrt{n}$ then $n$ is prime. The time required is $O(\sqrt{n})$, which is an exponential function of the size of $n$, and this algorithm is impractical for even moderately large $n$.

Fermat's Little Theorem says that if $n$ is a prime and $a$ is coprime to $n$ then $a^{n-1} \equiv 1 \bmod n$, which implies that the order of $a$ in $(\mathbb{Z}/n\mathbb{Z})^*$) divides $n - 1$. The condition $a^{n-1} \equiv 1 \bmod n$ is easy to check, using EXP in $(\mathbb{Z}/n\mathbb{Z})^*$.

REMARK 3. We take the opportunity to remind the reader that two integers are congruent modulo $n$ if their difference is divisible by $n$, that $\mathbb{Z}/n\mathbb{Z}$ denotes the set of $n$ classes under this equivalence relation, and that $\mathbb{Z}/n\mathbb{Z}$ is a ring under the natural addition and multiplication operations induced from operations on the integers. Moreover, $(\mathbb{Z}/n\mathbb{Z})^*$ denotes the group of units in this ring, i.e., the set of congruence classes containing integers that are coprime to $n$, under the operation of multiplication. This group has $\phi(n)$ elements, where $\phi(n)$ is the Euler-phi function—the number of positive integers less than $n$ that are coprime to $n$. Finally, we let $a \bmod n$ denote the class of $\mathbb{Z}/n\mathbb{Z}$ that contains $a$. We will tolerate the conflict with earlier usage because the meaning can be disambiguated from context: if $a \bmod n$ is an integer then the remainder is intended, and if $a \bmod n$ lies in $\mathbb{Z}/n\mathbb{Z}$ then the congruence class is intended.

In the favorable circumstance in which the prime factorization of $n-1$ is known, Fermat's Little Theorem can be turned on its head to give a proof of primality.

THEOREM 5. If $a$ and $n$ are integers such that $a^{n-1} \equiv 1 \bmod n$, and $a^{(n-1)/q} \not\equiv 1 \bmod n$ for all prime divisors $q$ of $n - 1$, then $n$ is prime.

PROOF. As noted above, the congruence $a^{n-1} \equiv 1 \bmod n$ implies that the order of $a \bmod n$ in $(\mathbb{Z}/n\mathbb{Z})^*$, which we will denote by $\mathrm{ord}_n(a)$, is a divisor of $n - 1$. Any proper divisor of $n - 1$ is a divisor of $(n - 1)/q$ for some prime $q$. The second condition of the theorem says that $\mathrm{ord}(a)$ does not divide $(n - 1)/q$ for any $q$, and we conclude that $\mathrm{ord}(a) = n - 1$. Thus $(\mathbb{Z}/n\mathbb{Z})^*$ has $n - 1$ elements and $n$ is a prime, as claimed. □

A generalization of this theorem due to Pocklington says that only a partial factorization of $n - 1$ is necessary: if $m$ is a divisor of $n - 1$ with $m > \sqrt{n}$, then

$n$ is a prime if $a^{n-1} \equiv 1 \bmod n$, and $a^{(n-1)/q} \not\equiv 1 \bmod n$ for prime divisors $q$ of $m$. Loosely, this says that primality is easy to test if $n-1$ is half-factored.

Unfortunately, this does not give an efficient primality test: For large $n$ no algorithm is known that efficiently factors or half-factors $n-1$.

As a first try, observe that if $n$ is composite then $a^{n-1}$ should be, intuitively, a random integer modulo $n$. Thus we could choose several random $a$ and report "probable prime" when Fermat's congruence $a^{n-1} \equiv 1 \bmod n$ holds for all $a$, and "composite" if it fails for any one of them. Unfortunately, there are positive composite integers called Carmichael numbers (e.g., $n = 561 = 3 \cdot 11 \cdot 17$; see [Crandall and Pomerance 2005]) such that the congruence holds for all $a$ that are relatively prime to $n$.

As a second try, we "take the square root" of the Fermat congruence. To explain this it is convenient to review Legendre and Jacobi symbols.

An integer $a$ is a *quadratic residue* modulo an odd prime $p$ if it is a nonzero square, i.e., if there is an $x$ (not divisible by $p$) such that $x^2 \equiv a \bmod p$. Nonsquares are said to be *quadratic nonresidues*. This information is encoded in the *Legendre symbol*

$$\left(\frac{a}{p}\right) = \begin{cases} 0 & \text{if } p \text{ divides } a, \\ 1 & \text{if } x \text{ is a quadratic residue mod } p, \\ -1 & \text{if } x \text{ is a quadratic nonresidue mod } p. \end{cases}$$

Euler's Criterion gives an explicit congruence for the Legendre symbol

$$\left(\frac{a}{p}\right) \equiv a^{(p-1)/2} \bmod p, \qquad (2\text{-}4)$$

from which the symbol can be computed efficiently using EXP in $(\mathbb{Z}/p\mathbb{Z})^*$. The *Jacobi symbol* generalizes the Legendre symbol and is defined, for an integer $a$ and a positive odd positive integer $b$, by reverting to the Legendre symbol when $b$ is an odd prime, and enforcing multiplicativity in the denominator; if $b = \prod b^{e_p}$ is the factorization of $b$ into prime powers, then

$$\left(\frac{a}{b}\right) = \prod_p \left(\frac{a}{p}\right)^{e_p}.$$

The Jacobi symbol is multiplicative in both the numerator and denominator, depends only on $a \bmod b$, and obeys the famous law of quadratic reciprocity

$$\left(\frac{a}{b}\right)\left(\frac{b}{a}\right) = (-1)^{((a-1)(b-1))/4}, \quad a, b \text{ odd}.$$

Moreover, two "supplementary laws" hold:

$$\left(\frac{-1}{b}\right) = (-1)^{(b-1)/2}, \qquad \left(\frac{2}{b}\right) = (-1)^{(b^2-1)/8}.$$

This leads to a natural recursive algorithm, using the identities

$$\left(\frac{a}{b}\right) = \left(\frac{a \bmod b}{b}\right),$$
$$\left(\frac{a}{b}\right) = (-1)^{(b^2-1)/8}\left(\frac{a/2}{b}\right),$$
$$\left(\frac{a}{b}\right) = (-1)^{(a-1)(b-1)/4}\left(\frac{b}{a}\right),$$

applicable, respectively, when $a < 0$ or $a \geq b$, $a$ is even, or $a$ is odd. The crucial point is that this is an efficient algorithm even if the factorization of $b$ is unknown. An actual implementation of this resembles Euclid's algorithm augmented with some bookkeeping.

REMARK 4. Let $n = F_k := 2^{2^k} + 1$ be the $k$th Fermat number. The reader may enjoy using Theorem 5 and quadratic reciprocity to show that if $n = F_k$ and $3^{(n-1)/2} \equiv -1 \bmod n$ then $n$ is prime, and to prove that if $n$ is prime then the congruence holds.

Now we return to the problem of giving an efficient algorithm for primality. We use Euler's congruence

$$\left(\frac{a}{p}\right) \equiv a^{(p-1)/2} \bmod p$$

together with one of the seminal ideas of twentieth century computer science: it can be beneficial to allow algorithms to make random moves!

A *probabilistic (randomized)* algorithm extends our earlier implicit notion of an algorithm in that such an algorithm is allowed to flip a coin as needed and make its next move depending on the result. Typically a coin flip is deemed to cost one unit of running time. It is possible to model this more formally by thinking of the sequence of random bits $b \in \{0, 1\}$ as a second input string to the algorithm. Saying that an algorithm has a property with probability $p$ means that it has the property for a fraction $p$ of the possible auxiliary input strings (coin flips). For example, if the algorithm returns a correct answer for two-thirds of all possible input bit strings, then we say that the probability of correctness is $p = 2/3$.

It may seem worse than useless to allow algorithms to make random moves unrelated to the problem at hand but, as we will see, this additional capability can be surprisingly powerful.

REMARK 5. Probabilistic algorithms are ubiquitous. In circumstances in which it is necessary to emphasize that an algorithm is not probabilistic it will be referred to as a *deterministic* algorithm.

The following primality test is a famous example of a probabilistic algorithm.

## SOLOVAY–STRASSEN PRIMALITY TEST

**Input:** A positive integer $k$, an odd integer $n > 1$
**Output:** "Prime" or "Composite"
1. For $i = 1, 2, \ldots, k$
   Choose $a$ randomly from $\{1, 2, \ldots, n-1\}$
   If $\left(\frac{a}{n}\right) = 0$ or $\left(\frac{a}{n}\right) \neq a^{(n-1)/2} \bmod n$ then
      Output "Composite" and halt.
2. Output "Prime"

REMARK 6. The algorithm chooses a uniformly random element of a finite set. The reader may enjoy the puzzle of figuring out how to do this using coin flips, i.e., events of probability $1/2$. The goal is to simulate an arbitrary probability, and to do so in an efficient manner, e.g., to use about $\log n$ coin flips on average to choose a uniformly random element from an $n$-element set.

There are different flavors of probabilistic algorithms, according to whether the output (or some possible outputs) are true or just highly likely to be true, whether running times are bounds, or merely expected running times, etc. The following theorem clarifies this in the case of Solovay–Strassen, e.g., showing that the answer "Composite" is always true, the answer "Prime" is highly likely to be true, and the running time is polynomial in the input size and absolute value of the logarithm of the error probability.

THEOREM 6. *The Solovay–Strassen algorithm returns "Prime" if $n$ is prime, and returns "Composite" with probability at least $1 - 1/2^k$ if $n$ is composite. Its expected running time is bounded by a polynomial in $\log n$ and $k$.*

PROOF. (Sketch.) The GCD and exponentiation steps can be done it at most time $O(\log^3 n)$, so each iteration takes polynomial time. The only aspect of the running time that depends on the input bits is the choice of a random element of the $(\mathbb{Z}/n\mathbb{Z})^*$.

If $n$ is prime, then Euler's congruence implies that the test returns "Prime."

If $n$ is composite then the algorithm will return "Composite" unless each of $k$ random choices of $a$ lies in the subgroup

$$E(n) := \left\{a \in (\mathbb{Z}/n\mathbb{Z})^* : \left(\frac{a}{n}\right) \equiv a^{(n-1)/2} \bmod n\right\}$$

of the multiplicative group $(\mathbb{Z}/n\mathbb{Z})^*$, sometimes called the group of "Euler liars" for the composite number $n$. The coin flips done by the algorithm allow it to choose a uniformly random element of $(\mathbb{Z}/n\mathbb{Z})^*$. It can be shown [Bach and Shallit 1996] that $|E(n)| \leq \phi(n)/2$, so that the chance of a composite number passing all $k$ of these tests is at most $1 - 1/2^k$. □

The technique of repeating a test to quickly drive the probability of bad behavior to near zero is common in practical probabilistic algorithms. For example, the probability of failure in one round of Solovay–Strassen is at most $1/2$, and that means that the probability of failure after $k$ independent tests is at most $1/2^k$, so assuring the failure probability of at most $\varepsilon$ requires $\log(1/\varepsilon)$ tests.

The Jacobi symbol compositeness test inside the Solovay–Strassen algorithm can be improved. If $n$ is odd, $a$ is coprime to $n$, and $n-1 = 2^k r$, where $r$ is odd, let $A := a^r \bmod n$. If $n$ is prime then, by Fermat's Theorem, $A^{2^k} \equiv 1 \bmod n$. Since the only square roots of 1 in a field are $\pm 1$, we know that if $n$ is prime then either $A = 1$, or else $-1$ occurs in the sequence $A, A^2, A^4, \ldots, A^{2^k} = 1 \bmod n$. If this happens we say that $n$ is an strong probable prime to the base $a$.

If this does not happen, i.e., either $A \neq 1 \bmod n$ and $A^{2^j} \neq -1 \bmod n$ for $0 \leq j < k$, then $n$ is definitely composite, and we say that $a$ is a *witness* of $n$'s compositeness.

The obvious generalization of Solovay–Strassen using this test is to choose a series of $a$'s, and see whether one is a witness of $n$'s compositeness or $n$ is a strong probable prime to all of those bases. This is sometimes called the strong Fermat test.

Here are several nontrivial results related to the above ideas; for a discussion and details see [Crandall and Pomerance 2005] and [Damgård et al. 1993].

(i) If an odd $n$ is composite then at least $3/4$ of the $a$'s coprime to $n$ are witnesses to $n$ being composite.

(ii) If a famous conjecture in number theory is true — the Extended Riemann Hypothesis (ERH) — then any odd composite $n$ has a witness that is less than $2 \log^2 n$. In particular, if we assume the ERH then there is a polynomial-time primality test (this also applies to the Solovay–Strassen test).

(iii) Let $n$ be an odd integer chosen uniformly randomly between $2^k$ and $2^{k+1}$, and let $a$ be chosen randomly among integers between 1 and $n$ that are coprime to $n$. If $n$ is a strong probable prime to the base $a$ then the probability that $a$ is composite is less than $k^2/4^{\sqrt{k}-2}$. Thus for large $n$ a single probable prime test can give high confidence of primality.

These ideas can provide overwhelming evidence that a given integer is prime, but no rigorous proof. If we want absolute proof, the story changes. Many proof techniques have been considered over the years, and this continues to be an active area for practical work. The central theoretical question is whether primality can be proved in polynomial time, and this was settled dramatically in 2002 when Manindra Agrawal, Neeraj Kayal, and Nitin Saxena [Agrawal et al. 2004] discovered a deterministic polynomial-time algorithm. For a description of this, and algorithms currently used in practice, see [Schoof 2008b].

PROBLEM 7. QUADRATIC NONRESIDUES: Given an odd prime $p$, find a quadratic nonresidue modulo $p$.

This simple problem illustrates stark differences between theory and practice, and deterministic and probabilistic algorithms.

If $p$ is an odd prime there are $(p-1)/2$ quadratic residues and $(p-1)/2$ nonresidues mod $p$, so if $a$ mod $p$ is nonzero, then $a$ has a 50/50 chance of being a nonresidue. Moreover, quadratic residuosity/nonresiduosity is easy to establish by calculating a Legendre symbol. Thus there is an obvious probabilistic algorithm: repeatedly choose $a$ at random until a nonresidue is found. With overwhelming likelihood a nonresidue will be found quickly, and thus quadratic nonresidues can be found quickly with a probabilistic polynomial-time algorithm.

However, no deterministic polynomial-time algorithm is known. Nothing seems to be better than testing $a = 2, 3, \ldots$ until arriving at a nonresidue. There are heuristic grounds to think that there is a nonresidue $a = O(\log^{1+\varepsilon} p)$. However, the best that can be proved is that one finds a nonresidue $a = O(p^{1/4})$. The simple-minded (deterministic) algorithm for finding a nonresidue could take exponential time. It is known that if the Extended Riemann Hypothesis is true, then there is a nonresidue $a < 2\log^2 p = O(\log^2 p)$ [Bach 1990].

PROBLEM 8. FACTORING: Given a positive integer $n > 1$, find a proper divisor $m$ of $n$, i.e., a divisor $m$ such that $1 < m < n$.

Factoring appears to be much harder than primality, both in theory and practice. Trial division again gives an obvious algorithm that is impractical unless $n$ has a small divisor. The problem has fascinated mathematicians for centuries, and a vast menagerie of algorithms are known. Details of two of the most important current algorithms are described elsewhere in this volume [Poonen 2008; Stevenhagen 2008b]. Both algorithms require the use of sophisticated mathematical ideas: one requires the use of elliptic curves, and the other relies extensively on algebraic number theory.

We now describe a striking factoring algorithm, called the Pollard $\rho$ algorithm, due to John Pollard [1978].

Let $n$ be a composite integer that is not a prime power. (It is easy to check whether or not $n$ is a perfect power by taking sufficiently accurate $k$th roots for $2 \le k \le \log_2 n$.) Let

$$f : \mathbb{Z}/n\mathbb{Z} \to \mathbb{Z}/n\mathbb{Z}, \quad x \mapsto f(x) = x^2 + 1,$$

and let

$$f^k(x) = f(f(\cdots f(x) \cdots))$$

denote the $k$th iterate of $f$ applied to $x$. The Pollard $\rho$ algorithm is:

POLLARD $\rho$ FACTORING ALGORITHM
**Input:** A composite $n > 1$, not a prime power
**Output:** A nontrivial factor of $n$
1. Choose a random $x \in \{0, 1, \ldots, n-1\}$
2. For $i := 1, 2, \ldots$
   $g := \gcd(f^i(x), f^{2i}(x))$
   If $g = 1$, go to the next $i$
   If $1 < g < n$ then output $g$ and halt
   If $g = n$ then go back to Step 1 and choose another $x$

What could possibly be going on here? The birthday paradox in elementary probability theory says that a collection of 23 or more people is more likely than not to have two people with the same birthday. More generally, if elements are chosen randomly from a set of size $n$, with replacement, a repeat is likely after $O(\sqrt{n})$ choices [Knuth 1981].

Let $p$ be an (unknown) prime divisor of $n$. Evidence suggests that $f^k(x)$ mod $p$ is indistinguishable from a random sequence. Assuming that this is the case, the birthday paradox implies that the sequence repeats a value after $O(\sqrt{p})$ steps, after which time it cycles. Thus the sequence of values mod $p$ has the form

$$y_1, \ldots, y_m, y_{m+1}, \ldots, y_{m+k} = y_m, y_{m+k+1} = y_{m+1}, \ldots.$$

A little calculation shows that if $i$ is the smallest multiple of $k$ that exceeds $m$ then $y_i = y_{2i}$. (This elegant idea is usually referred to as Floyd's cycle-finding algorithm, and it enables the algorithm to use only a very small amount of memory.) In the context of the Pollard $\rho$ algorithm this means that $p$ divides $\gcd(f^i(x), f^{2i}(x))$. Thus the GCDs in the algorithm will sooner or later be divisible by $p$. One catastrophe that can happen is that, by some strange coincidence, all primes dividing $n$ happen to divide the GCD at the same time, but this is unlikely in practice.

The complexity of this algorithm is $O(n^{1/4})$ in the worst case that $n$ is the product of two roughly equal primes. The Pollard $\rho$ algorithm does have the virtue that it finds smaller factors sooner, so that it is reasonable to apply the algorithm to a composite number when there is no knowledge of the size of the factors. Many further details and optimizations of this charming algorithm and variants can be found in [Cohen 1993; Knuth 1981; Teske 2001].

In the last 25 years, a number of algorithms have been proposed for factoring in subexponential time, i.e., in time less than $O(n^\varepsilon)$ for all $\varepsilon > 0$. All are probabilistic, and most rely, for this favorable complexity estimate, on highly plausible, but as yet unproved, assumptions. The conjectured run times are sometimes said to be the *heuristic complexity* of these algorithms.

To describe the complexity of these algorithms, let

$$L_n[a;c] = \exp\bigl((c+o(1))(\log n)^a(\log\log n)^{1-a}\bigr).$$

where $o(1)$ denotes a term that goes to 0 as $n$ goes to infinity. This function interpolates between polynomial time, for $a=0$, and exponential time, for $a=1$, and its peculiar shape arises from formulas for the density of smooth numbers [Granville 2008] (a number is "$y$-smooth" if all of its prime factors are less than or equal to $y$).

The first of the two new algorithms described elsewhere in this volume is the Elliptic Curve Method (ECM), due to Hendrik Lenstra [1987]; see also [Poonen 2008]. The algorithm relies on the theory of elliptic curves over finite fields and some eminently believable heuristics concerning them; the (heuristic) complexity of this probabilistic algorithm is $L[1/2;1]$. The ECM shares the advantage, with trial division and the Pollard $\rho$ algorithm, that it finds smaller factors faster. A number of other factoring algorithms are known that have complexity $L[1/2;1]$. An algorithm based on class groups of quadratic fields [Lenstra and Pomerance 1992] is "rigorous" in the sense that its $L[1/2;1]$ complexity does not rely on any heuristics.

A few years after the ECM appeared, John Pollard described an algorithm for factoring numbers of the form $n = a \cdot b^k + c$ for $k$ large. Shortly afterwards, this was generalized to arbitrary positive integers $n$. This algorithm is a successor to the so-called Quadratic Sieve algorithm (QS), and is called the Number Field Sieve (NFS); its heuristic complexity is $L_n[1/3; 4/3^{2/3}] = L_n[1/3; 1.92\ldots]$. The unproved assertions on which this complexity is based are natural statements concerning the proportion of polynomials values that are smooth; these assertions seem true in practice but their proof seems beyond current techniques. The basic idea of this so-called *index-calculus* algorithm is to find smooth integers, and smooth elements of algebraic number fields, by sieving, and then solve a system of linear equations over the field with two elements. The speed of the algorithm in practice derives from the fact that the basic sieving operation can be implemented efficiently on modern computers. Details can be found in [Crandall and Pomerance 2005; Lenstra and Lenstra 1993; Stevenhagen 2008b].

PROBLEM 9. DISCRETE LOGARITHMS: Given an element $x$ of a finite cyclic group $G$ and a generator $g$ of that group, find a nonnegative integer $k$ such that $x = g^k$.

The problem is not precise until the representation of elements of, and the operation in, the finite group is made explicit. The difficulty of the discrete logarithm problem (DLOG) depends on the size of the group and on its specific representation. The additive group $G = \mathbb{Z}/n\mathbb{Z} = \{0, 1, 2, \ldots, n-1\}$ is cyclic, and the discrete logarithm problem in $G$ is easy with the usual representation of elements. The discrete logarithm problem for the multiplicative group $(\mathbb{Z}/p\mathbb{Z})^*$ is hard, and its difficulty is comparable to the difficulty of factoring an integer of the size of $p$. There is a Pollard $\rho$ algorithm for solving this DLOG problem [Pomerance 2008; Teske 2001], but the best currently known algorithm for this group is a sub-exponential index calculus algorithm closely related to the NFS factoring algorithm, with heuristic complexity $L_p[1/3; 4 \cdot 3^{2/3}]$; see [Pomerance 2008; Schirokauer 2008].

REMARK 7. Interest in the DLOG problem has been stimulated by cryptographic applications, most notably the famous Diffie–Hellman protocol: If $A$ and $B$ want to have a public conversation which ends with them sharing a secret that no onlooker could reasonably discover, then they start by (publicly) agreeing on a suitable cyclic group $G$ of order $n$ together with a generator $g$ of that group. Then $A$ chooses a random integer $a < n$, and communicates $g^a$ to $B$, in public. Similarly $B$ chooses a random integer $b$ and transmits $g^b$ to $A$ over a public channel. They can then each privately compute a joint secret

$$s = (g^a)^b = (g^b)^a.$$

The most obvious way for an eavesdropper to defeat this is to intercept $g^a$, solve the implicit DLOG problem to find $a$, to intercept $g^b$ and compute $s = (g^b)^a$. More generally, the eavesdropper can succeed if he or she can find $g^{ab}$ knowing $g$, $g^a$, and $g^b$. No efficient algorithm to do this is known if $G$ is a large cyclic subgroup of prime order in $(\mathbb{Z}/p\mathbb{Z})^*$ for some prime $p$, or for a large cyclic subgroup $G$ of prime order of the group of points $E(\mathbb{F}_p)$ on a suitable elliptic curve $E$ over a finite field $\mathbb{F}_p := \mathbb{Z}/p\mathbb{Z}$. The representation of a cyclic group in the group of points of an elliptic curve seems particularly opaque, and in this case no sub-exponential discrete logarithm algorithms are known at all. For details on these cases, see [Pomerance 2008; Poonen 2008; Schirokauer 2008].

The difficulty of the abstract discrete logarithm problem in a group is dominated by the difficulty of the problem in the largest cyclic subgroup of prime order [Pomerance 2008]. This explains why in the cases of the groups $(\mathbb{Z}/p\mathbb{Z})^*$ and $E(\mathbb{F}_p)$ above, one usually considers cyclic subgroups of prime order.

The largest cyclic subgroup of prime order in a cyclic group $G$ of order $2^n$ has order 2, and there is a particularly efficient DLOG algorithm for $G$. Let $g$ be a generator of $G$. There is a chain of subgroups

$$1 = G_0 \subset G_1 \subset G_2 \subset \cdots \subset G_{n-1} \subset G_n = G,$$

where $G_m$ has $2^m$ elements. To find the logarithm of $a \in G$ with respect to $g$, note that $a^{2^{n-1}}$ is of order 1 or 2. In the first case, $a$ lies in $G_{n-1}$. In the second case, $ag$ lies in the subgroup. In either case we then use recursion.

PROBLEM 10. SQUARE ROOTS MODULO A PRIME: Given an odd prime $p$ and a quadratic residue $a$, find an $x$ such that $x^2 \equiv a \mod p$.

We start by showing how to efficiently reduce this problem to QUADRATIC NONRESIDUES. This means that modular square roots can be found efficiently once a quadratic nonresidue is known. Let $a$ be a quadratic nonresidue. Write $p-1 = 2^t q$, where $q$ is odd. The element $a^q$ lies in the subgroup $G$ of $(\mathbb{Z}/p\mathbb{Z})^*$ of order $2^t$, and $b = g^q$ is a generator of that group. By the observation above on 2-power cyclic groups, the discrete logarithm in the cyclic 2-group $G$ is easy and we can efficiently find a $k$ such that

$$a^q = b^k.$$

Note that $k$ is even since $a^{(p-1)/2} \equiv 1 \mod p$. Simple calculation shows that $x = a^{(p-q)/2} b^{k/2}$ is a square root of $a$:

$$x^2 = a^{(p-q)} b^k = a^{(p-q)} a^q = a^p \equiv a \mod p.$$

The running time of the this procedure depends on the power of 2 dividing $p-1$.

The conclusion is that square roots modulo $p$ can be found efficiently if quadratic nonresidues can be found efficiently; using this idea, taking square roots modulo a prime is easy in practice, and easy in theory if probabilistic algorithms are allowed.

More recently, René Schoof [1985] discovered a deterministic polynomial-time algorithm for finding square roots of a fixed integer $a$ modulo a prime, using elliptic curves. In practice, this algorithm is not competitive with the probabilistic algorithms above since the exponent on log $p$ is large. However, Schoof's paper has had an enormous impact on the study of elliptic curves over finite fields [Poonen 2008], since it also pioneered new techniques for finding the order of the group $E(\mathbb{F}_p)$ for large $p$.

PROBLEM 11. MODULAR SQUARE ROOTS: Given an integer $n$ and an integer $a$, determine whether $a$ is a square modulo $n$, and find an $x$ such that $x^2 \equiv a \mod n$ if $x$ exists.

If the prime factorization of $n$ is known, then the algorithm above can be combined with Hensel's Lemma and the Chinese Remainder Theorem (both discussed later) to find a square root of $a$ modulo $n$. However, the working algorithmic number theorist is often confronted with integers $n$ whose prime factorization is unknown. In this case, no efficient modular square root algorithm is known.

In fact, more is true: MODULAR SQUARE ROOTS is equivalent, in a sense to be made precise shortly, to FACTORING. This says that in addition to the above fact — that factoring $n$ enables the square root problem to be solved easily —

it is also the case that an algorithm for MODULAR SQUARE ROOTS enables $n$ to be factored efficiently. To factor $n$, first check that it is odd and not a perfect power. Then choose a random $y$ and apply the hypothetical MODULAR SQUARE ROOTS algorithm to $a = y^2$ mod $n$ to get an $x$ such that

$$x^2 \equiv y^2 \bmod n.$$

Any common divisor of $x$ and $y$ is a factor of $n$, so assume that $x$ and $y$ have no common divisor larger than 1. If $p$ is a factor of $n$ then $p$ divides $x^2 - y^2 = (x-y)(x+y)$. In addition, it divides exactly one of the factors $x - y$ or $x + y$ (if it divides both, it would divide their sum $2x$ and their difference $2y$).

If $y$ is random, then any odd prime that divides $x^2 - y^2$ has a 50/50 chance of dividing $x + y$ or $x - y$, and any two primes should presumably (i.e., heuristically) have a 50/50 chance of dividing different factors. In that case, greatest common divisor $\gcd(x - y, n)$ will be a proper factor of $n$. If this fails, try again by choosing another random $y$. After $k$ choices, the probability that $n$ remains unfactored is $2^{-k}$.

Thus FACTORING and MODULAR SQUARE ROOTS are in practice equivalent in difficulty.

REMARK 8. Replacing square roots by $e$th roots for $e > 2$ leads to a problem closely related to the RSA cryptosystem, perhaps the most famous of all public-key cryptographic systems. Let $n = pq$ be the product of two large primes, and $G = (\mathbb{Z}/n\mathbb{Z})^*$. The RSA cryptosystem uses exponentiation as a mixing transformation on $G$. A message of arbitrary length is broken a sequence of elements of $G$, and each element is encrypted separately. If $e > 1$ is an integer relatively prime to $(p-1)(q-1)$ then the encryption map $E\colon G \to G$ defined by

$$E(x \bmod n) = x^e \bmod n$$

is a bijection that can be computed efficiently using EXP. The decryption mapping is $D(y) = y^d$ where the integer $d$ is defined by the congruence

$$ed \equiv 1 \bmod (p-1)(q-1).$$

The decryption exponent $d$ can be found efficiently, if the factorization of $n$ is known, using the Extended Euclidean Algorithm described in the next section. If $n$ and $e$ are chosen suitably, then finding $D(y)$ without knowing $p$ and $q$ requires the solution of a modular $e$th roots problem. It is plausible that breaking this system is no easier than factoring.

PROBLEM 12. BOUNDED MODULAR SQUARE ROOTS: Given an integer $n$ and integers $a$ and $b$, determine whether there is an integer $x$ such that $x < b$ and $x^2 \equiv a \bmod n$.

It may be peculiar to ask for square roots below a given bound, but this additional condition turns out to make the problem much more difficult than MODULAR SQUARE ROOTS, assuming the widely believed $P \neq NP$ conjecture. Specifically, BOUNDED MODULAR SQUARE ROOTS is known to be $NP$-complete.

Briefly, a decision problem is in $P$ if there is a polynomial-time algorithm for it. A decision problem is in $NP$ if there is an algorithm $f(x, y)$ in $P$ that takes an instance $x$ of a problem and a "certificate" $y$ as input, such that for every "yes" instance $x$ there is a certificate $y$ such that $f(x, y)$ returns "yes." BOUNDED MODULAR SQUARE ROOTS is certainly in $NP$, as are almost all of the algorithms that we have considered. Indeed, given an instance $(a, b, n)$, the integer $x$ is a certificate: the verifications that $x < b$ and $x^2 \equiv a$ mod $n$ can be done in polynomial time. Membership in $NP$ doesn't imply that finding the certificate is easy, but merely that it exists; if certificates could be found easily the problem would be in $P$.

Finally, a decision problem is $NP$-complete if it at least as hard as any other problem in $NP$, in the sense that any other problem in $NP$ can be reduced to it. The notion of reduction needs to be defined precisely, and involves ideas similar to the equivalence of FACTORING and MODULAR SQUARE ROOTS sketched above. For details see [Garey and Johnson 1979].

A discussion of the complexity of number-theoretic problems would be incomplete without mentioning a problem for which no algorithm whatsoever exists.

PROBLEM 13. DIOPHANTINE EQUATIONS: Given a polynomial $f(x_1, \ldots, x_n)$ with integral coefficients in $n$ variables, is there $n$-tuple of integers $x$ such that $f(x) = 0$?

A famous result of Yuri Matijasevic, building on work of Julia Robinson, Martin Davis, and Hilary Putnam shows that this is an undecidable problem [Matiyasevich 1993; Davis 1973]. Although the problem might be easy for a specific $f$, there is no algorithm (efficient or otherwise) that takes $f$ as input and always determines whether $f(x) = 0$ is solvable in integers.

## 3. Euclid's algorithm

Euclid's algorithm, given above, has been an extraordinarily fertile source of algorithmic ideas, and can be viewed as a special case of famous modern algorithms, including Hermite normal form algorithms, lattice basis reduction algorithms, and Gröbner basis algorithms.

The GCD of integers $a$ and $b$ is the unique nonnegative integer $d$ such that $d\mathbb{Z} = a\mathbb{Z} + b\mathbb{Z}$. Here $d\mathbb{Z}$ denotes the principal ideal in the ring of integers consisting of all multiples of $d$, and $a\mathbb{Z} + b\mathbb{Z} := \{ax + by : x, y \in \mathbb{Z}\}$. To prove that $d$ exists, it suffices to consider nonzero $a$ and $b$, in which case $d$ can be taken to be the least positive element of $a\mathbb{Z} + b\mathbb{Z}$. Indeed, if $z = ax + by$ is an arbitrary element of $a\mathbb{Z} + b\mathbb{Z}$ then $z = qd + r$ where $r = z \bmod d$ so that $0 \le r < d$. Since $r = z - qd$ is a nonnegative element of $a\mathbb{Z} + b\mathbb{Z}$, it follows that $r = 0$, and $z$ is a multiple of $d$ as claimed.

This shows that for nonzero $a$ and $b$ the GCD $d = \gcd(a, b)$ is the smallest positive integral linear combination of $a$ and $b$. In addition, $d$ is the largest integer that divides both $a$ and $b$ and it is the only positive divisor that is divisible by all other divisors. Moreover, it is described in terms of the prime factorizations of positive $a$ and $b$ by

$$\gcd(a,b) = \prod_p p^{\min(a_p, b_p)}, \quad \text{if } a = \prod_p p^{a_p}, \; b = \prod_p p^{b_p}.$$

The Greeks would have been more comfortable with the following geometric definition of the GCD: if $a > b > 0$ consider a rectangle of width $a$ and height $b$. Remove a $b$-by-$b$ square from one end of the rectangle. Continue removing maximal subsquares (squares with one side being a side of the rectangle) until the remaining rectangle is a square. The side length of the remaining square is the "common unit of measure" of $a$ and $b$, i.e., their GCD.

For instance, if $a = 73$ and $b = 31$, then the removal of two maximal subsquares leaves an 11-by-31 rectangle, the removal of two further maximal squares leaves an 11-by-9 rectangle, the removal of one maximal subsquare leaves a 2-by-9 rectangle, the removal of four maximal subsquares leaves a 2-by-1 rectangle, and the removal of one maximal subsquare leaves a unit square.

This procedure makes sense for arbitrary real numbers, leading to the notion of a continued fraction; this process terminates for an $a$-by-$b$ rectangle if and only if $a/b$ is a rational number.

**3.1. Extended Euclidean algorithm.** In many applications of the Euclidean algorithm the identity $a\mathbb{Z} + b\mathbb{Z} = d\mathbb{Z}$ needs to be made explicit; i.e., in addition to finding $d$, both $x$ and $y$ must be found. To do this it suffices to augment Euclid's algorithm.

EXTENDED EUCLIDEAN ALGORITHM (EEA)
  **Input:** Positive integers $a$ and $b$
  **Output:** $x, y, z$ where $z = \gcd(a, b)$ and $z = ax + by$
  $\{X, Y, Z\} := \{1, 0, a\}$
  $\{x, y, z\} := \{0, 1, b\}$
  While $z > 0$
      $q := \lfloor Z/z \rfloor$
      $\{X, Y, Z, x, y, z\} := \{x, y, z, X - qx, Y - qy, Z - qz\}$
  Return $X, Y, Z$

Simple algebra shows that at every step

$$aX + bY = Z, \qquad ax + by = z, \qquad (3\text{-}1)$$

so that triples $(X, Y, Z)$ encode elements of the ideal $a\mathbb{Z} + b\mathbb{Z}$. Moreover, $Z - qz = Z \bmod z$ so that the $Z$-values recapitulate the ordinary Euclidean algorithm.

The execution of the algorithm on input $a = 73$ and $b = 31$ is summarized in the following table, where each row describes the state of the computation just after $q$ is computed.

| $X$ | $Y$ | $Z$ | $x$ | $y$ | $z$ | $q$ |
|---|---|---|---|---|---|---|
| 1 | 0 | 73 | 0 | 1 | 31 | 2 |
| 0 | 1 | 31 | 1 | −2 | 11 | 2 |
| 1 | −2 | 11 | −2 | 5 | 9 | 1 |
| −2 | 5 | 9 | 3 | −7 | 2 | 4 |
| 3 | −7 | 2 | −14 | 33 | 1 | 2 |
| −14 | 33 | 1 | 31 | −73 | 0 | |

Thus $\gcd(73, 31) = 1$ and $(-14) \cdot 73 + 33 \cdot 31 = 1$.

The reader should notice that from the third row onwards $X$ and $Y$ have opposite signs, and their values increase in absolute value in successive rows. It isn't hard to verify that (for positive $a$ and $b$) this is always the case.

What is the running time of this algorithm? It can be shown that the number of arithmetic operations required is linear in the input size. The idea of the argument is as follows. The maximal number of iterations are required when the quotients $q$ is always 1. An induction argument shows that if this is the case and $n$ iterations of the loop are required then $a \geq F_{n+2}$, $b \geq F_{n+1}$, where $F_n$ denotes the $n$th Fibonacci number ($F_1 = F_2 = 1$, $F_{n+1} = F_n + F_{n-1}$). On the other hand, the Euclidean algorithm starting with $a = F_{n+1}$, $b = F_n$ takes $n$ steps. Using $F_n = O(\phi^n)$, $\phi = (1 + \sqrt{5})/2$, it follows that $n = O(\log a)$, as desired. A careful accounting of the bit complexity, using the fact that the size of the integers decreases during the course of the algorithm, shows that the bit complexity is $O(\log^2 a)$.

On occasion it is useful to formulate the Euclidean algorithm in terms of 2-by-2 matrices. Associate the matrix

$$M := \begin{bmatrix} X & Y \\ x & y \end{bmatrix}.$$

to a row $X, Y, Z, x, y, z, q$. Then by (3-1)

$$M \begin{bmatrix} a \\ b \end{bmatrix} = \begin{bmatrix} Z \\ z \end{bmatrix}.$$

The matrix $M'$ of the next row is

$$M' = \begin{bmatrix} 0 & 1 \\ 1 & -q \end{bmatrix} M.$$

Iterating this in the specific calculation above gives

$$\begin{bmatrix} 1 \\ 0 \end{bmatrix} = \begin{bmatrix} 0 & 1 \\ 1 & -2 \end{bmatrix} \begin{bmatrix} 0 & 1 \\ 1 & -4 \end{bmatrix} \begin{bmatrix} 0 & 1 \\ 1 & -1 \end{bmatrix} \begin{bmatrix} 0 & 1 \\ 1 & -2 \end{bmatrix} \begin{bmatrix} 0 & 1 \\ 1 & -2 \end{bmatrix} \begin{bmatrix} 73 \\ 31 \end{bmatrix}, \qquad (3\text{-}2)$$

so that the Euclidean algorithm can be interpreted as a sequence of 2-by-2 matrix multiplications by matrices of a special form.

All of the ideas in the EEA apply to any commutative ring in which division with "smaller" remainder exists. For instance, the ring $F[x]$ of polynomials over a field $F$ admits a division algorithm for which the remainder polynomial has smaller degree than the divisor. The ring $\mathbb{Z}[i] := \{a + bi : a, b \in \mathbb{Z}\}$ of Gaussian integers inside the complex numbers has this property if size is measured as the absolute value, and we take the quotient in the ring to be the result of taking the quotient in the complex numbers and rounding to the nearest Gaussian integer.

One important application of the EEA is to find inverses in $(\mathbb{Z}/n\mathbb{Z})^*$. If $\gcd(a, n) = 1$ the EEA can be used to find $x, y$ such that $ax + ny = 1$, which implies that $ax \equiv 1 \bmod n$ and $(a \bmod n)^{-1} = x \bmod n$.

**3.2. Continued fractions.** The *partial quotients* of the continued fraction expansion of a real number $\alpha$ are the terms of the (finite or infinite) sequence $a_0, a_1, \ldots$ defined as follows.

CONTINUED FRACTION OF A REAL NUMBER
  **Input:** A real number $\alpha$
  **Output:** A sequence $a_i$ of integers
  $\alpha_0 := \alpha$
  For $i := 0, 1, \ldots$
      Output $a_i := \lfloor \alpha_i \rfloor$
      Halt if $\alpha_i = a_i$
      $\alpha_{i+1} := 1/(\alpha_i - a_i)$

EXAMPLE 1. If $\alpha = 73/31$ the partial quotients are $2, 2, 1, 4, 2$. If $\alpha = \sqrt{11}$ the partial quotients are ultimately periodic: $3, 3, 6, 3, 6, 3, 6, \ldots$.

The reader should verify that this procedure terminates if and only if $\alpha$ is a rational number.

Although written as if it was an algorithm, this is more of a mathematical construct than an algorithm for several reasons: (a) it may fail to terminate, (b) real numbers cannot be represented in finite terms suitable for a computer, and (c) testing real numbers for equality is (algorithmically) problematic.

The relationship between $\alpha_i$ and $\alpha_{i+1}$ can be written as

$$\alpha_i = a_i + \frac{1}{\alpha_{i+1}}. \tag{3-3}$$

Iterating this for $\alpha = 73/31$ gives the finite continued fraction

$$\frac{73}{31} = 2 + \cfrac{1}{2 + \cfrac{1}{1 + \cfrac{1}{4 + \cfrac{1}{2}}}} = [2; 2, 1, 4, 2],$$

where $[a_0; a_1, \ldots, a_n]$ denotes the value of a finite continued fraction with partial quotients $a_i$. The values of the successive partial continued fractions of a real number are called *convergents* of the continued fraction; e.g., the convergents of the continued fraction for $73/31$ are

$$[2] = 2, \quad [2; 2] = \frac{9}{2}, \quad [2; 2, 1] = \frac{11}{9}, \quad [2; 2, 1, 4] = \frac{31}{11}, \quad [2; 2, 1, 4, 2] = \frac{73}{31}.$$

Periodic continued fractions represent quadratic irrational numbers $a + b\sqrt{d}$, where $a, b, d$ are rational numbers (and $d$ is a nonsquare). For instance, if $\alpha$ denotes the purely periodic continued fraction $[3; 6, 3, 6, 3, 6, \ldots]$ then

$$\alpha = 3 + \cfrac{1}{6 + \cfrac{1}{\alpha}}.$$

Clearing fractions gives $2\alpha^2 - 6\alpha - 1 = 0$, so that $\alpha > 0$ implies that $\alpha = (3 + \sqrt{11})/2$.

PROPOSITION 14. Let $a_0, a_1, \ldots$ be a sequence of real numbers with $a_i > 0$ for $i > 0$. Define a real number $[a_0; a_1, \ldots, a_n]$ recursively by

$$[a_0;] = a_0, \qquad [a_0; a_1, \ldots, a_n, a_{n+1}] = [a_0; a_1, \ldots, a_n + 1/a_{n+1}].$$

Finally, define sequences $x_i, y_i$ recursively by

$$\begin{aligned} x_{-1} &= 0, & x_0 &= 1, & x_{n+1} &= a_{n+1} x_n + x_{n-1}, & n &\geq 0, \\ y_{-1} &= 1, & y_0 &= a_0, & y_{n+1} &= a_{n+1} y_n + y_{n-1}, & n &\geq 0. \end{aligned} \tag{3-4}$$

Then for nonnegative $n$,

$$y_n x_{n-1} - y_{n-1} x_n = (-1)^{n-1}, \qquad y_n/x_n = [a_0; a_1, \ldots, a_n].$$

Moreover, if the $a_i$ are the partial quotients of the continued fraction of a real number $\alpha$ then the integers $x_n$ and $y_n$ are coprime, and

$$\alpha = \frac{y_n \alpha_{n+1} + y_{n-1}}{x_n \alpha_{n+1} + x_{n-1}}. \tag{3-5}$$

REMARK. Several notations are commonly used for convergents; the choice here is dictated by a geometric interpretation to be given shortly.

PROOF. The coprimality of $x_n$ and $y_n$ follows once $x_n y_{n-1} - x_{n-1} y_n = (-1)^n$ is proved. This and all of the other assertions follow from straightforward induction arguments. For instance, assuming that $y_n/x_n = [a_0; a_1, \ldots, a_n]$ for a given $n$ and arbitrary $a_i$ we have

$$\frac{y_{n+1}}{x_{n+1}} = \frac{a_{n+1} y_n + y_{n-1}}{a_{n+1} x_n + x_{n-1}} = \frac{a_{n+1}(a_n y_{n-1} + y_{n-2}) + y_{n-1}}{a_{n+1}(a_n x_{n-1} + x_{n-2}) + x_{n-1}}$$

$$= \frac{(a_n + 1/a_{n+1}) y_{n-1} + y_{n-2}}{(a_n + 1/a_{n+1}) x_{n-1} + x_{n-2}}$$

$$= [a_0; a_1, \ldots, a_n + 1/a_{n+1}] = [a_0; a_1, \ldots, a_n, a_{n+1}]. \qquad \square$$

As with the Euclidean algorithm, it is sometimes convenient to formulate continued fractions in terms of 2-by-2 matrices, e.g., defining

$$M_n := \begin{bmatrix} a_n & 1 \\ 1 & 0 \end{bmatrix}, \qquad P_n := \begin{bmatrix} y_n & y_{n-1} \\ x_n & x_{n-1} \end{bmatrix} \qquad (3\text{-}6)$$

and observing that (3-4) implies that

$$P_n = M_0 M_1 \cdots M_n. \qquad (3\text{-}7)$$

A careful look at the convergents of the continued fraction of $73/31$ reveals that they recapitulate the Euclidean algorithm! This is easy to verify using

$$\begin{bmatrix} a & 1 \\ 1 & 0 \end{bmatrix}^{-1} = \begin{bmatrix} 0 & 1 \\ 1 & -a \end{bmatrix}.$$

Indeed, multiplying the relation $\begin{bmatrix} 73 \\ 31 \end{bmatrix} = P_4 \begin{bmatrix} 1 \\ 0 \end{bmatrix}$ repeatedly on the left by the $M_k^{-1}$ gives the earlier EEA formula (3-2); moreover, this process can be reversed to show that the EEA yields a continued fraction.

Since $\det(M_k) = -1$ and $\det(P_n) = y_n x_{n-1} - x_n y_{n-1}$ the product formula (3-7) gives another proof of the first statement of the proposition:

$$y_n x_{n-1} - y_{n-1} x_n = (-1)^{n+1}. \qquad (3\text{-}8)$$

**3.3. Rational approximation.** Let $\alpha, \alpha_n, a_n, y_n, x_n$ be as in the previous section. We wish to quantify the sense in which $y_n/x_n$ is a good approximation to $\alpha$. Dividing (3-8) by $x_n x_{n-1}$ gives

$$\frac{y_n}{x_n} - \frac{y_{n-1}}{x_{n-1}} = \frac{(-1)^{n+1}}{x_n x_{n-1}}.$$

Iterating this identity and using $y_0/x_0 = a_0$ gives the useful formula

$$\frac{y_n}{x_n} = a_0 + \frac{1}{x_0 x_1} - \frac{1}{x_1 x_2} + \cdots + (-1)^n \frac{1}{x_n x_{n+1}}. \tag{3-9}$$

Since the $x_n = a_n x_{n-1} + x_{n-2}$ are strictly increasing it follows that the sequence of convergents $y_n/x_n$ converges.

THEOREM 15. *The sequence $y_n/x_n$ converges to $\alpha$. For even $n$ the convergents increase and approach $\alpha$ from below; for odd $n$ the convergents decrease and approach $\alpha$ from above. Both $y_n/x_n - \alpha$ and $y_n - x_n \alpha$ alternate in sign, and decrease in absolute value. Moreover,*

$$\frac{1}{x_{n+2}} < |y_n - \alpha x_n| < \frac{1}{x_{n+1}}.$$

PROOF. From (3-5),

$$y_n - \alpha x_n = y_n - \frac{x_n (y_n \alpha_{n+1} + y_{n-1})}{x_n \alpha_{n+1} + x_{n-1}} = \frac{x_{n-1} y_n - x_n y_{n-1}}{x_n \alpha_{n+1} + x_{n-1}} = \frac{(-1)^{n+1}}{x_n \alpha_{n+1} + x_{n-1}}.$$

From $a_{n+1} = \lfloor \alpha_{n+1} \rfloor < \alpha_{n+1} < a_{n+1} + 1$ we get

$$x_{n+1} = x_n a_{n+1} + x_{n-1}$$
$$< x_n \alpha_{n+1} + x_{n-1} < x_n(a_{n+1} + 1) + x_{n-1} = x_{n+1} + x_n \leq x_{n+2},$$

and the desired inequalities in the theorem follow by taking reciprocals. It follows immediately that $\lim y_n/x_n = \alpha$ and the other statements follow from (3-9) and basic facts about alternating series. □

COROLLARY 16. *Any convergent $y_n/x_n$ in the continued fraction for $\alpha$ satisfies $|y_n/x_n - \alpha| < 1/x_n^2$.*

It is convenient to formalize the notion of a good approximation to $\alpha$ by rational numbers. If $(x, y)$ is a point in the $XY$-plane, then the distance from $(x, y)$ to the line $Y = \alpha X$ along the vertical line $X = x$ is $|y - \alpha x|$. Say that $(x, y)$, $x > 0$, is a *best approximator* to $\alpha$ if $x$ and $y$ are coprime and $(x, y)$ has the smallest vertical distance to $Y = \alpha X$ among all integer points with denominator at most $x$, i.e.,

$$|y - \alpha x| < |v - u\alpha| \quad \text{for all integers } u, v \text{ with } 0 < u < x.$$

The following theorem says that being a best approximator is equivalent to being a convergent, and it gives an explicit inequality for a rational number $y/x$ that is equivalent to being a convergent.

THEOREM 17. *Let $\alpha$ be an irrational real number and $x, y$ a pair of coprime integers with $x > 0$. Then the following are equivalent:*

(a) *$y/x$ is a convergent to $\alpha$.*

(b) If $x'$ is a multiplicative inverse of $y$ mod $x$ (in the sense that $yx' \equiv 1$ mod $x$ and $1 \le x' < x$), then

$$\frac{-1}{x(2x-x')} < \frac{y}{x} - \alpha < \frac{1}{x(x+x')}. \qquad (3\text{-}10)$$

(c) $y/x$ is a best approximator to $\alpha$.

COROLLARY 18. If $y/x$ is a rational number such that $\left|\frac{y}{x} - \alpha\right| < \frac{1}{2x^2}$, then $y/x$ is a convergent to $\alpha$.

PROOF OF THE COROLLARY. Without loss of generality, $x$ and $y$ are coprime. If $x'$ is as in the theorem, then $1 \le x' < x$, giving $2x - x' < 2x$ and $x + x' < 2x$. Multiplying by $x$ and taking reciprocals gives

$$\frac{-1}{x(2x-x')} < \frac{-1}{2x^2} < \frac{y}{x} - \alpha < \frac{1}{2x^2} < \frac{1}{x(x+x')},$$

and the corollary follows immediately from the theorem. $\square$

The convergents in the continued fraction of $\alpha$ determine lattice points (i.e., points with integer coordinates) in the plane, and they are unusually close to the line $Y = \alpha X$. The convergents alternate back and forth over the line, and each point $(x, y)$ is closer to the line than all points with smaller $X$. Since any best approximator $y_n/x_n$ comes from a convergent, it follows that the open parallelogram $\{(x, y) : 0 < x < x_{n+1}, \ |y - \alpha x| < |y_n - \alpha x_n|\}$ contains no lattice points other than the origin.

If you are standing at the origin of the plane looking in the direction of $Y = \alpha X$, then the lattice points $(x, y) \in \mathbb{Z}^2$ that are closest to the line occur alternately on both sides of the line, and become very close to the line. It is difficult to illustrate this graphically because the convergents approximate so well that they quickly appear to lie on the line. In Figure 1 an attempt has been made to convey this by greatly distorting the scale on the $y$-axis; specifically, a point at distance $d$ from the line $Y = \alpha X$ is displayed at a distance that is proportional to $d^{2/5}$. In that figure, $\alpha$ is the golden ratio $\phi = [1; 1, 1, 1, \ldots]$, and the coordinates of the convergents $(F_n, F_{n+1})$ are consecutive Fibonacci numbers. (The figure suggests a fact, brought to our attention by Bill Casselman, which the reader may enjoy proving: The even convergents $(x_n, y_n)$ are the vertices on the convex hull of the integer lattice points that lie below the line $Y = \alpha X$, and the odd convergents are the convex hull of the integer lattice points above the line.)

Before proving Theorem 17, it is convenient to present a lemma.

LEMMA 19. Let $x, x', y, y'$ be integers with $yx' - y'x = \pm 1$, and let $\alpha$ lie in between $y/x$ and $y'/x'$. Then there are no lattice points in the interior of the parallelogram bounded by the lines $X = x$, $X = x'$, the line through $(x, y)$ with slope $\alpha$, and the line through $(x', y')$ with slope $\alpha$.

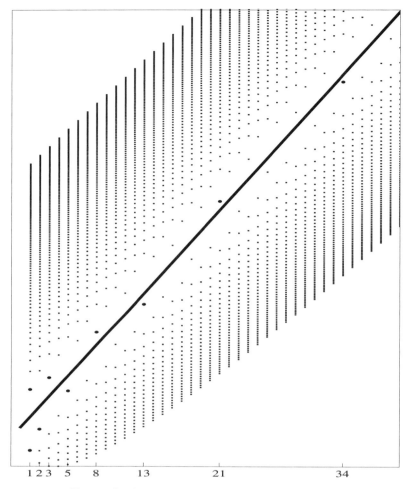

**Figure 1.** Approximations to the golden ratio $\phi$.

PROOF OF THE LEMMA. To fix the ideas, assume that $0 < x' < x$ and $y' - \alpha x' < 0 < y - \alpha x$ (any other case is similar). Let $(u, v)$ be a lattice point with $x' < u < x$. Then $(u, v)$ is in the stated parallelogram if and only if

$$y' - \alpha x < v - \alpha u < y - \alpha x. \tag{3-11}$$

The condition $yx' - y'x = \pm 1$ implies that there are integers $r, s$ such that

$$u = rx' + sx, \qquad v = ry' + sy.$$

The first equation implies that $r$ and $s$ are nonzero and have opposite sign; subtracting $\alpha$ times the first equation from the second gives

$$v - \alpha u = r(y' - \alpha x') + s(y - \alpha x).$$

Since $y - \alpha x$ and $y' - \alpha x'$ have opposite sign, the terms on the right hand side have the same sign. This implies that $|v - \alpha u| \geq \max(|y - \alpha x|, |y' - \alpha x'|)$ which contradicts (3-11), showing that $(u, v)$ does not lie in the parallelogram and proving the Lemma. □

PROOF OF THEOREM 17. (a) ⇒ (b), i.e., convergents satisfy the inequalities (3-10): Assume that $y/x = y_n/x_n$ is a convergent, and that $y_{n-1}/x_{n-1}$ is the preceding convergent. If $n$ is odd then $yx_{n-1} - y_{n-1}x = 1$ and $x' = x_{n-1}$. From Theorem 15,

$$0 < y - \alpha x = \frac{1}{x_n \alpha_{n+1} + x_{n-1}} < \frac{1}{x_n + x_{n-1}}$$

and the inequalities (3-10)

$$\frac{-1}{x(2x-x')} < 0 < \frac{y}{x} - \alpha < \frac{1}{x(x+x')}$$

follow. The case for even $n$ is similar except that $y/x - \alpha$ is negative, $yx_{n-1} - y_{n-1}x = -1$ requires us to take $x' = x - y$, and the left inequality in (b) is the nontrivial one.

(b) ⇒ (c), the inequalities (3-10) imply that $y/x$ is a best approximator: Assume that $x > 0$, $x$ and $y$ are coprime, and $y/x$ satisfies the inequalities. Let $(u, v)$ be a supposedly better approximator than $(x, y)$ so that $0 < u \leq x$. There are now two cases. First, suppose that $y - \alpha x > 0$. Choose $x'$ and $y'$ such that $yx' - xy' = 1$. Subtracting $y/x - y'/x' = 1/xx'$ from $y/x - \alpha < 1/(x(x+x'))$ gives

$$\frac{y'}{x'} - \alpha < \frac{1}{x(x + x')} - \frac{1}{xx'} = \frac{-1}{(x + x')x}, \qquad (3\text{-}12)$$

showing that $y' - x'\alpha$ is negative. The (proof of the) Lemma now shows that an inequality of the form (3-11) is impossible, which implies that $(x, y)$ is a best approximator.

The case $y - \alpha x < 0$ is similar and will be left to the reader.

(c) ⇒ (a): We show that if $(u, v)$ isn't a convergent then $v/u$ isn't a best approximator. If $(u, v)$ isn't a convergent there is an $n$ such that $x' = x_{n-1} < u \leq x = x_n$. Setting $y' = y_{n-1}$, $y = y_n$, the Lemma now applies, and $(u, v)$ isn't a best approximator unless $u = x_n$ in which case $v/u$ isn't a best approximator unless $v/u = y_n/x_n$. □

## 4. Modular polynomial equations

The goal of this section is to consider the problem of solving polynomial congruences $f(x) \equiv 0 \bmod n$, i.e., to finding roots of polynomials in $(\mathbb{Z}/n\mathbb{Z})[X]$.

We consider the successively more general cases in which $n$ is a prime, prime power, and arbitrary positive integer.

**4.1. Equations modulo primes.** The problem of finding integer solutions to polynomial equations modulo a prime $p$ is (essentially) the problem of finding roots to polynomials

$$f(X) \in \mathbb{F}_p[X]$$

with coefficients in the field with $p$ elements. Fermat's Little Theorem say that every element of $\mathbb{F}_p$ is a root of the polynomial $X^p - X$; comparing degrees and leading coefficients shows that

$$X^p - X = \prod_{a \in \mathbb{F}_p} (X - a) = \prod_{a=0}^{p-1} (X - a) \in \mathbb{F}_p[X].$$

It follows that if $f(X) \in \mathbb{F}_p[X]$ is an arbitrary polynomial, the gcd of $f(X)$ and $X^p - X$ is the product of all linear factors $X - a$, where $a$ is a root of $f$.

Similarly the roots of $X^{(p-1)/2} = 1$ are the quadratic residues modulo $p$, so

$$X^{(p-1)/2} - 1 = \prod_{a \in R_p} (X - a) \in \mathbb{F}_p[X],$$

where $R_p$ denotes the set of quadratic residues modulo $p$. This suggests the following elegant algorithm for finding all roots of a polynomial in $\mathbb{F}_p[X]$.

CANTOR–ZASSENHAUS $(f, p)$

  **Input:** A prime $p$ and a polynomial $f \in \mathbb{F}_p[X]$
  **Output:** A list of the roots of $f$
  1. $f := \gcd(f, X^p - X)$
  2. If $f$ has degree 1
      Return the root of $f$
  3. Choose $b$ at random in $\mathbb{F}_p$
      $g := \gcd(f(X), (X + b)^{(p-1)/2} - 1)$
      If $0 < \deg(g) < \deg(f)$ then
          Return CANTOR–ZASSENHAUS$(g,p)$
              $\cup$ CANTOR–ZASSENHAUS$(f/g,p)$
      Else, go to Step 3

The correctness follows from the earlier remarks — at Step 3, $f(X)$ is a product of distinct linear factors $X - a$, and then $g$ is the product of all $X - a$ such that $a + b$ is a quadratic residue. If $b$ is random then we would expect about half of the linear factors of $f$ to divide $g$.

Moreover, the algorithm takes (probabilistic) polynomial time since the input size for a (general) polynomial over $\mathbb{F}_p$ is $n \log p$ and all operations take time bounded by a polynomial in $n$ and $\log p$.

**4.2. Equations modulo prime powers.** Next we consider prime powers. The key idea is due to Kurt Hensel and gives an explicit construction that applies in sufficiently favorable situations to allow us to "lift" a solution to $f(X) \equiv 0 \bmod p^n$ to a solution modulo $p^{2n}$.

The idea is illustrated cleanly by the example of square roots modulo prime powers. Suppose that we have found a square root of $a \bmod p$, i.e., and integer $x$ such that $x^2 \equiv a \bmod p$. Let $y = (x^2+a)/(2x) \bmod p^2$. (Since $2x$ is invertible mod $p$ it is invertible mod $p^2$.) Some algebraic juggling shows that $y^2 \equiv a \bmod p^2$. The formula for $y$ comes from Newton's Lemma from calculus:

$$y = x - \frac{f(x)}{f'(x)} = x - \frac{x^2-a}{2x} = \frac{x^2+a}{2x}.$$

In any event, a root mod $p$ of $f(X) = X^2 - a \bmod p$ has been lifted to a root mod $p^2$. The general story is as follows.

THEOREM 20 (HENSEL'S LEMMA). *Let $p$ be a prime, $f(x) \in \mathbb{Z}[X]$ a polynomial with integer coefficients, and $a \in \mathbb{Z}$ an integer such that*

$$f(a) \equiv 0 \bmod p^k, \qquad f'(a) \not\equiv 0 \bmod p.$$

*Then $b \equiv a - f(a)/f'(a) \bmod p^{2k}$ is the unique integer modulo $p^{2k}$ that satisfies*

$$f(b) \equiv 0 \bmod p^{2k}, \qquad b \equiv a \bmod p^k.$$

REMARK 9. Since $u = f'(a)$ is relatively prime to $p$ it is relatively prime to any power of $p$ and has an inverse modulo that power, computable with the EEA. Thus the division in the formula in the theorem makes sense. Inverses modulo high powers can be found using Newton's method (for rational functions): find the inverse of $u$ modulo prime powers by finding roots of $f(X) = u - 1/X$. Specifically, if $au \equiv 1 \bmod p^k$ then $bu \equiv 1 \bmod p^{2k}$, where $b = a - f(a)/f'(a) = a(2 - au)$.

PROOF. Replacing $f(X)$ by $f(X+a)$, it suffices to prove the result when $a = 0$. Thus we are trying to find a root of $f(X) \equiv 0 \bmod p^{2k}$, given

$$f(X) = c_0 + c_1 X + c_2 X^2 + \cdots, \qquad c_0 = f(0), \; c_1 = f'(0).$$

By hypothesis, $c_0 \equiv 0 \bmod p^k$ and $c_1 \not\equiv 0 \bmod p$. It follows that $b = -c_0/c_1$ is the unique integer modulo $p^{2k}$ such that $b \equiv 0 \bmod p^k$ and $f(b) \equiv 0 \bmod p^{2k}$, finishing the proof. □

The reader will find it instructive to find a square root of, say, $-11$ modulo $3^8$.

The elementary form of Hensel's Lemma above is equivalent to a more elegant and general statement in the $p$-adic numbers. Since $p$-adic numbers (and, more generally, non-archimedean local fields) arise later in this volume, we digress to explain this idea. See [Cassels 1986], [Koblitz 1984], and [Serre 1973] for a much more thorough treatment.

Fix a prime $p$ and a constant $\rho$ such that $0 < \rho < 1$. An absolute value can be defined on the rational numbers $\mathbb{Q}$ by

$$|x|_p = \rho^n,$$

where $x = p^n y$, $n = v_p(x)$ is the unique integer (positive or negative) such that $y$ is a rational number whose numerator and denominator are coprime to $p$. By convention, $|0|_p = 0$.

Under this absolute value, powers of $p$ are "small." The absolute value satisfies the "non-archimedean" inequality $|x + y|_p \leq \max(|x|_p, |y|_p)$, which is stronger than the triangle inequality.

The absolute value $|x|_p$ gives the rational numbers $\mathbb{Q}$ the structure of a metric space by setting the distance between $x$ and $y$ to be $|x - y|_p$. The completion $\mathbb{Q}_p$ of this metric space is called the *p-adic numbers*, and it can be shown that the $p$-adic numbers are a locally compact topological field under natural extensions of the field operations and metric to the completion. The field is easily seen to be independent of the choice of $\rho$, though in some circumstances in algebraic number theory it is convenient to make the choice $\rho = 1/p$.

A nonzero element $a$ of $\mathbb{Q}_p$ can be represented concretely as a "Laurent series in $p$,", i.e.,

$$a = a_k p^k + a_{k+1} p^{k+1} + a_{k+1} p^{k+2} + \cdots,$$

where $k$ is an integer, the digits $a_n$ are integers chosen in the range $0 \leq a_n < p$, and $a_k \neq 0$. Moreover, $v_p(a) = k$ and $|a| = \rho^k$. Arithmetic operations are easy to visualize in this Laurent representation, at least approximately, by thinking of $p^n$ as being very small if $n$ is very large (analogous in some ways to the usual realization of real numbers as decimal expansions). The subset for which $v_p(a) \geq 0$ is the unit disk

$$\mathbb{Z}_p = \left\{ \sum_{n \geq 0} a_n p^n \right\} = \{x \in \mathbb{Q}_p : |x|_p \leq 1\},$$

which is a subring of $\mathbb{Q}_p$ and is called the ring of *p-adic integers*.

The $p$-adic numbers are a bit peculiar when first encountered, but it turns out that they beautifully capture the idea of calculating "modulo arbitrarily large

powers of $p$." The reader might enjoy proving that
$$-1 = 1 + 2 + 2^2 + 2^3 + \cdots \quad \text{in } \mathbb{Q}_2,$$
and the fact that infinite series in $\mathbb{Q}_p$ converge if and only if their terms go to zero.

Suppose that we are interested in roots of $f(X) \in \mathbb{Z}_p[X]$. (By multiplying by a power of $p$ the problem of finding roots for $f(X) \in \mathbb{Q}_p[X]$ readily reduces to $f(X) \in \mathbb{Z}_p[X]$.) Finding roots $x = a_0 + a_1 p + a_2 p^2 + \cdots$ to $f(x) = 0$ amounts to solving "$f(x) \equiv 0$ modulo $p^\infty$," and Hensel's Lemma can be translated to the statement that for $x \in \mathbb{Z}/p\mathbb{Z}$ with $|f(x)|_p < \varepsilon$ and $|f'(x)|_p = 1$, there is a unique $y \in \mathbb{Z}/p\mathbb{Z}$ such that
$$f(y) = 0, \qquad |y - x| < \varepsilon.$$
A more general form of Hensel's Lemma, which can be proved along the lines above, says that if $|f(x)| < |f'(x)|^2$ then there is a unique $y \in \mathbb{Z}_p$ such that $f(y) = 0$, $|y - x| \le |f(x)/f'(x)^2|$.

The problem of finding roots is a special case of finding factors of a polynomial, and an even more general form of Hensel's Lemma guarantees that approximate factors can, in suitable circumstances, be lifted to $p$-adic factors.

If $f$ is a polynomial with $p$-adic integer coefficients then let $f$ mod $p$ denote the polynomial obtained by looking at the coefficients modulo $p$.

THEOREM 21. *Assume that $f \in \mathbb{Z}_p[X]$ factors modulo $p$ as $f$ mod $p = gh \in (\mathbb{Z}/p\mathbb{Z})[x]$ and $g$ and $h$ are coprime in $(\mathbb{Z}/p\mathbb{Z})[x]$. Then there are unique $G$ and $H$ in $\mathbb{Z}_p[X]$ such that $f = GH$ and $G$ and $H$ lift $g$ and $h$ in the sense that $G$ mod $p = g$ and $H$ mod $p = h$.*

We sketch a proof of the theorem. Since $p$-adic integers are approximated arbitrarily closely by (ordinary) integers, it suffices to prove a statement about integer polynomials: if $f \in \mathbb{Z}[x]$ satisfies $f \equiv gh$ mod $p$ where $g$ and $h$ are coprime modulo $p$ (i.e., the elements that they determine in $\mathbb{F}_p[X]$ are coprime), then for arbitrarily large $n$ there are integer polynomials $G, H$ that are congruent to $g, h$, respectively, modulo $p$ and satisfy $f \equiv GH$ mod $p^n$.

To do this it turns out to be essential to simultaneously lift a certificate of the fact that $g$ and $h$ are relatively prime. Assume that we are given polynomials $f, g, h, r, s$ with integer coefficients such that
$$f \equiv gh \text{ mod } p^k, \qquad rg + sh \equiv 1 \text{ mod } p^k,$$
where $\deg(r) < \deg(h)$, $\deg(s) < \deg(g)$; our goal is to find $G, H, R, S$ satisfying a similar congruence modulo $p^{2k}$. Note that the hypotheses of the theorem imply that $r$ and $s$ exist for $k = 1$, using the Euclidean Algorithm for polynomials over $\mathbb{F}_p$. The pair $r, s$ is said to be a *coprimality certificate* of $g$ and $h$ mod $p^k$.

Before proving the theorem it is useful to show how to use a coprimality certificate mod $p^k$ to represent any polynomial as a linear combination of $g$ and $h$ modulo $p^k$.

LEMMA 22. If $rg + sh \equiv 1 \bmod p^k$, then for all $u \in \mathbb{Z}[X]$ there are $A, B \in \mathbb{Z}[X]$ such that
$$Ag + Bh \equiv u \bmod p^k.$$
If $\deg(u) < \deg(g) + \deg(h)$ then we can take $\deg(A) < \deg(h)$, $\deg(B) < \deg(g)$.

PROOF. Multiplying the assumed equation by $u$ gives $u \equiv rgu + shu = (ru)g + (su)h \bmod p^k$ so the first statement is immediate. To verify the assertion about degrees, we have to work harder. Let $A$ be the remainder when $ru$ is divided by $h$, i.e., $ru = q_1 h + A$, $\deg(A) < \deg(h)$. Similarly, let $B$ be the remainder when $su$ is divided by $g$, i.e., $su = q_2 g + B$, $\deg(B) < \deg(g)$. Then
$$Ag + Bh = (ru - q_1 h)g + (su - q_2 g)h$$
$$= (rg + sh)u - (q_1 + q_2)gh \equiv u + Qgh \bmod p^k,$$
where $Q = -q_1 - q_2$. Since $gh$ is monic of degree $\deg(g) + \deg(h)$ and all other terms in the congruence have degree strictly less than $\deg(g) + \deg(h)$ modulo $p^k$, it follows that $Q \equiv 0 \bmod p^k$, finishing the proof. □

Now we prove the theorem by showing how to lift the factorization, and lift the certificate, i.e., find the desired $G, H, R, S$ given $g, h, r, s$.

To lift the factorization write $f = gh + p^k u$, and use Lemma 22 to find $A, B$ with $u \equiv Ag + Bh \bmod p^k$; it is straightforward to check that $G = g + p^k B$, and $H = h + p^k A$ satisfy $f \equiv GH \bmod p^{2k}$.

To lift the certificate $rg + sh \equiv 1 \bmod p^k$, write $rg + sh = 1 + p^k u$, use Lemma 22 to find $u \equiv Ag + Bh \bmod p^k$, and check that $RG + SH \equiv 1 \bmod p^{2k}$ where $R = r + p^k A$ and $S = s + p^k B$.

**4.3. Chinese Remainder Theorem.** Finally, we come to the case of solving polynomial equations modulo arbitrary positive integers $n$, given that we can solve the equations modulo the prime power divisors of $n$. The Chinese Remainder Theorem gives an immediate way for doing this, by giving an explicit recipe for using solutions $f(x_1) \equiv 0 \bmod n_1$, $f(x_2) \equiv 0 \bmod n_2$ to produce an $x$ in $\mathbb{Z}/(n_1 n_2)\mathbb{Z}$ such that $f(x) \equiv 0 \bmod n_1 n_2$ when the $n_i$ are coprime.

THEOREM 23 (CHINESE REMAINDER THEOREM). If $m$ and $n$ are coprime, and $a$ and $b$ are arbitrary integers, then there is an integer $c$ such that
$$c \equiv a \bmod m, \qquad c \equiv b \bmod n.$$
Any two such integers $c$ are congruent modulo $mn$.

PROOF. Apply the EEA to $m$ and $n$ to find $x$ and $y$ such that $mx + ny = 1$. The integer
$$c = any + bmx$$
is obviously congruent to $a$ mod $m$ and $b$ mod $n$.

If $c'$ also satisfies these congruences then $d = c - c'$ is divisible by the coprime integers $m$ and $n$, and hence by $mn$. Thus $c$ and $c'$ are congruent modulo $mn$ as claimed. □

This can be stated more algebraically as follows.

COROLLARY 24. *If* $\gcd(m, n) = 1$ *then the rings* $\mathbb{Z}/(mn)\mathbb{Z}$ *and* $\mathbb{Z}/m\mathbb{Z} \times \mathbb{Z}/n\mathbb{Z}$ *are isomorphic via the map*
$$x \bmod mn \mapsto (x \bmod m, x \bmod n).$$
*Also,* $\phi(mn) = \phi(m)\phi(n)$ *where* $\phi$ *is the Euler* $\phi$-*function.*

Indeed, the first statement in the Chinese Remainder Theorem above says that the map is surjective, and the second says that it is injective. Units in the direct product of two rings are pairs $(u, v)$ where $u$ and $v$ are units in the respective rings, so the multiplicativity of $\varphi$ follows from the first statement.

As promised, the theorem shows how to combine modular solutions to polynomial equations: if $f(a) \equiv 0 \bmod m$ and $f(b) \equiv 0 \bmod n$ then apply the Chinese Remainder Theorem to find a $c$ such that $c \equiv a \bmod m$ and $c \equiv b \bmod n$; these congruences imply that $f(c) \equiv 0 \bmod mn$. By induction, this technique generalizes to more than 2 moduli: if $n_1, n_2, \ldots, n_k$ are pairwise coprime, and $a_i$ are given then there is an integer $x$, unique modulo the product of the $n_i$, such that $x \equiv a_i \bmod n_i$ for $1 \leq i \leq k$.

In fact, there are two natural inductive proofs. The product of $k$ moduli $n = \prod n_i$ could be reduced to the $k = 2$ case by $n = n_1 m$, where $m$ is the product of the $n_i$ for $i > 1$, or $n = m_1 m_2$ where each $m_i$ is the product of approximately half of the $n_i$. The latter method is significantly more efficient in practice: with reasonable assumptions, the first method takes $O(k^2)$ arithmetic operations, and the second takes $O(k \log k)$ arithmetic operations.

## 5. Quadratic extensions

A quadratic extension $K$ of a field $F$ is a field containing $F$ such that the dimension of $K$ as an $F$-vector space is 2. This means that $K$ is isomorphic to $F[X]/(f(X))$ where $f(X) \in F[X]$ is an irreducible quadratic polynomial. More concretely, $K = \{a + b\alpha : a, b \in F\}$ where $\alpha \in K$ satisfies an irreducible quadratic equation with coefficients in $F$. Finally, if $F = \mathbb{Q}$ then any quadratic extension is (isomorphic to) a subset of the complex numbers of the form $\mathbb{Q}(\sqrt{d}) = \{a + b\sqrt{d}\}$, where $d$ is a nonsquare in $\mathbb{Q}$.

In this section we consider four algorithms that are either illuminated by, or directly apply to, quadratic field extensions.

**5.1. Cipolla.** Let $p$ be an odd prime. A natural and direct algorithm, originally due to Cipolla [1902], finds square roots modulo $p$ by taking a single exponentiation in a quadratic extension of $\mathbb{F}_p = \mathbb{Z}/p\mathbb{Z}$. Unlike the modular square root algorithm presented earlier, it has a running time that is independent of the power of 2 that divides $p - 1$.

A polynomial $f(X) = X^2 - aX + b \in \mathbb{F}_p[X]$ is irreducible if and only if its discriminant $D := a^2 - 4b$ is a quadratic nonresidue. Let $\alpha = [X]$ be the coset of $X$ in $K := \mathbb{F}_p[X]/(f(X))$ so that $K$ can be viewed as the set of linear polynomials $\{a + b\alpha : a, b \in \mathbb{F}_p\}$ added in the usual way, and multiplied using the identity $\alpha^2 = a\alpha - b$.

In any field that contains $\mathbb{F}_p$ the map $\varphi(x) = x^p$ is a field automorphism of $K$, sometimes called the Frobenius map. Indeed, it is obvious that $\varphi$ is multiplicative, and $\varphi(x + y) = (x + y)^p = x^p + y^p$ follows from the fact that the interior binomial coefficients are divisible by $p$. Moreover $\varphi(x) = x$ if and only if $x \in \mathbb{F}_p$, since every element of $\mathbb{F}_p$ is a root of the polynomial $x^p - x$, and a polynomial of degree $p$ can have at most $p$ roots in a field.

CIPOLLA'S MODULAR SQUARE ROOT ALGORITHM

**Input:** An odd prime $p$ and a quadratic residue $b \in \mathbb{F}_p$
**Output:** $u \in \mathbb{F}_p$ such that $u^2 = b$
1. Select random $a$ until $a^2 - 4b$ is a nonresidue modulo $p$
2. Return $\alpha^{(p+1)/2}$, where $\alpha :=$ a root of $x^2 - ax + b$

Once an $a$ has been found in Step 1, then $f(X) = X^2 - aX + b$ is irreducible, and the above notation applies. Apply the Frobenius automorphism $\varphi$ to the equation $\alpha^2 - a\alpha + b = 0$ to find that $\beta := \alpha^p$ is also a root of $f$, so that $f(X) = (X - \alpha)(X - \beta)$. Comparing coefficients gives

$$b = \alpha \cdot \beta = \alpha \cdot \alpha^p = \alpha^{(p+1)}$$

so that $\alpha^{(p+1)/2}$ is a square root of $b$, proving correctness of the algorithm. Although the exponentiation is done in $K$, the final result lies in $\mathbb{F}_p$.

EXAMPLE 2. Let $p = 37$, $a = 34$. Easy Legendre symbol calculations show that 34 is a quadratic residue modulo 37, and $b = 1$ produces a nonresidue in Step 1, so that $f(x) = x^2 - x + 34$ is irreducible. We compute $\alpha^{(p+1)/2} = \alpha^{19}$ using EXP; at each step of the calculation we replace $\alpha^2$ by $\alpha - 34 = \alpha + 3$ and reduce the coefficients modulo 37. The following table summarizes the calculation:

| $k$ | $\alpha^k$ |
|---|---|
| 2 | $\alpha + 3$ |
| 4 | $\alpha^2 + 6\alpha + 9 = 7\alpha + 12$ |
| 8 | $49\alpha^2 + 168\alpha + 144 = 32\alpha + 32 = -5(\alpha + 1)$ |
| 9 | $-5(\alpha^2 + \alpha) = -5(2\alpha + 3)$ |
| 18 | $25(4\alpha^2 + 12\alpha + 9) = 30\alpha + 7$ |
| 19 | $30\alpha^2 + 7\alpha = 16$ |

As predicted, the final result lies in the field with 37 elements, and $16^2 \equiv 34$ mod 97.

**5.2. Lucas–Lehmer.** Theorem 5 gives a method for verifying primality of $n$ when the factorization of $n-1$ is known. A similar primality test, attributed to Lucas and Lehmer, enables the primality of $n$ to be established efficiently if the prime factorization of $n+1$ is known.

Let $f(X) = X^2 - aX + b \in \mathbb{Z}[x]$ be an irreducible polynomial with integer coefficients, $D := a^2 - 4b$ its discriminant, and $R := \mathbb{Z}[X]/(f(X))$. Let $\alpha := [X]$ and $\beta := a - \alpha$ be the two roots of $f$ in $R$.

For $k \geq 0$ define $V_k = \alpha^k + \beta^k \in R$. Both sequences $\{\alpha^k\}$ and $\{\beta^k\}$ satisfy the recursion

$$s_{k+1} = as_k - bs_{k-1},$$

so their sum $V_k$ also does, by linearity. Since $V_0 = 2$ and $V_1 = a$ are integers it follows that all of the $V_k$ are integers. More elaborate "doubling" recursions are easy to discover, allowing the $V_k$ to be computed in $O(\log k)$ arithmetic operations; these are equivalent to using EXP to calculate high powers of a matrix, noticing that

$$\begin{bmatrix} V_{k+2} & V_{k+1} \\ V_{k+1} & V_k \end{bmatrix} = \begin{bmatrix} a & -b \\ 1 & 0 \end{bmatrix}^k \begin{bmatrix} V_2 & V_1 \\ V_1 & V_0 \end{bmatrix}.$$

THEOREM 25. *Let $n$ be an odd positive integer, and $a, b$ integers such that $D = a^2 - 4b$ satisfies*

$$\left(\frac{D}{n}\right) = -1.$$

*Define $V_k$ as above, and let $m = (n+1)/2$. If $V_m \equiv 0$ mod $n$ and $\gcd(V_{m/q}, n) = 1$ for all prime divisors $q$ of $m$, then $n$ is prime.*

PROOF. Let $p$ be a prime divisor of $n$ with

$$\left(\frac{D}{p}\right) = -1.$$

Working modulo $p$, i.e., in

$$K := R/pR = \mathbb{Z}[x]/(p, f(x)) = \mathbb{F}_p[x]/(\overline{f}(x)),$$

we see that $\overline{f}(x) \in \mathbb{F}_p[x]$ is irreducible so that the ideas of the preceding section apply. Since the roots $\alpha$ and $\beta$ of an irreducible quadratic polynomial are nonzero, we can define

$$u = \beta/\alpha = \alpha^p/\alpha = \alpha^{p-1} \in K.$$

The equation $V_m \equiv 0 \mod p$ implies that $\alpha^m = -\beta^m$, i.e., $u^m = -1$ and $u^{2m} = 1$. Similarly, the equations $V_{m/q} \not\equiv 0 \mod p$ imply that $u^{2m/q} \neq 1$ for all prime divisors $q$ of $m$.

Therefore $u$ has order $2m = n + 1$ in $K$. Since

$$u^{p+1} = (\alpha^{p-1})^{p+1} = \alpha^{p^2-1} = 1,$$

it follows that $p + 1$ is divisible by $n + 1$, which implies that $n = p$, i.e., that $n$ is a prime as claimed. □

REMARK 10. As in the earlier primality test relying on the factorization of $n - 1$, it can be shown that it suffices to restrict attention to primes $q$ that divide a factor $F > \sqrt{n}$ of $n + 1$.

**5.3. Units in quadratic fields.** If $d$ is a rational number that is not a perfect square, then the quadratic extension $F = \mathbb{Q}(\sqrt{d}) = \{a + b\sqrt{d} : a, b \in \mathbb{Q}\}$ over $\mathbb{Q}$ is unchanged if we replace $d$ by $de^2$. We will from now on assume that $d$ is a squarefree integer, i.e., not divisible by any perfect square $n^2 > 1$. If $d < 0$ then $\mathbb{Q}(\sqrt{d})$ is said to be an *imaginary* quadratic field, and if $d > 0$ $\mathbb{Q}(\sqrt{d})$ is said to be a *real* quadratic field.

A *number field* is a field having finite degree over $\mathbb{Q}$. These fields are the core subject matter of algebraic number theory, and are of critical importance in several subsequent articles in this volume, e.g., [Stevenhagen 2008a; 2008b]. Quadratic fields $\mathbb{Q}(\sqrt{d})$ already illustrate many of the ideas that come up in studying number fields in general, and are a vast and fertile area of study.

Let $F = \mathbb{Q}(\sqrt{d})$. The *ring of integers* $\mathcal{O}_F$ in $F$ is the set of $v \in F$ that are roots of a monic polynomial with integer coefficients; this ring plays the role, in $F$, that the integers do in $\mathbb{Q}$.

PROPOSITION 26. With the above notation, $\mathcal{O}_F = \mathbb{Z}[\omega]$ where

$$\omega = \begin{cases} \sqrt{d} & \text{if } d \equiv 2 \text{ or } 3 \mod 4, \\ (1 + \sqrt{d})/2 & \text{if } d \equiv 1 \mod 4. \end{cases}$$

Thus $\mathcal{O}_F = \mathbb{Z}[\sqrt{d}]$ if $d \equiv 2$ or $3 \mod 4$, and

$$\mathcal{O}_F = \mathbb{Z}[\omega] = \left\{ \frac{x + y\sqrt{d}}{2} : x, y \in \mathbb{Z}, x \equiv y \mod 2 \right\}$$

if $d \equiv 1 \mod 4$.

The key idea of the proof is that if $v = x + y\sqrt{d} \in F$ is an algebraic integer then it satisfies the equation $X^2 - 2xX + (x^2 - dy^2) = 0$, which implies that $x$ and therefore $y$ are either integers or half-integers; we leave the details to the reader.

The algorithmically inclined reader will observe that the ring of integers is an unfriendly object. For instance, we have been assuming in this discussion that the integer $d$ is squarefree. In fact, it is difficult to detect or verify whether a large integer is squarefree, and therefore impossible to find the ring of integers in $F = \mathbb{Q}(\sqrt{d})$ for general $d$. See [Stevenhagen 2008a] for a discussion of how algorithmic number theorists deal with this issue for arbitrary number fields.

We now turn our attention to finding units in rings of integers of quadratic fields.

The only units (elements with multiplicative inverses) in $\mathbb{Z}$ are $\pm 1$. In order to find units in quadratic fields we introduce some standard terminology.

- If $v = x + y\sqrt{d} \in F$ then $v' = x - y\sqrt{d}$ is the *conjugate* of $v$.
- The *norm* of $v = x + y\sqrt{d} \in F$ is

$$N(v) = vv' = (x + y\sqrt{d})(x - y\sqrt{d}) = x^2 - dy^2 \in \mathbb{Q}.$$

The map $u \mapsto u'$ is easily checked to be an automorphism, and $(uv)' = u'v'$ implies that the norm map is multiplicative: $N(uv) = N(u)N(v)$.

Note that if $d \equiv 2$ or $3 \mod 4$ then $N(x + y\omega) = (x + y\omega)(x - y\omega) = x^2 - dy^2$, while if $d \equiv 1 \mod 4$ and $\omega = (1 + \sqrt{d})/2$ then

$$N(x + y\omega) = (x + y\omega)(x + y\omega') = x^2 - xy - \frac{d-1}{4}y^2.$$

LEMMA 27. An element $u$ in $\mathcal{O}_F$ is a unit if and only if $N(u) = \pm 1$.

PROOF. If $N(u) = \pm 1$ then $u(\pm u') = 1$ and $u$ is a unit. If $v$ is a unit then there is a $u \in \mathcal{O}_F$ such that $uv = 1$. Taking the norm and using the multiplicativity of the norm gives $N(u)N(v) = 1$, so that $N(u)$ is a unit in $\mathbb{Z}$, i.e., $N(u) = \pm 1$. □

In imaginary quadratic fields there are only finitely many units, and they are easy to determine by considering the explicit form of the norm mapping given above; the details of are again left to the reader, and the result is:

THEOREM 28. Units in an imaginary quadratic field $F = \mathbb{Q}(\sqrt{d})$, $d < 0$, are:

$$\mathcal{O}_F^* = \begin{cases} \{\pm 1, \pm i\} & \text{if } d = -1, \\ \{\pm 1, \pm \omega, \pm \omega^2\} & \text{if } d = -3, \\ \{\pm 1\} & \text{if } d < -3. \end{cases}$$

The units of the ring $\mathbb{Z}[\sqrt{d}]$ are $\{\pm 1\}$, unless $d = -1$.

Finding units in (the ring of integers of) real quadratic fields is considerably more interesting. If $d \equiv 2$ or $3 \mod 4$ then $\mathcal{O}_F = \mathbb{Z}[\sqrt{d}]$ and finding units is equivalent to solving Pell's equation [Lenstra 2008]

$$x^2 - dy^2 = \pm 1.$$

If $d \equiv 1 \mod 4$ then units $u = x + y\omega$ correspond to solutions to $x^2 - xy - y^2(d-1)/4 = \pm 1$. After multiplying by 4 and completing the square this is equivalent to integer solutions to $X^2 - dY^2 = \pm 4$.

In either case, the problem of finding units reduces to solving Pell's equation or a variant. This can be done by finding the continued fraction of $\omega$.

THEOREM 29. If $F = \mathbb{Q}(\sqrt{d})$, $d > 0$, and $u = x - y\omega$ is a unit in $\mathcal{O}_F^*$ with $x$ and $y$ positive, then $x/y$ is a convergent in the continued fraction expansion of $\omega$.

If $u = x + y\omega$ is a unit then $-u$, $u'$, and $-u'$ are all units. If $u$ is a unit, then one of $\pm u$, $\pm u'$ has the form $x - y\omega$ with $x$, $y$ positive, so the restriction to positive $x$, $y$ is unimportant: any unit can be obtained from a convergent by changing the sign and/or taking a conjugate.

PROOF. Assume that $u = x - y\omega$ is a unit with $x$ and $y$ positive. If $d \equiv 2$ or $3 \mod 4$, then $x^2 - dy^2 = \pm 1$. Moreover $d \geq 2$ and

$$\frac{x}{y} + \sqrt{d} = \sqrt{d \pm \frac{1}{y^2}} + \sqrt{d} \geq 1 + \sqrt{2} > 2. \tag{5-1}$$

From this equation and $(x/y - \sqrt{d})(x/y + \sqrt{d}) = \pm 1/y^2$ we get

$$\left| \frac{x}{y} - \sqrt{d} \right| < \frac{1}{2y^2},$$

so that, by Corollary 18, $x/y$ is a convergent in the continued fraction of $\omega$.

The case $d \equiv 1 \mod 4$ is similar, but slightly more involved. If $u = x - y\omega$ is a unit then $N(u) = x^2 - xy - (d-1)y^2/4 = \pm 1$. Solving the quadratic equation $(x/y)^2 - (x/y) - (d-1)/4 \pm 1/y^2 = 0$ for $x/y$ and doing some algebra shows that

$$\frac{x}{y} - \omega' \geq 2, \tag{5-2}$$

except possibly in the case $d = 5, y = 1$. Indeed, this inequality is equivalent to

$$\sqrt{d \pm 4/y^2} + \sqrt{d} \geq 4,$$

which is easy to verify if $d > 5$ (so that $d \geq 13$), or if $y \geq 2$. If either of these hold, proceed as above: $(x/y - \omega)(x/y - \omega') = \pm 1/y^2$ and (5-2) imply that $|x/y - \omega| < 1/2y^2$ and the theorem follows as before. If $d = 5$ and $y = 1$ then the theorem can be checked directly: both $2/1$ and $1/1$ are convergents. □

There is much more to say about continued fractions of arbitrary quadratic irrationals. We enumerate some of the most important facts.

(i) The continued fraction of a real number $\alpha$ is periodic if and only if $\alpha$ is a quadratic irrational.
(ii) The continued fraction of a quadratic irrational $\alpha$ is purely periodic if and only if $\alpha > 1$ and $-1 < \alpha' < 0$.
(iii) The continued fraction of a quadratic irrational $\alpha$ can be computed entirely with integer arithmetic; $\alpha_k$ has the form $(P_k + \sqrt{d})/Q_k$ for integers $P_k, Q_k$, where $Q_k$ divides $d - P_k^2$, and these integers are determined by the recursions

$$a_k = \lfloor \alpha_k \rfloor, \quad P_{k+1} = a_k Q_k - P_k, \quad Q_{k+1} = \frac{d - P_{k+1}^2}{Q_k}.$$

(iv) The period of the continued fraction of $\omega$ is $O(\sqrt{d} \log d)$. The continued fractions of $\sqrt{d}$ and $(1 + \sqrt{d})/2$ have the respective shapes

$$[a; \overline{P, 2a}], \quad [a, \overline{P, 2a - 1}],$$

where $a$ is a positive integer, the sequence $P = a_1, a_2, \ldots, a_2, a_1$ is a palindrome, and the $a_i$ are less than $a$.
(v) If the period of the continued fraction is $r$, then the $(r-1)$st convergent corresponds to a so-called fundamental unit $u$, and any other unit of the form $\pm u^n$ for integers $n$; i.e.,

$$\mathcal{O}_F^* = \{\pm u^n : n \in \mathbb{Z}\} \simeq \mathbb{Z} \times \mathbb{Z}/2\mathbb{Z}.$$

(vi) The index of $\mathbb{Z}[\sqrt{d}]^*$ in $\mathcal{O}_F^*$ is 1 or 3.
(vii) The fundamental generating unit of $\mathbb{Q}(\sqrt{d})$ has norm $-1$ if and only if $P$ has even length. If this is the case, the units of norm one are of the form $\pm u^{2k}$.

## 5.4. The Smith–Cornacchia algorithm.
Throughout this section integers $d$ and $n$ satisfy $0 < d < n$, and we consider integer solutions $x$, $y$ to

$$x^2 + dy^2 = n. \tag{5-3}$$

Note that the problem of solving this equation is equivalent to

$$(x + y\sqrt{-d})(x - y\sqrt{-d}) = n,$$

i.e., finding elements of $\mathbb{Z}[\sqrt{-d}]$ of norm $n$. Fix $d$ and let $R = \mathbb{Z}[\sqrt{-d}]$.

A coprime solution $x$, $y$ is said to be *primitive*. Any coprime solution determines a square root $t = x/y \bmod n$ of $-d$ modulo $n$; indeed,

$$t^2 + d = (x^2 + dy^2)/y^2 \equiv 0 \bmod n;$$

the unique such $t$ with $1 \leq t < n$ is sometimes called the *signature* of the solution.

A signature determines a map from $R$ to $\mathbb{Z}/n\mathbb{Z}$ defined by taking $\sqrt{-d}$ to $t$. The kernel of this ring homomorphism is the principal ideal

$$I_t = (x + y\sqrt{-d})R = nR + (t + \sqrt{-d})R.$$

The Smith–Cornacchia algorithm has two steps: determine all potential signatures $t$, and use the Euclidean algorithm to determine which of the ideals $I_t := nR + (t + \sqrt{-d})R$ are principal.

Given a primitive solution $(x, y)$ to $x^2 + dy^2 = n$ with signature $t$ define an integer $z$ by

$$z = \frac{ty - x}{n},$$

so that $x = ty - nz$. Dividing the inequality

$$|2xy| \leq x^2 + y^2 \leq x^2 + dy^2 = n$$

by $2ny^2$ gives

$$\left|\frac{x}{ny}\right| = \left|\frac{t}{n} - \frac{z}{y}\right| \leq \frac{1}{2y^2}.$$

Corollary 18 implies that $z/y$ is a convergent of the (finite) continued fraction of $t/n$.

Thus to solve (5-3) it suffices to compute the convergents in the continued fraction of $t/n$ and see whether any of the denominators give valid solutions $y$ in the equation $x^2 + dy^2 = n$.

In fact, it is slightly simpler to merely keep track of the remainders in the Euclidean algorithm applied to $t$ and $n$; indeed $|x|$ is a remainder, and the equation $x^2 + dy^2 = n$ can be solved for $y$. Since $x \leq \sqrt{n}$, the remainders have to be calculated at least to the point that they drop below $\sqrt{n}$.

EXAMPLE 3. If $d = 5$ and $n = 12829$ then we first have to find a square root $t$ of $-5$ modulo $12829$. Since $n$ is a prime, the square root is unique up to sign; in fact, Cipolla's algorithm gives $t = \pm 3705$ without difficulty. The sequence of remainders in the Euclidean algorithm is $12829, 3705, 1714, 277, 52, 17, 1$. The first $x$ below $\sqrt{12829}$ works:

$$x = 52, \qquad y = \sqrt{(n-x^2)/5} = 45, \qquad 52^2 + 5 \cdot 45^2 = 12829.$$

THEOREM 30. *If a primitive solution to $x^2 + dy^2 = n$ has signature $t$ then $|x|$ is the largest (i.e., first encountered) remainder that is less than $\sqrt{n}$ in the Euclidean algorithm applied to $n$ and $t$. Moreover, $(x, y)$ is (essentially) the unique solution with signature $t$: if $(u, v)$ is another such solution then $(x, y) = \pm(u, v)$ if $d > 1$, and if $d = 1$ reversal might occur, i.e., $(x, y) = \pm(u, v)$ or $(x, y) = \pm(v, u)$.*

PROOF. (See also [Nitaj 1995; Schoof 1995].) Let $x$ be the first remainder less than $\sqrt{n}$. Assume that the equation has some solution $(u, v)$ with signature $t$: $u^2 + dv^2 = n$, $u \equiv ty \bmod n$. Assume that $u \neq \pm x$. Then $|u|$ is a subsequent term in the remainder sequence for $t$ and $n$. By the Euclidean algorithm, there are $z, z'$ such that $x = yt - zn$ and $u = vt - z'n$. The remainders decrease in absolute value, so $x > u$, and the coefficients of $t$ increase in absolute value, so $|y| < |v|$. We know that $x^2 + dy^2 \equiv (t^2 + d)y^2 \equiv 0 \bmod n$. Moreover,

$$x^2 + dy^2 < n + dv^2 = n + d \cdot \frac{n - dv^2}{d} < 2n.$$

Since we also know that $x^2 + dy^2$ is divisible by $n$ it follows that $x^2 + dy^2 = n$.

Two different solutions $x, y$ and $u, v$ determine generators $x + y\sqrt{-d}$ and $u + v\sqrt{-d}$ of $I_t$ and hence there is a unit $a$ in $R$ such that $x + y\sqrt{-d} = a(u + v\sqrt{-d})$. If $d > 1$ the only units are $\pm 1$. If $d = 1$ then $a = \pm i$ is possible, and these account for all cases in the theorem. □

The algorithm that implements this is obvious.

SMITH–CORNACCHIA ALGORITHM
**Input:** Relatively prime positive integers $d$ and $n$
**Output:** All primitive solutions to $x^2 + dy^2 = n$ (possibly none)
1. Find all positive solutions (less than $n$) to $t^2 + d \equiv 0 \bmod n$
2. For each solution $t$, find the first remainder $x$ less than $\sqrt{n}$ in the Euclidean algorithm applied to $n$ and $t$; if $y := \sqrt{(n - x^2)/d}$ is an integer, output $(x, y)$

The second step of the algorithm is efficient. Unfortunately, the first step isn't in general, since it requires a modular square root. However, in the special case that $n$ is a prime, this step can be done efficiently by a probabilistic algorithm.

EXAMPLE 4. The special case $d = 1$ is interesting as it is the classic problem of writing an integer as the sum of two squares. By Euler's criterion, $-1$ is a quadratic residue modulo an odd prime $p$ if and only if $p$ is congruent to 1 modulo 4, and this algorithm gives a constructive proof of Euler's result that these are exactly the primes that are the sum of two squares. Moreover, it shows that if $z$ is a Gaussian integer then the factorization of $z$ in $\mathbb{Z}[i]$ reduces to the problem of factoring $N(z)$ in the integers. See [Bressoud and Wagon 2000] for further details.

## References

[Agrawal et al. 2004] M. Agrawal, N. Kayal, and N. Saxena, "PRIMES is in P", *Ann. of Math.* (2) **160**:2 (2004), 781–793.

[Bach 1990] E. Bach, "Explicit bounds for primality testing and related problems", *Math. Comp.* **55**:191 (1990), 355–380.

[Bach and Shallit 1996] E. Bach and J. Shallit, *Algorithmic number theory, I: Efficient algorithms*, MIT Press, Cambridge, MA, 1996.

[Bernstein 2008] D. Bernstein, "Fast multiplication and its applications", pp. 325–384 in *Surveys in algorithmic number theory*, edited by J. P. Buhler and P. Stevenhagen, Math. Sci. Res. Inst. Publ. **44**, Cambridge University Press, New York, 2008.

[Bressoud and Wagon 2000] D. Bressoud and S. Wagon, *A course in computational number theory*, Key College, Emeryville, CA, 2000.

[Cassels 1986] J. W. S. Cassels, *Local fields*, London Mathematical Society Student Texts **3**, Cambridge University Press, Cambridge, 1986.

[Cipolla 1902] M. Cipolla, "La determinazione assintotica dell'$n^{imo}$ numero primo", *Rend. Accad. Sci. Fis. Mat. Napoli* **8** (1902), 132–166.

[Cohen 1993] H. Cohen, *A course in computational algebraic number theory*, Graduate Texts in Mathematics **138**, Springer, Berlin, 1993.

[Cox 1989] D. A. Cox, *Primes of the form $x^2 + ny^2$: Fermat, class field theory and complex multiplication*, Wiley, New York, 1989.

[Crandall and Pomerance 2005] R. Crandall and C. Pomerance, *Prime numbers: A computational perspective*, 2nd ed., Springer, New York, 2005.

[Damgård et al. 1993] I. Damgård, P. Landrock, and C. Pomerance, "Average case error estimates for the strong probable prime test", *Math. Comp.* **61**:203 (1993), 177–194.

[Davis 1973] M. Davis, "Hilbert's tenth problem is unsolvable", *Amer. Math. Monthly* **80** (1973), 233–269.

[Flajolet and Vallée 1998] P. Flajolet and B. Vallée, "Continued fraction algorithms, functional operators, and structure constants", *Theoret. Comput. Sci.* **194**:1-2 (1998), 1–34.

[Garey and Johnson 1979] M. R. Garey and D. S. Johnson, *Computers and intractability: A guide to the theory of NP-completeness*, W. H. Freeman, San Francisco, 1979.

[von zur Gathen and Gerhard 2003] J. von zur Gathen and J. Gerhard, *Modern computer algebra*, 2nd ed., Cambridge University Press, Cambridge, 2003.

[Granville 2008] A. Granville, "Smooth numbers: computational theory and beyond", pp. 267–323 in *Surveys in algorithmic number theory*, edited by J. P. Buhler and P. Stevenhagen, Math. Sci. Res. Inst. Publ. **44**, Cambridge University Press, New York, 2008.

[Hopcroft and Ullman 1979] J. E. Hopcroft and J. D. Ullman, *Introduction to automata theory, languages, and computation*, Addison-Wesley, Reading, MA, 1979.

[Knuth 1981] D. E. Knuth, *The art of computer programming, II: Seminumerical algorithms*, 2nd ed., Addison-Wesley, Reading, MA, 1981.

[Koblitz 1984] N. Koblitz, *p-adic numbers, p-adic analysis, and zeta-functions*, 2nd ed., Graduate Texts in Mathematics **58**, Springer, New York, 1984.

[Kozen 2006] D. C. Kozen, *Theory of computation*, Springer, London, 2006.

[Lenstra 1987] H. W. Lenstra, Jr., "Factoring integers with elliptic curves", *Ann. of Math.* (2) **126**:3 (1987), 649–673.

[Lenstra 2008] H. W. Lenstra, Jr., "Solving the Pell equation", pp. 1–23 in *Surveys in algorithmic number theory*, edited by J. P. Buhler and P. Stevenhagen, Math. Sci. Res. Inst. Publ. **44**, Cambridge University Press, New York, 2008.

[Lenstra and Lenstra 1993] A. K. Lenstra and H. W. Lenstra, Jr. (editors), *The development of the number field sieve*, Lecture Notes in Mathematics **1554**, Springer, Berlin, 1993.

[Lenstra and Pomerance 1992] H. W. Lenstra, Jr. and C. Pomerance, "A rigorous time bound for factoring integers", *J. Amer. Math. Soc.* **5**:3 (1992), 483–516.

[Matiyasevich 1993] Y. V. Matiyasevich, *Hilbert's tenth problem*, MIT Press, Cambridge, MA, 1993.

[Nitaj 1995] A. Nitaj, "L'algorithme de Cornacchia", *Exposition. Math.* **13**:4 (1995), 358–365.

[Pollard 1978] J. M. Pollard, "Monte Carlo methods for index computation (mod $p$)", *Math. Comp.* **32**:143 (1978), 918–924.

[Pomerance 2008] C. Pomerance, "Elementary thoughts on discrete logarithms", pp. 385–396 in *Surveys in algorithmic number theory*, edited by J. P. Buhler and P. Stevenhagen, Math. Sci. Res. Inst. Publ. **44**, Cambridge University Press, New York, 2008.

[Poonen 2008] B. Poonen, "Elliptic curves", pp. 183–207 in *Surveys in algorithmic number theory*, edited by J. P. Buhler and P. Stevenhagen, Math. Sci. Res. Inst. Publ. **44**, Cambridge University Press, New York, 2008.

[van der Poorten 1986] A. J. van der Poorten, "An introduction to continued fractions", pp. 99–138 in *Diophantine analysis* (Kensington, 1985), edited by J. H. Loxton and A. J. van der Poorten, London Math. Soc. Lecture Note Ser. **109**, Cambridge Univ. Press, Cambridge, 1986.

[Schirokauer 2008] O. Schirokauer, "The impact of the number field sieve on the discrete logarithm problem in finite fields", pp. 397–420 in *Surveys in algorithmic number theory*, edited by J. P. Buhler and P. Stevenhagen, Math. Sci. Res. Inst. Publ. **44**, Cambridge University Press, New York, 2008.

[Schönhage 1971] A. Schönhage, "Schnelle Berechnung von Kettenbruchentwicklungen", *Acta Informatica* **1** (1971), 139–144.

[Schoof 1985] R. Schoof, "Elliptic curves over finite fields and the computation of square roots mod $p$", *Math. Comp.* **44**:170 (1985), 483–494.

[Schoof 1995] R. Schoof, "Counting points on elliptic curves over finite fields", *J. Théor. Nombres Bordeaux* **7**:1 (1995), 219–254.

[Schoof 2008a] R. Schoof, "Computing Arakelov class groups", pp. 447–495 in *Surveys in algorithmic number theory*, edited by J. P. Buhler and P. Stevenhagen, Math. Sci. Res. Inst. Publ. **44**, Cambridge University Press, New York, 2008.

[Schoof 2008b] R. Schoof, "Four primality testing algorithms", pp. 101–125 in *Surveys in algorithmic number theory*, edited by J. P. Buhler and P. Stevenhagen, Math. Sci. Res. Inst. Publ. **44**, Cambridge University Press, New York, 2008.

[Serre 1973] J.-P. Serre, *A course in arithmetic*, Graduate Texts in Mathematics **7**, Springer, New York, 1973.

[Shor 1997] P. W. Shor, "Polynomial-time algorithms for prime factorization and discrete logarithms on a quantum computer", *SIAM J. Comput.* **26**:5 (1997), 1484–1509.

[Shoup 2005] V. Shoup, *A computational introduction to number theory and algebra*, Cambridge University Press, Cambridge, 2005.

[Stevenhagen 2008a] P. Stevenhagen, "The arithmetic of number rings", pp. 209–266 in *Surveys in algorithmic number theory*, edited by J. P. Buhler and P. Stevenhagen, Math. Sci. Res. Inst. Publ. **44**, Cambridge University Press, New York, 2008.

[Stevenhagen 2008b] P. Stevenhagen, "The number field sieve", pp. 83–100 in *Surveys in algorithmic number theory*, edited by J. P. Buhler and P. Stevenhagen, Math. Sci. Res. Inst. Publ. **44**, Cambridge University Press, New York, 2008.

[Teske 2001] E. Teske, "Square-root algorithms for the discrete logarithm problem", pp. 283–301 in *Public-key cryptography and computational number theory* (Warsaw, 2000), edited by K. Alster et al., de Gruyter, Berlin, 2001.

[Yui and Top 2008] N. Yui and J. Top, "Congruent number problems in dimension one and two", pp. 613–639 in *Surveys in algorithmic number theory*, edited by J. P. Buhler and P. Stevenhagen, Math. Sci. Res. Inst. Publ. **44**, Cambridge University Press, New York, 2008.

JOE BUHLER
CENTER FOR COMMUNICATIONS RESEARCH
SAN DIEGO, CA, 92121
jpb@reed.edu

STAN WAGON
MACALESTER COLLEGE
ST. PAUL, MN 55105
wagon@macalester.edu

# Smooth numbers and the quadratic sieve

CARL POMERANCE

ABSTRACT. This article gives a gentle introduction to factoring large integers via the quadratic sieve algorithm. The conjectured complexity is worked out in some detail.

When faced with a large number $n$ to factor, what do you do first? You might say, "Look at the last digit," with the idea of cheaply pulling out possible factors of 2 and 5. Sure, and more generally, you can test for divisibility cheaply by all of the very small primes. So it may as well be assumed that the number $n$ has no small prime factors, say below $\log n$. Since it is also cheap to test for probable primeness, say through the strong probable prime test, and then actually prove primality as in [Schoof 2008] in the case that you become convinced $n$ is prime, it also may as well be assumed that the number $n$ is composite.

Trial division is a factoring method (and in the extreme, a primality test) that involves sequentially trying $n$ for divisibility by the consecutive primes. This method was invoked above for the removal of the small prime factors of $n$. The only thing stopping us from continuing beyond the somewhat arbitrary cut off of $\log n$ is the enormous time that would be spent if the smallest prime factor of $n$ is fairly large. For example, if $n$ were a modulus being used in the RSA cryptosystem, then as current protocols dictate, $n$ would be the product of two primes of the same order of magnitude. In this case, factoring $n$ by trial division would take roughly $n^{1/2}$ steps. This already is an enormous calculation if $n$ has thirty decimal digits, and for numbers only slightly longer, the calculation is not possible at this time by the human race and all of their computers.

**Difference of squares.** We have long known however that trial division is not the only game in town for factoring. Take the number 8051 for example. This number is composite and not divisible by any prime up to its logarithm. One can see instantly (if one looks) that it is $8100 - 49$, that is,

$$8051 = 90^2 - 7^2.$$

Thus, we can use algebra to factor 8051 as a difference of squares. It is $(90 - 7) \times (90 + 7)$, or $83 \times 97$. Every odd composite can be factored as a difference of squares (an easy exercise), so why don't we use this method instead of trial division?

Let us try again on the number $n = 1649$. Again, 1649 is composite, but not divisible by any prime up to its logarithm. What worked with 8051 was to take the first square above 8051, namely $90^2$, and then notice that $90^2 - 8051 = 49$, where 49 is recognized as a square. It would seem in general that one could walk through the sequence $x^2 - n$ with $x = \lceil n^{1/2} \rceil, \lceil n^{1/2} \rceil + 1, \ldots$, looking for squares. With $n = 1649$ we have

$$\begin{aligned} 41^2 - n &= 32, \\ 42^2 - n &= 115, \\ 43^2 - n &= 200, \end{aligned} \tag{1}$$

and so on, with no squares in immediate sight.

Despite this failure, the three equations (1) may in fact be already used to factor $n$. Note that while neither 32 nor 200 is a square, their product *is* a square: $6400 = 80^2$. Thus, since

$$\begin{aligned} 41^2 &\equiv 32 \pmod{n}, \\ 43^2 &\equiv 200 \pmod{n}, \end{aligned}$$

we have $41^2 \times 43^2 \equiv 80^2 \pmod{n}$, that is,

$$(41 \times 43)^2 \equiv 80^2 \pmod{n}. \tag{2}$$

We have found a solution to $a^2 \equiv b^2 \pmod{n}$. Is this interesting? There are surely plenty of uninteresting pairs $a, b$ with $a^2 \equiv b^2 \pmod{n}$. Namely, take any value of $a$ and let $b = \pm a$. Have we exhausted all solutions with this enumeration? Well no, since factoring $n$ as a difference of squares would give a pair $a, b$ with $b \neq \pm a$. Further, any pair $a, b$ with

$$a^2 \equiv b^2 \pmod{n}, \quad b \not\equiv \pm a \pmod{n} \tag{3}$$

must lead to a nontrivial factorization of $n$, via $\gcd(a - b, n)$. Indeed, (3) implies that $n$ divides $(a - b)(a + b)$, but divides neither factor, so both $\gcd(a - b, n), \gcd(a + b, n)$ must be nontrivial factors of $n$. Moreover, gcd's are simple to compute via Euclid's algorithm of replacing the larger member of $\gcd(u, v)$ by its residue modulo the smaller member, until one member reaches 0. Finally, if $n$ has at least two different odd prime factors, then it turns out that at least half of the solutions to $a^2 \equiv b^2 \pmod{n}$ with $ab$ coprime to $n$ also have $b \not\equiv \pm a$ $\pmod{n}$, that is, (3) is satisfied. The proof: For an odd prime power $p^u$, the congruence $y^2 \equiv 1 \pmod{p^u}$ has exactly 2 solutions, so since $n$ is divisible by at least two different odd primes, the congruence $y^2 \equiv 1 \pmod{n}$ has at least 4 solutions. Label these values of $y$ as $y_1, y_2, \ldots, y_s$, where $y_1 = 1, y_2 = -1$. Then a complete enumeration of the pairs of residues $a, b$ modulo $n$ that are coprime to $n$ and satisfy

$$a^2 \equiv b^2 \pmod{n}$$

consists of all pairs $a, y_i a$, where $a$ runs over residues coprime to $n$ and $i$ takes the values $1, \ldots, s$. Two of these pairs have $i = 1, 2$ and $s - 2$ (out of $s$) of these pairs have $i = 3, \ldots, s$. The latter pairs satisfy (3), and since $s \geq 4$, we are done.

So, I repeat the question, is our solution to (2) interesting? Here, we have $a = 114 = 41 \times 43 \bmod n$ and $b = 80$. Yes, we do have $a, b$ satisfying (3), so yes this is an interesting pair. We now compute $\gcd(114 - 80, 1649)$. The first step of Euclid gives $\gcd(34, 17)$, and the second step gives 17. That is, 17 is a divisor of 1649. Division reveals the cofactor, it is 97, and a couple of primality tests show that our factorization is complete. Actually, since we have previously checked that 1649 has no prime factors up to its logarithm, about 7.4, we already know that 17 and 97 have no prime factors below their square roots, and so are both prime.

If we had actually waded through the sequence in (1) looking for a square, we would have had to go all the way to $57^2 - n = 40^2$, and clearly in this example, trial division would have been superior. In general, trying to factor a number as a difference of squares is inferior to trial division for most numbers. But by altering $n$ by multiplying it by small odd numbers, one can then skew things so that in the worst case the time spent is only about $n^{1/3}$ steps, essentially a factoring method of R. S. Lehman. For example, if one tries to factor $5 \times 1649 = 8245$ by a difference of squares, the first try works, namely $8245 = 91^2 - 6^2 = 97 \times 85$. Taking the gcd of the factors with 1649 reveals the factorization we are looking for.

**A crucial lemma.** These thoughts lead down a different road. We would like to pursue what looked like perhaps fortuitous good luck with the equations in (1). If one were to try and describe what we did as an algorithm that works for a general number $n$, the part where we pick out the first and third of the equations to get the right sides to multiply to a square looks a bit suspect. In general we have this sequence of integers $x^2 - n$ as $x$ runs starting above $n^{1/2}$ and we wish to pick out a subsequence with product a square. Surely we should not be expected to examine all subsequences, since the number of these grows exponentially.

Let us look at a probabilistic model for this problem. We are presented with a random sequence of integers in the interval $[1, X]$ and we wish to stop the sequence as soon as some non-empty subsequence has product a square. After how many terms do we expect to stop, how do we recognize when to stop, and how do we find the subsequence? The answers to these questions involve *smooth numbers*.

A number $m$ is *smooth* if all of its prime factors are small. Specifically, we say $m$ is $B$-smooth if all of its prime factors are $\leq B$. The first observation is that if a number in our sequence is *not* smooth, then it is unlikely it will be used in a subsequence with product a square. Indeed, if the number $m$ is divisible by the large prime $p$, then if $m$ is to be used in the square subsequence, then there necessarily must be at least one other term $m'$ in the subsequence which also is divisible by $p$. (This other term $m'$ may be $m$ itself, that is, perhaps $p^2 | m$.) But given that $p$ is large, multiples of $p$ will be few and far between, and finding this mate for $m$ will not be easy. So, say we agree to choose some cut off $B$, and discard any number from the sequence that is not $B$-smooth. Let us look at the numbers that are not discarded, these $B$-smooth numbers. The following lemma is crucial.

LEMMA. If $m_1, m_2, \ldots, m_k$ are positive $B$-smooth integers, and if $k > \pi(B)$ (where $\pi(B)$ denotes the number of primes in the interval $[1, B]$), then some non-empty subsequence of $(m_i)$ has product a square.

PROOF. For a $B$-smooth number $m$, look at its *exponent vector* $\boldsymbol{v}(m)$. This is a simple concept. If $m$ has the prime factorization

$$m = \prod_{i=1}^{\pi(B)} p_i^{v_i},$$

where $p_i$ is the $i$th prime number and each exponent $v_i$ is a nonnegative integer, then $\boldsymbol{v}(m) = (v_1, v_2, \ldots, v_{\pi(B)})$. Then a subsequence $m_{i_1}, \ldots, m_{i_t}$ has product a square if and only if $\boldsymbol{v}(m_{i_1}) + \ldots + \boldsymbol{v}(m_{i_t})$ has all even entries. That is, if and only if this sum of vectors is the 0-vector mod 2. Now the vector space $\boldsymbol{F}_2^{\pi(B)}$, where $\boldsymbol{F}_2$ is the finite field with 2 elements, has dimension $\pi(B)$. And we have $k > \pi(B)$ vectors. So this sequence of vectors is linearly dependent in this vector space. However, a linear dependency when the field of scalars is $\boldsymbol{F}_2$ is exactly the same as a subsequence sum being the 0-vector. □

This proof suggests an answer to our algorithmic question of how to find the subsequence. The proof uses linear algebra, and this subject is rife with algorithms. Actually, there is really only one algorithm that is taught in beginning linear algebra classes, namely Gaussian reduction of a matrix, and then there are many, many applications of this one technique. Well, now we have another application. With a collection of smooth numbers, form their exponent vectors, reduce these modulo 2, and then use Gaussian reduction to find a nonempty subsequence with sum the 0-vector modulo 2.

In particular, we shall find the concept of an exponent vector very useful in the sequel. Note that knowledge of the complete exponent vector of a number $m$ is essentially equivalent to knowledge of the complete prime factorization of $m$. Also note that though in the proof of the lemma we wrote out exponent vectors in traditional vector notation, doing so in general may itself be an exponentially huge problem, even if we "know" what the vector is. Luckily, exponent vectors are sparse creatures with most entries being zero, so that one can work with them modulo 2 in a more compact notation that merely indicates where the odd entries are.

**A proto-algorithm.** So now we have a proto-algorithm. We are given a number $n$ which is composite, has no prime factors up to its logarithm, and is not a power. We insist that $n$ not be a power in order to ensure that $n$ is divisible by at least two different odd primes. It is easy to check if a number is a power by taking roots via Newton's method, and for close calls to integers, exponentiating that integer to see if $n$ is a power of it. Our goal is to find a nontrivial factorization of $n$.

1. Choose a parameter $B$, and examine the numbers $x^2 - n$ for $B$-smooth values, where $x$ runs through the integers starting at $\lceil n^{1/2} \rceil$.
2. When you have more than $\pi(B)$ numbers $x$ with $x^2 - n$ being $B$-smooth, form the exponent vectors of these $B$-smooth numbers, and use linear algebra to find a subsequence $x_1^2 - n, \ldots, x_t^2 - n$ which has product a square, say $A^2$.
3. From the exponent vectors of the numbers $x_i^2 - n$, we can produce the prime factorization of $A$, and thus find the least nonnegative residue of $A$ modulo $n$, call it $a$. Find too the least nonnegative residue of the product $x_1 \ldots x_t$ modulo $n$, call it $b$.
4. We have $a^2 \equiv b^2 \pmod{n}$. If $a \not\equiv \pm b \pmod{n}$, then compute $\gcd(a-b, n)$. Otherwise, return to step 1, find additional smooth values of $x^2 - n$, find a new linear dependency in step 2, and attempt once again to assemble congruent squares to factor $n$.

There are clearly a few gaps in this proto-algorithm. One is a specification of the number $B$. Another is a fuller description of how one examines a number for $B$-smoothness, namely how one recognizes a $B$-smooth number when presented with one.

**Recognizing smooth numbers.** Let us look at the second gap first, namely the recognition of $B$-smooth numbers. Trial division is a candidate for a good method, even though it is very slow as a worst-case factorization algorithm. The point is that $B$-smooth numbers are very far from the worst-case of trial division, in fact they approach the best case. There are fewer than $B$ primes up to $B$, so there are very few trials to make. In fact the number of trial divisions is at most the maximum of $\log_2 n$ and $\pi(B)$.

But we can do better. Let us review the sieve of Eratosthenes. One starts with a list of the numbers from 2 to some bound $X$. Then one recognizes 2, the first unmarked number, as prime, and crosses off every second number starting at 4, since these are all divisible by 2 and hence not prime. The next unmarked number is 3, which is the next prime, and then every third number is crossed off after 3, and so on. If one completes this with primes up to $X^{1/2}$, then every remaining unmarked number is prime. But we are not so interested in the unmarked numbers, rather the marked ones. In fact, the more marked a number the better, since it is then divisible by many small primes. This sieve procedure can rather easily be made completely rigorous. Indeed, interpret making a "mark" as taking the number in the relevant location and replacing it with its quotient by the prime being sieved. In addition, sieve by higher powers of the primes as well, again dividing by the underlying prime. If one does this procedure with the primes up to $B$ and their higher powers, the $B$-smooth numbers up to $X$ are detected by locations that have been transformed into the number 1. The time to do this is proportional to $X$ times the sum of the reciprocals of the primes up to $B$ and their powers up to $X$. This sum, as we will see in (4), is about $\log \log B$. Thus, in time proportional to $X \log \log B$ (for $B$ at least 16, say, to have $\log \log B > 1$) we can locate all of the $B$-smooth numbers up to $X$. That is, *per number*, we are spending only about $\log \log B$ steps on average. This count compares with about $B$ steps per number for trial division, a very dramatic comparison indeed. Further, there is a trick for reducing the complexity of the individual steps in the sieve involving the subtraction of low-precision logarithms to simulate the division. So it is a win-win for sieving.

However, we are not so interested in locating smooth numbers in the interval from 2 to $X$, but rather in the polynomial sequence $x^2 - n$ as $x$ runs. This is not a big hurdle, and essentially the same argument works. What makes the sieve of Eratosthenes work is the regular places in the sequence where we see a multiple of $p$. This holds as well for any polynomial sequence. One solves the quadratic congruence $x^2 - n \equiv 0 \pmod{p}$. For $p > 2$ there are either no solutions, 2 solutions, or in the case that $p | n$, just 1 solution. Of course the generic cases are 0 and 2 solutions, and each should occur roughly half the time. If there are no solutions, then there is no sieving at all. If there are 2 solutions, say $a_1, a_2$, where $a_i^2 - n \equiv 0 \pmod{p}$, then we find the multiples of the prime $p$ in the sequence for each $x \equiv a_i \pmod{p}$, where $i = 1, 2$. For example, say $n \equiv 4 \pmod{7}$, so that $a_1 = 2, a_2 = 5$. Then the values of $x$ where 7 divides $x^2 - n$ can be found in quite regular places, namely those values of $x$ that are congruent to 2 or 5 modulo 7. After an initial value is found that is 2 (mod 7), one can find the remaining places in this residue class by adding 7's sequentially to the location number, and similarly for the 5 (mod 7) residue class. One has a similar result for higher powers of $p$ and also for the special case $p = 2$. As with the sieve of Eratosthenes above, the number of steps we will spend per polynomial value is proportional to just $\log \log B$. That is, it takes about as much time to tell if a value is smooth as it does just to look at the value, as long as we amortize the time over many members of the polynomial sequence. So this is how we recognize $B$-smooth values of $x^2 - n$: we use a *quadratic sieve*.

**An optimization problem.** Next, in our proto-algorithm, there should be some guidance on the choice of the parameter $B$. Of course, we would like to choose $B$ *optimally*, that is, we should choose a value of $B$ which minimizes the time spent to factor $n$. This optimization problem must balance two principal forces. On the one hand, if $B$ is chosen very small, then we do not have to find very many $B$-smooths in the sieve, and the matrix we will be dealing with will be small. But numbers that are $B$-smooth with a very small value of $B$ are very sparsely distributed among the natural numbers, and so we may have to traverse a very long sequence of $x$ values to get even one $B$-smooth value of $x^2 - n$, much less the requisite number of them. On the other hand, if $B$ is chosen large, then $B$-smooth numbers are fairly common, and perhaps we will not have such a hard time finding them in the polynomial sequence $x^2 - n$. But, we will need to find a great many of them for large $B$, and the matrix we will be dealing with will be large.

So, the optimization problem must balance these conflicting forces. To solve this problem, we should have a measure of the likelihood that a value $x^2 - n$ is $B$-smooth. This in fact is a very hard problem in analytic number theory, one that is essentially unsolved in the interesting ranges. However, the number $n$ we are trying to factor may not be up on the latest research results! That is, perhaps we should be more concerned with what is *true* rather than what is *provable*, at least for the design of a practical algorithm. This is where heuristics enter the fray. Let us assume that a polynomial value is just as likely to be smooth as a random number of the same magnitude. This assumption has been roughly borne out in practice with the quadratic sieve algorithm, as well as other factorization methods such as the number field sieve; see the survey [Stevenhagen 2008].

What is the order of magnitude of our polynomial values? If $x$ runs in the interval

$$[n^{1/2}, n^{1/2} + n^\varepsilon],$$

where $0 < \varepsilon < 1/2$, then the numbers $x^2 - n$ are all smaller than approximately $2n^{1/2+\varepsilon}$. Let $X$ be this bound, and let us ask for the chance that a random number up to $X$ is $B$-smooth.

**An analytic tidbit.** In analytic number theory we use the notation $\psi(X, B)$ for the counting function of the $B$-smooth numbers in the interval $[1, X]$. That is,

$$\psi(X, B) = \#\{m : 1 \leq m \leq X, m \text{ is } B\text{-smooth}\}.$$

Let us try our hand at estimating this function in the special case that $B = X^{1/2}$. We can do this by an inclusion-exclusion, with a single exclusion, since no number up to $X$ is divisible by two primes $> X^{1/2}$. Thus,

$$\psi(X, X^{1/2}) = \lfloor X \rfloor - \sum_{X^{1/2} < p \leq X} \lfloor X/p \rfloor,$$

where $p$ runs over primes in the stated interval. This identity uses the fact that there are exactly $\lfloor X/p \rfloor$ multiples of $p$ in the interval $[1, X]$. And the multiples of $p$ are definitely not $x^{1/2}$-smooth, so must be excluded. By removing the floor functions in the above display, we create an error bounded by $\pi(X)$, so that the prime number theorem implies that

$$\psi(X, X^{1/2}) = X\left(1 - \sum_{X^{1/2} < p \leq X} 1/p\right) + O(X/\log X).$$

We now use a theorem of Mertens stating that we have

$$\sum_{p \leq t} 1/p = \log \log t + C + O(1/\log t), \tag{4}$$

for a particular constant $C$. This theorem predates the prime number theorem, but can also be derived from it. Using (4) we obtain

$$\sum_{X^{1/2} < p \leq X} 1/p = \sum_{p \leq X} 1/p - \sum_{p \leq X^{1/2}} 1/p$$
$$= \log \log X - \log \log(X^{1/2}) + O\bigl(1/\log(X^{1/2})\bigr)$$
$$= \log 2 + O(1/\log X).$$

We thus have

$$\psi(X, X^{1/2}) = (1 - \log 2)X + O(X/\log X),$$

so that

$$\frac{\psi(X, X^{1/2})}{X} \sim 1 - \log 2 \text{ as } X \to \infty.$$

For example, about 30% of all numbers have no prime factors above their square root. It may seem surprising that such a large proportion of numbers can be built out of so few primes.

**The $u^u$ philosophy.** In fact one can easily see that the exact same argument shows that
$$\frac{\psi(X, X^{1/u})}{X} \sim 1 - \log u$$
for each fixed value of $u$ in the interval $[1, 2]$. But what of larger values of $u$? Here we have the celebrated result of K. Dickman that
$$\frac{\psi(X, X^{1/u})}{X} \sim \rho(u) \tag{5}$$
for each fixed $u \geq 1$, where $\rho(u)$ is the Dickman–de Bruijn function. This function is defined as the continuous solution to the differential difference equation $u\rho'(u) = -\rho(u-1)$ for $u > 1$, with initial condition $\rho(u) \equiv 1$ on the interval $[0, 1]$. The function $\rho(u)$ is always positive, and as $u$ grows, it decays to 0 roughly like $u^{-u}$. A result of E. R. Canfield, P. Erdős, and myself is that even if $u$ is not fixed, we still have something like (5) holding. Namely, for $X \to \infty, u \to \infty$ subject to $X^{1/u} > (\log X)^{1+\varepsilon}$, we have
$$\frac{\psi(X, X^{1/u})}{X} = u^{-(1+o(1))u}, \tag{6}$$
for any fixed $\varepsilon > 0$.

**The choice of the smoothness bound.** Let $B = X^{1/u}$. Then $X/\psi(X, X^{1/u})$ is approximately equal to the reciprocal of the probability that a random integer up to $X$ is $B$-smooth (it is exactly this reciprocal if $X$ is an integer), and so $X/\psi(X, X^{1/u})$ is about equal to the expected number of random trials of choosing numbers in $[1, X]$ to find one which is $B$-smooth. However, we would like to have about $\pi(B)$ numbers that are $B$-smooth and sieving allows us to spend about $\log \log B$ steps per candidate, so the expected number of steps to find our requisite stable of $B$-smooths is about $\pi(B)(\log \log B)X/\psi(X, X^{1/u})$. Our goal is to minimize this expression. It is a bit ungainly, but if we make the rough approximations $\pi(B) \log \log B \approx X^{1/u}$, $X/\psi(X, X^{1/u}) \approx u^u$, we are looking at the simpler expression
$$X^{1/u} u^u.$$

We would like to choose $u$ so as to minimize this expression. Take logarithms: so we are to minimize
$$\frac{1}{u} \log X + u \log u.$$
The derivative is 0 when $u^2(\log u + 1) = \log X$. Taking the log of this equation, we find that $\log u \sim \frac{1}{2} \log \log X$, so that
$$u \sim (2 \log X / \log \log X)^{1/2}$$

and
$$B = \exp\bigl((2^{-1/2} + o(1))(\log X \log \log X)^{1/2}\bigr). \tag{7}$$
This calculation allows us not only to find the key parameter $B$, but also to estimate the running time. With $B$ given by (7), we have
$$X^{1/u} u^u = \exp\bigl((2^{1/2} + o(1))(\log X \log \log X)^{1/2}\bigr), \tag{8}$$
and so this expression stands as the number of steps to find the requisite number of $B$-smooth values. Recall that $X = 2n^{1/2+\varepsilon}$ in our proto-algorithm, so that the expression in (8) is of the form $n^{o(1)}$. That is, we do not need to take $\varepsilon$ as fixed in the expression for $X$; it may tend to 0 slowly. Letting $X = n^{1/2+o(1)}$, we get
$$B = \exp\bigl((1/2 + o(1))(\log n \log \log n)^{1/2}\bigr),$$
$$X^{1/u} u^u = \exp\bigl((1 + o(1))(\log n \log \log n)^{1/2}\bigr),$$
where the second expression is the number of steps to do the sieving to find the requisite number of $B$-smooth polynomial values.

**A general principle, a moral, and three bullets.** The heuristic analysis above is instructive in that it can serve, in an almost intact manner, for many factoring algorithms. What may change is the number $X$, which is the bound on the auxiliary numbers that are examined for smoothness. In the quadratic sieve algorithm, we have $X$ just a little above $n^{1/2}$. In the number field sieve, this bound on auxiliary numbers is much smaller, it is of the form $n^{o(1)}$. Smaller numbers are much more likely to be smooth than larger numbers, and this general principle "explains" the asymptotic superiority of the number field sieve over the quadratic sieve; see [Stevenhagen 2008].

Our story has a moral. Smooth numbers are not an artifact, they were forced upon us once we decided to combine auxiliary numbers to make a square. In fact for random numbers this is a theorem of mine: the bound in (8), of $X^{1/u} u^u$ for the depth of the search for random auxiliary numbers below $X$ to form a square, is tight. So the heuristic passage to $B$-smooth numbers is justified — one is unlikely to be able to assemble a square from random numbers below $X$ in fewer choices than the bound in (8), even if one does not restrict to $B$-smooth numbers with $B$ the bound in (7).

There are several important points about smooth numbers that make them indispensable in many number-theoretic algorithms:

- Smooth numbers have a simple multiplicative structure.
- Smooth numbers are easy to recognize.
- Smooth numbers are surprisingly numerous.

See [Granville 2008] in this volume for much more about smooth numbers.

**Gaussian reduction.** If Gaussian reduction is used on the final matrix, our complexity bound is ruined. In fact, our matrix will be about $B \times B$, and the bound for the number of steps to reduce such a matrix is about $B^3$. With $B$ given as in (8), the time for the matrix step is then about $\exp((3/2 + o(1))(\log n \log \log n)^{1/2})$, which then would be the dominant step in the algorithm, and would indicate that perhaps a smaller value of $B$ than in (8) would be in order.

There are several thoughts about the matrix issue. First, Gaussian reduction, though it may be the only trick in the beginning undergraduate linear algebra book, is not the only trick we have. There are in fact fast asymptotic methods, that are practical as well. I refer to the Wiedemann coordinate recurrence method, the Lanczos method, etc. These reduce the complexity to the shape $B^{2+o(1)}$, and so the total asymptotic, heuristic complexity of the quadratic sieve becomes, as first intimated above, $\exp((1 + o(1))(\log n \log \log n)^{1/2})$. Also, Gaussian reduction is not quite as expensive as its complexity estimate indicates. In practice, the matrix starts out as quite sparse, and so for awhile fill-in can be avoided. And, the arithmetic in the matrix is binary, so a programmer may exploit this, using say a 32-bit word size in the computer, and so process 32 matrix entries at once.

The matrix poses another problem as well, and that is the space that is needed to store it and process it. This space problem is mitigated by using a sparse encoding of the matrix, namely a list of where the 1's are in the matrix. This sparse encoding might be used at the start until the matrix can be cut down to size somewhat.

In practice, people have found that it pays to slightly deoptimize $B$ on the low side. This in essence is a concession to the matrix problem, both to the space required and the time required. While sieving can easily be distributed to many unextraordinary processors, no one knows how to do this efficiently with the matrix, and so this final step might well hog the memory and time of a large expensive computer.

**Conclusion.** The quadratic sieve is a deterministic factoring algorithm with conjectured complexity

$$\exp((1 + o(1))(\log n \log \log n)^{1/2}).$$

It is currently the algorithm of choice for "hard" composites with about 20 to 120 digits. (By "hard" I mean that the number does not have a small prime factor that could be discovered by trial division or the elliptic curve method, nor does the number succumb easily to other special methods such as the $p-1$ factoring method or the special number field sieve.) For larger numbers, the number field sieve moves to the front, but this "viability border" between the quadratic sieve and the number field sieve is not very well defined, and shifts as new computer architectures come on line and when new variations of the underlying methods are developed.

There are many variations of the quadratic sieve which speed it up considerably (but do not change the asymptotic complexity estimate of $\exp((1+o(1))(\log n \log \log n)^{1/2})$; that is, the variations only affect the "$o(1)$"). The most important of these variations is the idea of using multiple polynomials, due to J. Davis and P. Montgomery.

Essentially all of the practical factoring methods beyond trial division are heuristic, though the elliptic curve method is "almost" rigorous. The fastest, rigorous factoring algorithm is a probabilistic method of H. W. Lenstra and me, with expected complexity $\exp((1+o(1))(\log n \log \log n)^{1/2})$, namely the same as for the quadratic sieve (though here a "fatter" $o(1)$ makes the rigorous method inferior in practice). The fastest, rigorous deterministic factoring algorithm is due to J. Pollard and to V. Strassen, with a complexity of $n^{1/4+o(1)}$.

We refer to [Crandall and Pomerance 2005] and [Granville 2008] for further reading on this, and for references to original papers and other surveys.

**Acknowledgements.** I am indebted to John Voight for the careful notes he took at my lecture, and to Lancelot Pecquet, who offered numerous improvements for an earlier version of this article. This article was written while I was a member of the Technical Staff at Bell Laboratories.

# References

[Crandall and Pomerance 2005] R. Crandall and C. Pomerance, *Prime numbers*, 2nd ed., Springer-Verlag, New York, 2005.

[Granville 2008] A. Granville, "Smooth numbers: computational number theory and beyond", pp. 267–323 in *Surveys in algorithmic number theory*, edited by J. P. Buhler and P. Stevenhagen, Math. Sci. Res. Inst. Publ. **44**, Cambridge University Press, New York, 2008.

[Pomerance 1996] C. Pomerance, "A tale of two sieves", *Notices Amer. Math. Soc.* **43**:12 (1996), 1473–1485.

[Schoof 2008] R. J. Schoof, "Four primality testing algorithms", pp. 101–125 in *Surveys in algorithmic number theory*, edited by J. P. Buhler and P. Stevenhagen, Math. Sci. Res. Inst. Publ. **44**, Cambridge University Press, New York, 2008.

[Stevenhagen 2008] P. Stevenhagen, "The number field sieve", pp. 83–100 in *Surveys in algorithmic number theory*, edited by J. P. Buhler and P. Stevenhagen, Math. Sci. Res. Inst. Publ. **44**, Cambridge University Press, New York, 2008.

CARL POMERANCE
DEPARTMENT OF MATHEMATICS
DARTMOUTH COLLEGE
HANOVER, NH 03755-3551
(603) 646-2415
  carl.pomerance@dartmouth.edu

# The number field sieve

PETER STEVENHAGEN

ABSTRACT. We describe the main ideas underlying integer factorization using the number field sieve.

CONTENTS

| | |
|---|---|
| 1. Introduction | 83 |
| 2. Factoring by congruent squares | 84 |
| 3. Number rings | 86 |
| 4. Sieving for smooth elements | 88 |
| 5. Primes dividing $a - b\alpha$ | 90 |
| 6. Sieving and linear algebra | 91 |
| 7. Nonmaximality of $\mathbb{Z}[\alpha]$ | 93 |
| 8. Finiteness results from algebraic number theory | 95 |
| 9. Quadratic character columns | 96 |
| 10. Square root extraction | 97 |
| 11. Running time | 98 |
| References | 100 |

## 1. Introduction

The number field sieve is a factoring algorithm that tries to factor a hard composite number by exploiting factorizations of smooth numbers in a well-chosen algebraic number field. It is similar in nature to the quadratic sieve algorithm, but the underlying number theory is less elementary, and the actual implementation involves a fair amount of optimization of the various parameters.

The key idea of the algorithm, the use of smooth numbers in number rings different from $\mathbb{Z}$, was proposed in 1988 by Pollard. Many people have contributed theoretical and practical improvements since then. An excellent reference for many of the details left out in this paper is [Lenstra and Lenstra 1993]. It contains a complete bibliography of the early years of the number field sieve, as well as original contributions by most of the main developers of the algorithm.

Among the successes of the algorithm are the 2005 factorization of the 663-bit RSA challenge number

$$\text{RSA-}200 = p_{100} \cdot q_{100}$$

into a product of two primes of 100 decimal digits each, and the factorization in 2006 of the 275-digit Cunningham number

$$6^{353} - 1 = 5 \cdot p_{120} \cdot p_{155}$$

into a product of 5 and two primes of 120 and 155 digits, respectively. Unlike RSA-200, the second number has a special form that can be exploited by the number field sieve. No other algorithm is currently capable of factoring integers of this size.

For the quadratic sieve algorithm and the elliptic curve method, the conjectural asymptotic expected running time for factoring a large number $n$ is

$$\exp \sqrt{\log n \log \log n},$$

which is on a log-log scale halfway between exponential and polynomial. The number field sieve conjecturally improves this bound to

(1.1) $$\exp\bigl(c(\log n)^{1/3}(\log \log n)^{2/3}\bigr),$$

where the constant $c = (64/9)^{1/3} \approx 1.93$ can be lowered to $(32/9)^{1/3} \approx 1.53$ if we are dealing with numbers $n$ of the special form explained in section 3.

## 2. Factoring by congruent squares

The number field sieve is one of the algorithms that tries to factor $n$ by producing *congruent squares* modulo $n$, as explained in [Pomerance 2008]. For this we will assume from now on that $n$ is odd, composite and not a power of a prime number. Note that each of these conditions can easily be checked for large $n$. One tries to find integers $x$ and $y$ satisfying $x \not\equiv \pm y \bmod n$ and

(2.1) $$x^2 \equiv y^2 \bmod n.$$

In this case, $\gcd(x - y, n)$ is a non-trivial factor of $n$. As at least half of all pairs $(x, y)$ of invertible residue classes modulo $n$ satisfying (2.1) satisfy $x \not\equiv \pm y \bmod n$, we may expect to find a non-trivial factor of $n$ within a few tries if we can produce solutions $(x, y)$ to (2.1) in a pseudo-random way.

An old factoring algorithm based on this idea is the *continued fraction method*. It uses the convergents $x_i/y_i \in \mathbb{Q}$ ($i = 1, 2, \ldots$) occurring in the continued fraction expansion of $\sqrt{n}$ as defined in [Buhler and Wagon 2008]. These fractions, which can be computed from simple two-term recursive relations for the integers $x_i$ and $y_i$, provide rational approximations to the real number $\sqrt{n}$. The associated integers

$$Q_i = x_i^2 - ny_i^2$$

are of absolute value at most $2\sqrt{n}$, and we may hope to be able to find a fair number of these $Q_i$ which are *smooth*. As we saw in [Pomerance 2008], it is a matter of linear algebra over the field of two elements to construct a square from a sufficiently large set of integers that factor over a given factor base. From every *square* $y^2 = \prod_{i \in I} Q_i$, we find a solution

$$\left(\prod_{i \in I} x_i\right)^2 \equiv y^2 \bmod n$$

to the congruence (2.1).

The *quadratic sieve* replaces the integers $Q_i$ in the continued fraction algorithm by the values of the polynomial

$$Q(X) = X^2 - n.$$

For integers $x$ satisfying $|x - \sqrt{n}| < M$ for some small bound $M$, the absolute value of $Q(x)$ is not much larger than $2M\sqrt{n}$. As $M$ has to be large enough to allow for a reasonable supply of $x$-values, the numbers $Q(x)$ we encounter here are somewhat larger than the $Q_i$ above. However, the advantage of using values of the polynomial $Q$ is that the values of $x$ for which $Q(x)$ is smooth may be detected by *sieving*.

From the smooth values of $Q$, we construct a *square* $y^2 = \prod_{x \in S} Q(x)$ and a solution

$$\left(\prod_{s \in S} x\right)^2 \equiv y^2 \bmod n$$

to the basic congruence (2.1) exactly as for the continued fraction algorithm.

The algebraic description one may give of both methods is as follows. We have constructed squares $(x^2, y^2) \in \mathbb{Z} \times \mathbb{Z}$ whose images under the reduction map

$$\mathbb{Z} \times \mathbb{Z} \xrightarrow{\phi} \mathbb{Z}/n\mathbb{Z} \times \mathbb{Z}/n\mathbb{Z}$$
$$(x_i^2, x_i^2 - ny_i^2) \mapsto (x_i^2, x_i^2)$$
$$(x^2, x^2 - n) \mapsto (x^2, x^2)$$

lie in the *diagonal*. If we are lucky, $\phi(x, y)$ does *not* land in

$$D = \{(x, \pm x) : x \in \mathbb{Z}/n\mathbb{Z}\}$$

and we find a non-trivial factor of $n$. As $(x^2, y^2)$ is constructed in such a way that $\phi(x, y)$ has no obvious reason to always end up in $D$, we expect to be lucky in at least half of all cases.

The construction of squares in the continued fraction and quadratic sieve methods requires many auxiliary numbers $Q_i$ or $Q(x)$ of size $O(\sqrt{n})$ to be smooth. Asymptotically, the superior performance of the number field sieve stems from the fact that it is a sieving method that requires smaller auxiliary numbers to be smooth: they are of size

$$\exp\bigl(c'(\log n)^{2/3}(\log\log n)^{1/3}\bigr)$$

with $c' = (64/3)^{1/3} \approx 2.77$. Informally phrased, the length of these numbers is not *half* of the length of $n$, but only the *2/3-rd power* of the length of $n$. This improvement is obtained by replacing $\mathbb{Z} \times \mathbb{Z}$ by $\mathbb{Z} \times \mathbb{Z}[\alpha]$ for a suitable *number ring* $\mathbb{Z}[\alpha]$ and producing squares $(x^2, \gamma^2)$ with diagonal image under the reduction map

$$\mathbb{Z} \times \mathbb{Z}[\alpha] \xrightarrow{\phi} \mathbb{Z}/n\mathbb{Z} \times \mathbb{Z}/n\mathbb{Z}.$$

Exactly as before, this yields a solution

(2.2) $$x^2 \equiv \phi(\gamma)^2 \bmod n$$

to our basic congruence (2.1).

## 3. Number rings

A number field is a finite field extension of the field $\mathbb{Q}$ of rational numbers, and a *number ring* [Stevenhagen 2008] is by definition a subring of a number field. The basic type of number ring used in the number field sieve is the ring

$$\mathbb{Z}[\alpha] = \mathbb{Z}[X]/f\mathbb{Z}[X]$$

generated by a *formal zero* $\alpha = (X \bmod f\mathbb{Z}[X])$ of some *irreducible* polynomial $f \in \mathbb{Z}[X]$ of degree $d \geq 1$. The elements of this ring are finite expressions $\sum_{i \geq 0} a_i \alpha^i$ with $a_i \in \mathbb{Z}$. One may obtain an embedding $\mathbb{Z}[\alpha] \subset \mathbb{C}$ by taking $\alpha$ to be a *complex* zero of $f$. Note that even though the field of fractions of a number ring is always of the form $\mathbb{Q}(\alpha)$ for some root $\alpha$ of an irreducible polynomial in $\mathbb{Z}[X]$, there are many number rings that are of finite rank over $\mathbb{Z}$ but not of the form $\mathbb{Z}[\alpha]$.

We will take $f$ to be a *monic* irreducible polynomial in $\mathbb{Z}[X]$, such that

$$\mathbb{Z}[\alpha] = \mathbb{Z} \cdot 1 \oplus \mathbb{Z} \cdot \alpha \oplus \mathbb{Z} \cdot \alpha^2 \oplus \ldots \oplus \mathbb{Z} \cdot \alpha^{d-1}$$

is *integral* over $\mathbb{Z}$. It is an *order* in the field of fractions $\mathbb{Q}(\alpha)$ of $\mathbb{Z}[\alpha]$.

The *norm*

$$N : \mathbb{Q}(\alpha) \to \mathbb{Q}$$

takes $x \in \mathbb{Q}(\alpha)$ to the determinant of the multiplication-by-$x$ map on the $\mathbb{Q}$-vector space $\mathbb{Q}(\alpha)$. It is multiplicative, and for non-zero $x \in \mathbb{Z}[\alpha]$, the absolute value
$$|N(x)| = \#(\mathbb{Z}[\alpha]/x\mathbb{Z}[\alpha]) \in \mathbb{Z}$$
of the norm of $x$ measures the "size" of $x$.

(3.1) **Example.** The best known example of a number ring with $d = \deg(f) > 1$ is probably the ring $\mathbb{Z}[i]$ of *Gaussian integers* obtained by putting $f = X^2 + 1$ and $\alpha = i = \sqrt{-1}$. For this ring, the norm function is given by the simple formula
$$N(a+bi) = a^2 + b^2. \qquad //$$

More generally, one can find the norm of an element $x = a - b\alpha \in \mathbb{Q}(\alpha)$ from the irreducible polynomial $f = \sum_{i=0}^{d} c_i X^i$ of $\alpha$ as

(3.2) $$N(a - b\alpha) = b^d f(a/b) = \sum_{i=0}^{d} c_i a^i b^{d-i}.$$

For polynomial expressions $g(\alpha)$ in $\alpha$ of higher degree the norm can efficiently be computed from the resultant of $f$ and $g$, but we won't need this.

For a number ring $\mathbb{Z}[\alpha]$ to be useful in factoring $n$, it needs to come with a reduction homomorphism
$$\phi : \mathbb{Z}[\alpha] \to \mathbb{Z}/n\mathbb{Z}.$$
Giving such a homomorphism amounts to giving a zero $m = \phi(\alpha)$ of $f$ modulo $n$. In order to have a small number ring $\mathbb{Z}[\alpha]$, one tries to choose a polynomial $f$ of moderate degree — in practice $d$ is usually between 3 and 10, although its optimal value does slowly tend to infinity with $n$ — and having small coefficients. This is not an easy problem, but for certain *special* $n$ one can find very small $f$.

(3.3) **Example.** For the Fermat number
$$n = F_9 = 2^{2^9} + 1 = 2^{512} + 1$$
the polynomial $f = X^5 + 8$ is irreducible in $\mathbb{Z}[X]$ and satisfies
$$f(2^{103}) = 2^{515} + 8 = 8n \equiv 0 \bmod n.$$
Similarly, for the record factorization of the Cunningham number $n = 6^{353} - 1$ mentioned in the introduction, the polynomial $f = X^6 - 6$ is irreducible in $\mathbb{Z}[X]$ and satisfies
$$f(6^{59}) = 6^{354} - 6 = 6n \equiv 0 \bmod n. \qquad //$$

For numbers $n$ of the special form $n = r^e - s$, with $r$, $s$ and $e$ small, one can find a small polynomial $f$ as in the example. For *general* $n$ we cannot hope to be so lucky in finding $f$, and one has to deal with *large* number rings. The *special* and the *general* number field sieve stand for the versions of the algorithm corresponding to these two cases. As is to be expected, the special number field sieve has a somewhat better conjectural running time, and this is reflected by the size of the record factorizations for each of these versions.

We will mainly be concerned with the case of general integers $n$ to be factored. For such $n$, the *base m method* yields a polynomial $f$ of any desired degree $d > 1$ such that $m = m(d)$ is a zero of $f$ modulo $n$. One simply puts

$$m = \text{integer part of } n^{1/d}$$

and writes $n$ in base $m$ as

$$n = \sum_{i=0}^{d} c_i m^i.$$

Then $f = \sum_{i=0}^{d} c_i X^i$ is a polynomial in $\mathbb{Z}[X]$ satisfying $f(m) = n$. In realistic situations $n$ is much larger than $d$, which ensures that $f$ will be *monic*; one may further assume that $f$ is *irreducible*, as non-trivial factors of $f$ yield non-trivial factors of $n$.

From $|c_i| < m < n^{1/d}$ we deduce that the discriminant $\Delta(f)$ of $f$ satisfies

(3.4) $$|\Delta(f)| < d^{2d} n^{2-3/d}.$$

As $|\Delta(f)|$ often exceeds $n$, we cannot hope to be able to factor $\Delta(f)$.

## 4. Sieving for smooth elements

Having chosen $d$, $f$ and $m$ as above, we can combine the ordinary reduction map on $\mathbb{Z}$ with our reduction map on $\mathbb{Z}[\alpha]$ to obtain a ring homomorphism

$$\mathbb{Z} \times \mathbb{Z}[\alpha] \xrightarrow{\phi} \mathbb{Z}/n\mathbb{Z} \times \mathbb{Z}/n\mathbb{Z}$$
$$\left(x, \sum_{i=0}^{d-1} a_i \alpha^i\right) \longmapsto \left(x \bmod n, \sum_{i=0}^{d-1} a_i m^i \bmod n\right).$$

By construction, the elements $(a - bm, a - b\alpha)$ have $\phi$-image in the diagonal. In order to combine them into squares, we need to find sets $S$ of coprime integer pairs $(a, b)$ for which we have

(4.1) $$\prod_{(a,b) \in S} (a - bm) \quad \text{is a square in } \mathbb{Z};$$

(4.2) $$\prod_{(a,b) \in S} (a - b\alpha) \quad \text{is a square in } \mathbb{Z}[\alpha].$$

As in the case of the quadratic sieve, this is in principle done by sieving for *smooth* elements $(a - bm, a - b\alpha)$ and combining them into a square via linear algebra methods over $\mathbb{F}_2$. The details are however more involved.

Let us define an element $(a - bm, a - b\alpha) \in \mathbb{Z} \times \mathbb{Z}[\alpha]$ to be $y$-smooth if $a - bm$ is a $y$-smooth rational integer and $a - b\alpha$ is a $y$-smooth algebraic integer in $\mathbb{Z}[\alpha]$. The latter condition simply means that the norm $N(a - b\alpha) \in \mathbb{Z}$ is a $y$-smooth integer. On the rational side, the procedure to find a set $S$ for which (4.1) holds is more or less standard. We pick a universe

$$U = \{(a, b) : |a| \leq u, \quad 0 < b \leq u \quad \text{and} \quad \gcd(a, b) = 1\}$$

of coprime integer pairs $(a, b)$ depending on a parameter $u$.

Using the factor base $B_1$ consisting of primes $p \leq y$ and a sign-bit, we can determine the subset of pairs $(a, b) \in U$ for which $a - bm$ is $y$-smooth by sieving. here we have a 2-dimensional array of pairs $(a, b)$ over which the sieving with the primes in $B_1$ needs to be done. One may simply choose to sieve over $a$ for each value of $b$, but there exist other methods than this straightforward line-by-line sieving. Recent record factorizations have used a combination of different sieving methods.

On the algebraic side, the pairs $(a, b) \in U$ for which $N(a - b\alpha)$ is $y$-smooth can also be found by sieving with the primes in $B_1$, since we see from (3.2) that the norms

(4.3) $$N(a - b\alpha) = b^d f(a/b) = \sum_{i=0}^{d} c_i a^i b^{d-i}$$

are the $(a, b)$-values of the homogeneous polynomial $f(X, Y)$. But it is *not* sufficient to find elements $a - b\alpha$ whose norm factors over our factor base $B_1$. This information will only enable us to construct a product $\prod_{(a,b) \in S}(a - b\alpha)$ with square norm, which is far too weak to imply (4.2). A square in $\mathbb{Z}[\alpha]$ certainly has square norm, but the converse only holds in the trivial case $\mathbb{Z}[\alpha] = \mathbb{Z}$, where the norm of an element is the element itself.

(4.4) **Example.** In the ring of Gaussian integers $\mathbb{Z}[i]$ we have

$$N(3 + 4i) = 3^2 + 4^2 = 5^2 = N(5).$$

Now $3 + 4i = (2 + i)^2$ is indeed a square, but $5 = (2 + i)(2 - i)$ is not.  //

The problem we encounter is that *different* prime divisors of an element $x \in \mathbb{Z}[\alpha]$ can give rise to the *same* prime factor $p$ in its norm $N(x)$. This forces us to keep track of "prime factors" of $x$ *in the ring* $\mathbb{Z}[\alpha]$.

## 5. Primes dividing $a - b\alpha$

The theory of prime divisors in number rings lies at the very heart of algebraic number theory, and understanding the workings of the number field sieve is not possible without entering this area. Rather than assuming the more extensive exposition on the arithmetic of number rings in [Stevenhagen 2008], we use the concrete example of our number ring $\mathbb{Z}[\alpha]$ to illustrate and motivate the more general statements of algebraic number theory in that paper.

Let $R$ be any number ring. A *prime* in $R$ is a non-zero prime ideal $\mathfrak{p} \subset R$. The residue class ring $F = R/\mathfrak{p}$ of a prime is a finite field. We say that $\mathfrak{p}$ *divides* an element $x \in R$ if $x$ is contained in $\mathfrak{p}$. For the number ring $\mathbb{Z}$ the primes $p\mathbb{Z}$ correspond to the prime numbers $p \in \mathbb{Z}$. A prime $\mathfrak{p} \subset R$ *lies over* a unique prime $p\mathbb{Z} = \mathfrak{p} \cap \mathbb{Z}$ of $\mathbb{Z}$. The corresponding prime number $p$ is the characteristic of the field $F$. It is the unique prime number contained in $\mathfrak{p}$. The *degree* of $\mathfrak{p}$ is the degree of $F$ over its prime field $\mathbb{F}_p$.

A *prime of degree* 1 in $\mathbb{Z}[\alpha]$ is just the kernel $\mathfrak{p}$ of a ring homomorphism

$$\pi : \mathbb{Z}[\alpha] \longrightarrow \mathbb{F}_p$$

for some prime number $p$. As $\pi$ may be specified by giving the rational prime $p$ together with the zero $r_p = \pi(\alpha) \in \mathbb{F}_p$ of ($f \bmod p$) to which $\alpha$ is mapped, we use the ad hoc notation $(p; r_p)$ to denote $\mathfrak{p} = \ker \pi$.

Primes of degree 1 are the only primes we need for the number field sieve. Indeed, suppose that we have $(a, b) \in U$ as in the previous section, and that $\mathfrak{p}$ is a prime over $p$ dividing $a - b\alpha$. Then we have $p \nmid b$, since $p | b$ would imply $a \in \mathfrak{p} \cap \mathbb{Z} = p\mathbb{Z}$, contradicting the coprimality of $a$ and $b$. From $\bar{a} = \bar{b}\bar{\alpha} \in F = \mathbb{Z}[\alpha]/\mathfrak{p}$ we find that $r_p = \bar{\alpha} = ab^{-1} \bmod p$ is a zero of ($f \bmod p$), and that

$$\mathfrak{p} = p\mathbb{Z}[\alpha] + (a - b\alpha)\mathbb{Z}[\alpha]$$

is the kernel of the map

$$\mathbb{Z}[\alpha] \xrightarrow{\pi} \mathbb{F}_p$$
$$\alpha \longmapsto (ab^{-1} \bmod p).$$

Thus $\mathfrak{p}$ is the prime $(p; r_p)$ of degree 1 that is generated by $p$ and $r_p - \alpha$, with $r_p = ab^{-1} \bmod p$ a zero of ($f \bmod p$). Conversely, a prime $(p; r_p)$ of degree 1 divides $a - b\alpha$ if we have $r_p = ab^{-1} \bmod p$.

From our norm formula (3.2), we see that $ab^{-1} \bmod p$ is a zero of ($f \bmod p$) if and only if $N(a - b\alpha)$ is divisible by $p$. We conclude that for a rational prime number $p$, there is a prime divisor $\mathfrak{p}$ of $a - b\alpha$ in $\mathbb{Z}[\alpha]$ that lies over $p$ if and only if $p$ divides the norm $N(a - b\alpha)$. If $p$ divides $N(a - b\alpha)$, the prime $\mathfrak{p} = (p; r_p)$

with $r_p = (ab^{-1} \mod p)$ is the unique such prime, and we call

$$e_{\mathfrak{p}}(a - b\alpha) = \mathrm{ord}_p(N(a - b\alpha))$$

the *exponent* to which $\mathfrak{p}$ occurs in $a - b\alpha$. For the primes $\mathfrak{p}$ of $\mathbb{Z}[\alpha]$ that do not divide $a - b\alpha$, we put $e_{\mathfrak{p}}(a - b\alpha) = 0$. We then have the following fundamental fact.

(5.1) **Lemma.** *For each prime $\mathfrak{p}$ of degree 1, the exponent $e_{\mathfrak{p}}$ extends to a homomorphism*

$$e_{\mathfrak{p}} : \mathbb{Q}(\alpha)^* \to \mathbb{Z}.$$

This Lemma is slightly less innocent than it may appear at first sight, and we will define $e_{\mathfrak{p}}(x)$ for arbitrary $x \in \mathbb{Q}(\alpha)^*$ in (7.4). There is actually no need to restrict to primes of degree 1, but we do so as we have not defined the exponent at other primes. For our purposes, it suffices to know that we have $e_{\mathfrak{p}}(x) = 0$ whenever $x$ is a product of elements $a - b\alpha$ with $(a, b) \in U$ and $\mathfrak{p}$ is a prime of degree at least 2.

## 6. Sieving and linear algebra

On the rational side, we already chose a factor base $B_1$ consisting of the primes $p \leq y$ and a sign bit. For the factorization of our numbers $a - b\alpha$ in $\mathbb{Z}[\alpha]$, we choose a factor base $B_2$ consisting of all primes $(p; r_p)$ with $p \leq y$ prime and $r_p \in \mathbb{F}_p$ a root of ($f \mod p$). There may be several primes in $B_2$ lying over a given rational prime $p$, and the notation $(p; r_p)$ enables us to distinguish between such primes, and to identify the prime that accounts for the $p$-contribution (if any) to $N(a - b\alpha)$.

For each rational $p$, there are at most $d = \deg(f)$ values $r_p$. On average, there is 1 root of ($f \mod p$) in $\mathbb{F}_p$ if we let $y$ tend to infinity. This elegant result of Kronecker, which was generalized by Frobenius, is now often proved as a corollary of the Chebotarev density theorem [Stevenhagen and Lenstra 1996]. We deduce that both $B_1$ and $B_2$ are of size $y^{1+o(1)}$.

The combination of rational and algebraic sieving yields a subset $U' \subset U$ of pairs $(a, b) \in U$ that give rise to a $y$-smooth factorization of $a - bm$ in $\mathbb{Z}$ and a $y$-smooth factorization of $a - b\alpha$ in $\mathbb{Z}[\alpha]$. Such a pair $(a, b) \in U'$, together with the exponents of the rational primes $p \in B_1$ in $a - bm$ and the exponents of the algebraic primes $\mathfrak{p}$ in $a - b\alpha$, is usually referred to as a *relation*. All exponents are taken modulo 2, so they can be stored in a single bit.

In order to obtain dependencies between the exponent vectors of elements in $U'$, the number $\#U'$ of relations should exceed $\#B_1 + \#B_2$. For large factorizations, collecting sufficiently many relations may take several years of computer time. As different computers can independently test elements $(a, b) \in U$ for smoothness, distribution of the computation over a large number of computers is usually necessary to perform this step of the algorithm in practice.

The set $U'$, which may consist of millions of relations, is often so large that the linear algebra step over $\mathbb{F}_2$ needs to be performed on a computer that is equipped to handle huge amounts of data. It is important that the matrix of exponents is a very *sparse* matrix, which can be transformed into a much smaller dense matrix before it is given to the reduction algorithm that yields the desired dependencies. A practical reduction algorithm, such as the so-called block Lanczos method, may run for several days on a single large computer. In this case, distribution of the problem over more computers is not an easy matter.

Every dependency in the matrix of exponent vectors coming from the pairs $(a, b) \in U'$ corresponds to a subset $S \subset U'$ such that the following two conditions are satisfied:

(6.1) $\displaystyle\prod_{(a,b)\in S} (a - bm)$ is positive with even exponents at all primes $p \in \mathbb{Z}$;

(6.2) $\displaystyle\prod_{(a,b)\in S} (a - b\alpha)$ has even exponents at all primes $\mathfrak{p} \subset \mathbb{Z}[\alpha]$.

What we need is the validity of (4.1) and (4.2) in order to obtain the required square in $\mathbb{Z} \times \mathbb{Z}[\alpha]$. It is a simple and well known fact that (6.1) implies (4.1): requiring positivity is enough to produce true squares from integers having even exponents at all prime numbers. The situation is not so simple in $\mathbb{Z}[\alpha]$: several obstructions may prevent the validity of the implication (6.2) $\Rightarrow$ (4.2). Writing $\beta = \prod_{(a,b)\in S}(a - b\alpha)$ for the element in (6.2), they are the following.

(6.3) The ring $\mathbb{Z}[\alpha]$ is possibly not the *ring of integers* $\mathcal{O}$ of $\mathbb{Q}(\alpha)$. The ring of integers, which is the *maximal* order in $\mathbb{Q}(\alpha)$, is the "textbook ring" for which the theorem of unique prime ideal factorization holds. If we have $\mathbb{Z}[\alpha] \neq \mathcal{O}$, then (6.2) need not imply that $\beta\mathcal{O}$ is the square of an ideal.

(6.4) If $\beta\mathcal{O}$ is the square of some *ideal* $\mathfrak{c}$, then $\mathfrak{c}$ does not have to be a *principal* $\mathcal{O}$-ideal. This is exactly the reason why unique prime element factorization has to be replaced by unique prime ideal factorization in general number fields.

(6.5) If $\beta\mathcal{O}$ is the square of some principal ideal $\gamma\mathcal{O}$, we only have $\beta = \gamma^2$ up to multiplication by *units* in $\mathcal{O}$. This obstruction already occurs in the case for $\mathcal{O} = \mathbb{Z}$. Unlike $\mathbb{Z}$, the ring $\mathcal{O}$ usually has infinitely many units.

(6.6) If we do obtain an equality $\beta = \gamma^2$ in $\mathcal{O}$, we may have $\gamma \notin \mathbb{Z}[\alpha]$. If this happens, the reduction map $\phi$ is not defined on $\gamma$ and we do not obtain our final congruence (2.2).

Algebraic number theory provides the tools for dealing with all of these obstructions. In the next section, we will deal with the obstructions (6.3) and (6.6), which arise from the fact that $\mathbb{Z}[\alpha]$ may be strictly smaller than $\mathcal{O}$. Section 8 is devoted to the obstructions (6.4) and (6.5), which are classical and lie at the roots of algebraic number theory. It will be our aim to bound the index $[V:(V\cap\mathbb{Q}(\alpha)^{*2})]$ of the subgroup of true squares inside the group $V\subset\mathbb{Q}(\alpha)^*$ generated by the elements that meet condition (6.2).

## 7. Nonmaximality of $\mathbb{Z}[\alpha]$

The ring of integers $\mathcal{O}\subset\mathbb{Q}(\alpha)$, which consists by definition of all elements of $\mathbb{Q}(\alpha)$ that occur as the zero of some monic polynomial in $\mathbb{Z}[X]$, is the maximal order contained in $\mathbb{Q}(\alpha)$. It is free of rank $d=\deg(f)$ over $\mathbb{Z}$, and contains $\mathbb{Z}[\alpha]$ as a subring of *finite* index. There is the classical identity

(7.1) $$\Delta(f) = [\mathcal{O}:\mathbb{Z}[\alpha]]^2\cdot\Delta$$

relating the index $[\mathcal{O}:\mathbb{Z}[\alpha]]$ to the discriminant $\Delta(f)$ of the polynomial $f$ from (3.4) and the discriminant $\Delta$ of the number field $\mathbb{Q}(\alpha)$. As $\Delta$ is known to be a non-zero integer, we find that $[\mathcal{O}:\mathbb{Z}[\alpha]]$ is bounded by $|\Delta(f)|^{1/2}$. As we do not want to factor the possibly huge number $\Delta(f)$, we may not be able to determine $[\mathcal{O}:\mathbb{Z}[\alpha]]$ or $\mathcal{O}$. However, it is a standard fact that for any $x\in\mathcal{O}$, we have

$$f'(\alpha)\cdot x\in\mathbb{Z}[\alpha].$$

This is enough to deal with obstruction (6.6): we simply multiply our purported square in $\mathbb{Z}\times\mathbb{Z}[\alpha]$ by

$$(f'(m)^2, f'(\alpha)^2).$$

Then its square root gets multiplied by $(f'(m), f'(\alpha))$, so it will lie in $\mathbb{Z}\times\mathbb{Z}[\alpha]$. In order to keep an element that is invertible modulo $n$, we need to assume that $f'(m)$ is coprime to $n$. This is not a serious restriction as this condition is always satisfied in practice; if it isn't, we have found a factor of $n$ without applying the number field sieve!

(7.2) **Example.** Take $f=X^2+16$. Then the order $\mathbb{Z}[\alpha]=\mathbb{Z}[4i]$ has index 4 in the maximal order $\mathcal{O}=\mathbb{Z}[i]$ in $\mathbb{Q}(i)$. Example (4.4) shows that $3+\alpha=3+4i$ is a square in $\mathcal{O}$, but its square root $\gamma=2+\frac{1}{4}\alpha$ is not in $\mathbb{Z}[\alpha]$. However, the element $f'(\alpha)\cdot\gamma=2\alpha\cdot\gamma=4\alpha-8$ does lie in $\mathbb{Z}[\alpha]$. //

In order for an ideal $\mathfrak{c}\subset\mathcal{O}$ to be a square of some other ideal, it is necessary and sufficient that the exponents $\text{ord}_\mathfrak{q}(\mathfrak{b})$ are even at all primes $\mathfrak{q}$ of $\mathcal{O}$. This is an immediate corollary of the classical theorem of unique prime ideal factorization in $\mathcal{O}$. Now the primes $\mathfrak{q}$ of $\mathcal{O}$ coprime to the index $[\mathcal{O}:\mathbb{Z}[\alpha]]$ are "the same" as the ideals $\mathfrak{p}$ of $\mathbb{Z}[\alpha]$ coprime to the index. By this we mean that there is a natural bijection between the sets of such primes given by $\mathfrak{q} \mapsto \mathfrak{p} = \mathfrak{q} \cap \mathbb{Z}[\alpha]$. Moreover, if $\mathfrak{p}$ and $\mathfrak{q}$ are corresponding prime ideals, the inclusion map $\mathbb{Z}[\alpha] \subset \mathcal{O}$ induces an isomorphism of the local rings

$$\tag{7.3} \mathbb{Z}[\alpha]_{\mathfrak{p}} \xrightarrow{\sim} \mathcal{O}_{\mathfrak{q}}.$$

Both rings are discrete valuation rings, and the exponent $e_{\mathfrak{p}} : \mathbb{Q}(\alpha)^* \to \mathbb{Z}$ is in this case equal to the familiar prime ideal exponent $e_{\mathfrak{q}}$ for the ring of integers $\mathcal{O}$, which is multiplicative on the set of *all* non-zero $\mathcal{O}$-ideals, not just the principal ones.

For primes $\mathfrak{q}$ of $\mathcal{O}$ dividing the index $[\mathcal{O} : \mathbb{Z}[\alpha]]$, the situation is more complicated. There may be more primes $\mathfrak{q}$ lying above the same prime $\mathfrak{p} = \mathfrak{q} \cap \mathbb{Z}[\alpha]$, and even if $\mathfrak{p}$ has a single extension $\mathfrak{q}$ in $\mathcal{O}$, the natural map (7.3) need not be an isomorphism. If either of these happens, $\mathfrak{p}$ is said to be a *singular* prime of $\mathbb{Z}[\alpha]$. The other primes are the *regular* primes of $\mathbb{Z}[\alpha]$.

For a prime $\mathfrak{p}$ of $\mathbb{Z}[\alpha]$, we define the *exponent* at $\mathfrak{p}$ as the homomorphism $e_{\mathfrak{p}} : \mathbb{Q}(\alpha)^* \to \mathbb{Z}$ by

$$\tag{7.4} e_{\mathfrak{p}}(x) = \sum_{\mathfrak{q} \supset \mathfrak{p}} f(\mathfrak{q}/\mathfrak{p}) \, e_{\mathfrak{q}}(x),$$

where the sum ranges over the primes $\mathfrak{q} \subset \mathcal{O}$ lying over $\mathfrak{p}$, and $f(\mathfrak{q}/\mathfrak{p})$ is the degree of the residue field extension $\mathbb{Z}[\alpha]/\mathfrak{p} \subset \mathcal{O}/\mathfrak{q}$. This definition provides the extension of the homomorphism $e_{\mathfrak{p}}$ occurring in Lemma 5.1. For regular primes $\mathfrak{p}$, formula (7.4) reduces to $e_{\mathfrak{p}} = e_{\mathfrak{q}}$.

We now consider, inside the subgroup of $\mathbb{Q}(\alpha)^*$ that is generated by the elements $a - b\alpha \in \mathbb{Z}[\alpha]$ having $\gcd(a, b) = 1$, the group $V$ of those elements that have even exponents at the primes $\mathfrak{p}$ of $\mathbb{Z}[\alpha]$. We let $V_1 \subset V$ be the subgroup of elements $x \in V$ that have even exponents at all primes $\mathfrak{q}$ of $\mathcal{O}$, i.e., the elements $x \in V$ for which $x\mathcal{O}$ is the square of a $\mathcal{O}$-ideal. We have an injective homomorphism

$$V/V_1 \longrightarrow \bigoplus_{\mathfrak{q} \mid [\mathcal{O}:\mathbb{Z}[\alpha]]} \mathbb{Z}/2\mathbb{Z}$$

$$x \longmapsto (e_{\mathfrak{q}}(x) \bmod 2)_{\mathfrak{q}},$$

so $V/V_1$ is an $\mathbb{F}_2$-vector space of dimension bounded by the number of primes $\mathfrak{q}$ of $\mathcal{O}$ dividing the index $[\mathcal{O} : \mathbb{Z}[\alpha]]$. In view of (7.1), the number of rational primes dividing the index is no more than $\frac{1}{2} \log |\Delta(f)|$. For each of these primes there are at most $d = \deg(f)$ primes $\mathfrak{q}$ in $\mathcal{O}$ that divide it, so we find

$$\tag{7.5} \dim_{\mathbb{F}_2}(V/V_1) \leq \tfrac{1}{2} d \cdot \log \Delta(f).$$

This is a quantitative version of obstruction (6.3). Note that we have completely disregarded the fact that the elements of $V$ have even exponents at the singular primes of $\mathbb{Z}[\alpha]$. It is possible to obtain a slightly better upper bound for the index $[V : (V \cap \mathbb{Q}(\alpha)^{*2})]$ than that in 8.4 by taking this into account.

## 8. Finiteness results from algebraic number theory

Inequality (7.5) is the first step in bounding the successive $\mathbb{F}_2$-dimensions of the quotient spaces in the filtration

$$V \supset V_1 \supset V_2 \supset V_3 = V \cap \mathbb{Q}(\alpha)^{*2}.$$

Here $V_1$ is the subgroup from the previous section consisting of those $x \in V$ for which $x\mathcal{O}$ is an ideal square, and $V_2$ is the subgroup of those $x \in V$ for which $x\mathcal{O}$ is the square of a *principal* $\mathcal{O}$-ideal. Thus, the $\mathbb{F}_2$-spaces $V_1/V_2$ and $V_2/V_3$ measure the obstructions (6.4) and (6.5), respectively. We can bound their dimensions using two fundamental finiteness results from algebraic number theory.

The first result says that the class group of $\mathbb{Q}(\alpha)$, which is the group of all fractional $\mathcal{O}$-ideals modulo the subgroup of principal $\mathcal{O}$-ideals, is a finite abelian group. One can derive from [Lenstra 1992, Theorem 6.5] that its order $h$ can be bounded in terms of the degree $d$ and the discriminant $\Delta(f)$ of $f$ by

(8.1) $$h < |\Delta(f)|^{1/2} \cdot \frac{d-1 + \log|\Delta(f)|^{d-1}}{(d-1)!}.$$

We can map $V_1$ to the class group by sending $x \in V_1$ to the ideal class of the ideal $\mathfrak{a}$ satisfying $\mathfrak{a}^2 = x\mathcal{O}$. This map has kernel $V_2$, so we find the dimension of the $\mathbb{F}_2$-vector space $V_1/V_2$ to be bounded by $\log h$, yielding

(8.2) $$\dim_{\mathbb{F}_2}(V_1/V_2) \leq \frac{\log h}{\log 2}.$$

As the elements in $V_2$ are squares in $\mathbb{Q}(\alpha)^*$ up to multiplication by elements of the unit group $\mathcal{O}^*$, the order of $V_2/V_3$ does not exceed the order of $\mathcal{O}^*/\mathcal{O}^{*2}$. By the Dirichlet unit theorem [Stevenhagen 2008], the group $\mathcal{O}^*$ is the product of a finite cyclic group of roots of unity in $\mathbb{Q}(\alpha)$ with a free abelian group of rank at most $d-1$. It follows that $\mathcal{O}^*/\mathcal{O}^{*2}$ is finite of order at most $2^d$, and we find

(8.3) $$\dim_{\mathbb{F}_2}(V_2/V_3) \leq d.$$

Putting the estimates (3.4), (7.5), (8.1), (8.2) and (8.3) together, we arrive after a short computation at the following theorem for the values of $n$ and $d$ that we need.

(8.4) **Theorem.** *Let $V$ be as above, and suppose we have $n > d^{2d^2} > 1$. Then the subgroup $V_3 = V \cap \mathbb{Q}(\alpha)^{*2}$ of squares in $V$ satisfies*

$$\dim_{\mathbb{F}_2}(V/V_3) \leq (\log n)^{3/2}.$$

A more careful analysis using the information at the singular primes of $\mathbb{Z}[\alpha]$ as in [Lenstra and Lenstra 1993, Theorem 6.7, p. 61] shows that the exponent $3/2$ can be replaced by 1.

## 9. Quadratic character columns

The algorithm described so far is only able to produce elements in $\mathbb{Z} \times \mathbb{Z}[\alpha]$ for which the second component is in $V$, but not necessarily in the subgroup $V_3 = V \cap \mathbb{Q}(\alpha)^{*2}$ of squares. In order for an element $x \in V$ to be $V_3$, it is necessary and sufficient that all characters $\chi : V/V_3 \to \mathbb{F}_2$ vanish on $x$. At most $k = \dim(V/V_3)$ characters are needed to span the dual space $W = \text{Hom}(V/V_3, \mathbb{F}_2)$, and an element $x \in V$ is a square if and only if all these spanning characters assume the value 1 on $x$. As there is no easy way to produce a spanning set of characters, we will use *random* quadratic characters instead. An elementary calculation shows that if $W$ is any $k$-dimensional $\mathbb{F}_2$-vector space, a randomly chosen set of $k + e$ elements has probability at least $1 - 2^{-e}$ of generating $W$. As this probability converges exponentially to 1 in the number $e$ of extra random elements, we can be practically sure to generate $W$ for moderate values of $e$.

We are now faced with the problem of exhibiting suffciently many "quadratic characters" on $\mathbb{Z}[\alpha]$. On $\mathbb{Z}$, quadratic characters can be obtained from Legendre symbols $x \mapsto \left(\frac{x}{p}\right)$, which are easily evaluated. If $x \in \mathbb{Z}$ is *not* a square, we have, in a sense that is easily made precise,

$$\left(\tfrac{x}{p}\right) = -1$$

for about half of the primes $p$. More precisely, they are the odd primes $p$ that remain prime in the number ring $\mathbb{Z}[\sqrt{x}]$.

(9.1) **Example.** We have $\left(\frac{-16}{p}\right) = -1$ for all primes $p \equiv 3 \bmod 4$. //

Loosely speaking, we can say that an integer $x \neq 0$ that satisfies $\left(\tfrac{x}{p}\right) = 1$ for $t$ randomly chosen primes $p$ is a square with probability $1 - 2^{-t}$. We can use an analogue of this idea over $\mathbb{Z}[\alpha]$.

Every prime $\mathfrak{q} = \ker \pi \sim (q, r_q)$ of degree 1 of $\mathbb{Z}[\alpha]$ gives rise to a Legendre symbol

$$\left(\tfrac{\cdot}{\mathfrak{q}}\right) : \mathbb{Z}[\alpha] \xrightarrow{\pi} \mathbb{F}_q \xrightarrow{\left(\tfrac{\cdot}{q}\right)} \{\pm 1\} \cup \{0\}$$

such that for non-square $x \in \mathbb{Z}[\alpha]$, we have $\left(\tfrac{x}{\mathfrak{q}}\right) = -1$ about half the time. For $y$-smooth elements $x \in \mathbb{Z}[\alpha]$, we can avoid the character value 0 by restricting to Legendre symbols coming from primes $\mathfrak{q} \sim (q, r_q)$ of degree 1 with $q > y$. It is a consequence of the Chebotarev density theorem that the Legendre symbols coming from such $\mathfrak{q}$ are equidistributed over $\mathrm{Hom}(V/V_3, \{\pm 1\})$.

In the rational factor base $B_1$ consisting of primes $p \leq y$, we incorporated a sign bit to ensure that the integers with even prime exponents at all primes $p$ are actually squares. This sign bit for $\mathbb{Z}$ is nothing but the non-trivial chracter on the 1-dimensional $\mathbb{F}_2$-vector space that becomes $V/V_3$ if we replace $\mathbb{Z}$ by $\mathbb{Z}[\alpha]$.

In a similar way, we incorporate in our algebraic factor base $B_2$, which so far consisted of the primes $(p; r_p)$ with $p \leq y$, a sufficiently large number of $\mathbb{F}_2$-valued characters $\chi_\mathfrak{q} : V \to \mathbb{F}_2$ coming from the Legendre symbols of primes $\mathfrak{q} \sim (q, r_q)$ of degree 1 with $q > y$. The character $\chi_\mathfrak{q}$ is simply the Legendre symbol in additive notation, and the values $\chi_\mathfrak{q}(a-b\alpha)$ for $(a, b) \in U$ are treated exactly like the exponent values $e_\mathfrak{p}(a-b\alpha)$. In this way, we obtain a probabilistic algorithm for producing $y$-smooth elements $x \in V$ that do not only satisfy (6.2) but that are true squares. In this set-up the outcome of the linear algebra step, which reduces a matrix of approximate size $y \times y$, consists of subsets $S \subset U$ such that not only we have (6.1) and (6.2), but in addition

$$\left( f'(m)^2 \prod_{(a,b) \in S} (a-bm), \; f'(\alpha)^2 \prod_{(a,b) \in S} (a-b\alpha) \right)$$

is with very high probability a square $(x^2, \gamma^2) \in \mathbb{Z} \times \mathbb{Z}[\alpha]$.

## 10. Square root extraction

The element $(x^2, \gamma^2)$ just found yields a solution to our basic congruence (2.1). In order to obtain a factorization of $n$, we now need the values $(x \bmod n)$ and $\phi(\gamma)$ in $\mathbb{Z}/n\mathbb{Z}$. The gcd of $n$ with their difference is hopefully a non-trivial factor of $n$. Thus, we need to compute a square root $(x, \gamma)$ of our square $(x^2, \gamma^2) \in \mathbb{Z} \times \mathbb{Z}[\alpha]$. On the rational side, this is immediate since we know how to extract squares in $\mathbb{Z}$. It is even possible to avoid computing the large number

$$x^2 = f'(m)^2 \prod_{(a,b) \in S} (a-bm)$$

as we have a complete prime factorization of each of the elements $a - bm$ occurring in the product, and therefore a prime factorization of the product itself.

On the number field side, the situation is more complicated. The prime ideal factorization of

$$\prod_{(a,b) \in S} (a-b\alpha)$$

is easily determined, but this is not immediately useful as prime ideals may not have generators at all and, moreover, we most likely will be unable to compute generators for the unit group $\mathcal{O}^*$ in the large number field $\mathbb{Q}(\alpha)$. Only for the special number field sieve [Lenstra and Lenstra 1993, p. 21 ff.], which often yields rings of integers $\mathcal{O}$ with small units and trivial class group, one may be able to compute a square root of the element $\gamma^2 \in \mathbb{Z}[\alpha]$ using explicit generators of the primes in $\mathbb{Z}[\alpha]$.

For the general number field sieve, one can compute a root of the polynomial $X^2 - \gamma^2$ in $\mathbb{Q}(\alpha)$ by standard methods, such as successive approximation using Hensel's lemma [Buhler and Wagon 2008] at an appropriate prime. Theoretically, this can be done without affecting the expected asymptotic running time of the algorithm. In practice, it is feasible as well but rather cumbersome because of the size of the number $\gamma^2$, which necessitates the handling of very large numbers in the final iterations. Montgomery's method [1994] (see also [Nguyen 1998]), which uses complex approximations, has a better practical performance but has not yet been carefully analyzed.

## 11. Running time

From the analysis given in [Pomerance 2008], it follows that the conjectural asymptotic expected running time the quadratic sieve takes to factor $n$ is

$$\exp((1+o(1))\sqrt{\log n \log \log n})$$

for $n$ tending to infinity. The elliptic curve method has the same running time, which is halfway between exponential and polynomial.

For the number field sieve, we can do better if we carefully choose $d$ and $f$, and optimize the smoothness bound $y$ and the parameter $u$ for the size of the universe $U$ of pairs $(a, b)$ accordingly. We briefly sketch how to find heuristically the asymptotic optimal values, disregarding all lower order terms that occur along the way.

The basic cost of the algorithm, which is computed as in [Pomerance 2008], is $u^{2+o(1)} + y^{2+o(1)}$ as $n$ tends to infinity. The first term represents the sieving part of the algorithm, and equals the length of the sieve times a lower order factor. The second term is the matrix reduction part, which assumes that fast asymptotic methods are applied to a matrix of size at most $y \times y$. In order to balance these contributions, we will take $\log u \approx \log y$.

The numbers $a - b\alpha$ we consider are $y$-smooth if the integer $(a-bm) \cdot N(a-b\alpha)$ is, and using (4.3) and the size $n^{1/d}$ of $m$ and of the coefficients of $f$, we may bound this integer by

(11.1) $$un^{1/d} \cdot (d+1)u^d n^{1/d} \approx n^{2/d} u^{d+1},$$

Here we already take into account that $d$ will be chosen in (11.3) to be of much smaller order than the other factors. The "$u^u$-philosophy" in [Pomerance 2008] shows that a number $x \approx n^{2/d} u^{d+1}$ is $y$-smooth with probability $r^{-r}$, where $r = \log x / \log y$. In order to maximize this probability, we minimize the quantity

$$(11.2) \qquad r = \frac{\log x}{\log y} \approx \frac{\log x}{\log u} \approx \left(\frac{2 \log n}{\log u}\right) \frac{1}{d} + d + 1$$

by taking the degree of $f$ to be $d = (\frac{2 \log n}{\log u})^{1/2}$.

In order to obtain sufficiently many relations from our pairs $(a, b) \in U$ to create a dependent matrix, we need $u^2 \cdot r^{-r} \approx y$. Taking logarithms and replacing $\log y$ by $\log u$, we find $\log u \approx r \log r$ or, equivalently, $r \approx \log u / \log \log u$. Comparison with (11.2) for $d$ as above now leads to

$$\left(\frac{2 \log n}{\log u}\right)^{1/2} \approx \frac{\log u}{\log \log u},$$

and we take 2/3-rd powers to obtain $\log u (\log \log u)^{-2/3} \approx 2(\log n)^{1/3}$. In order to rewrite this, we observe that if we have real quantities $s, t$ satisfying $s = t(\log t)^a$ for some $a \in \mathbb{R}$, then, as $t$ tends to infinity, we have $t = (1 + o(1)) s (\log s)^{-a}$. Applying this for $t = \log u$ and $s = 2(\log n)^{1/3}$ with $a = -2/3$ we arrive at

$$\log y \approx \log u \approx 2(\log n)^{1/3} (\tfrac{1}{3} \log \log n)^{2/3} = (8/9)^{1/3} (\log n)^{1/3} (\log \log n)^{2/3}.$$

With this choice of the basic parameters $u$ and $y$, the asymptotic running time $u^{2+o(1)} + y^{2+o(1)}$ becomes

$$\exp\left(((64/9)^{1/3} + o(1))(\log n)^{1/3} (\log \log n)^{2/3}\right),$$

as claimed in (1.1). The optimal asymptotic value of the degree $d$ of $f$ comes out as

$$(11.3) \qquad d \approx \left(\frac{2 \log n}{\log u}\right)^{1/2} \approx \left(\frac{3 \log n}{\log \log n}\right)^{1/3},$$

and we find that the size in (11.1) of the integers we require to be smooth is

$$\exp\left(((64/3)^{1/3} + o(1))(\log n)^{2/3} (\log \log n)^{1/3}\right).$$

This bound, which we mentioned already in section 2, makes the number field sieve the fastest general purpose factoring algorithm that is currently known.

As with the quadratic sieve, there are various practical improvements to the basic number field sieve as we have described it here. The most important bells and whistles are mentioned in [Crandall and Pomerance 2001, Section 6.2.7]. Although they do not significantly change the asymptotic running time of the algorithm, they greatly enhance its practical performance, and they are instrumental in completing the record factorizations that mark the borderlines of what is currently feasible in factoring.

## References

[Buhler and Wagon 2008] J. P. Buhler and S. Wagon, "Basic algorithms in number theory", pp. 25–68 in *Surveys in algorithmic number theory*, edited by J. P. Buhler and P. Stevenhagen, Math. Sci. Res. Inst. Publ. **44**, Cambridge University Press, New York, 2008.

[Crandall and Pomerance 2001] R. Crandall and C. Pomerance, *Prime numbers: a computational perspective*, Springer, New York, 2001.

[Lenstra 1992] H. W. Lenstra, Jr., "Algorithms in algebraic number theory", *Bull. Amer. Math. Soc. (N.S.)* **26**:2 (1992), 211–244.

[Lenstra and Lenstra 1993] A. K. Lenstra and J. H. W. Lenstra, *The development of the number field sieve*, Lecture Notes in Mathematics **1554**, Springer, Berlin, 1993.

[Montgomery 1994] P. L. Montgomery, "Square roots of products of algebraic numbers", pp. 567–571 in *Mathematics of Computation 1943–1993: a half-century of computational mathematics* (Vancouver, 1993), edited by W. Gautschi, Proc. Sympos. Appl. Math. **48**, Amer. Math. Soc., Providence, RI, 1994.

[Nguyen 1998] P. Nguyen, "A Montgomery-like square root for the number field sieve", pp. 151–168 in *Algorithmic number theory, ANTS-III* (Portland, OR, 1998), edited by J. P. Buhler, Lecture Notes in Comput. Sci. **1423**, Springer, Berlin, 1998.

[Pomerance 2008] C. Pomerance, "Smooth numbers and the quadratic sieve", pp. 69–81 in *Surveys in algorithmic number theory*, edited by J. P. Buhler and P. Stevenhagen, Math. Sci. Res. Inst. Publ. **44**, Cambridge University Press, New York, 2008.

[Stevenhagen 2008] P. Stevenhagen, "The arithmetic of number rings", pp. 209–266 in *Surveys in algorithmic number theory*, edited by J. P. Buhler and P. Stevenhagen, Math. Sci. Res. Inst. Publ. **44**, Cambridge University Press, New York, 2008.

[Stevenhagen and Lenstra 1996] P. Stevenhagen and H. W. Lenstra, Jr., "Chebotarëv and his density theorem", *Math. Intelligencer* **18**:2 (1996), 26–37.

PETER STEVENHAGEN
MATHEMATISCH INSTITUUT
UNIVERSITEIT LEIDEN
POSTBUS 9512
2300 RA LEIDEN
THE NETHERLANDS
  psh@math.leidenuniv.nl

# Four primality testing algorithms

RENÉ SCHOOF

ABSTRACT. In this expository paper we describe four primality tests. The first test is very efficient, but is only capable of proving that a given number is either composite or 'very probably' prime. The second test is a deterministic polynomial time algorithm to prove that a given numer is either prime or composite. The third and fourth primality tests are at present most widely used in practice. Both tests are capable of proving that a given number is prime or composite, but neither algorithm is deterministic. The third algorithm exploits the arithmetic of cyclotomic fields. Its running time is almost, but not quite polynomial time. The fourth algorithm exploits elliptic curves. Its running time is difficult to estimate, but it behaves well in practice.

### CONTENTS

| | |
|---|---|
| 1. Introduction | 101 |
| 2. A probabilistic test | 102 |
| 3. A deterministic polynomial time primality test | 106 |
| 4. The cyclotomic primality test | 111 |
| 5. The elliptic curve primality test | 120 |
| References | 125 |

## 1. Introduction

In this expository paper we describe four primality tests.

In Section 2 we discuss the Miller–Rabin test. This is one of the most efficient probabilistic primality tests. Strictly speaking, the Miller–Rabin test is not a primality test but rather a 'compositeness test', since it does not prove the primality of a number. Instead, if $n$ is *not* prime, the algorithm proves this in all likelihood very quickly. On the other hand, if $n$ happens to be prime, the algorithm merely provides strong evidence for its primality. Under the assumption of the Generalized Riemann Hypothesis one can turn the Miller–Rabin algorithm into a deterministic polynomial time primality test. This idea, due to G. Miller, is also explained.

In Section 3 we describe the deterministic polynomial time primality test that was proposed by M. Agrawal, N. Kayal and N. Saxena in 2002 [Agrawal et al. 2004]. At the moment of this writing, this new test, or rather a more efficient probabilistic version of it, had not yet been widely implemented. In practice, therefore, for *proving* the primality of a given integer, one still relies on older tests that are either not provably polynomial time or not deterministic. In the remaining two sections we present the two most widely used such tests.

In Section 4 we discuss the cyclotomic primality test. This test is deterministic and is actually capable of *proving* that a given integer $n$ is either prime or composite. It does not run in polynomial time, but very nearly so. We describe a practical non-deterministic version of the algorithm. Finally in Section 5, we describe the elliptic curve primality test. This algorithm also provides a *proof* of the primality or compositeness of a given integer $n$. Its running time is hard to analyze, but in practice the algorithm seems to run in polynomial time. It is not deterministic. The two 'practical' tests described in Sections 4 and 5 have been implemented and fine tuned. Using either of them it is now possible to routinely prove the primality of numbers that have several thousands of decimal digits [Mihăilescu 1998; Morain 1998].

## 2. A probabilistic test

In this section we present a practical and efficient probabilistic primality test. Given a composite integer $n > 1$, this algorithm proves with high probability very quickly that $n$ is not prime. On the other hand, if $n$ passes the test, it is merely *likely* to be prime. The algorithm consists of repeating one simple step, a Miller–Rabin test, several times with different random initializations. The probability that a composite number is *not* recognized as such by the algorithm, can be made arbitrarily small by repeating the main step a number of times. The algorithm was first proposed by M. Artjuhov [1966/1967]. Later M. Rabin proposed the probabilistic version [1980]. Under assumption of the Generalized Riemann Hypothesis (GRH) one can actually *prove* that $n$ is prime by applying the test sufficiently often. This leads to G. Miller's *conditional* algorithm [1976]. Under assumption of GRH it runs in polynomial time. Our presentation follows the presentation of the algorithms in the excellent book [Crandall and Pomerance 2001].

Here is the key ingredient:

THEOREM 2.1. *Let $n > 9$ be an odd positive composite integer. We write $n - 1 = 2^k m$ for some exponent $k \geq 1$ and some odd integer $m$. Let*

$$B = \{x \in (\mathbb{Z}/n\mathbb{Z})^* : x^m = 1 \text{ or } x^{m 2^i} = -1 \text{ for some } 0 \leq i < k\}.$$

Then we have
$$\frac{\#B}{\varphi(n)} \leq \frac{1}{4}.$$

Here $\varphi(n) = \#(\mathbb{Z}/n\mathbb{Z})^*$ denotes Euler's $\varphi$-function.

PROOF. Let $2^l$ denote the largest power of 2 that has the property that it divides $p-1$ for every prime $p$ divisor of $n$. Then the set $B$ is contained in

$$B' = \{x \in (\mathbb{Z}/n\mathbb{Z})^* : x^{m2^{l-1}} = \pm 1\}.$$

Indeed, clearly any $x \in (\mathbb{Z}/n\mathbb{Z})^*$ satisfying $x^m = 1$ is contained in $B'$. On the other hand, if $x^{m2^i} = -1$ for some $0 \leq i < k$, we have $x^{m2^i} \equiv -1 \pmod{p}$ for every prime $p$ dividing $n$. It follows that for every $p$, the *exact* power of 2 dividing the order of $x$ modulo $p$, is equal to $2^{i+1}$. In particular, $2^{i+1}$ divides $p-1$ for every prime divisor $p$ of $n$. Therefore we have $l \geq i+1$. So we can write that $x^{m2^{l-1}} = (-1)^{2^{l-i-1}}$, which is $-1$ or $+1$ depending on whether $l = i+1$ or $l > i+1$. It follows that $B \subset B'$.

By the Chinese Remainder Theorem, the number of elements $x \in (\mathbb{Z}/n\mathbb{Z})^*$ for which we have $x^{m2^{l-1}} = 1$, is equal to the product over $p$ of the number of solutions to the equation $X^{m2^{l-1}} = 1$ modulo $p^{a_p}$. Here $p$ runs over the prime divisors of $n$ and $p^{a_p}$ is the exact power of $p$ dividing $n$. Since each of the groups $(\mathbb{Z}/p^{a_p}\mathbb{Z})^*$ is cyclic, the number of solutions modulo $p^{a_p}$ is given by

$$\gcd((p-1)p^{a_p-1}, m2^{l-1}) = \gcd(p-1, m)2^{l-1}.$$

The last equality follows from the fact that $p$ does not divide $m$. Therefore we have

$$\#\{x \in (\mathbb{Z}/n\mathbb{Z})^* : x^{m2^{l-1}} = 1\} = \prod_{p|n} \gcd(p-1, m)2^{l-1}.$$

Similarly, the number of solutions of the equation $X^{m2^l} = 1$ modulo $p^{a_p}$ is equal to $\gcd(p-1, m)2^l$, which is twice the number of solutions of $X^{m2^{l-1}} = 1$ modulo $p^{a_p}$. It follows that the number of solutions of the equation $X^{m2^{l-1}} = -1$ modulo $p^{a_p}$ is also equal to $\gcd(p-1, m)2^{l-1}$. Therefore we have

$$\#B' = 2 \prod_{p|n} \gcd(p-1, m)2^{l-1},$$

and hence

$$\frac{\#B'}{\varphi(n)} = 2 \prod_{p|n} \frac{\gcd(p-1, m)2^{l-1}}{(p-1)p^{a_p-1}}.$$

Suppose now that the propertion $\#B/\varphi(n)$ exceeds $\frac{1}{4}$. We want to derive a contradiction. Since we have $B \subset B'$, the inequality above implies that

$$\frac{1}{4} < 2 \prod_{p|n} \frac{\gcd(p-1,m)2^{l-1}}{(p-1)p^{a_p-1}}. \qquad (*)$$

We draw a number of conclusions from this inequality. First we note that $\gcd(p-1,m)2^{l-1}$ divides $(p-1)/2$ so that the right hand side of $(*)$ is at most $2^{1-t}$ where $t$ is the number of different primes dividing $n$. It follows that $t \le 2$.

Suppose that $t = 2$, so that $n$ has precisely two distinct prime divisors. If one of them, say $p$, has the property that $p^2$ divides $n$ so that $a_p \ge 2$, then the right hand side of $(*)$ is at most $2^{1-2}/3 = 1/6$. Contradiction. It follows that all exponents $a_p$ are equal to 1, so that $n = pq$ for two distinct primes $p$ and $q$. The inequality $(*)$ now becomes

$$\frac{p-1}{\gcd(p-1,m)2^l} \cdot \frac{q-1}{\gcd(q-1,m)2^l} < 2.$$

Since the factors on the left hand side of this inequality are positive integers, they are both equal to 1. This implies that $p-1 = \gcd(p-1,m)2^l$ and $q-1 = \gcd(q-1,m)2^l$. It follows that the exact power of 2 dividing $p-1$ as well as the exact power of 2 dividing $q-1$ are equal to $2^l$ and that the odd parts of $p-1$ and $q-1$ divide $m$. Considering the relation $pq = 1 + 2^k m$ modulo the odd part of $p-1$, we see that the odd part of $p-1$ divides the odd part of $q-1$. By symmetry, the odd parts of $p-1$ and $q-1$ are therefore equal. This implies $p-1 = q-1$ and contradicts the fact that $p \ne q$. Therefore we have $t = 1$ and hence $n = p^a$ for some odd prime $p$ and exponent $a \ge 2$. The inequality $(*)$ now says that $p^{a-1} < 4$, so that $p = 3$ and $a = 2$, contradicting the hypothesis that $n > 9$. This proves the theorem. □

When a random $x \in (\mathbb{Z}/n\mathbb{Z})^*$ is checked to be contained in the set $B$ of Theorem 2.1, we say that '$n$ passes a Miller–Rabin test'. Checking that $x \in B$ involves raising $x \in \mathbb{Z}/n\mathbb{Z}$ to an exponent that is no more than $n$. Using the binary expansion of the exponent, this takes no more that $O(\log n)$ multiplications in $\mathbb{Z}/n\mathbb{Z}$. Therefore a single exponentiation involves $O((\log n)^{1+\mu})$ elementary operations or bit operations. Here $\mu$ is a constant with the property that the multiplication algorithm in $\mathbb{Z}/n\mathbb{Z}$ takes no more than $O((\log n)^\mu)$ elementary operations. We have $\mu = 2$ when we use the usual multiplication algorithm, while one can take $\mu = 1 + \varepsilon$ for any $\varepsilon > 0$ by employing fast multiplication techniques.

By Theorem 2.1 the probability that a composite number $n$ passes a single Miller–Rabin test, is at most 25%. Therefore, the probability that $n$ passes $\log n$

such tests is smaller than $1/n$. The probability that a large composite $n$ passes $(\log n)^2$ tests is astronomically small: less than $n^{-\log n}$. Since for most composite $n$ the probability that $n$ passes a Miller–Rabin test is much smaller than $1/4$, one is in practice already convinced of the primality of $n$, when $n$ successfully passes a handful of Miller–Rabin tests. This is enough for most commercial applications.

Under assumption of the Generalized Riemann Hypothesis (GRH) for quadratic Dirichlet characters, the Miller–Rabin test can be transformed into a *deterministic* polynomial time primality test. This result goes back to [Miller 1976].

THEOREM 2.2 (GRH). *Let $n$ be an odd positive composite integer. Let $n-1 = 2^k m$ for some exponent $k \geq 1$ and some odd integer $m$. If for all integers $x$ between 1 and $2(\log n)^2$ one has*

$$x^m \equiv 1 \pmod{n} \quad \text{or} \quad x^{2^i m} \equiv -1 \pmod{n} \text{ for some } 0 \leq i < k,$$

*then $n$ is a prime number.*

PROOF. We first show that $n$ is squarefree. See also [Lenstra 1979]. Suppose that $p$ is a prime for which $p^2$ divides $n$. A special case of the result [Konyagin and Pomerance 1997, (1.45)] on the distribution of smooth numbers implies that for every odd integer $r \geq 5$ one has

$$\#\{a \in \mathbb{Z} : 1 \leq a \leq r \text{ and } a \text{ is product of primes} \leq (\log r)^2\} \geq \sqrt{r}.$$

We apply this with $r = p^2$. It follows that the subgroup $H$ of $(\mathbb{Z}/p^2\mathbb{Z})^*$ that is generated by the natural numbers $x \leq (\log n)^2$ has order at least $p$. On the other hand, the hypothesis of the theorem implies that every $x \in H$, being a product of numbers $a$ that satisfy $a^{n-1} \equiv 1 \pmod{p^2}$, satisfies $x^{n-1} \equiv 1 \pmod{p^2}$. Since the order of the group $(\mathbb{Z}/p^2\mathbb{Z})^*$ is $p(p-1)$ and $p$ does not divide $n-1$, we see that any $x \in H$ must satisfy $x^{p-1} \equiv 1 \pmod{p^2}$. But this is impossible, because the subgroup of $(\mathbb{Z}/p^2\mathbb{Z})^*$ that consists of elements having this property, has order $p-1$.

Therefore, if $n$ is composite, it is divisible by two odd distinct primes $p$ and $q$. Let $\chi$ denote the quadratic character of conductor $p$. By a result of E. Bach [1990], *proved under assumption of the GRH*, there exists a natural number $x \leq 2(\log p)^2 < 2(\log n)^2$ for which $\chi(x) \neq 1$. Since the condition of the theorem implies that we have $\gcd(x,n) = 1$, we must have $\chi(x) = -1$. Writing $p - 1 = 2^l \mu$ for some exponent $l \geq 1$ and some odd integer $\mu$, we have $x^{2^{l-1}\mu} \equiv \chi(x) = -1 \pmod{p}$. This implies that $-1$ is contained in the subgroup of $(\mathbb{Z}/p\mathbb{Z})^*$ generated by $x$. Since the 2-parts of the subgroups of $(\mathbb{Z}/p\mathbb{Z})^*$ generated by $x^m$ and by $x$ are the same, we have $x^m \not\equiv 1 \pmod{p}$ and hence $x^m \not\equiv 1 \pmod{n}$. Therefore the hypothesis of the theorem implies that $x^{2^i m} \equiv -1 \pmod{n}$ for some $0 \le i < k$. Since for this value of $i$ we also have $x^{2^i m} \equiv -1 \pmod{p}$, necessarily the equality $i = l - 1$ holds. It follows that we have $x^{2^{l-1} m} \equiv -1 \pmod{q}$, so that the order of $x^m \pmod{q}$ is equal to $2^l$. Writing $q - 1 = 2^{l'} \mu'$ for some exponent $l' \ge 1$ and some odd integer $\mu'$, we have therefore $l \le l'$.

Repeating the argument, but switching the roles of $p$ and $q$, we conclude that $l = l'$. Let $\chi'$ denote the quadratic character of conductor $q$. A second application of Bach's theorem, this time to the *non-trivial* character $\chi\chi'$, provides us with a natural number $y \le 2(\log n)^2$ for which $\chi\chi'(y) \ne 1$ and hence, say, $\chi(y) = -1$ while $\chi'(y) = 1$. The arguments given above, but this time applied to $y$, show that we cannot have $y^m \equiv -1 \pmod{n}$, so that necessarily $y^{2^i m} \equiv -1 \pmod{n}$ for some $0 \le i < k$. Moreover, the exponent $i$ is equal to $l - 1 = l' - 1$. It follows that $y^{2^{l'-1} m} \equiv -1 \pmod{q}$. This implies that the element $y^m \in (\mathbb{Z}/q\mathbb{Z})^*$ has order $2^{l'}$. Since the subgroups of $(\mathbb{Z}/q\mathbb{Z})^*$ generated by $y^m$ and $y^{\mu'}$ are equal, the order of $y^{\mu'} \in (\mathbb{Z}/q\mathbb{Z})^*$ is also $2^{l'}$. This contradicts the fact that $1 = \chi'(y) \equiv y^{2^{l'-1} \mu'} \pmod{q}$.

We conclude that $n$ is prime and the result follows. □

It is clear how to apply Theorem 2.2 and obtain a test that proves that $n$ is prime under condition of GRH: given an odd integer $n > 1$, we simply test the condition of Theorem 2.2 for all $a \in \mathbb{Z}$ satisfying $1 < a < 2(\log n)^2$. If $n$ passes all these tests and GRH holds, then $n$ is prime. Each test involves an exponentiation in the ring $\mathbb{Z}/n\mathbb{Z}$. Since the exponent is less than $n$, this can be done using only $O((\log n)^{1+\mu})$ elementary operations. Therefore this is a polynomial time primality test. Testing $n$ takes $O((\log n)^{3+\mu})$ elementary operations. As before, we have $\mu = 2$ when we use the usual multiplication algorithm, while we can take $\mu = 1 + \varepsilon$ for any $\varepsilon > 0$ by employing fast multiplication techniques.

## 3. A deterministic polynomial time primality test

In the summer of 2002 the three Indian computer scientists M. Agrawal, N. Kayal and N. Saxena presented a deterministic polynomial time primality test. We describe and analyze this extraordinary result in this section.

For any prime number $r$ we let

$$\Phi_r(X) = X^{r-1} + \cdots + X + 1$$

denote the $r$-th cyclotomic polynomial. Let $\zeta_r$ be a zero of $\Phi_r(X)$ and let $\mathbb{Z}[\zeta_r]$ denote the ring generated by $\zeta_r$ over $\mathbb{Z}$. For any $n \in \mathbb{Z}$ we write $\mathbb{Z}[\zeta_r]/(n)$ for the residue ring $\mathbb{Z}[\zeta_r]$ modulo the ideal $(n)$ generated by $n$. For $n \ne 0$, this is a finite ring.

THEOREM 3.1. *Let n be an odd positive integer and let r be a prime number. Suppose that*

(i) *n is not divisible by any of the primes $\leq r$;*
(ii) *the order of n (mod r) is at least $(\log n / \log 2)^2$;*
(iii) *for every $0 \leq j < r$ we have $(\zeta_r + j)^n = \zeta_r^n + j$ in $\mathbb{Z}[\zeta_r]/(n)$.*

*Then n is a prime power.*

PROOF. It follows from condition (ii) that we have $n \not\equiv 1 \pmod{r}$. Therefore there exists a prime divisor $p$ of $n$ that is not congruent to $1 \pmod{r}$. Let $A$ denote the $\mathbb{F}_p$-algebra $\mathbb{Z}[\zeta_r]/(p)$. It is a quotient of the ring $\mathbb{Z}[\zeta_r]/(n)$. For $k \in \mathbb{Z}$ coprime to $r$ we let $\sigma_k$ denote the ring automorphism of $A$ determined by $\sigma_k(\zeta_r) = \zeta_r^k$. The map $(\mathbb{Z}/r\mathbb{Z})^* \mapsto \Delta$ given by $k \mapsto \sigma_k$ is a well defined isomorphism. We single out two special elements of $\Delta$. One is the *Frobenius automorphism* $\sigma_p$ and the other is $\sigma_n$. Let $\Gamma$ denote the subgroup of $\Delta$ that is generated by $\sigma_p$ and $\sigma_n$.

Next we consider the subgroup $G$ of elements of the multiplicative group $A^*$ that are annihilated by the endomorphism $\sigma_n - n \in \mathbb{Z}[\Delta]$. In other words, we put

$$G = \{a \in A^* : \sigma_n(a) = a^n\}.$$

Pick a maximal ideal $\mathfrak{m}$ of $A$ and put $k = A/\mathfrak{m}$. Then $k$ is a finite extension of $\mathbb{F}_p$, generated by a primitive $r$-th root of unity. Let $H \subset k^*$ be the image of $G$ under the natural map $\pi : A \longrightarrow k$. The group $H$ is cyclic. Its order is denoted by $s$. We have the following commutative diagram.

$$\begin{array}{ccc} G & \subset & A^* \\ \downarrow \pi & & \downarrow \pi \\ H & \subset & k^* \end{array}$$

Since $\Delta$ is commutative, it acts on $G$. Since $\sigma_n$ and $\sigma_p$ act on $G$ by raising to the power $n$ and $p$ respectively, every $\sigma_m \in \Gamma$ acts by raising $g \in G$ to a certain power $e_m$ that is prime to $\#G$. The powers $e_m$ are well determined modulo the exponent $\exp(G)$ of $G$. Therefore the map $\Gamma \longrightarrow (\mathbb{Z}/\exp(G)\mathbb{Z})^*$, given by $\sigma_m \mapsto e_m$, is a well defined group homomorphism. Since $H$ is a cyclic quotient of $G$, its order $s$ divides the exponent of $G$ and the map $\sigma_m \mapsto e_m$ induces a homomorphism

$$\Gamma \longrightarrow (\mathbb{Z}/s\mathbb{Z})^*.$$

If $m \equiv p^i n^j \pmod{r}$, then it maps $\sigma_m \in \Gamma$ to $e_m \equiv p^i n^j \pmod{s}$.

It is instructive to see what all this boils down to when $n$ is prime. Then we have $n = p$ and $\sigma_n$ is equal to the Frobenius automorphism $\sigma_p$. The group $G$ is all of $A^*$ so that $H$ is equal to $k^*$. Writing $f$ for the order of $p$ modulo $r$, the group $\Gamma = \langle \sigma_p \rangle$ has order $f$ while the groups $H = k^*$ and its automorphism group $\text{Aut}(H)$ are *much larger*. Indeed, $H$ has order $s = p^f - 1 = n^{\#\Gamma} - 1$ and $\text{Aut}(H) \cong (\mathbb{Z}/s\mathbb{Z})^*$ is of comparable size

Under the conditions of the theorem, but *without assuming* that $n$ is prime, something similar can be shown to be true.

CLAIM. $s > n^{[\sqrt{\#\Gamma}]}$.

Using this inequality, we complete the proof of the theorem. Consider the homomorphism

$$\Gamma \longrightarrow (\mathbb{Z}/s\mathbb{Z})^*$$

constructed above. We first apply the box principle in the *small* group $\Gamma$ and then obtain a relation in $\mathbb{Z}$ from a relation in $(\mathbb{Z}/s\mathbb{Z})^*$ using the fact that the latter group is *very large*.

Let $q = n/p$. We consider the products $\sigma_p^i \sigma_q^j \in \Gamma$ for $0 \le i, j \le [\sqrt{\#\Gamma}]$. Since we have $(1 + [\sqrt{\#\Gamma}])^2 > \#\Gamma$, there are two pairs $(i, j) \ne (i', j')$ for which $\sigma_p^i \sigma_q^j$ and $\sigma_p^{i'} \sigma_q^{j'}$ are the same element in $\Gamma$. It follows that their images in the group $(\mathbb{Z}/s\mathbb{Z})^*$ are the same as well. Since $\sigma_q$ is mapped to $q \pmod{s}$, this means that $p^i q^j \equiv p^{i'} q^{j'} \pmod{s}$. The integer $p^i q^j$ does not exceed $n^{\max(i,j)} \le n^{[\sqrt{\#\Gamma}]} < s$. The same holds for $p^{i'} q^{j'}$. We conclude that $p^i q^j = p^{i'} q^{j'}$ in $\mathbb{Z}$! Since $(i, j) \ne (i', j')$ it follows that $n$ is a power of $p$. □

PROOF OF THE CLAIM. We first estimate $s = \#H$ in terms of $\#G$. Then we show that $G$ is large.

The first bound we show is

$$s \ge \#G^{1/[\Delta:\Gamma]}. \qquad (*)$$

Let $C$ denote a set of coset representatives of $\Gamma$ in $\Delta$ and consider the homomorphism

$$G \longrightarrow \prod_{i \in C} k^*$$

given by mapping $a \in G$ to the vector $(\sigma_i(a) \pmod{\mathfrak{m}})_{i \in C}$.

This map is injective. Indeed, if $a \in G$ has the property that $\sigma_i(a) = 1$ for some $i$, then we also have $\sigma_{in}(a) = \sigma_i(a^n) = \sigma_i(a)^n = 1$ and similarly $\sigma_{ip}(a) = 1$. In other words, we have $\sigma(a) = 1$ for all elements $\sigma$ in the coset of $\Gamma$ containing $\sigma_i$. Therefore, if $a \in G$ has the property that $\sigma_i(a) = 1$ for all $i \in C$, then automatically also $\sigma_i(a) = 1$ for all $i \in (\mathbb{Z}/r\mathbb{Z})^*$. It follows that $\sigma_i(a-1) = 0$ for all $i \in (\mathbb{Z}/r\mathbb{Z})^*$. Writing the element $a - 1$ as $f(\zeta_r)$ for some polynomial $f(X) \in \mathbb{F}_p[X]$, this implies that $f(\zeta_r^i) = 0$ for all $i \in (\mathbb{Z}/r\mathbb{Z})^*$. It follows that the cyclotomic polynomial $\Phi_r(X)$ divides $f(X)$ in $\mathbb{F}_p[X]$ and hence that $a - 1 = 0$, as required.

Since for every $i \in C$, the image of the map $G \longrightarrow k^*$ given by $a \mapsto \sigma_i(a) \pmod{\mathfrak{m}}$ is equal to $H$, the injectivity of the homomorphism implies that $\#G \leq s^{[\Delta:\Gamma]}$ as required.

The second estimate is
$$\#G \geq 2^{r-1}. \qquad (**)$$
Since $p \not\equiv 1 \pmod{r}$, the irreducible factors of $\Phi_r(X) = (X^r - 1)/(X - 1)$ in the ring $\mathbb{F}_p[X]$ have degree at least 2 and hence cannot divide any polynomial of degree 1. Therefore the elements $\zeta_r + j$ for $0 \leq j < r - 1$ are not contained in any maximal ideal of the ring $A$. It follows that they are *units* of $A$. By condition (iii), for each subset $J \subset \{0, 1, \ldots, r - 2\}$ the element
$$\prod_{j \in J}(\zeta_r + j)$$
is contained in $G$.

All these elements are *distinct*. Indeed, since the degree of the cyclotomic polynomial $\Phi_r$ is $r - 1$, the only two elements that could be equal to one another are the ones corresponding to the extreme cases $J = \varnothing$ and $J = \{0, 1, \ldots, r-2\}$. This can only happen when $\prod_{j=0}^{r-2}(X + j) - 1$ is divisible by $\Phi_r(X)$ in the ring $\mathbb{F}_p[X]$. Since both polynomials have the same degree, we then necessarily have $\prod_{j=0}^{r-2}(X + j) - 1 = \Phi_r(X)$. Inspection of the constant terms shows that $p = 2$. But this is impossible, because $n$ is odd.

Since there are $2^{r-1}$ subsets $J \subset \{0, 1, \ldots, r - 2\}$, we conclude that $\#G \geq 2^{r-1}$, as required.

Combining the inequalities $(*)$ and $(**)$ we find that
$$s \geq \#G^{1/[\Delta:\Gamma]} \geq 2^{(r-1)/[\Delta:\Gamma]} = 2^{\#\Gamma} > n^{\sqrt{\#\Gamma}} \geq n^{[\sqrt{\#\Gamma}]}.$$

Here we used the inequality $\#\Gamma > (\log n/\log 2)^2$. It follows from the fact that the order of $\sigma_n \in \Gamma$ is larger than $(\log n/\log 2)^2$. Indeed, this order is equal to the order of $n$ modulo $r$, which by condition (ii) is larger than $(\log n/\log 2)^2$. This proves the claim. $\qquad \square$

Theorem 3.1 leads to the following primality test.

ALGORITHM 3.2. Let $n > 1$ be an odd integer.

(i) First check that $n$ is not a proper power of an integer.
(ii) By successively trying $r = 2, 3, \ldots$, determine the smallest prime $r$ not dividing $n$ nor any of the numbers $n^i - 1$ for $0 \leq i \leq (\log n/\log 2)^2$.
(iii) For $0 \leq j < r - 1$ check that $(\zeta_r + j)^n = \zeta_r^n + j$ in the ring $\mathbb{Z}[\zeta_r]/(n)$.

If the number $n$ does not pass the tests, it is composite. If it passes them, it is a prime.

PROOF OF CORRECTNESS. If $n$ is prime, it passes the tests by Fermat's little theorem. Conversely, suppose that $n$ passes the tests. We check the conditions of Theorem 3.1. By the definition of $r$, the number $n$ has no prime divisors $\leq r$. Since $r$ does not divide any of the $n^i - 1$ for $1 \leq i \leq (\log n/\log 2)^2$, the order of $n$ modulo $r$ exceeds $(\log n/\log 2)^2$. This shows that the second condition of Theorem 3.1 is satisfied. Since test (iii) has been passed successfully, the third condition is satisfied. We deduce that $n$ is a prime power. Since $n$ passed the first test, it is therefore prime. □

RUNNING TIME ANALYSIS. The first test is performed by checking that $n^{1/m}$ is not an integer, for all integers $m$ between 2 and $\log n/\log 2$. This can be done in time $O((\log n)^4)$ by computing sufficiently accurate approximations to the real number $n^{1/m}$. The second test does not take more than $r$ times $O((\log n)^2)$ multiplications with modulus $\leq r$. This takes at most $O(r(\log r \log n)^2)$ bit operations. The third test takes $r$ times $O(\log n)$ multiplications in the ring $\mathbb{Z}[\zeta_r]/(n)$. The latter ring is isomorphic to $\mathbb{Z}[X]/(\Phi_r(X), n)$. If the multiplication algorithm that we use to multiply two elements of bit size $t$ takes no more than $O(t^\mu)$ elementary operations, then this adds up to $O((r \log n)^{1+\mu})$ elementary operations. Since $\mu \geq 1$ and since $r$ exceeds the order of $n$ mod $r$, we have $r > (\log n/\log 2)^2$. Therefore the third test is the dominating part of the algorithm.

We estimate how small we can take $r$. By the definition of $r$, the product $n \prod_i (n^i - 1)$ is divisible by all primes $l < r$. Here the product runs over values of $i \leq (\log n/\log 2)^2$. So

$$\sum_{l<r} \log l \leq \log n + \log n \sum_{1 \leq i \leq (\frac{\log n}{\log 2})^2} i = O((\log n)^5).$$

A weak and easily provable form of the prime number theorem says that there exists a constant $c > 0$ such that $\sum_{l<r} \log l \geq cr$ for every $r$. Therefore we have $r = O((\log n)^5)$. It follows that the algorithm takes $O((\log n)^{6(1+\mu)})$ elementary operations. When the usual multiplication algorithm is used, we have $\mu = 2$ and obtain an algorithm that takes at most $O((\log n)^{18})$ elementary operations. It takes $O((\log n)^{12+\varepsilon})$ elementary operations when fast multiplication techniques are employed. □

REMARK 1. Since the upper bound $\sqrt{\#\Gamma}$ is optimal for the box principle, the inequality $2^{\#\Gamma} > n^{\sqrt{\#\Gamma}}$ used above implies that $\#\Gamma = r - 1$ needs to be at least $(\log n/\log 2)^2$. This we know to be the case because the order of $\sigma_n \in \Gamma$, which is equal to the order of $n \in (\mathbb{Z}/r\mathbb{Z})^*$, exceeds $(\log n/\log 2)^2$. The argument involving the prime number theorem given above implies then that we cannot expect to be able to prove that the order of magnitude of the prime $r$

is smaller than $O((\log n)^5)$. Therefore this algorithm cannot be expected to be proved to run faster than $O((\log n)^{6(1+\mu)})$. On the other hand, in practice one easily finds a suitable prime of the smallest possible size $O((\log n)^2)$. Therefore the practical running time of the algorithm is $O((\log n)^{3(1+\mu)})$.

REMARK 2. One may replace the ring

$$\mathbb{Z}[\zeta_r]/(n) \cong (\mathbb{Z}/n\mathbb{Z})[X]/(\Phi_r(X))$$

by any Galois extension of $\mathbb{Z}/n\mathbb{Z}$ of the form

$$(\mathbb{Z}/n\mathbb{Z})[X]/(f(X))$$

that admits an automorphism $\sigma$ with the properties that

$$\sigma(X) = X^n \quad \text{and} \quad \sigma \text{ has order at least } (\log n/\log 2)^2.$$

This was pointed out by Hendrik Lenstra shortly after the algorithm described above came out. The running time of the resulting modified algorithm is then $O((d \log n)^{1+\mu})$ where $d$ is the degree of the polynomial $f(X)$. Since the order of $\sigma$ is at most $d$, one has

$$d > (\log n/\log 2)^2$$

and one cannot obtain an algorithm that runs faster than $O((\log n)^{3(1+\mu)})$. Since then Lenstra and Pomerance [≥ 2008] showed that for every $\varepsilon > 0$ one can construct suitable rings with $d = O((\log n)^{2+\varepsilon})$. This leads to a primality test that runs in time $O((\log n)^{(3+\varepsilon)(1+\mu)})$. This is essentially the same as the practical running time mentioned above.

## 4. The cyclotomic primality test

In this section we describe the cyclotomic primality test. This algorithm was proposed in 1981 by L. Adleman, C. Pomerance and R. Rumely [Adleman et al. 1983]. It is one of the most powerful practical tests available today (see [Mihăilescu 1998]). Our exposition follows H. Lenstra's lecture [1981]; see also [Cohen 1993, 9.1; Washington 1997, 16.1]. The actual computations involve *Jacobi sums*, but the basic idea of the algorithm is best explained in terms of *Gaussian sums*. See [Washington 1997; Lang 1978] for a more systematic discussion of the basic properties of Gaussian sums and Jacobi sums. For any positive integer $r$, we denote the subgroup of $r$-th roots of unity of $\overline{\mathbb{Q}}^*$ by $\mu_r$.

DEFINITION. Let $q$ be a prime and let $r$ be a positive integer prime to $q$. Let $\chi : (\mathbb{Z}/q\mathbb{Z})^* \longrightarrow \mu_r$ be a character and let $\zeta_q$ be a primitive $q$-th root of unity.

Then we define the *Gaussian sum* $\tau(\chi)$ by

$$\tau(\chi) = -\sum_{x \in (\mathbb{Z}/q\mathbb{Z})^*} \zeta_q^x \chi(x).$$

The Gaussian sum $\tau(\chi)$ is an algebraic integer, contained in the cyclotomic field $\mathbb{Q}(\zeta_r, \zeta_q)$. We have the diagram of fields

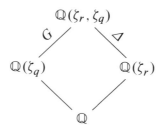

The Galois group of $\mathbb{Q}(\zeta_r, \zeta_q)$ over $\mathbb{Q}$ is isomorphic to $\Delta \times G$. Here we have $\Delta = \{\sigma_i : i \in (\mathbb{Z}/r\mathbb{Z})^*\}$, where $\sigma_i \in \Delta$ is the automorphism that acts trivially on $q$-th roots of unity, while its action of $r$-th roots of unity is given by $\sigma_i(\zeta_r) = \zeta_r^i$. The map $(\mathbb{Z}/r\mathbb{Z})^* \longrightarrow \Delta$ given by $i \mapsto \sigma_i$ is an isomorphism of groups. Similarly, we have $G = \{\rho_j : j \in (\mathbb{Z}/q\mathbb{Z})^*\}$ where $\rho_j \in \Delta$ is the automorphism given by $\rho_j(\zeta_r) = \zeta_r$ and $\rho_j(\zeta_q) = \zeta_q^j$. The map $(\mathbb{Z}/q\mathbb{Z})^* \longrightarrow G$ given by $j \mapsto \rho_j$ is an isomorphism of groups. We write the actions of the group rings $\mathbb{Z}[\Delta]$ and $\mathbb{Z}[G]$ on the multiplicative group $\mathbb{Q}(\zeta_r, \zeta_q)^*$ using exponential notation.

One easily checks the following relations.

$$\tau(\chi)^{\sigma_i} = \tau(\chi^i) \quad \text{for } i \in (\mathbb{Z}/r\mathbb{Z})^*.$$

and

$$\tau(\chi)^{\rho_j} = \chi(j)^{-1} \tau(\chi) \quad \text{for } j \in (\mathbb{Z}/q\mathbb{Z})^*.$$

We write $\overline{\tau(\chi)}$ for the complex conjugate of $\tau(\chi)$. For $\chi \neq 1$ one has

$$\tau(\chi)\overline{\tau(\chi)} = q,$$

showing that $\tau(\chi)$ is an algebraic integer that is only divisible by primes that lie over $q$.

For our purposes the key property of the Gaussian sums is the following.

PROPOSITION 4.1. *Let $q$ be a prime, let $r$ be a positive integer prime to $q$. Let $\chi : (\mathbb{Z}/q\mathbb{Z})^* \longrightarrow \mu_r$ be a character and let $\tau(\chi)$ be the corresponding Gaussian sum. Then, for every prime number $p$ not dividing $qr$ we have*

$$\tau(\chi)^{\sigma_p - p} = \chi^p(p) \quad \text{in the ring } \mathbb{Z}[\zeta_q, \zeta_r]/(p).$$

PROOF. We have $\tau(\chi)^p \equiv -\sum_{x \in (\mathbb{Z}/q\mathbb{Z})^*} \zeta_q^{px} \chi^p(x)$ modulo the ideal $p\mathbb{Z}[\zeta_q, \zeta_r]$. Multiplying by $\chi^p(p)$ and replacing the variable $x$ by $p^{-1}x$, we get that

$$\chi^p(p)\tau(\chi)^p \equiv -\chi^p(p) \sum_{x \in (\mathbb{Z}/q\mathbb{Z})^*} \zeta_q^x \chi^p(p^{-1}x) = \tau(\chi^p) \equiv \tau(\chi)^{\sigma_p} \pmod{p}$$

as required. $\square$

The cyclotomic primality test proceeds by checking the congruence of Proposition 4.1 for suitable characters $\chi : (\mathbb{Z}/q\mathbb{Z})^* \longrightarrow \mu_r$. The next theorem is the key ingredient for the cyclotomic primality test.

THEOREM 4.2. *Let $n$ be a natural number. Let $q$ be a prime not dividing $n$, let $r$ be a power of a prime number $l$ not dividing $n$ and let $\chi : (\mathbb{Z}/q\mathbb{Z})^* \longrightarrow \mu_r$ be a character. If*
*– for every prime $p$ dividing $n$ there exists $\lambda_p$ in the ring $\mathbb{Z}_l$ of $l$-adic integers such that*
$$p^{l-1} = n^{(l-1)\lambda_p} \quad \text{in } \mathbb{Z}_l^*;$$
*– the Gaussian sum $\tau(\chi)$ satisfies*
$$\tau(\chi)^{\sigma_n - n} \in \langle \zeta_r \rangle \quad \text{in the ring } \mathbb{Z}[\zeta_q, \zeta_r]/(n),$$
*then we have*
$$\chi(p) = \chi(n)^{\lambda_p}$$
*for every prime divisor $p$ of $n$.*

Note that $\lambda_p \in \mathbb{Z}_l$ in the first condition is well defined because both $n^{l-1}$ and $p^{l-1}$ are congruent to 1 (mod $l$). In addition, $\lambda_p$ is unique. When $l$ is odd, the first condition is equivalent to the fraction $(p^{l-1}-1)/(n^{l-1}-1)$ being $l$-integral. In the second condition, we denote by $\langle \zeta_r \rangle$ the cyclic subgroup of $(\mathbb{Z}[\zeta_r]/(n))^*$ of order $r$ generated by $\zeta_r$. Note that the group $\langle \zeta_r \rangle$ is not necessarily equal to the group of $r$-th roots of unity in the ring $\mathbb{Z}[\zeta_r]/(n)$.

PROOF OF THE THEOREM. We may assume that $\chi$ is a non-trivial character. By the second condition we have

$$\tau(\chi)^{\sigma_n^{-1} n} = \eta \tau(\chi) \quad \text{for some } \eta \in \langle \zeta_r \rangle \subset \mathbb{Z}[\zeta_q, \zeta_r]/(n).$$

Note that the operator $\sigma_n^{-1} n \in \mathbb{Z}[\Delta]$ has the property that $\eta^{\sigma_n^{-1} n} = \eta$. Therefore, for any integer $L \geq 0$, applying it $(l-1)L$ times leads to the relation

$$\tau(\chi)^{(\sigma_n^{-1} n)^{(l-1)L}} = \eta^{(l-1)L} \tau(\chi), \quad \text{in the ring } \mathbb{Z}[\zeta_q, \zeta_r]/(n).$$

On the other hand, Proposition 4.1 implies that for any prime divisor $p$ of $n$ we have $\tau(\chi)^{\sigma_p^{-1} p} = \chi(p)^{-1} \tau(\chi)$ and hence

$$\tau(\chi)^{(\sigma_p^{-1} p)^{l-1}} = \chi(p)^{1-l} \tau(\chi) \quad \text{in the ring } \mathbb{Z}[\zeta_q, \zeta_r]/(p).$$

Let $l^M$ be the order of the $l$-part of the finite multiplicative group $(\mathbb{Z}[\zeta_q,\zeta_r]/(n))^*$ and $A$ the group $(\mathbb{Z}[\zeta_q,\zeta_r]/(n))^*$ modulo $l^M$-th powers. Let $L$ be an integer between 0 and $l^M$ for which $L \equiv \lambda_p \pmod{l^M}$. Then we have $p^{l-1} \equiv n^{(l-1)L} \equiv n^{(l-1)\lambda_p} \pmod{l^M}$ and hence $(\sigma_n^{-1}n)^{(l-1)L} = \sigma_p^{-1}p$ in the ring $(\mathbb{Z}/l^M\mathbb{Z})[\Delta]$. It follows that the left hand sides of the two formulas above are equal in the group $A$. Then the same is true for the right hand sides. Since $\tau(\chi)$ is invertible modulo $p$, this means

$$\eta^{(l-1)L} = \chi(p)^{1-l} \quad \text{in the group } A.$$

Since $l-1$ is coprime to the order of $\mu_r$ and since the natural map $\langle \zeta_r \rangle \hookrightarrow A$ is injective, this implies

$$\chi(p)^{-1} = \eta^L = \eta^{\lambda_p},$$

in the group $\langle \zeta_r \rangle \subset (\mathbb{Z}[\zeta_q,\zeta_r]/(n))^*$. When we multiply the formulas of the first condition for the various prime divisors $p$ of $n$ together, we see that for every positive divisor $d$ of $n$ there exists $\lambda_d \in \mathbb{Z}_l$ for which $d^{l-1} = n^{(l-1)\lambda_d}$ in $\mathbb{Z}_l$. We have, of course, $\lambda_n = 1$. From the relation $\lambda_{dd'} = \lambda_d + \lambda_{d'}$, we deduce that $\eta^{\lambda_d} = \chi(d)^{-1}$ for every divisor $d$ of $n$. In particular, we have $\eta = \eta^{\lambda_n} = \chi(n)^{-1}$ and hence

$$\chi(p) = \chi(n)^{\lambda_p},$$

for every prime divisor $p$ of $n$, as required. □

ALGORITHM. The following algorithm is based on Theorem 4.2. Suppose we want to prove that a natural number $n$ is prime. First determine an integer $R > 0$ that has the property that

$$s = \prod_{\substack{q-1 \mid R \\ q \text{ prime}}} q$$

exceeds $\sqrt{n}$. At the end of this section we recall that there is a constant $c > 0$ such that for every natural number $n > 16$ there exists an integer $R < (\log n)^{c \log \log \log n}$ that has this property. Taking $R$ equal to the product of the first few small prime powers is a good choice. For all primes $q$ dividing $s$ and for each prime power $r$ that divides $q-1$ exactly, we make sure that $\gcd(n,qr) = 1$ and then check the two conditions of Theorem 4.2 for one character of conductor $q$ and order $r$. When $n$ passes all these tests, we check for $k = 1, \ldots, R-1$ whether the smallest positive residue of $n^k$ modulo $s$ divides $n$. If that never happens, then $n$ is prime.

PROOF OF CORRECTNESS. We first note that when $n$ is prime, Proposition 4.1 implies that it passes all tests. Conversely, suppose that $p \leq \sqrt{n}$ is a prime divisor of $n$. For every prime $l$ dividing $R$, let $\lambda_p$ be the $l$-adic number that occurs in the first condition of Theorem 4.2. Let $L \in \{0, 1, \ldots, R-1\}$ be the unique integer for which we have

$$L \equiv \lambda_p \pmod{r},$$

for the power $r$ of $l$ that exactly divides $R$. Theorem 4.2 implies therefore that $\chi(p) = \chi(n)^L$ for the set of characters of conductor $q$ and order $r$ for which the conditions of Theorem 4.2 have been checked. Since we have $s = \prod_{q-1 | R} q$, the exponent of the group $(\mathbb{Z}/s\mathbb{Z})^*$ divides $R$. Therefore our set of characters *generates* the group of *all* characters of $(\mathbb{Z}/s\mathbb{Z})^*$. It follows that

$$p \equiv n^L \pmod{s}.$$

Since we have $0 < p \leq \sqrt{n} < s$, this means that $p$ must actually be *equal* to the smallest positive residue of $n^k$ modulo $s$ for some $k = 1, 2, \ldots, R-1$. Since we checked that none of these numbers divide $n$, we obtain a contradiction. It follows that $p$ cannot exist, so that $n$ is necessarily prime. □

In practice, checking the first condition of Theorem 4.2 is easy. When $l \neq 2$, the number $\lambda_p \in \mathbb{Z}_l$ of the first condiction exists if and only if for any prime divisor $p$ of $n$, the rational number $(p^{l-1} - 1)/(n^{l-1} - 1)$ is $l$-integral. Since we have $p^{l-1} \equiv 1 \pmod{l}$, this is automatic when we have $n^{l-1} \not\equiv 1 \pmod{l^2}$. Given $n$, this usually holds true for various prime numbers $l$. Another useful criterion is the following. It can be checked for free when one checks the second condition of Theorem 4.2.

PROPOSITION 4.3. *Let $n > 1$ be an integer and let $l$ be a prime number not dividing $n$. Then there exists for every prime divisor $p$ of $n$ an exponent $\lambda_p \in \mathbb{Z}_l$ for which*

$$p^{l-1} = n^{(l-1)\lambda_p} \quad \text{in } \mathbb{Z}_l^*,$$

*whenever there exists a prime $q$ not dividing $n$ for which the following holds.*

(i) *($l \neq 2$) for some power $r > 1$ of $l$ and some character $\chi : (\mathbb{Z}/q\mathbb{Z})^* \longrightarrow \mu_r$ of order $r$ the number $\tau(\chi)^{\sigma_n - n}$ is a generator of the cyclic subgroup $\langle \zeta_r \rangle$ of $(\mathbb{Z}[\zeta_q, \zeta_r]/(n))^*$.*
(ii) *($l = 2$ and $n \equiv 1 \pmod{4}$) we have $\tau(\chi)^{\sigma_n - n} = -1$ for the quadratic character $\chi$ modulo $q$.*
(iii) *($l = 2$ and $n \equiv 3 \pmod{4}$) and for some character $\chi : (\mathbb{Z}/q\mathbb{Z})^* \longrightarrow \mu_r$ of 2-power order $r \geq 4$, the number $\tau(\chi)^{\sigma_n - n}$ is a generator of the cyclic subgroup $\langle \zeta_r \rangle$ of $(\mathbb{Z}[\zeta_q, \zeta_r]/(n))^*$. Moreover, the Gaussian sum associated to the quadratic character $\chi^{r/2}$ satisfies $\tau(\chi^{r/2})^{\sigma_n - n} = -1$ in the ring $\mathbb{Z}[\zeta_q]/(n)$.*

PROOF. Let $p$ be a prime divisor of $n$ and let $r$ be a power of $l$. As in the proof of Theorem 4.2, let $l^M$ denote the order of the $l$-part of the unit group

$(\mathbb{Z}[\zeta_q, \zeta_r]/(p))^*$ and let $A$ be the group $(\mathbb{Z}[\zeta_q, \zeta_r]/(p))^*$ modulo $l^M$-th powers. The latter is a module over the $l$-adic group ring $\mathbb{Z}_l[\Delta]$. The multiplicative subgroup $\{\sigma_m^{-1} m \in \mathbb{Z}_l[\Delta] : m \in \mathbb{Z}_l^*\}$ is naturally isomorphic to $\mathbb{Z}_l^*$. Therefore, when $l \neq 2$, its subgroup $G$ of $(l-1)$-th powers is isomorphic to the additive group $\mathbb{Z}_l$. When $l = 2$, this is not true, but in that case the subgroup $G^2$ of squares is isomorphic to $\mathbb{Z}_2$. By Proposition 4.1 for any prime $q$ and character $\chi : (\mathbb{Z}/q\mathbb{Z})^* \longrightarrow \mu_r$ of order $r$ we have

$$\tau(\chi)^{\sigma_p^{-1} p} = \chi(p)^{-1} \tau(\chi) \quad \text{in the group } A.$$

If $\tau(\chi)^{\sigma_n - n}$ is a generator of the group $\langle \zeta_r \rangle \subset (\mathbb{Z}[\zeta_q, \zeta_r]/(n))^*$, then we have

$$\tau(\chi)^{\sigma_n^{-1} n} = \eta \tau(\chi) \quad \text{in the group } A.$$

for some primitive $r$-th root of unity $\eta \in \langle \zeta_r \rangle \subset (\mathbb{Z}[\zeta_q, \zeta_r]/(n))^*$.

Now we prove (i). Since $\eta$ is a primitive root, the operator $(\sigma_n^{-1} n)^{l-1} \in \mathbb{Z}_l[\Delta]$ cannot be a 'proper' $l$-adic power of $(\sigma_p^{-1} p)^{l-1}$ in the sense that there cannot exist $\mu \in l\mathbb{Z}_l$ for which $(\sigma_n^{-1} n)^{l-1} = (\sigma_p^{-1} p)^{\mu(l-1)}$. Since both operators are contained in the pro-cyclic group $G \cong \mathbb{Z}_l$, the converse must therefore be true: we have $(\sigma_p^{-1} p)^{l-1} = (\sigma_n^{-1} n)^{(l-1)\lambda_p}$ and hence $p^{l-1} = n^{(l-1)\lambda_p}$ for some $\lambda_p \in \mathbb{Z}_l$.

To prove (ii), we observe that the values of $\chi$ are either $1$ or $-1$. Therefore we have $\tau(\chi)^{\sigma_n} = \tau(\chi)$. Since we have $\tau(\chi)^2 = \chi(-1)\tau(\chi)\overline{\tau(\chi)} = \chi(-1)q$, the condition $\tau(\chi)^{\sigma_n - n} = -1$ means precisely that

$$(\chi(-1)q)^{(n-1)/2} \equiv -1 \pmod{n}.$$

This shows that the 2-parts of the order of $\chi(-1)q \pmod{p}$ and of $n-1$ are equal. This means that $n-1$ divides $p-1$ in the ring of 2-adic integers $\mathbb{Z}_2$. Since $n \equiv 1 \pmod{4}$, this is equivalent to the statement that $p = n^{\lambda_p}$ for some $\lambda_p \in \mathbb{Z}_2$.

To prove (iii), we note that for $l = 2$, the group $G$ that we considered above is not isomorphic to $\mathbb{Z}_2$, but the subgroup $G^2$ is. Therefore the arguments of the proof of part (i) only show that $p^2 = n^{2\lambda_p}$ and hence $p = \pm n^{\lambda_p}$ for some $\lambda_p \in \mathbb{Z}_2$. We show that we have the plus sign. From the relation $p^2 = n^{2\lambda_p}$ we deduce that $\chi^{-1}(p)^2 = \eta^{2\lambda_p}$. Raising this relation to the power $-r/4$, we find

$$\left(\frac{p}{q}\right) = \chi^{r/2}(p) = \eta^{-r\lambda_p/2} = (-1)^{\lambda_p}.$$

Here we used the usual Legendre symbol to denote the quadratic character $\chi^{r/2}$. Since $q \equiv 1 \pmod{4}$, we have $\chi(-1) = 1$. Therefore the second condition

$\tau(\chi^{r/2})^{\sigma_n - n} \equiv -1 \pmod{n}$ says precisely that we have $q^{(n-1)/2} \equiv -1 \pmod{n}$. Since $(n-1)/2$ is odd, it follows that

$$\left(\frac{q}{p}\right) = \left(\frac{q^{(n-1)/2}}{p}\right) = \left(\frac{-1}{p}\right).$$

Since $\chi$ has order at least 4, we have $q \equiv 1 \pmod 4$ and hence, by quadratic reciprocity, $\left(\frac{p}{q}\right) = \left(\frac{q}{p}\right)$. The two formulas above imply that $\left(\frac{-1}{p}\right) = (-1)^{\lambda_p}$. This means precisely that $p \equiv n^{\lambda_p} \pmod 4$, so that we must have the plus sign, as required. □

If the number $n$ that is being tested for primality is actually prime, then in each instance the conditions of Proposition 4.3 are satisfied for a prime $q$ that has the property that $n$ is not an $l$-th power modulo $q$. Given $n$, one encounters in practice for every prime $l$ very quickly such a prime $q$, so that the first condition of Theorem 4.2 can be verified. In the unlikely event that for some prime $l$ none of the primes $q$ has this property, one simply tests the second condition of Theorem 4.2 for some more primes $q \equiv 1 \pmod l$.

Testing the second condition of Theorem 4.2 is a straightforward computation in the finite ring $\mathbb{Z}[\zeta_q, \zeta_r]/(n)$. In practice it is important to reduce this to a computation in the much smaller subring $\mathbb{Z}[\zeta_r]/(n)$. This is done by using Jacobi sums.

DEFINITION. Let $q$ be a prime and let $\chi, \chi' : (\mathbb{Z}/q\mathbb{Z})^* \longrightarrow \mu_r$ be two characters. Then we define the *Jacobi sum* $j(\chi, \chi')$ by

$$j(\chi, \chi') = - \sum_{x \in \mathbb{Z}/q\mathbb{Z}} \chi(x) \chi'(1-x).$$

Here we extend $\chi$ and $\chi'$ to $\mathbb{Z}/q\mathbb{Z}$ by putting $\chi(0) = \chi'(0) = 0$.

The Jacobi sum is an algebraic integer, contained in the cyclotomic field $\mathbb{Q}(\zeta_r)$. If the characters $\chi, \chi' : (\mathbb{Z}/q\mathbb{Z})^* \longrightarrow \mu_r$ satisfy $\chi\chi' \neq 1$, we have

$$j(\chi, \chi') = \frac{\tau(\chi)\tau(\chi')}{\tau(\chi\chi')}.$$

In particular, if $i > 0$ is prime to $r$ and less than the order of $\chi$, we have

$$\tau(\chi)^{i - \sigma_i} = \frac{\tau(\chi)^i}{\tau(\chi^i)} = \prod_{k=1}^{i-1} j(\chi, \chi^k).$$

The subgroup of the $l$-power order roots of unity in $\overline{\mathbb{Q}}^*$ is a $\mathbb{Z}[\Delta]$-module. Let $I \subset \mathbb{Z}[\Delta]$ be its annihilator. This ideal is generated by the elements of the form $\sigma_i - i$ with $i \in \mathbb{Z}$ coprime to $l$. Since we have $\tau(\chi)^{\rho_j - 1} \in \mu_r$ for all $j \not\equiv 0 \pmod{q}$, we have

$$1 = \tau(\chi)^{(\rho_j - 1)x} = \tau(\chi)^{x(\rho_j - 1)} \quad \text{for every } x \in I.$$

This shows that $\tau(\chi)^x$ and hence that $\tau(\chi)^x$ is contained in $\mathbb{Q}(\zeta_r)$ for every $x \in \mathbb{Z}[\Delta]$. This applies in particular to the element $x = \sigma_n - n \in I$. It turns out that it is possible to check the condition of Theorem 4.2 that $\tau(\chi)^{\sigma_n - n}$ is contained in $\langle \zeta_r \rangle$, without ever writing down the Gaussian sum $\tau(\chi) \in \mathbb{Z}[\zeta_r, \zeta_q]$, but by doing only computations with Jacobi sums in the ring $\mathbb{Z}[\zeta_r]/(n)$.

When $l$ is odd, the ideal $I$ generates a *principal* ideal in the $l$-adic group ring $\mathbb{Z}_l[\Delta]$. It is generated by any element of the form $\sigma_i - i$ for which $i^{l-1} \not\equiv 1 \pmod{l^2}$. We have $2^{l-1} \not\equiv 1 \pmod{l^2}$ for all primes $l < 3 \cdot 10^9$ except when $l = 1093$ or $3511$. Therefore we can in practice always use $i = 2$. In this case the relevant Jacobi sum is given by

$$\tau(\chi)^{\sigma_2 - 2} = \frac{\tau(\chi)\tau(\chi)}{\tau(\chi^2)} = j(\chi, \chi) = -\sum_{x \in \mathbb{Z}/q\mathbb{Z}} \chi(x(1-x)).$$

A computation [Cohen 1993, 9.1.5] shows that we have $\sigma_n - n = \alpha(\sigma_2 - 2)$ where $\alpha \in \mathbb{Z}_l[\Delta]$ is given by

$$\alpha = \sum_{\substack{1 \leq i < r \\ \gcd(i,r)=1}} \left[\frac{ni}{r}\right] \sigma_i^{-1}$$

times a unit in $\mathbb{Z}_l[\Delta]$. Here $[t]$ denotes the integral part of $t \in \mathbf{R}$. It follows that in order to verify that $\tau(\chi)^{\sigma_n - n}$ is contained in the group $\langle \zeta_r \rangle$ and to see whether it has order $r$, it suffices to evaluate the product

$$\prod_{\substack{1 \leq i < r \\ \gcd(i,r)=1}} j(\chi, \chi)^{[ni/r]\sigma_i^{-1}},$$

in the ring $\mathbb{Z}[\zeta_r]/(n)$ and check that it is contained in the group $\langle \zeta_r \rangle$ and see whether it has order $r$. Since the elements in the ring $\mathbb{Z}_l[\Delta]$ map the subgroup $\langle \zeta_r \rangle \subset (\mathbb{Z}[\zeta_r]/(n))^*$ to itself, the fact that we only know the element $\alpha$ up to multiplication by a unit in $\mathbb{Z}_l[\Delta]$ is of no importance.

When $l = 2$, the $\mathbb{Z}_l[\Delta]$-ideal generated by $I$ is *not* principal. It is generated by the elements $\sigma_3 - 3$ and $\sigma_{-1} + 1$. Suppose that the character $\chi : (\mathbb{Z}/q\mathbb{Z})^* \longrightarrow \mu_r$ has 2-power order $r \geq 8$.

When $n \equiv 1$ or $3 \pmod{8}$, the element $\sigma_n - n$ is contained in the $\mathbb{Z}_l[\Delta]$-ideal generated by $\sigma_3 - 3$ and we may proceed as above, replacing the Jacobi sum by the a product of two Jacobi sums: $\tau(\chi)^{\sigma_3 - 3} = j(\chi, \chi)j(\chi, \chi^2)$. We have $\sigma_n - n = \alpha(\sigma_3 - 3)$ where $\alpha \in \mathbb{Z}_l[\Delta]$ is given by $\alpha = \sum_{i \in E} \left[\frac{ni}{r}\right] \sigma_i^{-1}$ times a unit in $\mathbb{Z}_l[\Delta]$. Here $E$ denotes the set $\{i \in \mathbb{Z} : 1 \leq i < r \text{ and } i \equiv 1, 3 \pmod{8}\}$. Up to a $\mathbb{Z}_l[\Delta]$-automorphism we have

$$\tau(\chi)^{\sigma_n - n} = \prod_{i \in E} \left( j(\chi, \chi) j(\chi, \chi^2) \right)^{[ni/r]\sigma_i^{-1}},$$

and this expression involves only elements in the ring $\mathbb{Z}[\zeta_r]/(n)$.

When $n \equiv 5, 7 \pmod{8}$, we have $\sigma_n - n = -(\sigma_{-n} + n) + (\sigma_{-n} + \sigma_n)$. Now the element $\sigma_{-n} + n$ is contained in the ideal generated by $\sigma_3 - 3$, while we have $\tau(\chi)^{\sigma_{-n} + \sigma_n} = \tau(\chi^n)\tau(\chi^{-n}) = q\chi(-1)$. In this way one can express $\tau(\chi)^{\sigma_n - n}$ in a similar way in terms of elements of the subring $\mathbb{Z}[\zeta_r]/(n)$. See [Cohen 1993, 9.1.5] for the formulas.

When the order $r$ of the character is 2 or 4, it is easier to proceed dircetly. When $r = 2$, we have $\tau(\chi)^{\sigma_n - n} = (\chi(-1)q)^{(n-1)/2}$ and one should check that this is equal to $\pm 1$ in the ring $\mathbb{Z}/(n)$. Finally let $r = 4$. We have

$$\tau(\chi)^{n - \sigma_n} = \left( j(\chi, \chi)^2 \chi(-1) q \right)^{(n-1)/4}$$

when $n \equiv 1 \pmod{4}$, while

$$\tau(\chi)^{n - \sigma_n} = j(\chi, \chi) \left( j(\chi, \chi)^2 \chi(-1) q \right)^{(n-3)/4}$$

when $n \equiv 3 \pmod{4}$. In either case, in order to verify the second condition of Theorem 2.3, one should check that this number is a power of $i$ in the ring $\mathbb{Z}[i]/(n)$.

**Running time analysis.** All computations take place in finite rings of the form $\mathbb{Z}[\zeta_r]/(n)$, where $r$ divides $R$. The various summations range over the congruence classes modulo $r$ or $q$. Both $q$ and $r$ are less than $R$. The number of pairs $(q, r)$ involved in the computations is also at most $O(R)$. It follows that the number of elementary operations needed to perform the calculations is proportional to $R$ times a power of $\log n$. Therefore it is important that $R$ is small. On the other hand, the size of the $s$ should be at least $\sqrt{n}$.

By a result in analytic number theory [Crandall and Pomerance 2001, Theorem 4.3.5], there is a constant $c > 0$ so that for every natural number $n > 16$ there exists an integer $R < (\log n)^{c \log \log \log n}$ for which $s = \prod_{q-1 | R} q$ exceeds $\sqrt{n}$. It follows that the algorithm is almost polynomial time. It runs in time $O((\log n)^{c' \log \log \log n})$ for some constant $c' > 0$.

For instance, for $n$ approximately 880 decimal digits, a good choice is $R = 2^4 \cdot 3^2 \cdot 5 \cdot 7 \cdot 11 \cdot 13 \cdot 17 \cdot 19$, because then we have $s > 10^{441}$.

H. W. Lenstra proposed a slight modification of the cyclotomic test, which allows one to efficiently test integers satisfying $n < s^3$ rather than $n < s^2$, for primality. See [Lenstra 1981, Remark 8.7; Lenstra 1984] for this important practical improvement.

## 5. The elliptic curve primality test

The elliptic curve primality test, proposed by A. O. L. Atkin in 1988, is one of the most powerful primality tests that is used in practice [Morain 1998]. In order to explain its principle, we first consider a multiplicative group version of the test.

THEOREM 5.1. *Let $n > 1$ be a natural number and suppose that there is an element $a \in \mathbb{Z}/n\mathbb{Z}$ and an exponent $s > 0$ satisfying*

$$a^s = 1;$$
$$a^{s/q} - 1 \in (\mathbb{Z}/n\mathbb{Z})^* \quad \text{for every prime divisor } q \text{ of } s.$$

*Then any prime dividing $n$ is congruent to 1 (mod $s$). In particular, if $s > \sqrt{n}$, then $n$ is prime.*

PROOF. Let $p$ be a prime divisor of $n$. Then the image of $a$ in $\mathbb{Z}/p\mathbb{Z}$ is a unit of order $s$. Indeed, $a^s \equiv 1 \pmod{p}$ while $a^{s/q} \not\equiv 1 \pmod{p}$ for every prime divisor $q$ of $s$. Therefore $s$ divides the order of $(\mathbb{Z}/p\mathbb{Z})^*$. In other words, $p \equiv 1 \pmod{s}$, as required. Since a composite $n$ has a prime divisor $p \leq \sqrt{n}$, the second statement of the theorem is also clear. □

In applications, $s$ is a divisor of $n-1$ and the element $a$ of $\mathbb{Z}/n\mathbb{Z}$ is the $(n-1)/s$-th power of a randomly selected element. One tests the condition that $a^{s/q} - 1 \in (\mathbb{Z}/n\mathbb{Z})^*$ for every prime divisor $q$ of $s$, by evaluating the powers $b = a^{s/q}$ in the ring $\mathbb{Z}/n\mathbb{Z}$ and then checking that $\gcd(n, b-1) = 1$. In order to do this, one needs to know all prime divisors $q$ of $s$. In addition, $s$ needs to be large! Indeed, one needs that $s > \sqrt{n}$ in order to conclude that $n$ is prime. If $n$ is large, computing a divisor $s$ of $n - 1$ with these properties is usually very time consuming. Therefore only rarely a large number $n$ is proved prime by a direct application of this theorem.

Occasionally however, it may happen that one can compute a divisor $r > 1$ of $n - 1$ that has the property that $s = (n-1)/r$ is *probably* prime. In practice, $r$ is the product of the small prime divisors of $n - 1$ that one is able to find in a reasonable short time. The cofactor $s$ is much larger than $r$. If, by a stroke of luck, the number $s$ happens to pass some probabilistic primality test and one is confident that $s$ is prime, then one may reduce the problem of proving the primality of $n$ to proving the primality of $s$, which is at most as large as $n/2$ and usually quite a bit smaller. Indeed, pick a random $x \in \mathbb{Z}/n\mathbb{Z}$ and compute $a = x^r$. Almost certainly we have $a^s \equiv 1 \pmod{n}$ and $a-1 \in (\mathbb{Z}/n\mathbb{Z})^*$. Since $s > \sqrt{n}$, Theorem 5.1 implies then that $n$ is prime *provided that $s$* is prime. However, the chance that $n - 1$ factors like $r \cdot s$ this way is on the average $O(1/\log n)$. Therefore any attempt to proceed in some kind of inductive way, has only a very slight chance of succeeding.

Elliptic curves provide a way out of this situation. The main point is that for prime $n$ there are *many* elliptic curves $E$ over $\mathbb{Z}/n\mathbb{Z}$ and the orders of the groups $E(\mathbb{Z}/n\mathbb{Z})$ are rather uniformly distributed in the interval $(n+1-2\sqrt{n}, n+1+2\sqrt{n})$. S. Goldwasser and J. Kilian [1986] proposed a primality test based on the principle of Theorem 5.1 and on a deterministic polynomial time algorithm to determine the number of points on an elliptic curve over a finite field [Schoof 1985]. The running time of their probabilistic algorithm is polynomial time if one assumes a certain unproved assumption on the distribution of prime numbers in short intervals. Some years later, L. Adleman and M.-D. Huang [1992] eliminated the assumption, by proposing a probabilistic test involving abelian varieties of dimension 2. Both tests are of theoretical rather than practical value. From a theoretical point of view these algorithms have been superseded by the much simpler polynomial time deterministic algorithm explained in Section 3. However, the key idea leads to a powerful *practical* algorithm.

The main result is the following elliptic analogue of Theorem 5.1.

THEOREM 5.2. *Let $n > 1$ be a natural number and let $E$ be an elliptic curve over $\mathbb{Z}/n\mathbb{Z}$. Suppose that there is a point $P \in E(\mathbb{Z}/n\mathbb{Z})$ and an integer $s > 0$ for which*

$$sP = 0 \quad \text{in } E(\mathbb{Z}/n\mathbb{Z});$$

$$(s/q)P \neq 0 \quad \text{in } E(\mathbb{Z}/p\mathbb{Z}) \text{ for any prime divisor } p \text{ of } n.$$

*Then every prime $p$ dividing $n$ satisfies $\#E(\mathbb{Z}/p\mathbb{Z}) \equiv 0 \pmod{s}$. In particular, if $s > (\sqrt[4]{n}+1)^2$, then $n$ is prime.*

PROOF. Let $p$ be a prime divisor of $n$. Then the image of the point $P$ in $E(\mathbb{Z}/p\mathbb{Z})$ has order $s$. This implies that $\#E(\mathbb{Z}/p\mathbb{Z}) \equiv 0 \pmod{s}$. By Hasse's Theorem, we have $\#E(\mathbb{Z}/p\mathbb{Z}) \leq (\sqrt{p}+1)^2$. Therefore, if $s > (\sqrt[4]{n}+1)^2$, we have

$$(\sqrt{p}+1)^2 \geq \#E(\mathbb{Z}/p\mathbb{Z}) \geq s \geq (\sqrt[4]{n}+1)^2$$

and hence $p > \sqrt{n}$. If $n$ were composite, it would have a prime divisor $p \leq \sqrt{n}$. We conclude that $n$ is prime as required. □

The algorithm reduces the problem of proving the primality of $n$, to the problem of proving that a smaller number is prime as follows. Given a probable prime number $n$, one randomly selects elliptic curves $E$ over $\mathbb{Z}/n\mathbb{Z}$ and determines the order of the group $E(\mathbb{Z}/n\mathbb{Z})$ until one finds a curve for which $\#E(\mathbb{Z}/n\mathbb{Z})$ is of the form $r \cdot s$, where $r > 1$ and $s$ is a *probable* prime number satisfying $s > (\sqrt[4]{n}+1)^2$. In order to apply Theorem 5.2, one selects a random point $Q \in E(\mathbb{Z}/n\mathbb{Z})$ and computes $P = rQ$. One checks that $sP = 0$ in $E(\mathbb{Z}/n\mathbb{Z})$ and that $P \neq 0$ in $E(\mathbb{Z}/p\mathbb{Z})$ for every prime dividing $n$. If one works with projective coordinates $(x : y : z)$ satisfying a Weierstrass equation, then the latter simply means that the gcd of $n$ and the $z$-coordinate of $P$ is equal to 1. Theorem 5.2 implies then that $n$ is prime *if* $s$ is prime.

In practice, one computes $\#E(\mathbb{Z}/n\mathbb{Z})$ *under the assumption that $n$ is prime.* Then one attempts to factor the order of the group $E(\mathbb{Z}/n\mathbb{Z})$ by means of a simple trial divison algorithm or another method that finds small prime factors quicker than larger ones, like Lenstra's Elliptic Curve Method [1987]. Let $r$ be the product of these small prime factors. When $\#E(\mathbb{Z}/n\mathbb{Z})$ factors as a product $r \cdot s$ with $s$ a probable prime, it is in practice not a problem to verify the conditions of Theorem 5.2 for some randomly selected point $P$. That's because $n$ is probably prime. But we do not need to know this in order to apply Theorem 5.2.

Just as in the multiplicative case discussed above, this computation usually does not work out when $n$ is large. Typically one only succeeds in computing a small completely factored factor $r$ of $\#E(\mathbb{Z}/n\mathbb{Z})$ whose cofactor $s$ is *not* prime, but cannot be factored easily. In that case one discards the curve $E$, randomly selects another one and tries again. Since the curves $E$ are rather uniformly distributed with respect to the number of points in $\#E(\mathbb{Z}/n\mathbb{Z})$, the number of attempts one needs to make before one encounters a *prime* cofactor $s$, is expected to be $O(\log n)$. In the unlikely event that one is able to factor $\#E(\mathbb{Z}/n\mathbb{Z})$ completely or that one has $s < (\sqrt[4]{n} + 1)^2$, one is also satisfied. If this happens, one can switch the roles of $r$ and $s$ and almost certainly apply Theorem 5.2.

Atkin [1993] turns the test of Goldwasser and Kilian into a *practical* test by selecting the elliptic curves $E$ in the algorithm above more carefully. He considers suitable elliptic curves over the complex numbers with *complex multiplication* (CM) by imaginary quadratic orders of relatively small discriminant. He reduces the curves modulo $n$ and uses only these in his primality proof. The main point is that it is not only theoretically, but also *in practice* very easy to count the number of points on these elliptic curves modulo $n$. The resulting test is in practice very efficient, but its running time is very difficult to analyze rigorously, even assuming various conjectures on the distributions of smooth numbers and prime numbers. We sketch the algorithm and give a heuristic estimate of its running time.

Given $n$, Atkin first searches for imaginary quadratic integers $\varphi$ for which $N(\varphi) = n$ and $N(\varphi - 1) = r \cdot s$, where $r > 1$, $s > (\sqrt[4]{n} + 1)^2$ and $s$ is probably prime, in the sense that it passes a probabilistic primality test. Here $N(\alpha)$ denotes the *norm* of an imaginary quadratic number $\alpha$.

The theory of complex multiplication guarantees the existence of an elliptic curve $E$ over $\mathbf{C}$ with endomorphism ring isomorphic to the ring of integers of the imaginary quadratic field $\mathbb{Q}(\varphi)$. Moreover, if $n$ is prime, the characteristic polynomial of the Frobenius endomorphism of the reduced curve $E$ (mod $n$)

is equal to the minimum polynomial of $\varphi$. The number of points in $E(\mathbb{Z}/n\mathbb{Z})$ is equal to $N(\varphi - 1) = r \cdot s$. Therefore one may apply Theorem 5.2 to some randomly selected point and conclude that $n$ is prime when $s$ is. We first explain how to compute suitable imaginary quadratic integers $\varphi$ and then how to compute the corresponding elliptic curves.

If $n$ is prime, an imaginary quadratic field $F$ contains an element $\varphi$ with $N(\varphi) = n$ if and only if $n$ factors as a product of two *principal* prime ideals in the ring of integers $O_F$ of $F = \mathbb{Q}(\varphi)$. The probability that this happens is $1/2h$, where $h$ is the class number of $O_F$. Therefore in practice one first considers all imaginary quadratic fields with class number $h = 1$, then the ones with class number $h = 2, \ldots$, etc. First one checks whether or not $n$ splits in $F$. If $n$ is prime, this happens if and only if the discriminant $\Delta_F$ is a square modulo $n$. If $n$ splits, one sees whether it is a product of two prime *principal* ideals. To do this one computes a square root $z$ of $\Delta_F$ modulo $n$. Then the ideal $I$ generated by $n$ and $z - \sqrt{\Delta_F}$ is a prime divisor of $n$. To check that it is principal, one employs a lattice reduction algorithm and computes a shortest vector in the rank 2 lattice generated by $n$ and $z - \sqrt{\Delta_F}$ in $\mathbb{C}$. If the shortest vector has norm $n$, then we take it as our integer $\varphi$ and we know that $I = (\varphi)$ is principal. If the norm of the shortest vector is not equal to $n$, then the ideal $I$ is not principal and there does not exist an algebraic integer $\varphi \in F$ with $N(\varphi) = n$. In this case we cannot make use of the elliptic curves that have complex multiplication by the ring of integers of $F$.

We explain how to compute the elliptic curves $E$ over $\mathbb{Z}/n\mathbb{Z}$ from the quadratic integers $\varphi$. The $j$-invariants of elliptic curves over $\mathbb{C}$ that admit complex multiplication by the ring of integers of $F = \mathbb{Q}(\varphi)$ are algebraic integers contained in the Hilbert class field of $F$. The $j$-invariant of one such curve is

$$j(\tau) = \frac{\left(1 + 240 \sum_{k=1}^{\infty} \sigma_3(k) q^k\right)^3}{q \prod_{k=1}^{\infty} (1 - q^k)^{24}},$$

where $q = e^{2\pi i \tau}$, $\sigma_3(k) = \sum_{d|k} d^3$, and $\tau \in \mathbb{C}$ has positive imaginary part and has the property that the ring $\mathbb{Z} + \mathbb{Z}\tau$ is isomorphic to the ring of integers of $\mathbb{Q}(\varphi)$. The conjugates of $j(\tau)$ are given by $j\left(\frac{\tau+b}{a}\right)$ for suitable integers $a, b$. One computes approximations to these numbers and then the coefficients of the minimum polynomial of $j(\tau)$. This polynomial is contained in $\mathbb{Z}[X]$ and has huge coefficients. Therefore one rather works with modular functions that are contained in extensions of moderate degree $d$ (usually $d = 12$ or 24) of the function field $\mathbb{C}(j)$. The coefficients of these modular functions are much smaller. See [Cohen and Stevenhagen 2008, p. 532] for a precise statement.

If $n$ is prime, it splits by construction completely in the Hilbert class field $H$ of $F$. We compute a root of the minimal polynomial of $j(\tau)$ in $\mathbb{Z}/n\mathbb{Z}$ and call it $j$. From this we compute a Weierstrass equation of an elliptic curve $E$ over $\mathbb{Z}/n\mathbb{Z}$ with $j$-invariant equal to $j$. We perform all necessary computations as if $n$ were prime. Since $n$ probably is prime, they will be successful. If $n$ is prime, then we have $\#E(\mathbb{Z}/n\mathbb{Z}) = N(\zeta\varphi-1)$ for some root of unity $\zeta \in \mathbb{Q}(\varphi)$. If $\zeta \neq 1$, we 'twist' the curve $E$ so that we have $\#E(\mathbb{Z}/n\mathbb{Z}) = N(\varphi-1) = r \cdot s$. Usually, we have $\zeta \in \{\pm 1\}$. The exceptions are the fields $F = \mathbb{Q}(i)$ and $F = \mathbb{Q}(\sqrt{-3})$, in which cases $\zeta$ can be a fourth or sixth root of unity respectively.

It seems difficult to analyze this algorithm in a rigorous way. We present only a heuristic estimate of its running time. It is confirmed by the running times of actual implementations [Morain 2007].

In each step of the algorithm we reduce the proof of the primality of $n$ to the proof of the primality of a number that is at most $n/2$ and usually much smaller. Therefore the number of steps is bounded by $O(\log n)$. Each step consists of two phases.

First we search an imaginary quadratic number field $F$ with the property that $n$ is a norm of a principal ideal $(\varphi)$. Since the class number $h$ of $F$ is approximately the square root of the discriminant $|\Delta_F|$, the probability that this happens is $1/2h$. In addition, the probability that $N(\varphi-1)$ is equal to $r \cdot s$ where $r > 1$ is a small completely factored number and $s$ is a probable prime, is proportional to $1/\log n$. Therefore we expect to consider imaginary quadratic number fields $F$ with discriminants of size at most $O((\log n)^2)$.

As explained above, the search for $F$ involves computing square roots modulo $n$, lattice base reductions and Miller-Rabin primality tests. The cost of computing one square root modulo $n$ is $O((\log n)^{1+\mu})$. By making a list of square roots of small prime numbers $< \log n$, we can compute the square roots of a sufficiently large set of discriminants $\Delta_F$. This idea goes back to J. Shallit. It leads to an algorithm that takes $O((\log n)^{2+\mu})$ operations. The lattice base reduction is a gcd computation that can be set up to take no more than $O((\log n)^{\mu})$ operations. Since it is performed for $O((\log n)^2)$ fields $F$, this part of the algorithm also involves no more than $O((\log n)^{2+\mu})$ operations. Finally, the work involved in doing $O(\log n)$ Miller-Rabin tests is $O((\log n)^{2+\mu})$. It follows that the cost of the entire first phase is at most $O((\log n)^{2+\mu})$.

The second phase is a computation involving the 'lucky' quadratic number field $F$ that we found in the first phase: we compute an elliptic curve $E$ with CM by the ring of integers of $F$. Since the discriminant $\Delta_F$ is $O((\log n)^2)$, the amount of work to compute the minimum polynomial $g$ of its $j$-invariant by means of the high precision computations explained above is $O((\log n)^{1+\mu})$. See [Enge 2006]. The number of bits needed to write down $g$ is $O((\log n)^2)$.

The work to find a zero of $g$ in $\mathbb{Z}/n\mathbb{Z}$ is proportional to the effort to compute an $(n-1)/2$-th power modulo $g$, which is $O((\log n)^{1+2\mu})$. Finally we compute a large multiple of a random point on an elliptic curve modulo $n$. The work involved is $O((\log n)^{1+\mu})$, so that the cost of the entire second phase is $O((\log n)^{1+2\mu})$.

The total amount of work for a single step can therefore be estimated by $O((\log n)^{\max(2+\mu, 1+2\mu)}) = O((\log n)^{1+2\mu})$. The cost of the entire algorithm is therefore $O((\log n)^{2+2\mu})$. This is $O((\log n)^6)$ if one uses standard arithmetic in $\mathbb{Z}/n\mathbb{Z}$ and $O((\log n)^{4+\varepsilon})$ using fast multiplication techniques.

## References

[Adleman and Huang 1992] L. M. Adleman and M.-D. A. Huang, *Primality testing and abelian varieties over finite fields*, Lecture Notes in Mathematics **1512**, Springer, Berlin, 1992.

[Adleman et al. 1983] L. M. Adleman, C. Pomerance, and R. S. Rumely, "On distinguishing prime numbers from composite numbers", *Ann. of Math.* (2) **117**:1 (1983), 173–206.

[Agrawal et al. 2004] M. Agrawal, N. Kayal, and N. Saxena, "Primes is in P", *Annals of Math.* **160** (2004), 781–793.

[Artjuhov 1966/1967] M. M. Artjuhov, "Certain criteria for primality of numbers connected with the little Fermat theorem", *Acta Arith.* **12** (1966/1967), 355–364. In Russian.

[Atkin and Morain 1993] A. O. L. Atkin and F. Morain, "Elliptic curves and primality proving", *Math. Comp.* **61**:203 (1993), 29–68.

[Bach 1990] E. Bach, "Explicit bounds for primality testing and related problems", *Math. Comp.* **55**:191 (1990), 355–380.

[Cohen 1993] H. Cohen, *A course in computational algebraic number theory*, Graduate Texts in Mathematics **138**, Springer, Berlin, 1993.

[Cohen and Stevenhagen 2008] H. Cohen and P. Stevenhagen, "Computational class field theory", pp. 497–534 in *Surveys in algorithmic number theory*, edited by J. P. Buhler and P. Stevenhagen, Math. Sci. Res. Inst. Publ. **44**, Cambridge University Press, New York, 2008.

[Crandall and Pomerance 2001] R. Crandall and C. Pomerance, *Prime numbers: a computational perspective*, Springer, New York, 2001.

[Enge 2006] A. Enge, "The complexity of class polynomial computation via floating point approximations", preprint, 2006. Available at http://www.arXiv.org/abs/cs/0601104v1/2006.

[Goldwasser and Kilian 1986] S. Goldwasser and J. Kilian, "Almost all primes can be quickly certified", pp. 316–329 in *Proceedings of the eighteenth annual ACM Symposium on the theory of computing*, ACM, New York, 1986.

[Konyagin and Pomerance 1997] S. Konyagin and C. Pomerance, "On primes recognizable in deterministic polynomial time", pp. 176–198 in *The mathematics of Paul Erdős, I*, Algorithms Combin. **13**, Springer, Berlin, 1997.

[Lang 1978] S. Lang, *Cyclotomic fields*, Graduate Texts in Math. **59**, Springer, New York, 1978.

[Lenstra 1979] H. W. Lenstra, Jr., "Miller's primality test", *Inform. Process. Lett.* **8**:2 (1979), 86–88.

[Lenstra 1981] H. W. Lenstra, Jr., "Primality testing algorithms (after Adleman, Rumely and Williams)", pp. 243–257 in *Bourbaki Seminar* (1980/81), Lecture Notes in Math. **901**, Springer, Berlin, 1981.

[Lenstra 1984] H. W. Lenstra, Jr., "Divisors in residue classes", *Math. Comp.* **42**:165 (1984), 331–340.

[Lenstra 1987] H. W. Lenstra, Jr., "Factoring integers with elliptic curves", *Ann. of Math.* (2) **126**:3 (1987), 649–673.

[Lenstra and Pomerance $\geq$ 2008] H. W. Lenstra and C. Pomerance, "Primality testing with Gaussian periods", to appear.

[Mihăilescu 1998] P. Mihăilescu, "Cyclotomy primality proving: recent developments", pp. 95–110 in *Algorithmic number theory* (Portland, OR, 1998), Lecture Notes in Comput. Sci. **1423**, Springer, Berlin, 1998.

[Miller 1976] G. L. Miller, "Riemann's hypothesis and tests for primality", *J. Comput. System Sci.* **13**:3 (1976), 300–317.

[Morain 1998] F. Morain, "Primality proving using elliptic curves: an update", pp. 111–127 in *Algorithmic number theory* (Portland, OR, 1998), Lecture Notes in Comput. Sci. **1423**, Springer, Berlin, 1998.

[Morain 2007] F. Morain, "Implementing the asymptotically fast version of the elliptic curve primality proving algorithm", *Math. Comp.* **76**:257 (2007), 493–505.

[Rabin 1980] M. O. Rabin, "Probabilistic algorithm for testing primality", *J. Number Theory* **12**:1 (1980), 128–138.

[Schoof 1985] R. Schoof, "Elliptic curves over finite fields and the computation of square roots mod $p$", *Math. Comp.* **44**:170 (1985), 483–494.

[Washington 1997] L. C. Washington, *Introduction to cyclotomic fields*, 2nd ed., Graduate Texts in Mathematics **83**, Springer, New York, 1997.

RENÉ SCHOOF
DIPARTIMENTO DI MATEMATICA
UNIVERSITÀ DI ROMA 2 "TOR VERGATA"
VIA DELLA RICERCA SCIENTIFICA
I-00133 ROMA
ITALY
schoof@mat.uniroma2.it, schoof@science.uva.nl

# Lattices

HENDRIK W. LENSTRA, JR.

ABSTRACT. It occurs frequently in algorithmic number theory that a problem has both a discrete and a continuous component. A typical example is the search for a system of integers that satisfies certain inequalities. A problem of this nature can often be successfully approached by means of the algorithmic theory of lattices, a lattice being a discrete subgroup of a Euclidean vector space. This article provides an introduction to this theory, including a generous sample of applications.

## Contents

| | |
|---|---|
| 1. Introduction | 128 |
| 2. Lattices | 129 |
| 3. Examples in algebraic number theory | 131 |
| 4. Representing lattices | 132 |
| 5. The determinant | 134 |
| 6. The shortest vector problem | 137 |
| 7. Diophantine approximation | 139 |
| 8. The nearest vector problem | 143 |
| 9. Lattices of rank two | 145 |
| 10. Flags | 148 |
| 11. Finding a good flag | 152 |
| 12. Enumerating short vectors | 155 |
| 13. Factoring polynomials | 157 |
| 14. Linear algebra over the ring of integers | 162 |
| 15. Nonlinear problems | 167 |
| 16. Lattices over polynomial rings | 177 |
| Acknowledgments | 179 |
| References | 179 |

# 1. Introduction

A *lattice* is a discrete subgroup of a Euclidean vector space, and *geometry of numbers* is the theory that occupies itself with lattices. Since the publication of Hermann Minkowski's *Geometrie der Zahlen* in 1896, lattices have become a standard tool in number theory, especially in the areas of diophantine approximation, algebraic number theory, and the arithmetic theory of quadratic forms.

The theory of continued fractions, principally developed by Leonhard Euler (1707–1783), is in substance concerned with algorithmic aspects of lattices of rank 2. A significant advance in the algorithmic theory of lattices of general rank occurred in the early 1980's, with the development of the powerful lattice basis reduction algorithm that came to be called the LLL algorithm [Lenstra et al. 1982]. The LLL algorithm has found numerous applications in both pure and applied mathematics.

In algorithmic number theory, geometry of numbers now plays a role that is comparable to the role that linear programming plays in optimization theory, and that linear algebra plays throughout mathematics. This is due to a similar combination of circumstances: good algorithms are available for solving the basic problems, and many commonly encountered problems reduce to those basic problems. Just as a multitude of problems in mathematics can be linearized, so can many others be addressed by the introduction of a suitable lattice. Typically, this applies to problems that have both a discrete and a continuous component, such as the search for a system of integers that satisfies certain inequalities. Algorithmic number theory abounds in such problems.

The main purpose of the present introduction to the subject is to impart to the reader the ability to recognize situations in which a lattice basis reduction algorithm is useful. For this reason, all definitions and algorithms have been formulated in conceptual terms, appealing to the geometric rather than the algebraic intuition. At the same time, coordinates will be chosen when they have an actual role to play, which is unavoidably the case whenever the algorithms are to be translated into genuine computer programs. A generous sample of applications of lattice basis reduction to algorithmic number theory has been included; in many cases, the main point consists of recognizing a lattice behind a problem. For applications to integer programming, one may consult [Aardal and Eisenbrand 2005].

Complete proofs have not been provided for all results mentioned, though in many cases one will find a sketch of a proof or a 'convincing argument'. Generally, the subject matter is elementary enough that the readers can supply the details themselves, and in any case they can turn to the references at the end. The same applies to running time estimates of algorithms.

## 2. Lattices

*Euclidean vector spaces.* A *Euclidean vector space* is a finite-dimensional vector space $E$ over the field $\mathbb{R}$ of real numbers equipped with a map $\langle\,,\,\rangle\colon E\times E\to \mathbb{R}$ satisfying

$$\langle w+x, y\rangle = \langle w, y\rangle + \langle x, y\rangle, \qquad \langle rx, y\rangle = r\langle x, y\rangle,$$
$$\langle x, y\rangle = \langle y, x\rangle, \qquad \langle z, z\rangle > 0$$

for all $r \in \mathbb{R}$ and $w, x, y, z \in E$, $z \neq 0$. The map $\langle\,,\,\rangle$ is called the *inner product* on $E$. For $z \in E$, we write $\|z\| = \langle z, z\rangle^{1/2}$, and we refer to this number as the *length* of the vector $z$. Any Euclidean vector space $E$ is a *metric space* with distance function $d\colon E \times E \to \mathbb{R}$ defined by $d(x, y) = \|x - y\|$. If $E, E'$ are Euclidean vector spaces, then a map $\psi\colon E \to E'$ is an *isomorphism of Euclidean vector spaces* if it is an isomorphism of vector spaces over $\mathbb{R}$ that preserves inner products, in the sense that for all $x, y \in E$ one has $\langle \psi(x), \psi(y)\rangle = \langle x, y\rangle$. For each non-negative integer $n$, the vector space $\mathbb{R}^n$ is a Euclidean vector space with the *standard inner product* defined by

$$\langle (x_i)_{i=1}^n, (y_i)_{i=1}^n\rangle = \sum_{i=1}^n x_i y_i.$$

For each Euclidean vector space $E$ there is an isomorphism $\mathbb{R}^{\dim E} \cong E$ of Euclidean vector spaces, where $\dim E$ denotes the dimension of $E$ as a vector space over $\mathbb{R}$.

*Lattices.* A subset $L$ of a Euclidean vector space $E$ is *discrete* if the metric on $E$ defines the discrete topology on $L$; in other words, if for each $x \in L$ there is a positive real number $\varepsilon$ such that the only $y \in L$ with $d(x, y) < \varepsilon$ is given by $y = x$. A *lattice* is an additive subgroup $L$ of a Euclidean vector space $E$ such that $L$ is discrete as a subset of $E$; given that $L$ is a subgroup, discreteness is equivalent to the existence of a positive real number $\varepsilon$ such that the only vector $y \in L$ with $\|y\| < \varepsilon$ is given by $y = 0$.

A subset $L$ of a Euclidean vector space $E$ is a lattice if and only if there are $\mathbb{R}$-linearly independent vectors $b_1, \ldots, b_n \in E$ such that

$$L = \sum_{i=1}^n \mathbb{Z} b_i = \Big\{ \sum_{i=1}^n c_i b_i : c_i \in \mathbb{Z} \text{ for } i = 1, \ldots, n \Big\}.$$

If this is the case, then $b_1, \ldots, b_n$ are said to form a *basis* for $L$ (over $\mathbb{Z}$), and $L$ is isomorphic to $\mathbb{Z}^n$ as an abelian group; from $\#L/2L = 2^n$ one sees that $n$ is determined by the structure of $L$ as an abelian group, and it is called the *rank* of $L$, notation: $\mathrm{rk}\, L$.

One can also define lattices without reference to a Euclidean vector space. Namely, let $L$ be an abelian group, and let $q\colon L \to \mathbb{R}$ be a map. Then $L$ can be embedded as a lattice in a Euclidean vector space $E$ with $q(x) = \|x\|^2$ for all $x \in L$ if and only if $L$ is finitely generated and the following three conditions are satisfied:

$$q(x+y) + q(x-y) = 2q(x) + 2q(y) \text{ for all } x, y \in L,$$
$$q(x) \neq 0 \text{ for all } x \in L \text{ with } x \neq 0,$$
$$\{x \in L : q(x) \leq r\} \text{ is finite for each real number } r.$$

The proof of the 'if'-part (see [Lenstra 2001, Prop. 4.1]) shows that one may take $E = L \otimes_\mathbb{Z} \mathbb{R}$, the inner product being such that

$$\langle x, y \rangle = \bigl(q(x+y) - q(x) - q(y)\bigr)/2$$

for all $x, y \in L$. Thus, one can define a lattice to be a finitely generated abelian group $L$ equipped with a map $q: L \to \mathbb{R}$ satisfying the three conditions just listed. The first of these properties is called the *parallelogram law*, since it expresses that the sum of the squares of the lengths of the two diagonals of a parallelogram equals the sum of the squares of the lengths of its four sides. In general, if $L, q$ constitute a lattice, then one has $q(x) \geq 0$ for each $x \in L$, one thinks of $q(x)$ as the square of the length of $x$, and the function $d: L \times L \to \mathbb{R}$ defined by $d(x, y) = q(x - y)^{1/2}$ is a metric on $L$.

We shall often refer to a lattice as a pair $L, q$, emphasizing that all we need to know is the group $L$ and the lengths of all of its elements; when $q$ is clear from the context, it may be dropped. Often, it will tacitly be assumed that such a lattice is embedded in a Euclidean vector space $E$, and then it is always understood that $q(x) = \|x\|^2 = \langle x, x \rangle$ for all $x \in L$. The notation $q(x) = \langle x, x \rangle$ will also be used for other elements $x$ of $E$. Sometimes it is understood that $L$ is of *full rank* in $E$, which means that one has rk $L$ = dim $E$; one can always achieve this by replacing $E$ by the subspace of $E$ spanned by $L$.

*Isometries.* An isometry of a lattice $L, q$ to a lattice $L', q'$ is a bijection $f: L \to L'$ that preserves distances. One can compose each isometry with a translation to achieve that it maps 0 to 0, and each isometry mapping 0 to 0 is automatically a group isomorphism. One cares about lattices only up to isometry.

*Sublattices.* Let $L, q$ be a lattice. Every subgroup $M$ of $L$ becomes a lattice upon restricting $q$ to $M$; such a lattice is called a *sublattice* of $L$. A sublattice $M$ of $L$ is called *pure* if $L/M$ is *torsion-free* as an abelian group, which means that $L/M$ has no non-zero element of finite order. If $M$ is a pure sublattice of $L$, then $N = L/M$ acquires a natural lattice structure in the following way: embed $L$ in a Euclidean vector space $E$, let $E'$ be the subspace spanned by $M$, write $E'^\perp$ for the orthogonal complement $\{x \in E : \langle x, y \rangle = 0 \text{ for all } y \in E'\}$ of $E'$ in $E$, and $\pi: E \to E'^\perp$ for the orthogonal projection (so $\pi$ is $\mathbb{R}$-linear, zero on $E'$, and the identity on $E'^{\perp}$, and $\pi$ is uniquely determined by those properties); then $\pi L$ is a discrete subgroup of the Euclidean vector space $E'^{\perp}$ and therefore a lattice, and the natural isomorphism $N = L/M \to \pi L$ induced by $\pi$ identifies $\pi L$ with $N$, which therefore becomes a lattice as well.

*The dual lattice.* Let $L$ be a lattice of full rank in a Euclidean vector space $E$. Then $L^{\dagger} = \{x \in E : \langle x, L \rangle \subset \mathbb{Z}\}$ is also a lattice of full rank in $E$, the *dual* (or *polar*) of $L$. If $b_1, \ldots, b_n$ form a basis for $L$, then the unique elements $b_1^{\dagger}, \ldots, b_n^{\dagger} \in E$ satisfying $\langle b_i^{\dagger}, b_j \rangle = 1$ or $0$ according as $i + j = n+1$ or $i + j \ne n+1$ form a basis for $L^{\dagger}$. (This is the 'cobasis' of $E$ corresponding to the basis $b_1, \ldots, b_n$, numbered backwards for later convenience.) One has $\operatorname{rk} L^{\dagger} = \operatorname{rk} L$ and $L^{\dagger\dagger} = L$.

## 3. Examples in algebraic number theory

In this section we discuss three types of lattices that are naturally encountered in algebraic number theory. The examples are not typical of the examples that we shall encounter later on, and readers without an interest in algebraic number theory may safely skip this section.

*Additive groups of algebraic numbers.* Let $K$ be an algebraic number field, i.e., a field that is a finite extension of the field $\mathbb{Q}$ of rational numbers, and let $L$ be a finitely generated subgroup of the additive group of $K$; for example, one may take $L$ to be the ring $\mathbb{Z}_K$ of algebraic integers in $K$, or a fractional $\mathbb{Z}_K$-ideal. Then $L$ carries a natural lattice structure, which is defined by

$$q(x) = \sum_{\sigma} |\sigma x|^2$$

for $x \in L$, with $\sigma$ ranging over the set of field embeddings of $K$ in the field $\mathbb{C}$ of complex numbers, and where $|\ |$ denotes the usual absolute value on $\mathbb{C}$.

*Multiplicative groups of algebraic numbers.* One can deal with multiplicative subgroups in a similar manner. Let $K$ again be an algebraic number field, and denote by $\mu$ the set of roots of unity in $K$, which is a finite cyclic subgroup of the multiplicative group $K^*$ of $K$. Let now $L$ be a finitely generated subgroup of the quotient group $K^*/\mu$. Then $L$ has a natural lattice structure, which this time is defined by

$$q(x\mu) = \sum_{p} \sum_{\sigma} (\log |\sigma x|_p)^2$$

for $x\mu \in L \subset K^*/\mu$; here $p$ ranges over the set $\{\infty, 2, 3, 5, 7, \ldots\}$ of 'primes' of $\mathbb{Q}$, and $\sigma$ ranges, for fixed $p$, over the set of field embeddings of $K$ in an algebraic closure $\overline{\mathbb{Q}}_p$ of the $p$-adic completion $\mathbb{Q}_p$ of $\mathbb{Q}$; each $\overline{\mathbb{Q}}_p$ is chosen once and for all, and $|\ |_p$ denotes, for $p < \infty$, the $p$-adic absolute value on

$\overline{\mathbb{Q}}_p$ with $|p|_p = 1/p$, whereas on $\overline{\mathbb{Q}}_\infty = \mathbb{C}$ one takes $|\ |_\infty = |\ |$. If one takes $L = \mathbb{Z}_K^*/\mu$, where $\mathbb{Z}_K^*$ denotes the group of units of $\mathbb{Z}_K$, then all terms with $p \neq \infty$ vanish, and one obtains a lattice of which the rank is one less than the number of infinite places of $K$.

*Elliptic curves.* Consider an elliptic curve $\mathscr{E}$ over $\mathbb{Q}$, defined by a Weierstrass equation $y^2z + a_1xyz + a_3yz^2 = x^3 + a_2x^2z + a_4xz^2 + a_6z^3$, with all $a_i \in \mathbb{Q}$. It is well-known that the set $\mathscr{E}(\mathbb{Q})$ of points $(x:y:z)$ in the projective plane $\mathbf{P}^2(\mathbb{Q})$ that satisfy the equation, is in a natural way an abelian group, the *Mordell–Weil group* of $\mathscr{E}$ over $\mathbb{Q}$. Denote by $\mathscr{E}(\mathbb{Q})_{\mathrm{tor}}$ its subgroup of elements of finite order. Then $L = \mathscr{E}(\mathbb{Q})/\mathscr{E}(\mathbb{Q})_{\mathrm{tor}}$ is a lattice with

$$q(\overline{P}) = \frac{1}{2} \cdot \lim_{n \to \infty} \frac{h(2^n P)}{4^n}$$

for $P \in \mathscr{E}(\mathbb{Q})$, where $\overline{P}$ denotes the image of $P$ in $L$ and where for an element $(x:y:z) \in \mathbf{P}^2(\mathbb{Q})$, with $\mathbb{Z}x + \mathbb{Z}y + \mathbb{Z}z = \mathbb{Z}$, one defines $h(x:y:z) = \log\max\{|x|, |y|, |z|\}$; the number $q(\overline{P})$ is known as the *canonical height* of $P$.

## 4. Representing lattices

*Two different normalizations.* Suppose that $L$ is a lattice of full rank in a Euclidean vector space $E$. Writing $n = \mathrm{rk}\, L$, one has an isomorphism $L \cong \mathbb{Z}^n$ of groups as well as an isomorphism $E \cong \mathbb{R}^n$ of Euclidean vector spaces. However, these two isomorphisms are generally not compatible, and if, for whatever reason, one wishes to introduce coordinates, then one needs to choose between the two. Each option has its virtues, and the usefulness of the concept of lattices is in no small part due to the possibility of thinking about them in two different ways.

As we shall see, in many applications of lattices one takes $L$ equal to $\mathbb{Z}^n$ and $q$ equal to a function that reflects the problem at hand. On the other hand, when thinking about lattices one will often find it useful to imagine them as being embedded in ordinary Euclidean $n$-space, with $q(x)$ proportional to the square of the distance from $x$ to the origin. Here $n$ is bounded only by the limits of one's imagination. Experience shows that, even when the fourth dimension proves too hard to picture in one's mind, one can still avoid the common pitfall of implicitly assuming that the rank $n$ of $L$ is small, such as 2 or 3. Several subtle phenomena occur only for large $n$, and the fact that the LLL algorithm runs in polynomial time even when $n$ varies is one of the keys to its success.

*Representing lattices numerically.* If one wishes to run an algorithm on a lattice, one needs to specify the lattice and its elements in some numerical manner. There are many ways of doing this, and the two most important ones correspond to the two possibilities mentioned above. The first is to specify a lattice by writing down a real positive definite symmetric $n \times n$ matrix $\mathbf{A} = (a_{ij})_{1 \le i,j \le n}$; the lattice $L$ is then understood to be the abelian group $\mathbb{Z}^n$, its elements are represented as (column) vectors with $n$ integral entries, and $q$ is given by $q(x) = x^T \mathbf{A} x$ for $x \in L$, the superscript $T$ denoting passage to the transpose. In order to be able to write down $\mathbf{A}$ by means of a finite number of bits, one may require that all the $a_{ij}$ are rational, and that they are represented as $a_{ij} = a'_{ij}/d$ where $d$ and all $a'_{ij}$ are integers represented in binary, and $d > 0$.

The second way of specifying a lattice is by writing down a real $m \times n$ matrix $\mathbf{B} = (b_{ij})_{1 \le i \le m, 1 \le j \le n}$ of rank $n$; in this case, $L$ is understood to be the subgroup $\sum_{j=1}^{n} \mathbb{Z} b_j$ of $\mathbb{R}^m$, where $b_j = (b_{ij})_{i=1}^{m}$ and where $\mathbb{R}^m$ has the standard inner product. The elements of $L$ are then represented as real $m$-vectors. Again, one may require the entries of $\mathbf{B}$ to be rational, so that the coordinates of all elements of $L$ are rational as well.

*Conversion.* Whenever we discuss algorithms for lattices, it will always be assumed that lattices are specified in one of the two ways just described, by means of a matrix with rational entries. Which of the two one uses is immaterial, since there are polynomial-time algorithms for converting each type of presentation into the other. In one direction this is easy: the second type is converted into the first by the formula $\mathbf{A} = \mathbf{B}^T \cdot \mathbf{B}$. The conversion in the other direction is a little more laborious, and for lack of a suitable reference we give a quick sketch of a possible way to proceed. Given $\mathbf{A}$, one first uses the Gram–Schmidt process to diagonalize the induced quadratic form on $\mathbb{Q}^n$ (see Section 10). This has the effect of writing $\mathbf{A} = \mathbf{C}_1^T \cdot \mathbf{D} \cdot \mathbf{C}_1$, where $\mathbf{C}_1$ is an upper triangular $n \times n$ matrix over $\mathbb{Q}$, with 1's on the diagonal, and $\mathbf{D}$ is a diagonal matrix with $n$ positive rational diagonal entries $d_j$. Using a naive greedy algorithm, one writes each $d_j$ as the sum of $m_j = O(\log \max\{2, \log d'_j\})$ squares of non-zero rational numbers, where $d'_j$ denotes the product of the numerator and the denominator of $d_j$ (if one allows a probabilistic algorithm, as in [Rabin and Shallit 1986], one can take $m_j \le 4$). With $m = \sum_j m_j$, this leads to an $m \times n$ matrix $\mathbf{C}_2$ over $\mathbb{Q}$, with exactly one non-zero entry per row, such that $\mathbf{D} = \mathbf{C}_2^T \cdot \mathbf{C}_2$; and now the matrix $\mathbf{B} = \mathbf{C}_2 \cdot \mathbf{C}_1$ has rank $n$ and satisfies $\mathbf{A} = \mathbf{B}^T \cdot \mathbf{B}$, as desired. This procedure, while running in polynomial time, does give rise to a fairly large value for $m$, which is not bounded by a function of $n$ alone. The probabilistic algorithm from [Rabin and Shallit 1986] leads to $m \le 4n$. Theoretically, one can achieve $m \le n + 3$ (see [Cassels 1978, Chapter 6, Example 8]), but I do not know whether this can be done by means of an algorithm that is efficient in any sense of the word.

Whenever we assert that a lattice algorithm runs in polynomial time, then we mean that its run time is bounded by a polynomial function of the number of bits of the input, where all lattices forming part of the input or output of the algorithm are specified by a rational matrix **A** or **B** as above; that length will be at least the rank of the input lattice.

*Other representations.* There are other natural ways of specifying a lattice. For example, if $f: L \to L'$ is a group homomorphism from a lattice $L$ to a lattice $L'$, then the kernel and the image of $f$ are sublattices of $L$ and $L'$, respectively. If $L$ and $L'$ are specified by an $n \times n$ matrix **A** and an $n' \times n'$ matrix **A'** as above, so that $L = \mathbb{Z}^n$ and $L' = \mathbb{Z}^{n'}$, then the map $f: \mathbb{Z}^n \to \mathbb{Z}^{n'}$ is given by an $n' \times n$ matrix **F** over $\mathbb{Z}$; the three matrices **A**, **A'**, **F** can then serve to specify both the kernel and the image of $f$. One may convert this type of presentation into one of the earlier ones by means of the kernel and image algorithm presented in Section 14.

The examples from Section 3 show that sometimes lattices can be specified in ways that are very difficult to convert to any of our standard formats. For example, one can specify an algebraic number field $K$ by means of a defining equation over $\mathbb{Q}$, and this defining equation is then sufficient to specify the lattice $L = \mathbb{Z}_K$. However, no polynomial-time algorithm is known for actually finding a basis for $L = \mathbb{Z}_K$ over $\mathbb{Z}$ (even when one restricts to the case $[K : \mathbb{Q}] = 2$; see [Buchmann and Lenstra 1994]), and for typical fields $K$ with $[K : \mathbb{Q}] > 2$ the function $q$ is not $\mathbb{Q}$-valued. Similar comments apply to the unit lattice $L = \mathbb{Z}_K^*/\mu$, for which a $\mathbb{Z}$-basis appears to be even harder to compute, and to the Mordell–Weil lattices $L = \mathscr{E}(\mathbb{Q})/\mathscr{E}(\mathbb{Q})_{\text{tor}}$, for which $\mathbb{Z}$-bases are not even known to be computable.

## 5. The determinant

*Definition of the determinant.* After the rank, the most important numerical invariant attached to a lattice $L$ is its *determinant*, denoted by $d(L)$. It is defined by

$$d(L) = \lim_{r \to \infty} \frac{\text{vol } B(\sqrt{r})}{\#\{y \in L : q(y) \le r\}},$$

where for $n = \text{rk } L$ we define $B(\sqrt{r})$ to be the ball $\{x \in \mathbb{R}^n : \langle x, x \rangle \le r\}$ of radius $\sqrt{r}$ in $\mathbb{R}^n$, and vol denotes the standard $n$-dimensional volume. One has

$$\text{vol } B(\sqrt{r}) = r^{n/2} \cdot \text{vol } B(1) = r^{n/2} \cdot \pi^{n/2}/\tfrac{n}{2}!,$$

where $\tfrac{n}{2}!$ is inductively defined by $0! = 1$, $\tfrac{1}{2}! = \sqrt{\pi}/2$, and $\tfrac{n}{2}! = \tfrac{n}{2} \cdot \tfrac{n-2}{2}!$ for $n \ge 2$. (One has $\tfrac{n}{2}! = \Gamma(1 + \tfrac{n}{2})$.) To understand the definition of $d(L)$, and to show that the limit exists, one may assume $L$ to be embedded in the standard Euclidean vector space $\mathbb{R}^n$. Let **B** be a non-singular real $n \times n$ matrix such that the columns $b_j$ of $\mathbf{B}$ form a basis for $L$. Then the subset

$$F = \sum_{j=1}^{n} [0,1) b_j = \left\{ \sum_{j=1}^{n} c_j b_j : c_j \in \mathbb{R}, \ 0 \le c_j < 1 \text{ for } 1 \le j \le n \right\}$$

of $\mathbb{R}^n$ satisfies $\operatorname{vol} F = |\det \mathbf{B}|$, and $F$ is a fundamental domain for $L$ in the sense that each $x \in \mathbb{R}^n$ has a unique representation $x = y + z$ with $y \in L$ and $z \in F$. Restricting to the set of all $y \in L$ with $q(y) \le r$, one proves that the disjoint union $\bigcup_y (y + F)$, taken over those $y$, is a fair approximation to $B(\sqrt{r})$; more precisely, if one puts $s = \sup\{\langle z, z \rangle : z \in F\}$, then that union is contained in $B(\sqrt{r} + \sqrt{s})$, and for $r \ge s$ it contains $B(\sqrt{r} - \sqrt{s})$. Comparing volumes, one deduces

$$\lim_{r \to \infty} \frac{\#\{y \in L : q(y) \le r\} \cdot |\det \mathbf{B}|}{\operatorname{vol} B(\sqrt{r})} = 1.$$

It follows that $d(L)$ is well-defined, and that one has in fact $d(L) = |\det \mathbf{B}| = \operatorname{vol} F$. In particular, $\operatorname{vol} F$ is independent of the choice of the basis.

The zero lattice has determinant 1.

*Hadamard's inequality.* Let $L$, $b_1, \ldots, b_n$, $F$ be as above. The volume of the parallelepiped $F$ is at most the product of the lengths of the vectors $b_i$, so we have

$$d(L) \le \prod_{i=1}^{n} \|b_i\|.$$

This is *Hadamard's inequality*, which is valid for any basis $b_1, \ldots, b_n$ of a lattice $L$. Equality holds if and only if the vectors $b_i$ are pairwise orthogonal, in the sense that $\langle b_i, b_j \rangle = 0$ whenever $i \ne j$. In Section 10 we will see that if the basis $b_1, \ldots, b_n$ is *reduced* in a suitable sense, then one has the opposite inequality

$$\prod_{i=1}^{n} \|b_i\| \le c_n \cdot d(L),$$

where $c_n$ depends only on the rank $n$ of the lattice. Thus, a 'reduced' basis may be thought of as being 'nearly orthogonal'.

*Formulae for the determinant.* There are many formulae that can be used in the computation of $d(L)$, in addition to the formula $d(L) = |\det \mathbf{B}|$ mentioned above. If $L$ is given by means of a matrix $\mathbf{A}$ as in Section 4, then one has $d(L) = (\det \mathbf{A})^{1/2}$. These two formulae suffice for most algorithmic and numerical purposes. In a more theoretical context, they can be supplemented by the following rules. Let $L$ be a lattice. If $M$ is a sublattice of finite index $(L : M)$ of $L$, then one has $d(M) = (L : M) \cdot d(L)$. If $M$ is a pure sublattice of $L$ (see Section 2), then one has $d(L) = d(M) \cdot d(L/M)$. For the dual $L^\dagger$

of $L$, one has $d(L^\dagger) = 1/d(L)$. If $L$ is embedded as a lattice of full rank in a Euclidean vector space $E$, and $\tau\colon E \to E$ is a non-singular linear map, then $\tau L$ is a lattice, and one has $d(\tau L) = |\det \tau| \cdot d(L)$. The proofs are elementary and may be left to the reader.

*The volume discrepancy.* Let $E_1$ and $E_2$ be Euclidean vector spaces, and let $\tau\colon E_1 \to E_2$ be a linear map. We can associate to $\tau$ a positive real number $\gamma(\tau)$, the *volume discrepancy* of $\tau$, in the following way. Let $(\ker \tau)^\perp$ be the orthogonal complement of the kernel of $\tau$ in $E_1$. Then $\tau$ restricts to a vector space isomorphism $(\ker \tau)^\perp \to \tau E_1$. Identifying each of $(\ker \tau)^\perp$ and $\tau E_1$, as Euclidean vector spaces, with $\mathbb{R}^{\operatorname{rank} \tau}$, we obtain a non-singular linear map $\tau'\colon \mathbb{R}^{\operatorname{rank} \tau} \to \mathbb{R}^{\operatorname{rank} \tau}$, and we define $\gamma(\tau) = |\det \tau'|$; the independence of the choice of identifications with $\mathbb{R}^{\operatorname{rank} \tau}$ can either be shown directly, or be deduced from the formula $d(\tau L) = \gamma(\tau) \cdot d(L)$, which is valid for any lattice $L$ of full rank in $(\ker \tau)^\perp$. In the case $E_1 = E_2$ one has $\gamma(\tau) = |\det \tau|$ if $\tau$ is non-singular, but not if $\tau$ is singular, since one has $\gamma(\tau) > 0$.

Write $\tau^\dagger\colon E_2 \to E_1$ for the linear map that is *adjoint* to $\tau$; it is characterized by the property that $\langle x, \tau^\dagger y \rangle = \langle \tau x, y \rangle$ for all $x \in E_1$, $y \in E_2$. One has

$$\gamma(\tau) = \gamma(\tau^\dagger).$$

One can prove this by using that any square matrix and its transpose have the same determinant, or by considering dual lattices.

Some care is required with computing the volume discrepancy of a composed map. If $E_3$ is a third Euclidean vector space, and $\sigma\colon E_2 \to E_3$ is a linear map, then the formula $\gamma(\sigma\tau) = \gamma(\sigma)\gamma(\tau)$ is valid if one has $\tau E_1 = (\ker \sigma)^\perp$, but not in much greater generality.

The definition of the volume discrepancy given by Lang [1988, Chapter V, Section 2] generalizes the definition just given: the number $\gamma(\tau)$ defined above equals the volume discrepancy, as defined by Lang, of the exact sequence $0 \to \ker \tau \to E_1 \xrightarrow{\tau} E_2 \to E_2/\tau E_1 \to 0$. A still more general perspective is offered by de Smit [1996].

*Determinants of kernels and images.* Let $L_1$ and $L_2$ be lattices, and let $f\colon L_1 \to L_2$ be a group homomorphism. Embed $L_1$ and $L_2$ as lattices of full rank in Euclidean vector spaces $E_1$ and $E_2$, respectively, and write $f_\mathbb{R}$ for the $\mathbb{R}$-linear map $E_1 \to E_2$ induced by $f$. Then we have

$$d(\ker f) \cdot d(fL_1) = \gamma(f_\mathbb{R}) \cdot d(L_1)$$

with $\gamma(f_\mathbb{R})$ as defined above (cf. [Lang 1988, Chapter V, Theorem 2.1]). To prove this, one observes that $\ker f_\mathbb{R}$ is the $\mathbb{R}$-subspace of $E_1$ spanned by the pure sublattice $\ker f$ of $L_1$, and that $L = L_1/\ker f$ may be viewed as a lattice of full rank in $(\ker f_{\mathbb{R}})^{\perp}$ satisfying $f_{\mathbb{R}}L = fL_1$. Next one uses the formulae $d(L_1) = d(\ker f) \cdot d(L_1/\ker f)$ and $d(f_{\mathbb{R}}L) = \gamma(f_{\mathbb{R}}) \cdot d(L)$ that we encountered earlier.

The adjoint $f_{\mathbb{R}}^{\dagger}$ of $f_{\mathbb{R}}$ restricts to a map $f^{\dagger}: L_2^{\dagger} \to L_1^{\dagger}$. From $\gamma(f_{\mathbb{R}}) = \gamma(f_{\mathbb{R}}^{\dagger})$ and $d(L_1^{\dagger}) = d(L_1)^{-1}$ one obtains the *six lattices formula*

$$d(\ker f) \cdot d(fL_1) \cdot d(L_1^{\dagger}) = d(\ker f^{\dagger}) \cdot d(f^{\dagger} L_2^{\dagger}) \cdot d(L_2).$$

This formula is often helpful in computing determinants of lattices; see Sections 7 and 8 for illustrations.

## 6. The shortest vector problem

*Existence of short vectors.* The *shortest vector problem*, also known as the *homogeneous approximation problem*, is the following: given a lattice $L$ of positive rank, find a non-zero element $x \in L$ with $q(x)$ smallest possible. The formulation may be interpreted in several ways: writing $\lambda(L) = \min\{q(x) : x \in L, x \neq 0\}$, one may actually wish to find $x \in L$ with $q(x) = \lambda(L)$; or one may, in an algorithmic context, take 'smallest possible' to mean: smallest possible given the time that one is willing to spend.

The main theoretical result about the problem is the following.

THEOREM OF MINKOWSKI. *Each lattice $L$ of positive rank $n$ contains a non-zero element $x$ with $q(x) \leq \frac{4}{\pi} \cdot \frac{n}{2}!^{2/n} \cdot d(L)^{2/n} \leq n \cdot d(L)^{2/n}$.*

To see why this is true, assume again $L \subset \mathbb{R}^n$, and put $\lambda = \lambda(L) = \min\{q(x) : x \in L, x \neq 0\}$. Then no two distinct points of $L$ have distance smaller than $\sqrt{\lambda}$, so if one writes $B' = \{z \in \mathbb{R}^n : \langle z, z \rangle < \lambda/4\}$, then the open balls $y + B'$ of radius $\sqrt{\lambda}/2$ centered at the lattice points $y \in L$ are pairwise disjoint. Since the sets $y + F$ from the previous proof disjointly cover $\mathbb{R}^n$ as $y$ ranges over $L$, one deduces that $\operatorname{vol} B' \leq \operatorname{vol} F = d(L)$. Using that $\operatorname{vol} B' = (\sqrt{\lambda}/2)^n \cdot \operatorname{vol} B(1)$, one obtains the first inequality, and the second follows from the fact that $B(1)$ contains a cube with edge length $2/\sqrt{n}$. By Stirling's theorem, one actually has $\frac{4}{\pi} \cdot \frac{n}{2}!^{2/n} = \frac{2+o(1)}{\pi e} \cdot n$ for $n \to \infty$.

*The Hermite constant.* Both $\lambda(L)$ and $d(L)$ are *homogeneous functions* of $L$, of degrees 2 and $n$, respectively; that is, if inside $\mathbb{R}^n$ one replaces $L$ by $tL$ for some positive real number $t$ (or, equivalently, the function $q$ by $t^2 \cdot q$), then $\lambda$ is replaced by $t^2 \cdot \lambda$ and one has $d(tL) = t^n \cdot d(L)$. Hence, $d(L)^{2/n}$ is the only power of $d(L)$ that has the same degree as $\lambda(L)$, and therefore the only power of $d(L)$ that can possibly occur in a result like Minkowski's theorem. The supremum of $\lambda(L)/d(L)^{2/n}$, taken over all lattices $L$ of rank $n$, is called the *Hermite constant* and denoted by $\gamma_n$. Minkowski's theorem, as stated above, is equivalent to the inequalities $\gamma_n \leq \frac{4}{\pi} \cdot \frac{n}{2}!^{2/n} \leq n$. It is known that $n/(2\pi e) \leq \gamma_n \leq n/(\pi e + o(1))$

for $n \to \infty$; see [Conway and Sloane 1988, Chapter 1, Section 1] for more information and a slightly better result.

There is a sense in which, for a 'random' lattice of given positive rank $n$, the inequality $\lambda(L) \le \gamma_n \cdot d(L)^{2/n}$ is close to best possible. However, the lattices that occur in many applications are by no means random. As we shall see, one often constructs a lattice in such a manner that it has an 'exceedingly short' non-zero vector if and only if a certain problem has a solution, and that solution can then be read off from the short vector. In such cases, Minkowski's theorem plays at best a secondary role.

*Construction of short vectors.* A salient feature of the proof of Minkowski's theorem is its non-constructive character. The existence of $x$ is shown by a measure-theoretic version of the pigeon-hole principle, and no efficient algorithm for actually finding $x$ can be read from the proof. Indeed, all known algorithms for computing $\lambda(L)$, or for finding a lattice vector $x$ as in Minkowski's theorem, perform some sort of complete enumeration, and fail to run in polynomial time for varying $n$ (cf. Section 12).

In Section 11 we shall see that the construction of a 'fair' approximation to the shortest non-zero element of $L$ is a byproduct of so-called *lattice basis reduction* algorithms, such as the LLL algorithm. The LLL algorithm does run in polynomial time, but the non-zero vector $x \in L$ that it finds is not guaranteed to be the shortest one, or to be as short as in Minkowski's theorem. The quantity $q(x)/d(L)^{2/n}$ will be bounded by a function of $n$ alone, but this is an exponential function rather than a linear function as in Minkowski's theorem. For example, the standard variant of the LLL algorithm produces a non-zero element $x \in L$ with

$$q(x) \le 2^{n-1} \cdot \lambda(L), \qquad q(x) \le 2^{(n-1)/2} \cdot d(L)^{2/n}$$

(see Section 11). It is both fortunate and surprising that these exponential aberrations are small enough for most applications.

*Short vectors in the dual lattice.* Let $E$ be a Euclidean vector space. Write $' : E - \{0\} \to E - \{0\}$ for inversion in the unit sphere, so that $x' = x/\langle x, x \rangle$; note that $x'$ is a vector lying in the same direction from the origin as $x$, but with length equal to the inverse of the length of $x$. For each $x \in E - \{0\}$, one has $x'' = x$, and one also verifies easily that the subgroup $\{y \in E : \langle x, y \rangle \in \mathbb{Z}\}$ of $E$ is the (orthogonal) sum of the subgroup $\mathbb{Z}x'$ generated by $x'$ and the orthogonal complement $(\mathbb{R}x)^\perp = \{y \in E : \langle x, y \rangle = 0\}$ of the subspace spanned by $x$. Thus, $\{y \in E : \langle x, y \rangle \in \mathbb{Z}\}$ is the union, over $m \in \mathbb{Z}$, of the translates $(\mathbb{R}x)^\perp + mx'$ of the hyperplane $(\mathbb{R}x)^\perp$, and the successive distances between these translates are equal to $\|x'\| = 1/\|x\|$.

Next let $L \subset E$ be a lattice of full rank, and let $L^\dagger$ be its dual. For $x \in E - \{0\}$, one has $x \in L^\dagger$ if and only if $L$ is contained in the set $\{y \in E : \langle x, y \rangle \in \mathbb{Z}\} = (\mathbb{R}x)^\perp + \mathbb{Z}x'$. Since $x$ is 'short' if and only if $x'$ is 'long', and since every hyperplane in $E$ is of the form $(\mathbb{R}x)^\perp$, we conclude that the shortest vector problem for $L^\dagger$ is equivalent to the following problem posed in terms of $L$: given $L$, find a hyperplane $H$ in $E$ such that $L$ is contained in a collection of maximally widely spaced translates of $H$. The latter problem is useful in enumerating all lattice vectors that lie in a certain region, and it has applications in integer programming (see [Aardal and Eisenbrand 2005; Lenstra 1983]).

Minkowski's theorem now implies that for any lattice $L$ of positive rank $n$ there is a hyperplane $H$ as above, such that the distance between the successive translates of $H$ is at least $\gamma_n^{-1/2} \cdot d(L)^{1/n}$; and with the LLL algorithm one can find a hyperplane that is within a factor $2^{(n-1)/2}$ from optimal.

## 7. Diophantine approximation

This section and the next are devoted to some traditional applications of the shortest vector problem. For additional applications, see Sections 13–15 below.

*Continued fractions.* Suppose that $\alpha$ is a real number. Then the continued fraction expansion of $\alpha$ gives rise to a sequence $p_0/q_0, p_1/q_1, p_2/q_2, \ldots$ of rational numbers, with $\mathbb{Z}p_i + \mathbb{Z}q_i = \mathbb{Z}$ and $q_i > 0$ for all $i$, such that $|\alpha - p_i/q_i| < 1/q_i^2$ for all $i$ and such that any similarly written rational number $p/q$ satisfying $|\alpha - p/q| < 1/(2q^2)$ occurs in the sequence (see [Hardy and Wright 1938, Chapter X]). If $\alpha$ is rational then the sequence is finite; likewise, when $\alpha$ is irrational but known or given to finite precision only, as is often the case in an algorithmic context, then only finitely many terms of the sequence are meaningful.

Thus, the continued fraction expansion gives rise to a sequence of rational approximations $p/q$ to a given real number $\alpha$ that are 'good' in the sense that the error tends to 0 fairly quickly as a function of the denominator $q$ of the approximation.

It is instructive to see how one can achieve a similar purpose with the help of a lattice. Let again $\alpha$ be a real number, and define the lattice $L, q$ by $L = \mathbb{Z}^2$ and

$$q(x, y) = N \cdot (x - \alpha y)^2 + y^2 \qquad \text{for } x, y \in \mathbb{Z},$$

where $N$ is a suitably chosen 'large' real number. One verifies that rk $L = 2$ and $d(L) = N^{1/2}$, so there is a non-zero element $(x, y) \in L$ with $q(x, y) \le \gamma_2 N^{1/2}$, where $\gamma_2 = \sqrt{4/3}$ is the Hermite constant for $n = 2$ (see Section 9). Also, in algorithmic circumstances one can actually find such a vector efficiently (see Section 9). If $N > \gamma_2^2$ then from $(x - \alpha y)^2 \le q(x, y)/N \le \gamma_2/N^{1/2} < 1$ one deduces $y \ne 0$, and the inequality of the means implies $|N^{1/2} \cdot (x - \alpha y)| \cdot |y| \le$

$q(x, y)/2 \leq \gamma_2 N^{1/2}/2$, so that we have

$$\left|\alpha - \frac{x}{y}\right| \leq \frac{\gamma_2/2}{y^2}, \qquad 0 < |y| \leq \gamma_2^{1/2} N^{1/4}.$$

Thus one obtains a rational approximation to $\alpha$ that is of the same quality as what one obtains from the continued fraction algorithm. The main difference is that the continued fraction algorithm yields a whole sequence of approximations; to achieve this with lattices, one would need to vary $N$ and therefore consider a family of lattices. A discussion of techniques for doing this, and for deciding which values of $N$ are the crucial ones, falls outside the scope of the present introduction. In most circumstances where 'good' rational approximations to a real number $\alpha$ are required, a single well-chosen number $N$ will do.

*Higher-dimensional diophantine approximation.* The approximation problem just discussed allows several natural generalizations to higher dimensions, two of which will be discussed. Many corresponding higher-dimensional extensions of the continued fraction method have been proposed, but none appears to have all the properties that one desires. The translation into the shortest vector problem for a suitably constructed lattice generalizes readily to higher dimensions, and here again one encounters a proliferation of algorithms; that is, while in rank 2 there appears to exist only one reasonable lattice basis reduction algorithm (see Section 9), there is an entire family of them in rank greater than 2 (see Section 11).

*Simultaneous diophantine approximation.* Let $k$ real numbers $\alpha_1, \ldots, \alpha_k$, with $k \geq 1$, be given, and suppose that one is interested in finding simultaneous rational approximations $x_i/y$ to $\alpha_i$, all with the same denominator $y$; for $k = 1$ this is the problem discussed above. For general $k$, one can introduce the lattice $L, q$ defined by $L = \mathbb{Z}^{k+1}$ and

$$q(x_1, x_2, \ldots, x_k, y) = N \cdot \sum_{i=1}^{k}(x_i - \alpha_i y)^2 + y^2$$

for $(x_1, x_2, \ldots, x_k, y) \in \mathbb{Z}^{k+1}$, where $N$ plays the same role as above. One has rk $L = k + 1$ and $d(L) = N^{k/2}$. In the same manner as for $k = 1$ one now deduces that for $N > \gamma_{k+1}^{k+1}$ there is a integer vector $(x_1, x_2, \ldots, x_k, y)$ with $y \neq 0$ and

$$y^2 \leq \gamma_{k+1} N^{k/(k+1)}, \qquad \sum_{i=1}^{k}(x_i - \alpha_i y)^2 \leq \frac{k \cdot (\gamma_{k+1}/(k+1))^{1+1/k}}{|y|^{2/k}}.$$

In addition, with the LLL algorithm one can actually find such a vector, but with $2^{k/2}$ replacing $\gamma_{k+1}$.

Here is a possible algorithmic application of simultaneous diophantine approximation. Suppose one is given a $k \times k$ matrix $\mathbf{C}$ with integer entries and with $\det \mathbf{C} \neq 0$, as well as a column vector $b \in \mathbb{Z}^k$. Then there is a unique column vector $z \in \mathbb{R}^k$ with $\mathbf{C}z = b$, and the entries $z_1, \ldots, z_k$ of $z$ are rational; say $z_i = p_i/q$, where $q \in \mathbb{Z}$ is a common denominator. Further, suppose that one in interested in efficiently and *exactly* computing $z$, but that the only linear algebra package at one's disposal works in real precision. Then one can proceed as follows. First, use the linear algebra package to compute an approximate solution vector $\alpha$, so that the entries of $\mathbf{C}\alpha$ are very close to the entries of $b$ and the entries $\alpha_i$ of $\alpha$ are very close to $z_i$. If the approximations are good enough, then the lattice defined above will for large enough $N$ contain an exceptionally short vector, namely the (unknown!) vector $(p_1, \ldots, p_k, q)$. Next, one applies the LLL algorithm; it will find a non-zero vector $(x_1, \ldots, x_k, y) \in \mathbb{Z}^{k+1}$ that is at most $2^k$ times as long, and therefore still quite short; so short, that one estimates the integers entries of $\mathbf{C}x - yb$ (which is close to the tiny vector $y \cdot (\mathbf{C}\alpha - b)$) to be smaller than 1 in absolute value. Consequently, one has actually $\mathbf{C}x = yb$ and therefore $z = x/y$. The reader may enjoy filling in the details and working out explicit inequalities that make the argument valid.

There is a very similar but more complicated application of simultaneous rational approximations to linear programming (see [Schrijver 1986]).

*Approximate linear dependencies.* In a second higher-dimensional generalization of the approximation problem, one is given $k$ real numbers $\alpha_1, \ldots, \alpha_k$, with $k \geq 2$, and one is interested in finding an 'approximate' linear relation with integer coefficients among the $\alpha_i$, that is, a sequence $x_1, \ldots, x_k$ of integers, not all zero, such that $\left|\sum_i x_i \alpha_i\right|$ is small in relation to the sizes of the $x_i$ themselves. With $k = 2$, $\alpha_2 = 1$ this amounts to the problem of finding a good rational approximation to $\alpha_1$ that we considered above. Generally, one can take $L = \mathbb{Z}^k$ and define $q$ by

$$q(x_1, x_2, \ldots, x_k) = \sum_{i=1}^{k} x_i^2 + N \cdot \left(\sum_{i=1}^{k} x_i \alpha_i\right)^2 \quad (x_i \in \mathbb{Z}),$$

where $N$ is again a suitably large real number. We claim that one has

$$d(L) = \left(1 + N \cdot \sum_{i=1}^{k} \alpha_i^2\right)^{1/2}.$$

To prove this, consider the standard Euclidean vector spaces $E_1 = \mathbb{R}^k$ and $E_2 = E_1 \times \mathbb{R} = \mathbb{R}^{k+1}$, and define $\tau \colon E_1 \to E_2$ by

$$\tau\left((x_i)_{i=1}^{k}\right) = \left((x_i)_{i=1}^{k}, \sqrt{N} \cdot \sum_{i=1}^{k} \alpha_i x_i\right).$$

The lattices $L_1 = \mathbb{Z}^k$ and $L_2 = \tau L_1 + \mathbb{Z} \cdot (0, 1)$ are of full rank in $E_1$ and $E_2$, and $L$ may as a lattice be identified with $\tau L_1$. The six lattices formula from Section 5 now simplifies to the statement that $d(L)$ equals the determinant of the kernel of the map $\tau^\dagger \colon L_2^\dagger \to L_1^\dagger$; since the vector $((-\sqrt{N} \cdot \alpha_i)_{i=1}^k, 1)$ generates that kernel, its length $(1 + N \cdot \sum_{i=1}^k \alpha_i^2)^{1/2}$ equals $d(L)$.

One can now apply Minkowski's theorem to prove a general existence theorem for approximate linear dependencies. In addition, the LLL algorithm will find one.

In a typical practical application, one is interested in detecting a *true* linear dependency among certain numbers $\beta_i$, and each $\alpha_i$ is a good approximation to $\beta_i$. For example, with $\beta_i = \beta^{i-1}$ one may attempt to detect an algebraic number $\beta$ from a numerical approximation. A very similar application will be encountered in Section 13.

*Non-archimedean approximation.* The approximation problems discussed so far were concerned with *real* numbers, and the quality of the approximations was measured by means of the *real* absolute value. Sometimes it is felt that a different notion of lattice would be required if instead we are concerned with *p-adic numbers* and the *p-adic absolute value*. This is not true: both problems just considered, when transferred to a *p*-adic context, can still be addressed by means of suitably constructed lattices. The problem of finding approximate linear dependencies may serve as illustration.

Let $p$ be a prime number, denote by $\mathbb{Z}_p$ the ring of $p$-adic integers, by $\mathbb{Q}_p$ the field of fractions of $\mathbb{Z}_p$, and by $|\ |_p$ the $p$-adic absolute value on $\mathbb{Q}_p$ with $|p|_p = 1/p$. Given $k$ elements $\alpha_1, \ldots, \alpha_k$ of $\mathbb{Q}_p$, with $k \geq 2$, one looks for integers $x_1, \ldots, x_k$ that are not 'too large' in the usual absolute value, and not all zero, such that $\sum_{i=1}^k x_i \alpha_i$ is $p$-adically very close to 0. As in the case of real numbers, the $p$-adic numbers $\alpha_i$ will in an algorithmic context need to be specified to some finite precision; and in fact, if one wishes that $|\sum_{i=1}^k x_i \alpha_i|_p \leq p^{-m}$ for some given integer $m$, then it suffices to know the $\alpha_i$ modulo $p^m \mathbb{Z}_p$. Thus, we shall assume that the $\alpha_i$ are specified by means of approximations $\alpha'_i$ that belong to the ring $\mathbb{Z}[1/p]$ of rational numbers whose denominator is a power of $p$, and that are guaranteed to satisfy $|\alpha_i - \alpha'_i|_p \leq p^{-m}$. If that is the case, then for $x_i \in \mathbb{Z}$ one has $\left|\sum_i x_i \alpha_i\right|_p \leq p^{-m}$ if and only if $\sum_i x_i \alpha'_i \in p^m \mathbb{Z}$. We describe two constructions of lattices that one can use to find 'small' integers $x_i$, not all zero, with the latter property.

In the first construction, one simply takes $L$ to be the subgroup

$$\left\{ x = (x_i)_{i=1}^k \in \mathbb{Z}^k : \sum_i x_i \alpha'_i \in p^m \mathbb{Z} \right\}$$

of $\mathbb{Z}^k$, with $q(x) = \sum_i x_i^2$ for $x = (x_i)_{i=1}^k \in L$. One then has rk $L = k$ and

$d(L) = p^{m-m'}$, where $m'$ denotes the largest integer for which $p^{m'}\mathbb{Z}$ contains all $\alpha'_i$ as well as $p^m$. In many practical situations all $\alpha_i$ are in $\mathbb{Z}_p$ but not all are in $p\mathbb{Z}_p$, and $m \geq 0$; then one has $m' = 0$ and $d(L) = p^m$. A short non-zero vector in $L$, obtained with Minkowski's theorem or with LLL, gives rise to an approximate dependency as one requires. However, it should be observed that $L$ has not been specified in one of the standard formats from Section 4. Thus, before LLL can be applied, one needs to find a basis for $L$. One way of addressing this problem is found in Section 14. For now, we can achieve the same result by using the second construction instead.

In the second construction, one takes $L = \mathbb{Z}^{k+1}$ (so rk $L = k + 1$), with $q$ defined by

$$q(x_1, x_2, \ldots, x_k, y) = \sum_{i=1}^{k} x_i^2 + N \cdot \left(p^m y - \sum_{i=1}^{k} x_i \alpha'_i\right)^2,$$

where $N$ is a 'large' positive rational number. One has $d(L) = p^m N^{1/2}$. Suppose that $(x_1, x_2, \ldots, x_k, y)$ is a short non-zero lattice vector. Then the number $z = p^m y - \sum_{i=1}^{k} x_i \alpha'_i$ belongs to $p^{m'}\mathbb{Z}$, with $m'$ as defined above, and if $N$ is large enough then from the smallness of the vector and the inequality $z^2 \leq q(x_1, x_2, \ldots, x_k, y)/N$ one deduces $|z| < p^{m'}$. One concludes that $z = 0$, so that $\sum_{i=1}^{k} x_i \alpha'_i \in p^m \mathbb{Z}$. Therefore the $x_i$ do yield an approximate linear dependency, and from $\sum_i x_i^2 \leq q(x_1, x_2, \ldots, x_k, y)$ one sees that the $x_i$ are not too large.

As an interesting exercise, the reader may compare the quality of the approximations obtained from both constructions.

The $p$-adic absolute value that we just considered is a non-archimedean valuation of *mixed characteristic*, in the sense that the residue class field and the field on which the valuation is defined have different characteristics. One may also consider approximation problems for non-archimedean valuations of *equal characteristic*. These do give rise to a different notion of lattice, which we briefly treat in Section 16.

## 8. The nearest vector problem

*Inhomogeneous approximation.* The *nearest vector problem*, also known as the *inhomogeneous approximation problem*, is the following: given a lattice $L$ in a Euclidean vector space $E$, and an element $x \in E$, find $y \in L$ with smallest possible distance $d(x, y)$. By analogy with the case $L = \mathbb{Z} \subset \mathbb{R} = E$, one can think of this problem as a 'rounding' problem. As with the shortest vector problem in Section 6, the formulation allows for a strict and for a more relaxed interpretation.

For given $x \in E$, the set $\{x - y : y \in L\}$ equals the coset $x + L$ of $L$ in $E$, which is discrete in $E$; the nearest vector problem asks for an element of smallest possible length in this coset.

Let $E_0$ be the subspace of $E$ spanned by $L$, and denote the orthogonal projection of $x \in E$ on $E_0$ by $x_0$. Then for all $y \in L$ one has $d(x, y)^2 = d(x, x_0)^2 + d(x_0, y)^2$, so the nearest vector problem does not change if one replaces $E$, $x$ by $E_0$, $x_0$. Thus without loss of generality one may assume that $L$ spans $E$. In an algorithmic context one will usually also assume that the coordinates of $x$, when expressed on a basis for $L$, are rational numbers.

For $x = 0$ the nearest vector problem is solved by $y = 0$; so this special case is *not* the same as the shortest vector problem. Nevertheless, one thinks of the nearest vector problem as being harder than the shortest vector problem, and there are several observations that support this feeling. For one thing, the direct analogue of Minkowski's theorem is wrong; that is, if the rank $n$ is greater than 1, then one cannot guarantee the existence, for each $x \in E$, of an element $y \in L$ for which $d(x, y)$ is bounded by a function of $n$ and $d(L)$ alone (a suitable function of $n$, $d(L)$, and $\lambda(L)$ will do, however). There is also a formal result stating that the shortest vector problem reduces to no more than $n = \text{rk } L$ nearest vector problems, in the following manner (cf. [Goldreich et al. 1999]). Let $b_1, \ldots, b_n$ be a basis for $L$, and for each $j = 1, 2, \ldots, n$, let $L_j$ be the sublattice $\{\sum_i n_i b_i : n_i \in \mathbb{Z}, n_1, n_2, \ldots, n_j \text{ are } even\}$ of $L$. Then each set $b_j + L_j$ is a coset of $L_j$ in $L$. Their (disjoint) union, for $1 \le j \le n$, equals $L - 2L$, so if $x_j \in b_j + L_j$ has minimal length then the shortest among $x_1, \ldots, x_n$ will be a shortest non-zero element of $L$; and similarly one can reduce a relaxed version of the shortest vector problem to $n$ instances of a relaxed version of the nearest vector problem.

*The extended Euclidean algorithm.* Let $a_1, \ldots, a_k$ be positive integers, with $k \ge 2$, and put $d = \gcd(a_1, \ldots, a_k)$. If $k = 2$, the Euclidean algorithm can be used to compute $d$ when $a_1$ and $a_2$ are given, and with the extended Euclidean algorithm one can compute 'small' integers $x_1$ and $x_2$ with $x_1 a_1 + x_2 a_2 = d$ (see [Buhler and Wagon 2008, Section 3.1; Knuth 1981, Section 4.5.2]). Proceeding by induction on $k$, one can compute $d = \gcd(\gcd(a_1, \ldots, a_{k-1}), a_k)$ in polynomial time when $a_1, \ldots, a_k$ are given, and one can also inductively compute integers $x_1, \ldots, x_k$ with $\sum_i x_i a_i = d$; however, for $k > 2$ the integers $x_i$ computed in this manner will in general be very far from 'smallest possible'. Thus, one is faced with the question: given $a_1, \ldots, a_k$, as well as an integer solution $x = (x_i)_{i=1}^k$ to the equation $\sum_i x_i a_i = d$, find the smallest possible integer solution to the same equation. If we measure the 'size' of a solution by means of the Euclidean norm, then this is an instance of the nearest vector problem. Namely, let $L$ be the lattice in $\mathbb{R}^k$ (with the standard inner product)

defined by
$$L = \{y = (y_i)_{i=1}^k \in \mathbb{Z}^k : \sum_i y_i a_i = 0\}.$$
Then if $y \in L$ has smallest possible distance to $x$, the vector $x - y$ will be the smallest solution that one is looking for. One has $\operatorname{rk} L = k - 1$, and the six lattices formula from Section 5 readily implies $d(L) = (\sum_i (a_i/d)^2)^{1/2}$.

Note that $L$ is not given in one of the standard formats from Section 4, so before one can apply a lattice basis reduction algorithm one needs to find a basis for $L$. It is possible to obtain such a basis as a byproduct of the inductive computation that yields $d$ and the initial solution $x$. However, in Section 14 we shall see a much easier solution to the problem: if one works with the right lattice, then one can entirely forgo the inductive computation, and directly find both $d$ and a 'small' solution to $\sum_i x_i a_i = d$ by means of a lattice basis reduction algorithm.

*Finding the nearest vector.* As for the shortest vector problem, all known algorithms for solving the nearest vector problem perform some sort of complete enumeration, and they fail to run in polynomial time when the rank of $L$ varies (cf. Section 12). However, the LLL algorithm can be used to find an *approximate* solution. That is, the LLL algorithm computes a basis for a lattice $L$ that is 'reduced' in a suitable sense, and once a reduced basis is available one can, for given $x \in E$, efficiently compute an element $y \in L$ such that
$$d(x, y) \leq 2^n \cdot \min\{d(x, y') : y' \in L\},$$
where $n = \operatorname{rk} L$ (see Sections 10 and 11). An alternative formulation of the same algorithm is given in Section 14: given $L$ and $x$, a lattice $L'$ is constructed such that a 'reduced' basis for $L'$ immediately yields $y \in L$ as above.

## 9. Lattices of rank two

Lattices of rank two are easy to picture and to understand, and they play a pivotal role in lattice basis reduction algorithms.

*Reduced bases in rank two.* Let $L$ be a lattice with $\operatorname{rk} L = 2$, embedded in a two-dimensional Euclidean vector space $E$, and let $b_1, b_2 \in L$. Define the real numbers $a, b, c$ by
$$a = q(b_1), \qquad b = q(b_1 + b_2) - q(b_1) - q(b_2) = 2\langle b_1, b_2 \rangle, \qquad c = q(b_2).$$
Then for $x, y \in \mathbb{R}$ one has $q(xb_1 + yb_2) = ax^2 + bxy + cy^2$. We have $b^2 - 4ac \leq 0$, with strict inequality if and only if $b_1, b_2$ are linearly independent (over $\mathbb{R}$, or over $\mathbb{Z}$). The vectors $b_1, b_2$ form a basis for $L$ if and only if one has

$b^2 - 4ac = -4d(L)^2$. We call $b_1, b_2$ a *reduced basis* for $L$ if one has

$$q(b_1) = \lambda(L) = \min\{q(x) : x \in L, x \neq 0\},$$
$$b_2 \in L - \mathbb{Z}b_1, \qquad q(b_2) = \min\{q(x) : x \in L - \mathbb{Z}b_1\}.$$

It is automatic that any reduced basis for $L$ is a basis for $L$. Conversely, if $b_1$, $b_2$ form a basis for $L$, then they form a reduced basis if and only if one has $|b| \leq a \leq c$. It is clear from the definition that any lattice of rank 2 has a reduced basis.

*The shortest and nearest vector problems.* Let $L$ and $E$ be as above, and suppose that a reduced basis $b_1, b_2$ for $L$ is available. Let $a, b, c$ be defined as above. Then both the shortest vector problem and the nearest vector problem admit easy solutions. For the shortest vector problem this is obvious: $b_1$ is a shortest non-zero vector of $L$, and one has $\lambda(L) = q(b_1) = a \leq (4/3)^{1/2} d(L)$; the last inequality follows from $4d(L)^2 = 4ac - b^2 \geq 4a^2 - a^2 = 3a^2$. Considering the case $|b| = a = c > 0$ one proves that the Hermite constant $\gamma_2$ equals $(4/3)^{1/2}$.

The vector $-b_1$ is also a shortest non-zero vector of $L$, and the others, if any, are among $\pm b_2, \pm b_2 \pm b_1$.

For the nearest vector problem, assume $b = 2\langle b_1, b_2 \rangle \geq 0$, replacing $b_2$ by $-b_2$ if necessary. Define

$$F = \{z \in E : q(z) \leq q(z - y) \text{ for all } y \in \{\pm b_1, \pm b_2, \pm(b_1 - b_2)\}\}.$$

This is a hexagon if $b \neq 0$, and a rectangle if $b = 0$. Each $x \in E$ can be written as $x = y + z$ with $y \in L$ and $z \in F$, and in an algorithmic context such a representation is for given $x$ not hard to find. For 'most' $x$ it is unique, but whether or not it is unique, it is always true that $z$ is an element of the coset $x + L$ of minimal length, and that $y$ is a lattice element with minimal distance to $x$; so $y$ solves the nearest vector problem for $L$ and $x$.

It follows that the supremum, over all $x \in E$, of $\min\{q(x-y) : y \in L\}$ is equal to $\max\{q(z) : z \in F\}$. The latter number is given by the convenient formula

$$\max\{q(z) : z \in F\} = \frac{q(b_1) \cdot q(b_2) \cdot q(b_1 - b_2)}{4d(L)^2} = \frac{a \cdot c \cdot (a-b+c)}{-b^2 + 4ac},$$

where it is still assumed that $0 \leq b \leq a \leq c$. The reader may recognize the formula that expresses the circumradius of a plane triangle in terms of its area and the lengths of its sides.

*Lattice basis reduction in rank two.* Given a basis $b_1, b_2$ for a lattice $L$ of rank 2, the following iterative procedure replaces $b_1, b_2$ by a reduced basis. Let $m$ be an integer nearest to $\langle b_1, b_2 \rangle / \langle b_1, b_1 \rangle$, and replace $b_2$ by $b_2 - mb_1$. The new vector $b_2$ now satisfies $|\langle b_1, b_2 \rangle| \leq \frac{1}{2} \langle b_1, b_1 \rangle$. If it also satisfies $q(b_2) \geq q(b_1)$, then the basis $b_1$, $b_2$ is reduced, as desired; otherwise, interchange $b_1$ and $b_2$, and start all over again.

The procedure just described goes back to Gauss [1801], who used the language of binary quadratic forms. There is a strong analogy with the Euclidean algorithm for the computation of the greatest common divisor of two non-zero integers $a_1$, $a_2$: in a typical iteration step of the latter, one replaces $a_2$ by $a_2 - ma_1$, where $m$ equals $a_2/a_1$ rounded to an integer. The 'ideal' value $m = a_2/a_1$ would make the new value of $a_2$ equal to zero. Analogously, the ideal value $m = \langle b_1, b_2 \rangle / \langle b_1, b_1 \rangle$ would make the new vector $b_2$ orthogonal to $b_1$ in the sense that $\langle b_1, b_2 \rangle = 0$; one recognizes the Gram–Schmidt orthogonalization process. The actual choice of $m$ minimizes the value of $q(b_2 - mb_1)$ over $m \in \mathbb{Z}$; in particular, the new vector $b_2$ satisfies $q(b_2) \leq q(b_2 - b_1)$ and $q(b_2) \leq q(b_2 + b_1)$, which will be useful below.

Performing the procedure above for the sublattice $L$ of $\mathbb{Z}^2$ with basis $b_1 = (Na_1, 0)$, $b_2 = (Na_2, 1)$ (where $N$ is a suitably large integer) is in fact tantamount to the Euclidean algorithm for $a_1$, $a_2$.

*Termination.* The value of $q(b_1)$ decreases throughout the procedure just described. Since there are only finitely many vectors in $L$ whose length is bounded by the length of the initially given vector $b_1$, this implies that the procedure terminates in all cases.

To find a good bound for the number of iteration steps, we prove that in each step, except possibly the last two, the value of $q(b_1)$ decreases by a factor 3 or higher. That is to say, if in a certain step it occurs that, after the replacement of $b_2$ by $b_2 - mb_1$, the new vector $b_2$ satisfies $q(b_2) > q(b_1)/3$, then that step is either the last one or the next-to-last one. Namely, suppose it is not the last one; then one has $q(b_2) < q(b_1)$. The inequality $|\langle b_1, b_2 \rangle| \leq \frac{1}{2}\langle b_1, b_1 \rangle < \frac{3}{2}\langle b_2, b_2 \rangle$ then implies that the value for $m$ in the next step will be one of 0, 1, $-1$, and since all of the vectors $b_1$, $b_1 - b_2$, $b_1 + b_2$ are at least as long as $b_2$, that next step will be the last one, as asserted.

It follows that an upper bound for the number of iteration steps is given by $2 + (\log(q(b_{1,\text{initial}})/q(b_{1,\text{final}})))/\log 3$, where $b_{1,\text{initial}}$ and $b_{1,\text{final}}$ are the initially given basis vector $b_1$ and the basis vector $b_1$ as finally produced, respectively; here $q(b_{1,\text{final}}) = \lambda(L)$.

Suppose next that we are in an algorithmic context, and that $L$ and its basis are specified by means of a rational matrix $\mathbf{A}$ (or $\mathbf{B}$) as in Section 4. Then $q(L)$ is contained in $\mathbb{Z}\frac{1}{d}$ (or $\mathbb{Z}\frac{1}{d^2}$) if $d$ is a positive integer for which $\mathbb{Z}\frac{1}{d}$ contains the entries of $\mathbf{A}$ (or $\mathbf{B}$), and therefore one has $q(b_{1,\text{final}}) \geq \frac{1}{d}$ (or $\frac{1}{d^2}$). Combining this with the bound for the number of iteration steps just given, one now easily deduces that the entire algorithm runs in polynomial time.

## 10. Flags

*Flags.* It will be convenient to formulate lattice basis reduction algorithms for general rank not in terms of bases but in terms of flags. In this section, $L$ denotes a lattice embedded in a Euclidean vector space $E$, with $n = \text{rk } L = \dim E$. A *flag* of $L$ is a sequence $\mathcal{F} = (L_i)_{i=0}^n$ of pure sublattices $L_i$ of $L$ (as defined in Section 2) satisfying $\text{rk } L_i = i$ (for $0 \leq i \leq n$) and $L_{i-1} \subset L_i$ (for $0 < i \leq n$). Clearly one has $L_0 = \{0\}$ and $L_n = L$.

Every basis $b_1, \ldots, b_n$ for $L$ gives rise to the flag $\left(\sum_{j \leq i} \mathbb{Z} b_j\right)_{i=0}^n$. Conversely, each flag is of this form, but generally not for a unique basis; more precisely, two bases $a_1, \ldots, a_n$ and $b_1, \ldots, b_n$ for $L$ give rise to the same flag if and only if there are integers $c_{ij}$, for $1 \leq j \leq i \leq n$, such that $b_i = \sum_{j \leq i} c_{ij} a_j$ and $c_{ii} = \pm 1$ for all $i$. Thus, a flag may be said to carry a little less information than a basis.

*Successive distances and the Gram–Schmidt process.* Let $\mathcal{F} = (L_i)_{i=0}^n$ be a flag of $L$. For $1 \leq i \leq n$, the *$i$-th successive distance* $l_i(\mathcal{F})$ of $\mathcal{F}$ is defined by $l_i(\mathcal{F}) = d(L_i/L_{i-1})$.

The successive distances are related to the Gram–Schmidt orthogonalization process. Let $b_1, \ldots, b_n$ be a basis for $L$ that gives rise to $\mathcal{F}$. For each $i$, let $b_i^*$ be the unique vector in $b_i + \sum_{j<i} \mathbb{R} b_j$ that is orthogonal to $\sum_{j<i} \mathbb{R} b_j$. The vectors $b_i^*$ can be computed by means of the Gram–Schmidt orthogonalization process, that is, by an inductive application of the formula

$$b_i^* = b_i - \sum_{j<i} \mu_{ij} b_j^*, \qquad \text{where } \mu_{ij} = \frac{\langle b_i, b_j^* \rangle}{\langle b_j^*, b_j^* \rangle}.$$

One has $b_1^* = b_1$. With this notation, $l_i(\mathcal{F})$ is equal to the length $\|b_i^*\|$ of $b_i^*$ or, equivalently, to the distance of $b_i$ to the subspace $\sum_{j<i} \mathbb{R} b_j$ of $E$. In particular, one has $l_1(\mathcal{F}) = \|b_1\|$.

*The size of a flag.* Let $\mathcal{F} = (L_i)_{i=0}^n$ be a flag of $L$. The *size* $s(\mathcal{F})$ of $\mathcal{F}$ is defined by $s(\mathcal{F}) = \prod_{i=0}^n d(L_i)$. From $d(L_i) = \prod_{j \leq i} l_j(\mathcal{F})$ it follows that $s(\mathcal{F})$ can be expressed in terms of the successive distances by $s(\mathcal{F}) = \prod_{j=1}^n l_j(\mathcal{F})^{n+1-j}$.

It is not difficult to prove that a given lattice $L$ has, for each real number $r$, only finitely many flags of size at most $r$. Imprecisely speaking, a flag will be interesting for us if it has small size $s(\mathcal{F}) = \prod_{j=1}^n l_j(\mathcal{F})^{n+1-j}$, and this will be the case if the 'weight' in the product $\prod_{j=1}^n l_j(\mathcal{F})$, which assumes the constant value $d(L)$, is shifted towards the factors with large $j$. This may serve as a motivation for the following definition, which describes more precisely the property that one desires a flag to have.

*Reduced flags.* Let $c$ be a real number, $c \geq 1$, and let $\mathcal{F}$ be a flag for $L$. We say that $\mathcal{F}$ is *c-reduced* if for each $j$ with $0 < j < n$ one has $l_{j+1}(\mathcal{F})^2 \geq l_j(\mathcal{F})^2/c$; for $c = 1$ this is equivalent to the sequence of successive distances being non-decreasing, and the condition becomes weaker as $c$ gets larger. Not every lattice has a flag that is 1-reduced, but as we shall see, each lattice has a flag that is 4/3-reduced, and for each $c > 4/3$ a $c$-reduced flag can be quickly found. The standard choice is $c = 2$.

*The shortest vector problem.* Suppose $n > 0$. A $c$-reduced flag $\mathcal{F} = (L_i)_{i=0}^n$ gives rise to an approximate solution to the shortest vector problem, the quality of the approximation being measured by $c$. Namely, put $L_1 = \mathbb{Z}b_1$. Then $b_1$ is 'almost' the shortest non-zero vector of $L$ in the sense that

$$q(b_1) \leq c^{n-1} \min\{q(x) : x \in L - \{0\}\} = c^{n-1}\lambda(L).$$

To see this, let $x \in L - \{0\}$, and let $i$ be minimal with $x \in L_i$; then $\|x\|$ is at least the $i$-th successive distance $l_i(\mathcal{F})$, so

$$q(x) = \|x\|^2 \geq l_i(\mathcal{F})^2 \geq c^{1-i}l_1(\mathcal{F})^2 \geq c^{1-n}q(b_1),$$

as required. Combining the inequality just proved with Minkowski's theorem, we see that $q(b_1) \leq n \cdot c^{n-1} \cdot d(L)^{2/n}$, but this can be improved a little. Namely, multiplying together the inequalities $q(b_1) = l_1(\mathcal{F})^2 \leq c^{i-1}l_i(\mathcal{F})^2$ that we just proved, for $i = 1, \ldots, n$, and using that $\prod_{i=1}^n l_i(\mathcal{F}) = d(L)$, one finds

$$q(b_1) \leq c^{(n-1)/2} \cdot d(L)^{2/n}.$$

We also see from our inequalities that $b_1$ itself is actually a shortest non-zero vector of $L$ if one has $l_1(\mathcal{F}) = \min\{l_i(\mathcal{F}) : 1 \leq i \leq n\}$, which occurs if $c = 1$.

*The nearest vector problem.* A $c$-reduced flag $\mathcal{F} = \left(\sum_{j \leq i} \mathbb{Z}b_i\right)_{i=0}^n$ also gives rise to an approximate solution to the nearest vector problem, the quality of the approximation again being measured by $c$. To see this, let $b_i^*$ be as above, and write

$$F_i = \left\{\sum_{j=1}^i \mu_j b_j^* : \mu_i \in \mathbb{R}, -\tfrac{1}{2} < \mu_j \leq \tfrac{1}{2} \text{ for } 1 \leq j \leq i\right\}, \qquad F = F_n.$$

By induction on $i$ one checks that each $x \in \sum_{j \leq i} \mathbb{R}b_j$ admits a unique representation of the form $x = y + z$ with $y \in \sum_{j \leq i} \mathbb{Z}b_j$ and $z \in F_i$. In particular, each $x \in E$ can be written uniquely as $x = y + z$ with $y \in L$ and $z \in F$; moreover, in an algorithmic context this representation is easy to find. Thus, a $c$-reduced flag can be used to find, for every $x \in E$, an element $y \in L$ with

$$d(x, y)^2 \leq \max\{\langle z, z\rangle : z \in F\} = \tfrac{1}{4} \cdot \sum_{i=1}^n l_i(\mathcal{F})^2.$$

Also, the approximation of a given element $x \in E$ by an element $y \in L$ obtained in this way is not far from optimal, in the sense that for each other $y' \in L$ one has

$$d(x,y)^2 \leq (1+c+\cdots+c^{n-1}) \cdot d(x,y')^2.$$

To prove this, express $z = x - y$ and $z' = x - y'$ on the orthogonal basis $(b_j^*)_{j=1}^n$ of $E$:

$$z = \sum_{j=1}^n \mu_j b_j^*, \qquad z' = \sum_{j=1}^n \mu_j' b_j^*,$$

with $\mu_j, \mu_j' \in \mathbb{R}$, $-\frac{1}{2} < \mu_j \leq \frac{1}{2}$. From $z - z' \in L - \{0\}$ one deduces that the largest $i$ with $\mu_i \neq \mu_i'$ exists and satisfies $\mu_i - \mu_i' \in \mathbb{Z}$. Then one has the inequalities $|\mu_i'| \geq \frac{1}{2}$ and

$$q(z') = \sum_{j=1}^n \mu_j'^2 l_j(L)^2 \geq \tfrac{1}{4} l_i(L)^2 + \sum_{i<j\leq n} \mu_j^2 l_j(L)^2,$$

$$q(z) \leq \tfrac{1}{4} \cdot \sum_{j\leq i} l_j(L)^2 + \sum_{i<j\leq n} \mu_j^2 l_j(L)^2$$

$$\leq \tfrac{1}{4}(c^{i-1} + \cdots + c + 1)l_i(L)^2 + \sum_{i<j\leq n} \mu_j^2 l_j(L)^2,$$

which yield the desired inequality $q(z) \leq (1+c+\cdots+c^{n-1}) \cdot q(z')$.

*Specifying flags, size-reduced bases.* If one wishes to do computations with flags, one will need a way of specifying them numerically. Assuming that the lattice and its elements are specified in one of the standard formats of Section 4, one can specify a flag $\mathcal{F} = (L_i)_{i=0}^n$ by listing the elements of a basis $b_1, \ldots, b_n$ for $L$ that gives rise to $\mathcal{F}$. This representation is not unique, but it becomes unique, up to choosing $n$ signs, if one requires in addition that for each $i$ the vector $b_i - b_i^*$ belongs to the fundamental domain $F_{i-1}$ for $L_{i-1} = \sum_{j<i} \mathbb{Z} b_j$ in $\sum_{j<i} \mathbb{R} b_j$ defined above. A basis with this property is called *size-reduced*. To change a given basis for a lattice into a size-reduced one that gives rise to the same flag, it suffices to subtract a suitable element of $L_{i-1}$ from $b_i$, for each $i$.

In the course of computations, it may not be necessary to insist that no other bases than size-reduced ones be used for the purpose of specifying flags. However, size-reduced bases are important both in practice and in theory, because they help both in preventing excessive coefficient growth and in obtaining low run time estimates.

It will be convenient to say that a basis $b_1, \ldots, b_n$ for $L$ is *c-reduced*, for a real number $c \geq 1$, if it is size-reduced and the corresponding flag $\left(\sum_{j\leq i} \mathbb{Z} b_j\right)_{i=0}^n$ is $c$-reduced.

*Near-orthogonality of $c$-reduced bases.* Let $c$ be a real number, $c \geq 1$, and suppose that $b_1, \ldots, b_n$ is a $c$-reduced basis for a lattice $L$. With the notation as above, we have $b_i = b_i^* + \sum_{j<i} \mu_{ij} b_j^*$ for certain real numbers $\mu_{ij}$ with $-\frac{1}{2} < \mu_{ij} \leq \frac{1}{2}$, and this implies

$$q(b_i) \leq q(b_i^*) + \tfrac{1}{4} \sum_{j<i} q(b_j^*)$$

$$\leq q(b_i^*) + \tfrac{1}{4} \sum_{j<i} c^{i-j} q(b_i^*) = \left(1 + \tfrac{1}{4}(c^i - c)/(c-1)\right) \cdot q(b_i^*),$$

where $(c^i - c)/(c - 1) = i - 1$ if $c = 1$. Taking the product over $i$ and using the equality $\prod_i \|b_i^*\| = d(L)$ we find

$$\prod_{i=1}^n \|b_i\| \leq \prod_{i=1}^n \left(1 + \tfrac{1}{4}(c^i - c)/(c-1)\right)^{1/2} \cdot d(L).$$

Thus, for fixed $c$, a $c$-reduced basis is 'nearly orthogonal' in the sense of Section 5. If $c \geq 4/3$, then the inequalities just given can be simplified to

$$q(b_i) \leq c^{i-1} \cdot q(b_i^*) \quad \text{for } 1 \leq i \leq n,$$

$$\prod_{i=1}^n \|b_i\| \leq c^{n(n-1)/4} \cdot d(L).$$

*Successive minima and $c$-reduced bases.* For $1 \leq i \leq n$, the $i$-th successive minimum $\lambda_i(L)$ of $L$ is defined to be the infimum of the set of all real numbers $r$ with the property that $L$ contains at least $i$ linearly independent vectors $a$ with $q(a) \leq r$; equivalently, it is the *minimum* of that set of real numbers. Clearly, we have $\lambda_1(L) = \lambda(L)$. The following result shows that the successive minima can be approximately computed from a $c$-reduced basis.

PROPOSITION. *Let $c$ be a real number with $c \geq 4/3$, and let $b_1, \ldots, b_n$ be a $c$-reduced basis for a lattice $L$. Then we have*

$$c^{1-n} \cdot q(b_i) \leq \lambda_i(L) \leq \max\{q(b_j) : 1 \leq j \leq i\} \leq c^{i-1} \cdot q(b_i)$$

*for $1 \leq i \leq n$.*

*Proof.* Since $b_1, \ldots, b_i$ are $i$ linearly independent vectors, the middle inequality is immediate from the definition of $\lambda_i(L)$. For $1 \leq j \leq i$ we have

$$q(b_j) \leq c^{j-1} \cdot q(b_j^*) \leq c^{i-1} \cdot q(b_i^*) \leq c^{i-1} \cdot q(b_i),$$

which implies the third inequality. For the lower bound, let $\mathcal{F} = (L_j)_{j=0}^n$ be the flag of $L$ that $b_1, \ldots, b_n$ gives rise to. Choose $k$ minimal such that $L_k$ contains all $a \in L$ with $q(a) \leq \lambda_i(L)$. The set of such $a$ has rank at least $i$, so we have $k \geq i$, and therefore

$$l_k(\mathcal{F})^2 \geq c^{i-k} \cdot l_i(\mathcal{F})^2 = c^{i-k} \cdot q(b_i^*) \geq c^{1-k} \cdot q(b_i) \geq c^{1-n} \cdot q(b_i).$$

By definition of $k$, at least one $a$ does not belong to $L_{k-1}$, so we have $l_k(\mathcal{F})^2 \leq q(a) \leq \lambda_i(L)$. This proves the stated lower bound for $\lambda_i(L)$ and completes the proof of the Proposition.

*The dual flag.* Let $L$ be a lattice with dual $L^\dagger$ (see Section 2), and let $\mathcal{F} = (L_i)_{i=0}^n$ be a flag of $L$. For $M \subset L$, write $M^\perp = \{x \in L^\dagger : \langle x, y \rangle = 0$ for all $y \in M\}$; this is a pure sublattice of $L^\dagger$. Then $\mathcal{F}^\perp = (L_{n-i}^\perp)_{i=0}^n$ is a flag of $L^\dagger$, and one has $\mathcal{F}^{\perp\perp} = \mathcal{F}$. If $c$ is a real number with $c \geq 1$, then $\mathcal{F}$ is $c$-reduced if and only if $\mathcal{F}^\perp$ is $c$-reduced; this follows from the equality $l_i(\mathcal{F})l_j(\mathcal{F}^\perp) = 1$ for $i + j = n + 1$. If $(b_i)_{i=1}^n$ is a basis for $L$ that gives rise to $\mathcal{F}$, then the corresponding cobasis $(b_i^\dagger)_{i=1}^n$ (see Section 2) gives rise to $\mathcal{F}^\perp$. It is not generally true, for a real number $c \geq 1$, that $(b_i)_{i=1}^n$ is $c$-reduced if and only if $(b_i^\dagger)_{i=1}^n$ is $c$-reduced, though this is valid (up to a sign) for rk $L \leq 2$.

## 11. Finding a good flag

*Flags in rank two.* Suppose $L$ is a lattice of rank 2. Giving a flag $\mathcal{F} = (L_i)_{i=0}^2$ of $L$ is the same as giving a pure sublattice $L_1 = \mathbb{Z}b_1$ of rank 1 of $L$, since necessarily one has $L_0 = \{0\}$ and $L_2 = L$; the size $s(\mathcal{F})$ of such a flag is given by $s(\mathcal{F}) = l_1(\mathcal{F})d(L) = \|b_1\| \cdot d(L)$, so finding a flag of small size is equivalent to finding a non-zero vector of small length. Also, one has $l_2(\mathcal{F}) = d(L)/l_1(\mathcal{F})$, so if $c$ is a real number $\geq 1$ then $\mathcal{F}$ is $c$-reduced, as defined in the previous section, if and only if one has $q(b_1) \leq \sqrt{c} \cdot d(L)$. Since the Hermite constant $\gamma_2$ equals $\sqrt{4/3}$, it follows that $L$ has a $4/3$-reduced flag; and there is a lattice of rank 2 that does not have a $c$-reduced flag for any $c < 4/3$.

In Section 9 we saw a procedure for finding a $4/3$-reduced flag of $L$. If we rephrase one iteration step from that procedure in the language of flags, then we obtain the following: *if a flag $\mathcal{F}$ of $L$ is not $4/3$-reduced, then one can find a flag $\mathcal{F}'$ with smaller size: $s(\mathcal{F}') < s(\mathcal{F})$.* Namely, let $b_1, b_2$ be a size-reduced basis for $L$ giving rise to $\mathcal{F}$. Then one has $b_2 = b_2^* + \mu b_1$ with $|\mu| \leq \frac{1}{2}$, and therefore

$$q(b_2) = q(b_2^*) + \mu^2 q(b_1) \leq \left(\frac{l_2(\mathcal{F})^2}{l_1(\mathcal{F})^2} + \frac{1}{4}\right) \cdot q(b_1).$$

Since $\mathcal{F}$ is not $4/3$-reduced, we have $l_2(\mathcal{F})^2/l_1(\mathcal{F})^2 < 3/4$, and therefore $q(b_2) < q(b_1)$; so the flag $\mathcal{F}'$ corresponding to the basis $b_2, b_1$ is of smaller size than $\mathcal{F}$.

*Improving a given flag.* Suppose next that $L$ is a lattice of any rank $n$, and that $\mathcal{F} = (L_i)_{i=0}^n$ is a flag of $L$ that is not $4/3$-reduced. Then just as in the case of rank 2, one can find a flag $\mathcal{F}'$ of smaller size. To do this, first choose a *pivot*, i.e., an index $j$ with $0 < j < n$ for which $l_{j+1}(\mathcal{F})^2 < \frac{3}{4}l_j(\mathcal{F})^2$. Such an index exists, since by assumption the flag is not 4/3-reduced. Then $(L_i/L_{j-1})_{i=j-1}^{j+1}$ is a flag of the rank two lattice $L_{j+1}/L_{j-1}$, and that flag is not 4/3-reduced either. Thus, by the rank two case that we just did, one can replace it by a flag $(L'_i/L_{j-1})_{i=j-1}^{j+1}$ of smaller size; here one has $L'_{j-1}/L_{j-1} = \{0\}$ and $L'_{j+1}/L_{j-1} = L_{j+1}/L_{j-1}$. Writing $L'_i = L_i$ for all $i \neq j$, one now obtains a flag $\mathcal{F}' = (L'_i)_{i=0}^n$ of $L$ with $s(\mathcal{F}') < s(\mathcal{F})$. Notice that $\mathcal{F}'$ and $\mathcal{F}$ differ only in the rank $j$ sublattice.

Referring back to what we just proved for rank 2, we see that the inequality $s(\mathcal{F}') < s(\mathcal{F})$ can be sharpened to

$$s(\mathcal{F}') \leq \left(\frac{l_{j+1}(\mathcal{F})^2}{l_j(\mathcal{F})^2} + \frac{1}{4}\right)^{1/2} \cdot s(\mathcal{F}).$$

This will be useful below.

*Finding a 4/3-reduced flag.* Let $L$ be a given lattice, and let $\mathcal{F}$ be the flag of $L$ corresponding to a given basis $b_1, \ldots, b_n$ for $L$. If $\mathcal{F}$ is not 4/3-reduced, then as we just saw we can replace $\mathcal{F}$ by a flag $\mathcal{F}'$ that has smaller size. Since there are only finitely many flags of size smaller than the initially given flag, this procedure will, upon iteration, terminate with a flag of $L$ that is 4/3-reduced. This tells us, first, that each lattice has a 4/3-reduced flag and, second, how to find one in an algorithmic situation. Considering a size-reduced basis that gives rise to such a flag, we also conclude that each lattice has a 4/3-reduced basis.

*A basis reduction algorithm.* An algorithm that, given a lattice $L$ in one of the standard formats of Section 4, produces a basis for $L$ that is reduced in a certain sense, is called a *basis reduction algorithm*. For example, the procedure that we just sketched produces a basis that is 4/3-reduced. In the case $n = 2$, this procedure is nothing but the algorithm that we described in Section 9. For larger rank, the procedure becomes an actual basis reduction algorithm if it is supplemented with rules for choosing pivots and for deciding at which stages the basis corresponding to the current flag is to be replaced by a size-reduced basis.

It is an open problem whether, with appropriate rulings, the basis reduction algorithm obtained in this manner runs in polynomial time. As we saw in Section 9, it does run in polynomial time in the case $n = 2$, and in fact it runs in polynomial time for any fixed value of $n$ (see [Lenstra 2001]). The main obstacle towards proving such a result for varying $n$ is finding a good upper bound for the number of flags that the algorithm goes through.

It turns out that, in order to obtain a polynomial-time basis reduction algorithm, it suffices to be a little less demanding: if, instead of insisting on a flag or a basis that is 4/3-reduced, one allows a flag or a basis that is $c$-reduced with $c > 4/3$, then for any fixed value of $c$ such a flag or basis can be found in polynomial time. This is what we consider next.

*Finding a c-reduced flag.* Let a real number $c$ with $c > 4/3$ be fixed, and let $L$ be a lattice. The procedure that we indicated for finding a $4/3$-reduced flag can in an obvious way be shortened so as to find a flag that is merely $c$-reduced. One uses only pivots $j$ with $l_{j+1}(\mathcal{F})^2 < l_j(\mathcal{F})^2/c$, and at each step the improved flag $\mathcal{F}'$ satisfies

$$s(\mathcal{F}') \leq \left(\frac{l_{j+1}(\mathcal{F})^2}{l_j(\mathcal{F})^2} + \frac{1}{4}\right)^{1/2} \cdot s(\mathcal{F}) < \sqrt{1/c + 1/4} \cdot s(\mathcal{F}),$$

where $\sqrt{1/c + 1/4} < 1$. Starting from an initially given flag $\mathcal{F}_{\text{initial}}$, one terminates with a flag $\mathcal{F}_{\text{final}}$ that is $c$-reduced. Each time the flag is changed, its size gets multiplied by a factor smaller than $\sqrt{1/c + 1/4}$, so the number of times this happens is at most $(\log(s(\mathcal{F}_{\text{initial}})/s(\mathcal{F}_{\text{final}})))/|\log \sqrt{1/c + 1/4}|$. As in Section 9 one sees that in an algorithmic situation a good lower bound for $s(\mathcal{F}_{\text{final}})$ is available. This leads to an upper bound for the number of flags encountered in the course of the algorithm, an upper bound that is good enough to allow for a straightforward proof that the algorithm runs in polynomial time. The algorithm just described is the *LLL algorithm*. Properly speaking, the LLL algorithm is an entire family of algorithms, since there is considerable freedom in choosing $c$, in choosing the pivots, and in dealing with size-reduction.

*The LLL algorithm.* In summary, the LLL algorithm takes as input a lattice $L$, specified in one of the standard formats of Section 4, as well as a rational number $c > 4/3$; if no value for $c$ is specified, we assume that $c = 2$. For any fixed value of $c$, the algorithm runs in polynomial time. The output of the algorithm is a basis for $L$ that is $c$-reduced, as defined at the end of Section 10. If $n = \text{rk}\, L > 0$, then the first basis vector $b_1$ of that basis yields an approximate solution to the shortest vector problem for $L$, in the sense that one has

$$q(b_1) \leq c^{n-1} \cdot \min\{q(x) : x \in L - \{0\}\}, \qquad q(b_1) \leq c^{(n-1)/2} \cdot d(L)^{2/n}.$$

Further, such a basis being available, one can approximately solve the nearest vector problem for $L$, in the sense of having a polynomial-time algorithm that given a vector $x$ in the $\mathbb{Q}$-linear span of $L$ finds $y \in L$ such that

$$d(x, y) \leq (1 + c + \cdots + c^{n-1}) \cdot \min\{d(x, y') : y' \in L\}.$$

If $c = 2$, then the last inequality yields $d(x, y) \leq 2^n \cdot \min\{d(x, y') : y' \in L\}$.

## 12. Enumerating short vectors

In the present section we show how one can enumerate short vectors in a lattice with the help of a reduced basis. The method runs at best in polynomial time for fixed values of rk $L$. It relies on the following result, provided by R. J. Schoof, which gives upper bounds for the coefficients of a vector when expressed on a reduced basis, in terms of the length of the vector.

LEMMA. *Let $L$ be a lattice in a Euclidean vector space, and put $n = \text{rk } L$. Let $b_1, \ldots, b_n$ be a basis for $L$, and let $c$ be a real number with $c \geq 1$ such that $b_1, \ldots, b_n$ is c-reduced. For each $i = 1, \ldots, n$, denote by $b_i^*$ the unique vector in $b_i + \sum_{j<i} \mathbb{R} b_j$ that is orthogonal to $\sum_{j<i} \mathbb{R} b_j$. Let $r_1, \ldots, r_n \in \mathbb{R}$, and put $x = \sum_{i=1}^n r_i b_i$. Then one has*

$$|r_j| \leq (3\sqrt{c}/2)^{n-j} \cdot \frac{\|x\|}{\|b_j^*\|} \leq c^{(n-1)/2} \cdot (3/2)^{n-j} \cdot \frac{\|x\|}{\|b_1\|}$$

*for $j = 1, \ldots, n$.*

*Proof.* By the definition of $b_i^*$, we can write $b_i - b_i^* = \sum_{j<i} \mu_{ij} b_j^*$ with $\mu_{ij} \in \mathbb{R}$. The basis $b_1, \ldots, b_n$ being c-reduced is equivalent to the inequalities

$$\|b_j^*\| \leq c^{(i-j)/2} \|b_i^*\|, \qquad -\tfrac{1}{2} < \mu_{ij} \leq \tfrac{1}{2}$$

being valid for $1 \leq j < i \leq n$ (see the definition in Section 10). Substituting $b_i = b_i^* + \sum_{j<i} \mu_{ij} b_j^*$ into $x = \sum_{i=1}^n r_i b_i$ we find that we have $x = \sum_j r_j^* b_j^*$ for $r_j^* = r_j + \sum_{i>j} \mu_{ij} r_i$. The orthogonality of the $b_j^*$ implies $\|x\|^2 = \sum_j r_j^{*2} \|b_j^*\|^2$, so for each $j$ we have

$$|r_j^*| \cdot \|b_j^*\| \leq \|x\|.$$

We now prove the inequality $|r_j| \cdot \|b_j^*\| \leq (3\sqrt{c}/2)^{n-j} \cdot \|x\|$ by induction on $n - j$. From $r_j = r_j^* - \sum_{i>j} \mu_{ij} r_i$ and $|\mu_{ij}| \leq \tfrac{1}{2}$ we obtain

$$|r_j| \cdot \|b_j^*\| \leq |r_j^*| \cdot \|b_j^*\| + \sum_{i>j} \tfrac{1}{2} |r_i| \cdot \|b_j^*\| \leq \|x\| + \tfrac{1}{2} \sum_{i>j} c^{(i-j)/2} \cdot |r_i| \cdot \|b_i^*\|$$
$$\leq \left(1 + \tfrac{1}{2} \sum_{i=j+1}^n c^{(i-j)/2} (3\sqrt{c}/2)^{n-i}\right) \cdot \|x\|$$
$$= \left(1 + c^{(n-j)/2} \cdot ((3/2)^{n-j} - 1)\right) \cdot \|x\| \leq (3\sqrt{c}/2)^{n-j} \cdot \|x\|,$$

as required. This proves the first inequality in the Lemma. The second one follows from $\|b_1\| = \|b_1^*\| \leq c^{(j-1)/2} \|b_j^*\|$. This proves the Lemma.

*Computing $\lambda(L)$ and finding a shortest non-zero vector.* If, in the notation of the Lemma, the $r_i$ range independently over $\mathbb{Z}$, then $x$ ranges over $L$. If $x$ is a shortest non-zero vector of $L$, then one has $\|x\| \leq \|b_1\|$, so by the Lemma each $|r_i|$ is bounded by $c^{(n-1)/2} \cdot (3/2)^{n-i}$.

This suggests the following algorithm for computing $\lambda(L)$ for a given lattice $L$ of positive rank $n$. First, use the LLL algorithm to find a 2-reduced basis $b_1, \ldots, b_n$ for $L$. Next, compute $q(x)$ for each $x$ of the form $x = \sum_i r_i b_i$, where the $r_i$ range independently over all integers that are at most $2^{(n-1)/2} \cdot (3/2)^{n-i}$ in absolute value. Now $\lambda(L)$ is equal to the minimal non-zero value of $q(x)$ that is found. The algorithm can also be used to compute all shortest non-zero vectors of $L$; these are the vectors $x$ encountered that achieve the minimum.

Evidently, the number of systems of integers $r_i$ to be tried by the algorithm is bounded by a function of $n$ alone. Therefore, if $L$ is specified in one of the standard formats of Section 4, the algorithm just described runs in polynomial time for any fixed value of $n = \text{rk}\, L$.

*Enumerating all short vectors.* Suppose one is given a lattice $L$ of positive rank $n$, as well as a positive real number $r$, and one is interested in listing all $x \in L$ with $q(x) \leq r$. Then one can proceed in a similar fashion: apply the LLL algorithm with $c = 2$ (say), and try all $x$ of the form $\sum_i r_i b_i$, where each $r_i$ is an integer satisfying $|r_i| \leq 2^{(n-1)/2} \cdot (3/2)^{n-i} \cdot \sqrt{r}/\|b_1\|$. For 'small' values of $r$ — for example, no larger than $\lambda(L)$ multiplied by a function of $n$ alone — the resulting algorithm will for fixed $n$ run in polynomial time, as in the previous case.

In the case that $r$ is 'large', there is a special advantage in using the sharper upper bound $|r_i| \leq (3/\sqrt{2})^{n-i} \cdot \sqrt{r}/\|b_i^*\|$ from the Lemma. Namely, the number of vectors to be tried is in that case bounded by

$$\frac{r^{n/2}}{\prod_i \|b_i^*\|} = \frac{r^{n/2}}{d(L)}$$

multiplied by a function of $n$ alone. By what we saw in Section 5, this is a good approximation to the number of vectors $x \in L$ with $q(x) \leq r$ to be enumerated, again up to a factor depending on $n$ alone. In other words, for large enough $r$, the run time of the resulting algorithm is for fixed $n$ bounded by the length of the output of the algorithm multiplied by a polynomial function of the length of the input. This will in fact be true if $r$ is at least $1/\lambda(L^\dagger)$ times a suitable function of $n$.

*The nearest vector problem.* There is a similar enumeration algorithm for solving the nearest vector problem, which for any fixed value of $n = \text{rk}\, L$ runs in polynomial time. To see how this works, let $L$ be a lattice in a Euclidean vector space $E$ with $n = \dim E = \text{rk}\, L$, and let $x \in E$. We are interested in finding $y \in L$ with $q(x - y)$ minimal. One starts by applying the LLL algorithm, with any fixed $c > 4/3$. This gives rise to a $c$-reduced basis $b_1, \ldots, b_n$ for $L$, with

Gram–Schmidt orthogonalization $b_1^*, \ldots, b_n^*$ as in the Lemma. In Section 10 we saw how to use this basis in order to find $y_0 \in L$ such that

$$q(x - y_0) \le \tfrac{1}{4} \cdot \sum_{i=1}^{n} q(b_i^*) \le \tfrac{1}{4} \cdot (c^{n-1} + \cdots + c + 1) \cdot q(b_n^*).$$

Write $x = \sum_i r_i b_i$ with $r_i \in \mathbb{R}$, and let the vector $y \in L$ one is looking for be written $y = \sum_i m_i b_i$ with $m_i \in \mathbb{Z}$. Then one has

$$(r_n - m_n)^2 \cdot q(b_n^*) \le q(x - y) \le q(x - y_0).$$

In view of our bound for $q(x - y_0)$, this leaves a number of possibilities for the integer $m_n$ that is bounded by a function of $n$ alone. For each $m \in \mathbb{Z}$ satisfying $(r_n - m)^2 \cdot q(b_n^*) \le q(x - y_0)$, one now solves recursively the nearest vector problem for the lattice $L' = \sum_{i<n} \mathbb{Z} b_i$ of rank $n - 1$ and the element $x - m b_n - (r_n - m) b_n^*$ obtained by projecting $x - m b_n$ orthogonally to the subspace of $E$ spanned by $L'$; for each value of $m$, this gives rise to a nearest vector $y_m \in L'$, and one finds the solution to the nearest vector problem for $L$ and $x$ by putting $y = y_m + m b_n$, the value for $m$ being chosen so as to minimize $q(x - y_m - m b_n)$. One checks in a straightforward way that this correctly solves the nearest vector problem, and that for any fixed value of $n$ it does so in polynomial time. Its practical performance can be enhanced by a branch-and-bound technique.

## 13. Factoring polynomials

The present section is devoted to the earliest published application of the LLL algorithm, namely the construction of a polynomial-time algorithm for the problem of factoring non-zero polynomials in $\mathbb{Q}[X]$ into irreducible factors (see [Lenstra et al. 1982]).

*Summary description of the algorithm.* Let $f \in \mathbb{Q}[X]$ be a given non-constant polynomial, and write $n = \deg f$. One starts by choosing a 'prime' $p$ of the field $\mathbb{Q}$, and by finding an approximation $\beta$ to a zero $\alpha$ of $f$ in a finite extension of the completion $\mathbb{Q}_p$ of $\mathbb{Q}$ at $p$; for example, if one chooses $p = \infty$, then $\beta$ will be a complex number close to a complex zero $\alpha$ of $f$, and one can compute $\beta$ by means of techniques from numerical analysis. If $f$ is reducible, then $\alpha$ is a zero of a non-zero polynomial in $\mathbb{Q}[X]$ of degree smaller than $n$, so $1, \alpha, \ldots, \alpha^{n-1}$ are linearly dependent over $\mathbb{Q}$, and $1, \beta, \ldots, \beta^{n-1}$ are approximately linearly dependent. As we saw in Section 7, one can formulate the problem of finding an approximate linear dependence relation among the $\beta^i$ in lattice terms, and solve it by means of the LLL algorithm. If the vector found by LLL is short enough, then it will give rise to a non-trivial factor $g$ of $f$, and otherwise $f$ is irreducible. In the former case, one recursively applies the algorithm to $g$ and $f/g$, which leads to the full factorization of $f$ into irreducible factors in $\mathbb{Q}[X]$.

*Intermezzo on Berlekamp's algorithm.* In the more detailed description of the algorithm to be given below, we shall, instead of choosing $p = \infty$, take for $p$ a prime number depending on $f$. The role of the numerical analysis is then played by a combination of Berlekamp's algorithm and Hensel's algorithm. For the latter, see [Buhler and Wagon 2008, Section 4.2; von zur Gathen and Gerhard 1999, Section 15.4]; to the former we devote the present intermezzo.

Berlekamp's algorithm takes as input a prime number $p$ and a non-zero polynomial $f \in \mathbb{F}_p[X]$, and its output is the full factorization of $f$ into irreducible factors in $\mathbb{F}_p[X]$. The algorithm is deterministic, and its run time is $O(p \cdot (\log p + \deg f)^c)$ for a positive constant $c$.

For simplicity of description, we shall make the assumptions that the discriminant of $f$ is non-zero, that $f$ has positive degree, and that $f$ is *monic* in the sense of having leading coefficient 1; and in addition, instead of factoring $f$ completely, we shall find a single irreducible factor. It would be easy to remove these restrictions, but for the purposes of our application there is no need to do so.

Our assumptions imply that $f = \prod_i f_i$ for certain pairwise distinct monic irreducible polynomials $f_1, \ldots, f_t \in \mathbb{F}_p[X]$. There is a ring isomorphism

$$\mathbb{F}_p[X]/(f) \cong \prod_{i=1}^{t} \mathbb{F}_p[X]/(f_i),$$

where each $\mathbb{F}_p[X]/(f_i)$ is a field, with the subring $\{y \in \mathbb{F}_p[X]/(f_i) : y^p = y\}$ equal to its prime field $\mathbb{F}_p$. Hence one has $\{y \in \mathbb{F}_p[X]/(f) : y^p = y\} \cong \prod_{i=1}^{t} \mathbb{F}_p$. In particular, $f$ is irreducible if and only if $\{y \in \mathbb{F}_p[X]/(f) : y^p = y\}$ has dimension 1 as a vector space over $\mathbb{F}_p$; more generally, if $h$ is a non-constant factor of $f$, then $h$ is irreducible if and only if all $y \in \mathbb{F}_p[X]/(f)$ with $y^p = y$ reduce to a constant mod $h$.

To exploit these facts, Berlekamp's algorithm starts by finding a basis $g_1, g_2, \ldots, g_t$ of the $\mathbb{F}_p$-vector space $\{y \in \mathbb{F}_p[X]/(f) : y^p = y\}$. The latter space is the null-space of the linear map $\mathbb{F}_p[X]/(f) \to \mathbb{F}_p[X]/(f)$ sending $y$ to $y^p - y$, and a basis of this null-space can be computed by means of linear algebra. Next, the algorithm keeps track of a non-constant factor $h$ of $f$, starting with $h = f$, stopping when $h$ is irreducible, and replacing $h$ by a proper factor otherwise. This is done in the following manner.

If all $g_i$ are congruent to a constant modulo $h$, then $h$ is irreducible, and one stops. Otherwise, choose $i$ such that $g_i$ is not congruent to a constant modulo $h$. Then $h$ divides $g^p - g$, which equals the product $\prod_{j \in \mathbb{F}_p} (g_i - j)$, but $h$ does not divide any of the factors $g_i - j$. Hence, computing at most $p - 1$ greatest common divisors by means of the Euclidean algorithm, one finds $j \in \mathbb{F}_p$ with $0 < \deg \gcd(h, g_i - j) < \deg h$. Now replace $h$ by $\gcd(h, g_i - j)$, and iterate. This finishes the description of Berlekamp's algorithm. One checks in a straightforward way that it has the properties claimed.

For more information on factoring polynomials over finite fields, including the description of a probabilistic algorithm with polynomial expected run time, one may consult [von zur Gathen and Gerhard 1999, Chapter 14].

*Guaranteeing a common factor.* We prove a result that will be useful in proving the correctness of the factoring algorithm to be described. For a polynomial $e = \sum_i a_i X^i \in \mathbb{Z}[X]$, write $q(e) = \sum_i a_i^2$ and $\|e\| = q(e)^{1/2}$. For each positive integer $n$, write $\mathbb{Z}[X]_n$ for the set of polynomials in $\mathbb{Z}[X]$ of degree smaller than $n$; each $\mathbb{Z}[X]_n$ is, with the function $q$, a lattice of rank $n$ and determinant 1.

PROPOSITION. *Let $m$ be a positive integer, and let $h \in \mathbb{Z}[X]$ be a monic polynomial. Let $f$, $g$ be non-zero elements of the $\mathbb{Z}[X]$-ideal $(m, h)$ generated by $m$ and $h$, and suppose that we have*

$$\|f\|^{\deg g} \cdot \|g\|^{\deg f} < m^{\deg h}, \qquad \deg f + \deg g \geq \deg h.$$

*Then $f$ and $g$ have a common factor of positive degree in $\mathbb{Z}[X]$.*

*Proof.* First suppose that the only pair of polynomials $\lambda \in \mathbb{Z}[X]_{\deg g}$, $\mu \in \mathbb{Z}[X]_{\deg f}$ with $\lambda f + \mu g = 0$ is given by $\lambda = \mu = 0$. Then the set

$$M = \{\lambda f + \mu g : \lambda \in \mathbb{Z}[X]_{\deg g}, \mu \in \mathbb{Z}[X]_{\deg f}\}$$

is a sublattice of $\mathbb{Z}[X]_{\deg f + \deg g}$ of rank $\deg f + \deg g$, with basis

$$f, Xf, \ldots, X^{\deg g - 1} f, g, Xg, \ldots, X^{\deg f - 1} g.$$

By Hadamard's inequality, one has $d(M) \leq \|f\|^{\deg g} \cdot \|g\|^{\deg f}$. From $f, g \in (m, h)$ it follows that $M$ is contained in $L = (m, h) \cap \mathbb{Z}[X]_{\deg f + \deg g}$, which is also a sublattice of $\mathbb{Z}[X]_{\deg f + \deg g}$. From $\deg f + \deg g \geq \deg h$ it follows that $(m, h) + \mathbb{Z}[X]_{\deg f + \deg g} = \mathbb{Z}[X]$, and therefore

$$d(L) = \#\mathbb{Z}[X]_{\deg f + \deg g}/L = \#\mathbb{Z}[X]/(m, h) = m^{\deg h}.$$

Altogether we obtain

$$\|f\|^{\deg g} \cdot \|g\|^{\deg f} \geq d(M) = (L : M) \cdot d(L) \geq m^{\deg h},$$

contradicting our hypothesis. Thus, there do exist non-zero polynomials $\lambda \in \mathbb{Z}[X]_{\deg g}$ and $\mu \in \mathbb{Z}[X]_{\deg f}$ with $\lambda f = -\mu g$. This implies that $f$ and $g$ have a common factor of positive degree in $\mathbb{Z}[X]$, as required.

*Factoring polynomials.* We describe a polynomial-time algorithm that, given a non-constant polynomial $f \in \mathbb{Q}[X]$, finds the factorization of $f$ into irreducible factors in $\mathbb{Q}[X]$. Our description assumes that the discriminant $\Delta(f)$ of $f$ is non-zero, and that the coefficients of $f$ are in $\mathbb{Z}$; to achieve the first, one replaces $f$ by $f/\gcd(f, df/dX)$, and to achieve the second one multiplies the coefficients by a common denominator. We let $n = \deg f$.

(a) *Choose an auxiliary prime number.* Compute the least prime number $p$ not dividing the resultant $R(f, df/dX)$ of $f$ and its derivative. As $\pm R(f, df/dX)$ equals the product of the leading coefficient and the discriminant of $f$, the polynomial $(f \bmod p) \in \mathbb{F}_p[X]$ has degree $n$ and non-zero discriminant.

(b) *Find an irreducible factor mod $p$.* Apply Berlekamp's algorithm, as described above, to $(f \bmod p)$ divided by its leading coefficient. This leads to a monic irreducible factor $h_0 \in \mathbb{F}_p[X]$ of $(f \bmod p)$. If $\deg h_0 = \deg f$, then $f$ is irreducible in $\mathbb{Q}[X]$, and the algorithm stops. Assume now $\deg h_0 < \deg f$.

(c) *Determine the $p$-adic precision needed.* Compute the least integer $\mu$ with

$$p^{2\mu \deg h_0} > 2^{n(n-1)} \cdot \binom{2(n-1)}{n-1}^n \cdot q(f)^{2n-1}.$$

(d) *Find an approximate $p$-adic factor of $f$.* Use Hensel's algorithm, as described in [von zur Gathen and Gerhard 1999, Section 15.4], to find a monic polynomial $h \in \mathbb{Z}[X]$ such that $h_0 = (h \bmod p)$ and such that $(h \bmod p^\mu)$ divides $(f \bmod p^\mu)$ in $(\mathbb{Z}/p^\mu\mathbb{Z})[X]$; by Hensel's lemma and the fact that $\Delta(f) \not\equiv 0 \bmod p$, the polynomial $h$ exists and is unique modulo $p^\mu$. (*Note.* A formal zero of $h$ may be viewed as an approximate $p$-adic zero of $f$; so the computation of $h$ corresponds to the computation of $\beta$ in the summary description provided earlier.)

(e) *Apply lattice basis reduction.* Define $L$ to be the additive subgroup of $\mathbb{Z}[X]$ that has basis

$$p^\mu, \ p^\mu \cdot X, \ \ldots, \ p^\mu \cdot X^{(\deg h)-1}, \ h, \ X \cdot h, \ \ldots, \ X^{n-1-\deg h} \cdot h.$$

Viewing $L$ as a sublattice of the lattice $\mathbb{Z}[X]_n$ defined above, apply the LLL algorithm to find a 2-reduced basis $b_1, \ldots, b_n$ for $L$. (*Note.* The elements of $L$ are the polynomials of degree smaller than $n$ that assume $p$-adically small values at a zero $\beta$ of $h$, so they provide approximate linear dependencies among $1, \beta, \ldots, \beta^{n-1}$.)

(f) *Decide irreducibility or find a factor.* If

$$q(b_1) > 2^{n-1} \cdot \binom{2(n-1)}{n-1} \cdot q(f),$$

declare $f$ irreducible and stop. Otherwise, compute $g = \gcd(b_1, f)$ using the Euclidean algorithm in $\mathbb{Q}[X]$. Multiplying $g$ by a suitable scalar, we may assume that the coefficients of $g$ are in $\mathbb{Z}$ and generate the unit ideal of $\mathbb{Z}$. Factor $g$ and $f/g$ recursively into irreducible factors in $\mathbb{Q}[X]$, and combine their factorizations into the factorization of $f$.

*Correctness of the algorithm.* The proof that the algorithm, as described, runs in polynomial time, is largely routine. The only point worth emphasizing is that, by a very weak form of the prime number theorem, the prime number $p$ chosen in (a) is small enough for Berlekamp's algorithm to run in time polynomial in the length of the input data for our factoring algorithm. For more details on the run time analysis one may consult the original article [Lenstra et al. 1982].

The correctness of the algorithm, in particular of step (f), follows from the equivalence of the following statements: (i) $f$ is reducible; (ii) we have

$$q(b_1) \leq 2^{n-1} \cdot \binom{2(n-1)}{n-1} \cdot q(f);$$

(iii) $f$ and $b_1$ have a common factor of positive degree in $\mathbb{Z}[X]$. The implication (iii) $\Rightarrow$ (i) follows from $\deg b_1 < n = \deg f$. To prove (i) $\Rightarrow$ (ii), denote by $g$ the irreducible factor of $f$ in $\mathbb{Z}[X]$ for which $h_0$ divides $(g \bmod p)$; from $\Delta(f) \not\equiv 0 \bmod p$ it follows that $g$ exists and is unique up to sign. By Hensel's lemma, $(h \bmod p^\mu)$ divides $(g \bmod p^\mu)$ in $(\mathbb{Z}/p^\mu\mathbb{Z})[X]$. Also, if we assume (i), then we have $\deg g < n$, and therefore $g \in L$. A very general inequality of Mignotte [1974] on factors of polynomials implies

$$q(g) \leq \binom{2 \deg g}{\deg g} \cdot q(f) \leq \binom{2(n-1)}{n-1} \cdot q(f).$$

Since $b_1, \ldots, b_n$ is a 2-reduced basis for $L$ (see the end of Section 11), we have $q(b_1) \leq 2^{n-1} \cdot q(g)$, which leads to (ii). Finally, the inequalities in (ii) and (c) imply that the conditions of the Proposition are satisfied for $m = p^\mu$ and $g = b_1$, and this leads to a proof of (ii) $\Rightarrow$ (iii).

*Global fields.* The factoring algorithm in $\mathbb{Q}[X]$ described above admits a generalization to $K[X_1, \ldots, X_t]$, for any global field $K$ and any positive integer $t$. A significant special case is treated in [Lenstra 1985] by means of a different notion of lattice, as defined in Section 16 below. For a good general discussion with references, see [von zur Gathen and Gerhard 1999, Chapters 15 and 16].

*Van Hoeij's algorithm.* The reader may have noticed that, for practical purposes, the factoring algorithm as described allows many improvements. There is no need to care about these, since in virtually all practical situations there are other algorithms with a better performance. The chief one among these is *van Hoeij's algorithm*, which applies lattice basis reduction in an altogether different manner. We sketch the basic idea, without paying attention to refinements of practical value.

Let $f \in \mathbb{Z}[X]$ be a monic polynomial to be factored in irreducible factors in $\mathbb{Q}[X]$ or, equivalently, in $\mathbb{Z}[X]$. Put $n = \deg f$. As in the previous algorithm, one starts by choosing a prime $p$ of $\mathbb{Q}$, but next, instead of finding a good approximation to a single $p$-adic zero $\alpha$ of $f$, one finds good approximations $\beta_1, \ldots, \beta_n$ to *all* zeros $\alpha_1, \ldots, \alpha_n$ of $f$ in a suitable finite extension $K$ of the completion $\mathbb{Q}_p$ of $\mathbb{Q}$ at $p$. These approximations are found by means of techniques from classical or $p$-adic numerical analysis. Every monic factor $g$ of $f$ is of the form $\prod_{i \in I}(X - \alpha_i)$ for some subset $I \subset \{1, 2, \ldots, n\}$, and for $g$ to have coefficients in $\mathbb{Z}$ it is necessary that $\sum_{i \in I} \alpha_i$, $\sum_{i \in I} \alpha_i^2$, ... are in $\mathbb{Z}$, and hence that $\sum_{i \in I} \beta_i$, $\sum_{i \in I} \beta_i^2$, ... are $p$-adically very close to elements of $\mathbb{Z}$. Thus, van Hoeij's algorithm proceeds by choosing a positive integer $m$ and searching for an integer vector $(k_i)_{i=1}^n$ with the property that each of $\sum_{i=1}^n k_i \beta_i$, $\sum_{i=1}^n k_i \beta_i^2, \ldots, \sum_{i=1}^n k_i \beta_i^m$ is very close to an integer. This can be done by means of lattice basis reduction, the construction of the lattice being similar to the constructions shown in Section 7. If the only vectors that one finds have all $k_i$ equal, then one declares $f$ to be irreducible; if not all $k_i$ are equal, then for each $k$ that occurs among the $k_i$ one computes $\prod_{i, k_i = k}(X - \beta_i)$, and one hopes to be able to round its coefficients to integers and obtain a non-trivial factor of $f$. Using different vectors $(k_i)_{i=1}^n$ one may even hope to find the full factorization of $f$ into irreducible factors in $\mathbb{Z}[X]$ in this way. This strategy often works for very small values of $m$, such as $m = 1$ or $2$. If it doesn't work, then one increases the value of $m$ and tries again.

Van Hoeij's algorithm presents a number of interesting mathematical problems. The first is to give a version that can be rigorously analyzed and that runs in polynomial time. The second is to extend the algorithm from $\mathbb{Q}[X]$ to $K[X]$, for any global field $K$, including the case of positive characteristic. Neither of these problems is trivial, but they do admit solutions, see [Belabas et al. 2004]. The solution to the first problem uses an unrealistically large value for $m$, namely $m = n-1$. One may wonder whether smaller values of $m$ can be proved to work in all cases.

## 14. Linear algebra over the ring of integers

Lattice basis reduction is useful in solving linear algebra problems over $\mathbb{Z}$. Examples of such problems are: given an $m \times n$ matrix $\mathbf{F}$ with integral entries, find bases both for the kernel and for the image of the group homomorphism $\mathbb{Z}^n \to \mathbb{Z}^m$ mapping $x \in \mathbb{Z}^n$ to $\mathbf{F} \cdot x \in \mathbb{Z}^m$; and given such a matrix $\mathbf{F}$, and $b \in \mathbb{Z}^m$, determine all $x \in \mathbb{Z}^n$ with $\mathbf{F} \cdot x = b$.

The problems that we shall consider are purely linear, and their formulation does not refer to a lattice structure. Lattices are nevertheless useful in their solution, because they provide a natural way of coping with a difficulty that the more traditional approach, which depends on the *Hermite normal form* of an integer matrix (see [Cohen 1993, Section 2.4]), runs into. The straightforward algorithm for computing the Hermite normal form (see [Cohen 1993, Algorithm 2.4.4]) suffers from serious coefficient blow-up, and is therefore not expected to run in polynomial time. Preventing coefficient blow-up is tantamount to controlling the Euclidean length of the vectors that one works with, and that is what lattice algorithms are designed to do.

We shall in this section have occasion to endow groups of the form $\mathbb{Z}^k$, with $k$ a non-negative integer, with several different lattice structures; the notation $\| \ \|^2$ will always be reserved for the standard lattice structure, defined by $\|x\|^2 = \sum_{i=1}^{k} x_i^2$ for $x = (x_i)_{i=1}^{k} \in \mathbb{Z}^k$.

*Kernels, images, and reduced bases.* Let $n$ and $m$ be non-negative integers, and let $f\colon \mathbb{Z}^n \to \mathbb{Z}^m$ be a group homomorphism. Denote by $\mathbf{F}$ the $m \times n$ matrix over $\mathbb{Z}$ with the property that for all $x \in \mathbb{Z}^n$ one has $f(x) = \mathbf{F} \cdot x$; so the columns of $\mathbf{F}$ are the images of the standard basis vectors of $\mathbb{Z}^n$ under $f$. The following result shows how one can define a lattice with the property that bases for the kernel and the image of $f$ can be read off from a reduced basis for the lattice.

PROPOSITION. *Let $n$, $m$, $f$, $\mathbf{F}$ be as above, and write $r$ for the rank of $\mathbf{F}$. Let $F$ be a real number such that the absolute value of any entry of $\mathbf{F}$ is at most $F$, and let $c$ and $N$ be real numbers with*

$$c \geq 4/3, \qquad N > c^{n-1} \cdot (r+1) \cdot r^r \cdot F^{2r}.$$

*Let the lattice $L, q$ be defined by $L = \mathbb{Z}^n$ and*

$$q(x) = \|x\|^2 + N \cdot \|f(x)\|^2 \qquad \text{for } x \in \mathbb{Z}^n,$$

*and let $b_1, \ldots, b_n$ be a $c$-reduced basis for this lattice. Then we have:*

(a) $q(b_i) < N$ for $1 \leq i \leq n - r$;
(b) $b_1, \ldots, b_{n-r}$ *form a basis for* $\ker f$ *over* $\mathbb{Z}$;
(c) $q(b_i) \geq N$ for $n - r < i \leq n$;
(d) $f(b_{n-r+1}), \ldots, f(b_n)$ *form a basis for* $f(\mathbb{Z}^n)$ *over* $\mathbb{Z}$.

*Proof.* For notational convenience we may assume that the standard basis vectors of $\mathbb{Z}^n$ are numbered in such a way that the first $r$ columns of $\mathbf{F}$ are linearly independent. Let $r < h \leq n$. By Cramer's rule, there is a non-trivial linear dependency among the first $r$ columns and the $h$-th column of $\mathbf{F}$, with coefficients that are $r \times r$ minors of $\mathbf{F}$. This dependency gives rise to an element $x = (x_i)_{i=1}^{n}$ of $\ker f$ with $x_h \neq 0$ and $x_i = 0$ for all $i > r$ with $i \neq h$. By Hadamard's inequality we have $|x_i| \leq r^{r/2} F^r$ for all $i$, and therefore $q(x) = \|x\|^2 \leq (r+1) \cdot r^r \cdot F^{2r}$. The $n - r$ vectors obtained in this way for $h = r+1, \ldots, n$ are linearly independent, so for each $i \leq n - r$ the $i$-th successive minimum $\lambda_i(L)$, as defined in Section 10, satisfies $\lambda_i(L) \leq (r+1) \cdot r^r \cdot F^{2r}$. By the Proposition in Section 10, we now have

$$q(b_i) \leq c^{n-1} \cdot \lambda_i(L) \leq c^{n-1} \cdot (r+1) \cdot r^r \cdot F^{2r} < N \quad \text{for } i \leq n-r.$$

This proves (a). The definition of $q$ implies that every $x \in L$ with $q(x) < N$ belongs to ker $f$. Thus, from (a) we see that ker $f$ contains the linearly independent vectors $b_1, \ldots, b_{n-r}$. By linear algebra, the null space of $\mathbf{F}$ on $\mathbb{Q}^n$ has $\mathbb{Q}$-dimension equal to $n-r$ and is therefore spanned by $b_1, \ldots, b_{n-r}$. Consequently, inside $\mathbb{Q}^n$ we have

$$\ker f = \left(\sum_{i=1}^{n-r} \mathbb{Q} b_i\right) \cap \mathbb{Z}^n = \sum_{i=1}^{n-r} \mathbb{Z} b_i,$$

the latter equality because $b_1, \ldots, b_n$ form a basis for $\mathbb{Z}^n$ over $\mathbb{Z}$. This proves (b). It follows that for each $i > n-r$ we have $b_i \notin \ker f$ and therefore $q(b_i) \geq N$, which is (c). Finally, (d) follows from (b) and the homomorphism theorem from elementary group theory. This proves the Proposition.

*The kernel and image algorithm.* We describe an algorithm that, given nonnegative integers $n$ and $m$ and a group homomorphism $f: \mathbb{Z}^n \to \mathbb{Z}^m$, determines the kernel and the image of $f$. Here $f$ is specified by an $m \times n$ matrix $\mathbf{F}$ over $\mathbb{Z}$, as above. The kernel of $f$ is required to be specified by a sequence of vectors in $\mathbb{Z}^n$ that form a basis for ker $f$ over $\mathbb{Z}$, and likewise for the image of $f$ in $\mathbb{Z}^m$.

One starts by defining $F$ to be the maximum of the absolute values of the entries of $\mathbf{F}$, with $F = 0$ if $nm = 0$. One chooses $c = 2$, and one chooses $N$ to be an integer exceeding $2^{n-1} \cdot (r+1) \cdot r^r \cdot F^{2r}$, where $r$ denotes the rank of $\mathbf{F}$; if the value of $r$ is not known, one just uses the upper bound $r \leq \min\{n, m\}$. Next, one applies the LLL algorithm to find a $c$-reduced basis $b_1, \ldots, b_n$ for $L$. By the Proposition, the $b_i$ with $q(b_i) < N$ form a basis for ker $f$, and the images of the other $b_i$ under $f$ form a basis for the image of $f$. This completes the description of the algorithm. With a proper choice of $N$, this algorithm is readily shown to run in polynomial time.

*Ordered vector spaces.* We discuss a modification of the algorithm just described that both improves its practical performance and has theoretical interest. The modification consists of not choosing an actual value for $N$, but viewing it as an 'indefinitely large' symbol. More rigorously, one redefines the function $q$ on $L$ by $q(x) = (\|x\|^2, \|f(x)\|^2)$; its values are not in $\mathbb{R}$, but in the real vector space $\mathbb{R} \times \mathbb{R}$, which one endows with a total ordering by putting $(r_1, r_2) > (s_1, s_2)$ if and only if either $r_2 > s_2$, or $r_2 = s_2$ and $r_1 > s_1$ (the *anti-lexicographic ordering*). To capture the structure $L, q$ defined in this manner in a theoretical framework, one is led to define a generalized notion of Euclidean vector space, in which the real-valued inner product $\langle\,,\,\rangle$ defined on $E \times E$, as considered in Section 2, is replaced by one that takes values in a totally ordered real vector space; in addition to the axioms from Section 2, one requires that for any $x$, $y \in E$ there exists $r \in \mathbb{R}$ with $\langle x, y \rangle \le r \langle x, x \rangle$. It appears to be both worthwhile and feasible to define a correspondingly generalized notion of lattice, and to formulate conditions under which a natural extension of the LLL algorithm terminates in polynomial time. This theory, yet to be developed, should confirm that the modified kernel and image algorithm, and similar algorithms to be discussed below, run in polynomial time. The implications for diophantine approximation, where large weights $N$ are also encountered (see Section 7), are worth exploring as well.

*Solving a system of linear equations over* $\mathbb{Z}$. Let $m$ and $n$ be non-negative integers, let $\mathbf{F}$ be an $m \times n$ matrix over $\mathbb{Z}$, and let $b \in \mathbb{Z}^m$. We are interested in finding all $x \in \mathbb{Z}^n$ with $\mathbf{F} \cdot x = b$.

Define the group homomorphisms $g\colon \mathbb{Z}^n \times \mathbb{Z} = \mathbb{Z}^{n+1} \to \mathbb{Z}^m$ and $h\colon \mathbb{Z}^n \times \mathbb{Z} \to \mathbb{Z}$ by $g(x, z) = \mathbf{F} \cdot x - z \cdot b$ and $h(x, z) = z$, for $x \in \mathbb{Z}^n$, $z \in \mathbb{Z}$. Clearly, there exists $x \in \mathbb{Z}^n$ with $\mathbf{F} \cdot x = b$ if and only if 1 belongs to the image under $h$ of the kernel of $g$. Thus, one can decide whether the equation $\mathbf{F} \cdot x = b$ is solvable with $x \in \mathbb{Z}^n$ by performing the kernel and image algorithm twice. Actually, a single application of the LLL algorithm suffices, and the resulting algorithm does not only decide solvability, but in fact describes the set of all solutions. It runs as follows.

Let $N$ and $M$ be suitably chosen large integers with $N \gg M$, and make the group $L = \mathbb{Z}^n \times \mathbb{Z}$ into a lattice by putting

$$q(x, z) = \|x\|^2 + M \cdot z^2 + N \cdot \|\mathbf{F} \cdot x - z \cdot b\|^2 \qquad \text{for } x \in \mathbb{Z}^n,\ z \in \mathbb{Z}.$$

Use the LLL algorithm to determine a 2-reduced basis $b_1, \ldots, b_{n+1}$ for $L$. Then $\mathbf{F} \cdot x = b$ has a solution $x \in \mathbb{Z}^n$ if and only if there exists an index $j$ with $M \le q(b_j) < 4M$; moreover, if such an index exists, then it is unique, and the following is valid: each $b_i$ with $i < j$ is of the form $(b_i', 0)$ with $b_i' \in \mathbb{Z}^n$, the $z$-coordinate of $b_j$ equals $\pm 1$, and if $x_0 \in \mathbb{Z}^n$ is defined by $\pm b_j = (x_0, 1)$, then $x = x_0$ is a solution to $\mathbf{F} \cdot x = b$, whereas the general solution is given by $x = x_0 + \sum_{i=1}^{j-1} k_i b_i'$ with $k_1, \ldots, k_{j-1} \in \mathbb{Z}$.

One can show that the assertions just made are correct if $M > 2^n \cdot (r+1) \cdot r^r \cdot F^{2r}$ and $N > 2^n \cdot (r+M) \cdot r^r \cdot F^{2r}$, where $r$ equals the rank of $\mathbf{F}$ and $F \in \mathbb{Z}$ is an upper bound for the absolute values of all entries of $\mathbf{F}$ and $b$. As a consequence, one obtains a polynomial-time algorithm for solving $\mathbf{F} \cdot x = b$ over $\mathbb{Z}$. Alternatively, one may redefine $q$ to take values in the anti-lexicographically ordered real vector space $\mathbb{R} \times \mathbb{R} \times \mathbb{R}$, by putting $q(x, z) = \bigl(\|x\|^2, \|z\|^2, \|\mathbf{F} \cdot x - z \cdot b\|^2\bigr)$, and invoke the generalized algorithmic theory of lattices alluded to above.

*The Chinese remainder theorem.* Suppose one is given a positive integer $k$, a sequence $m_1, \ldots, m_k$ of pairwise coprime positive integers, as well as a sequence $r_1, \ldots, r_k$ of integers, and that one is interested in finding an integer $x$ satisfying the $k$ congruences $x \equiv r_i \bmod m_i$ ($1 \le i \le k$). The problem is equivalent to finding a vector $(x, y_1, \ldots, y_k) \in \mathbb{Z}^{k+1}$ satisfying the system of linear equations $x - y_i m_i = r_i$ ($1 \le i \le k$), and can thus be solved in polynomial time by the linear algebra algorithm just explained. There is also a more direct approach (see [Knuth 1981, Section 4.3.2]), and the reader is invited to make a comparison of run times.

*The generalized extended Euclidean algorithm.* We revisit a problem considered earlier. Let, slightly more generally than in Section 8, a non-negative integer $k$ as well as integers $a_1, \ldots, a_k$ be given; we want to compute an integer $d$ with $\sum_{i=1}^{k} \mathbb{Z} a_i = \mathbb{Z} d$, as well as 'small' integers $x_1, \ldots, x_k$ with $\sum_{i=1}^{k} x_i a_i = d$.

As in the linear algebra problem just considered, let $N$ and $M$ be suitably large positive integers with $N \gg M$, and make the group $\mathbb{Z}^{k+1}$ into a lattice by putting

$$q(x_1, \ldots, x_{k+1}) = \left(\sum_{i=1}^{k} x_i^2\right) + M \cdot x_{k+1}^2 + N \cdot \left(x_{k+1} - \sum_{i=1}^{k} x_i a_i\right)^2.$$

Let $b_1, \ldots, b_{k+1}$ be a 2-reduced basis for this lattice. If there is an index $j$ with $M \le q(b_j) < N$, and $b_j = (x_i)_{i=1}^{k+1}$, then for $d = x_{k+1}$ one has $\sum_{i=1}^{k} \mathbb{Z} a_i = \mathbb{Z} d$ and $\sum_{i=1}^{k} x_i a_i = d$. If no such index $j$ exists, then all $a_i$ are 0, and one can take $d$ and all $x_i$ to be 0 as well. The details, and the proof that the resulting algorithm runs in polynomial time, may again be left to the reader.

*The nearest vector problem.* The problem that we just discussed, was in Section 8 identified as a special case of the nearest vector problem. The general nearest vector problem admits a similarly direct solution by means of lattice basis reduction. Namely, suppose one is given a lattice $L$ in a Euclidean vector space $E$, as well as an element $x \in E$, and that one wants to find $y \in L$ with $q(x-y)$ small. Define a lattice $L'$, $q'$ by putting $L' = L \times \mathbb{Z}$ and $q'(y, z) = q(y - zx) + N \cdot z^2$ for $y \in L$, $z \in \mathbb{Z}$, where again $N$ is chosen large enough or indefinitely large. Only the last basis vector of a $c$-reduced basis $b_1, \ldots, b_{\mathrm{rk}\,L'}$ for $L'$ will then have a non-zero $z$-coordinate, and that $z$-coordinate will be $\pm 1$; if $\pm b_{\mathrm{rk}\,L'} = (y, 1)$, with $y \in L$, then $y$ is a 'good' solution to the nearest vector problem. This solution is essentially the same as the one constructed in Section 10.

*Operations on subgroups.* Let $n$ be a non-negative integer. The kernel and image algorithm can be used to perform several operations on subgroups of $\mathbb{Z}^n$. We give a number of examples; it is always assumed that, for algorithmic purposes, a subgroup $H \subset \mathbb{Z}^n$ is specified by means of a sequence of elements of $\mathbb{Z}^n$ that is a basis for $H$ over $\mathbb{Z}$. All algorithms to be described run in polynomial time, $n$ being viewed as part of the input.

Let $H_1$ and $H_2$ be two subgroups of $\mathbb{Z}^n$, and consider the group homomorphism $H_1 \times H_2 \to \mathbb{Z}^n$ sending $(x, y)$ to $x - y$. Its image is the subgroup $H_1 + H_2$ of $\mathbb{Z}^n$, and its kernel can in an obvious manner be identified with $H_1 \cap H_2$. Thus, from the kernel and image algorithm one obtains bases for both $H_1 + H_2$ and $H_1 \cap H_2$ over $\mathbb{Z}$. In fact, in the case of $H_1 \cap H_2$, one obtains *three* expressions for the same basis: one in terms of the given basis for $H_1$, one in terms of the given basis for $H_2$, and one in terms of the standard basis for $\mathbb{Z}^n$.

Let $H$ be a subgroup of $\mathbb{Z}^n$, and let **F** be an $n \times (\text{rk } H)$ matrix over $\mathbb{Z}$ of which the columns form a basis for $H$ over $\mathbb{Z}$. The *transpose* of **F** may be viewed as the matrix that describes the map $\varphi: \mathbb{Z}^n \to \text{Hom}(H, \mathbb{Z})$ defined by $\varphi(x)(y) = \langle x, y \rangle$ for $x \in \mathbb{Z}^n$, $y \in H$, where $\langle , \rangle$ denotes the standard inner product on $\mathbb{Z}^n$. Applying the kernel and image algorithm, one obtains a basis for $H^\perp = \ker \varphi = \{x \in \mathbb{Z}^n : \langle x, y \rangle = 0 \text{ for all } y \in H\}$. Doing this again, one obtains a basis for $H^{\perp\perp}$, which equals the subgroup $(\mathbb{Q} \cdot H) \cap \mathbb{Z}^n$ of $\mathbb{Z}^n$. Simultaneously, one obtains a basis for $\mathbb{Z}^n / H^{\perp\perp}$, which may be identified with the group $\mathbb{Z}^n / H$ modulo its torsion subgroup.

Define the *degree* $\deg x$ of a non-zero vector $x = (x_i)_{i=1}^n \in \mathbb{Z}^n$ to be $\max\{i : x_i \neq 0\}$. It is well-known that any subgroup $H \subset \mathbb{Z}^n$ has a basis $b_1, \ldots, b_{\text{rk } H}$ with the property that $\deg b_i$ is strictly increasing as a function of $i$. To compute such a basis from a given basis for $H$, it suffices to apply lattice basis reduction to the lattice $H, q$, where $q$ is defined by

$$q(x_1, \ldots, x_n) = \sum_{i=1}^n N_i x_i^2,$$

for suitable integers $N_i$ with $N_n \gg N_{n-1} \gg \cdots \gg N_2 \gg N_1 = 1$; again, the formalism involving ordered vector spaces would be applicable here. The same technique can be used to compute the Hermite normal form of an integer matrix by means of lattice basis reduction.

I do not know whether lattice basis reduction algorithms may assist in computing the *Smith normal form* of an integer matrix (see [Cohen 1993, Section 2.4.4]), or how useful they are in doing computations with finitely generated abelian groups that are allowed to have torsion.

## 15. Nonlinear problems

In Section 13 we saw that lattices can be used to solve the nonlinear problem of factoring in the ring $\mathbb{Q}[X]$. There is in fact a surprisingly large class of nonlinear problems that can be solved by means of lattices. In the present section we describe a general technique, and we illustrate it with three examples. Related methods are well-known in the area of diophantine approximation, where they are used to prove upper bounds for the number of integral solutions to certain systems of equations that satisfy certain inequalities (see [Heath-Brown 2002]). It is a more recent insight that in many cases these solutions can be efficiently enumerated by means of lattice basis reduction. One may consult [Bernstein 2008] for a different perspective, for references, and for a historical discussion, and [Elkies 2000] for an account of a very similar technique, with additional applications.

Let $V$ be an affine algebraic set defined over $\mathbb{R}$, embedded in affine $t$-space $\mathbf{A}_{\mathbb{R}}^t$, for some non-negative integer $t$; so the coordinate ring $\mathbb{R}[V]$ is equal to $\mathbb{R}[X_1, \ldots, X_t]/I$ for some ideal $I$ of the polynomial ring $\mathbb{R}[X_1, \ldots, X_t]$. The set $V(\mathbb{R})$ of real points of $V$ is defined by $\{x \in \mathbb{R}^t : f(x) = 0 \text{ for all } f \in I\}$. By abuse of notation, we write $V(\mathbb{Z}) = V(\mathbb{R}) \cap \mathbb{Z}^t$. Suppose in addition that $B$ is a subset of $\mathbb{R}^t$ for which $B \cap V(\mathbb{R})$ is bounded. Then the set $S = B \cap V(\mathbb{Z})$ is finite. We assume that one is interested in determining upper bounds for $\#S$ and, if $I$ and $B$ are given in some explicit manner, in algorithms for listing all elements of $S$.

The lattice-based technique that applies in this context, produces a non-zero element $g \in \mathbb{R}[V]$ that vanishes on $S$, so that $S$ remains unchanged if $V$ is replaced by the affine algebraic set $W$ defined by $\mathbb{R}[W] = \mathbb{R}[V]/(g)$, which can in principle be dealt with recursively.

In many situations of interest, the variety $V$ is an irreducible curve. In that case, the zero set of $g$ on $V$, which contains $S$, is finite; the lattice method gives an upper bound for its cardinality, and in algorithmic circumstances it is usually easy to first compute all zeros of $g$ in $V(\mathbb{Z})$ and next check them one by one for membership of $S$.

*Examples.* Rather than attempting to formulate general conditions under which the technique is useful, we describe three problems from algorithmic number theory to which it has been successfully applied. In each case, the efficiency of the resulting algorithm is contingent upon inequalities satisfied by the problem parameters.

(a) *Zeros of polynomials modulo n.* Suppose one is given integers $a$, $b$, and $n$ with $a < b$ and $n > 0$, as well as a monic polynomial $p \in \mathbb{Z}[X]$, and that one is interested in the set of all $x \in \mathbb{Z}$ with $a \leq x \leq b$ and $p(x) \equiv 0 \mod n$. Then one can take $t = 2$, and $V$ to be the algebraic subset of real affine 2-space defined by the equation $p(x) = n \cdot y$; that is, one has $\mathbb{R}[V] = \mathbb{R}[X, Y]/(p - nY)$. Note that the natural map $\mathbb{R}[X] \to \mathbb{R}[V]$ is a ring isomorphism, so that $V$ is actually isomorphic to the affine line over $\mathbb{R}$, which is an irreducible curve. With $B = \{(x, y) \in \mathbb{R}^2 : a \leq x \leq b\}$, the set $S = B \cap V(\mathbb{Z})$ defined above maps bijectively to the set $\{x \in \mathbb{Z} : a \leq x \leq b,\ p(x) \equiv 0 \bmod n\}$ that one is interested in, by the projection map $(x, y) \mapsto x$.

(b) *Divisors in residue classes.* Suppose one is given positive integers $u$, $v$, and $n$ with $\gcd(u, v) = 1$, and that one is interested in the set of divisors $x$ of $n$ that satisfy $x \equiv u \bmod v$. In this case, one can take $t = 3$ and define $V$ by $xy = n$, $x = u + vz$. Then one has $\mathbb{R}[V] = \mathbb{R}[X, Y, Z]/(XY - n, X - u - vZ)$, and there is an $\mathbb{R}$-algebra isomorphism from the ring $\mathbb{R}[X, X^{-1}]$ of Laurent polynomials in $X$ over $\mathbb{R}$ to the ring $\mathbb{R}[V]$ that maps $X$ to $X$ and $X^{-1}$ to $Y/n$. Hence, $V$ is isomorphic to the affine line with a single point removed, which is again an irreducible curve. With $B = \{(x, y, z) \in \mathbb{R}^3 : 1 \leq x \leq n\}$, the set $S = B \cap V(\mathbb{Z})$ may again be identified with the set one is interested in.

(c) *Diophantine approximation with restricted denominators.* Let $\alpha$ be a real number and let $n$ be a positive integer. We suppose that one is interested in 'good' rational approximations $y/z$ to $\alpha$, with $y, z \in \mathbb{Z}$, $z > 0$, of which the denominator $z$ is 'small' and satisfies the additional restriction that it divide $n$. Denote by $[a/n, b/n]$ the interval around $\alpha$ that one wishes $y/z$ to belong to, with the endpoints properly rounded to integer multiples of $1/n$, so that $a, b \in \mathbb{Z}$, $a < b$. We shall always assume $b - a < n$, since otherwise the interval $[a/n, b/n]$ contains rational numbers with any given denominator. Write $m$ for the desired upper bound on $z$. We can now take $t = 3$, define the surface $V$ by $xz = ny$, and put $B = \{(x, y, z) \in \mathbb{R}^3 : a \leq x \leq b,\ 1 \leq z \leq m\}$. One has $\mathbb{R}[V] = \mathbb{R}[X, Y, Z]/(XZ - nY)$, and the natural map $\mathbb{R}[X, Z] \to \mathbb{R}[V]$ is an isomorphism. The set $S = B \cap V(\mathbb{Z})$ maps bijectively to the set one is interested in, by $(x, y, z) \mapsto y/z$.

If two distinct rational numbers in $[a/n, b/n]$ each have denominator at most $m$, then their difference is a non-zero rational number of absolute value at most $(b - a)/n$ with denominator at most $m^2$, so that $(b - a)/n \geq 1/m^2$. Thus, for $m < \sqrt{n/(b-a)}$ the number $y/z$ is unique if it exists. One can find it using continued fractions or two-dimensional lattice basis reduction, as in Section 7. This approach, however, disregards the requirement that $z$ divide $n$. The approach of the present section does take that requirement into account, and it allows larger values for $m$ to be taken. More specifically, if $\varepsilon$ is such that $b - a = n^\varepsilon$, then Proposition C below shows that instead of $m < \sqrt{n/(b-a)} = n^{(1-\varepsilon)/2}$ we can allow $m < n^\eta$ for any $\eta < 1 - \sqrt{\varepsilon}$; note that one has $(1 - \varepsilon)/2 < 1 - \sqrt{\varepsilon}$.

The equation $xz = ny$ defining $V$ is homogeneous in $y$ and $z$, so it may also be thought of as defining a curve $V'$ in the product of the affine line $\mathbf{A}^1_\mathbb{R}$ parametrized by $x$ and the projective line $\mathbf{P}^1_\mathbb{R}$ parametrized by $y : z$. One may then view $V$ as a 'cone' over $V'$, the 'top' of the cone being the line in $\mathbf{A}^3_\mathbb{R}$ defined by $y = z = 0$. We will be careful to construct the non-zero element

$g \in \mathbb{R}[V]$ in such a way that it will likewise be homogeneous in $Y$ and $Z$, so that $g = 0$ defines a finite set of points in $V'$.

The following result shows the relevance of lattices for the type of problem we are considering. Let the notations $V$, $\mathbb{R}[V]$, $V(\mathbb{R})$, $V(\mathbb{Z})$, $B$, $S$ be as introduced at the beginning of this section.

LEMMA. *Let $L, q$ be a non-zero lattice and let $c$ be a positive real number such that*:
(i) *the group $L$ is a subgroup of the additive group of $\mathbb{R}[V]$ with the property that each $f \in L$ is integral-valued on $V(\mathbb{Z})$*,
(ii) *for each $x \in B \cap V(\mathbb{R})$ and each $f$ in the $\mathbb{R}$-linear span of $L$, one has $|f(x)| \le c \cdot q(f)^{1/2}$*,
(iii) *one has $c \cdot \sqrt{\operatorname{rk} L} \cdot d(L)^{1/\operatorname{rk} L} < 1$*.

*Then there exists a non-zero element $g \in L$ such that for all $x \in S$ one has $g(x) = 0$.*

PROOF. By the theorem of Minkowski (Section 6), we can choose a non-zero element $g \in L$ with $q(g) \le (\operatorname{rk} L) \cdot d(L)^{2/\operatorname{rk} L}$. Let $x \in S$. Applying (ii) to $f = g$ we obtain $|g(x)| \le c \cdot \sqrt{\operatorname{rk} L} \cdot d(L)^{1/\operatorname{rk} L}$, so by (iii) we have $|g(x)| < 1$. Since by (i) we have $g(x) \in \mathbb{Z}$, we obtain $g(x) = 0$. This proves the Lemma. □

In algorithmic circumstances, one replaces the theorem of Minkowski by a lattice basis reduction algorithm. This allows the actual construction of a non-zero element $g \in L$ that vanishes on $S$, provided that the condition (iii) is replaced by a slightly stronger one. Specifically, if one makes use of 2-reduced bases, then the factor $\sqrt{\operatorname{rk} L}$ in (iii) should be replaced by $2^{(\operatorname{rk} L - 1)/4}$.

The integrality condition (i) of the Lemma is satisfied if $L$ is chosen inside the image of the ring $\mathbb{Z}[X_1, \ldots, X_t]$ in $\mathbb{R}[X_1, \ldots, X_t]/I = \mathbb{R}[V]$. (Alternatively, the ring of integral-valued polynomials, which is generated by

$$\{\binom{X_i}{j} : 1 \le i \le t, j \in \mathbb{Z}_{\ge 0}\},$$

can be used.) Condition (ii) is, under weak conditions, probably automatic for *some* value of $c$; to keep $c$ small, with an eye on (iii), one adapts the choice of $q$ to the set $B$, as illustrated in the examples below. The inequality in (iii) expresses the condition under which the technique under discussion is useful.

Several strategies are available if (iii) is not satisfied. One strategy, which we shall follow in the proof of Proposition B below, is to cut up $B$ into several pieces, each piece having its own $L, q$ and a smaller value for $c$. Alternatively, one may decide to be satisfied with an element $g \in L$ with the weaker property that the zeros of $g - i$ cover all of $S$ when $i$ ranges over all integers with $|i|$ below a certain bound; to avoid the possibility that one of these $g - i$ is identically zero (that is, $g = i$ in $\mathbb{R}[V]$), one may have to find a non-zero element in the lattice $L/(L \cap \mathbb{Z})$ instead of in $L$ itself.

We return to our examples and illustrate how suitable lattices may be constructed.

PROPOSITION A. *There is a function* $\alpha: \mathbb{Z}_{>0} \to \mathbb{R}_{>0}$ *with* $\lim_{m \to \infty} \alpha(m) = 1/\log 2$ *such that for any integers* $a, b, n$ *and any polynomial* $p \in \mathbb{Z}[X]$ *with*

$$p \notin \mathbb{Z}, \qquad p \text{ monic}, \qquad n > 1, \qquad 0 < b - a \le n^{1/\deg p},$$

*the number of integers* $x$ *with* $a \le x \le b$ *and* $p(x) \equiv 0 \mod n$ *is at most* $\deg p + \alpha(n) \cdot \log n$. *In addition, there is a polynomial-time algorithm that given such* $a$, $b$, $n$, *and* $p$, *determines all those* $x$.

PROOF. We write $d = \deg p$, and we let $h$ be the least positive integer satisfying the inequality $2^{dh-1} > (dh)^2 \cdot n^{1-1/d}$. One readily checks that one has $dh < \deg p + \alpha(n) \cdot \log n$ for a function $\alpha$ as in the Proposition, so to prove the first statement it suffices to show that the number of desired values for $x$ is smaller than $dh$.

Define $L$ to be the additive group of polynomials in the subring $\mathbb{Z}[X, p/n]$ of $\mathbb{R}[X]$ that have degree smaller than $dh$. Then $L$ is a free abelian group of rank $dh$, with basis $\{X^i (p/n)^j : 0 \le i < d, 0 \le j < h\}$, and it contains $\sum_{i=0}^{dh-1} \mathbb{Z} \cdot X^i$ as a subgroup of index $n^{dh(h-1)/2}$. To endow $L$ with a lattice structure, write any polynomial $f \in \mathbb{R}[X]$ with $\deg f < dh$ in the form $f = \sum_{i=0}^{dh-1} c_i (X - \frac{b+a}{2})^i$ with $c_i \in \mathbb{R}$, and put $q(f) = \sum_i c_i^2 (\frac{b-a}{2})^{2i}$. This makes $L$ into a lattice, and a straightforward calculation gives

$$d(L) = \left(\frac{b-a}{2}\right)^{dh(dh-1)/2} \cdot n^{-dh(h-1)/2}.$$

For any real number $x$ with $a \le x \le b$ one has $\left\| \frac{x-(b+a)}{2} \right\| / \frac{b-a}{2} \le 1$, so the Cauchy–Schwarz inequality implies $|f(x)| \le (dh \cdot q(f))^{1/2}$ for any $f \in \mathbb{R}[X]$ with $\deg f < dh$. We can now apply the Lemma with $c = \sqrt{dh}$. Condition (iii) is

$$dh \cdot \left(\frac{b-a}{2}\right)^{(dh-1)/2} \cdot n^{-(h-1)/2} < 1.$$

From $b - a \le n^{1/d}$ and the choice of $h$ it follows that this condition is satisfied. The Lemma now implies that there is a non-zero polynomial $g \in \mathbb{Q}[X]$ of degree smaller than $dh$ that has all $x \in \mathbb{Z}$ with $a \le x \le b$ and $p(x) \equiv 0 \mod n$ among its zeros. It follows that the number of those $x$ is smaller than $dh$, as desired.

It is straightforward to convert the proof just given into a polynomial-time algorithm finding all desired values of $x$. Instead of the version of the Lemma that depends on Minkowski's theorem, one uses the algorithmic version, in which (iii) is replaced by a stronger condition. Thus, $h$ needs to be chosen somewhat larger, but one can still assure that $dh$ is small enough for the algorithm to run in polynomial time. Basis reduction yields a polynomial $g$ of degree smaller than $dh$ as above. All of its integral zeros can be determined by the method of Section 13, and these can be checked one by one. This proves Proposition A. $\square$

The exponent $1/\deg p$ in Proposition A is best possible as a function of $\deg p$. Namely, for any integer $d > 1$ and any real number $\eta > 1/d$, the number of $x \in \mathbb{Z}$ with $0 \leq x \leq n^\eta$ that are zeros of $p = X^d$ modulo an integer $n$ that is a $d$-th power, grows exponentially with $\log n$; thus, there does not exist a polynomial-time algorithm for enumerating all those $x$.

PROPOSITION B. *There is a positive real number $\beta$ such that for any three integers $u, v, n$ with*

$$\gcd(u, v) = 1, \qquad n > 1, \qquad v \geq n^{1/4},$$

*the number of positive divisors $x$ of $n$ with $x \equiv u \bmod v$ is at most $\beta \cdot (\log n)^2$. In addition, there is a polynomial-time algorithm that given such $u, v, n$, determines all those $x$.*

PROOF. Any divisor of $n$ that is congruent to $u \bmod v$ is coprime to $v$. Hence, replacing $n$ by the largest divisor of $n$ that is coprime to $v$ (and dealing separately with the case in which this divisor equals 1), we may assume $\gcd(n, v) = 1$. We shall do this throughout the proof.

Let $a, b, h$ be positive integers with $b > a$. We start by establishing, under suitable conditions, an upper bound for the number of divisors $x$ of $n$ with $a \leq x \leq b$ and $x \equiv u \bmod v$, the number $h$ being an auxiliary parameter.

The lattice to be used is of full rank in the $(2h+1)$-dimensional subspace $\sum_{i=-h}^{h} \mathbb{R} \cdot X^i$ of the ring $\mathbb{R}[X, X^{-1}]$ of Laurent polynomials over $\mathbb{R}$. On this vector space, we define a positive definite quadratic form $q$ by

$$q(f) = \sum_{i=0}^{h} c_i^2 \cdot \left(\frac{b-a}{2}\right)^{2i} + \sum_{i=1}^{h} d_i^2 \cdot \left(\frac{a^{-1}-b^{-1}}{2}\right)^{2i}$$

if

$$f = \sum_{i=0}^{h} c_i \cdot \left(X - \frac{b+a}{2}\right)^i + \sum_{i=1}^{h} d_i \cdot \left(X^{-1} - \frac{a^{-1}+b^{-1}}{2}\right)^i, \quad c_i, d_i \in \mathbb{R}.$$

As in the previous proof, for any such $f$ and any $x \in \mathbb{R}$ with $a \leq x \leq b$ one has $|f(x)| \leq ((2h+1) \cdot q(f))^{1/2}$, so that condition (ii) of the Lemma will be satisfied with $c = \sqrt{2h+1}$.

One checks that the lattice $L_0 = \sum_{i=-h}^{h} \mathbb{Z} \cdot X^i$ in $\sum_{i=-h}^{h} \mathbb{R} \cdot X^i$ has determinant

$$d(L_0) = \left(\frac{b-a}{2}\right)^{h(h+1)/2} \cdot \left(\frac{a^{-1}-b^{-1}}{2}\right)^{h(h+1)/2}.$$

Write $Y = n \cdot X^{-1}$. Then the elements of the sublattice $L_1 = \sum_{i=0}^{h} \mathbb{Z} \cdot X^i + \sum_{i=1}^{h} \mathbb{Z} \cdot Y^i$ of $L_0$ are integral-valued on the set of divisors of $n$. One has $(L_0 : L_1) = n^{h(h+1)/2}$ and therefore

$$d(L_1) = \left(\frac{b-a}{2}\right)^{h(h+1)/2} \cdot \left(\frac{n/a - n/b}{2}\right)^{h(h+1)/2}.$$

Write $Z = (X - u)/v$. Then all elements of the lattice $L = L_1 + \sum_{i=0}^{2h} \mathbb{Z} \cdot Y^h Z^i \subset \sum_{i=-h}^{h} \mathbb{R} \cdot X^i$ are integral-valued on the set of divisors $x$ of $n$ with $x \equiv u \bmod v$. From $\gcd(n, v) = 1$ one deduces $(L : L_1) = v^{h(2h+1)}$, so

$$d(L) = \left(\frac{b-a}{2}\right)^{h(h+1)/2} \cdot \left(\frac{n/a - n/b}{2}\right)^{h(h+1)/2} \cdot v^{-h(2h+1)}.$$

Now the Lemma shows: if $h$ satisfies the inequality

$$(2h+1)^2 \cdot \left(\frac{b-a}{2}\right)^{h(h+1)/(2h+1)} \cdot \left(\frac{n/a - n/b}{2}\right)^{h(h+1)/(2h+1)} \cdot v^{-2h} < 1,$$

then there exists a non-zero element $g \in L$ that has all divisors $x$ of $n$ with $x \equiv u \bmod v$ and $a \leq x \leq b$ among its zeros, so that the number of such $x$ is at most $2h$.

To investigate which values of $h$ satisfy the inequality, we restrict to the case $b = 2a$. Then one has $((b-a)/2) \cdot (n/a - n/b)/2 = n/8$. From $v \geq n^{1/4}$ one now deduces that the inequality for $h$ is satisfied if

$$(2h+1)^{2(2h+1)} \cdot n^{h/2} < 8^{h(h+1)}.$$

Such a value for $h$ can be chosen to satisfy $h \leq \delta \cdot \log n$ for some positive constant $\delta$. Thus, we have shown that for any positive integer $a$, the number of divisors $x$ of $n$ with $x \equiv u \bmod v$ and $a \leq x \leq 2a$ is at most $2\delta \log n$. We apply this to $a = 1, 2, 4, \ldots, 2^t$, where $t$ is maximal with $2^t < n$. It follows that the number of positive divisors $x$ of $n$ with $x \equiv u \bmod v$ is at most $(1 + (\log n)/\log 2) \cdot 2\delta \log n$. This implies the first statement of Proposition B.

The conversion of the proof just given into a polynomial-time algorithm follows the same lines as in the case of Proposition A. This proves Proposition B. □

The lattice $L$ used in the proof just given equals the intersection of $\sum_{i=-h}^{h} \mathbb{R} \cdot X^i$ with the subring $\mathbb{Z}[X, Y, Z]$ of $\mathbb{R}[X, X^{-1}]$. The reader may verify that use of the lattice $L_1 + \sum_{i=0}^{2h} \mathbb{Z} \cdot Y^h \binom{Z}{i}$ leads to a notably better result if $n$ has no small prime factors.

Choosing a different partition of $[1, n]$ into intervals $[a, b]$, and using the lattice $L = \mathbb{Z}[X, Y, Z] \cap \sum_{i=-h}^{k} \mathbb{R} \cdot X^i$ for suitable $h, k$ depending on $a, b$, one can improve the bound $\beta \cdot (\log n)^2$ given in Proposition B to $\beta \cdot (\log n)^{3/2}$. This result is due to D. J. Bernstein [2008, Theorem 6.4].

Pollard [1974] exhibited a deterministic and fully proved algorithm for factoring integers that runs in time $n^{1/4+o(1)}$ when the number $n$ to be factored tends to infinity. His result is still the best that is known. Pollard's algorithm depends on fast multiplication techniques. A different algorithm that proves the same result, and that has excellent parallelization properties, is obtained from Proposition B, as follows.

COROLLARY. *There exists, for some positive real number $c$, an algorithm that given a positive integer $n$, determines the complete prime factorization of $n$ in time at most $n^{1/4} \cdot (2 + \log n)^c$.*

*Proof.* We give a brief sketch of the algorithm. First, reduce to the case $n$ is odd. Next, let $v$ be the least power of 2 with $v > n^{1/4}$, and apply the algorithm from Proposition B to all odd values of $u$ with $0 < u < v$. This gives rise to a complete list of divisors of $n$, from which one easily assembles the prime factorization of $n$. This proves the Corollary.

PROPOSITION C. (a) *Let $a, b, n$ be integers with $0 < b - a < n$, let $\varepsilon$ be the real number with $b - a = n^\varepsilon$, and let $\eta \in \mathbb{R}$ satisfy $\eta < 1 - \sqrt{\varepsilon}$. Then there are at most $3/(1 - \sqrt{\varepsilon} - \eta)$ integers $x$ with $a \leq x \leq b$ for which the denominator of $x/n$ is at most $n^\eta$.*

(b) *There is an algorithm that, given integers $a, b, n, k, h$ with $h > k > 0$ and $0 < b - a \leq n^{k^2/h^2}$, determines, in time bounded by a polynomial function of $\log(|a| + |b|)$, $\log n$, and $h$, all integers $x$ with $a \leq x \leq b$ for which the denominator of $x/n$ is at most $n^{1-k/h-1/(2h)}$.*

PROOF. Let $a, b, n$ be as in (a). We let $m$ be a positive integer, to be thought of as an upper bound for the denominator of $x/n$. Further we let $h, k$ be integers satisfying $h > k > 0$; these are auxiliary parameters.

We consider full-rank lattices in the $h$-dimensional subspace $\sum_{i=0}^{h-1} \mathbb{R} \cdot X^i Z^k$ of the polynomial ring $\mathbb{R}[X, Z]$. For $f = \sum_{i=0}^{h-1} c_i (X - (b+a)/2)^i Z^k$ in that space ($c_i \in \mathbb{R}$), we write

$$q(f) = \sum_{i=0}^{h-1} c_i^2 \left(\frac{b-a}{2}\right)^{2i} \cdot m^{2k};$$

as in the earlier proofs in this section, we have

$$|f(x, z)| \leq (h \cdot q(f))^{1/2}$$

for all $x, z \in \mathbb{R}$ with $a \leq x \leq b$, $1 \leq z \leq m$.

The lattice $L_0 = \sum_{i=0}^{h-1} \mathbb{Z} \cdot X^i Z^k$ has rank $h$ and

$$d(L_0) = \left(\frac{b-a}{2}\right)^{h(h-1)/2} \cdot m^{kh}.$$

LATTICES                                         175

Write $Y = XZ/n$. The lattice $L = \sum_{i=0}^{k} \mathbb{Z} \cdot Y^i Z^{k-i} + \sum_{j=1}^{h-k-1} \mathbb{Z} \cdot X^j Y^k$ in $\sum_{i=0}^{h-1} \mathbb{R} \cdot X^i Z^k$ contains $L_0$ as a sublattice of index $n^{kh-k(k+1)/2}$, so one has

$$d(L) = \left(\frac{b-a}{2}\right)^{h(h-1)/2} \cdot m^{kh} \cdot n^{-kh+k(k+1)/2}.$$

All $f \in L$ are integral-valued on the set of pairs of integers $(x, z)$ for which $x/n$ has denominator dividing $z$.

Now the Lemma implies: if $m, h, k$ satisfy the inequality

$$\frac{h^2}{2^{h-1}} \cdot (b-a)^{h-1} \cdot m^{2k} \cdot n^{-2k+k(k+1)/h} < 1,$$

then there is a non-zero polynomial $g \in \mathbb{Q}[X]$ with $\deg g < h$ that has among its zeros all integers $x$ with $a \leq x \leq b$ for which $x/n$ has denominator at most $m$, so that the number of such $x$ is at most $h - 1$. For example, with $h = 2$, $k = 1$ this shows that $x$ is unique (if it exists) whenever $m < \sqrt{n/(b-a)}/\sqrt{2}$, which is slightly weaker than what we saw earlier.

To prove (a), put $\varepsilon = (\log(b-a))/\log n$ as in (a), and let $\eta < 1 - \sqrt{\varepsilon}$. Since we know that there is at most one $x$ as in (a) if $\eta < (1-\varepsilon)/2$, we may assume $\eta \geq (1-\varepsilon)/2$. Then we have $1 - \sqrt{\varepsilon} - \eta < 1/2$. Choose $h$ to be the unique integer with $1/h < (1 - \sqrt{\varepsilon} - \eta)/3 \leq 1/(h-1)$ and $k$ to be the least integer with $k \geq h\sqrt{\varepsilon}$. Then one verifies that we have $0 < k < h$ and

$$h \geq 7, \quad \frac{1}{2} \cdot \left(\frac{h-1}{k} \cdot \varepsilon + \frac{k+1}{h}\right) < 1 - \eta.$$

This implies that $h$, $k$, and $m = \lfloor n^\eta \rfloor$ satisfy the inequality above, so the number of $x$ is at most $h - 1$, which by the choice of $h$ is at most $3/(1 - \sqrt{\varepsilon} - \eta)$. This proves (a).

The proof of (b) follows the same lines as before. It depends on the inequality

$$\frac{1}{2} \cdot \left(\frac{h-1}{k} \cdot \frac{k^2}{h^2} + \frac{k+1}{h}\right) < \frac{k}{h} + \frac{1}{2h}.$$

Note that replacing $k$, $h$ by $4k$, $4h$, if necessary, one may assume $h \geq 7$. This proves Proposition C. □

REMARK. No particular effort has been spent on optimizing the constant 3 in the bound $3/(1 - \sqrt{\varepsilon} - \eta)$ in (a). A more pressing issue is to decide whether the number of $x$ in (a) may be bounded above by a continuous function of $\varepsilon$ alone.

*Error correction in $\mathbb{Z}/n\mathbb{Z}$.* The result just proved admits an attractive reformulation in the terminology of coding theory. Let $n$ be an integer with $n > 1$. We define an '$n$-adic' metric $d$ on the underlying set of the ring $\mathbb{Z}/n\mathbb{Z}$ by putting $d(r, s) = (\log \#J)/\log n$, where $J$ is the ideal of $\mathbb{Z}/n\mathbb{Z}$ generated by $r - s$; the reader may verify that $d$ is indeed a metric, and that the maximal value assumed by $d$ equals 1. This metric is closely related to the *Hamming metric* from coding theory (see [van Lint 1982]). To see this, assume momentarily that $n$ is squarefree, write $P$ for the set of prime factors of $n$, and identify $\mathbb{Z}/n\mathbb{Z}$ with $\prod_{p\in P} \mathbb{Z}/p\mathbb{Z}$ through the ring isomorphism sending $r$ to $(r \bmod p)_{p\in P}$. Two 'vectors' $(r_p)_{p\in P}$, $(s_p)_{p\in P}$ in $\prod_{p\in P} \mathbb{Z}/p\mathbb{Z}$ have Hamming distance $\#\{p : r_p \neq s_p\}$, whereas their newly defined distance equals $(\sum_{p, r_p \neq s_p} \log p)/\sum_{p\in P} \log p$; thus, $d$ is a weighted version of the Hamming distance, the weights having been normalized such that the maximum distance equals 1.

Note that, for general $n$ and all $x, x' \in \mathbb{Z}$, the denominator of $(x - x')/n$ equals $n^{d(x \bmod n, \, x' \bmod n)}$.

Next let, in addition to an integer $n > 1$, two integers $a, b$ with $0 < b - a < n$ be given, and write

$$C = \{(x \bmod n) : x \in \mathbb{Z}, a \leq x \leq b\}, \qquad \delta = 1 - \frac{\log(b-a)}{\log n}.$$

We think of the subset $C$ of $\mathbb{Z}/n\mathbb{Z}$ as a *code*, and, as in coding theory, we refer to $\delta$ as the *designed distance* of $C$. To justify this terminology, suppose that $x$, $x'$ are integers with $a \leq x < x' \leq b$. Then we have $d(x \bmod n, x' \bmod n) = 1 - (\log \gcd(x' - x, n))/\log n \geq 1 - (\log(b - a))/\log n = \delta$, so the 'distance' $\min\{d(v, w) : v, w \in C, v \neq w\}$ of $C$ is at least $\delta$. From $\delta > 0$ we also see that no two distinct integers $x, x' \in [a, b]$ are congruent modulo $n$, so we have $\#C = b - a + 1$.

For given $r \in \mathbb{Z}/n\mathbb{Z}$, one is now interested in the set of all $v \in C$ for which $d(v, r)$ is small; say, $d(v, r) \leq \eta$, where $\eta$ is a given real number. For $v, w \in C$, $v \neq w$, one has $d(v, r) + d(w, r) \geq d(v, w) \geq \delta$, so at most one $v \in C$ satisfies $d(v, r) < \delta/2$. If $u \in \mathbb{Z}$ is such that $r = (u \bmod n)$, then the set of all $v \in C$ with $d(v, r) \leq \eta$ is the same as the set of all $(x \bmod n) + r$, where $x$ ranges over those integers with $a - u \leq x \leq b - u$ for which $x/n$ has denominator at most $n^\eta$. Thus, the results of Proposition C can be transposed to the present setting. From (a) one sees that, for any $\eta < 1 - \sqrt{1 - \delta}$, the number of $v \in C$ with $d(v, r) \leq \eta$ is at most $3/(1 - \sqrt{1 - \delta} - \eta)$; note that $\delta/2 < 1 - \sqrt{1 - \delta}$. Similarly, (b) gives rise to an efficient 'decoding algorithm' past half the designed distance.

The analogue of Proposition C in non-zero characteristic, which may be based on the theory from Section 16 below, has applications to decoding Reed–Solomon and algebraic geometry codes from conventional coding theory, see [Guruswami and Sudan 1999; Bernstein 2008, Section 7].

# 16. Lattices over polynomial rings

There is an analogue of the notion of lattice in which the role of the ring $\mathbb{Z}$ of integers is played by the ring $k[t]$ of polynomials in one variable $t$ over a field $k$. The theory, to which we alluded in earlier sections, is in substance due to Mahler [1941]. Some of the main points are presented below, but we have good reasons to forgo a detailed treatment: from an algorithmic point of view, the theory has little to offer that one cannot obtain from linear algebra over $k$; and from a theoretical point of view the almost equivalent language of *vector bundles over the projective line* is more common.

Let $k$ and $k[t]$ be as above, and let $\deg: k[t] \to \{-\infty\} \cup \mathbb{R}$ map each non-zero polynomial to its degree and $0$ to $-\infty$. By a $k[t]$-*lattice* we mean a pair consisting of a finitely generated $k[t]$-module $L$ and a function $q: L \to \{-\infty\} \cup \mathbb{R}$ with the following properties:

$q(x+y) \leq \max\{q(x), q(y)\}$     for all $x, y \in L$,

$q(cx) = \deg c + q(x)$     for all $c \in k[t]$, $x \in L$,

$q(x) \neq -\infty$     for all $x \in L$, $x \neq 0$,

$\dim_k \{x \in L : q(x) \leq r\} < \infty$     for each $r \in \mathbb{R}$.

The first two properties imply that $\{x \in L : q(x) \leq r\}$ is a $k$-vector space for each $r \in \mathbb{R}$, so the dimension referred to in the last property is well-defined. To improve the resemblance to the definition given in Section 2, one may replace $q$ by the function $L \to \mathbb{R}$ sending $x$ to $\exp(q(x))$. One often restricts to lattices that are *integral-valued* in the sense that the image of $q$ is contained in $\{-\infty\} \cup \mathbb{Z}$.

*Examples.* (a) For each $\lambda \in \mathbb{R}$, an example of a $k[t]$-lattice is given by $L = k[t]$, $q(f) = \lambda + \deg f$; this lattice is denoted by $\mathcal{O}(-\lambda)$. If $L_1, q_1$ and $L_2, q_2$ are $k[t]$-lattices, then their *orthogonal sum* is the $k[t]$-lattice $L = L_1 \oplus L_2$ with $q(x_1, x_2) = \max\{q_1(x_1), q_2(x_2)\}$, for $x_1 \in L_1, x_2 \in L_2$. Somewhat surprisingly, there exists for every $k[t]$-lattice a finite sequence $\lambda_1, \ldots, \lambda_n$ of real numbers such that the lattice is, in an obvious sense, isomorphic to the orthogonal sum of the $n$ lattices $\mathcal{O}(-\lambda_i)$, $1 \leq i \leq n$; if we also require $\lambda_1 \leq \lambda_2 \leq \cdots \leq \lambda_n$, then the $\lambda_i$ are uniquely determined as the *successive minima* of the lattice, and all $\lambda_i$ are in $\mathbb{Z}$ if and only if the lattice is integral-valued. Thus, unlike usual lattices, $k[t]$-lattices admit a satisfactory classification.

(b) The reader acquainted with algebraic geometry (see [Hartshorne 1977]) can obtain $k[t]$-lattices from vector bundles over the projective line, as follows. Write $\mathbf{A}_k^1$ for $\operatorname{Spec} k[t]$, and let $\mathbf{P}_k^1 = \mathbf{A}_k^1 \cup \{\infty\}$ be the projective line over $k$. If $\mathcal{E}$ is a vector bundle over $\mathbf{P}_k^1$, then $L = \mathcal{E}(\mathbf{A}_k^1)$ is a $k[t]$-lattice, with $q(x) = \min\{m \in \mathbb{Z} : x \in t^m \mathcal{E}_\infty\}$ for $x \in L$, $x \neq 0$, and $q(0) = -\infty$. The $k[t]$-lattices obtained in this way are integral-valued, and conversely, each integral-valued

$k[t]$-lattice arises, up to isomorphism, from exactly one vector bundle over $\mathbf{P}_k^1$. The classification just referred to amounts in this case to Grothendieck's theorem describing all vector bundles over the projective line (see [Grothendieck 1957, Theorem 2.1]).

(c) Just as the ring $k[t]$ plays the role that in previous sections was played by $\mathbb{Z}$, so does the field $k(t)$ of fractions of $k[t]$ play the role of $\mathbb{Q}$. In Section 3 we obtained examples of lattices from algebraic number fields, and in a similar way one can obtain $k[t]$-lattices from fields that are finite extensions of $k(t)$. Let $K$ be such a field, and write $A$ for the integral closure of $k[t]$ in $K$. Consider the set $D$ of all maps $d \colon K \to \{-\infty\} \cup \mathbb{R}$ satisfying $d(xy) = d(x) + d(y)$ and $d(x + y) \leq \max\{d(x), d(y)\}$ for all $x, y \in K$, as well as $d(x) \neq -\infty$ for all $x \neq 0$ and $d(x) = \deg x$ for all $x \in k[t]$; so the maps $-d$, for $d \in D$, are the exponential valuations of $K$ that extend the 'infinite valuation' $-\deg$ of $k(t)$. From valuation theory it is well-known that the set $D$ is finite and non-empty. We may make $A$ into a $k[t]$-lattice by putting $q(x) = \max\{d(x) : d \in D\}$ for $x \in A$. If the infinite valuation is ramified in $K$, this is an example of a $k[t]$-lattice that is not integral-valued.

(d) The role of Euclidean vector spaces is in the current theory played by certain normed vector spaces over the completion $k(t)_\infty$ of $k(t)$ at the infinite prime. One may identify this completion with the field $k((t^{-1}))$ of formal Laurent series in $t^{-1}$ over $k$, and define $\deg \colon k(t)_\infty \to \{-\infty\} \cup \mathbb{R}$ by $\deg(\sum_{i \in \mathbb{Z}, i \leq m} a_i t^i) = m$ for $a_i \in k$, $a_m \neq 0$, and $\deg 0 = -\infty$. For integral-valued lattices, the only normed vector spaces one needs to consider are of the form $E = k(t)_\infty^n$, with $n \in \mathbb{Z}_{\geq 0}$, the norm $q \colon E \to \{-\infty\} \cup \mathbb{R}$ being defined by $q((c_i)_{i=1}^n) = \max\{\deg c_i : 1 \leq i \leq n\}$ (and $q(E) = \{-\infty\}$ if $n = 0$). For each basis $b_1, \ldots, b_n$ of $E$ over $k(t)_\infty$, the $k[t]$-module $L = \sum_{i=1}^n k[t] \cdot b_i$, together with the restriction of $q$ to $L$, is an integral-valued $k[t]$-lattice. This is the way in which integral-valued $k[t]$-lattices are often represented numerically. In order to specify the entries of the basis vectors $b_i$ by means of a finite number of elements of $k$, one may require them to be 'rational' in the sense that they belong to the subfield $k(t)$ of $k(t)_\infty$; in algorithmic circumstances, one will also need to place restrictions on the base field $k$.

To represent general $k[t]$-lattices in a similar way, it suffices to choose real numbers $\lambda_1, \ldots, \lambda_n$ and to redefine $q$ on $E$ by $q((c_i)_{i=1}^n) = \max\{\lambda_i + \deg c_i : 1 \leq i \leq n\}$.

*Basis reduction.* Let $L, q$ be a $k[t]$-lattice. Then $L$ has a *basis* as a $k[t]$-module, i.e., a sequence $b_1, \ldots, b_n$ of elements of $L$ such that the map $k[t]^n \to L$ sending $(c_i)_{i=1}^n$ to $\sum_{i=1}^n c_i b_i$ is bijective. A basis $b_1, \ldots, b_n$ is called *reduced* if for each $(c_i)_{i=1}^n \in k[t]^n$ one has $q(\sum_{i=1}^n c_i b_i) = \max\{q(c_i b_i) : 1 \leq i \leq n\}$. The classification theorem stated in Example (a) is readily seen to imply that each

$k[t]$-lattice has a reduced basis. One may wonder whether there is an algorithmic version of the classification theorem. In other words, is there an 'algorithm' that, given a $k[t]$-lattice $L$ as in Example (d), produces a reduced basis for $L$? In the case $k$ is finite and the lattice $L \subset E = k(t)_\infty^n$ from (d) is a sublattice of $k[t]^n$, such an algorithm, running in polynomial time, was exhibited by A. K. Lenstra [1985, Section 1]. It is not hard to adapt his algorithm to more general situations.

*Linear algebra.* The reader may enjoy developing the theory further, defining the determinant of a lattice and finding the analogue of Minkowski's theorem; but it is good to realize that almost anything that one can do with $k[t]$-lattices can also be done by means of linear algebra over $k$. In many applications, one is interested in the set $\{x \in L : q(x) \leq r\}$ for some $k[t]$-lattice $L$ and some $r \in \mathbb{R}$; that set is a finite-dimensional $k$-vector space, and one can usually compute a $k$-basis of that vector space using linear algebra over $k$ (see [Lenstra 1985, Section 1]). Over infinite fields, such as $k = \mathbb{Q}$, linear algebra has the distinct advantage of offering ready means for controlling coefficient blow-up. For finite $k$, however, the linear algebra approach is less efficient than the approach through $k[t]$-lattice basis reduction [Lenstra 1985, Section 1]. This algorithmic distinction may be the one redeeming feature of the theory of $k[t]$-lattices.

## Acknowledgments

I thank Karen Aardal, Jonathan Pila, Carl Pomerance, René Schoof, and Bart de Smit for their helpful assistance.

## References

[Aardal and Eisenbrand 2005] K. Aardal and F. Eisenbrand, "Integer programming, lattices, and results in fixed dimension", pp. 171–243 in *Discrete optimization*, edited by K. Aardal et al., Handbooks in Operations Research and Management Science **12**, Elsevier, Amsterdam, 2005.

[Belabas et al. 2004] K. Belabas, M. van Hoeij, J. Klüners, and A. Steel, "Factoring polynomials over global fields", Preprint, 2004. Available at http://arxiv.org/abs/math/0409510.

[Bernstein 2008] D. J. Bernstein, "Reducing lattice bases to find small-height values of univariate polynomials", pp. 421–446 in *Surveys in algorithmic number theory*, edited by J. P. Buhler and P. Stevenhagen, Math. Sci. Res. Inst. Publ. **44**, Cambridge University Press, New York, 2008.

[Buchmann and Lenstra 1994] J. A. Buchmann and H. W. Lenstra, Jr., "Approximating rings of integers in number fields", *J. Théor. Nombres Bordeaux* **6**:2 (1994), 221–260.

[Buhler and Wagon 2008] J. P. Buhler and S. Wagon, "Basic algorithms in number theory", pp. 25–68 in *Surveys in algorithmic number theory*, edited by J. P. Buhler and P. Stevenhagen, Math. Sci. Res. Inst. Publ. **44**, Cambridge University Press, New York, 2008.

[Cassels 1978] J. W. S. Cassels, *Rational quadratic forms*, London Mathematical Society Monographs **13**, Academic Press, London, 1978.

[Cohen 1993] H. Cohen, *A course in computational algebraic number theory*, Graduate Texts in Mathematics **138**, Springer, Berlin, 1993.

[Conway and Sloane 1988] J. H. Conway and N. A. Sloane, *Sphere packings, lattices and groups*, Grundlehren der Mathematischen Wissenschaften **290**, Springer, New York, 1988.

[Elkies 2000] N. D. Elkies, "Rational points near curves and small nonzero $|x^3 - y^2|$ via lattice reduction", pp. 33–63 in *Algorithmic number theory* (Leiden, 2000), edited by W. Bosma, Lecture Notes in Comput. Sci. **1838**, Springer, Berlin, 2000.

[von zur Gathen and Gerhard 1999] J. von zur Gathen and J. Gerhard, *Modern computer algebra*, Cambridge University Press, New York, 1999.

[Gauss 1801] C. F. Gauss, *Disquisitiones arithmeticae*, Fleischer, Leipzig, 1801.

[Goldreich et al. 1999] O. Goldreich, D. Micciancio, S. Safra, and J.-P. Seifert, "Approximating shortest lattice vectors is not harder than approximating closest lattice vectors", *Inform. Process. Lett.* **71**:2 (1999), 55–61.

[Grothendieck 1957] A. Grothendieck, "Sur la classification des fibrés holomorphes sur la sphère de Riemann", *Amer. J. Math.* **79** (1957), 121–138.

[Guruswami and Sudan 1999] V. Guruswami and M. Sudan, "Improved decoding of Reed–Solomon and algebraic-geometry codes", *IEEE Trans. Inform. Theory* **45**:6 (1999), 1757–1767.

[Hardy and Wright 1938] G. H. Hardy and E. M. Wright, *An introduction to the theory of numbers*, Oxford University Press, 1938.

[Hartshorne 1977] R. Hartshorne, *Algebraic geometry*, Graduate Texts in Mathematics **52**, Springer, New York, 1977.

[Heath-Brown 2002] D. R. Heath-Brown, "The density of rational points on curves and surfaces", *Ann. of Math.* (2) **155**:2 (2002), 553–595.

[Knuth 1981] D. E. Knuth, *The art of computer programming, II: Seminumerical algorithms*, 2nd ed., Addison-Wesley, Reading, MA, 1981.

[Lang 1988] S. Lang, *Introduction to Arakelov theory*, Springer, New York, 1988.

[Lenstra 1983] H. W. Lenstra, Jr., "Integer programming with a fixed number of variables", *Math. Oper. Res.* **8**:4 (1983), 538–548.

[Lenstra 1985] A. K. Lenstra, "Factoring multivariate polynomials over finite fields", *J. Comput. System Sci.* **30**:2 (1985), 235–248.

[Lenstra 2001] H. W. Lenstra, Jr., "Flags and lattice basis reduction", pp. 37–51 in *European Congress of Mathematics, I* (Barcelona, 2000), edited by C. Casacuberta et al., Progr. Math. **201**, Birkhäuser, Basel, 2001.

[Lenstra et al. 1982] A. K. Lenstra, H. W. Lenstra, Jr., and L. Lovász, "Factoring polynomials with rational coefficients", *Math. Ann.* **261**:4 (1982), 515–534.

[van Lint 1982] J. H. van Lint, *Introduction to coding theory*, Graduate Texts in Mathematics **86**, Springer, New York, 1982.

[Mahler 1941] K. Mahler, "An analogue to Minkowski's geometry of numbers in a field of series", *Ann. of Math.* (2) **42** (1941), 488–522.

[Mignotte 1974] M. Mignotte, "An inequality about factors of polynomials", *Math. Comp.* **28** (1974), 1153–1157.

[Pollard 1974] J. M. Pollard, "Theorems on factorization and primality testing", *Proc. Cambridge Philos. Soc.* **76** (1974), 521–528.

[Rabin and Shallit 1986] M. O. Rabin and J. O. Shallit, "Randomized algorithms in number theory", *Comm. Pure Appl. Math.* **39**:S, suppl. (1986), S239–S256.

[Schrijver 1986] A. Schrijver, *Theory of linear and integer programming*, John Wiley, Chichester, UK, 1986.

[de Smit 1996] B. de Smit, "Measure characteristics of complexes", *Cahiers Topologie Géom. Différentielle Catég.* **37**:1 (1996), 3–20.

HENDRIK W. LENSTRA, JR.
MATHEMATISCH INSTITUUT
UNIVERSITEIT LEIDEN
POSTBUS 9512
2300 RA LEIDEN
THE NETHERLANDS
hwl@math.leidenuniv.nl

# Elliptic curves

BJORN POONEN

ABSTRACT. This is an introduction to some aspects of the arithmetic of elliptic curves, intended for readers with little or no background in number theory and algebraic geometry. In keeping with the rest of this volume, the presentation has an algorithmic slant. We also touch lightly on curves of higher genus. Readers desiring a more systematic development should consult one of the references for further reading suggested at the end.

## CONTENTS

| | |
|---|---|
| 1. Plane curves | 183 |
| 2. Projective geometry | 185 |
| 3. Determining $X(\mathbb{Q})$: subdivision by degree | 186 |
| 4. Elliptic curves | 188 |
| 5. Structure of $E(k)$ for various fields $k$ | 190 |
| 6. Elliptic curves over the rational numbers | 192 |
| 7. The elliptic curve factoring method | 197 |
| 8. Curves of genus greater than 1 | 201 |
| 9. Further reading | 204 |
| References | 205 |

## 1. Plane curves

Let $k$ be a field. For instance, $k$ could be the field $\mathbb{Q}$ of rational numbers, the field $\mathbb{R}$ of real numbers, the field $\mathbb{C}$ of complex numbers, the field $\mathbb{Q}_p$ of $p$-adic numbers (see [Koblitz 1984] for an introduction), or the finite field $\mathbb{F}_q$ of $q$ elements (see Chapter I of [Serre 1973]). Let $\bar{k}$ be an algebraic closure of $k$.

A (geometrically integral, affine) *plane curve* $X$ over $k$ is defined by an equation $f(x, y) = 0$ where $f(x, y) = \sum a_{ij} x^i y^j \in k[x, y]$ is irreducible over $\bar{k}$.

---

The writing of this article was supported by NSF grant DMS-9801104, and a Packard Fellowship.

One defines the degree of $X$ and of $f$ by

$$\deg X = \deg f = \max\{i + j : a_{ij} \neq 0\}.$$

A *k-rational point* (or simply *k-point*) on $X$ is a point $(a, b)$ with coordinates in $k$ such that $f(a, b) = 0$. The set of all $k$-rational points on $X$ is denoted $X(k)$.

EXAMPLE. The equation $x^2 y - 6y^2 - 11 = 0$ defines a plane curve $X$ over $\mathbb{Q}$ of degree 3, and $(5, 1/2) \in X(\mathbb{Q})$.

Already at this point we can state an open problem, one which over the centuries has served as motivation for the development of a huge amount of mathematics.

QUESTION. Is there an algorithm, that given a plane curve $X$ over $\mathbb{Q}$, determines $X(\mathbb{Q})$, or at least decides whether $X(\mathbb{Q})$ is nonempty?

Although $X(\mathbb{Q})$ need not be finite, we will see later that it always admits a finite description, so this problem of determining $X(\mathbb{Q})$ can be formulated precisely using the notion of *Turing machine*: see [Hopcroft and Ullman 1969] for a definition. For the relationship of this question to Hilbert's Tenth Problem, see the survey [Poonen 2002].

The current status is that there exist computational methods that often answer the question for a particular $X$, although it has never been proved that these methods work in general. Even the following are unknown:

(1) Is there an algorithm that given a degree 4 polynomial $f(x) \in \mathbb{Q}[x]$, determines whether $y^2 = f(x)$ has a rational point?
(2) Is there an algorithm that given a polynomial $f(x, y) \in \mathbb{Q}[x, y]$ of degree 3, determines whether $f(x, y) = 0$ has a rational point?

In fact, problems (1) and (2) are equivalent, although this is by no means obvious! (For the experts: both (1) and (2) are equivalent to

(3) Is there an algorithm to compute the rank of an elliptic curve over $\mathbb{Q}$?

If the answer to (1) is yes, then one can compute the rank of any elliptic curve over $\mathbb{Q}$, by 2-descent. Conversely, if the answer to (3) is yes, the answer to (1) is also yes, since the only difficult case of (1) is when $y^2 = f(x)$ is a locally trivial principal homogeneous space of an elliptic curve $E$ over $\mathbb{Q}$, hence represented by an element of the 2-Selmer group of $E$, and knowledge of the rank of $E(\mathbb{Q})$ lets one decide whether its image in $\mathrm{III}(E)$ is nontrivial. Similarly (2) and (3) are equivalent, via 3-descent.)

## 2. Projective geometry

**2.1. The projective plane.** The affine plane $\mathbb{A}^2$ is the usual plane, with $\mathbb{A}^2(k) = \{(a,b) : a, b \in k\}$ for any field $k$. One "compactifies" $\mathbb{A}^2$ by adjoining some points "at infinity" to produce the projective plane $\mathbb{P}^2$. One of the main reasons for doing this is to make intersection theory work better: see Bézout's Theorem in Section 2.3.

The set of $k$-points on the projective plane $\mathbb{P}^2$ can be defined directly as $\mathbb{P}^2(k) := (k^3 - 0)/k^*$. In other words, a $k$-rational point on $\mathbb{P}^2$ is an equivalence class of triples $(a, b, c)$ with $a, b, c \in k$ not all zero, under the equivalence relation $\sim$, where $(a, b, c) \sim (\lambda a, \lambda b, \lambda c)$ for any $\lambda \in k^*$. The equivalence class of $(a, b, c)$ is denoted $(a : b : c)$. One can also identify $\mathbb{P}^2(k)$ with the set of lines through $0$ in $(x, y, z)$-space.

The injection $\mathbb{A}^2(k) \hookrightarrow \mathbb{P}^2(k)$ mapping $(a, b)$ to $(a : b : 1)$ is almost a bijection: the points of $\mathbb{P}^2(k)$ not in the image, namely those of the form $(a : b : 0)$, form a projective line $\mathbb{P}^1(k)$ of "points at infinity". Viewing $\mathbb{P}^2(k)$ as lines through $0$ in $(x, y, z)$-space, $\mathbb{A}^2(k)$ is the set of such lines passing through $(a, b, 1)$ for some $a, b \in k$, and the complement $\mathbb{P}^1(k)$ is the set of (horizontal) lines through $0$ in the $(x, y)$-plane.

Also, $\mathbb{P}^2$ can be covered by three copies of $\mathbb{A}^2$, namely $\{(x : y : z) \mid x \neq 0\}$, $\{(x : y : z) \mid y \neq 0\}$, and $\{(x : y : z) \mid z \neq 0\}$.

**2.2. Projective closure of curves.** The *homogenization* of a polynomial $f(x, y)$ of degree $d$ is $F(X, Y, Z) := Z^d f(X/Z, Y/Z)$. In other words, one changes $x$ to $X$, $y$ to $Y$, and then appends enough factors of $Z$ to each monomial to bring the total degree of each monomial to $d$. One can recover $f$ as $f(x, y) = F(x, y, 1)$.

If $f(x, y) = 0$ is a plane curve $C$ in $\mathbb{A}^2$, its *projective closure* is the curve $\widetilde{C}$ in $\mathbb{P}^2$ defined by the homogenized equation $F(X, Y, Z) = 0$. The curve $\widetilde{C}$ equals $C$ plus some points "at infinity".

EXAMPLE. If $f(x, y) = y^2 - x^3 + x - 7$, then

$$F(X, Y, Z) = Y^2 Z - X^3 + XZ^2 - 7Z^3$$

and

$$\begin{aligned}\widetilde{C}(\mathbb{Q}) &= \frac{\{\text{zeros of } F\}}{\mathbb{Q}^*} \\ &= \{\text{zeros of } F(X, Y, 1)\} \cup \frac{\{\text{zeros of } F(X, Y, 0)\} - 0}{\mathbb{Q}^*} \\ &= C(\mathbb{Q}) \cup \{P\},\end{aligned}$$

where $P$ is the point $(0 : 1 : 0)$ "at infinity".

## 2.3. Bézout's Theorem.
As mentioned earlier, one of the primary reasons for working in the projective plane is to obtain a good intersection theory. Let $F(X,Y,Z) = 0$ and $G(X,Y,Z) = 0$ be curves in $\mathbb{P}^2$ over $k$, of degree $m$ and $n$, respectively. Bézout's Theorem states that they intersect in exactly $mn$ points of $\mathbb{P}^2$, provided that

(i) $F$ and $G$ have no nontrivial common factor,
(ii) one works over an algebraically closed field, and
(iii) one counts intersection points with multiplicity (in case of singularities, or points of tangency).

We have given condition (i) in a form that yields a correct statement of Bézout's Theorem even without our assumption that plane curves are defined by irreducible polynomials. Of course, if $F$ and $G$ *are* irreducible, condition (i) states simply that the curves do not coincide.

## 3. Determining $X(\mathbb{Q})$: subdivision by degree

We return to the problem of determining the set of rational points $X(\mathbb{Q})$, where $X$ is an affine plane curve $f(x,y) = 0$ over $\mathbb{Q}$ or its projective closure. Let $d = \deg f$. We will look at the problem for increasing values of $d$.

### 3.1. $d = 1$: lines.
We know how to parameterize the rational points on $ax + by + c = 0$!

### 3.2. $d = 2$: conics.
Legendre proved that conics satisfy the *Hasse Principle*. This means: $X$ has a $\mathbb{Q}$-point if and only if $X$ has an $\mathbb{R}$-point and a $\mathbb{Q}_p$-point for each prime $p$. Since a projective conic is described by a quadratic form in three variables, Legendre's result can be viewed as a special case of the Hasse–Minkowski Theorem [Serre 1973, Chapter IV, §3.2], which states that a quadratic form in $n$ variables over $\mathbb{Q}$ represents 0 (in other words, takes the value 0 on some arguments in $\mathbb{Q}$ not all zero) if and only if it represents 0 over $\mathbb{R}$ and $\mathbb{Q}_p$ for all $p$.

Legendre's Theorem leads to an algorithm to determine the existence of a $\mathbb{Q}$-point on a conic $X$. Here is one such algorithm: complete the square, multiply by a constant, and absorb squares into the variables, to reduce to the case of $aX^2 + bY^2 + cZ^2 = 0$ in $\mathbb{P}^2$, where $a,b,c$ are nonzero, squarefree, pairwise relatively prime integers. Then one can show that there exists a $\mathbb{Q}$-point if and only if $a,b,c$ are not all of the same sign and the congruences

$$ax^2 + b \equiv 0 \pmod{c},$$
$$by^2 + c \equiv 0 \pmod{a},$$
$$cz^2 + a \equiv 0 \pmod{b}$$

are solvable in integers. Moreover, in this case, $aX^2 + bY^2 + cZ^2 = 0$ has a nontrivial solution in integers $X, Y, Z$ satisfying $|X| \le |bc|^{1/2}$, $|Y| \le |ac|^{1/2}$, and $|Z| \le |ab|^{1/2}$. See [Mordell 1969].

In the case where the conic $X$ has a $\mathbb{Q}$-point $P_0$, there remains the problem of describing the set of *all* $\mathbb{Q}$-points. For this there is a famous trick: for each $P \in X(\mathbb{Q})$ draw the line through $P_0$ and $P$, and let $t$ be its slope, which will be in $\mathbb{Q}$ (or maybe $\infty$). Conversely, given $t \in \mathbb{Q}$, Bézout's Theorem guarantees that the line through $P_0$ with slope $t$ will intersect the conic in one other point (provided that this line is not tangent to the conic at $P_0$), and this will be a rational point.

For example, if $X$ is the circle $x^2 + y^2 = 1$ and $P_0 = (-1, 0)$, then

$$t \longrightarrow \left( \frac{1-t^2}{1+t^2}, \frac{2t}{1+t^2} \right),$$

$$\frac{y}{x+1} \longleftarrow (x, y)$$

define *birational maps* from $\mathbb{A}^1$ to $X$ and back: this means that, ignoring finitely many subsets of smaller dimension (a few points), they are maps given by rational functions of the variables that induce a bijection between the $\overline{\mathbb{Q}}$-points on each side. These birational maps are defined over $\mathbb{Q}$; that is, the coefficients of the rational functions are in $\mathbb{Q}$, so they also induce a bijection between $\mathbb{Q}$-points (ignoring the same subsets as before). In particular, the complete set of rational solutions to $x^2 + y^2 = 1$ is

$$\left\{ \left( \frac{1-t^2}{1+t^2}, \frac{2t}{1+t^2} \right) : t \in \mathbb{Q} \right\} \cup \{(-1, 0)\}.$$

**3.3.** $d = 3$: **plane cubics.** Lind [1940] and Reichardt [1942] discovered that the Hasse Principle can fail for plane curves of degree 3. Here is a counterexample due to Selmer [1951; 1954]: the curve $3X^3 + 4Y^3 + 5Z^3 = 0$ in $\mathbb{P}^2$ has an $\mathbb{R}$-point $(((-4/3)^{1/3} : 1 : 0)$ is one) and a $\mathbb{Q}_p$-point for each prime $p$, but it has no $\mathbb{Q}$-point. (For $p > 5$, the existence of $\mathbb{Q}_p$-points can be proved by combining Hensel's Lemma [Koblitz 1984, Theorem 3] with a counting argument to prove the existence of solutions modulo $p$. For $p = 2, 3, 5$, a more general form of Hensel's Lemma can be used [Koblitz 1984, Chapter I, Exercise 6]. The nonexistence of $\mathbb{Q}$-points is more difficult to establish.)

As mentioned at the beginning of this article, deciding whether a plane cubic curve has a rational point is currently an unsolved problem. For the time being, we will restrict attention to those plane cubic curves that *do* have a rational point. These are called *elliptic curves*, and are birational to curves defined by a equation of a simple form. This leads to the first of the official definitions of elliptic curves that we present in the next section.

## 4. Elliptic curves

**4.1. Equivalent definitions.** Let $k$ be a perfect field. An *elliptic curve* over $k$ can be defined as any one of the following:

(i) The projective closure of a nonsingular curve defined by a "Weierstrass equation"
$$y^2 + a_1 xy + a_3 y = x^3 + a_2 x^2 + a_4 x + a_6$$
with $a_1, a_2, a_3, a_4, a_6 \in k$. If the characteristic of $k$ is not 2 or 3, one may restrict attention to projective closures of curves
$$y^2 = x^3 + Ax + B.$$
One can show that this is nonsingular if and only if $x^3 + Ax + B$ has distinct roots in $\bar{k}$, and that this holds if and only if the quantity $\Delta := -16(4A^3 + 27B^2)$ is nonzero.

(ii) A nonsingular projective genus 1 curve over $k$ equipped with a $k$-rational point $O$.

(iii) A one-dimensional projective group variety over $k$.

We need to define several of the terms occurring here. (Even then it will not be obvious that the definitions above are equivalent.)

**4.2. Singularities.** If $(0,0)$ is a point on the affine curve $f(x, y) = 0$ over $k$, then $(0,0)$ is a *singular* point if $\partial f/\partial x$ and $\partial f/\partial y$ both vanish at $(0,0)$. Equivalently, $(0,0)$ is singular if $f = f_2 + f_3 + \cdots + f_d$ where each $f_i \in k[x, y]$ is a homogeneous polynomial of degree $i$. For instance $(0,0)$ is singular on $y^2 = x^3$ and on $y^2 = x^3 + x^2$, but not on $y^2 = x^3 - x$. More generally, $(a, b)$ is singular on $f(x, y) = 0$ if and only if $(0,0)$ is singular on $f(X + a, Y + b) = 0$.

An affine curve is *nonsingular* if it has no singular points. A projective curve $F(X, Y, Z) = 0$ is nonsingular if its "affine pieces" $F(x, y, 1) = 0$, $F(x, 1, z) = 0$, $F(1, y, z) = 0$ are nonsingular. (One can generalize these notions to curves in higher dimensional affine or projective spaces. This is important because although every plane curve $X$, singular or not, is birational to some nonsingular projective curve $Y$, sometimes such a $Y$ cannot be found in the plane.)

*Smooth* is a synonym for nonsingular, at least for curves over a perfect field $k$.

**4.3. Genus.** Let $X$ be a nonsingular projective curve over a perfect field $k$. The *genus* of $X$ is a nonnegative integer $g$ that measures the geometric complexity of $X$. It has the following equivalent definitions:

(A) $g = \dim_k \Omega$ where $\Omega$ is the vector space of regular differentials on $X$. (Regular means "no poles". If $k = \mathbb{C}$, then regular is equivalent to holomorphic.)

(B) $g$ is the topological genus (number of handles) of the compact Riemann surface $X(\mathbb{C})$. (This definition makes sense only if $k$ can be embedded in $\mathbb{C}$.)
(C) $g = \frac{1}{2}(d-1)(d-2) - $ (terms for singularities of $Y$), where $Y$ is a (possibly singular) plane curve birational to $X$ and $d$ is the degree of $Y$. For example, a nonsingular plane cubic curve has genus 1.

We will not prove the equivalence of these definitions.

**4.4. Group law: definition.** To say that an elliptic curve $E$ over $k$ is a group variety means roughly that there is an "addition" map $E \times E \to E$, given by rational functions, that induces a group structure on $E(L)$ for any field extension $L$ of $k$. We now explain what the group law on $E(k)$ is, for an elliptic curve $E$ presented as a plane cubic curve in Weierstrass form.

The group law is characterized by the following two rules:

(i) The point $O = (0 : 1 : 0)$ at infinity is the identity of the group.
(ii) If a line $L$ intersects $E$ in three $k$-points $P, Q, R \in E(k)$ (taking multiplicities into account), then $P + Q + R = O$ in the group law.

From these one deduces:

(a) Given $P \in E(k)$ not equal to $0$, the vertical line through $P$ intersects $E$ in $P, O$, and a third point which is $-P$.
(b) Given $P, Q \in E(k)$ not equal to $O$, the line through $P$ and $Q$ (take the tangent to $E$ at $P$ if $P = Q$) intersects $E$ at $P, Q$, and a third point $R \in E(k)$. If $R = O$, then $P + Q = O$; otherwise $P + Q = -R$, where $-R$ can be constructed as in (a).

Note that $E(k)$ is an *abelian* group.

**4.5. Group law: formulas.** It is easy to see that, at least generically, the coordinates of $P + Q$ can be expressed as rational functions in the coordinates of $P$ and $Q$. Here we present explicit formulas that give an algorithm for computing $P + Q$. The existence of these formulas will be important in Section 7.5 as we develop the elliptic curve factoring method.

To compute the sum $R$ of points $P, Q \in E(k)$ on $y^2 = x^3 + Ax + B$ over $k$:

1. If $P = O$, put $R = Q$ and stop.
2. If $Q = O$, put $R = P$ and stop.
3. Otherwise let $P = (x_1 : y_1 : 1)$ and $Q = (x_2 : y_2 : 1)$. If $x_1 \neq x_2$, put

$$\lambda = (y_1 - y_2)(x_1 - x_2)^{-1},$$
$$x_3 = \lambda^2 - x_1 - x_2,$$
$$y_3 = \lambda(x_1 - x_3) - y_1,$$
$$R = (x_3 : y_3 : 1),$$

and stop.

4. If $x_1 = x_2$ and $y_1 = -y_2$, put $R = O$ and stop.
5. If $x_1 = x_2$ and $y_1 \ne -y_2$ (so $P = Q$), put

$$\lambda = (3x_1^2 + A)(y_1 + y_2)^{-1},$$
$$x_3 = \lambda^2 - x_1 - x_2,$$
$$y_3 = \lambda(x_1 - x_3) - y_1,$$
$$R = (x_3 : y_3 : 1),$$

and stop.

This requires $O(1)$ field operations in $k$. Using projective coordinates renders division unnecessary.

**4.6. Group law: examples.** Let $E$ be the elliptic curve $y^2 = x^3 - 25x$. (From now on, when we give a nonhomogeneous equation for an elliptic curve $E$, it is understood that we really mean for $E$ to be defined as the projective closure of this affine curve.) Since $x^3 - 25x$ has distinct roots, $E$ is nonsingular, so $E$ really is an elliptic curve. The line $L$ through $P := (-4, 6)$ and $Q := (0, 0)$ has the equation $y = (-3/2)x$. We compute $L \cap E$ by substitution:

$$((-3/2)x)^2 = x^3 - 25x$$
$$0 = (x+4)x(x - 25/4)$$

and find $L \cap E = \{P, Q, R\}$ where $R := (25/4, -75/8)$. Thus $P + Q + R = O$ in the group law, and $P + Q = -R = (25/4, 75/8)$.

The intersection of the line $X = 0$ in $\mathbb{P}^2$ with $E : Y^2 Z = X^3 - 25XZ^2$ is $X = 0 = Y^2 Z$, which gives $(0 : 1 : 0) = O$ and $(0 : 0 : 1) = Q$, the latter with multiplicity 2. (Geometrically, this corresponds to the vertical line $x = 0$ being tangent to $E$ at $Q$.) Thus $Q + Q + O = O$, and $2Q = O$; that is, $Q$ is a point of order 2, a *2-torsion point*. (In general, the nonzero 2-torsion points on $y^2 = x^3 + Ax + B$ are $(\alpha, 0)$ where $\alpha$ is a root of $x^3 + Ax + B$: they form a subgroup of $E(\bar{k})$ isomorphic to $\mathbb{Z}/2\mathbb{Z} \times \mathbb{Z}/2\mathbb{Z}$.)

# 5. Structure of $E(k)$ for various fields $k$

What kind of group is $E(k)$?

**5.1. Elliptic curves over the complex numbers.** On one hand $E(\mathbb{C})$ is a complex manifold, but on the other hand it is a group, and the coordinates of $P + Q$ are rational functions in the coordinates of $P$ and $Q$, so the group law is holomorphic. Hence $E(\mathbb{C})$ is a 1-dimensional Lie group over $\mathbb{C}$. Moreover, $E(\mathbb{C})$ is closed in $\mathbb{P}^2(\mathbb{C})$, which is compact, so $E(\mathbb{C})$ is compact. It turns out that it is also connected. By the classification of compact connected 1-dimensional Lie groups over $\mathbb{C}$, we must have $E(\mathbb{C}) \simeq \mathbb{C}/\Lambda$ as Lie groups over $\mathbb{C}$, for some lattice $\Lambda = \mathbb{Z}\omega_1 + \mathbb{Z}\omega_2$, where $\omega_1, \omega_2$ are an $\mathbb{R}$-basis of $\mathbb{C}$. The $\omega_i$ are called *periods*, because there is a meromorphic function $\wp(z)$ on $\mathbb{C}$ defined below such that $\wp(z + \omega) = \wp(z)$ for all $\omega \in \Lambda$. The lattice $\Lambda$ is not uniquely determined by $E$; scaling it by a nonzero complex number does not affect the isomorphism type of $\mathbb{C}/\Lambda$. But a particular $\Lambda$ can be singled out if a nonzero holomorphic differential on $E$ is also given (to be identified with $dz$ on $\mathbb{C}/\Lambda$).

Suppose conversely that we start with a discrete rank 2 lattice $\Lambda$ in $\mathbb{C}$. In other words, $\Lambda = \mathbb{Z}\omega_1 + \mathbb{Z}\omega_2$ for an $\mathbb{R}$-basis $\omega_1, \omega_2$ of $\mathbb{C}$. We will show how to reverse the previous paragraph to find a corresponding elliptic curve $E$ over $\mathbb{C}$. Set $g_2 = 60 \sum'_{\omega \in \Lambda} \omega^{-4}$ and $g_3 = 140 \sum'_{\omega \in \Lambda} \omega^{-6}$ where the $'$ means "omit the $\omega = 0$ term". Let

$$\wp(z) = z^{-2} + \sum\nolimits'_{\omega \in \Lambda}((z - \omega)^{-2} - \omega^{-2}).$$

Then one can prove the following: $\wp(z)$ is meromorphic on $\mathbb{C}$ with poles in $\Lambda$, $\wp(z + \omega) = \wp(z)$ for all $\omega \in \Lambda$, and $z \mapsto (\wp(z), \wp'(z))$ defines an analytic isomorphism $\mathbb{C}/\Lambda \to E(\mathbb{C})$ where $E$ is the elliptic curve

$$y^2 = 4x^3 - g_2 x - g_3$$

over $\mathbb{C}$. (The poles of $\wp(z)$ correspond under the isomorphism to the point $O \in E(\mathbb{C})$ at infinity.) The differential $dx/y$ on $E$ pulls back to $dz$ on $\mathbb{C}/\Lambda$; hence the inverse map $E(\mathbb{C}) \to \mathbb{C}/\Lambda$ is given by

$$(a, b) \mapsto \int_O^{(a,b)} \frac{dx}{y} = \int_\infty^a \frac{dx}{\sqrt{4x^3 - g_2 x - g_3}}$$

(with suitable choice of path and branch).

More generally, Riemann proved that any compact Riemann surface is isomorphic as complex manifold to $X(\mathbb{C})$ for some nonsingular projective curve $X$ over $\mathbb{C}$.

**5.2. Elliptic curves over the real numbers.** Let $E$ be an elliptic curve $y^2 = f(x)$ over $\mathbb{R}$, where $f(x) = x^3 + Ax + B \in \mathbb{R}[x]$ is squarefree. This time $E(\mathbb{R})$ is a 1-dimensional compact commutative Lie group over $\mathbb{R}$. Considering the intervals where $f$ is nonnegative, we find that $E(\mathbb{R})$ has one or two connected components, according as $f$ has one real root, or three real roots. Since the circle group

$$\mathbb{R}/\mathbb{Z} \simeq \{z \in \mathbb{C} : |z| = 1\}$$

is the only *connected* 1-dimensional compact commutative Lie group over $\mathbb{R}$, it follows that

$$E(\mathbb{R}) \simeq \begin{cases} \mathbb{R}/\mathbb{Z} & \text{if } f \text{ has 1 real root,} \\ \mathbb{R}/\mathbb{Z} \times \mathbb{Z}/2\mathbb{Z} & \text{if } f \text{ has 3 real roots.} \end{cases}$$

The group $E(\mathbb{R})$ can also be viewed as the subgroup of $E(\mathbb{C})$ fixed by complex conjugation. If $E$ is defined over $\mathbb{R}$, one can choose $\Lambda$ of the previous section to be stable under complex conjugation, and the coordinatewise action of complex conjugation on $E(\mathbb{C})$ corresponds to the natural action on $\mathbb{C}/\Lambda$. If a nonzero regular differential on $E$ (defined over $\mathbb{R}$) is fixed, then $\Lambda$ is determined, and in this situation one defines the *real period* as the positive generator of the infinite cyclic group $\Lambda \cap \mathbb{R}$.

**5.3. Elliptic curves over finite fields.** Let $E$ be an elliptic curve over the finite field $\mathbb{F}_q$ of $q$ elements. Since $E(\mathbb{F}_q)$ is a subset of $\mathbb{P}^2(\mathbb{F}_q)$, $E(\mathbb{F}_q)$ is a *finite* abelian group. Hasse proved

$$\#E(\mathbb{F}_q) = q + 1 - a,$$

where $|a| \leq 2\sqrt{q}$. This is a special case of the "Weil conjectures" (now proved). Moreover, an algorithm of Schoof [1985] computes $\#E(\mathbb{F}_q)$ in time $(\log q)^{O(1)}$ as follows: an algorithm we will not explain determines $\#E(\mathbb{F}_q) \bmod \ell$ for each prime $\ell$ up to about $\log q$, and then the Chinese Remainder Theorem recovers $\#E(\mathbb{F}_q)$.

EXAMPLE 1. Let $E$ be the elliptic curve $y^2 = x^3 - x + 1$ over $\mathbb{F}_3$. Hasse's Theorem implies $1 \leq \#E(\mathbb{F}_3) \leq 7$. In fact,

$$E(\mathbb{F}_3) = \{(0,1), (0,2), (1,1), (1,2), (2,1), (2,2), O\},$$

and $E(\mathbb{F}_3) \simeq \mathbb{Z}/7\mathbb{Z}$. Here is an exercise for the reader, an instance of what is called the *elliptic discrete logarithm problem*: which multiple of $(0,1)$ equals $(1,1)$?

EXAMPLE 2. Let $E$ be the elliptic curve $y^2 = x^3 - x$ over $\mathbb{F}_3$. Then

$$E(\mathbb{F}_3) = \{(0,0), (1,0), (2,0), O\},$$

and $E(\mathbb{F}_3) \simeq \mathbb{Z}/2\mathbb{Z} \times \mathbb{Z}/2\mathbb{Z}$.

# 6. Elliptic curves over the rational numbers

**6.1. Mordell's Theorem.** Let $E$ be an elliptic curve over $\mathbb{Q}$. Mordell proved that $E(\mathbb{Q})$ is a finitely generated abelian group:

$$E(\mathbb{Q}) \simeq \mathbb{Z}^r \times T,$$

where $r \in \mathbb{Z}_{\geq 0}$ is called the *rank*, and $T = E(\mathbb{Q})_{\text{tors}}$ is a finite abelian group called the *torsion subgroup*. (This is sometimes called the Mordell–Weil theorem, because Weil proved a generalization for *abelian varieties* over number fields. Abelian varieties are projective group varieties of arbitrary dimension.)

Although $T$ can be computed in polynomial time, it is not known whether $r$ is computable. We will say a little more about the latter at the end of Section 6.7.

EXAMPLE 1. Let $E$ be the elliptic curve $y^2 = x^3 - 25x$ over $\mathbb{Q}$. With work, one can show
$$E(\mathbb{Q}) \simeq \mathbb{Z} \times \mathbb{Z}/2\mathbb{Z} \times \mathbb{Z}/2\mathbb{Z},$$
where $E(\mathbb{Q})/E(\mathbb{Q})_{\text{tors}}$ is generated by $(-4, 6)$.

EXAMPLE 2. Let $E$ be the elliptic curve $y^2 + y = x^3 - x^2$ over $\mathbb{Q}$, also known as "the modular curve $X_1(11)$". Then
$$E(\mathbb{Q}) = \{(0, 0), (0, -1), (1, 0), (1, -1), O\} \simeq \mathbb{Z}/5\mathbb{Z}.$$

EXAMPLE 3. Let $E$ be the elliptic curve $1063y^2 = x^3 - x$. (This is not in Weierstrass form, but it is isomorphic to $y^2 = x^3 - 1063^2 x$.) Using "Heegner points on modular curves", Elkies [1994] computed that $E(\mathbb{Q}) \simeq \mathbb{Z} \times \mathbb{Z}/2\mathbb{Z} \times \mathbb{Z}/2\mathbb{Z}$, where $E(\mathbb{Q})/E(\mathbb{Q})_{\text{tors}}$ is generated by a point with $x$-coordinate $q^2/1063$, where

$$q = \frac{11091863741829769675047021635712281767382339667434645}{3173426575447721807352079773209000125228079367777887}.$$

EXAMPLE 4. Let $E$ be the elliptic curve $y^2 + xy + y = x^3 + ax + b$ where

$a = -12003982203699224530353461919116 6796374$, and

$b = 50422499248491067001080179916808272675944375622291 1415116$.

Martin and McMillen [2000] showed that $E(\mathbb{Q}) \simeq \mathbb{Z}^r$, where $r \geq 24$.

A folklore conjecture predicts that as E varies over all elliptic curves over $\mathbb{Q}$, the rank $r$ can be arbitrarily large. But the present author does not believe this.

**6.2. One-dimensional affine group varieties over $k$.** One way to begin studying an elliptic curve over $\mathbb{Q}$ or another number field is to reduce its coefficients modulo a prime and to study the resulting curve over a finite field. Unfortunately, the result can be singular even if the original curve was nonsingular. It turns out that upon deleting the singularity, we obtain a one-dimensional group variety, but it is no longer an elliptic curve: it is affine instead of projective due to the deletion.

The one-dimensional affine group varieties $G$ over $k$ can be classified. For simplicity, we assume that $k$ is a perfect field of characteristic not 2. The table on the next page lists all possibilities.

| $G$ | variety | group law | $G(k)$ |
|---|---|---|---|
| $\mathbb{G}_a$ | $\mathbb{A}^1$ | $x_1, x_2 \mapsto x_1 + x_2$ | $k$, under $+$ |
| $\mathbb{G}_m$ | $xy = 1$ | $(x_1, y_1), (x_2, y_2) \mapsto (x_1 x_2, y_1 y_2)$ | $k^*$, under $\cdot$ |
| $\mathbb{G}_m^{(a)}$ | $x^2 - ay^2 = 1$ | $(x_1, y_1), (x_2, y_2) \mapsto$ $(x_1 x_2 + a y_1 y_2, x_1 y_2 + x_2 y_1)$ | $\ker\left(k(\sqrt{a})^* \xrightarrow{\text{norm}} k^*\right)$ |

The first column gives the group variety $G$, which is either the *additive group* $\mathbb{G}_a$, or the *multiplicative group* $\mathbb{G}_m$ or one of its *twists* $\mathbb{G}_m^{(a)}$ for some $a \in k^*$. The isomorphism type of $\mathbb{G}_m^{(a)}$ as a group variety is determined by the image of $a$ in $k^*/k^{*2}$. If $a \in k^{*2}$, then $\mathbb{G}_m^{(a)}$ is isomorphic to $\mathbb{G}_m$.

The second column describes $G$ as a variety; in each case, $G$ is either $\mathbb{A}^1$ or a plane curve in $\mathbb{A}^2$. The third column expresses the group law morphism $G \times G \to G$ in coordinates. The final column describes the group of rational points $G(k)$.

**6.3. Singular Weierstrass cubics.** If $E$ is a *singular* curve defined as the projective closure of
$$y^2 = x^3 + ax^2 + bx + c$$
then there is at most one singularity. (Otherwise the formula for the genus would produce a negative integer.) Let $P_0$ be the singularity. By a change of variables, we may assume that $P_0 = (0, 0)$. The equation then has the form $(y^2 - ax^2) - x^3 = 0$. The tangent line(s) to the branches at $(0, 0)$ are $y = \pm\sqrt{a}x$. The singularity is called a *node* or a *cusp* according as $a \neq 0$ or $a = 0$.

In either case, $E_{\text{ns}} := E - \{P_0\}$ becomes a one-dimensional affine group variety using the same geometric construction as in the nonsingular case. (A line $L$ through two nonsingular points cannot pass through $P_0$, because the intersection multiplicity at $P_0$ would be at least 2, and Bézout's Theorem would be violated.) In fact,

$$E_{\text{ns}} \simeq \begin{cases} \mathbb{G}_a & \text{if } a = 0 \text{ (cusp)}, \\ \mathbb{G}_m & \text{if } a \in k^{*2} \text{ (node)}, \\ \mathbb{G}_m^{(a)} & \text{if } a \text{ is a nonsquare (node)}. \end{cases}$$

EXAMPLE. If $E$ is the projective closure of $y^2 = x^3$, which has a cusp at $(0, 0)$, then the isomorphism is given by

$$E_{\text{ns}} \longleftrightarrow \mathbb{G}_a,$$
$$(x, y) \longrightarrow x/y,$$
$$(t : 1 : t^3) = (t^{-2}, t^{-3}) \longleftarrow t.$$

The isomorphism respects the group structures: one can check for example that $(t : 1 : t^3)$, $(u : 1 : u^3)$, and $(v : 1 : v^3)$ are collinear in $\mathbb{P}^2$ whenever $t + u + v = 0$.

**6.4. Reduction mod $p$.** For any $u \in \mathbb{Q}^*$, the elliptic curve $E : y^2 = x^3 + Ax + B$ over $\mathbb{Q}$ is isomorphic to $Y^2 = X^3 + u^4 AX + u^6 B$. (Multiply the equation by $u^6$, and set $Y = u^3 y$ and $X = u^2 x$.) Therefore, we may assume that $A, B \in \mathbb{Z}$. Then one can reduce the equation modulo a prime $p$ to get a cubic curve $\overline{E}$ over $\mathbb{F}_p$. But $\overline{E}$ might be singular. This happens if and only if $p$ divides $\Delta$.

One says that $E$ has *good reduction* at $p$ if there is some Weierstrass equation for $E$ (obtained by change of coordinates) whose reduction modulo $p$ is nonsingular. Similarly if there is a Weierstrass equation for $E$ whose reduction modulo $p$ is a cubic curve with a node, one says that $E$ has *multiplicative reduction* at $p$; in this case one says that $E$ has *split multiplicative reduction* or *nonsplit multiplicative reduction* according as $\overline{E}_{\mathrm{ns}}$ is $\mathbb{G}_m$ or a nontrivial twist. Otherwise, if $E$ has neither good nor multiplicative reduction, then all Weierstrass equations for $E$ reduce modulo $p$ to a cubic curve with a cusp, and one says that $E$ has *additive reduction*. Some of this is summarized in the following table.

| singularity | $\overline{E}_{\mathrm{ns}}$ | terminology |
|---|---|---|
| none | $\overline{E}$ | good reduction |
| cusp | $\mathbb{G}_a$ | additive reduction |
| node | $\mathbb{G}_m$ or $\mathbb{G}_m^{(d)}$ | multiplicative reduction |

**6.5. Finiteness of $T := E(\mathbb{Q})_{\mathrm{tors}}$.** Suppose that an elliptic curve $E$ over $\mathbb{Q}$ has good reduction at $p$. By scaling, any point in $E(\mathbb{Q})$ can be written as $(a : b : c)$ with $a, b, c \in \mathbb{Z}$ such that $\gcd(a, b, c) = 1$, and then $a, b, c$ can be reduced modulo $p$ to obtain a point on $\overline{E}(\mathbb{F}_p)$. This defines a homomorphism $E(\mathbb{Q}) \to \overline{E}(\mathbb{F}_p)$.

THEOREM. *If $E$ has good reduction at a prime $p > 2$, then the torsion subgroup $T$ of $E(\mathbb{Q})$ injects into $\overline{E}(\mathbb{F}_p)$.*

COROLLARY. *$T$ is finite.*

EXAMPLE. Let $E$ be the elliptic curve $y^2 = x^3 - 4x + 4$ over $\mathbb{Q}$. Then
$$\Delta = -16(4(-4)^3 + 27 \cdot 4^2) = -2^8 \cdot 11,$$
so $E$ has good reduction at $p$ at least when $p \neq 2, 11$. We compute $\#\overline{E}(\mathbb{F}_3) = 7$ and $\#\overline{E}(\mathbb{F}_5) = 9$. The only group that can inject into groups of order 7 and 9 is the trivial group, so $T = \{O\}$. In particular, $(0, 2) \in E(\mathbb{Q})$ is of infinite order, and $E(\mathbb{Q})$ has positive rank.

**6.6. Other theorems about the torsion subgroup $T$.**

THEOREM (LUTZ, NAGELL). *Let $A, B \in \mathbb{Z}$ be such that $E : y^2 = x^3 + Ax + B$ is an elliptic curve. If $P \in T$ and $P \neq O$, then $P = (x_0, y_0)$ where $x_0, y_0 \in \mathbb{Z}$ and $y_0^2 \mid 4A^3 + 27B^2$.*

This gives a slow algorithm to determine $T$. (It requires factoring the integer $4A^3 + 27B^2$.)

THEOREM (MAZUR). *If $E$ is an elliptic curve over $\mathbb{Q}$, then either $T \simeq \mathbb{Z}/N\mathbb{Z}$ for some $N \leq 12$, $N \neq 11$, or $T \simeq \mathbb{Z}/2\mathbb{Z} \times \mathbb{Z}/2N\mathbb{Z}$ for some $N \leq 4$. In particular, $\#T \leq 16$.*

For each $m \geq 1$, one can use the group law to compute the polynomial $\phi_m(x) \in \mathbb{Q}[x]$ whose roots are the $x$-coordinates of the points of order $m$ in $E(\overline{\mathbb{Q}})$. Determining the points of order $m$ in $E(\mathbb{Q})$ is then a matter of finding the rational roots of $\phi_m$ and checking which of these give a rational $y$-coordinate. By Mazur's Theorem, only finitely many $m$ need be considered, so this gives a polynomial time algorithm for computing $T$.

**6.7. Height functions.** We next describe some of the ingredients which go into the proof of Mordell's Theorem. If $P = (a : b : c) \in \mathbb{P}^2(\mathbb{Q})$, we may scale to assume $a, b, c \in \mathbb{Z}$ and $\gcd(a, b, c) = 1$. Then define

$$H(P) := \max(|a|, |b|, |c|) \quad \text{and} \quad h(P) := \log H(P).$$

One calls $h(P)$ the *(logarithmic) height* of $P$. Roughly, $h(P)$ is the width of a sheet of paper needed to write down $P$.

It is easy to see that for any $B > 0$,

$$\#\{P \in \mathbb{P}^2(\mathbb{Q}) : H(P) \leq B\} \leq (2B+1)^3,$$

so

$$\{P \in \mathbb{P}^2(\mathbb{Q}) : h(P) \leq B\} \quad \text{is finite.} \tag{6-1}$$

The latter is a special case of Northcott's Theorem [Serre 1989, §2.4]. If $E \subset \mathbb{P}^2$ is an elliptic curve over $\mathbb{Q}$, then one can show that for $P, Q \in E(\mathbb{Q})$,

$$h(P+Q) + h(P-Q) = 2h(P) + 2h(Q) + O(1), \tag{6-2}$$

where the $O(1)$ depends on $E$ but not on $P$ or $Q$.

Define the *canonical height* or *Néron–Tate height* of a point $P \in E(\mathbb{Q})$ as $\hat{h}(P) := \lim_{n \to \infty} h(2^n P)/4^n$. The following are formal consequences of (6-1) and (6-2): for $P, Q \in E(\mathbb{Q})$ and $n \in \mathbb{Z}$,

(a) $h(2P) = 4h(P) + O(1)$;
(b) the limit defining $\hat{h}(P)$ exists;
(c) $\hat{h}(P) = h(P) + O(1)$;
(d) $\hat{h}(P+Q) + \hat{h}(P-Q) = 2\hat{h}(P) + 2\hat{h}(Q)$;
(e) $\hat{h}(nP) = n^2 \hat{h}(P)$;
(f) $\hat{h}(P) \geq 0$, with equality if and only if $P \in E(\mathbb{Q})_{\text{tors}}$.

In particular, $\hat{h}$ is a quadratic form on $E(\mathbb{Q})/E(\mathbb{Q})_{\text{tors}}$.

Moreover, (6-1) and (6-2) with the "Weak Mordell–Weil Theorem", which asserts the finiteness of $E(\mathbb{Q})/2E(\mathbb{Q})$, imply that $E(\mathbb{Q})$ is finitely generated. If generators of $E(\mathbb{Q})/2E(\mathbb{Q})$ could be found effectively, then the rank of $E(\mathbb{Q})$ and generators of $E(\mathbb{Q})$ could be found effectively. Unfortunately, the only known method for calculating $E(\mathbb{Q})/2E(\mathbb{Q})$, *descent*, requires a prime $p$ such that the $p$-primary part of a certain torsion group $\text{III}(E)$ associated to $E$ is finite. The group $\text{III}(E)$, called the *Shafarevich–Tate group*, is conjectured to be finite for all $E$, but this has been proved only in certain cases.

## 7. The elliptic curve factoring method

**7.1. An interpretation of factoring.** Suppose that $p$ and $q$ are unknown large primes, and $N = pq$ is given. One way to factor $N$ is to compute somehow an integer $m$ that is zero modulo $p$ but nonzero mod $q$. Then $\gcd(m, N)$, which can be computed quickly, yields $p$.

**7.2. Some factoring methods.** We outline various well-known factoring methods from the point of view of Section 7.1. (We use $\mathbb{Z}/N$ as an abbreviation for the quotient ring $\mathbb{Z}/N\mathbb{Z}$.)

<u>Trial division</u>: Try $m = 2, m = 3, m = 5 \ldots$.

<u>Pollard $\rho$</u>: Let $f$ be a function $\mathbb{Z}/N \to \mathbb{Z}/N$, compute a sequence $x_1, x_2, x_3, \ldots$ of elements of $\mathbb{Z}/N$ satisfying $x_{i+1} = f(x_i)$, and try $m = x_j - x_i$ for various $i, j$.

<u>Quadratic sieve, number field sieve</u>: After finding a nontrivial solution to $x^2 \equiv y^2 \pmod{n}$, try $m = x + y$.

<u>Pollard $p - 1$</u>: Choose random $a$ mod $N$, take $K = k!$ for some $k \geq 1$, and try $m = a^K - 1$.

<u>ECM (Lenstra's elliptic curve method)</u>: Instead of $a^K$ for $a \in (\mathbb{Z}/N)^*$, consider $K \cdot P$ for some $P \in E(\mathbb{Z}/N)$, for some elliptic curve $E$. Here $K \cdot P$ means $P + P + \cdots + P$, the sum of $K$ copies of $P$ in an abelian group $E(\mathbb{Z}/N)$ to be defined.

**7.3. The Pollard $p - 1$ method.** The elliptic curve method can be viewed as an analogue of the Pollard $p - 1$ method. As a warmup for the elliptic curve method, we describe the $p - 1$ method here more fully, but still ignoring details and practical improvements.

To factor $N$:

1. Choose an integer $K > 1$ with a lot of factors, for instance, $K = k!$ for some $k \geq 1$.

2. Choose an arbitrary integer $a$ satisfying $1 < a < N - 1$.
3. If $\gcd(a, N) > 1$, then we're done! Otherwise continue.
4. Use the binary expansion of $K$ to compute $a^K \bmod N$.
5. Compute $g = \gcd(a^K - 1, N)$. If $1 < g < N$, then we're done, since $g$ is a nontrivial factor of $N$. If $g = N$, try again with a different $a$, or with $K$ replaced by a divisor. If $g = 1$, try again with a larger $K$ (or if you're tired, give up).

If $K$ is a multiple of $p - 1$ for some prime $p$ dividing $N$, then in Step 4, $(a^K \bmod p)$ is a power of $(a^{p-1} \bmod p)$, which is $(1 \bmod p)$ by Fermat's Little Theorem. Then in Step 5, $a^K - 1$ is divisible by $p$, so $g = p$, (unless by chance $a^K - 1$ is divisible also by another factor of $N$).

The problem with this method is that it is not easy to arrange for $K$ to be divisible by $p - 1$, since we do not know what $p$ is! The best we can do is choose a $K$ with many factors, and hope that $p - 1$ will be among the factors. If we choose $K = k!$, we are essentially hoping that $p - 1$ will be *smooth*, that is, equal to a product of small primes. Thus the Pollard $p - 1$ method is effective only for finding factors $p$ of $N$ such that $p - 1$ is smooth.

**7.4. Variants of the $p-1$ method.** If instead of Fermat's Little Theorem in $\mathbb{F}_p^*$, one uses that every element of $\mathbb{F}_{p^2}^*/\mathbb{F}_p^*$ has order dividing $p + 1$, one can develop a $p + 1$ method by working in $A^*/(\mathbb{Z}/N)^*$, where $A = (\mathbb{Z}/N)[t]/(t^2 - b)$ for some $b \in \mathbb{Z}/N$. This will be effective for finding factors $p$ of $N$ such that $p + 1$ is smooth.

Similarly one can use subgroups of $\mathbb{F}_{p^r}^*$ for other small $r$ to develop methods that work well when $p^2 + p + 1$ is smooth, when $p^2 + 1$ is smooth, ..., when $\Phi_r(p)$ is smooth, where $\Phi_r(z)$ is the $r$-th cyclotomic polynomial. These rapidly become useless, because $p^2 + p + 1$ and so on are much larger than $p - 1$, and hence are much less likely to be smooth.

Lenstra's idea was instead to replace $\mathbb{F}_p^*$ by the group $E(\mathbb{F}_p)$ where $E$ is an elliptic curve! There are many different $E$ to try, of varying orders close to $p$.

**7.5. Elliptic curves over $\mathbb{Z}/N$.** The elliptic curve method requires working with elliptic curves over rings that are not fields. The theory of elliptic curves over rings is best expressed in the language of schemes, but this would take too many pages to develop. Fortunately, for the factoring application, we can make do with a more concrete development based on explicit formulas for the group law.

Let $N$ be a positive integer. To simplify the exposition, we assume that $\gcd(N, 6) = 1$. Define

$$\mathbb{P}^2(\mathbb{Z}/N) := \frac{\{(a : b : c) \mid a, b, c \in \mathbb{Z}/N, \gcd(a, b, c, N) = 1\}}{(\mathbb{Z}/N)^*}.$$

An *elliptic curve* $E$ over $\mathbb{Z}/N$ is given by an homogeneous equation

$$Y^2 Z = X^3 + AXZ^2 + BZ^3,$$

where $A, B \in \mathbb{Z}/N$ are such that the quantity $\Delta := -16(4A^3 + 27B^2)$ is in $(\mathbb{Z}/N)^*$. Then $E(\mathbb{Z}/N)$ denotes the subset of points $(a : b : c) \in \mathbb{P}^2(\mathbb{Z}/N)$ satisfying the cubic equation. For any prime $p$ dividing $N$, $E(\mathbb{Z}/p)$ denotes the set of points in $\mathbb{P}^2(\mathbb{Z}/p)$ satisfying the equation reduced modulo $p$.

For simplicity, suppose that $N = pq$ where $p$ and $q$ are distinct primes greater than 3. The Chinese Remainder Theorem implies

$$\frac{\mathbb{Z}}{N} \simeq \frac{\mathbb{Z}}{p} \times \frac{\mathbb{Z}}{q} \qquad \text{(as rings)},$$

$$\left(\frac{\mathbb{Z}}{N}\right)^* \simeq \left(\frac{\mathbb{Z}}{p}\right)^* \times \left(\frac{\mathbb{Z}}{q}\right)^* \qquad \text{(as groups)},$$

$$\mathbb{P}^2\left(\frac{\mathbb{Z}}{N}\right) \simeq \mathbb{P}^2\left(\frac{\mathbb{Z}}{p}\right) \times \mathbb{P}^2\left(\frac{\mathbb{Z}}{q}\right) \qquad \text{(as sets)},$$

$$E\left(\frac{\mathbb{Z}}{N}\right) \simeq E\left(\frac{\mathbb{Z}}{p}\right) \times E\left(\frac{\mathbb{Z}}{q}\right) \qquad \text{(as groups)}.$$

Hence the set $E(\mathbb{Z}/N)$ inherits the structure of an abelian group. Most pairs of points in $E(\mathbb{Z}/N)$ can be added using the formulas of Section 4.5. (Use the extended GCD algorithm to compute inverses modulo $N$.) In fact, the formulas fail only if some calculated quantity in $\mathbb{Z}/N$ is zero mod $p$ and nonzero mod $q$ or vice versa, in which case $N$ is factored!

**7.6. The elliptic curve method.** Assume that the integer $N$ to be factored satisfies $\gcd(N, 6) = 1$ and that $N \neq n^r$ for any integers $n, r \geq 2$. (The latter can be tested very quickly, in $(\log N)^{1+o(1)}$ time [Bernstein 1998].) To search for factors of $N$ of size less than about $P$:

1. Fix a "smoothness bound" $y$ much smaller than $P$, and let $K$ be the LCM of all $y$-smooth integers less than or equal to $P$.
2. Choose random integers $A, x_1, y_1 \in [1, N]$.
3. Let $B = y_1^2 - x_1^3 - Ax_1 \in \mathbb{Z}/N$ and let $E$ be $y^2 = x^3 + Ax + B$, so $P_1 := (x_1, y_1) \in E(\mathbb{Z}/N)$. If $\gcd(4A^3 + 27B^2, N) \neq 1$, go back to Step 2.
4. Use the binary expansion of the factors of $K$ to attempt to compute $K \cdot P_1 \in E(\mathbb{Z}/N)$ using the group law formula.
5. If at some point the formula fails (that is, we try to use the extended GCD to invert a nonzero non-unit in $\mathbb{Z}/N$), then we have found a factor of $N$. Otherwise, go back to Step 2 and try a different elliptic curve.

Note that in Steps 2 and 3 we choose the point first and then find an elliptic curve through it. This is because it is algorithmically difficult to find a "random" point on a given elliptic curve over $\mathbb{Z}/N$: doing this in the naive way, by choosing the $x$-coordinate first, would require taking a square root of an element of $\mathbb{Z}/N$, which is a problem known to be random polynomial time equivalent to factoring $N$!

### 7.7. Partial analysis of the elliptic curve method.
If $\#E(\mathbb{Z}/p)$ divides $K$, then $K \cdot P_1$ will reduce mod $p$ to $O$. In this case, it is probable that $K \cdot P_1$ is not also $O$ modulo $N$, or at least that at some point of the computation of $K \cdot P_1$, one reaches a subproduct $K'$ of $K$ such that $K' \cdot P_1$ is $O$ mod $p$ but not $O$ mod $N$. (This can be made precise.) Hence it is essentially true that factoring $N$ requires only being lucky enough to choose $E$ such that $\#E(\mathbb{Z}/p)$ divides $K$.

Suppose that $N$ has a prime factor $p$ such that $p + 1 + 2\sqrt{p} \le P$. Let $s$ be the probability that for a random $E$ constructed by the algorithm, the order of $E(\mathbb{Z}/p)$ is $y$-smooth. If $\#E(\mathbb{Z}/p)$ is $y$-smooth, then $\#E(\mathbb{Z}/p) \mid K$, by definition of $K$ and by Hasse's Theorem (Section 5.3) that $\#E(\mathbb{Z}/p) \le p + 1 + 2\sqrt{p}$. Hence the number of elliptic curves that need to be tried during the algorithm is $O(1/s)$. Each trial involves $O(\log K)$ group law operations, each requiring $(\log N)^{O(1)}$ bit operations, making the total running time

$$R = O(s^{-1}(\log K)(\log N)^{O(1)}).$$

Let $L(x) = \exp(\sqrt{(\log x)(\log \log x)})$, so $\log x \ll L(x) \ll x$ as $x \to \infty$. Take $y = L(P)^a$ for some $a > 0$ to be optimized later. In order to express the running time $R$ in terms of the parameters of the algorithm, namely $N$, $P$, and $a$, we first bound $K$:

$$K \le \prod_{\ell \le y} \ell^{\lfloor \log_\ell P \rfloor} \le \prod_{\ell \le y} P \le P^y,$$
$$\log K \le y \log P = L(P)^{a+o(1)}.$$

Next we need an estimate of the smoothness probability $s$. A theorem of Canfield, Erdős, and Pomerance [1983] states that the probability that a random integer in $[1, x]$ is $L(x)^a$-smooth is $L(x)^{-1/(2a)+o(1)}$ as $x \to \infty$. Using a formula of Deuring for the number of elliptic curves of given order over $\mathbb{Z}/p$, one can show that $\#E(\mathbb{Z}/p)$ is close to uniformly distributed over most of the Hasse range

$$[p + 1 - 2\sqrt{p}, p + 1 + 2\sqrt{p}].$$

To proceed, we assume the conjecture that the result of Canfield, Erdős, and Pomerance result holds for random integers in this much smaller range. Then $s = L(P)^{-1/(2a)+o(1)}$.

Under this assumption, the total running time for the elliptic curve method is

$$R = L(P)^{a+1/(2a)+o(1)}(\log N)^{O(1)}.$$

This is optimized at $a = 1/\sqrt{2}$, which makes $y = L(P)^{1/\sqrt{2}}$ and

$$R = L(P)^{\sqrt{2}+o(1)}(\log N)^{O(1)}.$$

To factor $N$ completely, we would take $P = \sqrt{N}$, which yields a running time of $L(N)^{1+o(1)}$. But one advantage of the elliptic curve method over most other factoring methods is that its running time depends on the size of the factor to be found: it is capable of finding small factors more quickly.

In practice, since the optimal choices of $y$ and hence $K$ depend on $P$, it is reasonable to run the elliptic curve method with $P$ small at first (to search for small factors) and then if no factor is found, try again and again, with an increasing sequence of values of $P$. Eventually, if no small factors are found, one should switch to the number field sieve, which is faster asymptotically if the factors are large.

**7.8. Elliptic curve method records.** The largest factor found by the elliptic curve method as of April 2008 is the 67-digit prime factor

4444349792156709907895752551798631908946180608768737946280238078881

(by Bruce Dodson in August 2006; see [Zimmerman 2008]).

Given Richard Brent's 2000 extrapolation [2000, §3.4] that the elliptic curve method record will be a $D$-digit factor in year $Y(D) := 9.3\sqrt{D} + 1932.3$, the value $Y(67) = 2008.4$ shows that the method performs well in practice.

# 8. Curves of genus greater than 1

**8.1. Divisors.** Divisors can be used to produce higher dimensional analogues of elliptic curves, attached to higher genus curves. They also give a natural proof of the associativity for the group law of elliptic curves.

Let $X$ be a nonsingular projective curve over a perfect field $k$. The *group of $\bar{k}$-divisors* $\text{Div}(X_{\bar{k}})$ is the set of formal sums of points in $X(\bar{k})$. The subgroup of $k$-*rational divisors* $\text{Div}(X)$ is the subgroup of $\text{Gal}(\bar{k}/k)$-stable divisors. The degree of a divisor $D = n_1(P_1) + \cdots + n_r(P_r)$ is the integer $n_1 + \cdots + n_r$. Then $\text{Div}^0(X_{\bar{k}})$ denotes the group of $\bar{k}$-divisors of degree zero. Similarly define $\text{Div}^0(X)$.

EXAMPLE. Let $E$ be the elliptic curve $y^2 = x^3 + 17$ over $\mathbb{Q}$. The points

$$P = (1+\sqrt{-7}, -2+\sqrt{-7}), \quad Q = (1-\sqrt{-7}, -2-\sqrt{-7}), \quad R = (-1, 4)$$

lie in $E(\overline{\mathbb{Q}})$. The divisor $D := 2(P) + 2(Q) - 7(R)$ is stable under $\text{Gal}(\overline{\mathbb{Q}}/\mathbb{Q})$ even though $P$ and $Q$ individually are not fixed, so $D \in \text{Div}(E)$. The degree of $D$ is $2 + 2 - 7 = -3$.

## 8.2. Principal divisors and the Picard group.
If $f$ is a nonzero rational function on $X$ (that is, a "function" given as a ratio of polynomials in the coordinates), and $P \in X(\bar{k})$, let $\operatorname{ord}_P(f)$ denote the "order of vanishing" of $f$ at $P$ (positive if $f(P) = 0$, negative if $f$ has a pole at $P$). Then the *divisor of* $f$ is

$$(f) := \sum_{P \in X(\bar{k})} \operatorname{ord}_P(f) \cdot P \quad \in \operatorname{Div}^0(X_{\bar{k}}),$$

and if the coefficients of $f$ are in $k$, then $(f) \in \operatorname{Div}^0(X)$. The set of such $(f)$ form the subgroup $\operatorname{Princ}(X)$ of *principal divisors*. If $D, D' \in \operatorname{Div}(X)$, one writes $D \sim D'$ if $D - D' \in \operatorname{Princ}(X)$. Define the *Picard group* or *divisor class group* of $X$ as $\operatorname{Pic}(X) := \operatorname{Div}(X)/\operatorname{Princ}(X)$. Also define $\operatorname{Pic}^0(X) := \operatorname{Div}^0(X)/\operatorname{Princ}(X)$.

EXAMPLE. If $E, P, Q, R$ are as in the example of the previous section and $f = y - x + 3$, one has $(f) = (P) + (Q) + (R) - 3(O)$, so $(P) + (Q) + (R) \sim 3(O)$ in $\operatorname{Pic}(E)$.

THEOREM. *If $E$ is an elliptic curve over $k$, the map $E(k) \to \operatorname{Pic}^0(E)$ sending $P$ to the class of $(P) - (O)$ is a bijection.*

The group structure on $\operatorname{Pic}^0(E)$ thus induces a group structure on $E(k)$, which is the same as the one we defined earlier.

## 8.3. Analogies.
It is helpful to keep in mind the following analogies when studying algebraic number fields or the geometry of curves (especially if they are over finite fields).

| Number field object | Function field analogue |
|---|---|
| $\mathbb{Z}$ | $k[t]$ |
| $\mathbb{Q}$ | $k(t)$ |
| $\mathbb{Q}_p$ | $k((t))$ |
| number field $F$ | finite extension of $k(t)$ |
| | (rational functions on curve $X$) |
| $\operatorname{Spec} \mathcal{O}_F$ | |
| (where $\mathcal{O}_F$ is the ring of integers of $F$) | smooth affine curve $X$ |
| finite extension $F'$ of $F$ | covering $X' \to X$ |
| fractional ideal | divisor |
| principal ideal | principal divisor |
| class group | $\operatorname{Pic}(X)$ |
| functional equation of $\zeta(s)$, | Weil conjectures |
| Riemann Hypothesis, | (all proved!) |
| and Generalized Riemann Hypothesis | |

**8.4. Rational points on curves.** We now turn briefly to the study of rational points on curves of arbitrary genus. (Recall that elliptic curves were curves of genus 1 equipped with a rational point.) A curve over $\mathbb{Q}$ of any genus can have $X(\mathbb{Q}) = \emptyset$: for example, if $X$ is birational to $y^2 = -(x^{2g+2} + 1)$, then $X$ has genus $g$ and has no rational points. In the table below, assume $X$ is a plane curve over $\mathbb{Q}$ with $X(\mathbb{Q}) \neq \emptyset$, and define

$$N_X(B) := \#\{P \in X(\mathbb{Q}) : h(P) \leq B\}.$$

The qualitative behavior of $X(\mathbb{Q})$, and in particular the rate of growth of the function $N_X(B)$, are roughly determined by the genus.

| Genus | $X(\mathbb{Q})$ | $N_X(B)$ |
|---|---|---|
| 0 | infinite | $(c_1 + o(1))e^{c_2 B}$ |
| 1 | finitely generated abelian group | $(c_3 + o(1))B^{c_4}$ |
| $\geq 2$ | finite | $O(1)$ |

In the third column, $c_1, c_2, c_3 > 0, c_4 \geq 0$ are constants depending on $X$, and the estimates hold as $B \to \infty$.

The fact that $X(\mathbb{Q})$ is finite when the genus is at least 2 was conjectured by Mordell [1922]. Proofs were given by Faltings [1983] and Vojta [1991], and a simplified version of Vojta's proof was given by Bombieri [1990]. All known proofs are ineffective: it is not known whether there exists an algorithm to determine $X(\mathbb{Q})$, although there probably is one.

**8.5. Jacobians: one tool for studying higher genus curves.** Recall that for elliptic curves $E$, we have a group isomorphism $E(k) \simeq \text{Pic}^0(k)$. But if $X$ is a nonsingular projective curve of genus $g > 1$ over a field $k$, then there is no natural bijection between $X(k)$ and $\text{Pic}^0(X)$, and in fact one can show that $X$ cannot be made into a group variety. (Here is one way to show this: if $X$ is a group variety, then one can create a global section of the tangent bundle of $X$ by choosing a nonzero tangent vector at the origin and moving it around by translations. But on a curve of genus $g > 1$, any nonzero meromorphic section of the tangent bundle has degree $2 - 2g < 0$, so it cannot be regular everywhere.)

Define an *abelian variety* to be a projective group variety, so that an elliptic curve is a one-dimensional abelian variety. Then for any $X$ of genus $g$ as above, there is an abelian variety $J$ of dimension $g$ called the *Jacobian* of $X$, with the property that $J(k) \simeq \text{Pic}^0(X)$, (at least under the technical assumption that $X(k) \neq \emptyset$). It is only when $X$ is an elliptic curve that $J$ coincides with $X$.

If $P_0 \in X(k)$, then there is an embedding of varieties $X \to J$ mapping each point $P$ on $X$ to the class of the divisor $(P) - (P_0)$ in $\text{Pic}^0(X)$.

EXAMPLE. Here we do a few computations in a Jacobian of a Fermat curve. Let $X$ be the projective closure of $x^{13} + y^{13} = 1$ over $\mathbb{Q}$. The genus of $X$ is $g = (13-1)(13-2)/2 = 66$. Let $P = (1, 0)$ and $Q = (0, 1)$, which are in $X(\mathbb{Q})$. Let $J$ be the Jacobian of $X$, so $J$ is a 66-dimensional abelian variety. The divisor of the function $(y-1)/(x-1)$ on $X$ is $13(Q) - 13(P)$, so the class of $13(Q) - 13(P)$ in $\text{Pic}^0(X) = J(\mathbb{Q})$ is trivial. Thus the class of $(Q) - (P)$ is a 13-torsion point (a point of order 13) in $J(\mathbb{Q})$. (One can use the fact that $X$ has positive genus to show that no function on $X$ has divisor $(Q) - (P)$, so this point of $J(\mathbb{Q})$ is nonzero.)

Suppose that $X$ is a nonsingular projective curve over $\mathbb{Q}$ of genus $g \geq 2$ with $X(\mathbb{Q}) \neq \varnothing$. By the Mordell–Weil theorem, $J(\mathbb{Q})$ is a finitely generated abelian group. Moreover, $J(\mathbb{Q})/J(\mathbb{Q})_{\text{tors}}$ can be equipped with a quadratic canonical height function $\hat{h}$. Vojta's proof of the finiteness of $X(\mathbb{Q})$ (following earlier ideas of Mumford) studies how points of $X(\mathbb{Q})$ can sit inside this lattice by applying diophantine approximation techniques.

Unfortunately the diophantine approximation techniques are ineffective: they give bounds on the number of rational points, but not bounds on their height. (Height bounds would reduce the problem of listing the points of $X(\mathbb{Q})$ to a finite computation.)

Nevertheless there exist techniques that often succeed in determining $X(\mathbb{Q})$ for particular curves $X$; some of these are discussed by Poonen [1996; 2002]. Also, there are many other conjectural approaches towards a proof that $X(\mathbb{Q})$ can be determined explicitly for all curves $X$ over $\mathbb{Q}$. (See [Hindry and Silverman 2000, Section F.4.2] for a survey of some of these.) But none have yet been successful. We need some new ideas!

## 9. Further reading

For the reader who wants more, we suggest a few other books and survey articles. Within each topic, references assuming less background are listed first. Of course, there are many other books on these topics; those listed here were chosen partly because they are the ones the author is most familiar with.

**9.1. Elliptic curves.** The book [Silverman and Tate 1992] is an introduction to elliptic curves at the advanced undergraduate level; among other things, it contains a treatment of the elliptic curve factoring method. The graduate level textbook [Silverman 1986] uses more algebraic number theory and algebraic geometry, but most definitions and theorems are recalled as they are used, so the book is readable even by those with minimal background. Cremona's book [1997] contains extensive tables of elliptic curves over $\mathbb{Q}$, and discusses many elliptic curve algorithms in detail.

**9.2. Algebraic geometry.** Fulton's text [1969] requires only a knowledge of abstract algebra at the undergraduate level; it develops the commutative algebra as it goes along. Shafarevich's text, now in two volumes [1994a; 1994b], is a extensive survey of the main ideas of algebraic geometry and its connections to other areas of mathematics. It is intended for mathematicians outside algebraic geometry, but the topics covered are so diverse that even specialists are likely to find a few things that are new to them. Finally, Hartshorne's graduate text [1977], although more demanding, contains a thorough development of the language of schemes and sheaf cohomology, as well as applications to the theory of curves and surfaces, mainly over algebraically closed fields.

**9.3. Surveys on arithmetic geometry.** Mazur's article [1986] is intended for a general mathematical audience: it begins with a discussion of various diophantine problems, and works its way up to a sketch of Faltings' proof of the Mordell conjecture. Lang's book [1991] is a detailed survey of the tools and results of arithmetic geometry, mostly without proofs (but it too sketches Faltings' proof).

# References

[Bernstein 1998] D. J. Bernstein, "Detecting perfect powers in essentially linear time", *Math. Comp.* **67**:223 (1998), 1253–1283.

[Bombieri 1990] E. Bombieri, "The Mordell conjecture revisited", *Ann. Scuola Norm. Sup. Pisa Cl. Sci. (4)* **17**:4 (1990), 615–640. Errata in "The Mordell conjecture revisited", *Ann. Scuola Norm. Sup. Pisa Cl. Sci. (4)* **18**:3 (1991), 473.

[Brent 2000] R. P. Brent, "Recent progress and prospects for integer factorisation algorithms", pp. 3–22 in *Computing and combinatorics* (Sydney, 2000), edited by D.-Z. Du et al., Lecture Notes in Comput. Sci. **1858**, Springer, Berlin, 2000.

[Canfield et al. 1983] E. R. Canfield, P. Erdős, and C. Pomerance, "On a problem of Oppenheim concerning "factorisatio numerorum"", *J. Number Theory* **17**:1 (1983), 1–28.

[Cremona 1997] J. E. Cremona, *Algorithms for modular elliptic curves*, 2nd ed., Cambridge University Press, Cambridge, 1997.

[Elkies 1994] N. D. Elkies, "Heegner point computations", pp. 122–133 in *Algorithmic number theory* (Ithaca, NY, 1994), edited by L. M. Adleman and M.-D. Huang, Lecture Notes in Comput. Sci. **877**, Springer, Berlin, 1994.

[Faltings 1983] G. Faltings, "Endlichkeitssätze für abelsche Varietäten über Zahlkörpern", *Invent. Math.* **73**:3 (1983), 349–366. Translated in *Arithmetic geometry*, edited by G. Cornell and J. Silverman, Springer, New York-Berlin, 1986, 9–27.

[Fulton 1969] W. Fulton, *Algebraic curves: An introduction to algebraic geometry*, Addison-Wesley, Reading, MA, 1969. With the collaboration of Richard Weiss.

[Hartshorne 1977] R. Hartshorne, *Algebraic geometry*, Graduate Texts in Mathematics **52**, Springer, New York, 1977.

[Hindry and Silverman 2000] M. Hindry and J. H. Silverman, *Diophantine geometry: An introduction*, Graduate Texts in Mathematics **201**, Springer, New York, 2000.

[Hopcroft and Ullman 1969] J. E. Hopcroft and J. D. Ullman, *Formal languages and their relation to automata*, Addison-Wesley, Reading, MA, 1969.

[Koblitz 1984] N. Koblitz, *p-adic numbers, p-adic analysis, and zeta-functions*, 2nd ed., Graduate Texts in Mathematics **58**, Springer, New York, 1984.

[Lang 1991] S. Lang, *Number theory. III*, Encyclopaedia of Mathematical Sciences **60**, Springer, Berlin, 1991. Diophantine geometry.

[Lind 1940] C.-E. Lind, *Untersuchungen über die rationalen Punkte der ebenen kubischen Kurven vom Geschlecht Eins*, thesis, University of Uppsala, 1940.

[Martin and McMillen 2000] R. Martin and W. McMillen, "An elliptic curve over $\mathbb{Q}$ with rank at least 24", electronic announcement on the NMBRTHRY list server, posted May 2, 2000. Available at http://listserv.nodak.edu/archives/nmbrthry.html.

[Mazur 1986] B. Mazur, "Arithmetic on curves", *Bull. Amer. Math. Soc. (N.S.)* **14**:2 (1986), 207–259.

[Mordell 1922] L. J. Mordell, "On the rational solutions of the indeterminate equations of the third and fourth degrees", *Proc. Cambridge Phil. Soc.* **21** (1922), 179–192.

[Mordell 1969] L. J. Mordell, "On the magnitude of the integer solutions of the equation $ax^2 + by^2 + cz^2 = 0$", *J. Number Theory* **1** (1969), 1–3.

[Poonen 1996] B. Poonen, "Computational aspects of curves of genus at least 2", pp. 283–306 in *Algorithmic number theory* (Talence, 1996), edited by H. Cohen, Lecture Notes in Comput. Sci. **1122**, Springer, Berlin, 1996.

[Poonen 2002] B. Poonen, "Computing rational points on curves", pp. 149–172 in *Number theory for the millennium, III* (Urbana, IL, 2000), edited by M. A. Bennett et al., A K Peters, Natick, MA, 2002.

[Reichardt 1942] H. Reichardt, "Einige im Kleinen überall lösbare, im Grossen unlösbare diophantische Gleichungen", *J. Reine Angew. Math.* **184** (1942), 12–18.

[Schoof 1985] R. Schoof, "Elliptic curves over finite fields and the computation of square roots mod $p$", *Math. Comp.* **44**:170 (1985), 483–494.

[Selmer 1951] E. S. Selmer, "The diophantine equation $ax^3 + by^3 + cz^3 = 0$", *Acta Math.* **85** (1951), 203–362.

[Selmer 1954] E. S. Selmer, "The diophantine equation $ax^3 + by^3 + cz^3 = 0$: Completion of the tables", *Acta Math.* **92** (1954), 191–197.

[Serre 1973] J.-P. Serre, *A course in arithmetic*, Graduate Texts in Mathematics **7**, Springer, New York, 1973.

[Serre 1989] J.-P. Serre, *Lectures on the Mordell–Weil theorem*, Aspects of Mathematics **E15**, Friedr. Vieweg & Sohn, Braunschweig, 1989.

[Shafarevich 1994a] I. R. Shafarevich, *Basic algebraic geometry, 1: Varieties in projective space*, 2nd ed., Springer, Berlin, 1994.

[Shafarevich 1994b] I. R. Shafarevich, *Basic algebraic geometry, 2: Schemes and complex manifolds*, 2nd ed., Springer, Berlin, 1994.

[Silverman 1986] J. H. Silverman, *The arithmetic of elliptic curves*, Graduate Texts in Mathematics **106**, Springer, New York, 1986.

[Silverman and Tate 1992] J. H. Silverman and J. Tate, *Rational points on elliptic curves*, Undergraduate Texts in Mathematics, Springer, New York, 1992.

[Vojta 1991] P. Vojta, "Siegel's theorem in the compact case", *Ann. of Math.* (2) **133**:3 (1991), 509–548.

[Zimmerman 2008] P. Zimmerman, "50 largest factors found by ECM", web page, 2008. Available at http://www.loria.fr/-zimmerma/records/top100.html.

BJORN POONEN
DEPARTMENT OF MATHEMATICS
UNIVERSITY OF CALIFORNIA
BERKELEY, CA 94720-3840
UNITED STATES
    poonen@math.berkeley.edu

# The arithmetic of number rings

PETER STEVENHAGEN

ABSTRACT. We describe the main structural results on number rings, that is, integral domains for which the field of fractions is a number field. Whenever possible, we avoid the algorithmically undesirable hypothesis that the number ring in question is integrally closed.

## CONTENTS

| | |
|---|---|
| 1. Introduction | 209 |
| 2. Number rings | 211 |
| 3. Localization | 214 |
| 4. Invertible ideals | 215 |
| 5. Ideal factorization in number rings | 217 |
| 6. Integral closure | 220 |
| 7. Linear algebra over $\mathbb{Z}$ | 224 |
| 8. Explicit ideal factorization | 228 |
| 9. Computing the integral closure | 233 |
| 10. Finiteness theorems | 236 |
| 11. Zeta functions | 242 |
| 12. Computing class groups and unit groups | 244 |
| 13. Completions | 253 |
| 14. Adeles and ideles | 255 |
| 15. Galois theory | 258 |
| Acknowledgments | 263 |
| References | 264 |

## 1. Introduction

The ring $\mathbb{Z}$ of 'ordinary' integers lies at the very root of number theory, and when studying its properties, the concept of *divisibility* of integers naturally leads to basic notions as primality and congruences. By the 'fundamental theorem of arithmetic', $\mathbb{Z}$ admits *unique prime factor decomposition* of nonzero integers. Though one may be inclined to take this theorem for granted, its proof is not completely trivial: it usually employs the Euclidean algorithm to show that the prime numbers, which are defined as *irreducible* elements having only 'trivial' divisors, are *prime elements* that only divide a product of integers if they divide one of the factors.

In the time of Euler, it gradually became clear that in solving problems concerning integers, it can be effective to pass from $\mathbb{Z}$ to larger rings that are not contained in the field $\mathbb{Q}$ of rational numbers but in number fields, that is, in finite field extensions of $\mathbb{Q}$. Such *number rings*, which occur everywhere in this volume, will be our objects of study. In [Lenstra 2008], we encounter the classical example of the Pell equation $x^2 - dy^2 = 1$, which can be viewed as the equation $(x + y\sqrt{d})(x - y\sqrt{d}) = 1$ in the quadratic ring $\mathbb{Z}[\sqrt{d}]$. In a similar way, writing an integer $n$ as a sum $n = x^2 + y^2$ of two squares amounts to decomposing $n$ as a product $n = (x + yi)(x - yi)$ of two conjugate elements in the ring $\mathbb{Z}[i]$ of Gaussian integers. The cyclotomic number ring $\mathbb{Z}[\zeta_p]$ obtained by adjoining a primitive $p$-th root of unity $\zeta_p$ has been fundamental in studying the Fermat equation $x^p + y^p = z^p$ since the first half of the nineteenth century, and it occurs center stage in Mihăilescu's recent treatment [2006] of the Catalan equation $x^p - 1 = y^q$. See also [Schoof 2008a].

Whereas the ring $\mathbb{Z}[i]$ is in many respects similar to $\mathbb{Z}$, an interesting property of the rings $\mathbb{Z}[\sqrt{d}]$ arising in the study of the Pell equation is that, unlike $\mathbb{Z}$, they have an *infinite* unit group. Kummer discovered around 1850 that the Fermat equation for prime exponent $p \geq 3$ has no solutions in nonzero integers if $\mathbb{Z}[\zeta_p]$ admits factorization into prime elements. As we now know [Washington 1997, Chapter 11], only the rings $\mathbb{Z}[\zeta_p]$ with $p \leq 19$ have this property. All other rings $\mathbb{Z}[\zeta_p]$, and in fact all number rings, admit factorization into irreducible elements, but this is not very useful as it is often not in any way unique. Kummer and others invented a theory of prime *ideal* factorization to salvage this situation. It lies at the heart of the algebraic number theory developed during the nineteenth century.

In the early twentieth century, Hensel showed how to *complete* the ring $\mathbb{Z}$ and other number rings at their prime ideals. This gives rise to rings in $p$-adic or *local* fields, which are algebraically simpler than number fields and in certain ways similar to the archimedean complete fields $\mathbb{R}$ and $\mathbb{C}$ of real and complex numbers. It led to the introduction of various 'analytic' techniques and gave rise to the insight that many questions in number rings can be studied *locally*, much like geometers study curves by focusing on neighborhoods of points. Precise formulations require the description of number theoretic objects in the language of 'abstract algebra', the language of groups, rings and fields that has become fundamental in many parts of mathematics.

Contrary to what is sometimes thought, the use of more abstract theory and language is not at all incompatible with algorithmic approaches to algebraic number theory. The recent advances with respect to 'down to earth' problems as integer factoring [Stevenhagen 2008] and primality testing [Schoof 2008c] rely on 'large' number rings and on Galois extensions of finite rings to achieve their goal. In the first case, the number rings $\mathbb{Z}[\alpha]$ one encounters are not necessarily the 'textbook rings' for which the nineteenth century theory was developed, whereas in the second case, Galois theory for rings rather than for fields is exploited. These examples show that number rings have 'concrete' algorithmic applications to problems not traditionally inside the domain of algebraic number theory, and that such applications require a slightly more general theory than is found in the classical textbooks. Fortunately, commutative algebra and, more in particular, ring theory provide us with the tools that are needed for this. In the case of number rings, the number field sieve alluded to above shows that it is undesirable to have a theory that only works for rings of integers in number fields, which may be computationally inaccessible, and that one needs to consider 'singular' number rings as well. In this paper, we impose no a priori restrictions on our number rings and define them as *arbitrary* subrings of number fields. As a consequence, the various localizations of number rings we encounter are in this same category. Special attention will be devoted to *orders* in number fields as defined in the next section, which play an important role in algorithmic practice. The analogy between number rings and algebraic curves explains the geometric flavor of much of our terminology, but we do not formally treat the case of subrings of function fields [Rosen 2002].

Number rings are the central objects in computational algebraic number theory, and algorithms in more specific areas as class field theory [Cohen and Stevenhagen 2008] assume that one can efficiently deal with them. In this paper, which is mostly a survey of more or less classical algebraic number theory, we include the modest amount of ring theory that is necessary to state and prove the results in the generality required by algorithmic practice. The next section introduces number rings and orders, and explains their relation to the classical textbook ring, the ring of integers. In addition, it outlines the further contents of this paper.

## 2. Number rings

A *number ring* is a domain $R$ for which the field of fractions $K = Q(R)$ is a *number field*, that is, a field of finite degree over $\mathbb{Q}$. Note that this is a rather general definition, and that already inside $\mathbb{Q}$ there are infinitely many number rings, such as $\mathbb{Q}$ itself and $\mathbb{Z}[\frac{1}{2}, \frac{1}{3}]$. In many ways, $\mathbb{Z}$ is the 'natural' number ring in $\mathbb{Q}$ to work with, as it governs the 'arithmetic behavior' of $\mathbb{Q}$ in a sense we will make precise. In a similar way, the arithmetic properties of an arbitrary number field $K$ are classically described in terms of the *ring of integers*

$$\mathcal{O}_K = \{x \in K : f_{\mathbb{Q}}^x \in \mathbb{Z}[X]\} \tag{2-1}$$

of $K$, which consists of the elements $x \in K$ for which the monic irreducible polynomial $f_{\mathbb{Q}}^x$ over $\mathbb{Q}$, also known as the *minimal polynomial* of $x$ over $\mathbb{Q}$, has integer coefficients. This *integral closure* of $\mathbb{Z}$ in $K$ is a natural algebraic notion and, as will be shown just before Theorem 6.5, a *ring*; it may however be inaccessible in computational practice.

Already in the case of a quadratic field $K = \mathbb{Q}(\sqrt{d})$ associated to a nonsquare integer $d$, we need to write $d$ as $d = m^2 \cdot d_0$ with $d_0$ squarefree in order to find $\mathcal{O}_K$, which is equal to $\mathbb{Z}[\sqrt{d_0}]$ or, in the case $d_0 \equiv 1 \bmod 4$, to $\mathbb{Z}[(1+\sqrt{d_0})/2]$. The only way we know to find $d_0$ proceeds by factoring $d$, which we may not be able to do for large $d$. However, even if we are unable to find a non-trivial square factor dividing $d$, we can often use the number ring $R = \mathbb{Z}[\sqrt{d}]$ (or $R = \mathbb{Z}[(1+\sqrt{d})/2]$) instead of $\mathcal{O}_K$ for our algorithmic purposes. Of course, we do need to know in which ways the subring $R$ will be 'just as good' as $\mathcal{O}_K$ itself, and in which ways it may fail to behave nicely. In this quadratic case, the subrings of $\mathcal{O}_K$ are well understood, and there is a classical description of their arithmetic in terms of binary quadratic forms that goes back to Gauss. Among their algorithmic 'applications', one finds a subexponential factoring algorithm for arbitrary integers $d$, the *class group method* [Seysen 1987].

Our potential inability to find the square divisors of large integers is also a fundamental obstruction [Buchmann and Lenstra 1994] to computing $\mathcal{O}_K$ in other number fields $K$. Indeed, let $K$ be given as $K = \mathbb{Q}(\alpha)$, with $\alpha$ the root of some monic irreducible polynomial $f = f_{\mathbb{Q}}^\alpha \in \mathbb{Q}[X]$ of degree $n = [K:\mathbb{Q}]$. Replacing $\alpha$ by $k\alpha$ for a suitable integer $k$ when necessary, we may assume that $f$ has integral coefficients. Then the index of $R = \mathbb{Z}[\alpha] = \mathbb{Z}[X]/(f)$ in $\mathcal{O}_K$ is finite, and we show in (7-7) that its square divides the discriminant $\Delta(f)$ of the polynomial $f$. Finding $\mathcal{O}_K$ starts with finding squares dividing $\Delta(f)$, which may not be feasible if the integer $\Delta(f)$ is too large to be factored. Such discriminants occur for the polynomials that are used in the number field sieve, and they force us to work with subrings of $K$ that are possibly smaller than the ring of integers $\mathcal{O}_K$.

The *simple integral extensions* $\mathbb{Z}[\alpha]$ obtained by adjoining to $\mathbb{Z}$ a root $\alpha$ of some monic irreducible polynomial $f \in \mathbb{Z}[X]$, also known as *monogenic* number rings, are in many ways computationally convenient to work with. The 'power basis' $1, \alpha, \alpha^2, \ldots, \alpha^{n-1}$ of $K = \mathbb{Q}(\alpha)$ as a vector space over $\mathbb{Q}$ is also a basis for $\mathbb{Z}[\alpha]$ as a module over $\mathbb{Z}$. More generally, a subring $R \subset K$ that is free of rank $n = [K:\mathbb{Q}]$ over $\mathbb{Z}$ is called an *order* in $K$. An element $x \in K$ is integral if and only if $\mathbb{Z}[x]$ is an order in $K$, so $\mathbb{O}_K$ is the union of all orders $\mathbb{Z}[x] \subset K$. The following will be proved in Section 7, as a direct corollary of formula (7-6).

THEOREM 2.2. *A number ring $R \subset K$ is an order in $K$ if and only if it is of finite index in $\mathbb{O}_K$.*

This shows that $\mathbb{O}_K$ is the *maximal order* in $K$. It need not be monogenic (Example 8.6).

In an arbitrary number ring $R$, the role played in $\mathbb{Z}$ by the prime numbers is taken over by the nonzero prime ideals or *primes* of $R$. Every ideal in $R$ containing $\alpha$ contains the multiples of the integer $f_\mathbb{Q}^\alpha(0)$, so nonzero ideals in $R$ 'divide' ordinary integers in the ideal-theoretic sense of the word. As $R/kR$ is finite of order at most $k^n = k^{[K:\mathbb{Q}]}$ for $k \in \mathbb{Z}_{\geq 1}$, with equality in the case that $R$ is an order, every nonzero ideal in a number ring is of finite index. In particular, all $R$-ideals are finitely generated, and all primes of $R$ are maximal. In ring-theoretic terms, number rings are *noetherian domains* of *dimension* at most 1. Every prime $\mathfrak{p} \subset R$ contains a unique prime number $p$, the characteristic of the finite field $R/\mathfrak{p}$. We say that $\mathfrak{p}$ *extends* $p$ or *lies over* $p$, and call the degree $f(\mathfrak{p}/p)$ of $R/\mathfrak{p}$ over the prime field $\mathbf{F}_p$ the *residue class degree* of $\mathfrak{p}$ over $p$.

In the next three sections, we describe ideal factorization in arbitrary number rings $R$ (Theorems 5.2 and 5.3), which turns out to be especially nice if $R$ equals or contains the ring of integers $\mathbb{O}_K$ of its field of fractions (Theorems 5.7 and 6.5). Some linear algebra over $\mathbb{Z}$ (Section 7) is involved in explicit factorization (Section 8), and the local computations involved in factoring rational primes in $R$ lead to algorithms to find $\mathbb{O}_K$ starting from an order $R$ (Section 9).

When using ideal factorization as a replacement for element factorization, the need arises to control the difference between elements and ideals. The problem of *nonprincipality* of ideals is quantified by the Picard group of the number ring introduced in Section 4. The problem of element identities 'up to units' arising from ideal arithmetic necessitates control of the unit groups of number rings. The classical *finiteness theorems* in Section 10 show that the Picard groups of number rings are finite (Corollary 10.6) and that the unit groups of many number rings are finitely generated (Theorem 10.9). The proofs of these theorems, which do not hold for arbitrary noetherian domains of dimension 1, exploit embeddings of number rings and their unit groups as lattices in Euclidean vector spaces. They are not directly constructive, and based on the *geometry of numbers*. Section 12 presents various explicit examples showing how the relevant groups may be computed using the explicit factorization techniques from Section 8. To guarantee that no units or Picard group relations have been overlooked, one uses the *analytic* information from Section 11 on the size of Picard and unit groups for $\mathbb{O}_K$; this information is encoded by the *zeta function* of the underlying number field. Relating the Picard group of $R$ to that of the ring of integers is made possible by the comparison statements Theorem 6.5 and 6.7.

Although actual computations are by their finite nature bound to process rational or algebraic numbers only, many mathematical *ideas* are most elegantly expressed in terms of 'limit objects' such as real numbers, which can only be approximated by computers. Number rings are naturally embedded in *local fields* (Section 13) and *adele rings* (Section 14), which are similar in nature to real numbers and provide a conceptual clarification of our local approach to number rings. The final Section 15 deals with Galois theoretic aspects of number rings.

## 3. Localization

When dealing algorithmically with a number field $K$ defined by some monic polynomial $f \in \mathbb{Z}[X]$, one often starts out with the simple order $R = \mathbb{Z}[\alpha]$ defined by $f$, and enlarges it to a bigger order whenever this is made possible by computations. This makes it important to carry over knowledge about, say, the primes over $p$ in $R$ to similar information in the larger ring. In this case, 'nothing changes' as long as the index of $R$ in the extension ring is coprime to $p$. Such statements are most conveniently made precise and proved by working 'locally', in a way that was already familiar to geometers in the nineteenth century. Algebraically, the corresponding process of localization of rings and modules [Atiyah and Macdonald 1969, Chapter 2] has become a standard procedure.

For a number ring $R$ with field of fractions $K$, one can form a localized ring

$$S^{-1}R = \{r/s \in K : r \in R, \ s \in S\} \subset K$$

whenever $S \subset R$ is a subset containing 1 that is closed under multiplication. There is a localization $K = Q(R)$ corresponding to $S = R \setminus \{0\}$, and, more generally, by taking $S = R \setminus \mathfrak{p}$, we have localizations

$$R_\mathfrak{p} = \{r/s \in K : r \in R, \ s \notin \mathfrak{p}\}$$

at all prime ideals $\mathfrak{p}$ of $R$. The number rings $R_\mathfrak{p}$ are *local number rings* in the sense that they have a unique maximal ideal

$$\mathfrak{p} R_\mathfrak{p} = \{r/s \in K : r \in \mathfrak{p}, \ s \notin \mathfrak{p}\}$$

consisting of the complement of the unit group $R_\mathfrak{p}^* = \{\frac{r}{s} \in K : r, s \notin \mathfrak{p}\}$. Conversely, a local number ring $R$ with maximal ideal $\mathfrak{p}$ is equal to its localization $R_\mathfrak{p}$ at $\mathfrak{p}$.

EXAMPLE 3.1. For $R = \mathbb{Z}$, the localization $\mathbb{Z}_{(p)} = \{r/s \in \mathbb{Q} : p \nmid s\}$ at the prime $p$ is a local ring with maximal ideal $\mathfrak{p} = p\mathbb{Z}_{(p)}$. Every fraction $x \in \mathbb{Q}^*$ can uniquely be written as $x = up^k$ with $u \in \mathbb{Z}_{(p)}^* = \{r/s \in \mathbb{Q} : p \nmid rs\}$ and $k \in \mathbb{Z}$. It follows that the ideals of $\mathbb{Z}_{(p)}$ are simply the powers of the principal ideal $p$, and this makes $\mathbb{Z}_{(p)}$ into the prototype of what is known as a *discrete valuation ring*. As will become clear in Proposition 5.4, these particularly simple rings arise as the localizations of a number ring $R$ at all of its 'regular' primes.

Localization often enables us to reduce the complexity of a number ring $R$ at hand by passing to a localized ring $S^{-1}R$. The ideals of $S^{-1}R$ are of the form $S^{-1}I = \{i/s : i \in I, s \in S\}$, with $I$ an ideal of the *global* ring $R$, and whenever $I \cap S$ is nonempty we have $S^{-1}I = S^{-1}R$. The primes of $S^{-1}R$ are the ideals $S^{-1}\mathfrak{p}$ with $\mathfrak{p}$ a prime of $R$ that does not meet $S$, and the natural map $R \to S^{-1}R$ induces an isomorphism between the local rings at $\mathfrak{p}$ and at $S^{-1}\mathfrak{p}$, respectively. If $R \subset R'$ is of finite index, we have $S^{-1}R = S^{-1}R' \subset K$ for all localizations for which the index is in $S$.

EXAMPLE 3.2. Taking $R$ an arbitrary number ring and $S = \{x \in \mathbb{Z} : p \nmid x\}$, as in Example 3.1, we obtain a *semilocal* number ring $R_{(p)}$ having only finitely many primes $\mathfrak{p}$, all containing $p$. The primes of $R_{(p)}$ correspond to the primes of $R$ lying over $p$. If $R$ is of finite index in $\mathcal{O}_K$ and $p$ is a prime number not dividing this index, the inclusion map $R \to \mathcal{O}_K$ becomes the identity when localized at $S$, and the local rings of $R$ and $\mathcal{O}_K$ are naturally isomorphic at primes over $p$. Informally phrased, $R$ and $\mathcal{O}_K$ are 'the same' at all primes that do not divide the index $[\mathcal{O}_K : R]$. Section 6 contains a more precise formulation of these statements.

## 4. Invertible ideals

As Kummer discovered, it is not in general possible to factor a nonzero element $x$ in a number ring $R$ into prime divisors, but something similar can be obtained when looking at 'ideal divisors' of $x$, that is, the ideals $I \subset R$ that satisfy $IJ = (x)$ for some $R$-ideal $J$.

For the modern reader, ideals are defined more generally as kernels of ring homomorphisms, and in this setting $I$ is said to *divide* $J$ whenever $I$ contains $J$. For a number ring $R$ with field of fractions $K$, it is convenient to slightly extend the concept of $R$-ideals and consider *fractional $R$-ideals*, that is, $R$-submodules $I \subset K$ with the property that $rI$ is a *nonzero* $R$-ideal for some $r \in R$. If we can take $r = 1$, then $I$ is an ordinary $R$-ideal, usually referred to as an *integral $R$-ideal*. For fractional ideals $I$ and $J$, we define the *ideal quotient* as

$$I : J = \{x \in K : xJ \subset I\}.$$

A standard verification shows that the sum, intersection, product, and quotient of two fractional ideals are again fractional ideals. For every fractional $R$-ideal $I$, the localized ideal $S^{-1}I$ is a fractional $S^{-1}R$-ideal. Moreover, localization respects the standard operations on ideals of taking sums, products, and intersections.

A fractional $R$-ideal $I$ is said to be *invertible* if there exists an $R$-ideal $J$ such that $IJ$ is a nonzero principal $R$-ideal. These 'ideals in Kummer's sense' are precisely the ones we need to 'factor' nonzero elements of $R$, and for which ideal multiplication gives rise to a *group operation*. If $I$ is invertible, we have $I \cdot I^{-1} = R$ for the fractional $R$-ideal

$$I^{-1} = R : I = \{x \in K : xI \subset R\},$$

and its *multiplier ring*

$$\Lambda(I) = \{x \in K : xI \subset I\},$$

which clearly contains $R$, is actually equal to $R$ as we have

$$\Lambda(I) = \Lambda(I) I \cdot I^{-1} \subset I \cdot I^{-1} = R.$$

The invertible fractional $R$-ideals form an abelian group $\mathcal{I}(R)$ under ideal multiplication. Clearly, all principal fractional $R$-ideals are invertible, and they form a subgroup $\mathcal{P}(R) = \{xR : x \in K^*\} = K^*/R^*$ of $\mathcal{I}(R)$.

For a principal ideal domain $R$ such as $R = \mathbb{Z}$, *all* fractional ideals are of the form $xR$ with $x \in K^*$, and we have $\mathcal{I}(R) = \mathcal{P}(R)$. For arbitrary number rings $R$, the quotient group $\mathrm{Pic}(R) = \mathcal{I}(R)/\mathcal{P}(R)$ measuring the difference between invertible and principal $R$-ideals is known as the *Picard group* or the *class group* of $R$. It fits in an exact sequence

$$1 \longrightarrow R^* \longrightarrow K^* \longrightarrow \mathcal{I}(R) \longrightarrow \mathrm{Pic}(R) \longrightarrow 1. \tag{4-1}$$

If $R = \mathcal{O}_K$ is the ring of integers of $K$, then $\mathrm{Pic}(\mathcal{O}_K)$, which only depends on $K$, is often referred to as the 'class group of $K$' and is denoted by $\mathrm{Cl}_K$. It is a fundamental invariant of the number field $K$.

The Picard group of a number ring $R$ vanishes if $R$ is a principal ideal domain. The converse statement does not hold in general: a number ring with trivial Picard group may have noninvertible ideals that are nonprincipal.

EXAMPLES 4.2. In the quadratic field $\mathbb{Q}(\sqrt{-3})$, the ring of integers $\mathcal{O} = \mathbb{Z}[\alpha]$ with $\alpha = (1 + \sqrt{-3})/2$ is a principal ideal domain as it admits a Euclidean 'division with remainder' with respect to the complex absolute value. In other words: for nonzero $\beta, \gamma \in \mathcal{O}$ there exist $q, r \in \mathcal{O}$ with $\beta/\gamma = q + r/\gamma$ and $|r| < |\gamma|$. In a picture, this boils down to the observation that the open disks of radius 1 around the points of $\mathcal{O}$ in the complex plane cover all of $\mathbb{C}$.

Taking $R = \mathbb{Z}[\sqrt{-3}]$ instead of $\mathcal{O}$, the $R$-translates of $\alpha$ are outside the open disks of radius 1 around $R$, and we find that every fractional $R$-ideal is either principal or of the form $xI$ with $I = \mathbb{Z} + \mathbb{Z} \cdot \alpha$. The nonprincipal ideals $xI$ have multiplier ring $\Lambda(xI) = \Lambda(I) = \mathbb{Z}[\alpha] \supsetneq R$, so they are not invertible, and we still have $\text{Pic}(R) = 0$. The prime ideal $\mathfrak{p} = 2I = \mathbb{Z} \cdot 2 + \mathbb{Z} \cdot (1 + \sqrt{-3})$ of $R$ of index 2 satisfies $\mathfrak{p}^2 = 2\mathfrak{p}$, which shows that $\mathfrak{p}$ is not invertible and that multiplication of arbitrary ideals in $R$ is not a 'group-like' operation.

For the order $\mathcal{O} = \mathbb{Z}[\sqrt{-5}]$, which is the ring of integers in $\mathbb{Q}(\sqrt{-5})$, the analogous picture shows that every fractional $\mathcal{O}$-ideal is either principal or of the form $xI$ with $I = \mathbb{Z} + \mathbb{Z} \cdot (1 + \sqrt{-5})/2$. In this case $\mathfrak{p} = 2I \subset \mathcal{O}$ satisfies $\mathfrak{p}^2 = 2\mathcal{O}$, which shows that *all* fractional $\mathcal{O}$-ideals are invertible and that $\text{Pic}(\mathcal{O})$ is cyclic of order 2.

If $R$ is a number ring and $I$ a fractional $R$-ideal, the localized ideals $I_\mathfrak{p}$ at the primes of $R$ are fractional $R_\mathfrak{p}$-ideals, and they are equal to $R_\mathfrak{p}$ for almost all $\mathfrak{p}$. The ideal $I$ can be recovered from its localizations as we have

$$I = \bigcap_\mathfrak{p} I_\mathfrak{p}. \tag{4-3}$$

To obtain the non-trivial inclusion $\supset$, note that for $x \in \bigcap_\mathfrak{p} I_\mathfrak{p}$, the ideal $\{r \in R : rx \in I\}$ equals $R$ as it is not contained in any prime of $R$.

PROPOSITION 4.4. *Let $R$ be a number ring and $I$ a fractional $R$-ideal. Then $I$ is invertible if and only if $I_\mathfrak{p}$ is a principal $R_\mathfrak{p}$-ideal for all primes $\mathfrak{p}$.*

PROOF. If $I$ is invertible, there exist $x_i \in I$ and $y_i \in I^{-1}$ with $\sum_{i=1}^n x_i y_i = 1$. Let $\mathfrak{p} \subset R$ be a prime. All terms $x_i y_i$ are in $R \subset R_\mathfrak{p}$, and they cannot all be in the maximal ideal of $R_\mathfrak{p}$. Suppose that we have $x_1 y_1 \in R_\mathfrak{p}^* = R_\mathfrak{p} \setminus \mathfrak{p} R_\mathfrak{p}$. Then any $x \in I$ can be written as $x = x_1 \cdot (x_1 y_1)^{-1} \cdot x y_1 \in x_1 R_\mathfrak{p}$. It follows that $I_\mathfrak{p}$ is principal with generator $x_1$.

For the converse, we argue by contradiction. If $I$ is not invertible, there exists a prime $\mathfrak{p}$ containing the ideal $II^{-1} \subset R$. Let $x \in I$ be an $R_\mathfrak{p}$-generator of $I_\mathfrak{p}$. If $I$ is generated over $R$ by $x_i$ for $i = 1, 2, \ldots, n$, we can write $x_i = x(r_i/s) \in R_\mathfrak{p}$, with $s \in R \setminus \mathfrak{p}$ independent of $i$. Then we have $sx^{-1} x_i = r_i \in R$ for all $i$, whence $sx^{-1} I \subset R$. We obtain $s = x \cdot sx^{-1} \in II^{-1} \subset \mathfrak{p}$, a contradiction. □

By Proposition 4.4, we may view the Picard group as a local-global obstruction group measuring the extent to which the locally principal $R$-ideals are globally principal.

## 5. Ideal factorization in number rings

It is not generally true that nonzero ideals in number rings, invertible or not, can be factored into a product of prime ideals. We can however use (4-3) to decompose $I \subset R$ into its $\mathfrak{p}$-*primary parts*

$$I_{(\mathfrak{p})} = I_\mathfrak{p} \cap R$$

at the various primes $\mathfrak{p}$ of $R$. We have $I_{(\mathfrak{p})} = R$ if $\mathfrak{p}$ does not divide $I$.

LEMMA 5.1. *Let $R_\mathfrak{p}$ be a local number ring. Then every nonzero ideal of $R_\mathfrak{p}$ contains some power of the maximal ideal of $R_\mathfrak{p}$.*

PROOF. As $R_\mathfrak{p}$ is noetherian, we can apply *noetherian induction*: if there are counterexamples to the lemma, the set of such ideals contains an element $I$ that is maximal with respect to the ordering by inclusion. Then $I$ is not prime, as the only nonzero prime ideal of $R_\mathfrak{p}$ is the maximal ideal. Let $x, y \in R \setminus I$ satisfy $xy \in I$. Then $I + (x)$ and $I + (y)$ strictly contain $I$, so they do satisfy the conclusion of the lemma and contain a power of the maximal ideal. The same then holds for $(I + (x))(I + (y)) \subset I$. Contradiction. □

By Lemma 5.1, the $\mathfrak{p}$-primary part $I_{(\mathfrak{p})}$ of a nonzero ideal $I \subset R$ contains some power of $\mathfrak{p}$. As there are no inclusions between different primes in number rings, this implies that $\mathfrak{p}$-primary parts at different $\mathfrak{p}$ are coprime.

THEOREM 5.2. *Let $R$ be a number ring. Then every nonzero ideal $I \subsetneq R$ has a primary decomposition $I = \prod_{\mathfrak{p} \supset I} I_{(\mathfrak{p})}$, and we have natural isomorphisms*

$$R/I \xrightarrow{\sim} \prod_{\mathfrak{p} \supset I} R/I_{(\mathfrak{p})} \xrightarrow{\sim} \prod_{\mathfrak{p} \supset I} R_\mathfrak{p}/I_\mathfrak{p}.$$

PROOF. We have $I = \bigcap_\mathfrak{p} I_\mathfrak{p} = \bigcap_\mathfrak{p} I_{(\mathfrak{p})}$ by (4-3), and we may take the intersections over those $\mathfrak{p}$ that contain $I$ only. By the coprimality of $\mathfrak{p}$-primary parts, the finite intersection obtained is actually a *product* $I = \prod_{\mathfrak{p} \supset I} I_{(\mathfrak{p})}$.

The isomorphism $R/I \xrightarrow{\sim} \prod_{\mathfrak{p} \supset I} R/I_{(\mathfrak{p})}$ is a special case of the Chinese remainder theorem for a product of pairwise coprime ideals. The localization map $R/I_{(\mathfrak{p})} \to R_\mathfrak{p}/I_\mathfrak{p}$ is injective by the definition of $I_{(\mathfrak{p})}$; for its surjectivity, we show that every $s \in R \setminus \mathfrak{p}$ is a unit in $R/I_{(\mathfrak{p})}$. By the maximality of $\mathfrak{p}$, there is an element $s' \in R \setminus \mathfrak{p}$ such that $ss' - 1$ is in $\mathfrak{p}$. As $I_{(\mathfrak{p})}$ contains $\mathfrak{p}^n$ for some $n$ by Lemma 5.1, the element $ss' - 1$ is nilpotent in $R/I_{(\mathfrak{p})}$, so $s$ is a unit. □

For *invertible* ideals, we can decompose the *group* $\mathcal{I}(R)$ of invertible fractional $R$-ideals in a similar way into $\mathfrak{p}$-primary components. By Proposition 4.4, invertible ideals are locally principal at each $\mathfrak{p}$, giving rise to elements of $\mathcal{P}(R_\mathfrak{p})$.

THEOREM 5.3. *Let $R$ be a number ring. Then we have an isomorphism*

$$\phi : \mathcal{I}(R) \xrightarrow{\sim} \bigoplus_{\mathfrak{p} \text{ prime}} \mathcal{P}(R_\mathfrak{p})$$

*that maps $I$ to its vector of localizations $(I_\mathfrak{p})_\mathfrak{p}$ at the primes $\mathfrak{p}$ of $R$.* □

To proceed from primary decomposition to prime ideal factorization, it is necessary that the localizations $I_\mathfrak{p}$ or, equivalently, the $\mathfrak{p}$-primary parts $I_{(\mathfrak{p})}$ of an ideal $I$ are powers of $\mathfrak{p}$. For the local rings $\mathbb{Z}_{(p)}$ of $R = \mathbb{Z}$, this is the case by

Example 3.1. For general number rings $R$, this depends on the nature of the local rings $R_\mathfrak{p}$.

PROPOSITION 5.4. *For a prime $\mathfrak{p}$ of a number ring $R$, the three following are equivalent*:

(1) $\mathfrak{p}$ *is an invertible $R$-ideal*;
(2) $R_\mathfrak{p}$ *is a principal ideal domain, and every $R_\mathfrak{p}$-ideal is a power of $\mathfrak{p} R_\mathfrak{p}$*;
(3) *there exists $\pi \in R_\mathfrak{p}$ such that every $x \in K^*$ can uniquely be written as $x = u \cdot \pi^k$ with $u \in R_\mathfrak{p}^*$ and $k \in \mathbb{Z}$.*

PROOF. For (1) $\Rightarrow$ (2), we use Proposition 4.4 to write $\mathfrak{p} R_\mathfrak{p} = \pi R_\mathfrak{p}$ and observe that all inclusions in the chain of principal $R_\mathfrak{p}$-ideals $R_\mathfrak{p} \supset \mathfrak{p} R_\mathfrak{p} = (\pi) \supset (\pi^2) \supset (\pi^3) \supset \ldots$ are strict: an equality $(\pi^n) = (\pi^{n+1})$ would imply $\pi^n = r\pi^{n+1}$ for some $r \in R_\mathfrak{p}$, whence $r\pi = 1$ and $\pi \in R_\mathfrak{p}^*$. We need to show there are no further $R_\mathfrak{p}$-ideals. Let $I \subset R_\mathfrak{p}$ be a nonzero ideal. As $I$ contains all sufficiently large powers of $(\pi)$ by Lemma 5.1, there is a largest value $n \geq 0$ for which we have $(\pi^n) \supset I$. Take any $r \in I \setminus (\pi^{n+1})$; then we have $r = a\pi^n$ with $a \notin (\pi)$. This implies that $a$ is a unit in $R_\mathfrak{p}$, so we have $(r) = (\pi^n) \subset I \subset (\pi^n)$ and $I = (\pi^n)$.

For (2) $\Rightarrow$ (3), take for $\pi$ a generator of $\mathfrak{p} R_\mathfrak{p}$. For every $x \in R_\mathfrak{p}$, we have $(x) = (\pi^k)$ for some uniquely determined integer $k \geq 0$, and $x = u \cdot \pi^k$ with $u \in R_\mathfrak{p}^*$. Taking quotients, we obtain (3).

For (3) $\Rightarrow$ (1), we note that we have $\pi \notin R_\mathfrak{p}^*$ and therefore

$$R_\mathfrak{p} = \{u \cdot \pi^k : u \in R_\mathfrak{p}^* \text{ and } k \geq 0\} \cup \{0\}.$$

This shows $R_\mathfrak{p}$ is local with principal maximal ideal $(\pi)$; so $\mathfrak{p}$ is invertible. $\square$

If the conditions in Proposition 5.4 are met, we call $\mathfrak{p}$ a *regular* prime of $R$ and the local number ring $R_\mathfrak{p}$ a *discrete valuation ring*. The exponent $k$ in (3) is then the *order* $\mathrm{ord}_\mathfrak{p}(x)$ to which $\mathfrak{p}$ occurs in $x \in K^*$, and the associated map $x \mapsto \mathrm{ord}_\mathfrak{p}(x)$ is the *discrete valuation* from which $R$ derives its name. It is a homomorphism $K^* \to \mathbb{Z}$ satisfying

$$\begin{aligned}\mathrm{ord}_\mathfrak{p}(xy) &= \mathrm{ord}_\mathfrak{p}(x) + \mathrm{ord}_\mathfrak{p}(y), \\ \mathrm{ord}_\mathfrak{p}(x+y) &\geq \min\{\mathrm{ord}_\mathfrak{p}(x), \mathrm{ord}_\mathfrak{p}(y)\}\end{aligned} \quad (5\text{-}5)$$

for all $x, y \in K$. We formally put $\mathrm{ord}_\mathfrak{p}(0) = +\infty$. With this convention, we have

$$R_\mathfrak{p} = \{x \in K : \mathrm{ord}_\mathfrak{p}(x) \geq 0\}. \quad (5\text{-}6)$$

If $x \in K$ has negative valuation, we have $R_\mathfrak{p}[x] = K$, so a discrete valuation ring in $K$ is a *maximal* subring of $K$ different from $K$.

If all primes of a number ring $R$ are regular, $R$ is said to be a *Dedekind domain*. As the next section will show, all rings of integers in number fields are

Dedekind domains. For such rings, the primary decomposition from Theorems 5.2 and 5.3 becomes a true prime ideal factorization, as all localizations $I_\mathfrak{p}$ of any fractional $R$-ideal $I$ are then of the form $\mathfrak{p}^k$ for some exponent $k = \mathrm{ord}_\mathfrak{p}(I) \in \mathbb{Z}$, the *valuation* of $I$ at $\mathfrak{p}$, that is equal to 0 for almost all $\mathfrak{p}$.

THEOREM 5.7. *For a number ring $R$ that is Dedekind, there is an isomorphism*

$$\mathcal{I}(R) \xrightarrow{\sim} \bigoplus_{\mathfrak{p}} \mathbb{Z}$$

$$I \longmapsto (\mathrm{ord}_\mathfrak{p}(I))_\mathfrak{p},$$

*and every $I \in \mathcal{I}(R)$ factors uniquely as a product $I = \prod_\mathfrak{p} \mathfrak{p}^{\mathrm{ord}_\mathfrak{p}(I)}$.* □

If $R$ is Dedekind, then *all* fractional $R$-ideals are invertible, and the Picard group $\mathrm{Pic}(R)$ is better known as the *class group* $\mathrm{Cl}(R)$ of $R$. It vanishes if and only if the Dedekind domain $R$ is a principal ideal domain.

## 6. Integral closure

A number ring $R$ is Dedekind if all of its localizations $R_\mathfrak{p}$ are discrete valuation rings, and in this case we have prime ideal factorization as in Theorem 5.7. Algorithmically, it may not be easy to test whether a given number ring is Dedekind. Theoretically, however, every number ring $R$ has a unique extension $R \subset \mathcal{O}$ inside its field of fractions $K$ that is of finite index over $R$, and is regular at all primes. This *normalization* of $R$ is the smallest Dedekind domain containing $R$, and represents what geometers would call a 'desingularization' of $R$.

The normalization of a number ring $R$ is defined as the *integral closure* of $R$ in its field of fractions $K$. It consists of those $x \in K$ that are *integral* over $R$, that is, for which there exists a monic polynomial $f \in R[X]$ with $f(x) = 0$. If $R$ equals its integral closure, it is said to be *integrally closed*. This is a 'local property'.

PROPOSITION 6.1. *A number ring $R$ is integrally closed if and only if all of the localizations $R_\mathfrak{p}$ at its primes $\mathfrak{p}$ are integrally closed.*

PROOF. Note that $R$ and its localizations have the same field of fractions $K$. If $x \in K$ is integral over $R$, it is obviously integral over all $R_\mathfrak{p}$. If all $R_\mathfrak{p}$ are integrally closed, we then have $x \in \bigcap_\mathfrak{p} R_\mathfrak{p} = R$ by (4-3), so $R$ is integrally closed.

Conversely, suppose $x \in K$ satisfies an integrality relation $x^n = \sum_{k=0}^{n-1} r_k x^k$ with $r_k \in R_\mathfrak{p}$ for some $\mathfrak{p}$. If $s \in R \setminus \mathfrak{p}$ is chosen such that we have $s r_k \in R$ for all $k$, multiplication by $s^n$ yields an integrality relation $(sx)^n = \sum_{k=0}^{n-1} r_k s^{n-k} (sx)^k$ for $sx$ with coefficients $r_k s^{n-k} \in R$. If $R$ is integrally closed, we have $sx \in R$ and therefore $x \in R_\mathfrak{p}$. Thus $R_\mathfrak{p}$ is integrally closed. □

PROPOSITION 6.2. *A local number ring is integrally closed if and only if it is a discrete valuation ring.*

COROLLARY 6.3. *A number ring is Dedekind if and only if it is integrally closed.*

Corollary 6.3 is immediate from Propositions 6.1 and 6.2. To prove 6.2, it is convenient to rephrase the definition of integrality in the following way.

LEMMA 6.4. *An element $x \in K$ is integral over $R$ if and only if there exists a finitely generated $R$-module $M \subset K$ with $M \neq 0$ and $xM \subset M$.*

PROOF. For integral $x$, the ring $R[x]$ is finitely generated as an $R$-module and yields a module $M$ of the required sort. For the converse, observe that the inclusion $xM \subset M$ for $M = Rm_1 + \cdots + Rm_n$ gives rise to identities $xm_i = \sum_{j=1}^{n} r_{ij} m_j$ for $j = 1, 2, \ldots, n$. As the $n \times n$ matrix $A = x \cdot \mathrm{id}_n - (r_{ij})_{i,j=1}^{n}$ with entries in $K$ maps the nonzero vector $(m_i)_i \in K^n$ to zero, we have $\det(A) = 0$, resulting in an integrality relation $x^n + \sum_{k=0}^{n-1} r_k x^k = 0$ for $x$. □

PROOF OF PROPOSITION 6.2. If $R \subset K$ is a discrete valuation ring, $\mathrm{ord}_\mathfrak{p}$ is the associated valuation, and $x \in K$ satisfies $\mathrm{ord}_\mathfrak{p}(x) < 0$, then a relation $x^n = \sum_{k=0}^{n-1} r_k x^k$ with $r_k \in R$ cannot hold since the valuation of the left hand side is by (5-5) and (5-6) smaller than that of the right hand side. This shows that $R$ is integrally closed.

Conversely, let $R$ be an integrally closed local number ring with maximal ideal $\mathfrak{p}$, and pick a nonzero element $a \in \mathfrak{p}$. By Lemma 5.1, there exists a smallest positive integer $n$ for which $\mathfrak{p}^n$ is contained in $aR$. Choose $b \in \mathfrak{p}^{n-1} \setminus aR$, and take $\pi = a/b$. By construction, we have $\pi^{-1} = b/a \notin R$ and $\pi^{-1}\mathfrak{p} \subset R$. As $\mathfrak{p}$ is a finitely generated $R$-module and $\pi^{-1} = b/a$ is not integral over $R$, we see from Lemma 6.4 that we cannot have $\pi^{-1}\mathfrak{p} \subset \mathfrak{p}$. It follows that $\pi^{-1}\mathfrak{p}$ equals $R$, so we have $\mathfrak{p} = \pi R$, and $R$ is a discrete valuation ring. □

Using Lemma 6.4, it is easy to see that the integral closure $\mathcal{O}$ of a number ring $R$ is indeed a ring: for $R$-integral $x, y \in K$, the finitely generated module $M = R[x, y]$ is multiplied into itself by $x \pm y$ and $xy$. Moreover, if $x \in K$ is integral over $\mathcal{O}$ and $M \subset \mathcal{O}$ is the $R$-module generated by the coefficients of an integrality relation for $x$ over $\mathcal{O}$, then $R[x] \cdot M$ is a finitely generated $R$-module that is multiplied into itself by $x$. Thus $x$ is integral over $R$ and contained in $\mathcal{O}$. This shows that $\mathcal{O}$, or, more generally, the integral closure in $K$ of *any* subring of $K$, is integrally closed. Clearly, $\mathcal{O}$ is the smallest Dedekind domain containing $R$.

THEOREM 6.5. *The integral closure $\mathcal{O}$ of a number ring $R$ in $K = Q(R)$ equals*

$$\mathcal{O} = R\mathcal{O}_K = \mathcal{O}_{K,T} = \{x \in K : \mathrm{ord}_\mathfrak{p}(x) \geq 0 \text{ for all } \mathfrak{p} \notin T\}$$

*for some set $T = T(R)$ of primes of $\mathcal{O}_K$.*

PROOF. It is clear that $\mathcal{O}$ contains $R$ and $\mathcal{O}_K$, and therefore $R\mathcal{O}_K$. To see that $R\mathcal{O}_K$ is a Dedekind domain and therefore equal to $\mathcal{O}$, we note that any of its localizations $(R\mathcal{O}_K)_\mathfrak{p}$ contains as a subring the localization of $\mathcal{O}_K$ at $\mathcal{O}_K \cap \mathfrak{p}$. This is a discrete valuation ring and, by the maximality of discrete valuation rings, it is equal to $(R\mathcal{O}_K)_\mathfrak{p}$.

We find that the primes of $\mathcal{O}$ correspond to a subset of the primes of $\mathcal{O}_K$, and that the local rings at corresponding primes coincide. Describing these as in (5-6), we arrive at the given description of $\mathcal{O}$ as the intersection of its localizations. □

We say that the ring $\mathcal{O}_{K,T}$ in Theorem 6.5 is obtained from $\mathcal{O}_K$ by 'inverting the primes in $T$'. The set $T = T(R)$ consists of those primes $\mathfrak{p}$ of $\mathcal{O}_K$ for which $R$ is not contained in the localization of $\mathcal{O}_K$ at $\mathfrak{p}$. It is empty if and only if $R$ is an order.

The class group of the Dedekind domain $\mathcal{O}_{K,T}$ can be obtained from $\text{Cl}(\mathcal{O}_K) = \text{Cl}_K$. The localization map $\mathcal{O}_K \to \mathcal{O}_{K,T}$ yields a natural map $I \mapsto I \cdot \mathcal{O}_{K,T}$ from $\mathcal{I}(\mathcal{O}_K)$ to $I(\mathcal{O}_{K,T})$ which maps principal ideals to principal ideals. This induces a homomorphism between the defining sequences (4-1) for their class groups, and the 'middle map' $\mathcal{I}(\mathcal{O}_K) \to I(\mathcal{O}_{K,T})$ is by Theorem 5.7 the natural surjection $\bigoplus_\mathfrak{p} \mathbb{Z} \to \bigoplus_{\mathfrak{p} \notin T} \mathbb{Z}$ with kernel $\bigoplus_{\mathfrak{p} \in T} \mathbb{Z}$. It follows that the natural map

$$\text{Cl}_K \to \text{Cl}(\mathcal{O}_{K,T})$$

is surjective, and an easy application of the snake lemma [Lang 2002, Section III.9] yields the exact sequence

$$1 \longrightarrow \mathcal{O}_K^* \longrightarrow \mathcal{O}_{K,T}^* \longrightarrow \bigoplus_{\mathfrak{p} \in T} \mathbb{Z} \xrightarrow{\varphi} \text{Cl}_K \longrightarrow \text{Cl}(\mathcal{O}_{K,T}) \longrightarrow 1. \quad (6\text{-}6)$$

Here $\varphi$ maps the generator corresponding to $\mathfrak{p} \in T$ to the class $[\mathfrak{p}] \in \text{Cl}_K$. Thus $\text{Cl}(\mathcal{O}_{K,T})$ is the quotient of $\text{Cl}_K$ modulo the subgroup generated by the ideal classes of the primes in $T$.

The inclusion map $R \to \mathcal{O}$ of a number ring $R$ in its normalization $\mathcal{O}$ also gives rise to an induced map $\text{Pic}(R) \to \text{Cl}(\mathcal{O})$ given by $[I] \mapsto [I\mathcal{O}]$. We conclude this section by analyzing it in a similar way. As the relation between primes in $R$ and $\mathcal{O}$ is of a different nature, the argument is slightly more involved.

At every prime $\mathfrak{p}$ of a number ring, the local ring $R_\mathfrak{p}$ is a subring of the ring $\mathcal{O}_\mathfrak{p}$ obtained by localizing the normalization $\mathcal{O}$ of $R$ at $S = R \setminus \mathfrak{p}$. If $\mathfrak{p}$ is regular, we have $R_\mathfrak{p} = \mathcal{O}_\mathfrak{p}$ by the maximality property of discrete valuation rings in $K$, and $R$ and $\mathcal{O}$ are 'locally the same' at $\mathfrak{p}$. If $\mathfrak{p}$ is noninvertible or *singular*, the inclusion $R_\mathfrak{p} \subset \mathcal{O}_\mathfrak{p}$ is strict as $\mathcal{O}_\mathfrak{p}$ is integrally closed by Proposition 6.1 and $R_\mathfrak{p}$ is not. Define the *conductor* of $R$ in its normalization $\mathcal{O}$ as

$$\mathfrak{f}_R = \{x \in \mathcal{O} : x\mathcal{O} \subset R\}.$$

This is an ideal in both $R$ and $\mathcal{O}$ and the largest $\mathcal{O}$-ideal contained in $R$. For $R = \mathbb{Z}[\sqrt{-3}] \subset \mathcal{O} = \mathbb{Z}[(1 + \sqrt{-3})/2]$ from 4.2, we have $R = \mathbb{Z} + 2\mathcal{O}$ and $\mathfrak{f}_R = 2\mathcal{O}$.

For an order $R$ the conductor is nonzero by Theorem 2.2, proved in the next section; for arbitrary $R$ the same is true as the index of the order $R \cap \mathcal{O}_K$ in $\mathcal{O}_K$ is an integer that multiplies $\mathcal{O} = R\mathcal{O}_K$ into $R$. From the comparison of local rings, we have

$$\mathfrak{p} \text{ is regular} \iff R_\mathfrak{p} = \mathcal{O}_\mathfrak{p} \iff \mathfrak{p} \nmid \mathfrak{f}_R,$$

so the conductor $\mathfrak{f}_R$ is a measure of the 'singularity' of $R$. Note that only finitely many primes of $R$ can be singular, just as algebraic curves only have finitely many singular points.

At the singular primes $\mathfrak{p}$, which divide $\mathfrak{f}_R$ and in particular the index $[\mathcal{O} : R]$, the local ring $R_\mathfrak{p}$ is a subring of finite index in $\mathcal{O}_\mathfrak{p}$. The primes of the *semilocal* ring $\mathcal{O}_\mathfrak{p}$ are the ideals $\mathfrak{q}\mathcal{O}_\mathfrak{p}$ coming from the primes $\mathfrak{q}$ of $\mathcal{O}$ extending $\mathfrak{p}$. As $\mathcal{O}_\mathfrak{p}$ is a Dedekind domain with only finitely many primes, it has trivial Picard group by the Chinese remainder theorem, and by Theorem 5.3 we can write its ideal group as

$$\mathcal{I}(\mathcal{O}_\mathfrak{p}) = \bigoplus_{\mathfrak{q} \supset \mathfrak{p}} \mathcal{P}(\mathcal{O}_\mathfrak{q}) = \mathcal{P}(\mathcal{O}_\mathfrak{p}) = K^*/\mathcal{O}_\mathfrak{p}^*.$$

THEOREM 6.7. *Let $R \subset \mathcal{O}$ be a number ring of conductor $\mathfrak{f}$ in its normalization $\mathcal{O}$, and $v : \mathrm{Pic}(R) \to \mathrm{Cl}(\mathcal{O})$ the natural map defined by $v([I]) = [I \cdot \mathcal{O}]$. Then we have a natural exact sequence*

$$1 \longrightarrow R^* \longrightarrow \mathcal{O}^* \longrightarrow (\mathcal{O}/\mathfrak{f})^*/(R/\mathfrak{f})^* \longrightarrow \mathrm{Pic}(R) \xrightarrow{v} \mathrm{Cl}(\mathcal{O}) \longrightarrow 1.$$

PROOF. Write the Picard groups of $R$ and $\mathcal{O}$ in terms of their defining exact sequence (4-1), and express $\mathcal{I}(R)$ and $\mathcal{I}(\mathcal{O})$ as in Theorem 5.3. Using the identity for $K^*/\mathcal{O}_\mathfrak{p}^*$ preceding the theorem, we obtain a diagram

$$\begin{array}{ccccccccc} 1 & \longrightarrow & K^*/R^* & \longrightarrow & \bigoplus_\mathfrak{p} K^*/R_\mathfrak{p}^* & \longrightarrow & \mathrm{Pic}(R) & \longrightarrow & 1 \\ & & \downarrow & & \downarrow & & \downarrow & & \\ 1 & \longrightarrow & K^*/\mathcal{O}^* & \longrightarrow & \bigoplus_\mathfrak{p} K^*/\mathcal{O}_\mathfrak{p}^* & \longrightarrow & \mathrm{Cl}(\mathcal{O}) & \longrightarrow & 1 \end{array}$$

with exact rows. Again, the middle vertical map is surjective, this time with kernel $\bigoplus_\mathfrak{p} \mathcal{O}_\mathfrak{p}^*/R_\mathfrak{p}^*$. By the snake lemma, $\mathrm{Pic}(R) \to \mathrm{Cl}(\mathcal{O})$ is surjective and its kernel $N$ fits in an exact sequence $1 \to \mathcal{O}^*/R^* \to \bigoplus_\mathfrak{p} \mathcal{O}_\mathfrak{p}^*/R_\mathfrak{p}^* \to N \to 1$. We are thus reduced to giving a natural isomorphism $(\mathcal{O}/\mathfrak{f})^*/(R/\mathfrak{f})^* \xrightarrow{\sim} \bigoplus_\mathfrak{p} \mathcal{O}_\mathfrak{p}^*/R_\mathfrak{p}^*$. Note that we may restrict the direct sum above to the singular primes $\mathfrak{p} \mid \mathfrak{f}$, since at regular primes we have $\mathcal{O}_\mathfrak{p} = R_\mathfrak{p}$, and therefore $\mathcal{O}_\mathfrak{p}^*/R_\mathfrak{p}^* = 1$.

We first apply Theorem 5.2 to $I = \mathfrak{f}$ to obtain localization isomorphisms $R/\mathfrak{f} \cong \bigoplus_\mathfrak{p} R_\mathfrak{p}/\mathfrak{f} R_\mathfrak{p}$ and $\mathcal{O}/\mathfrak{f} \cong \bigoplus_\mathfrak{p} \bigoplus_{\mathfrak{q} \supset \mathfrak{p}} \mathcal{O}_\mathfrak{q}/\mathfrak{f}\mathcal{O}_\mathfrak{q} \cong \bigoplus_\mathfrak{p} \mathcal{O}_\mathfrak{p}/\mathfrak{f}\mathcal{O}_\mathfrak{p}$. Taking unit groups, we arrive at a natural isomorphism

$$(\mathcal{O}/\mathfrak{f})^*/(R/\mathfrak{f})^* \cong \bigoplus_{\mathfrak{p} \mid \mathfrak{f}} (\mathcal{O}_\mathfrak{p}/\mathfrak{f}\mathcal{O}_\mathfrak{p})^*/(R_\mathfrak{p}/\mathfrak{f} R_\mathfrak{p})^*. \tag{6-8}$$

For $\mathfrak{p} \mid \mathfrak{f}$, an element $x \in \mathcal{O}_\mathfrak{p}$ is invertible modulo $\mathfrak{f}\mathcal{O}_\mathfrak{p}$ if and only if it is not contained in any maximal ideal $\mathfrak{q}\mathcal{O}_\mathfrak{p} \supset \mathfrak{p}\mathcal{O}_\mathfrak{p} \supset \mathfrak{f}\mathcal{O}_\mathfrak{p}$, that is, if and only if it is in $\mathcal{O}_\mathfrak{p}^*$. It follows that the natural map $\mathcal{O}_\mathfrak{p}^* \to (\mathcal{O}_\mathfrak{p}/\mathfrak{f}\mathcal{O}_\mathfrak{p})^*$ is surjective. From the equality $\mathfrak{f}\mathcal{O} = \mathfrak{f} R$ we obtain $\mathfrak{f}\mathcal{O}_\mathfrak{p} = \mathfrak{f} R_\mathfrak{p}$, so $(R_\mathfrak{p}/\mathfrak{f} R_\mathfrak{p})^*$ is a subgroup of $(\mathcal{O}_\mathfrak{p}/\mathfrak{f}\mathcal{O}_\mathfrak{p})^*$, and the image of $x \in \mathcal{O}_\mathfrak{p}^*$ lies in it exactly when we have $x \in R_\mathfrak{p}^*$. We obtain natural maps $\mathcal{O}_\mathfrak{p}^*/R_\mathfrak{p}^* \xrightarrow{\sim} (\mathcal{O}_\mathfrak{p}/\mathfrak{f}\mathcal{O}_\mathfrak{p})^*/(R_\mathfrak{p}/\mathfrak{f} R_\mathfrak{p})^*$ at all $\mathfrak{p}$. Combining this with (6-8), we obtain the desired isomorphism. □

EXAMPLE 6.9. Let $K$ be a quadratic field with ring of integers $\mathcal{O} = \mathcal{O}_K = \mathbb{Z}[\omega]$. For each positive integer $f$, there is a unique subring $R = R_f = \mathbb{Z}[f\omega] = \mathbb{Z} + \mathbb{Z} \cdot f\omega$ of index $f$ in $\mathcal{O}$. It has conductor $\mathfrak{f}_R = f\mathcal{O}$, and its Picard group $\text{Pic}(R)$ is the *ring class group* of the order of conductor $f$ in $\mathcal{O}$. This class group can be described as a *form class group* of binary quadratic forms, and it has an interpretation in class field theory [Cohen and Stevenhagen 2008] as the Galois group of the *ring class field* of conductor $f$ over $K$. By Theorem 6.7, it is the extension

$$1 \longrightarrow (\mathcal{O}/f)^*/\text{im}[\mathcal{O}^*](\mathbb{Z}/f\mathbb{Z})^* \longrightarrow \text{Pic}(R) \longrightarrow \text{Cl}_K \longrightarrow 1$$

of $\text{Cl}_K$ by a finite abelian group that is easily computed, especially for imaginary quadratic $K$, which have $\mathcal{O}^* = \{\pm 1\}$ in all but two cases.

For the order $R = \mathbb{Z}[\sqrt{-3}]$ of index 2 in $\mathcal{O} = \mathbb{Z}[\omega]$ with $\omega = (1 + \sqrt{-3})/2$, the group $\mathbb{F}_4^*/\langle\omega\rangle\mathbb{F}_2^*$ vanishes, and we find as in Examples 4.2 that, just like $\text{Cl}(\mathcal{O})$, the Picard group $\text{Pic}(R)$ is trivial.

## 7. Linear algebra over $\mathbb{Z}$

Before we embark on the algorithmic approach to the ring theory of the preceding sections, we discuss the computational techniques from linear algebra that yield finiteness statements such as Theorem 2.2, and more.

Let $A$ be a ring, and $B$ an $A$-algebra that is free of finite rank $n$ over $A$. For $x \in B$, let $M_x : B \to B$ denote the $A$-linear multiplication map $b \mapsto xb$. With respect to an $A$-basis of $B = \bigoplus_{i=1}^n A \cdot x_i$, the map $M_x$ can be described by an $n \times n$ matrix with coefficients in $A$, and we define the *norm* and the *trace* from $B$ to $A$ by

$$N_{B/A}(x) = \det M_x \quad \text{and} \quad \text{Tr}_{B/A}(x) = \text{trace } M_x.$$

It follows immediately from this definition that the norm is a multiplicative map, whereas the trace $\text{Tr}_{B/A} : B \to A$ is a homomorphism of the additive groups.

The notions of norm and trace are stable under *base change*. This means that if $f : A \to A'$ is *any* ring homomorphism and $f_* : B \to B' = B \otimes_A A'$ is the induced map from $B$ to the free $A'$-algebra $B' = \bigoplus_{i=1}^n A' \cdot (x_i \otimes 1)$, the diagrams

$$\begin{array}{ccc} B & \xrightarrow{f_*} & B' = B \otimes_A A' \\ {\scriptstyle N_{B/A}} \downarrow {\scriptstyle \text{Tr}_{B/A}} & & \downarrow {\scriptstyle N_{B'/A'}} {\scriptstyle \text{Tr}_{B'/A'}} \\ A & \xrightarrow{f} & A' \end{array}$$

describing the 'base change' $A \to A'$ for norm and trace commute. Indeed, for an element $x \in B$ the multiplication matrix of $f_*(x)$ on $B'$ with respect to the $A'$-basis $x_1 \otimes 1, x_2 \otimes 1, \ldots, x_n \otimes 1$ is obtained by applying $f$ to the entries of $M_x$ with respect to the $A$-basis $x_1, x_2, \ldots, x_n$.

Base changing a domain $A$ to its field of fractions suffices to recover the classical 'linear algebra fact' that norms and traces do not depend on the choice of a basis for $B$ over the domain $A$. In fact, the issue of dependency on a basis does not even arise if one uses *coordinatefree* definitions for the determinant and the trace of an endomorphism $M \in \text{End}_A(B)$ of a free $A$-module $B$ of rank $n$. For the determinant, one notes [Bourbaki 1989, Section III.8.1] that the $n$-th exterior power $\bigwedge^n B$ is a free $A$-module of rank 1 on which $M$ induces scalar multiplication by $\det M \in A$. For the trace [Bourbaki 1989, Section II.4.1], one views $\text{End}_A(B) = B \otimes_A B^*$ as the tensor product of $B$ with its dual module $B^* = \text{Hom}_A(B, A)$ and defines $\text{Tr}_{B/A}(\sum b \otimes f) = \sum f(b)$.

For an order $B = R$ in $K$, base changing from $A = \mathbb{Z}$ to $\mathbb{Q}$ and to $\mathbb{F}_p$, respectively, shows that the norm and trace maps $R \to \mathbb{Z}$ are the restrictions to $R$ of the 'field maps' $K \to \mathbb{Q}$, and that their reductions modulo $p$ are the norm and trace maps $R/pR \to \mathbb{F}_p$ for the $\mathbb{F}_p$-algebra $R/pR$.

For $B = K$ a number field of degree $n$ over $A = \mathbb{Q}$, we can use the $n$ distinct embeddings $\sigma_i : K \to \mathbb{C}$ and the base change $\mathbb{Q} \to \mathbb{C}$ to diagonalize the matrix for $M_x$ as $M_x = (\sigma_i(x))_{i=1}^n$, since we have an isomorphism

$$\begin{aligned} K \otimes_{\mathbb{Q}} \mathbb{C} &\xrightarrow{\sim} \mathbb{C}^n \\ x \otimes y &\longmapsto (\sigma_i(x)y)_{i=1}^n. \end{aligned} \quad (7\text{-}1)$$

This yields the formulas $N_{K/\mathbb{Q}}(x) = \prod_{i=1}^n \sigma_i(x)$ and $\text{Tr}_{K/\mathbb{Q}}(x) = \sum_{i=1}^n \sigma_i(x)$ for the norm and trace from $K$ to $\mathbb{Q}$.

In a free $A$-algebra $B$ of rank $n$, the *discriminant* of $x_1, x_2, \ldots, x_n \in B$ is defined as

$$\Delta(x_1, x_2, \ldots, x_n) = \det(\text{Tr}_{B/A}(x_i x_j))_{i,j=1}^n,$$

and the *discriminant* $\Delta(B/A)$ of $B$ over $A$ is the discriminant of any $A$-basis of $B$. We have $(\text{Tr}_{B/A}(y_i y_j))_{i,j=1}^n = T \cdot (\text{Tr}_{B/A}(x_i x_j))_{i,j=1}^n \cdot T^t$ for any $n \times n$ matrix $T \in \text{GL}_n(A)$ over $A$ transforming $\{x_i\}_{i=1}^n$ to $\{y_i\}_{i=1}^n$, so the discriminant of a free $A$-algebra does depend on the choice of the basis, but only up to the square of a unit in $A$. Over a field $A$, it is usually the vanishing of $\Delta(B/A)$ that has intrinsic significance, and the algebras of non-vanishing discriminant are known as *separable A-algebras*. Over $A = \mathbb{Z}$, the discriminant

$$\Delta(R) = \det(\text{Tr}_{R/\mathbb{Z}}(x_i x_j))_{i,j=1}^n \in \mathbb{Z} \tag{7-2}$$

of an order $R = \bigoplus_{i=1}^n \mathbb{Z} \cdot x_i$ of rank $n$ is a well-defined integer. Considering arbitrary $\mathbb{Z}$-linear transformations of bases, one obtains, for an inclusion $R' \subset R$ of orders in $K$, the index formula

$$\Delta(R') = [R : R']^2 \cdot \Delta(R). \tag{7-3}$$

The discriminant of an order $R = \bigoplus_{i=1}^n \mathbb{Z} \cdot x_i$ in $K$ can also be defined in terms of the embeddings $\sigma_i : K \to \mathbb{C}$ from (7-1) as

$$\Delta(R) = [\det(\sigma_i(x_j))_{i,j=1}^n]^2. \tag{7-4}$$

To see that (7-4) agrees with (7-2), one multiplies the matrix $X = (\sigma_i(x_j))_{i,j=1}^n$ by its transpose and uses the description of the trace map following (7-1) to find

$$X^t \cdot X = \left(\sum_{k=1}^n \sigma_k(x_i x_j)\right)_{i,j=1}^n = (\text{Tr}_{L/K}(x_i x_j))_{i,j=1}^n.$$

Taking determinants, the desired equality follows.

For an order $\mathbb{Z}[\alpha]$ in $K = \mathbb{Q}(\alpha)$, the elements $\alpha_i = \sigma_i(\alpha)$ are the roots of $f_\mathbb{Q}^\alpha$, and by (7-4) its discriminant can be evaluated as the square of a Vandermonde determinant and equals the polynomial discriminant $\Delta(f_\mathbb{Q}^\alpha)$:

$$\begin{aligned}\Delta(\mathbb{Z}[\alpha]) &= \Delta(1, \alpha, \alpha^2, \ldots, \alpha^{n-1}) = [\det(\sigma_i(\alpha^{j-1}))_{i,j=1}^n]^2 \\ &= [\det(\alpha_i^{j-1})_{i,j=1}^n]^2 = \prod_{i>j}(\alpha_i - \alpha_j)^2 = \Delta(f_\mathbb{Q}^\alpha).\end{aligned} \tag{7-5}$$

In the same way, one shows that the discriminant over a field $A$ of a simple field extension $A \subset A(\alpha)$ is up to squares in $A^*$ equal to $\Delta(f_A^\alpha)$. The extension $A \subset A(\alpha)$ is separable (as a field extension of $A$, or as an $A$-algebra) if and only if $f_\mathbb{Q}^\alpha$ is a separable polynomial.

For an order $R$ in $K$ containing $\mathbb{Z}[\alpha]$, the identities (7-3) and (7-5) yield

$$\Delta(f_\mathbb{Q}^\alpha) = [R : \mathbb{Z}[\alpha]]^2 \cdot \Delta(R), \tag{7-6}$$

so $\Delta(R)$ is nonzero. We also see that we have $\mathbb{Z}[\alpha] \subset R \subset d^{-1}\mathbb{Z}[\alpha]$ for every order $R$ containing $\alpha$, with $d$ the largest integer for which $d^2$ divides $\Delta(f_\mathbb{Q}^\alpha)$. It follows that $\mathcal{O}_K$ itself is an order contained in $d^{-1}\mathbb{Z}[\alpha]$, and in principle $\mathcal{O}_K$ can be found by a finite computation starting from $\mathbb{Z}[\alpha]$: there are finitely many residue classes in $d^{-1}\mathbb{Z}[\alpha]/\mathbb{Z}[\alpha]$, and for each class one decides whether it is in $\mathbb{O}_K/\mathbb{Z}[\alpha]$ by computing the irreducible polynomial of an element from the class and checking whether it is integral. Finally, (7-6) implies that any order in $K$ is of finite index in $\mathbb{O}_K$; so we have proved Theorem 2.2.

The discriminant $\Delta_K = \Delta(\mathbb{O}_K)$ is often referred to as the *discriminant of $K$*. It is a fundamental invariant of $K$, and by (7-6) it satisfies

$$\Delta(f_\mathbb{Q}^\alpha) = [\mathbb{O}_K : \mathbb{Z}[\alpha]]^2 \cdot \Delta_K \qquad (7\text{-}7)$$

for every $\alpha \in \mathbb{O}_K$ of degree $n$. In cases where $\Delta(f_\mathbb{Q}^\alpha)$ can be factored, (7-7) is used as the starting point for the computation of the extension $\mathbb{Z}[\alpha] \subset \mathbb{O}_K$. If we are lucky and $\Delta(f_\mathbb{Q}^\alpha)$ can be shown to be squarefree, then we know immediately that $\mathbb{Z}[\alpha]$ is the full ring of integers $\mathbb{O}_K$, and that we have $\Delta_K = \Delta(f_\mathbb{Q}^\alpha)$.

EXAMPLE 7.8. Let $K$ be imaginary quadratic of discriminant $\Delta_K$, and let $\tau \in K \setminus \mathbb{Q}$ be a zero of the irreducible polynomial $f_\mathbb{Z}^\tau = aX^2 + bX + c \in \mathbb{Z}[X]$. Then $I = \mathbb{Z} + \mathbb{Z} \cdot \tau$ is an invertible ideal for the order

$$R_\tau = \mathbb{Z}[a\tau] = \Lambda(I) = \{x \in K : xI \subset I\}.$$

As $a\tau$ is a zero of $X^2 + bX + ac \in \mathbb{Z}[X]$, the order $R_\tau$ has discriminant

$$\Delta(R_\tau) = b^2 - 4ac = f^2 \Delta_K$$

by (7-5) and (7-7), with $f$ the index of $R_\tau$ in $\mathbb{O}_K$.

As any quadratic order $R$ has field of fractions $K = \mathbb{Q}(\sqrt{\Delta(R)})$, it is determined up to isomorphism by its discriminant, which can uniquely be written as $\Delta(R) = f^2 \Delta_K$. In higher degree, there exist non-isomorphic *maximal* orders having the same discriminant.

To compute polynomial discriminants, one makes use of the *resultant*. The resultant of nonzero polynomials $g = b \prod_{i=1}^r (X - \beta_i)$ and $h = c \prod_{j=1}^s (X - \gamma_j)$ with coefficients and zeros in some field is defined as

$$R(g, h) = b^s c^r \prod_{i=1}^r \prod_{j=1}^s (\beta_i - \gamma_j).$$

It can be expressed [Lang 2002, Section IV.8] as a determinant in terms of the coefficients of $g$ and $h$, but computations are usually based on the following obvious properties:

(R1) $R(g, h) = (-1)^{rs} R(h, g)$;
(R2) $R(g, h) = b^s \prod_{i=1}^r h(\beta_i)$;
(R3) $R(g, h) = b^{s-s_1} R(g, h_1)$ if $h_1 \neq 0$ satisfies $h_1 \equiv h \bmod g$ and $s_1 = \deg h_1$.

For an element $g(\alpha) \in K = \mathbb{Q}(\alpha)$, property (R2) yields

$$N_{K/\mathbb{Q}}(g(\alpha)) = R(f, g).$$

For $f = f_\mathbb{Q}^\alpha$ as above and $\alpha_i = \sigma_i(\alpha)$, one has $f'(\alpha_1) = \prod_{i \geq 2}(\alpha_1 - \alpha_i)$. Taking $g$ to be the derivative $f'$ of $f$, one can write the discriminant of $f$ as

$$\Delta(f) = \prod_{i<j}(\alpha_i - \alpha_j)^2 = (-1)^{n(n-1)/2} N_{K/\mathbb{Q}}(f'(\alpha)) = (-1)^{n(n-1)/2} R(f, f').$$

This reduces the computation of norms and polynomial discriminants to the computation of resultants, which can be performed inside the field containing the coefficients of the polynomials.

EXAMPLE 7.9. The discriminant of the polynomial $X^n + a$ equals

$$(-1)^{n(n-1)/2} R(X^n + a, nX^{n-1}) = (-1)^{n(n-1)/2} n^n a^{n-1}.$$

For $f = X^3 - X^2 - 15X - 75$, long division shows that the remainder of $f$ upon division by its derivative $f' = 3X^2 - 2X - 15$ equals $f - \frac{1}{9}(3X-1)f' = -\frac{9}{88}(X - \frac{15}{2})$. This is a linear polynomial with zero $\frac{15}{2}$, so we find

$$\Delta(f) = -R(f', f) = 3^2 \cdot R(f', -\tfrac{88}{9}(X - \tfrac{15}{2}))$$
$$= 3^2 \cdot R(-\tfrac{88}{9}(X - \tfrac{15}{2}), f')$$
$$= -3^2 \cdot (-\tfrac{88}{9})^2 \cdot f'(\tfrac{15}{2}) = -2^4 \cdot 3 \cdot 5^2 \cdot 11^2.$$

## 8. Explicit ideal factorization

In order to factor an ideal $I$ in a number ring $R$ in the sense of Theorem 5.2, we have to determine for all primes $\mathfrak{p} \supset I$ the $\mathfrak{p}$-primary part $I_{(\mathfrak{p})}$ of $I$. Every prime $\mathfrak{p} \supset I$ divides the index $[R:I]$, so a first step towards factoring $I$ consists of factoring $[R:I]$ in $\mathbb{Z}$ to determine the rational primes $p$ over which the primes $\mathfrak{p} \mid I$ lie.

The index map $I \mapsto [R:I]$ for integral $R$-ideals extends to a *multiplicative* map $\mathcal{I}(R) \to \mathbb{Q}^*$ on invertible ideals known as the *ideal norm*. Its multiplicativity follows from Theorems 5.2 and 5.3 and the observation that for *principal* $R_\mathfrak{p}$-ideals, the index is a multiplicative function. At regular primes $\mathfrak{p}$, all ideals in $R_\mathfrak{p}$ are principal by Proposition 5.4. At singular primes this is not the case, and the behavior of the singular prime $\mathfrak{p} = (2, 1 + \sqrt{-3}) \subset R = \mathbb{Z}[\sqrt{-3}]$ in Examples 4.2 is typical: $[R:\mathfrak{p}^2] = [R:2\mathfrak{p}] = 2^3 > 2^2 = [R:\mathfrak{p}]^2$.

If $R$ is an order, the ideal norm of a principal ideal $xR \subset R$ equals $|N_{R/\mathbb{Z}}(x)|$ as the element norm is by definition the determinant of the multiplication map $M_x : R \to R$, and we have $[R:M_x[R]] = |\det M|$ for the $\mathbb{Z}$-module $R$. By multiplicativity, this compatibility of element and ideal norms in orders extends to all $x \in K^*$.

If singular primes are encountered in $R$, one usually replaces $R$ by an extension ring in which the primes over $p$ are all regular, and then $\mathrm{ord}_{\mathfrak{p}}(I)$ can be determined to complete the factorization of $I$. If possible, one tries to take $R$ to be the ring of integers $\mathcal{O}_K$, which has no singular primes at all.

Given a number field $K = \mathbb{Q}(\alpha)$ generated by an element $\alpha$ with irreducible monic polynomial $f \in \mathbb{Z}[X]$, we have the simple order $\mathbb{Z}[\alpha] \subset \mathcal{O}_K$ as a first approximation to $\mathcal{O}_K$, and factoring $p$ in $\mathcal{O}_K$ or $\mathbb{Z}[\alpha]$ is 'the same' as long as $p$ does not divide the index $[\mathcal{O}_K : \mathbb{Z}[\alpha]]$ from (7-7). For such $p$ we have an isomorphism $\mathbb{Z}[\alpha]/p\mathbb{Z}[\alpha] \xrightarrow{\sim} \mathcal{O}_K/p\mathcal{O}_K$, and the order $\mathbb{Z}[\alpha]$ is called *p-maximal* or *regular* above $p$.

The primes over $p$ in $\mathbb{Z}[\alpha]$ are the kernels of the ring homomorphisms

$$\varphi : \mathbb{Z}[\alpha] = \mathbb{Z}[X]/(f) \to \overline{\mathbf{F}}_p$$

from $\mathbb{Z}[\alpha]$ to an algebraic closure $\overline{\mathbf{F}}_p$ of the field $\mathbf{F}_p$ of $p$ elements. As $\varphi$ factors via $\mathbb{Z}[\alpha]/p\mathbb{Z}[\alpha] = \mathbf{F}_p[X]/(\overline{f})$, such kernels correspond to the irreducible factors $\overline{g} \in \mathbf{F}_p[X]$ of $\overline{f} = f \bmod p$. Pick monic polynomials $g_i \in \mathbb{Z}[X]$ such that $\overline{f}$ factors as $\overline{f} = \prod_{i=1}^{s} \overline{g}_i^{e_i} \in \mathbf{F}_p[X]$. Then the ideals in $\mathbb{Z}[\alpha]$ lying over $p$ are the ideals

$$\mathfrak{p}_i = (p, g_i(\alpha)) \subset \mathbb{Z}[\alpha]. \tag{8-1}$$

From the isomorphism $\mathbb{Z}[\alpha]/\mathfrak{p}_i \cong \mathbf{F}_p[X]/(\overline{g}_i)$, we see that the residue class degree of $\mathfrak{p}_i$ over $p$ equals $f(\mathfrak{p}_i/p) = \deg(g_i)$. For any polynomial $t \in \mathbb{Z}[X]$, the element $t(\alpha) \in \mathbb{Z}[\alpha]$ is in $\mathfrak{p}_i$ if and only if $\overline{g}_i$ divides $\overline{t}$ in $\mathbf{F}_p[X]$.

THEOREM 8.2 (KUMMER–DEDEKIND). *Let $p$ and $\mathbb{Z}[\alpha]$ be as above, and define $\mathfrak{p}_i = (p, g_i(\alpha)) \subset \mathbb{Z}[\alpha]$ corresponding to the factorization $\overline{f} = \prod_{i=1}^{s} \overline{g}_i^{e_i} \in \mathbf{F}_p[X]$ as in (8-1). Then the inclusion*

$$\prod_{i=1}^{s} \mathfrak{p}_i^{e_i} \subset (p)$$

*of $\mathbb{Z}[\alpha]$-ideals is an equality if and only if all $\mathfrak{p}_i$ are invertible. If $r_i \in \mathbb{Z}[X]$ is the remainder of $f$ upon division by $g_i$ in $\mathbb{Z}[X]$, say $f = q_i g_i + r_i$, then we have*

$$\mathfrak{p}_i \text{ is regular} \iff e_i = 1 \text{ or } p^2 \nmid r_i \in \mathbb{Z}[X].$$

*If $\mathfrak{p}_i$ is singular, then $\frac{1}{p} q_i(\alpha) \notin \mathbb{Z}[\alpha]$ is an integral element of $\mathbb{Q}(\alpha)$.*

PROOF. Write $R = \mathbb{Z}[\alpha]$. As $\prod_{i=1}^{s} g_i(\alpha)^{e_i}$ is in $f(\alpha) + pR = pR$, the inclusion $\prod_{i=1}^{s} \mathfrak{p}_i^{e_i} \subset pR + \prod_{i=1}^{s} g_i(\alpha)^{e_i} \subset pR$ is immediate. If it is an equality, all $\mathfrak{p}_i$ are clearly invertible. Conversely, if all $\mathfrak{p}_i$ are invertible, then the invertible ideal $\prod_{i=1}^{s} \mathfrak{p}_i^{e_i}$ has index $\prod_{i=1}^{s} p^{e_i \deg(g_i)} = p^{\deg f} = [R : pR]$, so equality holds.

As the remainder $r_i$ of $f$ upon division by $g_i$ in $\mathbb{Z}[X]$ is divisible by $p$, there are polynomials $q_i, s_i \in \mathbb{Z}[X]$ satisfying $f = q_i \cdot g_i + ps_i$ and $\deg(s_i) < \deg(g_i)$. Substitution of $\alpha$ yields the relation

$$ps_i(\alpha) = -q_i(\alpha)g_i(\alpha) \in \mathfrak{p}_i \qquad (*)$$

between the two $R$-generators $p$ and $g_i(\alpha)$ of $\mathfrak{p}_i$. If $\bar{g}_i$ occurs with exponent $e_i = 1$ in $\bar{f}$, we have $\bar{g}_i \nmid \bar{q}_i \in \mathbb{F}_p[X]$, so $q_i(\alpha)$ is not in $\mathfrak{p}_i$. In this case $q_i(\alpha)$ is a unit in $R_{\mathfrak{p}_i}$, and $(*)$ shows that $\mathfrak{p}_i R_{\mathfrak{p}_i}$ is principal with generator $p$. Similarly, the hypothesis $r_i = ps_i \notin p^2 \mathbb{Z}[X]$ means that $\bar{s}_i \in \mathbb{F}_p[X]$ is nonzero of degree $\deg(\bar{s}_i) < \deg(\bar{g}_i)$. This implies $\bar{g}_i \nmid \bar{s}_i$, so now $s_i(\alpha)$ is a unit in $R_{\mathfrak{p}_i}$ and $\mathfrak{p}_i R_{\mathfrak{p}_i}$ is principal with generator $g_i(\alpha)$. In either case $\mathfrak{p}_i$ is regular by Proposition 5.4.

If we have $e_i > 1$ and $p^2$ divides $r_i$ in $\mathbb{Z}[X]$, then $\mathfrak{p}_i$ is singular, as the identity

$$\tfrac{1}{p} q_i(\alpha)\mathfrak{p}_i = \tfrac{1}{p} q_i(\alpha) \cdot pR + \tfrac{1}{p} q_i(\alpha) \cdot g_i(\alpha) R = q_i(\alpha) R + s_i(\alpha) R \subset \mathfrak{p}_i$$

shows that its multiplier ring $\Lambda(\mathfrak{p}_i)$ contains the integral element $\tfrac{1}{p} q_i(\alpha) \notin R_{\mathfrak{p}_i}$, so $R_{\mathfrak{p}_i}$ is not integrally closed. □

If $\mathbb{Z}[\alpha]$ is regular above $p$, we can factor $p$ in $\mathbb{Z}[\alpha]$ or $\mathcal{O}_K$ using the Kummer–Dedekind theorem, whereas if $\mathbb{Z}[\alpha]$ is singular above $p$, then $p$ divides the index $[\mathcal{O}_K : \mathbb{Z}[\alpha]]$, and $p^2$ divides $\Delta(f_\mathbb{Q}^\alpha)$ by (7-7). For every singular prime $\mathfrak{p}_i \mid p$ of $\mathbb{Z}[\alpha]$ we encounter, an element $p^{-1} q_i(\alpha)$ is provided by Theorem 8.2 that can be adjoined to the order $\mathbb{Z}[\alpha]$ to obtain an order of smaller index in $\mathcal{O}_K$.

EXAMPLE 8.3. Let $\alpha$ be a zero of $f = X^3 + 44 = X^3 + 2^2 \cdot 11 \in \mathbb{Z}[X]$, and $\mathbb{Z}[\alpha]$ the associated cubic order in $K = \mathbb{Q}(\alpha)$. Then $f$ is separable modulo the primes $p \neq 2, 3, 11$ coprime to $\Delta(X^3 + 44) = -2^4 \cdot 3^3 \cdot 11^2$, and for these $p$ we can factor $(p)$ into prime ideals in $\mathbb{Z}[\alpha]$ as

$$(p) = \begin{cases} \mathfrak{p}_1 \mathfrak{p}_2 \mathfrak{p}_3 & \text{if } p \equiv 1 \bmod 3 \text{ and } 44 \text{ is a cube modulo } p, \\ (p) & \text{if } p \equiv 1 \bmod 3 \text{ and } 44 \text{ is not a cube modulo } p, \\ \mathfrak{p}\mathfrak{P} & \text{if } p \equiv 2 \bmod 3. \end{cases}$$

In the first case, the primes $\mathfrak{p}_i = (p, \alpha - k_i)$ corresponding to the three cube roots $k_i$ of $-44$ modulo $p$ are of degree 1. In the second case, the rational prime $p$ is called *inert* as it remains prime in $\mathbb{Z}[\alpha]$ (but becomes of degree 3). For $p \equiv 2 \bmod 3$, the element $-44 \in \mathbb{F}_p^*$ has a unique cube root $k$ giving rise to a prime $\mathfrak{p} = (p, \alpha - k)$ of degree 1, and the irreducible quadratic polynomial $g = (X^3 + 44)/(X - k) \in \mathbb{F}_p[X]$ yields the other prime $\mathfrak{P} = (p, g(\alpha))$ of degree 2 lying over $p$.

For $p = 2, 11$, the triple factor $X$ of $f$ mod $p$ leaves as remainder $44 = 2^2 \cdot 11$ upon division in $\mathbb{Z}[X]$. For $p = 11$, this yields the factorization $(11) = (11, \alpha)^3$.

For $p=2$, it implies that the unique prime $\mathfrak{p}_2 = (2,\alpha)$ above 2 in $\mathbb{Z}[\alpha]$ is singular, and that $\beta = \alpha^2/2$ is an integral element outside $\mathbb{Z}[\alpha]$. We have

$$f_{\mathbb{Q}}^{\beta} = X^3 - 2 \cdot 11^2,$$

and $\mathbb{Z}[\beta]$ is a 2-integral ring in which we have $(2) = (2,\beta)^3$ by Theorem 8.2. This factorization also holds in $\mathbb{Z}[\alpha,\beta]$ and in $\mathcal{O}_K$.

For $p=3$, the triple factor $X-1$ of $f$ mod 3 leaves remainder $45 = 3^2 \cdot 5$ upon division in $\mathbb{Z}[X]$, so the unique prime ideal $\mathfrak{p}_3 = (3, \alpha-1)$ over 3 in $\mathbb{Z}[\alpha]$ is singular, and from $X^3 + 44 = (X-1)(X^2+X+1) + 45$ we see that $\gamma = \frac{1}{3}(\alpha^2 + \alpha + 1)$ is integral. Its irreducible polynomial is the polynomial $f_{\mathbb{Q}}^{\gamma} = X^3 - X^2 + 15X - 75$ of discriminant $-2^4 \cdot 3 \cdot 5^2 \cdot 11^2$ from Example 7.9, so $\mathbb{Z}[\gamma]$ is regular above 3 and Theorem 8.2 gives us the factorization $(3) = (3,\gamma)^2(3,\gamma-1)$ in any $K$-order containing $\gamma$.

In this small example, $\mathbb{Z}[\alpha]$ has index $6 = 2 \cdot 3$ in $\mathcal{O}_K = \mathbb{Z}[\alpha, \beta, \gamma]$, and we have $\Delta(\mathcal{O}_K) = \Delta_K = 6^{-2} \cdot \Delta(f) = -2^2 \cdot 3 \cdot 11^2$. The multiplier rings of the singular primes $\mathfrak{p}_2 = (2,\alpha)$ and $\mathfrak{p}_3 = (3, \alpha-1)$ of $\mathbb{Z}[\alpha]$ contain $\beta$ and $\gamma$, respectively, so $\mathfrak{f}_{\mathbb{Z}[\alpha]} = \mathfrak{p}_2\mathfrak{p}_3$ is a $\mathbb{Z}[\alpha]$-ideal of index 6 that multiplies $\mathcal{O}_K$ into $\mathbb{Z}[\alpha]$. As an $\mathcal{O}_K$-ideal, it is the regular ideal $(2,\beta)^2(3,\gamma)(3,\gamma-1)$ of norm 36.

Having computed $\mathcal{O}_K = \mathbb{Z}[\beta,\gamma]$, one may verify that $\beta - \gamma = (\alpha^2 - 2\alpha - 2)/6$ has irreducible polynomial $X^3 + X^2 - 7X - 13$ and generates $\mathcal{O}_K$ over $\mathbb{Z}$, so in this case $\mathcal{O}_K$ is actually a simple extension of $\mathbb{Z}$. However, finding such a generator starting from $X^3 + 44$ is not immediate.

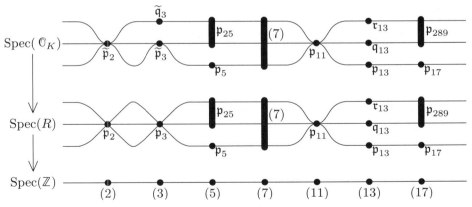

The picture above is a 'geometric' rendering of the cubic order $R = \mathbb{Z}[\sqrt[3]{44}]$ and its normalization $\mathcal{O}_K$. The inclusions $\mathbb{Z} \subset R \subset \mathcal{O}_K$ correspond to 'covering maps' $\operatorname{Spec} \mathcal{O}_K \to \operatorname{Spec} R \to \operatorname{Spec} \mathbb{Z}$. Here the *spectrum* of a number ring (see [Eisenbud and Harris 2000] for more details) is represented as a 'curve' having the primes of the ring as its points, and the covering maps intersect primes in the larger ring with the smaller ring. As suggested by the picture, $\operatorname{Spec} R$ is a

3-to-1 cover of Spec $\mathbb{Z}$. The fiber above any rational prime different from the prime divisors 2, 3, and 11 consists of exactly three extension primes, provided that we count a prime $\mathfrak{p} \mid p$ with 'weight' $f(\mathfrak{p}/p)$. Primes of weight 2 and 3 are represented by vertical dashes intersecting 2 or 3 of the lines rather than points. The primes $\mathfrak{p}_2$ and $\mathfrak{p}_3$ over 2 and 3 in $R$ are singular 'triple points', whereas the unique prime $\mathfrak{p}_{11} = (11, \alpha)$ in $R$ over 11 is a regular prime in which the three lines are tangent to each other, illustrating the identity $\mathfrak{p}_{11}^3 = (11)$. The normalization $\mathcal{O}_K$ of $R$ is locally isomorphic to $R$ at all primes not dividing $\mathfrak{f}_{\mathbb{Z}[\alpha]} = \mathfrak{p}_2 \mathfrak{p}_3$. The singular prime $\mathfrak{p}_2$ has a unique extension $\widetilde{\mathfrak{p}}_2$ in $\mathcal{O}_K$, for which we have $\mathfrak{p}_2 \mathcal{O}_K = \widetilde{\mathfrak{p}}_2^2$. The singular prime $\mathfrak{p}_3$ factors in $\mathcal{O}_K$ as a product $\mathfrak{p}_3 \mathcal{O}_K = \widetilde{\mathfrak{p}}_3 \widetilde{\mathfrak{q}}_3$ of two primes, and we have $3\mathcal{O} = \widetilde{\mathfrak{p}}_3^2 \widetilde{\mathfrak{q}}_3$ with $\widetilde{\mathfrak{p}}_3 = (3, \gamma)$.

If $R$ is regular above $p$, we can factor the rational prime $p$ in $R$ as $pR = \prod_{\mathfrak{p} \mid p} \mathfrak{p}^{e(\mathfrak{p}/p)}$. The exponent $e(\mathfrak{p}/p) = \mathrm{ord}_\mathfrak{p}(pR)$ is known as the *ramification index* of $\mathfrak{p}$ over $p$. We say that $p$ is *unramified* in $R$ if all $e(\mathfrak{p}/p)$ equal 1, and *ramified* in $R$ if we have $e(\mathfrak{p}/p) > 1$ for some $\mathfrak{p}$. Thus, the primes 2, 3 and 11 are ramified in the ring $\mathcal{O}_K$ in Example 8.3, and all other primes are unramified. This example also shows the validity of the following relation for orders, which, for orders of the form $\mathbb{Z}[\alpha]$, is immediate from the identity $f(\mathfrak{p}_i/p) = \deg g_i$ in Theorem 8.2.

THEOREM 8.4. *Let $R$ be an order of rank $n$, and suppose that $R$ is regular above $p$. Then we have $\sum_{\mathfrak{p} \mid p} e(\mathfrak{p}/p) f(\mathfrak{p}/p) = n$.*

PROOF. The ideal norm of $pR$, which is $\#(R/pR) = p^n$, is also equal to

$$\prod_{\mathfrak{p} \mid p} \#(R/\mathfrak{p}^{e(\mathfrak{p}/p)}) = \prod_{\mathfrak{p} \mid p} \#(R/\mathfrak{p})^{e(\mathfrak{p}/p)} = p^{\sum_{\mathfrak{p} \mid p} e(\mathfrak{p}/p) f(\mathfrak{p}/p)}$$

by the multiplicativity of the ideal norm for powers of regular primes. □

In the situation of Theorem 8.4, we call $p$ *totally split* (or just *split*) in $R$ if there are $n$ extension primes over $p$, which then have $e(\mathfrak{p}/p) = f(\mathfrak{p}/p) = 1$. If there is a single prime $\mathfrak{p} \mid p$ with $e(\mathfrak{p}/p) = n$, we call $p$ *totally ramified* in $R$. If $\mathfrak{p} = pR$ is a prime ideal of $R$, we have $f(\mathfrak{p}/p) = n$, and we say that $p$ is *inert* in $R$.

For singular primes $\mathfrak{p} \mid p$, the ramification index of $\mathfrak{p}$ over $p$ is not defined.

THEOREM 8.5. *Let $R$ be an order and $p$ a prime number. If $R$ is singular above $p$, then $p^2$ divides $\Delta(R)$. If $R$ is regular above $p$, we have*

$$p \text{ divides } \Delta(R) \iff p \text{ is ramified in } R.$$

PROOF. If $R$ is singular above $p$, then $p$ divides the index of $R$ in its integral closure $\mathcal{O}$, so $p^2$ divides $\Delta(R) = [\mathcal{O} : R]^2 \cdot \Delta(\mathcal{O})$ by (7-3).

We have $\Delta(R) \equiv 0 \bmod p$ if and only if the discriminant $\Delta(R/pR) \in \mathbb{F}_p$ of the $\mathbb{F}_p$-algebra $R/pR$ vanishes. If $p$ is unramified in $R$, then $R/pR$ is a product $\prod_{\mathfrak{p} \mid p} R/\mathfrak{p}$ of finite fields and $\Delta(R/pR)$ is a product of discriminants of field extensions of $\mathbb{F}_p$, each of which is nonzero by the remark following (7-5). If $p$ is ramified in $R$, the $\mathbb{F}_p$-algebra $R/pR$ has a nonzero nilradical. Taking a basis containing nilpotent elements, which have trace zero, we see that its discriminant vanishes. □

If $R$ is an order of rank $n$ in which some prime $p < n$ splits completely, then $R$ is not monogenic as a monogenic ring admits at most $p$ homomorphisms to $\mathbb{F}_p$. This makes it easy to construct examples of number fields $K$ for which $\mathcal{O}_K$ is not monogenic, and for which every order $\mathbb{Z}[\alpha]$ has index in $\mathcal{O}_K$ divisible by $p$.

EXAMPLE 8.6. The ring of integers of $K = \mathbb{Q}(\sqrt{-7}, \sqrt{17})$ is generated by $\beta = (1+\sqrt{-7})/2$ and $\gamma = (1+\sqrt{17})/2)$, and $f_\mathbb{Q}^\beta = X^2 - X + 2$ and $f_\mathbb{Q}^\gamma = X^2 - X - 4$ each have two roots in $\mathbb{F}_2$. This yields *four* different maps $\mathcal{O}_K = \mathbb{Z}[\beta, \gamma] \to \mathbb{F}_2$, so 2 splits completely in $\mathcal{O}_K$. The discriminant $\Delta_K = 7^2 \cdot 17^2$ is odd, but every order $R = \mathbb{Z}[\alpha]$ in $K$ has $\Delta(R) \equiv 0 \bmod 4$.

## 9. Computing the integral closure

If $R$ is a number ring, the integral closure $\mathcal{O}$ of $R$ in $K = Q(R)$ contains $R$ as a subring of finite index. As the example $R = \mathbb{Z}[\sqrt{d}]$ in Section 2 shows, efficient computation of $\mathcal{O}$ from $R$ is hampered by our inability to factor integers or, more precisely, to determine the largest squarefree divisor of a given integer. The algorithms we do have to compute $\mathcal{O}$ from $R$ are mostly 'local at $p$' for a rational prime number $p$. They work inside the $\mathbb{F}_p$-algebra $R/pR$, which is finite of rank at most $[K : \mathbb{Q}]$ even in case $R$ is not assumed to be an order. As they only use linear algebra over $\mathbb{F}_p$, they are fairly efficient. However, it may not be easy to find the primes $p$ dividing $[\mathcal{O} : R]$ at which these computations need to be performed.

In the case of an order of discriminant $\Delta(R)$, the 'critical' primes are the primes that divide $\Delta(R)$ more than once. For such $p$, one wants to find a $p$-maximal extension

$$R \subset O_p = \{x \in \mathcal{O}_K : p^k x \in R \text{ for some } k \in \mathbb{Z}_{\geq 0}\}$$

of $R$ inside $\mathcal{O}_K$, for which the index in $\mathcal{O}_K$ is coprime to $p$. The index $[O_p : R]$ is a $p$-power, and its square $[O_p : R]^2$ divides $\Delta(R)$. Taken together, the rings $O_p$ with $p^2 \mid \Delta(R)$ generate $\mathcal{O}_K$ over $R$.

In practice one starts with a simple order $\mathbb{Z}[\alpha]$, which has $\Delta(\mathbb{Z}[\alpha]) = \Delta(f_\mathbb{Q}^\alpha)$, and applies the Kummer–Dedekind theorem to determine for which critical primes $p$ the order $\mathbb{Z}[\alpha]$ is singular above $p$ and the inclusion $\mathbb{Z}[\alpha] \subset O_p$ is strict. In case singular primes over $p$ are encountered, the elements $p^{-1}q_i(\alpha) \in O_\mathfrak{p} \setminus \mathbb{Z}[\alpha]$ provided by the theorem are adjoined to $\mathbb{Z}[\alpha]$ to obtain an extension ring $R$ for which the index $[R : \mathbb{Z}[\alpha]]$ is a power of $p$, and which has by (7-6) a discriminant $\Delta(R)$ having fewer factors $p$ than $\Delta(f_\mathbb{Q}^\alpha)$. If $p^2$ still divides $\Delta(R)$, it may be necessary to further extend the ring $R$ to obtain $O_p$, and as $R$ will not in general be simple, we now need an algorithm that is not restricted to the monogenic setting of the Kummer–Dedekind theorem to obtain the extension $R \subset O_p$. There are two ways to proceed.

The first method considers the individual primes $\mathfrak{p} \mid p$ and is also useful to compute valuations at $\mathfrak{p}$ in case $\mathfrak{p}$ is regular. It tries to compute the fractional $R$-ideal

$$\mathfrak{p}^{-1} = R : \mathfrak{p} = \{x \in K : x\mathfrak{p} \subset R\},$$

which clearly satisfies $R \subset \mathfrak{p}^{-1} \subset \frac{1}{p}R$. In fact, $\mathfrak{p}^{-1}$ *strictly* contains $R$. To see this, one picks a nonzero element $x \in \mathfrak{p}$ and notices that, by Theorem 5.2 and Lemma 5.1, the ideal $(x)$ contains a product of prime ideals of $R$. We can write this product as $\mathfrak{p} \cdot I \subset (x)$ and we may assume $I \subsetneq (x)$ by taking the minimal number of primes in the product. For $y \in I \setminus (x)$, the element $a = y/x$ is in $\mathfrak{p}^{-1} \setminus R$.

Finding an element $a = \frac{1}{p}r$ in $\mathfrak{p}^{-1} \setminus R$ amounts to finding a nonzero element $\bar{r} \in R/pR$ that annihilates the ideal $\mathfrak{p}/pR \subset R/pR$, and this is a matter of linear algebra in $R/pR$. If we have $\mathfrak{p}$ in its 'standard form' $\mathfrak{p} = (p, \beta) \subset R$ on 2 generators, then we only need to find a nonzero element annihilating $\bar{\beta} \in R/pR$.

An element $a \in \mathfrak{p}^{-1} \setminus R$ tells us all about $\mathfrak{p}$. As $(R + Ra)\mathfrak{p} \subset R$ is an $R$-ideal containing the maximal ideal $\mathfrak{p}$, we have two possibilities. If $(R + Ra)\mathfrak{p}$ equals $\mathfrak{p}$, then $a$ is in the multiplier ring $\Lambda(\mathfrak{p})$ but not in $R$, so $\mathfrak{p}$ is singular, and we have found an element of $\mathcal{O} \setminus R$ that we use to enlarge $R$. If it equals $R$, then $R + Ra = \mathfrak{p}^{-1}$ is the inverse of the regular ideal $\mathfrak{p}$ in $\mathcal{I}(R)$, and we can use $a$ to determine valuations at $\mathfrak{p}$.

PROPOSITION 9.1. *Let $\mathfrak{p}$ be an invertible $R$-ideal, and $\mathfrak{p}^{-1} = R + Ra$ its inverse. Then the $\mathfrak{p}$-adic valuation of an ideal $I \subset R$ equals*

$$\mathrm{ord}_\mathfrak{p}(I) = \max\{k \geq 0 : a^k I \subset R\}.$$

PROOF. The valuation $\mathrm{ord}_\mathfrak{p}(I)$ of an integral ideal $I$ is the largest integer $k$ for which $\mathfrak{p}^{-k}I = (R + Ra)^k I$ is contained in $R$. □

The second method to enlarge a number ring $R$ to a $p$-maximal extension is similar in nature, but does not find the individual primes over $p$ first. Assuming that $p$ is not a unit in $R$, it defines the *$p$-radical* of $R$ as the intersection or, equivalently, the product

$$I_p = \bigcap_{\mathfrak{p} \in \mathfrak{p}} \mathfrak{p} = \prod_{\mathfrak{p} \in \mathfrak{p}} \mathfrak{p} \supset pR \tag{9-2}$$

of the primes of $R$ lying over $p$. Then $I_p/pR$, being the intersection of all prime ideals of $R/pR$, is the *nilradical* $\text{nil}(R/pR)$ of the finite ring $R/pR$. To compute it, we let $F_p : R/pR \to R/pR$ be the Frobenius map defined by $F_p(x) = x^p$. Then $F_p$ is an $\mathbb{F}_p$-linear map that can be described by a matrix with respect to a basis of the finite $\mathbb{F}_p$-algebra $R/pR$. We have $I_p/pR = \text{nil}(R/pR) = \ker F_p^k$ if $k$ is chosen so that $p^k$ exceeds $\dim_{\mathbb{F}_p}(R/pR) \leq [K:\mathbb{Q}]$. This makes the computation of $I_p/pR \subset R/pR$ a standard matter of linear algebra over $\mathbb{F}_p$.

If we find $I_p/pR = 0$, then $R$ is regular and unramified at $p$, and finding the primes over $p$ amounts to splitting the separable $\mathbb{F}_p$-algebra $R/pR$ into a product of finite fields. This is done by finding the idempotents in $R/pR$, which is easy as these span the kernel of the linear map $F_p - \text{id}$.

The more interesting case arises when the inclusion in (9-2) is strict, that is, when $R$ is singular or ramified above $p$. To find out whether $R$ is singular above $p$ and the inclusion $R \subset O_p$ strict, one now considers the multiplier ring

$$R' = \Lambda(I_p) = \{x \in K : xI_p \subset I_p\}$$

of the $p$-radical of $R$. From $p \in I_p$ we obtain $R \subset R' \subset \frac{1}{p}R$, and as $I_p$ is a finitely generated $R$-ideal, we have $R' \subset O_p$ by Lemma 6.4.

PROPOSITION 9.3. *Define the extensions $R \subset R' \subset O_p$ as above. If the inclusion $R \subset O_p$ is strict, then so is the inclusion $R \subset R'$.*

PROOF. Suppose we have $[O_p : R] = p^r > 1$. As all sufficiently high powers of $I_p$ contain $pR$, we have $I_p^k O_p \subset R$ for large $k$. Let $m \geq 0$ be the largest integer for which we have $I_p^m O_p \not\subset R$, and pick $x \in I_p^m O_p \setminus R \subset O_p \setminus R$. For $y \in I_p$ we now have $xy \in I_p^{m+1} O_p \subset R \cap I_p O_p = I_p$, so $x$ is in $R' \setminus R$. □

By Proposition 9.3, we can find $O_p$ by repeatedly replacing $R$ by $R'$ until we have $R = R' = O_p$. As we can work 'modulo $p$' all the time, the resulting *Pohst–Zassenhaus algorithm* reduces to linear algebra over $\mathbb{F}_p$. One starts by computing $\mathbb{F}_p$-bases for $I_p/pR$ and $I_p/pI_p$ from a basis of $R/pR$. As $I_p/pI_p$ is an $R$-module, we have a structure map

$$\varphi : R \to \text{End}(I_p/pI_p),$$

and we find $R' = \frac{1}{p}N$ for $N = \ker \varphi$. Note that $N = \ker \varphi$ contains $pR$ since $\varphi$ factors via $R/pR$, and that $R'$ is generated over $R$ by $1/p$ times the lifts to $R$ of an $\mathbb{F}_p$-basis for $N/pR$. Computing $N$ is again a matter of linear algebra. A slight drawback of the method is that for $I_p/pI_p$ of $\mathbb{F}_p$-dimension $n \leq [K:\mathbb{Q}]$, the endomorphism ring $\text{End}(I_p/pI_p)$ is a matrix ring of dimension $n^2$ over $\mathbb{F}_p$. Thus, the relevant map $\overline{\varphi} : R/pR \to \text{End}(I_p/pI_p)$ is described by a matrix of size $n^2 \times n$ over $\mathbb{F}_p$.

Once a $p$-regular ring $R$ has been obtained, the extension primes $\mathfrak{p} \mid p$ are found by finding the idempotents of the $\mathbb{F}_p$-algebra $R/pR$. We refer to [Cohen 1993, Chapter 6] for further details.

## 10. Finiteness theorems

In order to use ideal factorization in a number ring $R$ to establish divisibility results between *elements* as one does in $\mathbb{Z}$, there are two obstacles one has to deal with. The first is the obstruction to invertible ideals being principal, which is measured by the Picard group $\mathrm{Pic}(R)$. The second is the problem that generators of principal ideals in $R$ are only unique up to multiplication by *units* in $R$. As the example of the number ring $R = \mathbb{Z}[\sqrt{d}]$ occurring in Pell's equation shows, the unit group $R^*$ may be infinite. We do however have two basic finiteness theorems: the Picard group $\mathrm{Pic}(R)$ of a number ring $R$ is a *finite* abelian group, and for orders $R$, the unit group $R^*$ is a *finitely generated* abelian group. These are not algebraic properties of one-dimensional Noetherian domains, and the proofs use the fact that number rings allow embeddings in Euclidean vector spaces, in which techniques from the geometry of numbers can be applied.

Let $V$ be an $n$-dimensional real vector space equipped with a scalar product $\langle \cdot , \cdot \rangle : V \times V \to \mathbb{R}$, that is, a positive definite bilinear form on $V \times V$. Then we define the *volume* of a parallelepiped $B = \{r_1 x_1 + r_2 x_2 + \cdots + r_n x_n : 0 \leq r_i < 1\}$ spanned by $x_1, x_2, \ldots, x_n$ as

$$\mathrm{vol}(B) = |\det(\langle x_i, x_j \rangle)_{i,j=1}^n|^{1/2}.$$

Thus, the 'unit cube' spanned by an orthonormal basis for $V$ has volume 1, and the image of this cube under a linear map $T$ has volume $|\det(T)|$. If the vectors $x_i$ are written with respect to an orthonormal basis for $V$ as $x_i = (x_{ij})_{j=1}^n$, then we have

$$|\det(\langle x_i, x_j \rangle)_{i,j=1}^n|^{1/2} = |\det(M \cdot M^t)|^{1/2} = |\det(M)|$$

for $M = (x_{ij})_{i,j=1}^n$. The volume function on parallelepipeds can be extended to a Haar measure on $V$ that, under the identification $V \cong \mathbb{R}^n$ via an orthonormal basis for $V$, is the well-known Lebesgue measure on $\mathbb{R}^n$.

A subgroup $L = \mathbb{Z} \cdot x_1 + \mathbb{Z} \cdot x_2 + \cdots + \mathbb{Z} \cdot x_k \subset V$ spanned by $k$ linearly independent vectors $x_i \in V$ is called a *lattice of rank $k$* in $V$. We clearly have $k \leq n$, and all discrete subgroups of $V$ are of this form. If $L \subset V$ has maximal rank $n$, the *covolume* $\mathrm{vol}(V/L)$ of $L$ in $V$ is the volume of a parallelepiped $F$ spanned by a basis of $L$. Such a parallelepiped is a *fundamental domain* for $L$ as every $x \in V$ has a unique representation $x = f + l$ with $f \in F$ and $l \in L$. In fact, $\mathrm{vol}(V/L)$ is the volume of $V/L$ under the induced Haar measure on the factor group $V/L$.

All finiteness results in this section are applications of Minkowski's 'continuous version' of Dirichlet's box principle. His theorem is simple to state and amazingly effective in the sense that *many* results can be derived from it. However, like the box principle itself, it has little algorithmic value since its proof is a pure existence proof that suggests no efficient algorithm.

THEOREM 10.1 (MINKOWSKI). *Let $L$ be a lattice of maximal rank $n$ in $V$. Then every closed bounded subset of $V$ that is convex and symmetric and has volume $\mathrm{vol}(X) \geq 2^n \cdot \mathrm{vol}(V/L)$ contains a nonzero lattice point.*

PROOF. Suppose first that we have $\mathrm{vol}(X) > 2^n \cdot \mathrm{vol}(V/L)$. Then the set $\frac{1}{2}X = \{\frac{1}{2}x : x \in X\}$ has volume $\mathrm{vol}(\frac{1}{2}X) > \mathrm{vol}(V/L)$, so the map $\frac{1}{2}X \to V/L$ cannot be injective. Pick distinct points $x_1, x_2 \in X$ with $\frac{1}{2}x_1 - \frac{1}{2}x_2 = \omega \in L$. As $X$ is symmetric, $-x_2$ is contained in $X$. By convexity, we find that the convex combination $\omega$ of $x_1$ and $-x_2 \in X$ is in $X \cap L$.

Under the weaker assumption $\mathrm{vol}(X) \geq 2^n \mathrm{vol}(V/L)$, we observe that each of the sets $X_\varepsilon = (1+\varepsilon)X$ with $0 < \varepsilon \leq 1$ contains a nonzero lattice point $\omega_\varepsilon \in L$. There are only finitely many distinct lattice points $\omega_\varepsilon \in L \cap 2X$, and a point occurring for infinitely many is in the closed set $X = \cap_\varepsilon X_\varepsilon$. □

If $K$ is a number field of degree $n$ over $\mathbb{Q}$, the base change $\mathbb{Q} \to \mathbb{C}$ provides us with a canonical embedding of $K$ in the $n$-dimensional complex vector space $K_\mathbb{C} = K \otimes_\mathbb{Q} \mathbb{C}$:

$$\Phi_K : \quad K \longrightarrow K_\mathbb{C} \cong \mathbb{C}^n$$
$$x \longmapsto (\sigma(x))_\sigma.$$

Here the isomorphism $K_\mathbb{C} \cong \mathbb{C}^n$ is as in (7-1), with $\sigma$ ranging over the $n$ embeddings $K \to \mathbb{C}$. Note that $\Phi_K$ is a ring homomorphism, and that the norm and trace on the free $\mathbb{C}$-algebra $K_\mathbb{C}$ extend the norm and the trace of the field extension $K/\mathbb{Q}$. The image of $K$ under the embedding lies in the $\mathbb{R}$-algebra

$$K_\mathbb{R} = \{(z_\sigma)_\sigma \in K_\mathbb{C} : z_{\bar{\sigma}} = \bar{z}_\sigma\}$$

consisting of the elements of $K_\mathbb{C}$ invariant under the involution $F : (z_\sigma)_\sigma \mapsto (\bar{z}_\sigma)_\sigma$. Here $\bar{\sigma}$ denotes the embedding of $K$ in $\mathbb{C}$ that is obtained by composition of $\sigma$ with complex conjugation.

On $K_\mathbb{C} \cong \mathbb{C}^n$, we have the standard hermitian scalar product, which satisfies $\langle Fz_1, Fz_2 \rangle = \overline{\langle z_1, z_2 \rangle}$. Its restriction to $K_\mathbb{R}$ is a *real* scalar product that equips $K_\mathbb{R}$ with a Euclidean structure and a *canonical* volume function.

It is customary to denote the real embeddings of $K$ by $\sigma_1, \sigma_2, \ldots, \sigma_r$ and the pairs of complex embeddings of $K$ by $\sigma_{r+1}, \bar{\sigma}_{r+1}, \sigma_{r+2}, \bar{\sigma}_{r+2}, \ldots, \sigma_{r+s}, \bar{\sigma}_{r+s}$.

We have $r + 2s = n$ and an isomorphism of $\mathbb{R}$-algebras

$$K_\mathbb{R} \xrightarrow{\sim} \mathbb{R}^r \times \mathbb{C}^s \qquad (10\text{-}2)$$
$$(z_\sigma)_\sigma \longmapsto (z_{\sigma_i})_{i=1}^{r+s}.$$

The inner product on $K_\mathbb{R}$ is taken componentwise, with the understanding that at a 'complex' component $(\sigma, \bar\sigma)$, the inner product of $z_1 = x_1 + iy_1$ and $z_2 = x_2 + iy_2$ equals

$$\langle \begin{pmatrix}z_1\\ \bar z_1\end{pmatrix}, \begin{pmatrix}z_2\\ \bar z_2\end{pmatrix} \rangle = z_1 \bar z_2 + \bar z_1 z_2 = 2 \operatorname{Re}(z_1 \bar z_2) = 2(x_1 x_2 + y_1 y_2).$$

This differs by a factor 2 from the inner product under the identification of $\mathbb{C}$ with the 'complex plane' $\mathbb{R}^2$, so volumes in $K_\mathbb{R}$ are $2^s$ times *larger* than they are in $\mathbb{R}^r \times \mathbb{C}^s$ with the 'standard' Euclidean structure. The following theorem shows that the 'canonical' volume is indeed canonical.

LEMMA 10.3. *Let $R$ be an order in a number field $K$. Then $\Phi_K[R]$ is a lattice of covolume $|\Delta(R)|^{1/2}$ in $K_\mathbb{R}$.*

PROOF. Choose a $\mathbb{Z}$-basis $\{x_1, x_2, \ldots, x_n\}$ for $R$. Then $\Phi_K[R]$ is spanned by the vectors $(\sigma x_i)_\sigma \in K_\mathbb{R}$. In terms of the matrix $X = (\sigma_i(x_j))_{i,j=1}^n$ following (7-4), the covolume of $\Phi_K[R]$ equals

$$|\det(\langle (\sigma x_i)_\sigma, (\sigma x_j)_\sigma \rangle)_{i,j=1}^n|^{1/2} = |\det(X^t \cdot \bar X)|^{1/2} = |\Delta(R)|^{1/2}. \qquad \square$$

For $I \in \mathcal{I}(R)$, the lattice $\Phi_K[I]$ has covolume $N(I) \cdot |\Delta(R)|^{1/2}$ in $K_\mathbb{R}$. Define the closed convex symmetric subset $X_t \subset K_\mathbb{R}$ by $X_t = \{(z_\sigma)_\sigma \in K_\mathbb{R} : \sum_\sigma |z_\sigma| \leq t\}$, and choose $t$ such that its volume $\operatorname{vol}(X_t) = 2^r \pi^s t^n$ equals $2^n N(I) \cdot |\Delta(R)|^{1/2}$. Using the arithmetic-geometric-mean inequality, we find that $X_t$ contains the $\Phi_K$-image of a nonzero element $x \in I$ of absolute norm

$$|N_{K/\mathbb{Q}}(x)| = \prod_\sigma |\sigma(x)| \leq \left(\frac{1}{n} \sum_\sigma |\sigma(x)|\right)^n \leq \frac{t^n}{n^n} = M_R \cdot N(I),$$

where the Minkowski constant of the order $R$ is defined as

$$M_R = \left(\frac{4}{\pi}\right)^s \frac{n!}{n^n} \cdot |\Delta(R)|^{1/2}. \qquad (10\text{-}4)$$

It follows that $xI^{-1}$ is integral and of norm at most $M_R$. As $R$ has only finitely many ideals of norm at most $M_R$, we obtain our first finiteness result.

THEOREM 10.5. *Let $R$ be an order and $M_R$ its Minkowski constant. Then every ideal class in the Picard group $\operatorname{Pic}(R)$ contains an integral ideal of norm at most $M_R$, and $\operatorname{Pic}(R)$ is a finite abelian group.* $\square$

COROLLARY 10.6. *The Picard group of a number ring is finite.*

PROOF. If $\mathcal{O}$ is Dedekind, this is clear from Theorem 6.5 and (6-6). The case of a general number ring $R$ of conductor $\mathfrak{f}$ in its normalization $\mathcal{O}$ then follows from Theorem 6.7, as $(\mathcal{O}/\mathfrak{f})^*$ is a finite group. □

As the Minkowski constant $M_R$ in (10-4) is at least equal to 1, the absolute value of the discriminant of an order of rank $n$ satisfies

$$|\Delta(R)| \geq \left(\frac{\pi}{4}\right)^{2s}\left(\frac{n^n}{n!}\right)^2.$$

This lower bound grows exponentially with $n$, and for $R \neq \mathbb{Z}$ we have $n > 1$ and $|\Delta(R)| \geq \pi^2/4 > 2$. By Theorem 8.5, it follows that *every* order $R \neq \mathbb{Z}$ has singular or ramifying primes. In particular, the ring of integers of a number field $K \neq \mathbb{Q}$ is ramified at the primes $p$ dividing the integer $|\Delta_K| > 1$.

If we fix the value of the discriminant $\Delta(R)$ in Lemma 10.3, this puts a bound on its rank $n$, and one can use Theorem 10.1 to generate $R$ by elements lying in small boxes of $K_\mathbb{R}$. This leads to *Hermite's theorem*: up to isomorphism, there are only finitely many orders of given discriminant $D$. It gives rise to the problem of finding asymptotic expressions for $x \to \infty$ for the number of number fields $K$, say with $r$ real and $s$ complex primes, for which $|\Delta_K|$ is at most $x$. As the suggested proof of the theorem, based on Theorem 10.1, is not at all constructive, this is a non-trivial problem for $n > 2$. For $n > 3$ there has only recently been substantial progress [Bhargava 2005; ≥ 2008].

In a more geometric direction, the finiteness of the number of curves (up to isomorphism) of given genus and 'bounded ramification' that are defined over $\mathbb{Q}$ is a 1962 conjecture of Shafarevich that was proved by Faltings [Cornell and Silverman 1986] in 1983. The *ineffective* proof yields no explicit cardinalities of any kind.

The unit group of a number ring $R$ has a finite cyclic torsion subgroup $\mu_R$ consisting of the roots of unity in $R$. As $R$ is countable, $R^*/\mu_R$ is countably generated. Not much more can be said in general, as the case $R = K$ shows, but for orders we can be more precise by considering the restriction of the ring homomorphism $\Phi_K$ to $R^*$:

$$\Phi_K : R^* \to K_\mathbb{R}^* = \{(z_\sigma)_\sigma \in K_\mathbb{C}^* : \bar{z}_\sigma = z_{\bar\sigma}\}.$$

In order to produce lattices, we apply the logarithm $z \mapsto \log|z|$ componentwise on $K_\mathbb{C}^* = (\mathbb{C}^*)^n$ to obtain a homomorphism $K_\mathbb{C}^* \to \mathbb{R}^n$ that sends $(z_\sigma)_\sigma$ to $(\log|z_\sigma|)_\sigma$.

THEOREM 10.7 (DIRICHLET UNIT THEOREM). *Let $R$ be an order of maximal rank $n = r + 2s$ in $K$, with $r$ and $s$ as in (10-2). Then the homomorphism*

$$L : R^* \xrightarrow{L} \mathbb{R}^n, \quad x \mapsto (\log|\sigma x|)_\sigma$$

*has kernel $\mu_R$ and maps $R^*$ onto a lattice of rank $r + s - 1$ in $\mathbb{R}^n$.*

PROOF. For a bounded set $B = [-M, M]^n \subset \mathbb{R}^n$, the inverse image in $K_\mathbb{R}^*$ under the logarithmic map is the bounded set $\{(z_\sigma)_\sigma \in K_\mathbb{R} : e^{-M} \leq |z_\sigma| \leq e^M\}$, which has finite intersection with the lattice $\Phi(R)$. Thus $L^{-1}[B] \subset R^*$ is finite, and the discrete subgroup $L[R^*] \subset \mathbb{R}^n$ is a lattice in $\mathbb{R}^n$. Taking $M = 0$, we see that $\ker L$ is finite and equal to $\mu_R$.

We have $\log |\sigma(x)| = \log |\bar\sigma(x)|$ for every $x \in K^*$, so $L[R^*]$ lies in the $(r+s)$-dimensional subspace $\{(x_\sigma)_\sigma \in \mathbb{R}^n : x_\sigma = x_{\bar\sigma}\} \subset \mathbb{R}^n$, and we lose no information if we replace $L$ by its composition $L' : R^* \to \mathbb{R}^n = \mathbb{R}^{r+2s} \to \mathbb{R}^{r+s}$ of $L$ with the linear map that *adds* the components at each of the $s$ pairs of complex conjugate embeddings $(\sigma, \bar\sigma)$ into a single component.

For $\eta \in R^*$ we have $[R : \eta R] = |N_{R/\mathbb{Z}}(\eta)| = \prod_\sigma |\sigma(x)| = 1$, so $L'[R^*] \cong R^*/\mu_R$ is a lattice in the 'trace-zero-hyperplane'

$$H = \{(x_i)_{i=1}^{r+s} : \sum_i x_i = 0\} \subset \mathbb{R}^{r+s}. \tag{10-8}$$

Showing that $L'[R^*]$ has maximal rank $r + s - 1$ in $H$ is done using Theorem 10.1. Let $E = \{(z_\sigma)_\sigma : \prod_\sigma z_\sigma = \pm 1\} \subset K_\mathbb{R}^*$ be the 'norm-$\pm 1$-subspace' that is mapped onto $H$ under the composition $\varphi : K_\mathbb{R}^* \xrightarrow{\log} \mathbb{R}^n \to \mathbb{R}^{r+s}$, and choose $t$ such that the box $X = \{(z_\sigma)_\sigma \in K_\mathbb{R} : |z_\sigma| \leq t \text{ for all } \sigma\}$ has $\mathrm{vol}(X) = 2^n \cdot |\Delta_K|^{1/2}$. For every $e = (e_\sigma)_\sigma \in E$, the set

$$eX = \{ex : x \in X\} = \{(z_\sigma)_\sigma \in K_\mathbb{R} : |z_\sigma| < |e_\sigma|t\}$$

is a box around the origin with volume $\mathrm{vol}(eX) = \mathrm{vol}(X)$, so it contains an element $\Phi(x_e) \in \Phi[R]$ by Theorem 10.1. The norm $N(x_e)$ of $x_e \in R$ is bounded by $\prod_\sigma |e_\sigma| t = t^n$ for each $e$, so the set of *ideals* $\{x_e R : e \in E\}$ is finite, say equal to $\{a_i R\}_{i=1}^k$. Now

$$Y = E \cap \left(\bigcup_{i=1}^k \Phi(a_i^{-1}) X\right)$$

is a bounded subset of $E$ as all boxes $\Phi(a_i^{-1})X$ are bounded in $K_\mathbb{R}$. By the norm condition on the elements of $Y \subset E$, the absolute values $|y_\sigma|$ of $y = (y_\sigma)_\sigma \in Y$ are bounded away from zero, so $\varphi[Y]$ is a *bounded* subset of $H$.

To show that $L'[R^*]$ has maximal rank in $H$, it now suffices to show that we have $L'[R^*] + \varphi[Y] = H$ or, equivalently, $\Phi[R^*] \cdot Y = E$. For the non-trivial inclusion $\supset$, pick $e \in E$. Then there exist a nonzero element $a \in R$ such that $\Phi(a)$ is contained in $e^{-1}X$ and an element $a_i \in R$ as defined above satisfying $a_i a^{-1} = u \in R^*$. It follows that $e$ is contained in $\Phi(a^{-1})X = \Phi(u)\Phi(a_i^{-1})X$, whence in $\Phi[R^*] \cdot Y$. $\square$

Less canonically, Theorem 10.7 states that there exists a finite set $\eta_1, \eta_2, \ldots, \eta_{r+s-1}$ of *fundamental units* in $R$ such that we have

$$R^* = \mu_R \times \langle\eta_1\rangle \times \langle\eta_2\rangle \times \cdots \times \langle\eta_{r+s-1}\rangle.$$

Such a system of fundamental units, which forms a $\mathbb{Z}$-basis for $R^*/\mu_R$, is only unique up to $\mathrm{GL}_{r+s-1}(\mathbb{Z})$-transformations and multiplication by roots of unity.

The regulator of a set $\{\varepsilon_1, \varepsilon_2, \ldots, \varepsilon_{r+s-1}\}$ of elements of norm $\pm 1$ in $K^*$ is defined as

$$\mathrm{Reg}(\varepsilon_1, \varepsilon_2, \ldots, \varepsilon_{r+s-1}) = \left|\det(n_i \log |\sigma_i \varepsilon_j|)_{i,j=1}^{r+s-1}\right|.$$

Here the integer $n_i \in \{1, 2\}$ equals 1 if $\sigma_i$ is a real embedding and 2 otherwise. The *regulator* $\mathrm{Reg}(R)$ of an order $R$ in $K$ is the regulator of a system of fundamental units for $R^*$, with $\mathrm{Reg}(R) = 1$ if $R^*$ is finite. Its value is the covolume of the lattice $L'[R^*]$ in the trace-zero-hyperplane $H$ from (10-8) after a projection $H \xrightarrow{\sim} \mathbb{R}^{r+s-1}$ obtained by leaving out one of the coordinates.

The regulator of the ring of integers of $K$ is simply referred to as the *regulator* $R_K$ of $K$. Unlike the discriminant $\Delta_K$, which is an integer, $R_K$ is a positive real number which is usually transcendental, as it is an expression in terms of *logarithms* of algebraic numbers. For an order $R$ in $K$, we have $[\mathcal{O}_K^* : \mu_K R^*] = \mathrm{Reg}(R^*)/R_K$. For a subring $R \subset K$ that is not an order, we can extend Theorem 10.7 to describe $R^*$, even though $L[R^*] \subset H$ is a dense subset.

THEOREM 10.9. *Let $R$ be a number ring with field of fractions $K$, and define $r$ and $s$ as in* (10-2). *Write $T$ for the set of primes $\mathfrak{p}$ of $\mathcal{O}_K$ for which $R$ contains elements of negative valuation. Then we have an isomorphism*

$$R^* \cong \mu_R \oplus \mathbb{Z}^{r+s-1} \oplus \mathbb{Z}^T,$$

*and $R^*$ is finitely generated if and only if $T$ is finite.*

PROOF. The unit group $R^*$ of $R$ is by Theorem 6.7 of finite index in the unit group $\mathcal{O}^*$ of the normalization $\mathcal{O}$ of $R$. By Theorem 6.5, we have $\mathcal{O} = \mathcal{O}_{K,T}$ for our set $T$, and (6-6) provides us with an exact sequence $1 \to \mathcal{O}_K^* \to \mathcal{O}_{K,T}^* \to \mathbb{Z}^T \to \mathrm{Cl}_K$. As subgroups of finite index in $\mathbb{Z}^T$ are free of the same rank, we have a split exact sequence $1 \to \mathcal{O}_K^* \to \mathcal{O}_{K,T}^* \to \mathbb{Z}^{\#T} \to 1$, and the result follows from Theorem 10.7. □

The $r$ real embeddings and the $s$ complex conjugate pairs of embeddings of $K$ are often referred to as the *real* and *complex* primes of $K$, a point of view on which we will elaborate in Section 13. It is customary to include the set $T_\infty$ of these *infinite primes* in the set $T$ we use in Theorem 6.5 to define the ring $\mathcal{O}_{K,T}$. With this convention, Theorem 10.9 states that the group $\mathcal{O}_{K,T}^*$ of $T$-*units* is the product of $\mu_K$ and a free abelian group of rank $\#T - 1$.

The group $\mu_K$ of roots of unity is easily found. For $r > 0$ we simply have $\mu_K = \mu_\mathbb{R} = \{\pm 1\}$, and for totally complex $K$ the group $\mu_K$ reduces injectively modulo all odd unramified primes of $K$. This implies that the order $w_K$ of $\mu_K$ is an integer dividing $\#(\mathcal{O}_K/\mathfrak{p})^* = p^{f(\mathfrak{p}/p)} - 1$ for all primes $\mathfrak{p} \nmid 2\Delta_K$.

As $w_K$ is actually the greatest common divisor of these orders, a few well-chosen primes are usually enough to determine $w_K$ and the maximal cyclotomic subfield $\mathbb{Q}(\mu_K) \subset K$.

## 11. Zeta functions

Although our approach to number rings has been mostly algebraic, we do need a few results from analytic number theory that play an important role in the verification of the *correctness* of any computation of Picard and unit groups. We do not give the proofs of these results.

For a number field $K$, the *Dedekind zeta function* $\zeta_K$ is the complex analytic function defined on the half plane $\text{Re}(t) > 1$ by

$$\zeta_K(t) = \sum_{I \neq 0} N_{K/\mathbb{Q}}(I)^{-t}, \tag{11-1}$$

where the sum ranges over all nonzero ideals $I \subset \mathcal{O}$ of the ring of integers $\mathcal{O}_K$ of $K$. For $K = \mathbb{Q}$, this is the well-known Riemann zeta function $\zeta(t) = \sum_{n=1}^{\infty} n^{-t}$. The sum defining $\zeta_K(t)$ converges absolutely and uniformly on compact subsets of $\text{Re}(t) > 1$, and the holomorphic limit $\zeta_K$ can be expanded into an Euler product

$$\zeta_K(t) = \prod_{\mathfrak{p}} (1 - N_{K/\mathbb{Q}}(\mathfrak{p})^{-t})^{-1} = \prod_{\mathfrak{p}} (1 - p^{-f(\mathfrak{p}/p)t})^{-1} \tag{11-2}$$

over the primes of $\mathcal{O}_K$; this shows that $\zeta_K$ is zero-free on $\text{Re}(t) > 1$. To see this, note first that for each rational prime number $p$, there are at most $n = [K : \mathbb{Q}]$ primes $\mathfrak{p} \mid p$ by Theorem 8.4, and each of these has $N_{K/\mathbb{Q}}(\mathfrak{p}) = p^{f(\mathfrak{p}/p)} \geq p$. The resulting estimate

$$\sum_{N_{K/\mathbb{Q}}(\mathfrak{p}) \leq X} |N_{K/\mathbb{Q}}(\mathfrak{p})^{-t}| \leq n \sum_{p \leq X} p^{-\text{Re}(t)}$$

shows that $\sum_{\mathfrak{p}} N_{K/\mathbb{Q}}(\mathfrak{p})^{-t}$ converges absolutely and uniformly in every half plane $\text{Re}(t) > 1 + \varepsilon$, so the same is true for the right hand side of (11-2). Multiplication of the geometric series

$$(1 - N_{K/\mathbb{Q}}(\mathfrak{p})^{-t})^{-1} = \sum_{k=0}^{\infty} N_{K/\mathbb{Q}}(\mathfrak{p})^{-kt}$$

for all primes $\mathfrak{p}$ reduces (11-2) to (11-1), as every ideal $I$ has a unique factorization as a product of prime ideal powers.

Hecke proved that $\zeta_K$ can be extended to a holomorphic function on $\mathbb{C} \setminus \{1\}$, and that it has particularly nice properties when Euler factors are added in (11-2) for the $r$ real and $s$ complex primes of $K$. More precisely, the function

$$Z(t) = |\Delta_K|^{t/2} \left( \Gamma(t/2) \pi^{-t/2} \right)^r \left( \Gamma(t)(2\pi)^{-t} \right)^s \zeta_K(t) \tag{11-3}$$

satisfies the simple functional equation $Z(t) = Z(1-t)$. In [Lang 1994, Chapter XIII and XIV], one can find both Hecke's classical proof and Tate's 1959 adelic proof of this result. More details on Tate's approach are found in [Ramakrishnan and Valenza 1999].

Hecke's techniques have been used by Zimmert to show that not only $\Delta_K$ but also the regulator $R_K$ grows exponentially with the degree. An explicit lower bound [Skoruppa 1993] is

$$R_K/w_K \geq .02 \cdot \exp(.46r + .1s), \tag{11-4}$$

with $w_K = \#\mu_K$ the number of roots of unity in $K$. Depending on the degree and the size of the discriminant, there are better lower bounds [Friedman 1989]. For all but nine explicitly known fields, one has $R_K/w_K \geq 1/8$.

The functional equation shows that the meromorphic extension of $\zeta_K$ to $\mathbb{C}$ has 'trivial zeros' at all negative integers $k \in \mathbb{Z}_{<0}$: for $k$ odd the multiplicity of the zero equals $s$, whereas for $k$ even the multiplicity equals $r + s$. All other zeros $\rho$ satisfy $0 < \text{Re}(\rho) < 1$, and one of the deepest open problems in number theory, the *Generalized Riemann Hypothesis (GRH)*, predicts that we actually have $\text{Re}(\rho) = 1/2$ for these non-trivial zeros.

At $t = 1$, the Dedekind zeta function $\zeta_K$ has a simple pole. Its residue

$$2^r (2\pi)^s \frac{h_K R_K}{w_K |\Delta_K|^{1/2}}$$

at $t = 1$ can be found using extensions of the techniques in the previous section [Lang 1994, Section VIII.2, Theorem 5]. It gives us information on the fundamental invariants $h_K$, $R_K$, and $\Delta_K$ of $K$ we have defined before. For $K = \mathbb{Q}$ the residue equals 1, and for general $K$ we can approximate it by evaluating the limit $\lim_{t \to 1} \zeta_K(t)/\zeta_\mathbb{Q}(t)$ using (11-2) to obtain

$$\frac{h_K R_K}{w_K} = 2^{-r} (2\pi)^{-s} |\Delta_K|^{1/2} \prod_p E(p), \tag{11-5}$$

where the Euler factor $E(p)$ at the rational prime $p$ is defined by

$$E(p) = \frac{1 - p^{-1}}{\prod_{\mathfrak{p} \mid p}(1 - p^{-f(\mathfrak{p}/p)})}.$$

Identity (11-5) allows us to approximate $h_K R_K$ by multiplying the Euler factors $E(p)$ for sufficiently many $p$. Convergence is slow, but we will only need single digit precision. In fixed degree $n$, it suggests that $h_K R_K$ is a quantity of order of magnitude $|\Delta_K|^{1/2}$. In an asymptotic sense, this is made precise by the *Brauer–Siegel theorem* [Lang 1994, Section XIII.4, Theorem 4], which states that the quotient of the *logarithms* $\log(h_K R_K)$ and $\frac{1}{2} \log |\Delta_K|$ tends to 1 in any sequence of pairwise non-isomorphic normal number fields $K$ of some fixed degree. The condition of normality can be dispensed with under assumption of GRH, and for the degree it actually suffices to assume that $[K : \mathbb{Q}]/\log|\Delta_K|$ tends to 0. Unfortunately, the theorem is not *effective*.

At $t = 0$, the functional equation shows that $\zeta_K$ has a zero of order $r + s - 1$ with leading coefficient $-h_K R_K/w_K$ in the Taylor expansion. The idea that such a coefficient should 'factor over $\chi$-eigenspaces', much like zeta functions themselves factor into $L$-functions, underlies the largely conjectural theory of *Stark units* [Tate 1984].

## 12. Computing class groups and unit groups

The actual computations of class groups and units groups are inextricably linked, in the sense that one uses a single algorithm yielding both the class group and the unit group. The complexity of the calculation is exponential in any reasonable measure for the size of the number field, and for number rings that do allow explicit calculations of the type discussed in this section, factoring discriminants and computing normalizations are not expected to pose great difficulties. For this reason, we will focus on the computation of the class group and unit group of a number field $K$. For other number rings $R \subset K$, the results from Section 5 can be used to relate $\mathrm{Pic}(R)$ and $R^*$ to $\mathrm{Cl}_K$ and $\mathcal{O}_K^*$.

The kind of computation that yields $\mathrm{Cl}_K$ and $\mathcal{O}_K^*$ has become standard in algorithmic number theory: it factors smooth elements over a suitably chosen factor base and produces relations using linear algebra over $\mathbb{Z}$. In this volume, it occurs in [Lenstra 2008; Pomerance 2008a; 2008b; Stevenhagen 2008; Schirokauer 2008; Schoof 2008b].

Suppose we are given a number field

$$K = \mathbb{Q}[X]/(f)$$

of degree $n = r + 2s$ by means of a defining monic polynomial $f \in \mathbb{Z}[X]$ having $r$ real and $s$ complex conjugate pairs of roots. Computing $r$ and $s$ from $f$ is classical and easy: one counts sign changes in a *Sturm sequence* that can be obtained as a by-product of the computation of the discriminant $\Delta(f)$ from the resultant $R(f, f')$ as in Example 7.9 [Cohen 1993, Theorem 4.1.10]. We take the order $\mathbb{Z}[\alpha]$ defined by $f = f_\mathbb{Q}^\alpha$, and extend it to $\mathcal{O} = \mathcal{O}_K$ using the methods of Sections 8 and 9. Note that this requires factoring $\Delta(f)$. We then select a smoothness bound $B$ such that $\mathrm{Cl}_K$ is generated by the primes $\mathfrak{p}$ in $\mathcal{O}$ of norm at most $B$. One can take for $B$ the Minkowski constant

$$M_K = \left(\frac{4}{\pi}\right)^s \frac{n!}{n^n} \cdot |\Delta_K|^{1/2}$$

from (10-4) or use the asymptotically *much* smaller *Bach bound*

$$B_K = 12(\log|\Delta_K|)^2,$$

which is good enough if one is willing to assume GRH [Bach 1990]. In practice even the Bach bound is usually overly pessimistic, and correct results may be obtained using even smaller values of $B$.

EXAMPLE 12.1. For $f = X^3 + 44$ from Example 8.3, we have $\Delta_K = -2^2 \cdot 3 \cdot 11^2$ and $r = s = 1$. In this case the Minkowski constant $M_K < 11$ is so small that all calculations can be done by hand. As 44 as not a cube modulo 7, the prime 7 is inert in $K$, and $\mathrm{Cl}_K$ is generated by the primes over 2, 3, and 5. We factored these primes explicitly in Theorem 8.4 as $(2) = \mathfrak{p}_2^3$, $(3) = \mathfrak{p}_3^2\mathfrak{q}_3$ and $(5) = \mathfrak{p}_5\mathfrak{p}_{25}$, where the subscripts denote the norms of the primes involved. Using $g = X^3 + X^2 - 7X - 13$, the irreducible polynomial of the generating element $\delta = \beta - \gamma = (\alpha^2 - 2\alpha - 2)/6$ of $\mathcal{O}_K$ we found in Example 8.3, it suffices to tabulate a few smooth values of $g$ and compute the factorization of those principal ideals $(k - \gamma)$ of norm $N_{K/\mathbb{Q}}(k - \gamma) = g(k)$ that are 5-smooth.

| $k$ | $-3$ | $-2$ | $-1$ | $0$ | $1$ | $2$ | $3$ |
|---|---|---|---|---|---|---|---|
| $g(k)$ | $-2 \cdot 5$ | $-3$ | $-2 \cdot 3$ | $13$ | $-2 \cdot 3^2$ | $-3 \cdot 5$ | $2$ |
| $(k-\delta)$ | $\mathfrak{p}_2\mathfrak{p}_5$ | $\mathfrak{q}_3$ | $\mathfrak{p}_2\mathfrak{p}_3$ | — | $\mathfrak{p}_2\mathfrak{q}_3^2$ | $\mathfrak{p}_3\mathfrak{p}_5$ | $\mathfrak{p}_2$ |

Note that $g$ has a double zero modulo 3 at $k = -1$ mod 3 giving rise to the ramified prime $\mathfrak{p}_3 = (3, \delta + 1)$ dividing $(-1 - \delta)$, whereas the unramified prime $\mathfrak{q}_3 = (3, \delta + 2)$ divides $(k - \delta)$ at the values $k \equiv 1$ mod 3. The table shows that $\mathfrak{p}_2 = (3 - \delta)$ and $\mathfrak{q}_3 = (2 + \delta)$ are principal, and by the entries for $k = -1$ and $k = -3$, the other generators $\mathfrak{p}_3 = (1 + \delta)/(3 - \delta)$ and $\mathfrak{p}_5 = (3 + \delta)/(3 - \delta)$ of $\mathrm{Cl}_K$ are principal as well, so without any further computation we have $\mathrm{Cl}_K = 0$. The unused entries $k = 1, 2$ now yield *different* generators for $(1 - \delta)$ and $(2 - \delta)$, namely $(3 - \delta)(2 + \delta)^2$ and $(1 + \delta)(3 + \delta)/(3 - \delta)^2$. Their quotients are the unit $(1 - \delta)(3 - \delta)^{-1}(2 + \delta)^{-2} = -1$ and the non-trivial unit

$$\varepsilon = (2 - \delta)(1 + \delta)^{-1}(3 + \delta)^{-1}(3 - \delta)^2 = -5\delta^2 + 17\delta - 7.$$

In terms of the root $\alpha$ of $f = X^3 + 44$, we have $\varepsilon = \frac{1}{6}(17\alpha^2 - 4\alpha - 226)$. The unit $\varepsilon$ is actually fundamental, so we have $\mathcal{O}_K^* = \langle -1 \rangle \times \langle \varepsilon \rangle$. To *prove* this fact, we have several options.

One can generate more units using the prime ideal factorizations of 5-smooth ideals such as (2), (3), (5), and $(8 + \delta) = \mathfrak{q}_3^4\mathfrak{p}_5$, find that they are all up to sign a power of $\varepsilon$, and decide on *probabilistic* grounds that $\varepsilon$ must be fundamental. Alternatively, one can divide the regulator $\mathrm{Reg}(\varepsilon) \approx 8.3$ by the absolute lower bound for $R_K$ following from (11-4) to obtain $[\mathcal{O}_K^* : \langle -1 \rangle \times \langle \varepsilon \rangle] \geq 33$, and show that $-\varepsilon$, which has norm 1, is not a $p$-th power for any prime number $p \leq 31$.

For this, it suffices to find a prime $\mathfrak{p}$ of norm $N\mathfrak{p} \equiv 1 \bmod p$ in $\mathcal{O}_K$ and to show that we have $\varepsilon^{(N\mathfrak{p}-1)/p} \neq 1 \in \mathcal{O}_K/\mathfrak{p}$. If $\varepsilon$ is *not* a $p$-th power in $K$, such $\mathfrak{p}$ are usually abundant. A third possibility consists in approximating the Euler product (11-5) to precision sufficient to convince oneself that $R_K$ is not equal to $\text{Reg}(\varepsilon)/k$ for some $k > 1$. We will do this in the case of a larger example in Example 12.4.

For a 'small' number field $K = \mathbb{Q}[X]/(f)$ defined by a monic polynomial $f \in \mathbb{Z}[X]$, the basic invariants of $K$ can be found from a small table of factored values of $f$.

EXAMPLE 12.2. We take $f = X^3 + X^2 + 5X - 16$, which gives rise to the values in Table 1. As $f$ has no zeros modulo 11, 13, and 17, it is irreducible modulo these primes, and in $\mathbb{Z}[X]$. The discriminant $\Delta(f) = -R(f, f')$ can be computed as in Example 7.9, and equals

$$-R(X^3 + X^2 + 5X - 16, 3X^2 + 2X + 5) = -3^2 R(\tfrac{28}{9}X - \tfrac{149}{9}, 3X^2 + 2X + 5)$$
$$= -3^2 \cdot \left(\tfrac{28}{9}\right)^2 \cdot \left(3\left(\tfrac{149}{28}\right)^2 + 2\left(\tfrac{149}{28}\right) + 5\right)$$
$$= -8763$$
$$= -3 \cdot 23 \cdot 127.$$

As $\Delta(f)$ is squarefree, $K = \mathbb{Q}[X]/(f)$ has $\Delta_K = -8763$ and $\mathcal{O} = \mathcal{O}_K = \mathbb{Z}[\alpha]$. As $f$ has a single real root, we have $r = s = 1$, and Minkowski's constant equals

$$M_K = \frac{3!}{3^3} \frac{4}{\pi} \sqrt{8763} \approx 26.5.$$

The primes in $\mathcal{O}$ of norm at most 25 can be found by factoring the rational primes up to 23 in $\mathcal{O}$. This is an application of Theorem 8.2 using the values of $f$ in

| $k$ | $f(k)$ | $k$ | $f(k)$ |
|---|---|---|---|
| $-10$ | $-2 \cdot 3 \cdot 7 \cdot 23$ | 0 | $-2^4$ |
| $-9$ | $-709$ | 1 | $-3^2$ |
| $-8$ | $-2^3 \cdot 3^2 \cdot 7$ | 2 | $2 \cdot 3$ |
| $-7$ | $-3 \cdot 5 \cdot 23$ | 3 | $5 \cdot 7$ |
| $-6$ | $-2 \cdot 113$ | 4 | $2^2 \cdot 3 \cdot 7$ |
| $-5$ | $-3 \cdot 47$ | 5 | $3 \cdot 53$ |
| $-4$ | $-2^2 \cdot 3 \cdot 7$ | 6 | $2 \cdot 7 \cdot 19$ |
| $-3$ | $-7^2$ | 7 | $3 \cdot 137$ |
| $-2$ | $-2 \cdot 3 \cdot 5$ | 8 | $2^3 \cdot 3 \cdot 5^2$ |
| $-1$ | $-3 \cdot 7$ | 9 | $839$ |

**Table 1.** Values of $f = X^3 + X^2 + 5X - 16$.

our table. Leaving out the inert primes 11, 13, and 17, we obtain factorizations

$$20 = \mathfrak{p}_2\mathfrak{p}_4 = (2,\alpha)\cdot(2,\alpha^2+\alpha+1)$$
$$30 = \mathfrak{p}_3^2\mathfrak{q}_3 = (3,\alpha+1)^2\cdot(3,\alpha-1)$$
$$50 = \mathfrak{p}_5\mathfrak{p}_{25} = (5,\alpha+2)\cdot(5,x_5)$$
$$70 = \mathfrak{p}_7\mathfrak{q}_7\mathfrak{r}_7 = (7,\alpha+1)(7,\alpha-3)(7,\alpha+3)$$
$$190 = \mathfrak{p}_{19}\mathfrak{p}_{361} = (19,\alpha-4)\cdot(19,x_{19})$$
$$230 = \mathfrak{p}_{23}^2\mathfrak{q}_{23} = (23,\alpha+7)^2(23,\alpha+10)$$

in which $x_5$ and $x_{19}$ denote elements that we do not bother to compute. This shows that $\mathrm{Cl}_K$ is generated by the classes of the primes $\mathfrak{p}_2$, $\mathfrak{p}_3$, $\mathfrak{p}_5$, $\mathfrak{p}_{19}$, $\mathfrak{p}_{23}$, and two of the primes over 7. We can express the classes of the large primes in those of smaller primes using the factorizations of principal ideals $(k-\alpha)$ resulting from the values of $f(k)$ in our table.

The entry $k=-7$ yields $(-7-\alpha) = \mathfrak{p}_3\mathfrak{p}_5\mathfrak{p}_{23}$, so we can omit $[\mathfrak{p}_{23}]$ from our list of generators. Similarly, we can omit $[\mathfrak{p}_{19}]$ as the entry $k=6$ gives $(6-\alpha) = \mathfrak{p}_2\mathfrak{p}_7\mathfrak{p}_{19}$. The primes over 7 can be dealt with using the identities $(-1-\alpha) = \mathfrak{p}_3\mathfrak{p}_7$ and $(3-\alpha) = \mathfrak{p}_5\mathfrak{q}_7$. The relation $(-2-\alpha) = \mathfrak{p}_2\mathfrak{q}_3\mathfrak{p}_5 = 3\mathfrak{p}_2\mathfrak{p}_3^{-2}\mathfrak{p}_5$ takes care of $[\mathfrak{p}_5]$, and finally $(2-\alpha) = \mathfrak{p}_2\mathfrak{p}_3$ shows that the class group of $K$ is generated by $[\mathfrak{p}_2]$. The order of this class divides 4 since we have $(\alpha) = \mathfrak{p}_2^4$, and further relations do not indicate that it is smaller.

To show that $\mathfrak{p}_2^2$ is not principal and that $\mathrm{Cl}_K = \langle[\mathfrak{p}_2]\rangle$ is cyclic of order 4, we need to know the group $\mathcal{O}^*/(\mathcal{O}^*)^2$. As in Example 12.1, we can produce a non-trivial unit from the fact that the factorizations of 3, $(\alpha)$, $(\alpha-1)$, and $(\alpha-2)$ involve only $\mathfrak{p}_2$ and the primes over 3. One deduces that

$$\eta = \frac{(\alpha-1)(\alpha-2)^4}{9\alpha} = 4\alpha^2 + \alpha - 13$$

is a unit of norm $N(\eta) = 1$. From the Dirichlet unit theorem (with $r = s = 1$) we have $\mathcal{O}^* \cong \langle -1 \rangle \times P$, where $P \cong \mathbb{Z}$ can be taken to be the group of units of norm 1. In order to prove that $\eta$ generates $P/P^2$, it suffices to show that $\eta$ is not a square in $\mathcal{O}^*$. This is easy: reducing modulo $\mathfrak{p}_3$ we find $\eta \equiv 4 - 1 - 13 \equiv -1$, and $-1$ is not a square in $\mathcal{O}/\mathfrak{p}_3 = \mathbb{F}_3$.

Suppose now that $\mathfrak{p}_2^2 = (y)$ is principal. Then $y^2$ and $\alpha$ are both generators of $\mathfrak{p}_2^4$, so there exists a unit $\varepsilon$ with $y^2 = \varepsilon\cdot\alpha$. As the norm $N(\varepsilon\cdot\alpha) = 16N(\varepsilon) = N(y)^2$ is positive, we have $\varepsilon \in P$. If $\varepsilon$ is in $P^2$, then $\alpha = \varepsilon^{-1}y^2$ is a square, contradicting that we have $\alpha \equiv -2 \bmod \mathfrak{p}_5$. If $\varepsilon$ is in $\eta P^2$, then $\eta\cdot\alpha$ is a square, and this is contradicted by the congruence

$$\eta\cdot\alpha \equiv (4(-2)^2 + (-2) - 13)\cdot -2 \equiv 3 \bmod \mathfrak{p}_5.$$

We conclude that no unit $\varepsilon$ exists, and that $\mathfrak{p}^2$ has order 2 in $\mathrm{Cl}_K \cong \mathbb{Z}/4\mathbb{Z}$. As in Example 12.1, we can show in various ways that $\eta$ is actually a fundamental unit, and that we have $\mathcal{O}^* = \langle -1 \rangle \times \langle \eta \rangle$.

In number fields $K = \mathbb{Q}(\alpha)$ that are bigger than the two baby examples we just did, more work is involved, but the underlying idea remains the same. Having chosen a factor bound $B$, one explicitly obtains the factorization of all primes $p \leq B$ using the Kummer–Dedekind theorem for $\mathbb{Z}[\alpha]$. For those primes $p$ dividing the index of $\mathbb{Z}[\alpha]$ in $\mathcal{O}_K$, some extra work is needed as $\mathcal{O}/p\mathcal{O}$ may not be monogenic over $\mathbb{F}_p$. However, everything can be determined using the techniques of Section 9, using linear algebra over $\mathbb{F}_p$. One is left with a factor base consisting of the set $T$ of primes of norm at most $B$.

Next, one tries to factor sufficiently many principal ideals $(x) \subset \mathcal{O}_K$ over the chosen factor base. The $B$-smooth elements $x \in \mathcal{O}_K \setminus \{0\}$ for which this is possible generate the subgroup $\mathcal{O}_{K,T}^* \subset K^*$ of $T$-units, and we have an exact sequence

$$1 \to \mathcal{O}_K^* \longrightarrow \mathcal{O}_{K,T}^* \xrightarrow{F} \mathbb{Z}^T \xrightarrow{C} \mathrm{Cl}_K \to 0.$$

Here $F$ is the 'factorization map' that sends $x \in \mathcal{O}_{K,T}^*$ to its exponent vector $(\mathrm{ord}_{\mathfrak{p}}(x\mathcal{O}_K))_{\mathfrak{p} \in T}$, and $C$ is the natural map sending the characteristic function of $\mathfrak{p} \in T$ to the ideal class of $[\mathfrak{p}]$. As $\mathcal{O}_{K,T}^*$ is a free abelian group of rank $\#T + r + s - 1$ by Theorem 10.9, we expect it to be generated by any 'random' subset of its elements of cardinality substantially larger than its rank.

EXAMPLES 12.3. In $K = \mathbb{Q}(\sqrt[3]{44})$ from Example 12.1, the set $T$ of primes of norm at most 5 consists of 4 primes, and $\#T + r + s - 1$ equals 5. The 6 elements $k - \delta$ with $|k| \in \{1, 2, 3\}$ already generate $\mathcal{O}_{K,T}^* \cong \langle -1 \rangle \times \mathbb{Z}^5$.

In Example 12.2, with $T$ consisting of the 3 primes of norm at most 3, the elements $3$, $(\alpha)$, $(\alpha - 1)$, and $(\alpha - 2)$ are independent in $\mathcal{O}_{K,T}^* \cong \langle -1 \rangle \times \mathbb{Z}^4$.

For a subset $X \subset \mathcal{O}_{K,T}^*$ generating a subgroup of maximal rank, it is a matter of linear algebra over $\mathbb{Z}$ to reduce the matrix of exponent vectors $(F(x))_{x \in X}$ and to compute the group $\mathbb{Z}^T / F[X]$, which will be of finite order

$$h' = [F[\mathcal{O}_{K,T}^*] : F[X]] \cdot h_K$$

if $X$ is sufficiently large. The dependencies found between the vectors give rise to elements $u \in \mathcal{O}_K^* = \ker(F)$ generating a subgroup $U \subset \mathcal{O}_K^*$. For these elements we compute their log-vectors $L'(u) \in H \subset \mathbb{R}^{r+s}$ as in Theorem 10.7. Linear algebra over $\mathbb{R}$ will give us $r + s - 1$ independent units generating the lattice $L[U]$ if $X$ is sufficiently large, and the associated regulator is

$$R' = [\mathcal{O}_K^* : L[U]] \cdot R_K.$$

If our set $X$ truly generates $\mathcal{O}_{K,T}^*$, then we have $h' = h_K$ and $R' = R_K$, so

$$\mathrm{Cl}_K \cong \mathbb{Z}^T / F[X]$$

and $\mathcal{O}_K^* = U$. If this is not the case, $h'R' > h_K R_K$ will be an integral multiple of $h_K R_K$, and we discover this by comparing our value $h'R'$ with the analytic estimate of $h_K R_K$ obtained by approximating (11-5) with a truncated Euler product. The factors $E(p)$ in (11-5) are computed from the 'factorization type' of $p$ in $\mathcal{O}_K$, which we know already for $p < B$, and which follows for more $p$, if desired, from the factorization type of the defining polynomial $f$ modulo $p$.

The description of the fundamental units furnished by the algorithm is a *power product representation* in terms of $T$-units in $X$. Although the set $X$ we pick to generate $\mathcal{O}_{K,T}^*$ usually consists of elements that are relatively small, the units obtained from them can be huge when written out on a basis of $K$ over $\mathbb{Q}$. This is unavoidable in view of the order of magnitude $|\Delta_K|^{1/2}$ of $h_K R_K$: in many cases $h_K$ appears to be rather small, and this means that the regulator measuring the *logarithmic* size of the unit group will be of size $|\Delta_K|^{1/2}$. In such cases the units themselves require a number of bits that is exponential in $\log |\Delta_K|$. Already in the simplest non-trivial case of real quadratic fields, the phenomenon of the smallest solution of the Pell equation $x^2 - dy^2 = 1$ being very large in comparison to $d > 0$ was noticed 350 years ago by Fermat.

This paper does not intend to present cutting-edge examples of the performance of the algorithm above, which involve serious linear algebra to reduce large matrices over $\mathbb{Z}$. The final example below is small and 'hands-on' like the cubic Examples 12.1 and 12.2. It illustrates the use of log-vectors in the determination of the unit group, and the analytic confirmation of the algebraically obtained output.

EXAMPLE 12.4. Let $K$ be the quartic field generated by a root $\alpha$ of the polynomial

$$f = X^4 - 2X^2 + 3X - 7 \in \mathbb{Z}[X]$$

of prime discriminant $\Delta(f) = -98443$. Then we have $\mathcal{O}_K = \mathbb{Z}[\alpha]$, an order with $r = 2$, $s = 1$ and Minkowski constant $M_K \approx 37.4$. To deal with all primes of norm up to 37, we tabulate consecutive values of $f$ in Table 2 on the next page.

This shows that $f$ has no roots modulo the primes $p = 2, 3, 17, 23, 29$, and also modulo 37 once we check $37 \nmid f(18)$. In fact, $f$ is irreducible modulo 2 and 3, and the factorization $(5) = \mathfrak{p}_5 \mathfrak{q}_5 \mathfrak{p}_{25}$ shows that $\mathrm{Cl}_K$ is generated by the ideals of prime norm $p \in [5, 31]$, which all 'occur' in Table 2.

| $n$ | $f(n)$ | $n$ | $f(n)$ | $n$ | $f(n)$ |
|---|---|---|---|---|---|
| $-18$ | $127 \cdot 821$ | $-6$ | $11 \cdot 109$ | $6$ | $5 \cdot 13 \cdot 19$ |
| $-17$ | $5 \cdot 11^2 \cdot 137$ | $-5$ | $7 \cdot 79$ | $7$ | $7 \cdot 331$ |
| $-16$ | $64969$ | $-4$ | $5 \cdot 41$ | $8$ | $5 \cdot 797$ |
| $-15$ | $50123$ | $-3$ | $47$ | $9$ | $7^2 \cdot 131$ |
| $-14$ | $5^2 \cdot 7^2 \cdot 31$ | $-2$ | $-5$ | $10$ | $11 \cdot 19 \cdot 47$ |
| $-13$ | $19 \cdot 1483$ | $-1$ | $-11$ | $11$ | $5^2 \cdot 577$ |
| $-12$ | $5 \cdot 7 \cdot 11 \cdot 53$ | $0$ | $-7$ | $12$ | $20477$ |
| $-11$ | $83 \cdot 173$ | $1$ | $-5$ | $13$ | $5 \cdot 5651$ |
| $-10$ | $13 \cdot 751$ | $2$ | $7$ | $14$ | $7 \cdot 5437$ |
| $-9$ | $5 \cdot 19 \cdot 67$ | $3$ | $5 \cdot 13$ | $15$ | $149 \cdot 337$ |
| $-8$ | $31 \cdot 127$ | $4$ | $229$ | $16$ | $5 \cdot 7 \cdot 11 \cdot 13^2$ |
| $-7$ | $5^2 \cdot 7 \cdot 13$ | $5$ | $11 \cdot 53$ | $17$ | $31 \cdot 2677$ |

**Table 2.** Values of $f = X^4 - 2X^2 + 3X - 7 \in \mathbb{Z}[X]$.

In case $f(n) = \pm p$ is prime, the prime ideal $(p, \alpha - n)$ is principal and generated by $\alpha - n$. The following list of ideals of prime norm $p \leq 31$ results.

$$\mathfrak{p}_5 = (\alpha - 1) \qquad \mathfrak{q}_5 = (\alpha + 2)$$
$$\mathfrak{p}_7 = (\alpha) \qquad \mathfrak{q}_7 = (\alpha - 2)$$
$$\mathfrak{p}_{11} = (\alpha + 1) \qquad \mathfrak{q}_{11} = (11, \alpha - 5)$$
$$\mathfrak{p}_{13} = (13, \alpha - 3) \qquad \mathfrak{q}_{13} = (13, \alpha - 6)$$
$$\mathfrak{p}_{19} = (19, \alpha - 6) \qquad \mathfrak{q}_{19} = (19, \alpha + 9)$$
$$\mathfrak{p}_{31} = (31, \alpha + 8) \qquad \mathfrak{q}_{31} = (31, \alpha + 14)$$

The primes lying over 5 and 7 are all principal, and so is $\mathfrak{p}_{11}$. This suggests strongly that $\mathrm{Cl}(\mathcal{O})$ is trivial. In order to prove this, we try to express all primes in the table in terms of the principal ideals. From the entry with $k = 3$ in our table we obtain $(3 - \alpha) = \mathfrak{p}_{13}\mathfrak{q}_5$, showing that $\mathfrak{p}_{13}$ is principal. The relation $(16 - \alpha) = \mathfrak{p}_5 \mathfrak{q}_7 \mathfrak{q}_{11} \mathfrak{p}_{13}^2$ then shows that $\mathfrak{q}_{11}$ is also principal. Similarly, we have principality of $\mathfrak{q}_{13}$ from $(-7 - \alpha) = \mathfrak{q}_5^2 \mathfrak{p}_7 \mathfrak{q}_{13}$ and of $\mathfrak{p}_{19}$ from $(6 - \alpha) = \mathfrak{p}_5 \mathfrak{q}_{13} \mathfrak{p}_{19}$. Finally, we use $(-14 - \alpha) = \mathfrak{p}_5^2 \mathfrak{p}_7^2 \mathfrak{q}_{31}$ to eliminate $\mathfrak{q}_{31}$. This exploits all useful relations from our table, leaving us with the primes $\mathfrak{q}_{19}$ and $\mathfrak{p}_{31}$. In order to prove that these primes are also principal, we factor a small element in them. Modulo $\mathfrak{q}_{19} = (19, \alpha + 9)$ we have $\alpha = -9 \in \mathbb{F}_{19}$, and $1 - 2\alpha$ is therefore a small element in the ideal. Similarly, we have $\alpha = -8 \in \mathbb{F}_{31}$ when working modulo $\mathfrak{p}_{31} = (31, \alpha + 8)$, so $1 + 4\alpha$ is in $\mathfrak{p}_{31}$. The norms of these elements are $N(1 - 2\alpha) = 2^4 f(1/2) = -5 \cdot 19$ and $N(1 + 4\alpha) = (-4)^4 f(-1/4) = 5 \cdot 13 \cdot 31$, which implies that $\mathfrak{q}_{19}$ and $\mathfrak{p}_{31}$ are principal. The corresponding explicit factorizations are $(1 - 2\alpha) = \mathfrak{q}_5 \mathfrak{q}_{19}$ and $(1 + 4\alpha) = \mathfrak{p}_5 \mathfrak{p}_{13} \mathfrak{p}_{31}$. This proves that $\mathrm{Cl}(\mathcal{O})$ is trivial.

At this stage, we have produced explicit generating elements for all prime ideals of norm below the Minkowski bound. Although we do not need all of these generators, we list them for completeness sake.

$$\mathfrak{p}_5 = (\alpha - 1) \qquad \mathfrak{q}_5 = (\alpha + 2)$$
$$\mathfrak{p}_7 = (\alpha) \qquad \mathfrak{q}_7 = (\alpha - 2)$$
$$\mathfrak{p}_{11} = (\alpha + 1) \qquad \mathfrak{q}_{11} = (32\alpha^3 + 53\alpha^2 + 25\alpha + 138)$$
$$\mathfrak{p}_{13} = (\alpha^3 - 2\alpha^2 - 2\alpha - 2) \qquad \mathfrak{q}_{13} = (2\alpha^3 - 4\alpha^2 + 5\alpha - 6)$$
$$\mathfrak{p}_{19} = (\alpha^3 + \alpha^2 + \alpha + 8) \qquad \mathfrak{q}_{19} = (\alpha^3 - 2\alpha^2 + 2\alpha - 3)$$
$$\mathfrak{p}_{31} = (2\alpha^3 + 3\alpha^2 + 2\alpha + 11) \qquad \mathfrak{q}_{31} = (\alpha^3 + \alpha^2 + 6)$$

These generators are not necessarily the smallest or most obvious generators of the ideals in question, they happen to come out of the arguments by which we eliminated all generators of the class group. The search for units that is to follow will provide other generators, and one can for instance check that the large coefficients of our generator for $\mathfrak{q}_{11}$ are not necessary as we have $\mathfrak{q}_{11} = (\alpha^2 - 3)$.

From now on every further factorization of a principal ideal $(x)$ as a product of primes in this table will give us a unit in $\mathcal{O}$: since both $x$ and a product of generators from our table generate $(x)$, this means that their quotient is a unit. Trying some elements $a + b\alpha$, for which we can easily compute the norm $N_{K/\mathbb{Q}}(a + b\alpha) = b^4 f(-a/b)$, one quickly generates a large number of units. The rank of the unit group $\mathcal{O}^*$ for our field $K$ equals $r + s - 1 = 2$, so some administration is needed to keep track of the subgroup of $\mathcal{O}^*$ generated by these units. As in the proof of the Dirichlet unit theorem, one looks at the lattice in $\mathbb{R}^2$ generated by the 'log-vectors' $L(u) = (\log|\sigma_1(u)|, \log|\sigma_2(u)|)$ for each unit $u$. Here $\sigma_1$ and $\sigma_2$ are taken to be the real embeddings $K \to \mathbb{R}$, so they send $\alpha$ to the real roots $\alpha_1 \approx -2.195$ and $\alpha_2 \approx 1.656$ of $f$.

The table below lists a couple of units obtained from small elements $a + b\alpha$.

| relation | $u$ | $L(u)$ |
|---|---|---|
| $(2\alpha + 1) = \mathfrak{q}_{11}\mathfrak{q}_{13}$ | $\alpha^3 - 2\alpha^2 + 3\alpha - 4$ | $(3.4276, -3.7527)$ |
| $(2\alpha - 3) = \mathfrak{q}_{31}$ | $\alpha^3 - 2\alpha^2 + 3\alpha - 4$ | $(3.4276, -3.7527)$ |
| $(2\alpha + 3) = \mathfrak{p}_5^2\mathfrak{q}_7$ | $-3\alpha^3 - 5\alpha^2 - 2\alpha - 12$ | $(-3.4276, 3.7527)$ |
| $(3\alpha + 1) = \mathfrak{q}_5\mathfrak{q}_7\mathfrak{p}_{19}$ | $5\alpha^3 - 11\alpha^2 + 14\alpha - 16$ | $(5.0281, -1.2731)$ |
| $(3\alpha - 5) = \mathfrak{q}_{13}$ | $\alpha^3 - 4\alpha + 2$ | $(-1.6005, -2.4796)$ |
| $(3\alpha - 4) = \mathfrak{q}_5^2\mathfrak{q}_{11}$ | $-4743\alpha^3 + 10412\alpha^2 - 13371\alpha + 15124$ | $(11.8833, -8.7785)$ |
| $(4\alpha - 7) = \mathfrak{q}_5\mathfrak{p}_7\mathfrak{p}_{11}$ | $-\alpha^3 + 2\alpha^2 - 3\alpha + 4$ | $(3.4276, -3.7527)$ |

We see that

$$\eta_1 = \alpha^3 - 2\alpha^2 + 3\alpha - 4 \quad \text{and} \quad \eta_2 = \alpha^3 - 4\alpha + 2$$

are likely to be fundamental. From the log-vectors, the units in the fourth and sixth lines of the table are easily identified (up to sign) as being equal to $\eta_1 \eta_2^{-1}$ and $\eta_1^3 \eta_2^{-1}$.

In order to show that $\mathcal{O}^*$ is equal to $\langle -1 \rangle \times \langle \eta_1 \rangle \times \langle \eta_2 \rangle$, we have to check that the regulator of $K$ is equal to

$$\text{Reg}(\eta_1, \eta_2) = \begin{vmatrix} \log |\sigma_1(\eta_1)| & \log |\sigma_1(\eta_2)| \\ \log |\sigma_2(\eta_1)| & \log |\sigma_2(\eta_2)| \end{vmatrix} \approx \begin{vmatrix} 3.4276 & -1.6005 \\ -3.7527 & -2.4796 \end{vmatrix} \approx 14.506.$$

If this is the case, the residue in $t = 1$ of the zeta function $\zeta_K(t)$ of $K$ should equal

$$\frac{2^r (2\pi)^s h_K R(\eta_1, \eta_2)}{w_K \sqrt{|\Delta|}} \approx \frac{2^2 (2\pi) \cdot 1 \cdot 14.506}{2 \cdot \sqrt{98443}} \approx 0.5810.$$

We can approximate this residue using (11-5), using the Euler product $\prod_p E(p)$, with

$$E(p)^{-1} = \frac{\prod_{\mathfrak{p} \mid p} (1 - N_{K/\mathbb{Q}}(\mathfrak{p})^{-1})}{1 - p^{-1}}.$$

The factor $E(p)^{-1}$ is a polynomial expression in $p^{-1}$ that depends only on the residue class degrees of the primes $\mathfrak{p} \mid p$, that is, on the degrees of the irreducible factors of the defining polynomial $f$ modulo $p$. If we disregard the single ramified prime 98443, there are 5 possible factorization types of $f$ modulo $p$. If the number $n_p$ of zeros of $f$ mod $p$ equals 4, 2, or 1, we immediately know the degree of all irreducible factors of $f$ mod $p$. For $n_p = 0$, the polynomial $f$ is either irreducible modulo $p$ or a product of two quadratic irreducibles, and we can use the fact that the *parity* of the number $g$ of irreducible factors of $f$ mod $p$ can be read off from $\left(\frac{\Delta(f)}{p}\right) = (-1)^{n-g}$ for $p \nmid \Delta(f)$. It follows that we have

$$E(p)^{-1} = \begin{cases} (1 - p^{-1})^3 & \text{if } n_p = 4; \\ (1 - p^{-1})(1 - p^{-2}) & \text{if } n_p = 2; \\ 1 - p^{-3} & \text{if } n_p = 1; \\ (1 + p^{-1})(1 - p^{-2}) & \text{if } n_p = 0 \text{ and } \left(\frac{\Delta(f)}{p}\right) = 1; \\ 1 + p^{-1} + p^{-2} + p^{-3} & \text{if } n_p = 0 \text{ and } \left(\frac{\Delta(f)}{p}\right) = -1. \end{cases}$$

The following data indicate the speed of convergence of this product.

| $N$ | $\prod_{p<N} E(p)$ | $N$ | $\prod_{p<N} E(p)$ |
|---|---|---|---|
| 100 | 0.625211 | 5000 | 0.579408 |
| 200 | 0.595521 | 10000 | 0.579750 |
| 500 | 0.581346 | 20000 | 0.581892 |
| 1000 | 0.584912 | 50000 | 0.581562 |
| 2000 | 0.585697 | 100000 | 0.581423 |

We see that the convergence is non-monotonous and slow, but all values are close to the expected value 0.5810. If our units $\eta_1$ and $\eta_2$ were not fundamental, the Euler product should be at least twice as small as 0.5810, which is highly unlikely. Under GRH, one can effectively bound the error of a finite approximation [Buchmann and Williams 1989, Theorem 3.1] and *prove* the correctness of the result obtained.

## 13. Completions

We have seen in Section 5 that a regular prime $\mathfrak{p}$ of a number ring $R$ gives rise to a discrete valuation ring $R_\mathfrak{p} \subset R$ with maximal ideal containing $\mathfrak{p}$. If $\mathfrak{p}$ is singular, there is a discrete valuation ring with this property for every prime $\widetilde{\mathfrak{p}}$ over $\mathfrak{p}$ in the normalization of $R$ from Theorem 6.5. Thus, the discrete valuation rings having a given number field $K$ as their field of fractions correspond bijectively to the primes of the ring of integers $\mathcal{O}_K$ of $K$. One may even follow the example of the geometers in their definition of *abstract non-singular curves* [Hartshorne 1977, Section I.6] and say that these discrete valuation rings *are* the *primes* or *places* of $K$.

For each prime $\mathfrak{p}$ of $\mathcal{O}_K$, one can use the discrete valuation $\text{ord}_\mathfrak{p} : K \to \mathbb{Z} \cup \{\infty\}$ to define a $\mathfrak{p}$-adic *absolute value* or *exponential valuation* on $K$ by

$$|x-y|_\mathfrak{p} = N_{K/\mathbb{Q}}(\mathfrak{p})^{-\text{ord}_\mathfrak{p}(x-y)}. \tag{13-1}$$

By (5-5), it satisfies $|xy|_\mathfrak{p} = |x|_\mathfrak{p}|y|_\mathfrak{p}$ for $x, y \in K$ and the *ultrametric inequality*

$$|x+y|_\mathfrak{p} \leq \max\{|x|_\mathfrak{p}, |y|_\mathfrak{p}\}. \tag{13-2}$$

Instead of $N_{K/\mathbb{Q}}(\mathfrak{p}) = \#(\mathcal{O}_K/\mathfrak{p})$, we could have taken any real number $c > 1$ in (13-1) to get an equivalent metric inducing the same topology on $K$, but our normalization will be natural in view of (13-4) and the remark following it.

One may now *complete* the number field $K$ as in [Weiss 1963, Section I.7] with respect to the metric in (13-1) to obtain a field $K_\mathfrak{p}$ that is complete with respect to the $\mathfrak{p}$-adic absolute value, using a process similar to the construction of the field of real numbers $\mathbb{R}$ as consisting of *limits* of Cauchy sequences of rational numbers. If we choose a *uniformizer* $\pi \in K$ of order $\text{ord}_\mathfrak{p}(\pi) = 1$ as in Proposition 5.4(3) and some finite set $S \subset \mathcal{O}$ of representatives of the cosets of $\mathfrak{p}$ in $\mathcal{O}$, then every $x \in K_{\mathfrak{p}}$ can *uniquely* be written as a converging Laurent series

$$x = \sum_{k=k_0}^{\infty} a_k \pi^k \in K_{\mathfrak{p}}$$

with 'digits' $a_k \in S$. The field operations can be performed as for real numbers represented in terms of their decimal expansions, so effective computations in $K_{\mathfrak{p}}$ are possible to any given $\mathfrak{p}$-adic *precision*.

Topologically, the fields $K_{\mathfrak{p}}$ are locally compact fields, but their topology is different from that of the more familiar *archimedean* locally compact fields $\mathbb{R}$ and $\mathbb{C}$. As a result of the *non-archimedean* ultrametric inequality (13-2), small quantities do not become large when repeatedly added to themselves in $K_{\mathfrak{p}}$, so all open disks $\{x \in K_{\mathfrak{p}} : |x|_{\mathfrak{p}} < \varepsilon\}$ around $0 \in K_{\mathfrak{p}}$ are additive subgroups of $K_{\mathfrak{p}}$. By the discreteness of the absolute value, these open disks are also closed in $K_{\mathfrak{p}}$, and $K_{\mathfrak{p}}$ is a *totally disconnected* topological space.

The closure $O_{\mathfrak{p}}$ of the ring of integers $\mathcal{O}_K$ in $K_{\mathfrak{p}}$, which is equal to the closed unit disk of radius 1 in $K_{\mathfrak{p}}$, is a compact *ring* consisting of the $\mathfrak{p}$-*adic integers*, that is, elements of the form $\sum_{k \geq 0} a_k \pi^k$.

For $K = \mathbb{Q}$, the completion at primes leads as in [Gouvêa 1993; Koblitz 1984] to the rings $\mathbb{Z}_p$ of $p$-adic integers and the $p$-adic fields $\mathbb{Q}_p$ from [Buhler and Wagon 2008, Section 4.3]. Just like the field of real numbers, the field $\mathbb{Q}_p$ is algebraically 'simpler' than $\mathbb{Q}$: the number of extensions of $\mathbb{Q}_p$ of fixed degree $n$ (inside an algebraic closure) is a *finite* number. In fact, defining such extensions by a monic polynomial from $\mathbb{Z}_p[X]$, one can derive this from the compactness of $\mathbb{Z}_p$ by showing (*Krasner's lemma* [Lang 1994, Proposition II.2.3]) that irreducible polynomials in $\mathbb{Z}_p[X]$ that are coefficientwise sufficiently close define the same extension of $\mathbb{Q}_p$. In particular, all finite extensions of $\mathbb{Q}_p$ arise as completions of number fields at primes over $p$.

For a number field $K = \mathbb{Q}(\alpha)$ defined by a monic polynomial $f = f_{\mathbb{Q}}^{\alpha} \in \mathbb{Z}[X]$, finding the primes of $K$ lying over a rational prime $p$ amounts to factoring $f$ over $\mathbb{Q}_p$, as a factorization $f = \prod_{i=1}^{s} f_i \in \mathbb{Q}_p[X]$ yields an isomorphism

$$K \otimes_{\mathbb{Q}} \mathbb{Q}_p = \mathbb{Q}_p[X]/(f) \xrightarrow{\sim} \prod_{i=1}^{s} \mathbb{Q}_p[X]/(f_i) = \prod_{i=1}^{s} K_{\mathfrak{p}_i} \qquad (13\text{-}3)$$

that maps the subring $\mathbb{Z}[\alpha] \otimes_{\mathbb{Z}} \mathbb{Z}_p = \mathbb{Z}_p[X]/(f)$ into $\prod_{i=1}^{s} O_{\mathfrak{p}_i}$, with equality if and only if $\mathbb{Z}[\alpha]$ is regular over $p$.

By Hensel's lemma [Buhler and Wagon 2008, Section 4.3], the factorization of $f$ over $\mathbb{Q}_p$ can be found by factoring $f$ modulo a sufficiently high power of $p$ and lifting the factors by a Newton-type algorithm. Finite precision is enough to determine the nature of the extension field $K_{\mathfrak{p}_i}$ corresponding to a factor $f_i$ in (13-3), and in this light the Kummer–Dedekind Theorem 8.2 is a first step that exploits the factorization of $(f \bmod p) \in \mathbb{F}_p[X]$. In case $f \bmod p$ is separable,

Hensel's lemma yields the 'unramified case' of the Kummer–Dedekind theorem (Theorem 8.2), and $K_{\mathfrak{p}_i} = \mathbb{Q}_p[X]/(\bar{f}_i)$ is *the* unramified extension of $\mathbb{Q}_p$ of degree $f(\mathfrak{p}_i/p) = \deg(\bar{f}_i)$.

Apart from the completions at the primes $\mathfrak{p}$ of the ring of integers, we also have completions of $K$ arising from the embeddings $\sigma : K \to \mathbb{C}$ occurring in (10-2). Up to complex conjugation, there are $r$ real and $s$ complex embeddings known as the *infinite* primes of $K$, and the completion is either $\mathbb{R}$ or $\mathbb{C}$ for these primes. We normalize the absolute values at an infinite prime $\mathfrak{p}$ corresponding to $\sigma$ in each of these cases by putting $|x|_{\mathfrak{p}} = |\sigma(x)|$ if $\mathfrak{p}$ is real, and $|x|_{\mathfrak{p}} = |\sigma(x)|^2$ if $\mathfrak{p}$ is complex.

The normalization of infinite absolute values is actually *the same* as for the finite absolute values in (13-1). To explain this, we note that each *local field* $K_{\mathfrak{p}}$ obtained by completing $K$ at a prime $\mathfrak{p}$ is locally compact, and comes with a translation invariant measure [Ramakrishnan and Valenza 1999, Theorem 1.8] that is unique up to a scalar factor; this is the *Haar measure* $\mu_{\mathfrak{p}}$. For the infinite primes, $\mu_{\mathfrak{p}}$ is a multiple of the familiar Lebesgue measure on $\mathbb{R}$ or $\mathbb{C}$ that we used implicitly in the volume computations in the $\mathbb{R}$-algebra $K_{\mathbb{R}} = \prod_{\mathfrak{p}\mid\infty} K_{\mathfrak{p}}$ from (10-2). The normalization of the infinite absolute values is inspired by the fact that for $x \in K_{\mathfrak{p}}$, we have $\mu_{\mathfrak{p}}(xB_{\mathfrak{p}}) = |x|_{\mathfrak{p}}\mu_{\mathfrak{p}}(B_{\mathfrak{p}})$ for all measurable subsets $B_{\mathfrak{p}} \subset K_{\mathfrak{p}}$. For finite $\mathfrak{p}$, the very same identity gives rise to the normalization (13-1): multiplication by $x$ increases all volumes in $K_{\mathfrak{p}}$ by a factor $|x|_{\mathfrak{p}}$.

With our normalization, the product $\prod_{\mathfrak{p}\mid\infty} |x|_{\mathfrak{p}}$ of the infinite absolute values of $x$ equals $|N_{K/\mathbb{Q}}(x)|$ by the remark following (7-1). In view of (13-1) and the compatibility of element and ideal norm, we arrive at the *product formula*

$$\prod_{\mathfrak{p}} |x|_{\mathfrak{p}} = 1 \qquad \text{for } x \in K^*, \tag{13-4}$$

where the product is taken over *all* primes of $K$, both finite and infinite. It is the arithmetic analogue of the complex geometric fact that functions on curves have 'as many zeros as they have poles' when we count them with multiplicity and all 'points at infinity' are included in our (projective) curves.

## 14. Adeles and ideles

Despite the intrinsic differences between finite and infinite primes, the product formula already indicates that it is often useful to treat them equally, in order to obtain a closer analogy with the geometric situation, where all 'places' of a curve are of the same finite nature. This has given rise to the concept of the *adele ring* $\mathbf{A}_K = \prod'_{\mathfrak{p}} K_{\mathfrak{p}}$ of $K$, a global object obtained by taking a restricted direct product of *all* completions $K_{\mathfrak{p}}$, both finite and infinite. The restriction means that we deal only with elements that are in the local ring of integers $O_{\mathfrak{p}}$

at almost all finite $\mathfrak{p}$, that is,

$$\mathbf{A}_K = \{(x_\mathfrak{p})_\mathfrak{p} \in \prod_\mathfrak{p} K_\mathfrak{p} : |x_\mathfrak{p}|_\mathfrak{p} \leq 1 \text{ for almost all } \mathfrak{p}\}.$$

With this restriction, the adele ring is naturally a locally compact ring whose topology is generated by sets of the form $\prod_{\mathfrak{p} \in T} X_\mathfrak{p} \times \prod_{\mathfrak{p} \notin T} O_\mathfrak{p}$. Here $T$ is a finite set of primes containing the infinite primes, $X_\mathfrak{p}$ is some open subset of the locally compact field $K_\mathfrak{p}$, and $O_\mathfrak{p}$ is the compact ring of integers in $K_\mathfrak{p}$ at finite $\mathfrak{p}$. The diagonal embedding $K \to \mathbf{A}_K$ defined by $x \mapsto (x)_\mathfrak{p}$ makes $K$ into a subring of $\mathbf{A}_K$.

For $A_\mathbb{Q} = \mathbb{R} \times \prod'_p \mathbb{Q}_p$, the compact open neighborhood $W = [-\frac{1}{2}, \frac{1}{2}] \times \prod_p \mathbb{Z}_p$ of 0 satisfies $W \cap \mathbb{Q} = \{0\}$ and $\mathbb{Q} + W = \mathbf{A}_\mathbb{Q}$. It follows that $\mathbb{Q}$ is discrete in $\mathbf{A}_\mathbb{Q}/\mathbb{Q}$ and that the quotient group $\mathbf{A}_\mathbb{Q}/\mathbb{Q}$ is compact. The same statements hold for $K \subset \mathbf{A}_K$ and the quotient group $\mathbf{A}_K$, as (13-3) shows that the adele ring $\mathbf{A}_K \cong \mathbf{A}_\mathbb{Q} \otimes_\mathbb{Q} K$ is obtained from $\mathbf{A}_\mathbb{Q}$ by applying the base change $\mathbb{Q} \to K$. One can show that the locally compact group $\mathbf{A}_K$ is naturally isomorphic to its Pontryagin dual, and that isomorphism makes $K$ into its own annihilator. This lies at the basis of the adelic proof of the functional equation of the zeta function alluded to after (11-3).

The unit group $\mathbf{A}_K^* = \prod'_\mathfrak{p} K_\mathfrak{p}^*$ of the adele ring of $K$ is the *idele group*

$$\mathbf{A}_K^* = \{(x_\mathfrak{p})_\mathfrak{p} \in \prod_\mathfrak{p} K_\mathfrak{p}^* : |x_\mathfrak{p}|_\mathfrak{p} = 1 \text{ for almost all } \mathfrak{p}\}. \tag{14-1}$$

This is naturally a locally compact group, when its topology is generated as above by sets of the form $\prod_{\mathfrak{p} \in T} Y_\mathfrak{p} \times \prod_{\mathfrak{p} \notin T} O_\mathfrak{p}^*$. The idele group contains $K^*$ diagonally as a discrete subgroup of *principal ideles*. We define the *norm* of an idele by

$$\|(x_\mathfrak{p})_\mathfrak{p}\| = \prod_\mathfrak{p} |x_\mathfrak{p}|_\mathfrak{p}, \tag{14-2}$$

which is well defined by (14-1), and we note that the norm map factors by the product formula (13-4) via the *idele class group* $C_K = \mathbf{A}_K^*/K^*$. The idele class group is of fundamental importance in class field theory, which describes the Galois group of the maximal abelian extension of $K$ over $K$ as a quotient of $C_K$ under the *Artin map* [Cohen and Stevenhagen 2008]. In the most classical case where $K$ is imaginary quadratic and the class field theory goes under the name of *complex multiplication*, ideles have proved to be a most convenient tool even in a computational setting [Gee and Stevenhagen 1998].

For every prime $\mathfrak{p}$, the local unit group $U_\mathfrak{p} = \{x_\mathfrak{p} \in K_\mathfrak{p}^* : |x_\mathfrak{p}|_\mathfrak{p} = 1\}$ is a maximal compact subgroup of $K_\mathfrak{p}^*$. It is equal to $O_\mathfrak{p}^*$ if $\mathfrak{p}$ is finite, and to the group $\{z : |z| = 1\}$ in $\mathbb{R}^*$ or $\mathbb{C}^*$ if $\mathfrak{p}$ is infinite and $K_\mathfrak{p}$ is isomorphic to $\mathbb{R}$ or $\mathbb{C}$. The subgroup $U_K = \prod_\mathfrak{p} U_\mathfrak{p}$ is a maximal compact subgroup of $\mathbf{A}_K^*$ that intersects $K^*$ in the group $\mu_K$ of roots of unity of $K$. It is the kernel of the surjective homomorphism

$$A_K^* \xrightarrow{\delta} \text{Div}_K = \bigoplus_{\mathfrak{p}<\infty} \mathbb{Z} \times \bigoplus_{\mathfrak{p}|\infty} \mathbb{R}$$
$$(x_\mathfrak{p})_\mathfrak{p} \longmapsto ((\text{ord}_\mathfrak{p}(x_\mathfrak{p}))_\mathfrak{p}, (-\log|x_\mathfrak{p}|)_\mathfrak{p})$$

to the *Arakelov divisor group* $\text{Div}_K$ of $K$. The elements of $\text{Div}_K$ are usually represented as finite formal sums $D = \sum_\mathfrak{p} n_\mathfrak{p} \mathfrak{p}$, with $n_\mathfrak{p} \in \mathbb{Z}$ if $\mathfrak{p}$ is finite and $n_\mathfrak{p} \in \mathbb{R}$ if $\mathfrak{p}$ is infinite. One can think of $\text{Div}_K$ as a 'completion' of the group $\mathcal{I}(\mathcal{O}_K) = \bigoplus_{\mathfrak{p}<\infty} \mathbb{Z}$ of finite divisors from Theorem 5.7 by $\bigoplus_{\mathfrak{p}|\infty} \mathbb{R} = \mathbb{R}^{r+s}$, the group of infinite divisors occurring in (10-8). In these terms, every $x \in K^*$ gives rise to a *principal* Arakelov divisor $\delta(x) = (x\mathcal{O}_K, -L'(x)) \in \text{Div}_K$, with $L'$ the homomorphic extension to $K^*$ of the logarithmic map $L': R^* \to \mathbb{R}^{r+s}$ from the proof of the Dirichlet unit theorem (Theorem 10.7). The quotient $\text{Pic}_K$ of $\text{Div}_K$ modulo its subgroup of principal divisors fits in an exact sequence

$$1 \to \mu_K \longrightarrow K^* \xrightarrow{\delta} \text{Div}_K \longrightarrow \text{Pic}_K \to 1$$

that is analogous to (4-1), and much closer to the definition of the Picard group of a *complete* algebraic curve.

Just as for functions on an algebraic curve, we define a *degree map* deg : $A_K^* \to \mathbb{R}$ in terms of the norm (14-2) by $\deg(x) = -\log \|x\|$. As ideles in $U_K$ and principal ideles have degree 0, this gives rise to a homomorphism deg : $\text{Pic}_K \to \mathbb{R}$. Its kernel $\text{Pic}_K^0$ is the *Arakelov class group* of $K$ that occurs center stage in [Schoof 2008b]. As in [Proposition 2.2] there, it is an extension

$$0 \to H/L'[\mathcal{O}_K^*] \to \text{Pic}_K^0 \to \text{Cl}_K \to 0 \qquad (14\text{-}3)$$

of the ordinary class group $\text{Cl}_K$ of $K$ by the 'unit torus' obtained by taking the trace-zero-hyperplane $H \subset \mathbb{R}^{r+s}$ from (10-8) modulo the unit lattice $L'[\mathcal{O}_K^*]$ of covolume the regulator $R_K$ of $K$. The algorithm in Section 12 for computing class groups and unit groups in $K$ can be viewed as an algorithm for computing the Arakelov class group $\text{Pic}_K^0$, and it is in these terms that a proper analysis of the algorithm can be given; see [Lenstra 1992, Section 6] and [Schoof 2008b, Section 12].

By (14-3), the compactness of the Arakelov class group $\text{Pic}_K^0$ is tantamount to the finiteness of $\text{Cl}_K$ and the compactness of $H/L'[\mathcal{O}_K^*]$ expressed by the Dirichlet unit theorem (Theorem 10.7). As $\text{Pic}_K^0$ is the quotient of the group $C_K^1 = \ker[\deg : C_K \to \mathbb{R}]$ of idele classes of norm 1 (and degree zero) by the compact group $U_K$, the compactness of $\text{Pic}_K^0$ implies the compactness of $C_K^1$. One can also go in the reverse direction, prove the compactness of $C_K^1$ directly as in [Cassels and Fröhlich 1967, Section II.16], and derive from this the finiteness results of Corollary 10.6 and Theorem 10.7.

## 15. Galois theory

Let $K$ be a number field that is Galois over $\mathbb{Q}$ with group $G = \text{Gal}(K/\mathbb{Q})$. Then $G$ acts on every object that is 'intrinsically defined' in terms of $K$. Examples of such objects are the ring of integers $\mathcal{O}_K$, its unit group $\mathcal{O}_K^*$, and its class group $\text{Cl}_K$, and in each of these examples the natural problem of determining their *Galois module structure* over $\mathbb{Z}[G]$ was and is an area of active research [Fröhlich 1983; Weiss 1996].

A number ring with field of fractions $K$ does not necessarily have an action of $G$, but the number ring $R$ generated by all $G$-conjugates of the original ring is a *Galois number ring* that does. For an order $\mathbb{Z}[\alpha]$, this amounts to passing to the 'Galois order' $R$ generated by *all* roots of $f = f_\mathbb{Q}^\alpha$. The invariant ring

$$R^G = \{x \in R : \sigma(x) = x \text{ for all } \sigma \in G\} = R \cap \mathbb{Q}$$

is equal to $\mathbb{Z}$ for such an order, and the fundamental observation that we have a *transitive* $G$-action on the primes of $R$ extending a given rational prime is true in great generality.

LEMMA 15.1. *Let $A$ be a commutative ring and $G \subset \text{End}(A)$ a finite group of automorphisms. If $\varphi, \psi : A \to k$ are homomorphisms to a domain $k$ that coincide on the invariant ring $A^G$, then $\varphi$ equals $\psi \circ \sigma$ for some $\sigma \in G$.*

PROOF. Extend $\varphi$ and $\psi$ coefficientwise to homomorphisms $\varphi, \psi : A[X] \to k[X]$. An element $a \in A$ is a zero of the polynomial $f = \prod_{\sigma \in G}(X - \sigma a) \in A^G[X]$ on which $\varphi$ and $\psi$ coincide, so $\varphi(a)$ is a zero of $\varphi(f) = \psi(f) = \prod_{\sigma \in G}(X - \psi \sigma a) \in k[X]$, and we have $\varphi(a) = (\psi \sigma)(a)$ for some $\sigma \in G$ since $k$ is a domain. Now $\sigma$ depends on $a$, but our argument shows that the union over $\sigma \in G$ of

$$A_\sigma = \{a \in A : \varphi(a) = (\psi \sigma)(a)\}$$

equals $A$. To show that we have $A = A_\sigma$ for some $\sigma$, as stated by the lemma, we repeat the previous argument starting with the maps $\varphi, \psi : A[X] \to k[X]$ to obtain

$$A[X] = \bigcup_{\sigma \in G}(A[X])_\sigma = \bigcup_{\sigma \in G} A_\sigma[X].$$

If $a_\sigma \in A \setminus A_\sigma$ exists for all $\sigma \in G$, all polynomials $\sum_{\sigma \in G} a_\sigma X^{n_\sigma}$ that are sums of monomials of *different* degrees $n_\sigma$ are in $A[X]$ but not in $A_\sigma[X]$ for any $\sigma \in G$. □

The primes over $p$ in a Galois number ring $R$ are kernels of homomorphisms $R \to k$, with $k = \overline{\mathbb{F}}_p$ an algebraic closure of $\mathbb{F}_p$, and they extend the homomorphism $R^G = \mathbb{Z} \to \mathbb{F}_p$. By Lemma 15.1, they are transitively permuted by the Galois group. As a consequence, the residue class degree $f_p = f(\mathfrak{p}/p)$ for a prime $\mathfrak{p} \mid p$ does not depend on the choice of the extension prime $\mathfrak{p}$ in $R$. If

$R$ is regular above $p$, the same is true for the ramification index $e_\mathfrak{p} = e(\mathfrak{p}/p)$, and, with $g_p$ the number of primes in $R$ lying over $p$, Theorem 8.4 for Galois number rings becomes

$$e_p f_p g_p = [K : \mathbb{Q}]. \qquad (15\text{-}2)$$

Identity (15-2) is actually an identity for the primes over $p$ in a Galois number field $K$. It also holds for the infinite prime $p = \infty$ of $\mathbb{Q}$, as does Theorem 8.4, if we set $f_\infty = 1$ and $e_\infty = [K_\mathfrak{p} : \mathbb{R}] \in \{1, 2\}$.

EXAMPLE 15.3. Let $f = X^3 + 44$ be as in Example 8.3, and $K = \mathbb{Q}(\zeta_3, \sqrt[3]{44})$ a splitting field of $f$ over $\mathbb{Q}$. Then $K$ is Galois over $\mathbb{Q}$ with nonabelian Galois group of order 6. The prime 3 ramifies in the quadratic subfield $\mathbb{Q}(\zeta_3) = \mathbb{Q}(\sqrt{-3})$, so the primes in $K$ over 3 have even ramification index $e_3$. As 3 has two extensions in $\mathbb{Q}(\sqrt[3]{44})$ by Example 8.3, we have $g_3 \geq 2$. From $e_3 f_3 g_3 = 6$ we find $e_3 = 2$, $f_3 = 1$, and $g_3 = 3$. This shows without any explicit computation that the primes occurring in the factorization $(3) = \mathfrak{p}_3^2 \mathfrak{q}_3$ in $\mathbb{Q}(\sqrt[3]{44})$ factor in the quadratic extension $K$ of $\mathbb{Q}(\sqrt[3]{44})$ as $\mathfrak{p}_3 \mathcal{O}_K = \mathfrak{P}_3 \mathfrak{P}_3'$ and $\mathfrak{q}_3 \mathcal{O}_K = \mathfrak{Q}_3^2$. A similar argument leads to the same values of $e$, $f$, and $g$ for describing the 'ramification' of the infinite prime $p = \infty$.

The prime 5 is inert in $\mathbb{Q}(\zeta_3)$ and splits as $(5) = \mathfrak{p}_5 \mathfrak{p}_{25}$ in $\mathbb{Q}(\sqrt[3]{44})$. For this prime, $f_5$ is even and $g_5$ is at least 2, so we have $e_5 = 1$, $f_5 = 2$, and $g_5 = 3$. We conclude that $\mathfrak{p}_5$ is inert in $\mathbb{Q}(\sqrt[3]{44}) \subset K$, giving rise to a prime $\mathfrak{P}_{25}$ of norm 25, and that $\mathfrak{p}_{25}$ splits into two primes $\mathfrak{Q}_{25}$ and $\mathfrak{R}_{25}$ of norm 25 each.

Let $K$ be Galois over $\mathbb{Q}$, and $\mathfrak{p} \mid p$ a prime of $K$. Then the stabilizer

$$G_\mathfrak{p} = \{\sigma \in G : \sigma\mathfrak{p} = \mathfrak{p}\} \subset G$$

of $\mathfrak{p}$ is the *decomposition group* of $\mathfrak{p}$. It is the subgroup of automorphisms in $G$ that leave the $\mathfrak{p}$-adic absolute value on $K$ invariant, and it may be identified with the Galois group $\mathrm{Gal}(K_\mathfrak{p}/\mathbb{Q}_p)$ of the $\mathfrak{p}$-adic completion over $\mathbb{Q}_p$. For $p = \infty$, read $\mathbb{Q}_\infty = \mathbb{R}$.

As $G$ acts transitively on the primes over $p$, all decomposition groups of primes over $p$ are conjugate in $G$. The $G$-set $G/G_\mathfrak{p}$ of left cosets of $G_\mathfrak{p}$ in $G$ may be identified with the set of extensions of $p$ to $K$. As $G/G_\mathfrak{p}$ has cardinality $g_p$, the order of $G_\mathfrak{p}$ equals $e_p f_p$ by (15-2).

For finite primes, the decomposition group $G_\mathfrak{p}$ acts naturally as a group of automorphisms on the residue class field extension $\mathbb{F}_p \subset k_\mathfrak{p} = \mathcal{O}_K/\mathfrak{p}$, which is cyclic of degree $f_p$ and has a canonical generator of its Galois group in the *Frobenius automorphism* $\mathrm{Frob}_\mathfrak{p} : x \mapsto x^p$ on $k_\mathfrak{p}$.

LEMMA 15.4. *For every prime $\mathfrak{p} \mid p$ in $K$, there exists $\sigma_\mathfrak{p} \in \mathrm{Gal}(K/\mathbb{Q})$ inducing the Frobenius automorphism $\mathrm{Frob}_\mathfrak{p}$ on $k_\mathfrak{p}$.*

PROOF. Applying Lemma 15.1 with $\psi : \mathcal{O}_K \to k_\mathfrak{p}$ the reduction map and $\varphi = \mathrm{Frob}_\mathfrak{p} \circ \psi$, we find that there exists $\sigma_\mathfrak{p} \in G$ that induces $\mathrm{Frob}_\mathfrak{p}$. □

Denoting the kernel of reduction modulo $\mathfrak{p}$ in $G_\mathfrak{p}$ by $I_\mathfrak{p}$, we obtain an exact sequence
$$1 \longrightarrow I_\mathfrak{p} \longrightarrow G_\mathfrak{p} \longrightarrow \mathrm{Gal}(k_\mathfrak{p}/\mathbb{F}_p) \longrightarrow 1. \tag{15-5}$$
The *inertia group* $I_\mathfrak{p}$ is a normal subgroup of $G_\mathfrak{p}$ of order $e_p$. Its invariant field $K^{I_\mathfrak{p}}$ is the largest subfield of $K$ on which the $\mathfrak{p}$-adic valuation is an *unramified* prime over $p$. The invariant subfield $K^{G_\mathfrak{p}}$ of the decomposition group itself is the largest subfield of $K$ for which the completion under the $\mathfrak{p}$-adic valuation is equal to $\mathbb{Q}_p$. For infinite primes $\mathfrak{p}$, we use the convention $I_\mathfrak{p} = G_\mathfrak{p}$ to make this correct.

EXAMPLE 15.6. Take $K = \mathbb{Q}(\zeta_3, \sqrt[3]{44})$ as in Example 15.3. The primes 2 and 11 are unramified and inert in $\mathbb{Q} \subset \mathbb{Q}(\zeta_3)$, with extensions that are totally ramified in $K/\mathbb{Q}(\zeta_3)$. Their decomposition groups are equal to $G$ itself, and their inertia groups are equal to the normal subgroup of index 2 in $G \cong S_3$.

The decomposition groups at the three primes over 3 in $G$ are the three subgroups of order 2. Note that these are conjugate subgroups, and that $G_{\mathfrak{Q}_3} = I_{\mathfrak{Q}_3}$ is the subgroup with invariant field $\mathbb{Q}(\sqrt[3]{44})$. For the three unramified primes over 5, we have the same three decomposition groups of order 2, and $G_{\mathfrak{P}_{25}}$ is the one with invariant field $\mathbb{Q}(\sqrt[3]{44})$.

The prime 7 splits in $\mathbb{Q}(\zeta_3)$ into two primes that remain inert in $K/\mathbb{Q}(\zeta_3)$. Their decomposition group is the normal subgroup of order 3 in $G$. The decomposition groups of the six extension primes of the totally splitting prime 13 are trivial.

For unramified primes $\mathfrak{p}$, the decomposition group $G_\mathfrak{p} \cong \mathrm{Gal}(k_\mathfrak{p}/\mathbb{F}_p)$ in (15-5) is cyclic of order $f_p$ with *canonical* generator $\sigma_\mathfrak{p}$, the *Frobenius at* $\mathfrak{p}$ in $G$. The elements $\mathrm{Frob}_\mathfrak{p} \in G$ for the primes $\mathfrak{p} \mid p$ form a conjugacy class $C_p \subset G$ in case $p$ is unramified in $K$. In the case of Example 15.6, the three conjugacy classes of $G \cong S_3$ are realized by the smallest unramified primes 5, 7, and 13. It is even true that every conjugacy class occurs as the *Frobenius class* for infinitely many $p$, in the following precise sense.

THEOREM 15.7 (CHEBOTAREV DENSITY THEOREM). *Let $\mathbb{Q} \subset K$ be Galois with group $G$ and $C \subset G$ a conjugacy class. Then the set of rational primes $p$ that are unramified in $K$ and have $C$ as their Frobenius class is infinite and has natural density $\#C/\#G$ in the set of all primes.*

EXAMPLE 15.8. In the case of the degree 6 field $K$ from Example 15.6, the totally splitting primes $p$ having 'trivial' Frobenius class $C_p = \{\mathrm{id}\} \subset G$ form a set of density $1/6$, whereas the unramified primes $p \equiv 2 \bmod 3$ having the three elements of order 2 in $G$ in their Frobenius class form, as expected, a set of density $1/2$. The primes $p$ modulo which $X^3 + 44$ is irreducible have the two elements of order 3 in $G$ in their Frobenius class and form a set of density $1/3$.

The cyclotomic field $\mathbb{Q}(\zeta_n)$ is abelian over $\mathbb{Q}$ with Galois group

$$\mathrm{Gal}(\mathbb{Q}(\zeta_n)/\mathbb{Q}) \cong (\mathbb{Z}/n\mathbb{Z})^*$$
$$(\sigma_a : \zeta_n \mapsto \zeta_n^a) \leftrightarrow (a \bmod n). \tag{15-9}$$

Here $\{\sigma_p\}$ is the Frobenius class of $p$, and Theorem 15.7 reduces to Dirichlet's theorem on primes in arithmetic progressions: the primes $p \nmid n$ are equidistributed over $(\mathbb{Z}/n\mathbb{Z})^*$. Chebotarev's original proof (1924) of Theorem 15.7 reduces the general case by a clever trick (see [Stevenhagen and Lenstra 1996, Appendix]) to the cyclotomic case, and forms a key ingredient in the proof of *Artin's reciprocity law* in class field theory [Cohen and Stevenhagen 2008], a far reaching generalization of Example 15.8 to arbitrary abelian extensions of number fields. Assuming class field theory, there are shorter proofs [Lang 1994, Theorem VIII.4.10] of Theorem 15.7.

If $K$ is not Galois over $\mathbb{Q}$, the absolute Galois group $G_\mathbb{Q} = \mathrm{Gal}(\overline{\mathbb{Q}}/\mathbb{Q})$ of $\mathbb{Q}$ acts not on $K$ itself but on its *fundamental set* $X_K = \mathrm{Hom}(K, \overline{\mathbb{Q}})$ of embeddings of $K$ in $\overline{\mathbb{Q}}$. Choosing $\overline{\mathbb{Q}}$ as a subfield of $\mathbb{C}$, the elements of $X_K$ are the $n$ embeddings $\sigma : K \to \mathbb{C}$ considered in Section 10. The images $\sigma[K] \subset \overline{\mathbb{Q}}$ for $\sigma \in X_K$ generate the *normal closure* $L$ of $K$ in $\overline{\mathbb{Q}}$, which is Galois over $\mathbb{Q}$. The natural left action of $G_\mathbb{Q}$ on $X_K$ by composition factors via the finite quotient $\mathrm{Gal}(L/\mathbb{Q})$. Writing $K = \mathbb{Q}(\alpha)$, one may, more classically, identify $X_K$ with the $G_\mathbb{Q}$-set of roots of $f_\mathbb{Q}^\alpha$ in $\overline{\mathbb{Q}}$ under $\sigma \mapsto \sigma(\alpha)$, and view $L$ as the splitting field over $\mathbb{Q}$ of the polynomial $f_\mathbb{Q}^\alpha$. The splitting of a prime $p$ in $K$ can be described in terms of the Galois action on $X_K$ of the decomposition and inertia groups $G_\mathfrak{P}$ and $I_\mathfrak{P}$ in $\mathrm{Gal}(L/\mathbb{Q})$ of a prime $\mathfrak{P} \mid p$ in $L$.

THEOREM 15.10. *Let $K$ be a number field with fundamental set $X_K$ and normal closure $L$ over $\mathbb{Q}$. Given a prime $p$ and integers $e_i, f_i > 0$ for $i = 1, 2, \ldots, t$ with $\sum_{i=1}^t e_i f_i = [K : \mathbb{Q}]$, the following are equivalent:*

(1) *there are $t$ different primes $\mathfrak{p}_1, \mathfrak{p}_2, \ldots, \mathfrak{p}_t$ over $p$ in $K$ having $e(\mathfrak{p}_i/p) = e_i$ and $f(\mathfrak{p}_i/p) = f_i$;*

(2) *for any prime $\mathfrak{P}$ over $p$ in $L$, there are $t$ different $G_\mathfrak{P}$-orbits $X_i \subset X_K$ of length $\#X_i = e_i f_i$; under the action of $I_\mathfrak{P}$ on $X_i$, there are $f_i$ orbits of length $e_i$.*

We will merely sketch the proof, stressing once more the analogy between finite and infinite primes. For the infinite prime $p = \infty$, we embed $\overline{\mathbb{Q}}$ in the algebraic closure $\mathbb{C}$ of the completion $\mathbb{R}$ of $\mathbb{Q}$ at $p$ to view $X_K$ as the set of embeddings of $K$ in $\mathbb{C}$. Then the absolute Galois group $G_\mathbb{R} = \mathrm{Gal}(\mathbb{C}/\mathbb{R})$ of $\mathbb{R}$ acts on $X_K$, and embeddings in $\mathbb{C}$ give rise to the same infinite prime on $K$ if and only if they are complex conjugate, as in Section 10, and we see that the $G_\mathbb{R}$-orbits of $X_K$ of length 1 and 2 correspond to the real and complex primes of $K$.

For finite $p$, we embed $\overline{\mathbb{Q}}$ in an algebraic closure $\overline{\mathbb{Q}}_p$ of $\mathbb{Q}_p$, to which the $p$-adic absolute value extends uniquely [Weiss 1963, Corollary 2-2-11], and we have $G_{\mathbb{Q}_p} = \mathrm{Gal}(\overline{\mathbb{Q}}_p/\mathbb{Q}_p)$ act on $X_K$. Again, two embeddings of $K$ in $\overline{\mathbb{Q}}_p$ give rise to the same p-adic value on $K$ if and only if they are in the same $G_{\mathbb{Q}_p}$-orbit, and the length of such an orbit is

$$\mathrm{Hom}_{\mathbb{Q}_p}(K_\mathfrak{p}, \overline{\mathbb{Q}}_p) = [K_\mathfrak{p} : \mathbb{Q}_p] = e(\mathfrak{p}_i/p) f(\mathfrak{p}_i/p).$$

A concrete application of Theorem 15.10 is the following classical method to obtain $\mathrm{Gal}(L/\mathbb{Q})$ for the normal closure $L$ of a field $K = \mathbb{Q}(\alpha)$ generated by the root $\alpha$ of a monic irreducible polynomial $f \in \mathbb{Z}[X]$.

COROLLARY 15.11. *Let $f$, $K$ and $L$ be as above. Then the following are equivalent*:

(1) *there exists a prime $p$ for which $f$ mod $p$ factors as a product of $t$ distinct irreducible factors of degrees $d_1, d_2, \ldots, d_t$.*
(2) $\mathrm{Gal}(L/\mathbb{Q})$, *viewed as a permutation group on $X_K$, contains a permutation that is the product of $t$ disjoint cycles of lengths $d_1, d_2, \ldots, d_t$.*

PROOF. For a prime as in (1), apply the Kummer–Dedekind theorem (Theorem 8.2) to the number ring $\mathbb{Z}[\alpha] = \mathbb{Z}[X]/(f)$ to deduce that the primes over $p$ in $K = \mathbb{Q}(\alpha)$ are unramified with residue class degrees $d_1, d_2, \ldots, d_t$. The Frobenius of such a prime, which generates the decomposition group, will act on $X_K$ as a product of $t$ disjoint cycles of lengths $d_1, d_2, \ldots, d_t$, as these are the orbit lengths under the action of Frobenius by Theorem 15.10. Conversely, every element of $\mathrm{Gal}(L/\mathbb{Q})$ is the Frobenius of some prime over $p \nmid \Delta(f)$ by Theorem 15.7, so all cycle types can be obtained from the factorization of $f$ mod $p$ for a suitable prime $p$ in 1. □

EXAMPLES 15.12. An irreducible cubic polynomial $f \in \mathbb{Z}[X]$ has $S_3$ as the Galois group of its splitting field if and only if it splits as a product of a linear and an irreducible quadratic factor modulo some prime $p$.

For the quartic polynomial $f = X^4 - 2X^2 + 3X - 7$ from Example 12.4, we noticed that it was irreducible modulo 2 and 3, and had exactly two zeros modulo 5. The Galois group of its splitting field is therefore a subgroup of $S_4$ containing a 4-cycle and a 2-cycle. As the value $f(-4) = 5 \cdot 41$ is the only root of $f$ modulo 41, the factorization modulo 41 shows that the Galois group contains a 3-cycle as well, and is therefore equal to the full symmetric group $S_4$.

More generally, one can show that the Galois group Gal($f$) over $\mathbb{Q}$ of the splitting field of *any* polynomial $f \in \mathbb{Z}[X]$ of squarefree discriminant is the full symmetric group. This is because a prime $p$ that divides $\Delta(f)$ only once has a unique ramified extension $\mathfrak{p}$ to $K = \mathbb{Q}[X]/(f)$, with $e(\mathfrak{p}/p) = 2$. By Theorem 15.10, we deduce that all non-trivial inertia groups in Gal($f$) are generated by a single 2-cycle in their action on the fundamental set $X_K$. Now the Galois group of a number field over $\mathbb{Q}$ is generated by its inertia groups, as every proper extension of $\mathbb{Q}$ ramifies at some finite prime. We can then apply the group theoretical fact that Gal($f$), as a transitive subgroup of the symmetric group that is generated by transpositions, is equal to the full symmetric group.

EXAMPLE 15.13. From Theorem 15.7 and Corollary 15.11, it follows that the *factorization type* of an irreducible polynomial $f \in \mathbb{Z}[X]$ modulo rational primes $p$ occurs with a frequency that depends on the group Gal($f$) of its splitting field over $\mathbb{Q}$. We illustrate this for the quartic polynomial $f = X^4 - 2X^2 + 3X - 7$ from Example 12.4, which has group $S_4$.

| $X$ | # primes | 4 | 1-3 | 2-2 | 1-1-2 | 1-1-1-1 |
|---|---|---|---|---|---|---|
| $10^3$ | 194 | .27976 | .35714 | .10119 | .23810 | .02381 |
| $10^4$ | 1229 | .26688 | .34255 | .11391 | .23759 | .03905 |
| $10^5$ | 9592 | .25378 | .33208 | .12407 | .24617 | .04390 |
| $10^6$ | 78498 | .25063 | .33377 | .12448 | .25048 | .04064 |
| $10^7$ | 664579 | .24962 | .33366 | .12517 | .25003 | .04152 |
| $\infty$ | $\infty$ | .25000 | .33333 | .12500 | .25000 | .04167 |

There are five (separable) factorization types of polynomials of degree 4, just like there are five cycle types in $S_4$. They correspond to the partitions of 4. In the table above, we have counted the fractions of the primes up to some bound $X$ yielding a given factorization type. For increasing $X$, the fractions tend to the limit fractions $\frac{1}{4}$, $\frac{1}{3}$, $\frac{1}{8}$, $\frac{1}{4}$ and $\frac{1}{24}$ in the bottom line that come from the five conjugacy classes in Gal($f$) $\cong S_4$. In fact, the general density result for 'factorization types' of a polynomial in $\mathbb{Z}[X]$ modulo primes is a weak version of Theorem 15.7 that is due to Frobenius, and Theorem 15.7 can be seen as a common generalization of this result and Dirichlet's theorem on primes in arithmetic progressions.

## Acknowledgments

Useful comments on earlier versions of this paper were provided by Reinier Bröker, Joe Buhler, Capi Corrales, Eduardo Friedman, René Schoof and William Stein.

## References

[Atiyah and Macdonald 1969] M. F. Atiyah and I. G. Macdonald, *Introduction to commutative algebra*, Addison-Wesley, MA–London–Don Mills, Ont., 1969.

[Bach 1990] E. Bach, "Explicit bounds for primality testing and related problems", *Math. Comp.* **55**:191 (1990), 355–380.

[Bhargava 2005] M. Bhargava, "The density of discriminants of quartic rings and fields", *Ann. of Math.* (2) **162**:2 (2005), 1031–1063.

[Bhargava ≥ 2008] M. Bhargava, "The density of discriminants of quintic rings and fields", *Ann. of Math.* (2). To appear.

[Bourbaki 1989] N. Bourbaki, *Algebra. I. Chapters 1–3*, Elements of Mathematics (Berlin), Springer, Berlin, 1989.

[Buchmann and Lenstra 1994] J. A. Buchmann and H. W. Lenstra, Jr., "Approximating rings of integers in number fields", *J. Théor. Nombres Bordeaux* **6**:2 (1994), 221–260.

[Buchmann and Williams 1989] J. Buchmann and H. C. Williams, "On the computation of the class number of an algebraic number field", *Math. Comp.* **53**:188 (1989), 679–688.

[Buhler and Wagon 2008] J. P. Buhler and S. Wagon, "Basic algorithms in number theory", pp. 25–68 in *Surveys in algorithmic number theory*, edited by J. P. Buhler and P. Stevenhagen, Math. Sci. Res. Inst. Publ. **44**, Cambridge University Press, New York, 2008.

[Cassels and Fröhlich 1967] J. W. S. Cassels and A. Fröhlich (editors), *Algebraic number theory: Proceedings of an instructional conference* (Brighton, 1965), Academic Press, London, 1967.

[Cohen 1993] H. Cohen, *A course in computational algebraic number theory*, Graduate Texts in Mathematics **138**, Springer, Berlin, 1993.

[Cohen and Stevenhagen 2008] H. Cohen and P. Stevenhagen, "Computational class field theory", pp. 497–534 in *Surveys in algorithmic number theory*, edited by J. P. Buhler and P. Stevenhagen, Math. Sci. Res. Inst. Publ. **44**, Cambridge University Press, New York, 2008.

[Cornell and Silverman 1986] G. Cornell and J. H. Silverman (editors), *Arithmetic geometry* (Storrs, CT, 1984), Springer, New York, 1986.

[Eisenbud and Harris 2000] D. Eisenbud and J. Harris, *The geometry of schemes*, Graduate Texts in Mathematics **197**, Springer, New York, 2000.

[Friedman 1989] E. Friedman, "Analytic formulas for the regulator of a number field", *Invent. Math.* **98**:3 (1989), 599–622.

[Fröhlich 1983] A. Fröhlich, *Galois module structure of algebraic integers*, vol. 1, Ergebnisse der Mathematik und ihrer Grenzgebiete (3), Springer, Berlin, 1983.

[Gee and Stevenhagen 1998] A. Gee and P. Stevenhagen, "Generating class fields using Shimura reciprocity", pp. 441–453 in *Algorithmic number theory* (Portland, OR,

1998), edited by J. P. Buhler, Lecture Notes in Comput. Sci. **1423**, Springer, Berlin, 1998.

[Gouvêa 1993] F. Q. Gouvêa, *p-adic numbers: An introduction*, Universitext, Springer, Berlin, 1993.

[Hartshorne 1977] R. Hartshorne, *Algebraic geometry*, Graduate Texts in Mathematics **52**, Springer, New York, 1977.

[Koblitz 1984] N. Koblitz, *p-adic numbers, p-adic analysis, and zeta-functions*, 2nd ed., Graduate Texts in Mathematics **58**, Springer, New York, 1984.

[Lang 1994] S. Lang, *Algebraic number theory*, 2nd ed., Graduate Texts in Mathematics **110**, Springer, New York, 1994.

[Lang 2002] S. Lang, *Algebra*, 3rd ed., Graduate Texts in Mathematics **211**, Springer, New York, 2002.

[Lenstra 1992] H. W. Lenstra, Jr., "Algorithms in algebraic number theory", *Bull. Amer. Math. Soc. (N.S.)* **26**:2 (1992), 211–244.

[Lenstra 2008] H. W. Lenstra, Jr., "Solving the Pell equation", pp. 1–23 in *Surveys in algorithmic number theory*, edited by J. P. Buhler and P. Stevenhagen, Math. Sci. Res. Inst. Publ. **44**, Cambridge University Press, New York, 2008.

[Mihăilescu 2006] P. Mihăilescu, "On the class groups of cyclotomic extensions in presence of a solution to Catalan's equation", *J. Number Theory* **118**:1 (2006), 123–144.

[Pomerance 2008a] C. Pomerance, "Elementary thoughts on discrete logarithms", pp. 385–396 in *Surveys in algorithmic number theory*, edited by J. P. Buhler and P. Stevenhagen, Math. Sci. Res. Inst. Publ. **44**, Cambridge University Press, New York, 2008.

[Pomerance 2008b] C. Pomerance, "Smooth numbers and the quadratic sieve", pp. 69–81 in *Surveys in algorithmic number theory*, edited by J. P. Buhler and P. Stevenhagen, Math. Sci. Res. Inst. Publ. **44**, Cambridge University Press, New York, 2008.

[Ramakrishnan and Valenza 1999] D. Ramakrishnan and R. J. Valenza, *Fourier analysis on number fields*, Graduate Texts in Mathematics **186**, Springer, New York, 1999.

[Rosen 2002] M. Rosen, *Number theory in function fields*, Graduate Texts in Mathematics **210**, Springer, New York, 2002.

[Schirokauer 2008] O. Schirokauer, "The impact of the number field sieve on the discrete logarithm problem in finite fields", pp. 397–420 in *Surveys in algorithmic number theory*, edited by J. P. Buhler and P. Stevenhagen, Math. Sci. Res. Inst. Publ. **44**, Cambridge University Press, New York, 2008.

[Schoof 2008a] R. Schoof, *Catalan's Conjecture*, Springer, New York, 2008.

[Schoof 2008b] R. J. Schoof, "Computing Arakelov class groups", pp. 447–495 in *Surveys in algorithmic number theory*, edited by J. P. Buhler and P. Stevenhagen, Math. Sci. Res. Inst. Publ. **44**, Cambridge University Press, New York, 2008.

[Schoof 2008c] R. J. Schoof, "Four primality testing algorithms", pp. 101–125 in *Surveys in algorithmic number theory*, edited by J. P. Buhler and P. Stevenhagen, Math. Sci. Res. Inst. Publ. **44**, Cambridge University Press, New York, 2008.

[Seysen 1987] M. Seysen, "A probabilistic factorization algorithm with quadratic forms of negative discriminant", *Math. Comp.* **48**:178 (1987), 757–780.

[Skoruppa 1993] N.-P. Skoruppa, "Quick lower bounds for regulators of number fields", *Enseign. Math.* (2) **39**:1-2 (1993), 137–141.

[Stevenhagen 2008] P. Stevenhagen, "The number field sieve", pp. 83–100 in *Surveys in algorithmic number theory*, edited by J. P. Buhler and P. Stevenhagen, Math. Sci. Res. Inst. Publ. **44**, Cambridge University Press, New York, 2008.

[Stevenhagen and Lenstra 1996] P. Stevenhagen and H. W. Lenstra, Jr., "Chebotarëv and his density theorem", *Math. Intelligencer* **18**:2 (1996), 26–37.

[Tate 1984] J. Tate, *Les conjectures de Stark sur les fonctions L d'Artin en $s = 0$*, Progress in Mathematics **47**, Birkhäuser, Boston, 1984.

[Washington 1997] L. C. Washington, *Introduction to cyclotomic fields*, 2nd ed., Graduate Texts in Mathematics **83**, Springer, New York, 1997.

[Weiss 1963] E. Weiss, *Algebraic number theory*, McGraw-Hill, New York, 1963.

[Weiss 1996] A. Weiss, *Multiplicative Galois module structure*, Fields Institute Monographs **5**, American Mathematical Society, Providence, RI, 1996.

PETER STEVENHAGEN
MATHEMATISCH INSTITUUT
UNIVERSITEIT LEIDEN
POSTBUS 9512
2300 RA LEIDEN
THE NETHERLANDS
psh@math.leidenuniv.nl

# Smooth numbers: computational number theory and beyond

ANDREW GRANVILLE

ABSTRACT. The analysis of many number theoretic algorithms turns on the role played by integers which have only small prime factors; such integers are known as "smooth numbers". To be able to determine which algorithm is faster than which, it has turned out to be important to have accurate estimates for the number of smooth numbers in various sequences. In this chapter, we will first survey the important estimates for application to computational number theory questions, results as well as conjectures, before moving on to give sketches of the proofs of many of the most important results. After this, we will describe applications of smooth numbers to various problems in different areas of number theory. More complete surveys, with many more references, though with a different focus, were given by Norton [1971] and Hildebrand and Tenenbaum [1993a].

### CONTENTS

| | |
|---|---|
| Notation and references | 268 |
| 1. The basic estimates for practical applications | 268 |
| 2. Applications to computational number theory | 275 |
| 3. Estimates: more details | 281 |
| 4. Smooths in short intervals, in arithmetic progressions, and as values of polynomials | 294 |
| 5. Understanding, computing, and playing with smooth numbers | 302 |
| 6. Applications to other areas of number theory and beyond | 309 |
| Acknowledgments | 315 |
| Appendix: Notation | 315 |
| References | 316 |

L'auteur est partiellement soutenu par une bourse du Conseil de recherches en sciences naturelles et en génie du Canada.

## Notation and references

This article has two target audiences. For those primarily interested in computational number theory, I have tried to write this paper so that they can better understand the main tools used in analyzing algorithms. For those primarily interested in analytic problems, I have tried to give concise introductions to simplified versions of various key computational number theory algorithms, and to highlight applications and open counting questions. Besides the danger of never quite getting it right for either reader, I have had to confront the difficulty of the differences in notation between the two areas, and to work with some standard concepts in one area that might be puzzling to people in the other. Please consult the appendix for notation that is non-standard for one of the two fields.

This article is not meant to be a complete survey of all progress in this very active field. Thus I have not referred to many excellent works that are not entirely pertinent to my view of the subject, nor to several impressive works that have been superseded in the aspects in which I am interested.

## 1. The basic estimates for practical applications

Let $S(x, y)$ be the set of integers up to $x$, all of whose prime factors are $\leq y$ (such integers are called "$y$-smooth"), and let $\Psi(x, y)$ be the number of such integers. Throughout we will let $p_1 = 2 < p_2 = 3 < p_3 < \cdots$ be the sequence of primes, and select $k$ so that $p_k$ is the largest prime $\leq y$ (and thus $k = \pi(y)$, the number of primes $\leq y$).

Dickman [1930] showed the remarkable result that for any fixed $u \geq 1$, the proportion of the integers up to $x$, that only have prime factors $\leq x^{1/u}$, tends to a nonzero limit as $x \to \infty$. This limit, denoted by $\rho(u)$, is known as the Dickman–de Bruijn $\rho$-function, and we shall describe it more fully below. Dickman's result may be stated more precisely as

$$\Psi(x, y) \sim x\rho(u) \quad \text{as } x \to \infty, \quad \text{where } x = y^u. \tag{1.1}$$

It is obvious that

$$\rho(u) = 1 \quad \text{for } 0 \leq u \leq 1, \tag{1.2}$$

and we shall prove below that

$$\rho(u) = 1 - \log u \quad \text{for } 1 \leq u \leq 2. \tag{1.3}$$

One cannot write down a useful, simple function that gives the value of $\rho(u)$ for all $u$. The neatest way to define $\rho(u)$ in general is via the *integral delay equation*

$$\rho(u) = \frac{1}{u} \int_{u-1}^{u} \rho(t) \, dt \quad \text{for all } u > 1. \tag{1.4}$$

Note that by differentiating this expression, we obtain

$$\rho'(u) = -\frac{\rho(u-1)}{u}. \tag{1.5}$$

As we shall see later, $\rho(u)$ decays to 0 extremely rapidly as a function of $u$. A crude but very useful estimate is

$$\rho(u) = 1/u^{u+o(u)} \quad \text{as } u \to \infty, \tag{1.6}$$

and even a little more precisely

$$\rho(u) = \left(\frac{e+o(1)}{u \log u}\right)^u \quad \text{as } u \to \infty. \tag{1.7}$$

In many applications we need to take $y$ to be smaller than a given, fixed positive power of $x$, and so we might ask in what range the asymptotic formula $\Psi(x, y) \sim x\rho(u)$ actually holds? De Bruijn [1951a; 1951b; 1966] showed that

$$\Psi(x, y) = x\rho(u)\left\{1 + O\left(\frac{\log(u+1)}{\log y}\right)\right\}, \quad \text{where } x = y^u, \tag{1.8}$$

holds for

$$1 \le u \le (\log y)^{3/5-\varepsilon}, \quad \text{that is, } y > \exp\left((\log x)^{5/8+\varepsilon}\right). \tag{1.9}$$

Hildebrand [1986] improved this substantially to the range

$$1 \le u \le \exp\left((\log y)^{3/5-\varepsilon}\right), \quad \text{that is, } y > \exp\left((\log \log x)^{5/3+\varepsilon}\right). \tag{1.10}$$

How much further is possible? Hildebrand [1984a] showed that such an estimate holds uniformly for

$$1 \le u \le y^{1/2-\varepsilon}, \quad \text{that is, } y \ge (\log x)^{2+\varepsilon}, \tag{1.11}$$

if and only if the Riemann Hypothesis is true. We shall sketch the ideas of these proofs later.

Some authors work (for example, Tenenbaum [1990]) with the more accurate though (what I find to be) more unwieldy approximation

$$\Psi(x, y) \approx x \int_0^x \rho\left(u - \frac{\log t}{\log y}\right) d\left(\frac{[t]}{t}\right)$$

(due to de Bruijn). We shall not pursue this here.

Canfield, Erdős, and Pomerance [1983] proved a weaker result than (1.8) but which is applicable in a much wider range and so is very important for computational number theorists: We have

$$\Psi(x, y) = \frac{x}{u^{u+o(u)}}, \tag{1.12}$$

for
$$u \leq y^{1-\varepsilon} \quad \text{with } u \to \infty, \quad \text{that is, } y \geq (\log x)^{1+\varepsilon}.$$

There seems to be no hope of proving $\Psi(x, y) \sim x\rho(u)$ in such a wide range (since this would imply the Riemann Hypothesis, as we saw above!), although Hildebrand [1986] did improve (1.12) to

$$\Psi(x, y) = x\rho(u) \exp\bigl(O\bigl(u \exp(-(\log u)^{3/5-o(1)})\bigr)\bigr) \qquad (1.13)$$

in the same range. Note that $x\rho(u) < 1$ if $y < \{e - o(1)\}\log x$ by (1.7), so $\Psi(x, y) \sim x\rho(u)$ certainly cannot hold in this range.

Results discussed below do tell us, for instance, that

$$\Psi(x, \log^A x) = x^{1-1/A + o(1)} \quad \text{for any } A > 1. \qquad (1.14)$$

Moreover if $0 < \alpha < 1$ then

$$\Psi(x, e^{c(\log x)^\alpha (\log \log x)^\beta}) = xe^{-((1-\alpha)/c + o(1))(\log x)^{1-\alpha}(\log \log x)^{1-\beta}}, \qquad (1.15)$$

the most important special case being

$$\Psi(x, L(x)^c) = x/L(x)^{1/2c + o(1)} \qquad (1.16)$$

where, here and henceforth, we adopt the commonly used notation

$$L(x) := \exp(\sqrt{\log x \log \log x}).$$

What about for smaller $y$? In the next section we shall see that if $y$ is very small compared to $x$ then one can obtain an asymptotic formula for $\Psi(x, y)$ which looks quite different from the formulae above, since it now depends very much on the primes $\leq y$: For $y \leq \sqrt{\log x \log \log x}$, we have

$$\Psi(x, y) = \frac{1}{\pi(y)!} \prod_{p \leq y}\Bigl(\frac{\log x}{\log p}\Bigr)\Bigl(1 + O\Bigl(\frac{y^2}{\log x \log y}\Bigr)\Bigr). \qquad (1.17)$$

For $y = o(\log x)$ with $y \to \infty$, we have

$$\Psi(x, y) = \Bigl(\frac{\log x}{y}\Bigr)^{(1+o(1))\pi(y)}. \qquad (1.18)$$

In fact for any $x \geq y \geq 2$, we have

$$\log \Psi(x, y) = ug\Bigl(\frac{y}{\log x}\Bigr)\Bigl(1 + O\Bigl(\frac{1}{\log y} + \frac{1}{\log \log x}\Bigr)\Bigr), \qquad (1.19)$$

where $g(\kappa) = \log(1 + \kappa) + \kappa \log(1 + 1/\kappa)$. Notice that the estimates (1.14) as $\alpha \to 1$, and (1.18) as $y \to \log x$, take a rather different shape. For the rather delicate region in-between, where $y$ is a constant multiple of $\log x$, say $y = \kappa \log x$, we get from (1.19) that

$$\Psi(x, \kappa \log x) = \exp\left(g(\kappa)\frac{\log x}{\log \log x}\left(1 + O\left(\frac{1}{\log \log x}\right)\right)\right).$$

This is a typical "phase transition" type function. The reasons for such a change in behavior are explored in detail in [Granville 1989].

**1.1. The supposed universality of Dickman's density function.** Computational number theory abounds with examples of sequences $\mathcal{N}$ of integers from which we need to extract $y$-smooth numbers. As we saw in the previous section, if $y = x^{1/u}$, then the proportion of $y$-smooth integers up to $x$ is $\rho(u)$, in the surprisingly wide range (1.10) for $y$. However, in most examples that arise, the sequence $\mathcal{N}$ is rarely so simple as a random sample of integers up to $x$. Nonetheless, we typically assume for the sequences $\mathcal{N}$ which arise, that more-or-less the same proportion of them are $y$-smooth as for randomly chosen numbers of roughly the same size. In Section 1.5, we will attempt to formulate appropriate hypotheses to precisely understand what we are assuming in the most important algorithms. These hypotheses appear to be fairly ad hoc, tied in to the algorithms, but as an analytic number theorist, I prefer to formulate more natural conjectures from which our hypotheses may follow: Greg Martin [2002] made an in-depth study of smooth values of polynomials and made a remarkable universal prediction:

CONJECTURE. *Suppose that $f(x) \in \mathbb{Z}[x]$ has distinct irreducible factors over $\mathbb{Z}[x]$ of degrees $d_1, d_2, \ldots, d_k \geq 1$, respectively, and fix $u > 0$. There are*

$$\sim \rho(d_1 u)\rho(d_2 u) \cdots \rho(d_k u) x \tag{1.20}$$

*integers $n \leq x$ for which $|f(n)|$ is $y$-smooth, where $x = y^u$, as $x \to \infty$.*

The case $k = 1$ implies that for irreducible polynomials, $f(n)$ is as likely to be $y$-smooth as random integers of the same size. The general conjecture implies that the property of being $y$-smooth for the various irreducible factors of $f$ is statistically independent. The jury is out on this as a conjecture: Under rather strong assumptions about prime values of polynomials, Martin proves this in a very limited range ($y > x^{d-1/k+\varepsilon}$), but it is plausible that rather different behavior emerges for (say) $y = \sqrt{x}$. The conjecture is true for $k = 1$ with $f$ of degree 1, but that is the only case we know for sure. See Section 4.3 for more on what is known.

It is also true that what we have conjectured here is not really what is needed for applications to algorithmic number theory issues. What we really need to understand is far more difficult: In what range of values for $x$ and $y$ does (1.20) hold? Such a range is certainly dependent on the coefficients of $f$, but we need results in which this dependence is simply stated and easily applicable. As far as I know, no-one has thought through appropriate general conjectures of this nature, though I hope some reader will accept this as a challenge. In Section 4.3 we state what current results imply about such ranges in the degree one case.

One should not be seduced into thinking that these proportions (as in (1.20)) hold up for all naturally defined sequences. Although it is true that the proportion of $y$-smooth values of $\{a^2 + b^2 \leq x : a, b \geq 1\}$ is $\rho(u)$, we are counting here our numbers with multiplicity (that is, how often each $n$ is represented as a sum of two squares). However, if we don't count with multiplicities, that is, we look at the proportion of $y$-smooth values of $\{n \leq x : n = a^2 + b^2$ for some $a, b \in \mathbb{Z}\}$, then the proportion changes to $\sigma(u)$, where $\sigma(u) = 1$ for $0 \leq u \leq 1$ and $\sigma'(u) = -\sigma(u-1)/2\sqrt{u^2-u}$ for $u > 1$ (see [Moree 1993]). In fact, $\sigma(u) = \rho(u)/(2 + o(1))^u$ is quite a bit smaller than $\rho(u)$.

**1.2. Beliefs, in cryptography and in estimates.** There are many cryptographic schemes around, mostly based on ideas in number theory and combinatorics. For public key cryptography, the goal is to produce a truly "one-way" function in which a practical decryption method cannot be deduced from the (publicly available) encryption method. Most such schemes rely on the difficulty of solving a particular mathematics problem, be it factoring, discrete logs on some group, or an atrociously convoluted linear algebra problem. None of these are provably difficult to solve but, correctly formulated, they certainly seem difficult (indeed we know of no sensible mathematics problem at all that is provably difficult to solve; finding one is an outstanding question in theoretical computer science). Some of the mathematical problems used are ancient chestnuts, like factoring (see Sections 2.3, 2.4, 2.5, 2.6 below), and it perhaps gives one some faith when a cryptoscheme is based on a problem that has remained unscathed through two centuries of attacks by the finest minds from Gauss onwards. By the same token I have less faith in cryptoschemes that rely on convoluted problems a few years old and that are too ugly to attract the finest minds. For some applications one must trade such a feeling of security for speed, so I would advise the reader to remain wary of what they say into their cellphone!

There is now an extensive literature on counting smooth numbers (much of which we are reviewing in this chapter). In some situations there are sharp estimates in wide ranges, and yet in other seemingly tractable situations, there does seem to be serious difficulty in extending the range of what is known. In particular it is intriguingly difficulty to prove that there are smooth numbers in all intervals of length $\sqrt{x}$, close to $x$. It is unclear whether our inability to prove strong results in this problem is due to some intrinsic difficulty in the problem, or to our own incompetence. I write this because most work on smooth numbers uses tools that were designed for other questions and modified to fit here, in particular, tools used in understanding the distribution of primes where there are certain natural barriers (like proving that there are primes in every interval of length $\sqrt{x}$ close to $x$). In the case of smooths in arithmetic progressions, I overcame such barriers that restrict the range in which one can estimate primes in arithmetic progressions (see Section 4.2 for the range (4.6)) by applying an elementary idea of Hildebrand (although even this has its roots in Selberg's elementary proof of the prime number theorem); this success with "smooths in arithmetic progressions" had long made me suspect that these "smooths in short intervals" problems might succumb to a clever combinatorial argument, rather than sophisticated technique.

However, I am now pessimistic about solving the main "smooths in short intervals" problem, that is, that every interval of length $x^\varepsilon$, close to sufficiently large $x$, contains an $x^\varepsilon$-smooth integer. The reason for my pessimism, as we shall see in Section 4.4, is that solving this problem will allow us to solve an old well-tested chestnut of analytic number theory, which one feels certain lies deep. Let me explain. Obviously if an estimate implies some version of the Riemann Hypothesis — like (1.8) in the range $y > (\log x)^{2+\varepsilon}$ — then we expect that it will be hard to prove! There are many other problems in analytic number theory which have been intensively studied for a long time with little success, and the one we need is one of my favorites:

VINOGRADOV'S CONJECTURE. *Fix $\varepsilon > 0$. If $p$ is a sufficiently large prime, then the least quadratic nonresidue (mod $p$) is $< p^\varepsilon$.*

As we will discuss in Section 6.4, Burgess [1962; 1963] proved this conjecture for any $\varepsilon > 1/(4\sqrt{e})$. There has been no improvement in forty years, though we do now understand how improvements are intimately tied in to deep questions on the zeros of $L$-functions. In Section 4.4 we prove that Vinogradov's conjecture for $\varepsilon$ for primes $p \equiv 3 \pmod 4$ does follow if every interval of length $x^\varepsilon$ close to sufficiently large $x$ contains an $x^\varepsilon$-smooth integer. Given my belief that Vinogradov's conjecture is an intrinsically difficult problem, I am pessimistic that researchers will prove such a "smooths in short intervals" result for some $\varepsilon < 1/(4\sqrt{e})$ in the near future, though I would be delighted to be wrong!

**1.3. Smooths in number fields.** For a given number field $K$, define $\Psi_K(x, y)$ to be the number of ideals in the ring of integers of $K$ which have norm $\leq x$ and whose prime ideal factors have norm $\leq y$. It is easy to imitate methods from $K = \mathbb{Q}$ to prove that

$$\Psi_K(x, y) = \Psi_K(x, x)\rho(u)\left(1 + O_K\left(\frac{\log(u+1)}{\log y}\right)\right) \quad (1.21)$$

in the range (1.10) and also to prove results analogous to (1.12) and (1.13).

There is a much harder and more mysterious problem: When $K$ has units of infinite order, estimate asymptotically the number of algebraic integers in $K$ of height $\leq x$ whose prime ideal factors have norm $\leq y$. There are several substantial technical problems in solving this, particularly because of the involvement of the class and unit groups in such estimates. This question is pertinent to better analyzing the number field sieve, as well as discrete logarithms in finite fields when the field is presented as a ring of integers modulo a prime ideal.

**1.4. Entirely explicit results.** There are very few results in the literature with precise inequalities where every constant is explicit. However, these can be very useful. I will list a few here: Konyagin and Pomerance [1997] showed that if $x \geq y \geq 2$ and $x \geq 4$, then

$$\Psi(x, y) \geq x/(\log x)^u, \qquad (1.22)$$

which implies Lenstra's [1979] result that $\Psi(x, \log^2 x) \geq \sqrt{x}$. In Section 3.1, we will see that

$$\binom{\left[\frac{\log x}{\log 2}\right] + \pi(y)}{\pi(y)} \geq \Psi(x, y) \geq \binom{[u] + \pi(y)}{\pi(y)} \geq \left(\frac{\pi(y)}{[u]}\right)^{[u]}; \qquad (1.23)$$

and in Section 3.2, we will see that

$$\frac{1}{k!} \prod_{p \leq y} \frac{\log x}{\log p} \leq \Psi(x, y) \leq \frac{1}{k!} \prod_{p \leq y} \frac{\log X}{\log p}, \qquad (1.24)$$

where $X = x \prod_{p \leq y} p$.

One can show (using (3.18)) that the lower bound in (1.23) is $> 2x\rho(u)$ for $y \ll (\log x \log \log x)/(\log \log \log x)$; hence the asymptotic formula $\Psi(x, y) \sim x\rho(u)$ evidently cannot hold in this range.

Pomerance notes that the asymptotic expansion for $\Psi(x, y)$ is

$$\Psi(x, y) = x\rho(u) + (1 - \gamma)\frac{x\rho(u-1)}{\log x} + O\left(x\rho(u)\frac{\log^2 u}{\log^2 y}\right),$$

where $\gamma$ is Euler's constant, so that

$$\Psi(x, y) \geq x\rho(u) \qquad (1.25)$$

if $u$ or $y$ is sufficiently large. He asks, is this true for all $x \geq 2y \geq 2$?

Hildebrand [1985a] gave a gorgeous upper bound for smooths in short intervals:

$$\Psi(x + z, y) \leq \Psi(x, y) + \Psi(z, y)$$

for all $x, z > y$ if $y$ is sufficiently large. Does this hold for all $x, z > y \geq 2$?

## 1.5. Useful conjectures.
We conjecture that for some fixed $c$ with $0 < c < 4$ and sufficiently small $c' > 0$, we have

$$\Psi(x + c\sqrt{x}, y) - \Psi(x, y) \gg \sqrt{x}/u^{u+o(u)}, \quad \text{where } y = x^{1/u}, \quad (1.26)$$

for $y > L(x)^{c'}$. Several known methods might, if pushed to their extreme, allow us to prove such an estimate if $c \to \infty$ slowly, perhaps like a power of $\log x$. Unfortunately, in the essential application to the elliptic curve factoring method (Section 2.6), we must have $c < 4$; results for larger $c$ have no such consequences!

The elliptic curve primality test (Section 2.8) can be proved to always run in random polynomial time provided

$$\sum_{n \in S(x,y)} \left( \pi\left(\frac{x + 4\sqrt{x}}{n}\right) - \pi\left(\frac{x}{n}\right) \right) \gg \frac{\sqrt{x}}{\log^A x} \quad (1.27)$$

for $y = \exp(O((\log \log x)^2 / \log \log \log x))$. Unfortunately we can only prove this estimate for considerably larger $y$.

To unconditionally prove that the basic quadratic sieve algorithm (Section 2.4) factors $n$ in the time claimed, we need to show that for a given quadratic polynomial $f(t) = t^2 + 2bt - c$ with $1 \le c < 2b$ and $y = L(b)^{1/\sqrt{2}}$, there are $\gg y^{1+o(1)}$ values of $m \le y^2$ for which $f(m)$ is $y$-smooth. (Note that $f$ has discriminant $4n$).

## 1.6. The dual problem.
Define $\Phi(x, y)$ to be the number of integers up to $x$ whose prime factors are all $> y$. Buhštab [1949] showed that

$$\Phi(x, y) \sim \omega(u) \frac{x}{\log y}, \quad (1.28)$$

where $\omega(u) = 1/u$ for $1 \le u \le 2$ and

$$u\omega(u) = 1 + \int_1^{u-1} \omega(t) dt \quad \text{for all } u \ge 2. \quad (1.29)$$

Note that $(u\omega(u))' = \omega(u-1)$. In fact $\lim_{u \to \infty} \omega(u) = e^{-\gamma}$ and

$$\max_{u+2 \ge v \ge u} |\omega(u) - e^{-\gamma}| = \rho(u) e^{O(u)}. \quad (1.30)$$

## 2. Applications to computational number theory

In this section, we shall survey the use of smooth numbers in computational number theory, without too detailed descriptions of the algorithms (which may be found elsewhere in this volume). Further considerations of the rôle played by smooth numbers in computational number theory can be found in Pomerance's beautiful article [1995].

## 2.1. Why are smooth numbers so often involved?

As we shall see, a significant step in many algorithms that we encounter is to quickly determine a non-empty subset of a sequence $m_1, m_2, \ldots$ of integers whose product is a square. Often elements of the sequence, although explicitly determined, seem like they have more-or-less the same multiplicative properties as randomly chosen integers in $[1, x]$.

Pomerance [1996a] showed that if $m_1, \ldots, m_N, \ldots$ are selected randomly and independently from $[1, x]$ then, with probability 1, the smallest $N$ for which there is such a non-empty subset of $m_1, \ldots, m_N$ whose product is a square, satisfies $N = L(x)^{\sqrt{2}+o(1)}$. Moreover if $N = L(x)^{\sqrt{2}+o(1)}$ then with probability 1 there is a non-empty subset of $m_1, \ldots, m_N$ consisting only of $L(x)^{1/\sqrt{2}}$-smooth integers whose product is a square; and these lead us to a very simple linear algebra algorithm to determine such subsets (see Section 2.2). Similar remarks may also be made when looking for a multiplicatively dependent finite subsequence of the $m_i$'s.

Smooths also appear in other contexts: for example, integers that we know we can factor in polynomial time are typically smooth, or a smooth times a prime (see Section 2.9).

## 2.2. Products that are a square.

If $u_1, u_2, \ldots$ are $y$-smooth integers, factor each $u_j$ as

$$u_j = 2^{a_{j,1}} 3^{a_{j,2}} \cdots p_k^{a_{j,k}}.$$

Then $\prod_{j \in J} u_j$ is a square if and only if

$$\sum_{j \in J} (a_{j,1}, a_{j,2}, \ldots, a_{j,k}) = 0 \quad \text{as a vector in } (\mathbb{Z}/2\mathbb{Z})^k.$$

Thus such a nontrivial subset is guaranteed amongst $u_1, u_2, \ldots, u_{k+1}$. To determine the appropriate subset one can use Gaussian elimination or other algorithms.

Using this we can justify at least part of Pomerance's result (with an argument due, earlier, to Schroeppel): If we randomly choose $N$ integers from $[1, x]$, we expect approximately $N\Psi(x, y)/x$ of them to be $y$-smooth. Once this is $> \pi(y)$, then we are guaranteed a subset whose product is a square. Thus we pick $y$ so as to minimize $N_x := x\pi(y)/\Psi(x, y)$. By (1.16), this occurs when

$$u = (\sqrt{2} + o(1))\sqrt{\log x / \log \log x} \quad \text{and} \quad y = L(x)^{1/\sqrt{2}+o(1)},$$

so that $N_x = L(x)^{\sqrt{2}+o(1)}$, as claimed in Section 2.1. Recently Croot, Granville, Pemantle, and Tetali [2008] improved Pomerance's result by showing that the smallest $N$ for which there is a non-empty subset of $m_1, \ldots, m_N$ whose product is a square, satisfies $\frac{1}{4} N_x \leq N \leq \{e^{-\gamma} + o(1)\} N_x$ with probability $1 + o(1)$. We conjecture that this transition is quite sudden, in that $N = \{e^{-\gamma} + o(1)\} N_x$ with probability $1 + o(1)$: This prediction is borne out by the practical experience of computers.

**2.3. Dixon's random squares factoring method.** Our goal is to determine "random" integers $a$ and $b$ such that

$$a^2 \equiv b^2 \pmod{n} \tag{2.1}$$

and then, with a little luck, $(a \pm b, n)$ is a nontrivial factor of $n$.

In Dixon's method, one randomly selects $r_1, r_2, \ldots$ and determines $m_j \equiv r_j^2 \pmod{n}$ with $|m_j| \leq n$. By Pomerance's result in Section 2.1, we expect that there is a subset $J$ of $\{1, \ldots, N\}$ once $N = L(n)^{\sqrt{2}+o(1)}$ such that $a^2 = \prod_{j \in J} m_j$ for some integer $a$. Taking $b = \prod_{j \in J} r_j$, we have candidates for (2.1). This argument can be modified to rigorously prove that the expected running time is $L(n)^{\sqrt{2}+o(1)}$ (see the remarks at the end of Section 2.6).

**2.4. Smaller squares.** In 1640, Fermat suggested to Frenicle a method of factoring $n$: Take $r_j = [\sqrt{n}] + j$ for $j = 1, 2, \ldots$ so that $m_j = r_j^2 - n$. If any $m_j$ is a square, then $n = (r_j - \sqrt{m_j})(r_j + \sqrt{m_j})$. Calculation of consecutive $m_j$'s is easy since $m_{j+1} = m_j + 2[\sqrt{n}] + 2j + 1$; determining whether $m_j$ is not a square is easy by successively testing whether it is a quadratic residue mod $8, 3, 5, 7, \ldots$. In this way Fermat factored some impressively large numbers.

In the most primitive version of the quadratic sieve one uses the same values of $m_j$ with $j < n^{o(1)}$, so that each $m_j < n^{1/2+o(1)}$. Then the result of Section 2.1 indicates that the running time is $L(n^{1/2+o(1)})^{\sqrt{2}} = L(n)^{1+o(1)}$, faster than the random squares method. In fact Brillhart and Morrison's "continued fractions method" and Lenstra and Pomerance's "class group method" both attain the same speed up for the same reason, but the first has the advantage of being very practical, whereas the second is rigorously analyzable. However the flexibility of the quadratic sieve allows various, very effective, speed-up strategies.

**2.5. Spectacular savings with smaller squares: the number field sieve.** Let $d$ be a large integer, define $m = [n^{1/d}]$, and write $n$ in base $m$ as $n = m^d + a_1 m^{d-1} + a_2 m^{d-2} + \cdots + a_d$, where each $a_i$ is an integer with $0 \leq a_i \leq m-1$. Define $f(T) = T^d + a_1 T^{d-1} + \cdots + a_d$.

If $f$ factors as $g(T)h(T)$ then $n = f(m) = g(m)h(m)$ provides a nontrivial factorization of $n$, as follows from a clever theorem of Brillhart, Filaseta, and Odlyzko [1981]. Thus we can assume $f$ is irreducible and let $\alpha$ be a root of $f$. Let $I$ be the ideal $(\alpha - m, n)$ so that $\text{Norm}(I) = n$. The idea is to find $u, v \in \mathbb{Q}(\alpha)$ so that $u^2 \equiv v^2 \pmod{I}$, and hopefully $\text{Norm}(u \pm v, \alpha - m, n)$ will provide nontrivial factors of $n$.

Now for any integers $a$ and $b$, we have $a + b\alpha \equiv a + bm \pmod{I}$, and we proceed, as in the quadratic sieve, keeping only pairs $a, b$ where $a + bm$ is $y$-smooth and $a + b\alpha$ is $y$-smooth in $\mathbb{Q}(\alpha)$.

We need a product of terms, $(a + b\alpha)(a + bm)$, to be a square: To use Pomerance's estimate, we consider trying to make a product of terms, $(a + bm)\operatorname{Norm}(a + b\alpha)$, a square. Note though that even if this is a square, this calculation is insufficient to guarantee things work, for several reasons: Most important is that different primes of our field can have the same norm — however, each unramified prime ideal that arises is of the form $(p, \alpha - w)$, and this divides $\operatorname{Norm}(a + b\alpha)$ if and only if $a + bw \equiv 0 \pmod{p}$ so we can easily distinguish, in our calculation, between unramified prime ideals of the same norm. There are several other obstructions: ramified primes of the same norm, the difference between $\mathbb{Z}[\alpha]$ and the ring of integers of $\mathbb{Q}(\alpha)$, and various considerations of the 2-parts of the unit and class groups of $\mathbb{Q}(\alpha)$. All of these difficulties can be handled (see [Stevenhagen 2008] in this volume or [Crandall and Pomerance 2001, Chapter 6.2]. If we take all $0 < a, b \leq y$, then

$$|(a + bm)\operatorname{Norm}(a + b\alpha)| = |(a + bm)b^d f(-a/b)| \leq 2(d + 1)m^2 y^{d+1}.$$

By Pomerance's result we thus want $y^2 = L(dm^2 y^{d+1})^{\sqrt{2}+o(1)}$, which implies that $(\log^2 y)/(\log\log y) \sim (2\log n)/d + d\log y$ assuming $d \to \infty$. To minimize the left side we take $d = \sqrt{(2\log n)/(\log y)}$, leading to the choices $\log y \approx ((8/9)\log n(\log\log n)^2)^{1/3}$ and $d = ((3\log n)/(\log\log n))^{1/3} + O(1)$. This gives a running time of

$$\exp\big(\{1 + o(1)\}(\tfrac{64}{9}\log n(\log\log n)^2)^{1/3}\big), \tag{2.2}$$

an amazing speed up over $L(n)^{1+o(1)}$ from the previous section.

The constant $(64/9)^{1/3}$ can be slightly improved, though at the price of complicating the algorithm.

**2.6. Factoring using smooth group orders.** Pollard's $p - 1$ method: If $p$ is a prime factor of $n$ and the order of $2 \pmod{p}$ is $y$-smooth, and this is not so for all other prime factors of $n$, then $\gcd(2^\ell - 1, n) = p$, where $\ell$ is any multiple of the order of $2 \pmod{p}$ yet still a $y$-smooth integer. In practice, one takes $\ell = \operatorname{lcm}[1, 2, \ldots, y]$ for each successive integer $y$ and computes $2^\ell \pmod{n}$. This is an efficient algorithm if the structure of the factorization of $n$ is just right, though this will not be so for many integers $n$.

Lenstra suggested replacing the group $(\mathbb{Z}/p\mathbb{Z})^*$ in this calculation by the group of $\mathbb{F}_p$-points on an elliptic curve. The advantage is that these groups have orders between $p - 2\sqrt{p} + 1$ and $p + 2\sqrt{p} + 1$, and in fact for any integer in-between, there is an elliptic curve group of that order. It seems far more likely that some, even many, of these numbers are smooth, and so with an algorithm analogous to Pollard's $p-1$ method, Lenstra provided an efficient general purpose factoring method.

Lenstra's "elliptic curve factoring method" [1987] proceeds by first randomly choosing elliptic curves (mod $n$) together with a point $R$ — first select $R = (x_0, y_0)$ and $a$, and then pick $b$ so that $y_0^2 \equiv x_0^3 + ax_0 + b \pmod{n}$ — and then compute $\ell R$ on $E : y^2 = x^3 + ax + b$ over $\mathbb{Z}/n\mathbb{Z}$. If $p$ is a prime factor of $n$ such that $\#E(\mathbb{F}_p)$ divides $\ell$, then this procedure is likely to factor $n$. The algorithm takes expected time $L(p)^{\sqrt{2}+o(1)}$ to find $p$ provided it is true that there are roughly the number of $y$-smooths in $(p - \sqrt{p}, p + \sqrt{p})$ as we might guess (see Section 1.5). The proof uses Deuring's theory of CM-elliptic curves which implies that there are $H(t^2 - 4p)$ elliptic curves (mod $p$) with $p + 1 - t$ points, where $H(*)$ is the Kronecker class number; it also uses Siegel's work in analytic number theory which shows that $H(d) = \sqrt{d}(\log d)^{O(1)}$ with very few exceptions. The great advantage of Lenstra's idea (over Pollard's $p-1$ method, and variants like the $p+1$ method) is that it should always work and that its success does not depend on properties of the factorization of $n$.

Analysis of the elliptic curve factoring method hinges on an unproved assumption, that there are roughly as many $y$-smooths in $(p - \sqrt{p}, p + \sqrt{p})$ as we might guess ; thus for any particular $n$, we cannot rigorously prove that the algorithm has the claimed expected running time. However we can prove that the algorithm does have the claimed expected running time for all but a small set of exceptional $n$: In certain circumstances this is very useful, for we may have an algorithm that applies the method to any of a large set of randomly chosen $n$ so that the overall running time can be proved rigorously. For example, Dixon's algorithm, described in barest outline in Section 2.3, was made rigorous in this way by Pomerance [1987]. Other examples are the hyperelliptic smoothness test as proposed by Lenstra, Pila, and Pomerance [1993], and Lenstra and Pomerance's [1992] rigorous time bound for factoring.

### 2.7. Smooths solving the discrete log problem.
One wishes to find, for given elements $g$ and $b$ of a group $G$, an integer $m$ for which $g^m = b$ in $G$. Typically one takes $G = (\mathbb{Z}/p\mathbb{Z})^*$. This can be solved in (provable) expected running time $L(p)^{\sqrt{2}+o(1)}$ using smooth numbers. The idea is to start by computing $u_j \equiv g^{m_j} \pmod{p}$ for random integers $m_j$ with $|u_j| < p$, and keeping $u_j$ if it is $y$-smooth. Once we find enough such $u_j$, we should be able to solve for each prime $q \leq y$, writing $q = g^{v_q} \pmod{p}$. Once this is done, we compute $u \equiv bg^{-k} \pmod{p}$ for random $k$. If $u$ is ever $y$-smooth, then we know how to write it as a power of $g$, and therefore also $bg^{-k}$; thus we determine $m$.

This can be generalized to finite fields $\mathbb{F}_{p^d}$: One way is by representing the field as $\mathbb{F}_p[x]/(f(x))$, where $f$ is irreducible over $\mathbb{F}_p[x]$ of degree $d$ and then developing a theory of smooth $\mathbb{F}_p[x]$-polynomials (see Section 5.2).

A similar strategy can be used to efficiently determine the structure of the class group of an imaginary quadratic field.

There is no obvious analogy of this solution in the discrete logarithm problem for elliptic curves over finite fields, which makes these groups tempting to use for cryptographic protocols. It seems to me, however, foolish to view this as the basis for a belief that such protocols are more secure!

**2.8. Goldwasser and Kilian's elliptic curve primality test.** This generalizes Pocklington's test and almost certainly works in random polynomial time even providing a certificate of primality for $p$. To prove the primality of $p$, we find an elliptic curve $E$ defined over $\mathbb{F}_p$ such that a completely factored integer $m > (p^{1/4}+1)^2$ divides the order of a point on $E(\mathbb{F}_p)$. (See [Schoof 2008] in this volume for details on how to do this in practice.) For simplicity authors often restrict $\#E(\mathbb{F}_p)$ to be a prime or twice a prime, and so a probabilistic primality test follows provided, for $x = (\sqrt{p}-1)^2$,

$$\pi(x+4\sqrt{x}) - \pi(x) \gg \sqrt{x}/\log^A x \qquad (2.3)$$

for some fixed $A(>1)$. Although this holds for "almost all $x$", we cannot prove it for all $x$ even assuming the Riemann Hypothesis. However, one does get a random polynomial time algorithm provided (1.27) holds, since then $\#E(\mathbb{F}_p)$ is an easily factorable number (see the next section).

Adleman and Huang's [1992] hyperelliptic curve primality test does provably work in random polynomial time. The idea is that since such a curve has between $p^2 - cp^{3/2}$ and $p^2 + cp^{3/2}$ points, we can find a curve with a prime $q$ number of points in this interval. Then with probability 1, we can prove $q$ is prime by the elliptic curve test.

In light of the recent deterministic polynomial time primality testing algorithm of Agrawal, Kayal and Saxena [2004] (see also Granville [2005]), the test of Adleman and Huang is now of mostly historical interest.

**2.9. Easily factorable numbers.** In practice, what numbers can be factored in polynomial time? Certainly, almost all primes can be identified as primes in probabilistic polynomial time, for example, by the Goldwasser–Kilian test or, for $\log^C x$-smooth numbers, by trial division. Lenstra's "elliptic curve factoring method" completely factors $x$ in expected running time

$$(L(y)\log x)^{O(1)},$$

where $y$ is the second largest prime factor of $x$. Thus an integer that is a prime times a $y$-smooth, with $y = \exp(O((\log\log x)^2/\log\log\log x))$, can be factored in practice in polynomial time. There are

$$\sim e^\gamma x/u \asymp x(\log\log x)^2/(\log x \log\log\log x)$$

such integers $\leq x$. In fact, one can rigorously prove that this number of integers can be factored in expected polynomial time by this method, despite the fact that Lenstra's elliptic curve factoring method is only guaranteed to work for some but not all $n$, since in this calculation we are averaging over many possibilities.

**2.10. Lenstra's polynomial time test as to whether an integer that is conjecturally prime is rigorously squarefree.** If $n > 32$ and $a^{n-1} \equiv 1 \pmod{n}$ for all $a < \log^2 n$, then $n$ is squarefree: This algorithm is only of interest if we already suspect that $n$ is prime; otherwise it is extremely unlikely that the hypothesis will be satisfied (note that if the Generalized Riemann Hypothesis holds and $a^{n-1} \equiv 1 \pmod{n}$ for all $a < 2\log^2 n$, then $n$ is indeed prime). To see that Lenstra's algorithm works, note that if $p^2$ divides $n$, then for any $a < 4\log^2 p$ $(< \log^2 n)$ we have $a^{n-1} \equiv 1 \equiv a^{p(p-1)} \pmod{p^2}$, and since $(p, n-1) = 1$ we have $a^{p-1} \equiv 1 \pmod{p^2}$. Therefore every $(4\log^2 p)$-smooth integer $n$ satisfies $n^{p-1} \equiv 1 \pmod{p^2}$, and there are just $p - 1$ of these $\leq p^2$, so that

$$p - 1 \geq \Psi(p^2, 4\log^2 p) \geq p$$

by (1.22), giving a contradiction.

## 3. Estimates: more details

In this section, we introduce many of the important techniques used in estimating $\Psi(x, y)$. These ideas were introduced by various authors, though de Bruijn and later, Hildebrand, certainly deserve the lion's share of the credit.

**3.1. Upper and lower bounds: elementary combinatorics.** Evidently $n \in S(x, y)$ if and only if we can write $n$ in the form $n = p_1^{a_1} p_2^{a_2} \cdots p_k^{a_k}$, where the $a_k$ are nonnegative integers with $n \leq x$, that is,

$$a_1 \log p_1 + a_2 \log p_2 + \cdots + a_k \log p_k \leq \log x. \tag{3.1}$$

Since each $\log p_i \geq \log 2$, this implies that

$$a_1 + a_2 + \cdots + a_k \leq \left\lceil \frac{\log x}{\log 2} \right\rceil = A,$$

say, so that

$$\Psi(x, y) \leq \binom{A+k}{k},$$

the upper bound in (1.23). Similarly, because each $\log p_i \leq \log y$, any solution to

$$a_1 + a_2 + \cdots + a_k \leq \left\lceil \frac{\log x}{\log y} \right\rceil = [u]$$

gives a solution to (3.1) where $x = y^u$. This implies

$$\Psi(x, y) \geq \binom{[u]+k}{k},$$

the lower bound in (1.23). We can use these bounds, in practice, since

$$k = \pi(y) \sim \frac{y}{\log y}, \quad u = \frac{\log x}{\log y}, \quad \text{and} \quad A = \frac{\log x}{\log 2} + O(1)$$

by the prime number theorem. If $k > c \log x$, then $\binom{A+k}{k} > x$; so the upper bound is useless! But

$$\Psi(x, y) \geq \frac{k^{[u]}}{[u]!} \approx \left(\frac{ey}{\log x}\right)^u = \frac{x}{((1/e)\log x)^u} \quad (3.2)$$

(which should be compared to (1.22)). If $k < \varepsilon \log x/(\log\log x)$, then

$$\frac{A^k}{k!} \gtrsim \Psi(x, y) \gtrsim \frac{[u]^k}{k!} \approx \frac{u^k}{k!}. \quad (3.3)$$

**3.2. Upper and lower bounds: lattices.** In general, if we wish to determine the number of lattice points inside an $n$-dimensional tetrahedron, then we can get good estimates using a little geometry. For if we represent the number of lattice points inside the triangle $\{x_1, x_2 \geq 0, \ w_1 x_1 + w_2 x_2 \leq \tau\}$ by shading in the unit square to the right and above each lattice point, we get a shaded region whose area equals the number of such lattice points. However the original triangle is entirely in the shade, and so has area less than or equal to the number of lattice points, so that there are $\geq \tau^2/(2! w_1 w_2)$ lattice points. On the other hand, the triangle $\{x_1, x_2 \geq 0, \ w_1(x_1-1) + w_2(x_2-1) \leq \tau\}$ contains the shaded region, and so the number of lattice points is no more than this triangle's area, which is $(\tau + w_1 + w_2)^2/(2! w_1 w_2)$. More generally, the boxes "to the right and above" each lattice point in

$$\{x_1, \ldots, x_k \geq 0 : x_1 w_1 + x_2 w_2 + \cdots + x_k w_k \leq \tau\}$$

contain this $k$-dimensional tetrahedron and are contained inside the tetrahedron

$$\{x_1, \ldots, x_k \geq 0 : (x_1 - 1)w_1 + (x_2 - 1)w_2 + \cdots + (x_k - 1)w_k \leq \tau\}.$$

Now $\Psi(x, y)$ equals the number of solutions to (3.1), and so

$$\frac{1}{k!} \prod_{p \leq y} \frac{\log x}{\log p} \leq \Psi(x, y) \leq \frac{1}{k!} \prod_{p \leq y} \frac{\log X}{\log p}, \quad (1.24)$$

where $\log X = \log x + \sum_{p \leq y} \log p$. The ratio of the upper to the lower bound is $((\log X)/(\log x))^k$, and since $\sum_{p \leq y} \log p \sim y$ by the prime number theorem, we get

$$\left(\frac{\log X}{\log x}\right)^k \approx \left(\frac{\log x + y}{\log x}\right)^{y/\log y} \approx e^{y^2/(\log x \log y)},$$

which implies (1.17) and (1.18); if $y \leq \log x$, we get

$$\Psi(x, y) = \frac{1}{\pi(y)!} \prod_{p \leq y} \left(\frac{\log x}{\log p}\right) e^{O(y^2/(\log x \log y))}. \tag{3.4}$$

**3.3. Sieving.** Begin with the numbers $1, 2, \ldots, [x]$. If we remove those integers divisible by 2 or 3, then the number of integers left is

$$x\left(1 - \tfrac{1}{2}\right)\left(1 - \tfrac{1}{3}\right)$$

plus or minus 2, since each prime $p$ "removes" about $1/p$ of the remaining integers, leaving a proportion of about $1 - 1/p$ of them. Therefore after sieving with primes from a finite set $P$, we expect to have approximately

$$x \prod_{p \in P} \left(1 - \tfrac{1}{p}\right) \tag{3.5}$$

integers left. Sieve theory tells us that this is a good guess if $p < x^{1/4}$ for all $p \in P$. If $P$ is the set of primes between $y$ and $x$ where $x = y^u$, then, by the above heuristic, we might expect

$$\Psi(x, y) \approx x \prod_{y < p \leq x} \left(1 - \tfrac{1}{p}\right) \sim \frac{x}{u}$$

integers left, but this is very wrong. We will now see this by determining $\Psi(x, x^{1/u})$ for $1 \leq u \leq 2$. Suppose that $n$ is a positive integer $\leq x$ that does not belong to $S(x, x^{1/u})$. Then $n$ must have a prime factor $p > x^{1/u}$. Evidently, $n$ cannot have two such prime factors, else $x \geq n \geq pq > x^{2/u} \geq x$. Therefore, any such $n$ can be written as $n = pm$, where $p$ is a prime in $[x^{1/u}, x]$ and $m \leq x/p$. Moreover, any $n$ that can be written in this form is $\leq x$ and $\notin S(x, x^{1/u})$. Therefore

$$\Psi(x, x^{1/u}) = [x] - \sum_{x^{1/u} < p \leq x} \#\{m \leq x/p\} = x - \sum_{x^{1/u} < p \leq x} (x/p + O(1))$$

$$= x\left(1 - (\log \log x - \log \log(x^{1/u}) + o(1))\right) + O(x/\log x)$$

$$= x\left(1 - \log\left(\frac{\log x}{(1/u) \log x}\right) + o(1)\right) = x(1 - \log u + o(1)).$$

(Here we use that $\sum_{p \leq x} 1/p = \log \log x + C + o(1)$ for some constant $C$, as well as that there are $\ll x/\log x$ primes $\leq x$.) We thus see that for $1 \leq u \leq 2$, we actually get the proportion $1 - \log u$ of the integers left, rather than the proportion $1/u$ as had been expected. More generally, (1.6) states that the proportion left is $\rho(u) = 1/u^{u+o(u)}$, a far cry from $1/u$. This big difference can be explained by the proof above: (3.5) is valid as long as divisibility for two different primes $p, q \in P$ is essentially independent. However, we saw in our proof that no $n \leq x$

can ever be divisible by two such primes $p$, so such divisibility is certainly not statistically independent.

### 3.4. Estimates for larger $u$. The *inclusion-exclusion principle* gives

$$\Psi(x, y) = [x] - \sum_{y<p\leq x} \left[\frac{x}{p}\right] + \sum_{y<p<q\leq x} \left[\frac{x}{pq}\right] - \sum_{y<p<q<r\leq x} \left[\frac{x}{pqr}\right] + \cdots$$

$$\stackrel{?}{\approx} x\left(1 - \sum_{y<p\leq x} \frac{1}{p} + \sum_{y<p<q\leq x} \frac{1}{pq} - \sum_{y<p<q<r\leq x} \frac{1}{pqr} + \cdots\right)$$

$$= x \prod_{y<p\leq x} (1 - 1/p).$$

This looks unlikely to provide a good approximation since there are $\pi(x) - \pi(y)$ terms here to worry about for the first sum, and far more for the second sum, usually more than $x$. However, if $p_1 \cdots p_k > x$, then $[x/(p_1 \cdots p_k)] = 0$ so we may ignore this term; in particular, any term with at least $u$ prime divisors is $> y^u = x$ so that $[x/(p_1 p_2 \ldots p_k)] = 0$. Thus we may refine the above formula to

$$\Psi(x, x^{1/u}) = [x] - \sum_{y<p\leq x} \left[\frac{x}{p}\right] + \sum_{\substack{y<p<q\leq x, \\ pq\leq x}} \left[\frac{x}{pq}\right] - \sum_{\substack{y<p<q<r\leq x, \\ pqr\leq x}} \left[\frac{x}{pqr}\right] + \cdots$$

$$\approx x\left(1 - \sum_{y<p\leq x} \frac{1}{p} + \sum_{\substack{y<p<q\leq x, \\ pq\leq x}} \frac{1}{pq} - \sum_{\substack{y<p<q<r\leq x, \\ pqr\leq x}} \frac{1}{pqr} + \cdots\right) \quad (3.6)$$

$$= x \sum_{\substack{d\leq x, \\ p|d \Rightarrow y<p\leq x}} \frac{\mu(d)}{d}. \quad (3.7)$$

Notice that the error term here is

$$\leq \sum_{\substack{n\leq x, \\ p|n \Rightarrow p>y}} 1 \ll \frac{x}{\log y} = u \frac{x}{\log x},$$

where the second inequality follows from the "fundamental lemma of the sieve" (which implies that if $m$ is an $x$-smooth integer, then there are $\ll \varphi(m)x/m$ integers up to $x$ that are coprime to $m$). However, it is complicated to evaluate (3.7). We note though that combining the above estimates with (1.1) in the range (1.9) implies

$$\sum_{\substack{d\leq x, \\ p|d \Rightarrow y<p\leq x}} \frac{\mu(d)}{d} = \rho(u) + O\left(\frac{1}{\log y}\right) \quad (3.8)$$

## 3.5. How about for larger $u$? (II)

We now proceed by induction: We shall "prove" that there exists a constant $\rho(u)$ for each $u > 0$ such that

$$\Psi(x, x^{1/u}) \sim x\rho(u). \tag{3.9}$$

As we saw above, this is true for $0 \le u \le 2$, with

$$\rho(u) = \begin{cases} 1 & \text{for } 0 < u \le 1; \\ 1 - \log u & \text{for } 1 \le u \le 2. \end{cases}$$

We will use the Buchstab–de Bruijn identity

$$\Psi(x, y) = 1 + \sum_{p \le y} \Psi\left(\frac{x}{p}, p\right), \tag{3.10}$$

which may be proved by writing each $n \in S(x, y)$ with $n > 1$ as $n = pm$, where $p$ is the largest prime factor of $n$.

Suppose (3.9) holds for $0 \le u \le N$, and consider values of $u \in (N, N+1]$: Subtracting (3.10) with $y = x^{1/N}$ from the same equation with $y = x^{1/u}$, we obtain

$$\Psi(x, x^{1/u}) = \Psi(x, x^{1/N}) - \sum_{x^{1/u} < q \le x^{1/N}} \Psi\left(\frac{x}{q}, q\right)$$

$$\approx x\left(\rho(N) - \sum_{x^{1/u} < q \le x^{1/N}} \frac{1}{q} \rho\left(\frac{\log(x/q)}{\log q}\right)\right).$$

Note that

$$\frac{\log(x/q)}{\log q} = \frac{\log x}{\log q} - 1 < \frac{\log x}{\log(x^{1/u})} - 1 = u - 1 \le N,$$

so we can apply the induction hypothesis. The prime number theorem states, in one form, that $\theta(T) := \sum_{p \text{ prime}, p \le T} \log p = T + O(T/\log^A T)$ for any fixed $A > 0$. Taking $T = x^{1/t}$, we obtain

$$\sum_{x^{1/u} < q < x^{1/N}} (1/q) \rho\left(\frac{\log(x/q)}{\log q}\right) = \int_{x^{1/u}}^{x^{1/N}} \rho\left(\frac{\log x}{\log T} - 1\right) \frac{d\theta(T)}{T \log T}$$

$$\approx \int_{x^{1/u}}^{x^{1/N}} \rho\left(\frac{\log x}{\log T} - 1\right) \frac{dT}{T \log T}$$

$$= \int_N^u \rho(t-1) \frac{dt}{t},$$

so we deduce that

$$\rho(u) = \rho(N) - \int_N^u \rho(t-1) \frac{dt}{t}. \tag{3.11}$$

This implies (1.5), and so (1.4). Such an approach to (1.1) involves a double induction and has led to the proof of (1.1) in the range (1.9) in the skillful hands of de Bruijn. We shall now see how a different identity, based on ideas of Chebyshev, allowed Hildebrand to prove (1.1) in the far wider range (1.10).

**3.6. Hildebrand's identity [1984a].** Another, easier, approach uses the (Chebyshev)–Hildebrand identity

$$\Psi(x, y) \log x = \int_1^x \Psi(t, y) \frac{dt}{t} + \sum_{\substack{p^m \leq x, \\ p \leq y}} \Psi\left(\frac{x}{p^m}, y\right) \log p. \tag{3.12}$$

As we shall see, the (Chebyshev)–Hildebrand identity has an advantage over the Buchstab–de Bruijn identity in that one of the parameters is held fixed; this perhaps explains the exponential difference in the ranges (1.9) and (1.10).

The (Chebyshev)–Hildebrand identity is proved in two steps: First write

$$\sum_{n \in S(x,y)} \log n = \int_1^x \log t \, d\Psi(t, y) = [\Psi(t, y) \log t]_1^x - \int_1^x \frac{\Psi(t, y)}{t} dt$$

$$= \Psi(x, y) \log x - \int_1^x \frac{\Psi(t, y)}{t} dt;$$

then write

$$\sum_{n \in S(x,y)} \log n = \sum_{n \in S(x,y)} \sum_{p^a | n} \log p$$

$$= \sum_{\substack{p^a \leq x, \\ p \leq y}} \log p \sum_{\substack{n \in S(x,y), \\ p^a | n}} 1 = \sum_{\substack{p^a \leq x, \\ p \leq y}} \log p \, \Psi\left(\frac{x}{p^a}, y\right)$$

and compare. Note that $\Psi(x, y) \log x \approx x\rho(u) \log x$. So,

$$\int_1^x \frac{\Psi(t, y)}{t} dt \leq x \quad \text{and} \quad \sum_{\substack{p \leq y, \\ a \geq 2}} \log p \, \Psi\left(\frac{x}{p^a}, y\right) \leq \sum_{p \leq y} \frac{x \log p}{p(p-1)} \leq cx$$

can be safely ignored in a (narrow) range for $u$ (though see (3.20) below for a more careful estimate), leading to

$$\Psi(x, y) \log x = \sum_{p \leq y} \log p \, \Psi\left(\frac{x}{p}, y\right) + O(x). \tag{3.13}$$

Therefore, using the prime number theorem again, taking $t = y^v$,

$$x \log y \cdot u\rho(u) \approx \int_1^y \Psi\left(\frac{x}{t}, y\right) d\theta(t) \approx \int_1^y \frac{x}{t} \rho\left(u - \frac{\log t}{\log y}\right) dt$$

$$= x \int_0^1 \rho(u-v)\, dv \log y = x \log y \int_{u-1}^u \rho(w)\, dw,$$

which gives (1.4).

**3.7. The value of $\rho$.** From (1.4) one can compute values of the function $\rho$, but it is appealing to find a non-self-referential expression to describe $\rho$, perhaps in terms of "simple" functions. By iterating (1.4) one determines the values

$$\rho(u) = \begin{cases} 1 & \text{for } 0 < u \leq 1, \\ 1 - \log u & \text{for } 1 \leq u \leq 2, \\ 1 - \log u + \int_2^u \log(v-1)\,(dv/v) & \text{for } 2 \leq u \leq 3. \end{cases} \quad (3.14)$$

One can continue in this way, or one can refer back to our inclusion-exclusion argument (3.6), to deduce

$$\rho(u) = 1 - \int_{t_1=1}^u \frac{dt_1}{t_1} + \frac{1}{2!} \int_{t_1=1}^u \int_{\substack{t_2=1 \\ t_1+t_2 \leq u}}^u \frac{dt_1\, dt_2}{t_1 t_2} - \cdots$$

$$+ \frac{(-1)^k}{k!} \int_{t_1=1}^u \int_{t_2=1}^u \cdots \int_{\substack{t_k=1 \\ t_1+t_2+\cdots+t_k \leq u}}^u \frac{dt_1\, dt_2 \cdots dt_k}{t_1 t_2 \cdots t_k} + \cdots \quad (3.15)$$

One can use these formulae to approximate $\rho(u)$, as we do in Tables 1 and 2.

**3.8. The Laplace transform.** A more sophisticated way to evaluate $\rho(u)$ is to take the integral equation just derived, multiply by $e^{-su}$, and integrate over $u$:

$$\int_{u \geq 0} u\rho(u)e^{-su} = \int_{u \geq 0} \int_{t=u-1}^u \rho(t)e^{-st} \cdot e^{-s(u-t)}\, dt.$$

If $\mathcal{L}(\sigma, s) = \int_0^\infty \sigma(t)e^{-st}\, dt$, then

$$-\mathcal{L}'(\rho, s) = \mathcal{L}(\rho, s) \int_0^1 e^{-sv}\, dv = \mathcal{L}(\rho, s)\left(\frac{1-e^{-s}}{s}\right),$$

so that

$$\mathcal{L}(\rho, s) = \mathcal{L}(\rho, 0) \exp\left(-\int_0^s \frac{1-e^{-t}}{t}\, dt\right). \quad (3.16)$$

| $u$ | $\rho(u)$ | $u$ | $\rho(u)$ |
|---|---|---|---|
| 1.0 | 1 | 4.6 | $1.0514448555 \times 10^{-3}$ |
| 1.1 | $9.0468982020 \times 10^{-1}$ | 4.7 | $8.0455864484 \times 10^{-4}$ |
| 1.2 | $8.1767844321 \times 10^{-1}$ | 4.8 | $6.1395732200 \times 10^{-4}$ |
| 1.3 | $7.3763573553 \times 10^{-1}$ | 4.9 | $4.6727987480 \times 10^{-4}$ |
| 1.4 | $6.6352776338 \times 10^{-1}$ | 5.00 | $3.5472470048 \times 10^{-4}$ |
| 1.5 | $5.9453489190 \times 10^{-1}$ | 5.25 | $1.7608050363 \times 10^{-4}$ |
| 1.6 | $5.2999637076 \times 10^{-1}$ | 5.50 | $8.6018611125 \times 10^{-5}$ |
| 1.7 | $4.6937174895 \times 10^{-1}$ | 5.75 | $4.1401923703 \times 10^{-5}$ |
| 1.8 | $4.1221333511 \times 10^{-1}$ | 6.00 | $1.9649696355 \times 10^{-5}$ |
| 1.9 | $3.5814611384 \times 10^{-1}$ | 6.25 | $9.1989056666 \times 10^{-6}$ |
| 2.0 | $3.0685281945 \times 10^{-1}$ | 6.50 | $4.2503555174 \times 10^{-6}$ |
| 2.1 | $2.6040578017 \times 10^{-1}$ | 6.75 | $1.9396328773 \times 10^{-6}$ |
| 2.2 | $2.2035713792 \times 10^{-1}$ | 7.00 | $8.7456699538 \times 10^{-7}$ |
| 2.3 | $1.8579946160 \times 10^{-1}$ | 7.25 | $3.8977236841 \times 10^{-7}$ |
| 2.4 | $1.5599126388 \times 10^{-1}$ | 7.50 | $1.7178674921 \times 10^{-7}$ |
| 2.5 | $1.3031956184 \times 10^{-1}$ | 7.75 | $7.4903397724 \times 10^{-8}$ |
| 2.6 | $1.0827244298 \times 10^{-1}$ | 8.00 | $3.2320693044 \times 10^{-8}$ |
| 2.7 | $8.9418565728 \times 10^{-2}$ | 8.25 | $1.3806442282 \times 10^{-8}$ |
| 2.8 | $7.3391580766 \times 10^{-2}$ | 8.50 | $5.8405695633 \times 10^{-9}$ |
| 2.9 | $5.9878115989 \times 10^{-2}$ | 8.75 | $2.4474945382 \times 10^{-9}$ |
| 3.0 | $4.8608388294 \times 10^{-2}$ | 9.00 | $1.0162482828 \times 10^{-9}$ |
| 3.1 | $3.9322969543 \times 10^{-2}$ | 9.25 | $4.1822758017 \times 10^{-10}$ |
| 3.2 | $3.1703444514 \times 10^{-2}$ | 9.50 | $1.7063527387 \times 10^{-10}$ |
| 3.3 | $2.5464723875 \times 10^{-2}$ | 9.75 | $6.9034598009 \times 10^{-11}$ |
| 3.4 | $2.0371779062 \times 10^{-2}$ | 10.00 | $2.7701718379 \times 10^{-11}$ |
| 3.5 | $1.6229593244 \times 10^{-2}$ | 10.50 | $4.3559526093 \times 10^{-12}$ |
| 3.6 | $1.2875434187 \times 10^{-2}$ | 11.00 | $6.6448090707 \times 10^{-13}$ |
| 3.7 | $1.0172837816 \times 10^{-2}$ | 11.50 | $9.8476421050 \times 10^{-14}$ |
| 3.8 | $8.0068721888 \times 10^{-3}$ | 12.00 | $1.4197131651 \times 10^{-14}$ |
| 3.9 | $6.2803730622 \times 10^{-3}$ | 12.50 | $1.9934633331 \times 10^{-15}$ |
| 4.0 | $4.9109256480 \times 10^{-3}$ | 13.00 | $2.7291890306 \times 10^{-16}$ |
| 4.1 | $3.8285861740 \times 10^{-3}$ | 13.50 | $3.6468386519 \times 10^{-17}$ |
| 4.2 | $2.9754747898 \times 10^{-3}$ | 14.00 | $4.7606300143 \times 10^{-18}$ |
| 4.3 | $2.3050505145 \times 10^{-3}$ | 14.50 | $6.0765096099 \times 10^{-19}$ |
| 4.4 | $1.7799424649 \times 10^{-3}$ | 15.00 | $7.5899080047 \times 10^{-20}$ |
| 4.5 | $1.3701177412 \times 10^{-3}$ | 15.50 | $9.2840614064 \times 10^{-21}$ |

**Table 1.** These tabulated values of the $\rho$-function were kindly supplied by Dan Bernstein.

| $u$ | $\rho(u)$ | $u$ | $\rho(u)$ |
|---|---|---|---|
| 16.00 | $1.1129193527 \times 10^{-21}$ | 36 | $1.2186971835 \times 10^{-63}$ |
| 16.50 | $1.3082753696 \times 10^{-22}$ | 37 | $6.2216867863 \times 10^{-66}$ |
| 17.00 | $1.5090797501 \times 10^{-23}$ | 38 | $3.0739529917 \times 10^{-68}$ |
| 17.50 | $1.7090489298 \times 10^{-24}$ | 39 | $1.4711270490 \times 10^{-70}$ |
| 18.00 | $1.9013542117 \times 10^{-25}$ | 40 | $6.8254908515 \times 10^{-73}$ |
| 18.50 | $2.0790325732 \times 10^{-26}$ | 41 | $3.0725325059 \times 10^{-75}$ |
| 19.00 | $2.2354265872 \times 10^{-27}$ | 42 | $1.3429776221 \times 10^{-77}$ |
| 19.50 | $2.3646133399 \times 10^{-28}$ | 43 | $5.7038156797 \times 10^{-80}$ |
| 20 | $2.4617828289 \times 10^{-29}$ | 44 | $2.3555177956 \times 10^{-82}$ |
| 21 | $2.5480499999 \times 10^{-31}$ | 45 | $9.4649292957 \times 10^{-85}$ |
| 22 | $2.4863827200 \times 10^{-33}$ | 46 | $3.7028093193 \times 10^{-87}$ |
| 23 | $2.2937113098 \times 10^{-35}$ | 47 | $1.4112017836 \times 10^{-89}$ |
| 24 | $2.0054951700 \times 10^{-37}$ | 48 | $5.2425207999 \times 10^{-92}$ |
| 25 | $1.6658044238 \times 10^{-39}$ | 49 | $1.8994303041 \times 10^{-94}$ |
| 26 | $1.3172582250 \times 10^{-41}$ | 50 | $6.7153344971 \times 10^{-97}$ |
| 27 | $9.9360680532 \times 10^{-44}$ | 55 | $2.6127284053 \times 10^{-109}$ |
| 28 | $7.1621362879 \times 10^{-46}$ | 60 | $5.8980293741 \times 10^{-122}$ |
| 29 | $4.9417994435 \times 10^{-48}$ | 65 | $8.0954516406 \times 10^{-135}$ |
| 30 | $3.2690443253 \times 10^{-50}$ | 70 | $7.0280992226 \times 10^{-148}$ |
| 31 | $2.0762615316 \times 10^{-52}$ | 75 | $3.9915358890 \times 10^{-161}$ |
| 32 | $1.2678257178 \times 10^{-54}$ | 80 | $1.5268607441 \times 10^{-174}$ |
| 33 | $7.4525736262 \times 10^{-57}$ | 85 | $4.0351170225 \times 10^{-188}$ |
| 34 | $4.2222207383 \times 10^{-59}$ | 90 | $7.5340256724 \times 10^{-202}$ |
| 35 | $2.3080811963 \times 10^{-61}$ | 95 | $1.0137476011 \times 10^{-215}$ |

**Table 2.** More values of the $\rho$-function.

Using the formula for the inverse Laplace transform gives

$$\rho(u) = \frac{1}{2i\pi} \int_{\mathrm{Re}(s)=\alpha} \mathcal{L}(\rho, s) e^{us} \, ds$$

$$= c \int_{\mathrm{Re}(s)=\alpha} \exp\left(us - \int_0^s \frac{1-e^{-t}}{t} dt\right) ds, \quad (3.17)$$

where $c = \mathcal{L}(\rho, 0)/(2i\pi)$ $(= e^\gamma/(2i\pi))$. This formula is not easy to work with, though one can deduce that

$$\rho(u) = \left(1 + O\left(\frac{1}{u}\right)\right) \sqrt{\frac{\xi'(u)}{2\pi}} \exp\left(\gamma - u\xi + \int_0^\xi \frac{e^t - 1}{t} dt\right), \quad (3.18)$$

where $\xi = \xi(u)$ is the unique positive solution to the equation $e^\xi = 1 + \xi u$.

## 3.9. More about $\rho$.
One can deduce

$$\rho(u) = \left(\frac{e+o(1)}{u \log u}\right)^u \tag{1.7}$$

from (3.18), though this can be derived more simply as follows: A simple calculation gives

$$\int_{u-1}^{u} \left(\frac{k}{t \log t}\right)^t dt \sim \frac{eu}{k}\left(\frac{k}{u \log u}\right)^u.$$

Fix $k < e$ and select $u_0$ sufficiently large. Now select $c > 0$ so that $\rho(t) > c(k/(t \log t))^t$ for all $t \leq u_0$. We claim that this inequality holds for all $t$. If not, select the smallest $u$ for which it fails so that, since $\rho$ is continuous,

$$c\left(\frac{k}{u \log u}\right)^u = \rho(u) = \frac{1}{u}\int_{u-1}^{u} \rho(t)\,dt > \frac{c}{u}\int_{u-1}^{u}\left(\frac{k}{t \log t}\right)^t dt \sim \frac{ce}{k}\left(\frac{k}{u \log u}\right)^u,$$

which gives a contradiction since $k < e$.

The upper bound can be proved analogously, but a more elegant upper bound can be derived as follows: Since $\rho(u) > 0$ for all $u$ by (1.2) and (1.4), but is nonincreasing by (1.5), we see that $\rho(t) \leq \rho(u-1)$ for all $t$ such that $u-1 \leq t \leq u$, and so $\rho(u) \leq \rho(u-1)/u$ by (1.4). Therefore, if $m$ is that integer for which $m+1 \geq u > m$ then, by induction,

$$\rho(u) \leq \frac{\rho(u-m)}{u(u-1)\cdots(u-m+1)} = \frac{\rho(u-m)\Gamma(u-m+1)}{\Gamma(u+1)} \leq \frac{1}{\Gamma(u+1)},$$

since $\rho(t) \leq 1$ for all $t$ and $\Gamma(t) \leq 1$ for $t \in [1,2]$.

Amusing identities involving the $\rho$ function include

$$e^\gamma = \int_0^\infty \rho(t)\,dt = \delta + \sum_{n \geq 1}(n+\delta)\rho(n+\delta) \quad \text{for any } 0 \leq \delta \leq 1.$$

## 3.10. Rankin's (clever) upper bound method (1938).
Fix any $\sigma > 0$. Then, since $(x/n)^\sigma \geq 1$ if $n \leq x$, and is $> 0$ if $n > x$,

$$\Psi(x,y) \leq \sum_{\substack{n \leq x, \\ p|n \Rightarrow p \leq y}} \left(\frac{x}{n}\right)^\sigma = x^\sigma \prod_{p \leq y}\left(1 - \frac{1}{p^\sigma}\right)^{-1}.$$

To minimize the right hand side (RHS) we can use calculus:

$$\log(\text{RHS}) = \sigma \log x - \sum_{p \leq y} \log(1 - p^{-\sigma}).$$

Differentiating gives

$$\log x = \sum_{p \leq y} \frac{\log p}{p^\sigma - 1}. \tag{3.19}$$

Not easy to explicitly solve! Since the right side is monotone decreasing, continuous as a function of $\sigma$, and decreasing from $\infty$ to 0, there is a unique solution $\sigma = \alpha(x, y)$. In fact,

$$\alpha(x, y) = \frac{\log(1 + y/\log x)}{\log y}\left(1 + O\left(\frac{\log\log(1+y)}{\log y}\right)\right) \approx 1 - \frac{u \log u}{\log y},$$

where the last approximation is valid if it is $> 1/2$. This formula for $\alpha$ is tricky and technical to substitute in, and comes out bigger than the correct answer by a small factor (see below (3.23)). However, it has the great advantage of being a relatively simple method for obtaining a good upper bound in all ranges for $x$ and $y$, and so has been very useful for applications. In fact this method has featured in many of Erdős and Pomerance's works on counting numbers of interest in computational number theory; see the survey [Pomerance 1989], for example.

*Special case*: $y = (\log x)^A$ for $A > 1$. Let $\sigma = 1 - 1/A$ to get

$$\log \prod_{p \leq y}\left(1 - \frac{1}{p^\sigma}\right)^{-1} \ll \sum_{p \leq y} \frac{1}{p^\sigma} = \sum_{p \leq y} \frac{p^{1/A}}{p} \ll \frac{y^{1/A}}{\log y} \ll \frac{\log x}{\log\log x}.$$

Therefore $\Psi(x, (\log x)^A) \leq x^{1 - 1/A + O(1/\log\log x)}$. Combining this with (3.3), we obtain

$$\Psi(x, (\log x)^A) = x^{1 - 1/A + O(1/\log\log x)}.$$

**3.11. Iterating the identities more carefully.** With more care, the error term in (3.13) can be improved to, for all $y \geq \log^{2+\varepsilon} x$,

$$\Psi(x, y) \log x = \sum_{p \leq y} \Psi\left(\frac{x}{p}, y\right) \log p + O(\Psi(x, y)). \tag{3.20}$$

From this we can deduce the following result ( much as in the proof of [Granville 1993a, Proposition 1]), which explains Hildebrand's result (1.10) (and after):

THEOREM. *Define $\rho_y(u)$ as follows*:

$$\rho_y(u) = 1 \quad \text{for } 0 \leq u \leq 1$$

and

$$\rho_y(u) = \frac{1}{u \log y} \sum_{p \leq y} \rho_y\left(u - \frac{\log p}{\log y}\right)\frac{\log p}{p}$$

*for $u > 1$. Then*

$$\Psi(x, y) = x\rho_y(u)\left(1 + O_\varepsilon\left(\frac{\log(u+1)}{\log y}\right)\right)$$

*for all $x \geq y \geq (\log x)^{2+\varepsilon}$, where $x = y^u$.*

Moreover
$$\rho_y(u) = \rho(u)\left(1 + O_\varepsilon\left(\frac{\log(u+1)}{\log y}\right)\right)$$
uniformly in a range for $u$ and $y$ if and only if
$$|\theta(y) - y| \ll \frac{y}{u^{1+o(1)}} \qquad (3.21)$$
uniformly in the same range for $u$ and $y$, where $\theta(y) := \sum_{p \leq y} \log p$. The strongest form of the prime number theorem known gives the range (1.10). The Riemann Hypothesis is equivalent to: One can take $u = y^{1/2-\varepsilon}$ for all $\varepsilon > 0$ in (3.21), which gives us (1.11).

**3.12. The saddle point method.** Hildebrand and Tenenbaum [1986], developing an old approach of de Bruijn, used the saddle point method to get an asymptotic for $\Psi(x, y)$ in all ranges that are not easily handled by other methods. They started with Perron's formula: Fix $\alpha > 0$ real. Then

$$\int_{\text{Re}(s)=\alpha} \frac{y^s}{s}\,ds = \begin{cases} 1 & \text{if } y > 1, \\ 1/2 & \text{if } y = 1, \\ 0 & \text{if } 0 < y < 1. \end{cases}$$

If we want those $n \geq 1$ for which $n < x$, we can recognize them as those with $x/n > 1$. Therefore

$$\Psi(x, y) = \sum_{\substack{n \leq x, \\ p|n \Rightarrow p \leq y}} 1 = \sum_{\substack{n \geq 1, \\ p|n \Rightarrow p \leq y}} \int_{\text{Re}(s)=\alpha} \frac{(x/n)^s}{s}\,ds + O(1)$$

$$= \int_{\text{Re}(s)=\alpha} \left(\sum_{n \geq 1,\, p|n \Rightarrow p \leq y} \frac{1}{n^s}\right) \frac{x^s}{s}\,ds + O(1)$$

$$= \int_{\text{Re}(s)=\alpha} \xi(s, y) \frac{x^s}{s}\,ds + O(1), \qquad (3.22)$$

where $\xi(s, y) := \prod_{p \leq y}(1 - 1/p^s)^{-1}$.

Select $\alpha = \alpha(x, y)$ (the optimization point in Rankin's upper bound). Define $\varphi_k = \varphi_k(s, y) = (d/ds)^k \log \xi(s, y)$ with $\varphi = \varphi_0$, so that $\log x + \varphi_1(\alpha, y) = 0$ by (3.19). One shows that the main contribution to this integral comes from a very short segment close to $\alpha(x, y)$, the "saddle point", so that

$$\Psi(x, y) = \frac{1}{2i\pi} \int_{\alpha(x,y)-i/\log y}^{\alpha(x,y)+i/\log y} \xi(s, y) x^s \frac{ds}{s} + \text{small error}.$$

Now if $s = \alpha + it$, then
$$\frac{x^s}{s} = \frac{x^\alpha}{\alpha} \cdot \frac{x^{it}}{1+it/\alpha} = \frac{x^\alpha}{\alpha} e^{it \log x}\left(1 - \frac{it}{\alpha} + \text{error}\right);$$

and, developing the Taylor series,

$$\log\left(\frac{\xi(s, y)}{\xi(\alpha, y)}\right) = \varphi(\alpha + it, y) - \varphi(\alpha, y) = it\varphi_1 - \frac{t^2}{2}\varphi_2 - \frac{it^3}{6}\varphi_3 + O(t^4\varphi_4),$$

so that

$$\frac{x^s}{s}\xi(s, y) = \frac{x^\alpha}{\alpha}\xi(\alpha, y)e^{it(\log x + \varphi_1(\alpha, y))} \times e^{-t^2\varphi_2/2}\left(1 - \frac{it}{\alpha} - \frac{it^3}{6}\varphi_3 + \text{error}\right).$$

Therefore, since $\log x + \varphi_1(\alpha, y) = 0$,

$$\frac{1}{2i\pi}\int_{\alpha - i/\log y}^{\alpha + i/\log y} \xi(s, y)\frac{x^s}{s}\,ds$$

$$= \frac{x^\alpha}{\alpha}\xi(\alpha, y)\frac{1}{2\pi}\int_{-1/\log y}^{1/\log y} e^{-t^2\varphi_2/2}\left(1 - \frac{it}{\alpha} - \frac{it^3}{6}\varphi_3 + \text{error}\right)dt$$

$$= \frac{x^\alpha}{\alpha}\xi(\alpha, y)\frac{1}{\sqrt{2\pi\varphi_2}}(1 + \text{error})$$

(since $\int_{-\infty}^{\infty} e^{-at^2}\,dt = \sqrt{\pi/a}$). Therefore, for all $x \geq y \geq 2$,

$$\Psi(x, y) = \frac{x^\alpha \xi(\alpha, y)}{\alpha\sqrt{2\pi\varphi_2(\alpha, y)}}\left(1 + O\left(\frac{1}{u} + \frac{\log y}{y}\right)\right). \qquad (3.23)$$

Note that this is smaller than the upper bound $\Psi(x, y) \leq x^\alpha \xi(\alpha, y)$, given by Rankin's method, by a factor $\asymp \alpha\sqrt{\varphi_2(\alpha, y)} \ll \log x$. Evaluating the esoteric expression on the right side of (3.23) is in general a difficult problem (indeed it may be argued that we have traded in one intractable problem for another!). However one can make some interesting deductions; for example, one can deduce that if $1 \leq c \leq y$ then

$$\Psi(cx, y) = \Psi(x, y)c^{\alpha(x,y)}\left(1 + O\left(\frac{1}{u} + \frac{\log y}{y}\right)\right), \qquad (3.24)$$

so that one can solve an old conjecture of Erdős:

$$\Psi(2x, y)/\Psi(x, y) \sim \left(1 + \frac{y}{\log x}\right)^{(\log 2)/(\log y)}$$

$$\sim \begin{cases} 1 & \text{if and only if } y \leq (\log x)^{1+o(1)}, \\ 2 & \text{if and only if } y > (\log x)^\infty. \end{cases}$$

In between is a transition:

$$\Psi(2x, y) \sim 2^{1-1/\alpha}\Psi(x, y) \quad \text{when} \quad y = (\log x)^{\alpha + o(1)} \quad \text{with } \alpha > 1.$$

This, and a function field analogue, inspired Soundararajan to find an easy deduction of (1.8) in the range (1.10) from (3.23), based on the idea that, in a wide range,

$$\frac{\Psi(x^2, y^2)}{x^2} \sim \frac{\Psi(x, y)}{x}.$$

## 4. Smooths in short intervals, in arithmetic progressions, and as values of polynomials

**4.1. Smooths in short intervals.** One might guess that smooth numbers are "well-distributed" in short intervals, that is, that roughly the same proportion of integers in a short interval near $x$ are smooth as amongst all of the integers up to $x$. More precisely we expect that, usually, we have

$$\Psi(x+z, y) - \Psi(x, y) \sim \frac{z}{x}\Psi(x, y) \sim z\rho(u). \tag{4.1}$$

Hildebrand [1986] showed this for $x \geq z \geq x/y^{5/12}$ in the range (1.10); this can be improved to the range $x \geq z \geq x/y^{1-o(1)}$ (using identities involving integers with exactly two prime factors in short intervals as at the end of Section 4.2). Friedlander and I [1993] proved (4.1) in the range

$$\exp((\log x)^{5/6+o(1)}) \leq y \leq x \quad \text{and} \quad \sqrt{x}y^2 \exp((\log x)^{1/6}) \leq z \leq x.$$

The most elusive goal in the subject is to show that (4.1) holds when $y$ and $z$ are arbitrary powers of $x$. Specifically if $1 > \beta, \alpha > 0$, then one wants to show that

$$\Psi(x+x^\beta, x^\alpha) - \Psi(x, x^\alpha) \sim x^\beta \rho(1/\alpha); \tag{4.2}$$

our result implies this for $\beta > 1/2 + 2\alpha$. Remarks in Section 4.4 suggest that (4.2) is inaccessible if $\alpha < 1/(4\sqrt{e}) = .15163\ldots$ and $\beta < \alpha e^{\rho(1/\alpha)}$ (for if we can prove this then we can improve what is known on Vinogradov's conjecture; see Section 1.2). Note that for $\alpha$ in this range, $1 < e^{\rho(1/\alpha)} < 1 + 3/10^5$, so the most accessible inaccessible cases have $\alpha, \beta \approx 5/33$.

A slight improvement from [Balog 1987] gives that

$$\Psi(x+x^\beta, x^\alpha) - \Psi(x, x^\alpha) \gg_{\alpha,\beta} x^\beta \tag{4.3}$$

for all $\beta > 1/2$ and $\alpha > 0$. Harman [1991] extended Balog's result by showing

$$\Psi(x+x^\beta, y) - \Psi(x, y) > 0$$

for any fixed $\beta > 1/2$ and $x \geq y \geq \exp((\log x)^{2/3+o(1)})$. Lenstra, Pila, and Pomerance [1993] slightly strengthened this result and gave an explicit lower bound of the correct order of magnitude. Xuan [1999] showed, under the assumption of the Riemann Hypothesis, that there is an $x^\varepsilon$-smooth integer in any interval $[x, x + \sqrt{x}(\log x)^{1+o(1)}]$.

Obtaining results like (4.2) and (4.3) with $\beta = 1/2$ and $\alpha > 0$ fixed but small seems to be rather difficult, and there are few results. I believe that this is the outstanding problem in the whole area of the distribution of smooth numbers:

CHALLENGE PROBLEM 2000. *Prove that, for all $\alpha > 0$, if $x$ is sufficiently large then*

$$\Psi(x + \sqrt{x}, x^\alpha) - \Psi(x, x^\alpha) > 0. \qquad (4.4)$$

The "$\sqrt{x}$" barrier for the length of the interval has only been broken in certain special circumstances: In 1987, Friedlander and Lagarias showed, by ingeniously constructing such an integer, that there is always an $x^{1/2}$-smooth integer in any interval of length $x^{1/4+\varepsilon}$ near to $x$, and also an $x^{1/3}$-smooth integer in any interval of length $x^{5/12+\varepsilon}$ near to $x$. Harman [1999] proved (4.3) for $1/2 \geq \beta \geq 3/7$ with $\alpha > (3 - 5\beta)/(2\sqrt{e})$. In particular this shows one can take any $\alpha > 1/(4\sqrt{e})$ for $\beta = 1/2$ in (4.3). This was recently improved by Croot [2001], who showed that for any fixed $\varepsilon > 0$,

$$\Psi(x + c\sqrt{x}, x^{3/(14\sqrt{e})+\varepsilon}) - \Psi(x, x^{3/(14\sqrt{e})+\varepsilon}) \gg_\varepsilon \sqrt{x}/(\log x)^{\log 4 + o(1)}$$

for some constant $c = c(\varepsilon) > 0$. This is far from what we need for applications: In Section 1.5, we saw that we would like to prove a rather better lower bound than that given in (4.4), with "$x^\alpha$" replaced by "$L(x)^c$".

Following Conjecture 4.4.2, we will justify our belief that an estimate like $\Psi(x + x^\beta, x^\alpha) - \Psi(x, x^\alpha) > 0$ is inaccessible whenever $\alpha, \beta < 1/(4\sqrt{e})$; in this context the results of Harman and Croot seem all the more remarkable for having been obtained with such small $\alpha$.

Friedlander and Lagarias [1987] also proved that one can "break the $\sqrt{x}$ barrier" most of the time. The best result to date of this type, due to Hildebrand and Tenenbaum [1993a], states that for any fixed $\varepsilon > 0$, (4.1) holds when

$$\exp((\log X)^{5/6+\varepsilon}) \leq y \leq X \quad \text{and} \quad y \exp((\log X)^{1/6}) \leq z \leq X,$$

for all but at most $X/\exp((\log X)^{1/6-\varepsilon})$ integers $x \leq X$. Assuming the Riemann Hypothesis, Hafner [1993] showed such a result for $L(X) \leq y \leq X$ with $\sqrt{L(X)} \leq z \leq X$.

**4.2. Smooths in arithmetic progressions.** Let $\Psi_q(x, y)$ be the number of integers in $S(x, y)$ that are coprime to $q$. As one might guess,

$$\Psi_q(x, y) \sim \frac{\varphi(q)}{q} \Psi(x, y)$$

in a very wide range: Tenenbaum [1993] showed this, provided there are at most $y^{o(1/\log u)}$ prime factors of $q$ that are $\leq y$. In fact Xuan [1995] showed that if $q$

has no more than $y^{1/2}$ prime factors $\leq y$ then

$$\Psi_q(x,y) \sim \prod_{p|q,\, p\leq y} \left(1 - \frac{1}{p^{\alpha(x,y)}}\right) \Psi(x,y)$$

for $(\log x)^{1+o(1)} < y < x^{o(1)}$.

Let $\Psi(x,y;a,q)$ be the number of integers in $S(x,y)$ that are equivalent to $a$ modulo $q$. We would expect

$$\Psi(x,y;a,q) \sim \frac{1}{\varphi(q)} \Psi_q(x,y) \quad \text{whenever } (a,q) = 1. \tag{4.5}$$

I showed that this is true for

$$x \geq y \geq q^{1+\varepsilon} \quad \text{as } (\log x)/(\log q) \to \infty. \tag{4.6}$$

For $y \geq q^{3/4+\varepsilon}$ and $x \geq y^2$ we get (using results of Balog and Pomerance [1992])

$$\Psi(x,y;a,q) \asymp \frac{1}{\varphi(q)} \Psi_q(x,y) \quad \text{whenever } (a,q) = 1. \tag{4.7}$$

Harman [Harman 1999] remarkably proves that if $q$ is cube free then

$$\Psi(x,y;a,q) \gg \frac{1}{\varphi(q)} \Psi_q(x,y) \quad \text{whenever } (a,q) = 1$$

for $y > q^{1/(4\sqrt{e})+\varepsilon}$ and $x > q^{9/4+\varepsilon}$. No wider range for $y$ is feasible with the current state of knowledge; see Section 4.4. Recently Soundararajan [2006] has developed a new analytic method which is likely to give asymptotics like (4.5) in a range like $y > q^{1/(4\sqrt{e})+\varepsilon}$ for a wide range of values of $x$. Also the method is liable to prove that the smooth numbers are equidistributed in the subgroup of $(\mathbb{Z}/q\mathbb{Z})^*$ generated by the primes $\leq y$ in an even wider range for $y$.

Fouvry and Tenenbaum [1996] showed that (4.5) holds for almost all $x \leq X$ and for almost all $q \leq \min\{x^{3/5-o(1)}, \exp(c(\log y \log\log y)/(\log\log\log y))\}$; and for almost all $q \leq \sqrt{x}/\exp((\log x)^{1/3})$ provided $y > \exp((\log x)^{2/3+o(1)})$.

For a given sequence $\mathcal{N} = n_1 < n_2 < \cdots$ of integers it is often relatively easy to give an asymptotic estimate for

$$\sum_{q \leq Q} \sum_{\substack{a=1 \\ (a,q)=1}}^{q} \left| \sum_{\substack{n \in \mathcal{N},\, n \leq x, \\ n \equiv a \pmod{q}}} 1 - \frac{1}{\varphi(q)} \sum_{\substack{n \in \mathcal{N},\, n \leq x, \\ (n,q)=1}} 1 \right|^2 \tag{4.8}$$

when $Q = x/\log^A x$: For example, when $\mathcal{N}$ is the sequence of primes, this is $\sim Qx \log x$. Bob Vaughan and I have noted that we can get a nontrivial upper bound, but have had difficulties obtaining an asymptotic, when $\mathcal{N}$ is the sequence of $y$-smooth numbers, for various ranges of values of $y$.

One proves results like (4.5) and (4.7) in a way similar to Hildebrand's method in Section 3.6. By summing

$$\sum_{\substack{n \leq x,\, p\mid n \Rightarrow p \leq y, \\ n \equiv a \pmod{q}}} \log n$$

in two ways, we get the analogy of the Chebyshev–Hildebrand identity (3.12), namely

$$\Psi(x, y; q, a) \log x = \int_1^x \Psi(t, y; q, a) \frac{dt}{t}$$
$$+ \sum_{\substack{p^m \leq x, \\ p \leq y,\, p \nmid q}} \Psi\!\left(\frac{x}{p^m}, y; q, \frac{a}{p^m}\right) \log p. \quad (4.9)$$

Now sum this over all $a$ with $1 \leq a \leq q$ and $(a, q) = 1$, and divide by $\varphi(q)$, to get

$$\frac{\Psi_q(x, y)}{\varphi(q)} \log x = \int_1^x \frac{\Psi_q(t, y)}{\varphi(q)} \frac{dt}{t} + \sum_{\substack{p^m \leq x \\ p \leq y,\, p \nmid q}} \frac{\Psi_q(x/p^m, y)}{\varphi(q)} \log p, \quad (4.10)$$

and we have functional equations for $\Psi(x, y; q, a)$ and $\Psi_q(x, y)/\varphi(q)$ that are the same. We can use this to show that (4.5) holds for all $x \geq y$ provided it holds for all $x$ in the range

$$y^{2+\delta}/q > x \geq y.$$

To get a very sharp range like (4.6) we first write our functional equation as

$$\Psi(x, y; q, a) \log x = \sum_{p \leq y,\, p \nmid q} \Psi\!\left(\frac{x}{p}, y; q, \frac{a}{p}\right) \log p + O\!\left(\frac{\Psi_q(x, y)}{\varphi(q)}\right). \quad (4.11)$$

"Iterate this", that is, substitute in

$$\Psi\!\left(\frac{x}{p}, y; q, \frac{a}{p}\right) \log\!\left(\frac{x}{p}\right) = \cdots,$$

as determined by (4.11), to get

$$\Psi(x, y; q, a) \log x = \sum_{\substack{p_1, p_2 \leq y, \\ p_1, p_2 \nmid q}} \frac{\log p_1 \log p_2}{\log(x/p_1)} \Psi\!\left(\frac{x}{p_1 p_2}, y; q, \frac{a}{p_1 p_2}\right)$$
$$+ O\!\left(\frac{\Psi_q(x, y)}{\varphi(q)}\right).$$

To use this functional equation, we need to know that integers with exactly two prime factors are well-distributed in arithmetic progressions, rather than primes. Such a result is provided by [Mikawa 1989]: If $x \geq y \geq x^{1-\varepsilon}$ and $q = o(y/\log^5 x)$, then

$$\sum_{\substack{x-y \leq n < x, \\ n \equiv a \pmod{q}, \\ n = p_1 p_2}} \log p_1 \log p_2 \gg \frac{y \log x}{\varphi(q)}$$

for almost all $(a, q) = 1$. Therefore (4.5) holds in the range (4.6).

**4.3. Polynomial values.** For a given polynomial $f(x) \in \mathbb{Z}[x]$, define

$$\Psi_f(x, y) = \#\{n \leq x : p \mid f(n) \Rightarrow p \leq y\}.$$

For $f$ of degree one, the results above imply $\Psi_f(x, x^{1/u}) \sim x\rho(u)$ for fixed $u$ as $x \to \infty$. Actually they even imply that $\Psi_f(x, y) \sim x\rho(u)$ uniformly for such $f$ when $u \to \infty$ and $v/\log u \to \infty$, where $x = y^u$ and $y = H(f)^v$, and $H(f)$, the height of $f$, equals the maximum of the absolute values of the coefficients of $f$.

For irreducible $f$ of degree $d$ we expect

$$\Psi_f(x, x^{1/u}) \sim \rho(du)x \quad \text{as } x \to \infty, \tag{4.12}$$

for fixed $u > 0$, as we discussed in Section 1.1. Hmyrova [1966] gave an upper bound of similar order of magnitude[1] for irreducible polynomials by showing $\Psi_f(x, x^{1/u}) \ll_f x(e/u)^u$ if $y = x^{1/u} \geq \log x$.

Balog and Ruzsa [1997] showed that

$$\Psi_f(x, x^{1/u}) \asymp_{f,u} x \tag{4.13}$$

when $f$ is the product of two linear polynomials in $\mathbb{Z}[x]$. Hildebrand [1989] proved (4.13) for $k/u > e^{-1/(k-1)}$ when $f$ splits completely into $k$ distinct linear factors over $\mathbb{Z}[x]$. Dartyge [1996] proved (4.13) for $f(t) = t^2 + 1$ when $u < 179/149$.

For general $f$ of higher degree, let $d_1, d_2, \ldots, d_k$ be the degrees of the distinct irreducible factors of $f$, where $d$ is the maximum of the $d_j$ and $\ell$ is the number of $j$ for which $d_j = d$. Very recently Dartyge, Martin and Tenenbaum [2001] proved (4.12) for fixed $1/u > d - 1/\ell$; and Martin [2002] proved

$$\Psi_f(x, x^{1/u}) \sim \rho(d_1 u)\rho(d_2 u) \cdots \rho(d_k u)x \tag{1.20}$$

in the same range, assuming a suitable and plausible uniform version of hypothesis $H$. (Hypothesis $H$ is a grand generalization of the prime twins conjecture which, in its simplest form, states that if $f_1(t), \cdots, f_k(t) \in \mathbb{Z}[t]$ are irreducible polynomials, which have the property that for every prime $p$ there exists an

---

[1] Wolke [1971] claimed to have given an upper bound with a factor $1/u^{du}$ in both these problems but Friedlander points out that the deduction of [(24)] there, and thus the whole proof, seems flawed.

integer $a_p$ such that $p$ does not divide $f_1(a_p)f_2(a_p)\cdots f_k(a_p)$, then there are infinitely many distinct integers $n$ for which $|f_1(n)|, |f_2(n)|, \ldots, |f_k(n)|$ are all prime.)

Dartyge, Martin and Tenenbaum [2001] also showed that

$$\pi_f(x, x^{1/u}) \gg \pi(x),$$

where $\pi_f(x, y) = \#\{q \leq x : q \text{ prime and } p \mid f(q) \Rightarrow p \leq y\}$ for $1/u > d - 1/(2\ell)$. Hmyrova [1966] gave the general upper bound (see again our footnote on page 298) of $\pi_f(x, x^{1/u}) \ll_f \pi(x)/u^{\{1+o(1)\}u}$ for $y = x^{1/u} > \log x$.

**4.4. Limitations on what we might prove.** In our current state of knowledge it seems inaccessible to improve upon what is known about Vinogradov's conjecture, or the following generalization:

VINOGRADOV'S CONJECTURE $(+\varepsilon)$. *Fix an integer $k \geq 2$. For each prime $p$, the least $k$-th power nonresidue (mod $p$) is $\ll_{\varepsilon,k} p^\varepsilon$.*

In Section 6.4 we give the proof that this holds for any $\varepsilon > 1/(4u_k)$ where $u_k$ is defined so that $\rho(u_k) = 1/k$ (and thus $u_k \sim (\log k)/(\log \log k)$). Given how long this result has remained unimproved, one must surely regard any estimate on smooth numbers that implies an improvement to be "inaccessible" for now, since it would have wide ramifications, not only affecting problems in computational number theory but also such an old chestnut of analytic number theory; see Section 1.2 for further discussion.

If $\Psi(x, y; a, q) > 0$ for sufficiently large $x$ then every residue class (mod $q$) is generated by the primes $\leq y$ implying Vinogradov's Conjecture $(+\varepsilon)$ if we can take $y$ to be an arbitrarily small power of $x$. Thus we assume that such a result is inaccessible. This is why Harman's result, mentioned in Section 4.2, seems so remarkable in the context of the $k = 2$ case of Vinogradov's conjecture; see Section 6.4.

In fact, if $\Psi(x, y; a, p) > 0$ for sufficiently large $x$ for more than $(p-1)/k$ residue classes $a \pmod{p}$ with $(a, p) = 1$, then the primes $\leq y$ generate more than just the $k$-th power residues (mod $p$). Such a result would be implied if $\Psi(x, y; a, p) < k\Psi_p(x, y)/(p-1)$. Thus we believe that the following conjecture is inaccessible for any $\varepsilon < 1/(4\sqrt{e})$:

CONJECTURE 4.4.1. *Fix $\varepsilon > 0$ sufficiently small. If $y = q^\varepsilon$, then for sufficiently large $x$ we have*

$$\Psi(x, y; a, q) < \frac{2\Psi_q(x, y)}{\varphi(q)}. \quad (4.14)$$

These observations are all due to Friedlander who was motivated by them to give the first bounds of type (4.7).

The following conjecture is certainly a goal of researchers into smooths in short intervals (see Section 4.1):

CONJECTURE 4.4.2. *Fix $\varepsilon > 0$. If $x$ is sufficiently large, then*

$$\Psi(x, x^\varepsilon) - \Psi(x - x^\varepsilon, x^\varepsilon) > 0. \tag{4.15}$$

If this conjecture is true then Vinogradov's Conjecture $(+\varepsilon)$ holds whenever $-1$ is not a $k$-th power residue (mod $p$): Let $q$ be the least positive $k$-th power nonresidue (mod $p$), which we can assume is $> p^\varepsilon$. By the above conjecture, there exists a $y$-smooth integer $p - m$ in $(p - q, p - 1)$, where $y = q - 1$, which is thus a $k$-th power residue (mod $p$); but then so is $-1 \equiv (p - m)/m$ (mod $p$), which gives a contradiction.

Thus Conjecture 4.4.2 seems inaccessible even for $\varepsilon < 1/(4\sqrt{e})$, since this would imply an improvement on what is known about Vinogradov's conjecture for primes $p \equiv 3$ (mod 4) (though Croot and Harman have recently given results that approach this boundary, as reported in Section 4.1).

This argument still works if

$$\Psi(x^\beta, x^\alpha) + \{\Psi(x, x^\alpha) - \Psi(x - x^\beta, x^\alpha)\} > x^\beta$$

for sufficiently large $x$. Such an estimate would be guaranteed if (4.2) holds, for $\beta < \alpha e^{\rho(1/\alpha)}$.

Even a rather weaker conjecture of the form (4.3) seems inaccessible for similar reasons:

CONJECTURE 4.4.3. *Fix $\varepsilon > 0$. For $\delta > 0$ sufficiently small, we have*

$$\Psi(x + z, y) - \Psi(x, y) \gtrsim \frac{\delta}{\varepsilon} z \quad \text{for } x \geq z \geq x^\delta \quad \text{and } x \geq y \geq x^\varepsilon, \tag{4.16}$$

*whenever $x$ is sufficiently large.*

Note that this follows from (4.2) for any fixed $0 < \delta < (1/2)\,\varepsilon\rho(1/\varepsilon)$.

Vinogradov's Conjecture $(+\varepsilon)$ does follow, in general, from this conjecture: Let $q$ be the least positive $k$-th power nonresidue (mod $p$), which we can assume is $> p^\varepsilon$. We may also assume that $\varepsilon < 1/\sqrt{e}$ (since Vinogradov's conjecture is known to be true for larger $\alpha$).

Let $y = q - 1$. Define $t_j = [jp/q]$. Note that at most one of $t_j + i$ and $q(t_j + i) - jp$ is a $k$-th power residue (mod $p$) since their ratio is $q$ (mod $p$). Thus, taking $0 \leq j \leq q - 1$ and $1 \leq i \leq I = [p^\delta]$, we deduce that there are at most $(q - 1)I$ $k$-th power residues (mod $p$) amongst the union of

$$\bigcup_{j=0}^{q-1} (t_j, t_j + I] \quad \text{and} \quad \bigcup_{j=0}^{q-1} \bigcup_{i=1}^{I} \{q(t_j + i) - jp\} = (0, qI].$$

Since every $y$-smooth integer amongst these is a $k$-th power residue, we deduce that

$$qI > \Psi(qI, y) + \sum_{j=0}^{q-1}\{\Psi(t_j + I, y) - \Psi(t_j, y)\}. \tag{4.17}$$

By (4.2) we expect the above to be $\sim qI(\rho(1 + \delta/\varepsilon) + \rho(1/\varepsilon))$. Now $1/\varepsilon > \sqrt{e}$, so that $\rho(1/\varepsilon) < 1/2$, and thus to get a contradiction we certainly need $\rho(1+\delta/\varepsilon) > 1/2$, so that $\delta/\varepsilon < 1$ and therefore $\rho(1+\delta/\varepsilon) = 1 - \log(1+\delta/\varepsilon)$ by (1.3). This implies we get a contradiction (so that Vinogradov's Conjecture ($+\varepsilon$) holds) provided (4.2) holds uniformly for $\delta < \varepsilon(e^{\rho(1/\varepsilon)} - 1)$.

In the hypothesis of Conjecture 4.4.3 we allow "$\delta > 0$ sufficiently small"; in particular, assume $\delta < \varepsilon$. Then the right side of (4.17) is

$$\gtrsim qI\left(\rho\left(1 + \frac{\delta}{\varepsilon}\right) + \frac{\delta}{\varepsilon}\right) = qI\left(1 + \left(\frac{\delta}{\varepsilon} - \log\left(1 + \frac{\delta}{\varepsilon}\right)\right)\right)$$

by Conjecture 4.4.3, giving a contradiction. Thus Vinogradov's Conjecture ($+\varepsilon$) also follows from Conjecture 4.4.3.

**4.5. The distribution of $y$-smooth integers and their prime divisors.** It is evident that if $y$ is sufficiently small, then most $y$-smooth integers up to $x$ will be divisible by high powers of primes, whereas if $y$ is large (say $y = x$), then few $y$-smooth integers are divisible by high powers. To understand the change in nature of $\Psi(x, y)$ at $y$ near $\log x$, from (1.14) if $y > (\log x)^{1+\varepsilon}$ to (1.18) if $y = o(\log x)$, we need to better understand what the elements of $S(x, y)$ look like.

A proportion $6/\pi^2$ of the $y$-smooth integers up to $x$ are squarefree, whenever $y$ is larger than any fixed power of $\log x$ (that is, $(\log y)/(\log\log x) \to \infty$). That proportion drops to 0 once $y < (\log x)^{2+o(1)}$; and, since there are only $2^{\pi(y)}$ $y$-smooth, squarefree integers, this is $\Psi(x, y)^{o(1)}$ for $y = o(\log x)$ by (1.18).

Alladi [1982] noted that for $u > 2$ but fixed, one has

$$\sum_{n \in S(x,y)} \mu(n) \sim \omega'(u)\frac{x}{\log^2 y},$$

where $\omega$ is as in Section 1.5.

In 1940, Erdős and Kac showed that the values of $\Omega(n)$, the total number of prime factors of $n$, for $n$ up to $x$, satisfy the normal distribution with mean and variance $\sim \log\log x$. Alladi [1987], Hensley [1987], and Hildebrand [1987a; 1987b] showed that, for $n \in S(x, y)$, the values of $\Omega(n)$ also satisfy the normal distribution whenever $y \gg \log x$: The mean and variance are $\sim \log\log x$ when $u = o(\log\log x)$; the mean is $\approx u$ and the variance is $\approx u/\log^2 u$ for

$$\log x \ll y \ll \exp((\log x)^{1/21}).$$

Fouvry and Tenenbaum [1991] considered exponential sums over smooth numbers, getting the upper bound

$$\sum_{n \in S(x,y)} e^{2i\pi an/q} \ll x(\log qx)^3 \left( \frac{\sqrt{y}}{x^{1/4}} + \frac{1}{\sqrt{q}} + \sqrt{\frac{qy}{x}} \right). \tag{4.18}$$

I particularly like their result on Riemann's summation theorem: Of course if $\theta \in (0, 2\pi)$, then

$$\sum_{n \geq 1} e^{i\theta n}/n = -\log(1 - e^{i\theta}).$$

However, ordering $n$ in the sum by their largest prime factor, we get

$$\lim_{y \to \infty} \sum_{n \in S(x,y)} e^{i\theta n}/n = -\log(1-e^{i\theta}) + \begin{cases} \log p/\varphi(p^k) & \text{if } \theta = a/p^k, \ p \text{ prime,} \\ 0 & \text{otherwise.} \end{cases}$$

The types of results discussed in this subsection are beautifully developed by de la Bretéche and Tenenbaum [2005b].

## 5. Understanding, computing, and playing with smooth numbers

**5.1. Gaps in the sequence of smooth numbers.** What can we say about gaps in the sequence $S(x, y) = \{1 = n_1 < n_2 < \cdots\}$? Tijdeman [1973; 1974] showed there exist constants $c_1(y), c_2(y) > 0$ such that

$$\frac{n_i}{(\log n_i)^{c_1(y)}} \ll_y n_{i+1} - n_i \ll_y \frac{n_i}{(\log n_i)^{c_2(y)}}.$$

By (1.17), the average gap is $\asymp_y n_i/(\log n_i)^{\pi(y)}$, so that $c_2(y) \leq \pi(y) \leq c_1(y)$.

Erdős [1955] showed there are no more than $cy/\log y$ consecutive integers $n_i$ with each $n_i \gg y$; this was later improved to $c(y \log \log \log y)/(\log y \log \log y)$ by Shorey [1973/74].

**5.2. Smooth polynomials in finite fields (related to discrete log problem).** We define $N_q(n, m)$ to be the number of polynomials $f \in \mathbb{F}_q[t]$ of degree $n$, all of whose irreducible polynomial factors in $\mathbb{F}_q[t]$ have degree $\leq m$.

Manstavičius [1992] gave an asymptotic formula for $n \geq m$ in the range $m/\sqrt{n \log n} \to \infty$, namely, that

$$N_q(n, m) \sim \rho(n/m) q^n,$$

which is not hard to prove using the Buchstab–de Bruijn identity

$$N_q(n, m) - N_q(n, m-1) = I_q(m) N_q(n - m, m),$$

where $I_q(m) = (1/m) \sum_{d \mid m} \mu(d) q^{m/d}$, the number of irreducible polynomials in $\mathbb{F}_q[t]$ of degree $m$. He also showed that for

$$\sqrt{n \log n} \geq m \geq (1 + o(1))(\log n)/(\log q),$$

one can prove an asymptotic formula for $N_q(n, m)$ analogous to Hildebrand and Tenenbaum's (3.23): note that since $f_m(z) = \prod_{k \leq m}(1 - z^k)^{-I_q(k)} = \sum_{n \geq 1} N_q(n, m) z^n$, we have, for any $0 < r < 1$, the Rankin-type upper bound

$$N_q(n, m) = \frac{1}{2i\pi} \int_{|z|=r} \frac{f_m(z)}{z^{n+1}} dz \leq \frac{f_m(r)}{r^n},$$

which, when optimized at the saddle point $r(n, m)$, is too large by only a small factor. Manstavičius's estimate is difficult to work with, but by comparing $r(n, m)$ with $r(2n, 2m)$, Soundararajan [2001] deduces that

$$N_q(n, m) = \rho(n/m) q^n \exp\left(O\left(\frac{n \log n}{m^2}\right)\right)$$

for $n \geq m \geq \log(n \log^2 n)/(\log q)$ (and a similar method works for $\Psi(x, y)$). As another consequence, Soundararajan shows that if $n^{2/3} \geq m \geq 1$, then

$$\frac{N_q(n+1, m)}{N_q(n, m)} \sim \max(1, qn^{-1/m}).$$

A consequence of this is that $N_q(n, m) = q^n/u^{u+o(u)}$, where $u = n/m$, provided $q \geq (n \log^2 n)^{1/m}$. Bender and Pomerance [1998] showed $N_q(n, m) \geq q^n/n^u$ whenever $m \leq \sqrt{n}$.

Soundararajan also noted that for $m \leq (\log n)/(2 \log q)$, one has the analogy to (1.17) and for $m \leq (\log n)/(\log q)$ the analogy to (1.18), so that in this range $N_q(n, m)$ is "polynomial in $n$", whereas for $m \gtrsim (\log n)/(\log q)$ it is "exponential in $n$". The transition is analogous to (1.19): For all $n \geq m \geq 1$ we have

$$\log N_q(n, m) = \frac{n}{m} g\left(\frac{1}{n} \sum_{k \leq m} k I_q(k)\right)\left(1 + O\left(\frac{1}{m} + \frac{1}{\log n}\right)\right).$$

Soundararajan's title "Smooth polynomials: analogies and asymptotics" is thus very fitting.

The referee has justifiably complained that I have not adequately discussed the contributions of Adleman, ElGamal, Odlyzko, Bender and Pomerance, and Schirokauer, who, using various clever and difficult techniques, give subexponential bounds for the discrete logarithm problem over finite fields. However this is one subject I will leave the interested reader to check up on independently!

## 5.3. Primes $p$ where $p-1$ is smooth. Define

$$\pi(x,y) = \#\{p \le x : p-1 \in S(x,y)\}.$$

One would guess that

$$\pi(x,y) \sim \pi(x)\rho(u) \quad \text{when } y = x^{1/u}; \tag{5.1}$$

and, indeed, this follows for all $u \ge 1$ from a weak form of the Elliott–Halberstam conjecture, a well-believed conjecture in analytic number theory. For some applications $\pi(x,y) \gg_u \pi(x)$ is enough and is known [Friedlander 1989] to hold for $u \le 2\sqrt{e} = 3.297442542\ldots$; for other applications even $\pi(x,y) \gg_u x/\log^A x$ is enough and is known [Baker and Harman 1998] for $u \le 3.3772\ldots$ and some value of $A$. One quite surprising application is due to Erdős [1935]:

If (5.1) holds uniformly for given $u$, then there exist arbitrarily large integers $n$ for which there are more than $n^{1-1/u-\varepsilon}$ solutions $m$ to $\varphi(m) = n$.

PROOF. Let $P$ be the set of primes $p \le (\log N)^u$ such that all prime factors of $p-1$ are $\le y := \log N$. Consider the set $A$ of integers $m$ that are the product of $k = [(\log N)/(u \log \log N)]$ distinct primes in $P$: There are

$$\binom{\pi((\log N)^u, \log N)}{k} \ge \left(\frac{\pi((\log N)^u, \log N)}{k}\right)^k$$

$$\ge \left(\frac{(1+o(1))\rho(u)\pi((\log N)^u)}{k}\right)^k = N^{1-1/u+o(1)}$$

such integers. However, if $m \in A$ then $\varphi(m) < m \le N$, and $q$ divides $\varphi(m)$ which implies that $q \mid p-1$ for some $p \in P$, which in turn implies that $q \le y$, so that there are $\le \Psi(N,y) = N^{o(1)}$ such values. Therefore some such value of $\varphi(m)$ is taken by $\ge N^{1-1/u-o(1)}$ different $m \in A$. □

Assuming (5.1) holds uniformly with $y = e^{\sqrt{\log x}}$, Pomerance [1980] showed

$$\max_{n \le x} \#\{m : \varphi(m) = n\} = x/\exp((1+o(1))(\log x \log \log \log x)/(\log \log x)).$$

Using the ideas of the proof of Theorem 4 and of the second part of Corollary 3 of [Fouvry and Tenenbaum 1996], one can show that

$$\pi(x,y) \ll \pi(x)\rho(u)$$

for $x \ge y \ge \exp((\log x)^{2/3+o(1)})$. Pomerance and Shparlinski [2002] gave the slightly weaker upper bound $\pi(x,y) \ll \pi(x)u\rho(u)$ in the extended range $x \ge y \ge \exp(\sqrt{\log x \log \log x})$.

Adleman, Pomerance and Rumely [1983] developed the fastest known deterministic primality test using primes $p$ where $p-1$ is $y$-smooth; the test was made practical by Cohen and Lenstra.

Alford, Pomerance, and I [1994] gave another application showing that there are more than $x^{c(1-1/u)-\varepsilon}$ Carmichael numbers $\leq x$. We believe one can take $u$ arbitrarily large in (5.1) and take $c$ arbitrarily close to 1 (which also follows from a weak form of the Elliott–Halberstam conjecture), so that there are $\gg x^{1-\varepsilon}$ Carmichael numbers $\leq x$. From the current state of these analytic quantities, one can then deduce that there are $\gg x^{2/7}$ Carmichael numbers $\leq x$; recently Harman [2005] (and subsequently) improved the method to obtain $\gg x^{1/3}$ Carmichael numbers $\leq x$.

**5.4. As a sieve, again.** Again consider sieving $[1, x]$ with a set of primes $P$ such that $\prod_{p \in P}(1-1/p) \sim 1/u$. We wish to determine the (best possible) upper and lower bounds for the number of integers left unsieved. In Section 3.3 we saw that the expected number is $\sim x/u$. Hildebrand [1984a; 1984b] showed

$$\rho(u)x \lesssim \text{ the number of integers left unsieved } \leq \left(e^\gamma - \frac{c}{u^u}\right)\frac{x}{u}$$

for some constant $c > 0$. Soundararajan and I [2004] recently improved this upper bound to

$$\leq \left(e^\gamma - \frac{1}{u^{1+o(1)}}\right)\frac{x}{u}$$

which is "best possible". Note that the bound $\rho(u)x$ cannot be reduced since smooth numbers give such an example. From this perspective, one sees that there are remarkably few smooth numbers compared to what one might expect from the sieve; on the other hand, Pomerance [1995; 2008] again and again makes the opposite assertion, that smooths are "fairly numerous", at least in the contexts that arise in computational number theory. The point is that in sieve theory one is interested in the number of integers composed of primes from the given set using the Euler product as the measure of the size of the set, whereas in computational number theory the correct measure is simply the number of elements of the set, and thus one reaches such different conclusions.

**5.5. Smooth twins and the smooth Goldbach problem.** Hildebrand [1985b] showed that there are infinitely many pairs $(n, n+1)$ of $n^\varepsilon$-smooth integers. This has been improved to $n^{c(\log \log \log n)/(\log \log n)}$-smooth. Konyagin (unpublished) and Balog and Wooley [1998] showed for any fixed $\varepsilon > 0$ and integer $k$, there are infinitely many $k$-tuples of consecutive $n^\varepsilon$-smooth integers. Note that the conjecture in Section 1.1 implies that there are $\sim \rho(1/\varepsilon)^k x$ such $k$-tuples up to $x$. The construction of Konyagin and of Balog and Wooley is remarkably clever yet simple:

Find coprime integers $m_1, m_2, \ldots, m_k$ with each $\varphi(m_i)/m_i \leq \varepsilon/2$, and let $M = m_1 m_2 \ldots m_k$. For each prime $p \leq k$, determine the integer $a_p$ with $a_p \equiv v_p(j) \pmod{m_j}$ for each $1 \leq j \leq k$ and $0 \leq a_p \leq M-1$ (here $v_p(j)$ is the exact power of $p$ dividing $j$). Let $b = \prod_{p \le k} p^{a_p}$ and then $N = bn^M$ for any large integer $n$. We claim $N-1, \ldots, N-k$ are all $N^\varepsilon$-smooth for

$$N - j = j(N/j - 1) = j(n_j^{m_j} - 1) = j \prod_{d \mid m_j} \varphi_d(n_j),$$

where $n_j = n^{M/m_j} \prod_{p \le k} p^{(a_p - v_p(j))/m}$ and each

$$\varphi_d(n_j) \ll n_j^{\varphi(d)} \le n_j^{\varphi(m_j)} \le n_j^{\varepsilon m_j/2} = (N/j)^{\varepsilon/2} \le N^{\varepsilon/2}.$$

Balog [1989] showed that every integer $N$ can be written as the sum of two $N^{.2695}$-smooths. Balog and Sárközy [1984a; 1984b] showed that every integer $N$ can be written as the sum of three $L(N)^{3+o(1)}$-smooths.

**5.6. Average sizes of factors.** Dickman's original motivation for studying smooth numbers was to gain a better understanding of the distribution of the largest prime factor $p(n)$ of integers $n$. One can easily deduce from his result that the average size of $(\log p(n))/(\log n)$ is

$$\sim \int_0^\infty \frac{\rho(t)}{(1+t)^2} dt = 0.624\ldots,$$

which is called Golomb's constant.

Knuth and Trabb Pardo [1976/77] studied the distribution of the $k$-th largest prime factor of an integer, proving that there are $\sim \rho_k(u)x$ integers[2] $\le x$ with $k$-th largest prime factor $\le x^{1/u}$, where $\rho_k(u) = 1$ for $u \le 1$, and

$$\rho_k(u) = 1 - \int_1^u (\rho_k(t-1) - \rho_{k-1}(t-1)) \frac{dt}{t}$$

for $u > 1$. They gave the lovely inclusion-exclusion formula (analogous to (3.15) above)

$$\rho_k(u) = 1 - \sum_{n \ge 0} \binom{n+k-1}{k-1} (-1)^n \log_{n+k}(u),$$

where $\log_1(u) = \log(u)$ and $\log_m(u) = \int_1^u (\log_{m-1}(t-1)/t)dt$. It turns out $\rho_k(u) \sim e^\gamma (\log u)^{k-2}/((k-2)!u)$ for all $k \ge 2$, a very different behavior from the $k = 1$ case.

Using the above, they show that the average order of the logarithm of the second largest prime factor is $\sim .20958\ldots \log n$, and of the third largest is $\sim .08831\ldots \log n$.

Suggestively, they show that $\rho_k(u)$ is also the probability that the $k$-th longest cycle in a random permutation of $N$ letters has length $\le N/u$.

---

[2]Note that this function $\rho_k$ is not related to $\rho_y$ of Section 3.11.

Billingsley [1972] generalized Dickman's result to the following: For any $1 \geq \alpha_1 \geq \alpha_2 \geq \cdots \geq \alpha_k > 0$ there are $\sim \rho(\alpha_1, \alpha_2, \ldots, \alpha_k) x$ integers $n$ up to $x$ such that the $j$-th largest prime factor of $n$ is $\leq x^{\alpha_j}$ for $j = 1, 2, \ldots, k$, where

$$\rho(\alpha_1, \ldots, \alpha_k) = \int \cdots \int_{\substack{v_1 \geq v_2 \geq \cdots \geq v_k \geq 0, \\ v_1 + v_2 + \cdots + v_k \leq 1, \\ \text{each } v_j \leq \alpha_j}} \rho\left(\frac{1 - v_1 - v_2 - \cdots - v_k}{v_k}\right) \frac{dv_1 \cdots dv_k}{v_1 \cdots v_k}.$$

As discussed in Arratia, Barbour, and Tavaré [1997], this is also the probability that the $j$-th longest cycle in a random permutation of $N$ letters has length $\leq \alpha_j N$ for $j = 1, 2, \ldots, k$. In fact, they show that $\rho(\alpha_1, \ldots, \alpha_k)$ is the distribution function for several interesting combinatorial problems.

De Koninck [1994] showed that the prime $p$ that is most likely to be the largest prime factor of an integer up to $x$, satisfies $p = L(x)^{1/\sqrt{2}+o(1)}$. He went on to make the delightful observation that, for any $k \geq 2$, the prime which is most often the $k$-th largest prime factor of $n$ is $p = 3$ (where we range over integers $n$ up to sufficiently large $x$).

Another related combinatorial problem was found by Chamayou [1973]: If $X_1, X_2, \ldots$ are independent random variables each uniformly distributed on $(0, 1)$, then the probability that $X_1 + X_1 X_2 + X_1 X_2 X_3 + X_1 X_2 X_3 X_4 + \cdots \leq u$ is $e^{-\gamma} \int_0^u \rho(t) dt$.

**5.7. Finding smooth numbers computationally.** The obvious way to find $y$-smooth numbers in $(x, x + z)$ with $z \leq x$ is to initialize an array $a[i] := 0$ for $1 \leq i \leq z$ (where $a[i]$ corresponds to $x + i$). For each successively larger prime power $p^j \leq x + z$ with $p \leq y$, determine the smallest $i$ such that $p^j$ divides $x + i$ and then add $\log p$ to $a[i], a[i + p^j], a[i + 2p^j]$ and so on, up until the end of the array. When we've finished, if any $a[i] \geq \log x$, then $x + i$ is $y$-smooth. This algorithm has running time $\ll z \log \log y + uy$.

A very similar idea can be used when we wish to find smooth values of polynomials, since then $p \mid f(n)$ if and only if $p \mid f(n + p)$.

It seems likely that any such algorithm must have a running time $\geq \pi(y)$, since the primes up to $y$ are part of the input in the problem. However, one might guess that it is not necessary to have $z$ in the running time (if $z$ is part of the running time, it suggests that the algorithm examines most integers in the interval) since one might be able to find smooths by a more constructive approach. Boneh [2002] did this in an ingenious way, obtaining an algorithm with running time $(y \log x)^{O(1)}$, as follows.

The goal in Boneh's method is to determine a polynomial $f(t) \in \mathbb{Z}[t]$ for which every integer $m \leq z$ with $(x + m, L) \geq x$ is a root, where $L = \prod_{p \leq y} p$. Now, if $g = (x + m, L)$, then $g^k$ divides $L^j (x + m)^{k-j}$ and $(x + m)^k x^j$ for

$j = 0, 1, 2, \ldots, k-1$, and so $g^k$ also divides any linear combination of these polynomials. By the LLL-algorithm, one can rapidly find a polynomial $f(t)$ with small coefficients which is a linear combination of these polynomials, so that $f(m) \equiv 0 \pmod{g^k}$. Now, since $f$ has small coefficients we deduce that $|f(m)| < x^k \le g^k$, which implies that $f(m) = 0$ since $f(m) \equiv 0 \pmod{g^k}$.

As a consequence, Boneh deduces that if $(1/5) \log^2 x > y > \log x$, then there are $\ll y/\log x$ integers in $[x, x + x^{(\log x)/(4y)}]$ that divide LCM$[1, \ldots, y]$.

### 5.8. Computational upper and lower bounds on $\Psi(x, y)$.
Computing the precise value of $\Psi(x, y)$ is likely to be impractical once $x$ and $y$ are large, since it seems likely that one would have to do at least $\Psi(x, y)$ bit operations. Perhaps I am too pessimistic.

However, good approximations for $\Psi(x, y)$ may be all that are needed in certain applications: Hunter and Sorenson [1997] noted that one could estimate $\alpha$ and then use (3.23) to approximate $\Psi(x, y)$ up to a factor $1 + O(1/u + (\log y)/y)$ in time $\ll (y \log \log x)/\log y + y/\log \log y$.

Bernstein [2002] has indicated how lattice point arguments allow one to quickly obtain good upper and lower bounds: For a large integer $N$, select $m_j$ to be the smallest integer with $m_j \ge N \log p_j$, so that any solution to

$$a_1 m_1 + \cdots + a_k m_k \le d, \tag{5.2}$$

where $d = [N \log x]$, gives rise to a solution of (3.1) and thus a $y$-smooth number $\prod_{j=1}^{k} p_j^{a_j} \le x$. Therefore, the number of solutions to (5.2) satisfying $d = [N \log x]$ provides a lower bound for $\Psi(x, y)$. To obtain an upper bound by similar methods, note that if (3.1) holds, then

$$\sum_{i=1}^{k} a_i m_i < \sum_{i=1}^{k} a_i (1 + N \log p_i) \le \left(N + \frac{1}{\log 2}\right) \log x = N \log X,$$

where $X = x^{1+1/(N \log 2)}$. Thus the number of solutions to (5.2) satisfying $d = [N \log X]$ provides the desired upper bound.

Now, the number of solutions to (5.2) equals the coefficient of $T^d$ in

$$\prod_{i=0}^{k} (1 + T^{m_i} + T^{2m_i} + T^{3m_i} + \cdots) \pmod{T^{d+1}},$$

where $m_0 = 1$. The obvious algorithm for doing these multiplications takes time $\ll kd^2 \log^2 x$, and this can be sped up in several ways; see [Bernstein 2002].

Note these bounds will get more accurate the larger $N$ is; for example, combining both arguments gives the upper bound $\le \Psi(x^{1+1/(N \log 2)}, y)$.

## 5.9. Determining the smooth part of each integer in a large set.
In several algorithms we wish to rapidly determine the $y$-smooth parts of many integers. One can use trial division if that is not significant in the running time of the algorithm; otherwise one can use Lenstra's elliptic curve factoring algorithm or even some sort of sieving technique if, for instance, the integers are the consecutive values of a polynomial. Recently Bernstein [2000] has come up with a remarkably clever procedure that allows one to find all the small prime factors of $\gg y$ integers, each with $(\log y)^{O(1)}$ bits, in $(\log y)^{O(1)}$ bit operations per integer on average. The central idea is to multiply many of these integers together, determine the $y$-smooth part of that product, and then gradually dismantle this into smaller and smaller subproducts.

# 6. Applications to other areas of number theory and beyond

## 6.1. Smooth numbers and character sums.
A central problem in analytic number theory is to determine when

$$\left|\sum_{n \leq x} \chi(n)\right| = o(x) \tag{6.1}$$

for a primitive character $\chi \pmod{q}$. Burgess [1957] used ingenious combinatorial methods together with the "Riemann Hypothesis for hyperelliptic curves" to establish (6.1) whenever $x > q^{1/4+o(1)}$, and $q$ is cubefree. Soundararajan and I [2001a] investigated the idea that $\sum_{n \leq x} \chi(n)$ is well approximated by

$$\Psi(x, y; \chi) := \sum_{n \in S(x,y)} \chi(n),$$

for "small" $y$. We conjecture that there exists $A > 0$ for which

$$\sum_{n \leq x} \chi(n) = \Psi(x, y; \chi) + o(\Psi(x, y; \chi_0)) \tag{6.2}$$

holds uniformly for $y = (\log q + \log^2 x)(\log \log q)^A$. Here $\chi_0$ is the principal character $(\bmod\, q)$, that is, $\chi_0(n) = 1$ if $(n, q) = 1$, and $\chi_0(n) = 0$ otherwise. This implies that

$$\max_{\chi \neq \chi_0} \left|\sum_{n \leq x} \chi(n)\right| \sim \Psi(x, \log q) \tag{6.3}$$

whenever $\log x = o(((\log \log q)/(\log \log \log q))^2)$, for any prime $q$. Smooth numbers thus appear naturally in a central question in analytic number theory. Assuming the Generalized Riemann Hypothesis, we showed that (6.2) holds with $y = \log^2 q \log^2 x (\log \log q)^{O(1)}$, which implies that

$$\left|\sum_{n \leq \log^u q} \chi(n)\right| \leq (1 + o(1))\rho(u/2) \log^u q.$$

If we assume, as is now widely believed, that the imaginary part of the zeros of zeta functions follow the same distributions as the eigenvalues of (the classical) groups of random matrices (see Katz and Sarnak [1999]), then we can improve this to the upper bound implicit in (6.3). Unconditionally we showed that (6.2) holds for "almost all" characters $\chi$ (mod $q$) with $y = \log q \log x (\log \log q)^{O(1)}$.

We also unconditionally derived lower bounds for character sums. For example, we proved the lower bound implicit in (6.3), that is, that for any fixed $A > 0$ and for any given angle $\theta$, there are many $\chi$ modulo prime $q$ for which $\sum_{n \leq \log^u q} \chi(n) = (e^{i\theta} + o(1))\rho(u) \log^u q$.

In the more useful case of real characters, we showed that there are infinitely many fundamental discriminants $-D < 0$ for which $\sum_{n \leq \log^u D}(-D/n) \geq (\rho(u) + o(1))x$. Montgomery and Vaughan [1977] showed that character sums are always $\ll \sqrt{q} \log \log q$ assuming the Generalized Riemann Hypothesis; Paley [1932] had shown that this is best possible other than the constant. Smooth numbers have something to say about this problem: There are infinitely many fundamental discriminants $-D < 0$ for which

$$\sum_{n \leq D/\log^u D}(-D/n) \gg (\rho(u)/\log(u+2))\sqrt{D} \log \log D.$$

Recently Soundararajan and I [2007] further developed the idea of (6.2) showing, under the Generalized Riemann Hypothesis that if $\chi$ (mod $q$) is a primitive character and $x < q^{3/2}$, then

$$\sum_{n \leq x} \chi(n) e^{2i\pi n\alpha} = \sum_{\substack{n \leq x, \\ n \in \mathcal{S}(y)}} \chi(n) e^{2i\pi n\alpha} + O(xy^{-1/6} \log q)$$

for any $\alpha \in [0, 1)$. We next apply this result to give new upper bounds on character sums.

### 6.2. The proportion of integers that are quadratic residues, and a generalization.
We can describe $\Psi(x, y)/x$ as a "mean value" of a multiplicative function, it being the $\alpha = 1$ case of

$$F_\alpha(x) = \sum_{n \leq x} f_\alpha(n) \quad \text{where} \quad f_\alpha(p) = \begin{cases} 1 & \text{for } p \leq y, \\ 1 - \alpha & \text{for } p > y. \end{cases}$$

One can show that $F_\alpha(x) \sim x\rho_\alpha(u)$, where $\rho_\alpha(u) = 1$ for $0 \leq u \leq 1$ and

$$u\rho_\alpha(u) = \int_{u-1}^{u} \rho_\alpha(t) dt + (1-\alpha) \int_0^{u-1} \rho_\alpha(t) dt$$

for each $u > 1$. This implies that $\rho'_\alpha(u) = -\alpha \rho_\alpha(u-1)/u$.

Goldston and McCurley [1988] showed there are $x\rho_\alpha(u)(1 + O(1/\log y))$ integers all of whose prime factors are either $\leq y$ or from a set of primes that has density $1 - \alpha$. The asymptotic behavior of $\rho_\alpha$ is surprising: For $\text{Re}(\alpha) > 0$ we have

$$\rho_\alpha(u) \sim \frac{e^{\gamma\alpha}}{\Gamma(1-\alpha)} u^{-\alpha}$$

if $\alpha$ is not a positive integer. However, if $\alpha$ is a positive integer, then the behavior of $\rho_\alpha$ is quite different: For large $u$,

$$\rho_\alpha(u) = (-1)^{1-\alpha}\left(\frac{e\alpha + o(1)}{u \log u}\right)^u,$$

and, for small $u$, the function $\rho_\alpha(u)$ oscillates about the real axis if $\alpha \geq 2$.

Now $\rho_2(u)$ can get negative, and the minimum of $\rho_2(u)$ occurs at $u = 1 + \sqrt{e}$ where

$$\rho_2(1 + \sqrt{e}) = 1 - \frac{\pi^2}{3} - 2\log(1 + \sqrt{e})\log\left(\frac{e}{1+\sqrt{e}}\right) + 4\sum_{n=1}^{\infty} \frac{1}{n^2} \frac{1}{(1+\sqrt{e})^n}.$$

Soundararajan and I [2001b] deduced that for every sufficiently large integer $N$ and every prime $p$, more than 17.15% of the integers $\leq N$ are quadratic residues (mod $p$). (Note $(1 + \rho_2(1 + \sqrt{e}))/2 \approx .1715$.)

In a further generalization, suppose that $f(p) = 1$ for all $p \leq y$ and that $f(p)$ lies inside or on the unit circle for $p > y$. Define $\chi(t) = 1$ for $0 \leq t \leq 1$, and let $\chi(t) := y^{-t} \sum_{p \leq y^t} \log p$ for $1 < t \leq u$. Then define $\sigma(t) = 1$ for $0 \leq t \leq 1$, and let

$$u\sigma(u) = \int_0^u \chi(t)\sigma(u-t)dt$$

for all $u > 1$. Soundararajan and I showed that $\sum_{n \leq x} f(n) = (\sigma(u) + o(1))x$ for $x = y^u$.

**6.3. Large gaps between primes.** Rankin [1938] and Erdős [1940] showed how to construct large gaps between consecutive primes by finding long sequences of consecutive integers which each have a small prime factor, say $\leq x$. Suppose $[r+1, r+V]$ is such an interval. For each prime $p \leq x$ let $a_p \equiv -r \pmod{p}$. Then $p \mid r+n$ if and only if $p \mid n - a_p$, that is, $n \equiv a_p \pmod{p}$. On the other hand, if every $n \in [1, V]$ belongs to some arithmetic progression $a_p \pmod{p}$, then select $r \equiv -a_p \pmod{p}$ for every prime $p \leq x$; so every number in $[r+1, r+V]$ is composite. Therefore finding such an interval is equivalent to selecting arithmetic progressions $a_p \pmod{p}$ to sieve out the integers in $[1, V]$. Now, in Section 5.4, we saw that very efficient sieving is given by using smooth numbers appropriately. That is how everyone proceeded... In fact, using the primes up to $x$, we know how to sieve out an interval of length $(2e^\gamma + o(1))x \log x (\log \log \log x)/(\log \log x)^2$. Iwaniec [1978] showed that we cannot sieve out an interval of length $\gg x^2$.

## 6.4. Least quadratic nonresidue (mod $p$).

From the work of Littlewood, Ankeny, Montgomery, and then Bach [1985], we have the following: Assuming the Riemann Hypothesis for $L(s, (\cdot/q))$, there is an integer $n \leq 2\log^2 q$ such that $(n/q) = 0$ or $-1$. (Here $q$ may be composite.)

Burgess [1957] showed unconditionally, via an argument of Vinogradov, that there is a value of $n \leq p^{1/(4\sqrt{e})+\varepsilon}$ such that $(n/p) = -1$ if $p$ is a sufficiently large prime.

PROOF. Burgess had already proved (6.1) for $x > q^{1/4+o(1)}$. Taking $\chi = (\cdot/p)$ with $q = p$ in (6.1), this implies that if $N > p^{1/4+\varepsilon}$, then there are $\sim N/2$ integers $n \leq N$ for which $(n/p) = 1$. He then applied an old argument of Vinogradov as follows. If $(q/p) = 1$ for all primes $q \leq y$, then $(n/p) = 1$ for all $n \in S(N, y)$. Therefore $N\rho(u) \sim \Psi(N, y) \lesssim N/2$, where $N = y^u$, so that $\rho(u) \leq 1/2$, which holds if and only if $u \geq 1/\sqrt{e}$. □

*Remark*: Vinogradov observed that one can proceed analogously for $k$-th power residues for any integer $k \geq 2$: Define $u_k$ so that $\rho(u_k) = 1/k$ and $u_k \sim (\log k)/(\log \log k)$ by (1.6). By an analogous proof, for every sufficiently large prime $p \equiv 1 \pmod{k}$, there is an integer $n \leq p^{1/(4u_k)+\varepsilon}$ which is not a $k$-th power (mod $p$).

*The large sieve inequality:* Let $\mathcal{N}$ be a sequence of positive integers with

$$N(x) := \sum_{\substack{n \in \mathcal{N}, \\ n \leq x}} 1 \quad \text{and} \quad N(x; p, a) := \sum_{\substack{n \in \mathcal{N}, n \leq x, \\ n \equiv a \pmod{p}}} 1.$$

Then

$$\sum_{p \leq \sqrt{x}} p \sum_{a=1}^{p} (N(x; p, a) - N(x)/p)^2 \leq 2xN(x).$$

Linnik [1941] showed there are no more than $4/\rho(2u)$ primes $\leq z$ for which the least quadratic nonresidue is $> y$, where $z = y^u$.

PROOF. Let $\mathcal{N}$ be the sequence of integers whose prime factors are all $\leq y$. Let $x = z^2$. If $p \leq z \, (= \sqrt{x})$ is a prime for which the least quadratic nonresidue $q_p$ is $> y$, then $(n/p) = 1$ for all $n \in \mathcal{N}$: In particular $N(x; p, a) = 0$ if $(a/p) = -1$ or $0$. Therefore, by the large sieve inequality above,

$$\sum_{\substack{p \leq z \\ q_p > y}} p \cdot \frac{(p+1)}{2} \left(\frac{N(x)}{p}\right)^2 \leq 2xN(x),$$

so that the number of such primes is $\leq 4x/N(x)$. Note that $N(x) = \Psi(x, y) = \Psi(z^2, y) \sim x\rho(2u)$. □

Linnik's result implies that there are $\ll_\varepsilon \log\log x$ counterexamples to Vinogradov's conjecture up to $x$.

**6.5. Fermat's Last Theorem (first case).** Long before the famous work of Andrew Wiles, there were other, simpler approaches! The first case of Fermat's Last Theorem (FLT I) states that there are no solutions to

$$x^p + y^p = z^p \quad \text{for } x, y, z > 0 \text{ and } p \nmid xyz.$$

In 1910, Wieferich showed that a solution implies $2^{p-1} \equiv 1 \pmod{p^2}$. Others then proved a solution implies $3^{p-1} \equiv 1 \pmod{p^2}$ — it's unlikely that these two congruences ever happen simultaneously — then $5^{p-1}, 7^{p-1}, \ldots \equiv 1 \pmod{p^2}$. Hendrik Lenstra proved $\Psi(x, (\log x)^2) > \sqrt{x}$ (which also follows from (1.22)), which implies that there is a $q < 4\log^2 p$ for which $q^{p-1} \not\equiv 1 \pmod{p^2}$.

PROOF. Otherwise take $x = p^2$ above to get $\Psi(p^2, 4\log^2 p) > p$. Now if $n \in S(p^2, 4\log^2 p)$ then $q^{p-1} \equiv 1 \pmod{p^2}$ for every prime $q$ dividing $n$ so that $n^{p-1} \equiv 1 \pmod{p^2}$. But there are only $p-1$ such values of $n < p^2$. □

In my PhD thesis, I showed how to systematically deduce $q^{p-1} \equiv 1 \pmod{p^2}$ for each successive prime $q$ from a supposed solution to FLT I, developing an approach of Frobenius. A Maple implementation gave each $q \leq 89$. If my algorithm never degenerated, one would deduce that $q^{p-1} \equiv 1 \pmod{p^2}$ for every prime $q \leq (\log p)^{1/4}$, which is unfortunately not quite enough to deduce FLT I!

**6.6. Important applications to Waring's problem, and beyond.** In 1941 Vinogradov used the circle method to prove that every sufficiently large integer is the sum of $\leq (2+\varepsilon)k \log k$ terms, each of which is the $k$-th power of an integer. Vaughan's work [1989] suggested that one might do better by working with $k$-th powers of smooth numbers. Wooley [1992] did this in improving Vinogradov's result to $\leq k \log k + k \log \log k + Ck$ $k$-th powers of integers, the biggest breakthrough in this well-explored problem in fifty years! Moreover, Vaughan and Wooley have developed this into a powerful tool, not only for Waring's problem, but for various other questions: For instance, Harman [1993] uses an exponential sum estimate in the circle method, involving only smooth numbers in the exponents, to study when there are solutions to

$$\|a_1 n_1^k + \cdots + a_r n_r^k\| \ll 1/N^k,$$

where $1 \leq n_i \leq N$, the $a_i$'s are given real algebraic numbers, and $\|t\|$ is the distance from the nearest integer to $t$. See Vaughan's [1993] survey and Vaughan and Wooley [2002] for a thorough discussion of related questions.

**6.7. Egyptian fractions.** In his PhD thesis, Croot [2001] solved an old ($500) problem of Erdős and Graham by showing that for any $r$-coloring of the integers in $[2, e^{167000r}]$, there is a monochromatic subset $S$ for which

$$\sum_{n \in S} \frac{1}{n} = 1.$$

The proof reduces the problem to looking over smooth subsets of the integers. The "167000" comes from an estimate involving $\rho(u)$. Clearly, one needs an interval longer than $[2, e^{(1-\varepsilon)r}]$. What is the correct upper limit for the interval?

**6.8. The abc-conjecture.** Suppose $c > a > 0$ are both $y$-smooth. The "abc-conjecture" (see [Granville and Tucker 2002]) tells us that

$$c^{1-\varepsilon} \ll_\varepsilon \prod_{p \mid ac(c-a)} p \leq (c-a) \prod_{p \leq y} p.$$

Thus if $a < c$ are the first two $y$-smooth integers $\geq x$, then

$$c - a \gg_\varepsilon x^{1-\varepsilon}/e^{(1+o(1))y}.$$

So for $y = o(\log x)$ we see that $\Psi(x + x^{1-o(1)}, y) - \Psi(x, y) = 0$ or $1$.

**6.9. $S$-unit equations with lots of solutions.** Let $S$ be a set of $s$ primes. Evertse [1984] showed that there are $\leq 3 \cdot 7^{2s+3}$ solutions to $a+b=c$ in coprime integers $a, b, c$ whose prime divisors all come from the set $S$ (that is, $a, b, c$ are "$S$-units"). Erdős, Stewart and Tijdeman [1988] showed that there exist sets $S$ with at least $\exp((4 + o(1))(s/\log s)^{1/2})$ solutions. Recently Konyagin and Soundararajan [2007] improved this to at least $\exp(s^{2-\sqrt{2}-o(1)})$ solutions. They also prove that there exist sets $S$ with at least $\exp(s^{1/16})$ solutions to $a + 1 = c$ in $S$-units, and as a consequence that there exist integers $N$ with more than $\exp((\log N)^{1/16})$ consecutive divisors.

More generally, it is known that there are only finitely many solutions to

$$a_1 + a_2 + \cdots + a_n = b \tag{6.4}$$

in positive integers $a_1, \ldots, a_n, b$ with $\gcd(a_1, \ldots, a_n, b) = 1$, where all prime factors of $a_1 \cdots a_n b$ come from $S$. One can easily show that there are sets with at least

$$\exp\bigl((n^2/(n-1) + o(1))s/(s \log s)^{1/n}\bigr) \tag{6.5}$$

solutions: Let $y$ be large, and let $u = y^{1-1/n}/((1-1/n) \log y)$. If $a_1, \ldots, a_n \in S(y^u, y)$, then $1 \leq b \leq ny^u$. Therefore some value of $b$ is taken in (6.4) at least $\Psi(y^u, y)^n/(ny^u)$ times, which is (6.5) by (1.13) and (1.7). To assume coprimeness, we divide each solution through by $\gcd(a_1, \ldots, a_n, b)$. (Here $S = \{p : p \mid b\} \cup \{p \leq y\}$.)

One might guess that each such value of $b$ is taken roughly equally often, in which case the number of solutions is at least (6.5) with $n+1$ replacing $n$.

**6.10. Ramanujan–Nagell equations with lots of solutions.** Evertse [1984] showed that if $F(x, y) \in \mathbb{Z}[x, y]$ is homogenous of degree $d$, with at least three distinct linear factors, then $F(m, n)$ is an $S$-unit for at most $2 \cdot 7^{(2s+3)d^3}$ pairs of coprime integers $m, n$, which Bombieri (1987) improved to $(12(d+5))^{12(s+1)}$. On the other hand, Erdős, Stewart, and Tijdeman [1988] showed there exist such $F$ with at least

$$\exp((d^2 + o(1))(s \log s)^{1/d} / \log s). \tag{6.6}$$

solutions. The proof is similar to that in the previous section. Fix large $y$, and let $u = dy^{1/d}/\log y$. Consider all vectors of the form

$$(n - a_1, n - a_2, \ldots, n - a_d) \quad \text{with } n \le x \text{ and each } a_i \in S(x, y). \tag{6.7}$$

Each entry of each vector is in $(-x, x)$, and there are $x\Psi(x, y)^d$ vectors, so some vector, say $(r_1, \ldots, r_d)$, is attained $\ge x\Psi(x, y)^d/(2x)^d$ times, which equals (6.6) with $s = \pi(y)$. Hence if $F(x, y) = \prod_{i=1}^{d}(x - r_i y)$, then for each such vector in (6.7) we get $F(n, 1) = a_1 \cdots a_d \in S(x, y)$.

## Acknowledgments

I would like to thank Dan Bernstein, John Friedlander, Adolf Hildebrand, the referee, Carl Pomerance, K. Soundararajan, and Gerald Tenenbaum for their help in preparing this article.

## Appendix: Notation

Whenever $\varepsilon$ is used in the text, it means some arbitrarily small positive constant; whenever $c$ or $C$ is used, it means some fixed, usually positive, constant that we have not determined. When we write $f(x) := \ldots$ this means that $f(x)$ is defined by the quantity on the right side of the equation.

As in analytic number theory:

- $f(x) \ll g(x)$ and $f(x) = O(g(x))$ both mean that there exists a constant $c > 0$ such that $|f(x)| \le cg(x)$ for all $x$ in the domain. If the domain is not specified, then we usually mean for all sufficiently large $x$.
- $f(x) \ll_{\varepsilon,k} g(x)$ means that the constant $c$ above depends on the values of $\varepsilon$ and $k$, but nothing else.
- $f(x) \gg g(x)$ means that there exists a constant $c > 0$ such that $f(x) \ge c|g(x)|$ for all $x$ in the domain.

- $f(x) \asymp g(x)$ means that $f(x) \ll g(x)$ and $f(x) \gg g(x)$; in other words, there exist constants $C > c > 0$ such that $Cg(x) > f(x) > cg(x) > 0$ for all $x$ in the domain.
- $f(x) = o(g(x))$ means that $f(x)/g(x) \to 0$ as $x \to \infty$.
- $f(x) \sim g(x)$ means that $f(x)/g(x) \to 1$ as $x \to \infty$.
- $f(x) \lesssim g(x)$ means that $\limsup_{x \to \infty} f(x)/g(x) \le 1$. We define $f(x) \gtrsim g(x)$ analogously.
- Finally, I have abused the notation $f(x) \approx g(x)$ to mean "it is true that $f$ is more-or-less equal to $g$ other than a small error for most values of $x$, but one needs to take care for extreme values of $x$ as this equality might then be false. Similarly $f(x) \gtrapprox g(x)$ means "it is more-or-less true that $f(x) \ge g(x)$", and we give an analogous definition to $f(x) \lessapprox g(x)$. My reason for doing this is that in some of the more difficult analytic arguments I have chosen to emphasize the main ideas, thereby neglecting what are sometimes very difficult error terms.

## References

[Adleman and Huang 1992] L. M. Adleman and M.-D. A. Huang, *Primality testing and abelian varieties over finite fields*, Lecture Notes in Mathematics **1512**, Springer, Berlin, 1992.

[Adleman et al. 1983] L. M. Adleman, C. Pomerance, and R. S. Rumely, "On distinguishing prime numbers from composite numbers", *Ann. of Math.* (2) **117**:1 (1983), 173–206.

[Agrawal et al. 2004] M. Agrawal, N. Kayal, and N. Saxena, "PRIMES is in P", *Ann. of Math. (2)* **160**:2 (2004), 781–793.

[Alford et al. 1994] W. R. Alford, A. Granville, and C. Pomerance, "There are infinitely many Carmichael numbers", *Ann. of Math.* (2) **139**:3 (1994), 703–722.

[Alladi 1982] K. Alladi, "Asymptotic estimates of sums involving the Moebius function. II", *Trans. Amer. Math. Soc.* **272**:1 (1982), 87–105.

[Alladi 1987] K. Alladi, "An Erdős–Kac theorem for integers without large prime factors", *Acta Arith.* **49**:1 (1987), 81–105.

[Arratia et al. 1997] R. Arratia, A. D. Barbour, and S. Tavaré, "Random combinatorial structures and prime factorizations", *Notices Amer. Math. Soc.* **44**:8 (1997), 903–910.

[Bach 1985] E. Bach, *Analytic methods in the analysis and design of number-theoretic algorithms*, ACM Distinguished Dissertations, MIT Press, Cambridge, MA, 1985.

[Baker and Harman 1998] R. C. Baker and G. Harman, "Shifted primes without large prime factors", *Acta Arith.* **83**:4 (1998), 331–361.

[Balog 1987] A. Balog, "On the distribution of integers having no large prime factor", pp. 27–31, 343 in *Journées arithmétiques de Besançon* (1985), Astérisque **147–148**, Soc. math. de France, Paris, 1987.

[Balog 1989] A. Balog, "On additive representation of integers", *Acta Math. Hungar.* **54**:3-4 (1989), 297–301.

[Balog and Pomerance 1992] A. Balog and C. Pomerance, "The distribution of smooth numbers in arithmetic progressions", *Proc. Amer. Math. Soc.* **115**:1 (1992), 33–43.

[Balog and Ruzsa 1997] A. Balog and I. Z. Ruzsa, "On an additive property of stable sets", pp. 55–63 in *Sieve methods, exponential sums, and their applications in number theory* (Cardiff, 1995), edited by G. R. H. Greaves et al., London Math. Soc. Lecture Note Ser. **237**, Cambridge Univ. Press, Cambridge, 1997.

[Balog and Sárközy 1984a] A. Balog and A. Sárközy, "On sums of integers having small prime factors, I", *Studia Sci. Math. Hungar.* **19**:1 (1984), 35–47.

[Balog and Sárközy 1984b] A. Balog and A. Sárközy, "On sums of integers having small prime factors, II", *Studia Sci. Math. Hungar.* **19**:1 (1984), 81–88.

[Balog and Wooley 1998] A. Balog and T. D. Wooley, "On strings of consecutive integers with no large prime factors", *J. Austral. Math. Soc. Ser. A* **64**:2 (1998), 266–276.

[Bender and Pomerance 1998] R. L. Bender and C. Pomerance, "Rigorous discrete logarithm computations in finite fields via smooth polynomials", pp. 221–232 in *Computational perspectives on number theory* (Chicago, IL, 1995), edited by D. A. Buell and J. T. Teitelbaum, AMS/IP Stud. Adv. Math. **7**, Amer. Math. Soc., Providence, RI, 1998.

[Bernstein 2000] D. Bernstein, "How to find small factors of integers", Preprint, 2000.

[Bernstein 2002] D. J. Bernstein, "Arbitrarily tight bounds on the distribution of smooth integers", pp. 49–66 in *Number theory for the millennium, I* (Urbana, IL, 2000), edited by M. A. Bennett et al., A K Peters, Natick, MA, 2002.

[Billingsley 1972] P. Billingsley, "On the distribution of large prime divisors", *Period. Math. Hungar.* **2** (1972), 283–289.

[Boneh 2002] D. Boneh, "Finding smooth integers in short intervals using CRT decoding", *J. Comput. System Sci.* **64**:4 (2002), 768–784.

[de la Bretèche and Tenenbaum 2005a] R. de la Bretèche and G. Tenenbaum, "Entiers friables: inégalité de Turán–Kubilius et applications", *Invent. Math.* **159**:3 (2005), 531–588.

[de la Bretèche and Tenenbaum 2005b] R. de la Bretèche and G. Tenenbaum, "Propriétés statistiques des entiers friables", *Ramanujan J.* **9**:1-2 (2005), 139–202.

[Brillhart et al. 1981] J. Brillhart, M. Filaseta, and A. Odlyzko, "On an irreducibility theorem of A. Cohn", *Canad. J. Math.* **33**:5 (1981), 1055–1059.

[de Bruijn 1951a] N. G. de Bruijn, "The asymptotic behaviour of a function occurring in the theory of primes", *J. Indian Math. Soc. (N.S.)* **15** (1951), 25–32.

[de Bruijn 1951b] N. G. de Bruijn, "On the number of positive integers $\leq x$ and free of prime factors $> y$", *Nederl. Acad. Wetensch. Proc. Ser. A.* **54** (1951), 50–60.

[de Bruijn 1966] N. G. de Bruijn, "On the number of positive integers $\leq x$ and free prime factors $> y$. II", *Nederl. Akad. Wetensch. Proc. Ser. A 69=Indag. Math.* **28** (1966), 239–247.

[Buhštab 1949] A. A. Buhštab, "On those numbers in an arithmetic progression all prime factors of which are small in order of magnitude", *Doklady Akad. Nauk SSSR (N.S.)* **67** (1949), 5–8. In Russian.

[Burgess 1957] D. A. Burgess, "The distribution of quadratic residues and non-residues", *Mathematika* **4** (1957), 106–112.

[Burgess 1962] D. A. Burgess, "On character sums and $L$-series, I", *Proc. London Math. Soc. (3)* **12** (1962), 193–206.

[Burgess 1963] D. A. Burgess, "On character sums and $L$-series, II", *Proc. London Math. Soc. (3)* **13** (1963), 524–536.

[Canfield et al. 1983] E. R. Canfield, P. Erdős, and C. Pomerance, "On a problem of Oppenheim concerning "factorisatio numerorum"", *J. Number Theory* **17**:1 (1983), 1–28.

[Chamayou 1973] J.-M.-F. Chamayou, "A probabilistic approach to a differential-difference equation arising in analytic number theory", *Math. Comp.* **27** (1973), 197–203.

[Crandall and Pomerance 2001] R. Crandall and C. Pomerance, *Prime numbers: a computational perspective*, Springer, New York, 2001.

[Croot 2001] E. S. Croot, *Smooth numbers in short intervals*, PhD thesis, University of Georgia, 2001.

[Croot et al. 2008] E. S. Croot, A. Granville, R. Pemantle, and P. Tetali, "Sharp transitions in making squares", Preprint, 2008. Available at http://www.dms.umontreal.ca/~andrew/PDF/Transition.pdf.

[Dartyge 1996] C. Dartyge, "Entiers de la forme $n^2 + 1$ sans grand facteur premier", *Acta Math. Hungar.* **72**:1-2 (1996), 1–34.

[Dartyge et al. 2001] C. Dartyge, G. Martin, and G. Tenenbaum, "Polynomial values free of large prime factors", *Period. Math. Hungar.* **43**:1-2 (2001), 111–119.

[Dickman 1930] K. Dickman, "On the frequency of numbers containing prime factors of a certain relative magnitude", *Ark. Mat. Astr. Fys.* **22** (1930), 1–14.

[Erdős 1935] P. Erdős, "On the normal number of prime factors of $p-1$ and some other related problems concerning Euler's $\Phi$-function", *Quart. J. Math. (Oxford)* **6** (1935), 205–213.

[Erdős 1940] P. Erdős, "The difference of consecutive primes", *Duke Math. J.* **6** (1940), 438–441.

[Erdős 1955] P. Erdős, "On consecutive integers", *Nieuw Arch. Wisk. (3)* **3** (1955), 124–128.

[Erdős et al. 1988] P. Erdős, C. L. Stewart, and R. Tijdeman, "Some Diophantine equations with many solutions", *Compositio Math.* **66**:1 (1988), 37–56. Available at http://www.numdam.org/item?id=CM_1988__66_1_37_0.

[Evertse 1984] J.-H. Evertse, "On equations in $S$-units and the Thue–Mahler equation", *Invent. Math.* **75**:3 (1984), 561–584.

[Fouvry and Tenenbaum 1991] É. Fouvry and G. Tenenbaum, "Entiers sans grand facteur premier en progressions arithmetiques", *Proc. London Math. Soc. (3)* **63**:3 (1991), 449–494.

[Fouvry and Tenenbaum 1996] E. Fouvry and G. Tenenbaum, "Répartition statistique des entiers sans grand facteur premier dans les progressions arithmétiques", *Proc. London Math. Soc.* (3) **72**:3 (1996), 481–514.

[Friedlander 1973] J. B. Friedlander, "Integers without large prime factors", *Nederl. Akad. Wetensch. Proc. Ser. A* **76**=*Indag. Math.* **35** (1973), 443–451.

[Friedlander 1989] J. B. Friedlander, "Shifted primes without large prime factors", pp. 393–401 in *Number theory and applications* (Banff, AB, 1988), edited by R. A. Mollin, NATO Adv. Sci. Inst. Ser. C Math. Phys. Sci. **265**, Kluwer, Dordrecht, 1989.

[Friedlander and Granville 1993] J. B. Friedlander and A. Granville, "Smoothing "smooth" numbers", *Philos. Trans. Roy. Soc. London Ser. A* **345**:1676 (1993), 339–347.

[Friedlander and Lagarias 1987] J. B. Friedlander and J. C. Lagarias, "On the distribution in short intervals of integers having no large prime factor", *J. Number Theory* **25**:3 (1987), 249–273.

[Goldston and McCurley 1988] D. A. Goldston and K. S. McCurley, "Sieving the positive integers by large primes", *J. Number Theory* **28**:1 (1988), 94–115.

[Granville 1989] A. Granville, "On positive integers $\leq x$ with prime factors $\leq t \log x$", pp. 403–422 in *Number theory and applications* (Banff, AB, 1988), edited by R. A. Mollin, NATO Adv. Sci. Inst. Ser. C Math. Phys. Sci. **265**, Kluwer, Dordrecht, 1989.

[Granville 1993a] A. Granville, "Integers, without large prime factors, in arithmetic progressions. I", *Acta Math.* **170**:2 (1993), 255–273.

[Granville 1993b] A. Granville, "Integers, without large prime factors, in arithmetic progressions. II", *Philos. Trans. Roy. Soc. London Ser. A* **345**:1676 (1993), 349–362.

[Granville 2005] A. Granville, "It is easy to determine whether a given integer is prime", *Bull. Amer. Math. Soc. (N.S.)* **42**:1 (2005), 3–38.

[Granville and Soundararajan 2001a] A. Granville and K. Soundararajan, "Large character sums", *J. Amer. Math. Soc.* **14**:2 (2001), 365–397.

[Granville and Soundararajan 2001b] A. Granville and K. Soundararajan, "The spectrum of multiplicative functions", *Ann. of Math.* (2) **153**:2 (2001), 407–470.

[Granville and Soundararajan 2004] A. Granville and K. Soundararajan, "The number of unsieved integers up to $x$", *Acta Arith.* **115**:4 (2004), 305–328.

[Granville and Soundararajan 2007] A. Granville and K. Soundararajan, "Large character sums: pretentious characters and the Pólya–Vinogradov theorem", *J. Amer. Math. Soc.* **20**:2 (2007), 357–384.

[Granville and Tucker 2002] A. Granville and T. J. Tucker, "It's as easy as $abc$", *Notices Amer. Math. Soc.* **49**:10 (2002), 1224–1231.

[Hafner 1993] J. L. Hafner, "On smooth numbers in short intervals under the Riemann Hypothesis", Preprint, 1993.

[Halberstam and Richert 1974] H. Halberstam and H.-E. Richert, *Sieve methods*, London Mathematical Society Monographs **4**, Academic Press, London–New York, 1974.

[Harman 1991] G. Harman, "Short intervals containing numbers without large prime factors", *Math. Proc. Cambridge Philos. Soc.* **109**:1 (1991), 1–5.

[Harman 1993] G. Harman, "Small fractional parts of additive forms", *Philos. Trans. Roy. Soc. London Ser. A* **345**:1676 (1993), 327–338.

[Harman 1999] G. Harman, "Integers without large prime factors in short intervals and arithmetic progressions", *Acta Arith.* **91**:3 (1999), 279–289.

[Harman 2005] G. Harman, "On the number of Carmichael numbers up to $x$", *Bull. London Math. Soc.* **37**:5 (2005), 641–650.

[Hensley 1987] D. Hensley, "The distribution of $\Omega(n)$ among numbers with no large prime factors", pp. 247–281 in *Analytic number theory and Diophantine problems* (Stillwater, OK, 1984), edited by A. C. Adolphson et al., Progr. Math. **70**, Birkhäuser, Boston, 1987.

[Hildebrand 1984a] A. Hildebrand, "Integers free of large prime factors and the Riemann hypothesis", *Mathematika* **31**:2 (1984), 258–271.

[Hildebrand 1984b] A. Hildebrand, "Quantitative mean value theorems for nonnegative multiplicative functions. I", *J. London Math. Soc.* (2) **30**:3 (1984), 394–406.

[Hildebrand 1985a] A. Hildebrand, "Integers free of large prime divisors in short intervals", *Quart. J. Math. Oxford Ser.* (2) **36**:141 (1985), 57–69.

[Hildebrand 1985b] A. Hildebrand, "On a conjecture of Balog", *Proc. Amer. Math. Soc.* **95**:4 (1985), 517–523.

[Hildebrand 1986] A. Hildebrand, "On the number of positive integers $\leq x$ and free of prime factors $> y$", *J. Number Theory* **22**:3 (1986), 289–307.

[Hildebrand 1987a] A. Hildebrand, "On the number of prime factors of integers without large prime divisors", *J. Number Theory* **25**:1 (1987), 81–106.

[Hildebrand 1987b] A. Hildebrand, "Quantitative mean value theorems for nonnegative multiplicative functions. II", *Acta Arith.* **48**:3 (1987), 209–260.

[Hildebrand 1989] A. Hildebrand, "On integer sets containing strings of consecutive integers", *Mathematika* **36**:1 (1989), 60–70.

[Hildebrand and Tenenbaum 1986] A. Hildebrand and G. Tenenbaum, "On integers free of large prime factors", *Trans. Amer. Math. Soc.* **296**:1 (1986), 265–290.

[Hildebrand and Tenenbaum 1993a] A. Hildebrand and G. Tenenbaum, "Integers without large prime factors", *J. Théor. Nombres Bordeaux* **5**:2 (1993), 411–484. Available at http://jtnb.cedram.org/item?id=JTNB_1993_5_2_411_0.

[Hildebrand and Tenenbaum 1993b] A. Hildebrand and G. Tenenbaum, "On a class of differential-difference equations arising in number theory", *J. Anal. Math.* **61** (1993), 145–179.

[Hmyrova 1966] N. A. Hmyrova, "On polynomials with small prime divisors. II", *Izv. Akad. Nauk SSSR Ser. Mat.* **30** (1966), 1367–1372. In Russian.

[Hunter and Sorenson 1997] S. Hunter and J. Sorenson, "Approximating the number of integers free of large prime factors", *Math. Comp.* **66**:220 (1997), 1729–1741.

[Iwaniec 1978] H. Iwaniec, "On the problem of Jacobsthal", *Demonstratio Math.* **11**:1 (1978), 225–231.

[Katz and Sarnak 1999] N. M. Katz and P. Sarnak, "Zeroes of zeta functions and symmetry", *Bull. Amer. Math. Soc. (N.S.)* **36**:1 (1999), 1–26.

[Knuth and Trabb Pardo 1976/77] D. E. Knuth and L. Trabb Pardo, "Analysis of a simple factorization algorithm", *Theoret. Comput. Sci.* **3**:3 (1976/77), 321–348.

[de Koninck 1994] J.-M. de Koninck, "On the largest prime divisors of an integer", pp. 447–462 in *Extreme value theory and its applications*, edited by J. Galambos et al., Kluwer, Dordrecht, 1994.

[Konyagin and Pomerance 1997] S. Konyagin and C. Pomerance, "On primes recognizable in deterministic polynomial time", pp. 176–198 in *The mathematics of Paul Erdős, I*, edited by R. L. Graham and J. Nešetřil, Algorithms Combin. **13**, Springer, Berlin, 1997.

[Konyagin and Soundararajan 2007] S. Konyagin and K. Soundararajan, "Two $S$-unit equations with many solutions", *J. Number Theory* **124**:1 (2007), 193–199.

[Lenstra 1979] H. W. Lenstra, Jr., "Miller's primality test", *Inform. Process. Lett.* **8**:2 (1979), 86–88.

[Lenstra 1987] H. W. Lenstra, Jr., "Factoring integers with elliptic curves", *Ann. of Math.* (2) **126**:3 (1987), 649–673.

[Lenstra and Pomerance 1992] H. W. Lenstra, Jr. and C. Pomerance, "A rigorous time bound for factoring integers", *J. Amer. Math. Soc.* **5**:3 (1992), 483–516.

[Lenstra et al. 1993] H. W. Lenstra, Jr., J. Pila, and C. Pomerance, "A hyperelliptic smoothness test. I", *Philos. Trans. Roy. Soc. London Ser. A* **345**:1676 (1993), 397–408.

[Linnik 1941] U. V. Linnik, "The large sieve", *C. R. (Doklady) Acad. Sci. URSS (N.S.)* **30** (1941), 292–294.

[Manstavičius 1992] E. Manstavičius, "Remarks on elements of semigroups that are free of large prime factors", *Liet. Mat. Rink.* **32**:4 (1992), 512–525. In Russian; translated in *Lithuanian Math. J.* **32**:4 (1992), 400–409.

[Martin 2002] G. Martin, "An asymptotic formula for the number of smooth values of a polynomial", *J. Number Theory* **93**:2 (2002), 108–182.

[Mikawa 1989] H. Mikawa, "Almost-primes in arithmetic progressions and short intervals", *Tsukuba J. Math.* **13**:2 (1989), 387–401.

[Montgomery and Vaughan 1977] H. L. Montgomery and R. C. Vaughan, "Exponential sums with multiplicative coefficients", *Invent. Math.* **43**:1 (1977), 69–82.

[Moree 1993] P. Moree, *Psixyology and Diophantine equations*, Thesis, Leiden University, 1993.

[Moree and Stewart 1990] P. Moree and C. L. Stewart, "Some Ramanujan–Nagell equations with many solutions", *Indag. Math. (N.S.)* **1**:4 (1990), 465–472.

[Norton 1971] K. K. Norton, *Numbers with small prime factors, and the least $k$th power non-residue*, Memoirs of the American Mathematical Society, No. 106, American Mathematical Society, Providence, R.I., 1971.

[Paley 1932] R. E. A. C. Paley, "A theorem on characters", *J. London Math. Soc.* **7** (1932), 28–32.

[Pomerance 1980] C. Pomerance, "Popular values of Euler's function", *Mathematika* **27**:1 (1980), 84–89.

[Pomerance 1987] C. Pomerance, "Fast, rigorous factorization and discrete logarithm algorithms", pp. 119–143 in *Discrete algorithms and complexity* (Kyoto, 1986), edited by D. S. Johnson et al., Perspect. Comput. **15**, Academic Press, Boston, 1987.

[Pomerance 1989] C. Pomerance, "Two methods in elementary analytic number theory", pp. 135–161 in *Number theory and applications* (Banff, AB, 1988), edited by R. A. Mollin, NATO Adv. Sci. Inst. Ser. C Math. Phys. Sci. **265**, Kluwer, Dordrecht, 1989.

[Pomerance 1995] C. Pomerance, "The role of smooth numbers in number-theoretic algorithms", pp. 411–422 in *Proceedings of the International Congress of Mathematicians, I* (Zürich, 1994), edited by S. D. Chatterji, Birkhäuser, Basel, 1995.

[Pomerance 1996a] C. Pomerance, "Multiplicative independence for random integers", pp. 703–711 in *Analytic number theory, II* (Allerton Park, IL, 1995), edited by B. C. Berndt et al., Progr. Math. **139**, Birkhäuser, Boston, 1996.

[Pomerance 1996b] C. Pomerance, "A tale of two sieves", *Notices Amer. Math. Soc.* **43**:12 (1996), 1473–1485.

[Pomerance 2008] C. Pomerance, "Smooth numbers and the quadratic sieve", pp. 69–81 in *Surveys in algorithmic number theory*, edited by J. P. Buhler and P. Stevenhagen, Math. Sci. Res. Inst. Publ. **44**, Cambridge University Press, New York, 2008.

[Pomerance and Shparlinski 2002] C. Pomerance and I. E. Shparlinski, "Smooth orders and cryptographic applications", pp. 338–348 in *Algorithmic number theory* (Sydney, 2002), edited by C. Fieker and D. R. Kohel, Lecture Notes in Comput. Sci. **2369**, Springer, Berlin, 2002.

[Rankin 1938] R. A. Rankin, "The difference between consecutive prime numbers", *J. London Math. Soc.* **13** (1938), 242–247.

[Schoof 2008] R. J. Schoof, "Computing Arakelov class groups", pp. 447–495 in *Surveys in algorithmic number theory*, edited by J. P. Buhler and P. Stevenhagen, Math. Sci. Res. Inst. Publ. **44**, Cambridge University Press, New York, 2008.

[Shorey 1973/74] T. N. Shorey, "On gaps between numbers with a large prime factor. II", *Acta Arith.* **25** (1973/74), 365–373.

[Soundararajan 2001] K. Soundararajan, "Smooth polynomials: analogies and asymptotics", Preprint, 2001.

[Soundararajan 2006] K. Soundararajan, Smooth numbers in arithmetic progressions, 2006.

[Stevenhagen 2008] P. Stevenhagen, "The number field sieve", pp. 83–100 in *Surveys in algorithmic number theory*, edited by J. P. Buhler and P. Stevenhagen, Math. Sci. Res. Inst. Publ. **44**, Cambridge University Press, New York, 2008.

[Tenenbaum 1990] G. Tenenbaum, *Introduction à la théorie analytique et probabiliste des nombres*, Publ. Inst. Elie Cartan **13**, 1990.

[Tenenbaum 1993] G. Tenenbaum, "Cribler les entiers sans grand facteur premier", *Philos. Trans. Roy. Soc. London Ser. A* **345**:1676 (1993), 377–384.

[Tijdeman 1973] R. Tijdeman, "On integers with many small prime factors", *Compositio Math.* **26** (1973), 319–330.

[Tijdeman 1974] R. Tijdeman, "On the maximal distance between integers composed of small primes", *Compositio Math.* **28** (1974), 159–162.

[Vaughan 1989] R. C. Vaughan, "A new iterative method in Waring's problem", *Acta Math.* **162**:1-2 (1989), 1–71.

[Vaughan 1993] R. C. Vaughan, "The use in additive number theory of numbers without large prime factors", *Philos. Trans. Roy. Soc. London Ser. A* **345**:1676 (1993), 363–376.

[Vaughan and Wooley 2002] R. C. Vaughan and T. D. Wooley, "Waring's problem: a survey", pp. 301–340 in *Number theory for the millennium, III* (Urbana, IL, 2000), edited by M. A. Bennett et al., A K Peters, Natick, MA, 2002.

[Wolke 1971] D. Wolke, "Polynomial values with small prime divisors", *Acta Arith.* **19** (1971), 327–333.

[Wooley 1992] T. D. Wooley, "Large improvements in Waring's problem", *Ann. of Math.* (2) **135**:1 (1992), 131–164.

[Xuan 1995] T. Z. Xuan, "Integers with no large prime factors", *Acta Arith.* **69**:4 (1995), 303–327.

[Xuan 1999] T. Z. Xuan, "On smooth integers in short intervals under the Riemann hypothesis", *Acta Arith.* **88**:4 (1999), 327–332.

ANDREW GRANVILLE
DÉPARTMENT DE MATHÉMATIQUES ET STATISTIQUE
UNIVERSITÉ DE MONTRÉAL
CP 6128 SUCC CENTRE-VILLE
MONTRÉAL, QC H3C 3J7
CANADA
andrew@dms.umontreal.ca

# Fast multiplication and its applications

DANIEL J. BERNSTEIN

ABSTRACT. This survey explains how some useful arithmetic operations can be sped up from quadratic time to essentially linear time.

## 1. Introduction

This paper presents fast algorithms for several useful arithmetic operations on polynomials, power series, integers, real numbers, and 2-adic numbers.

Each section focuses on one algorithm for one operation, and describes seven features of the algorithm:

- Input: What numbers are provided to the algorithm? Sections 2, 3, 4, and 5 explain how various mathematical objects are represented as inputs.
- Output: What numbers are computed by the algorithm?
- Speed: How many coefficient operations does the algorithm use to perform a polynomial operation? The answer is at most $n^{1+o(1)}$, where $n$ is the problem size; each section states a more precise upper bound, often using the function $\mu$ defined in Section 4.
- How it works: What is the algorithm? The algorithm may use previous algorithms as subroutines, as shown in (the transitive closure of) Figure 1.
- The integer case (except in Section 2): The inputs were polynomials (or power series); what about the analogous operations on integers (or real numbers)? What difficulties arise in adapting the algorithm to integers? How much time does the adapted algorithm take?

---

*Mathematics Subject Classification:* Primary 68–02. Secondary 11Y16, 12Y05, 65D20, 65T50, 65Y20, 68W30.

Permanent ID of this document: 8758803e61822d485d54251b27b1a20d. Date: 2008.05.15.

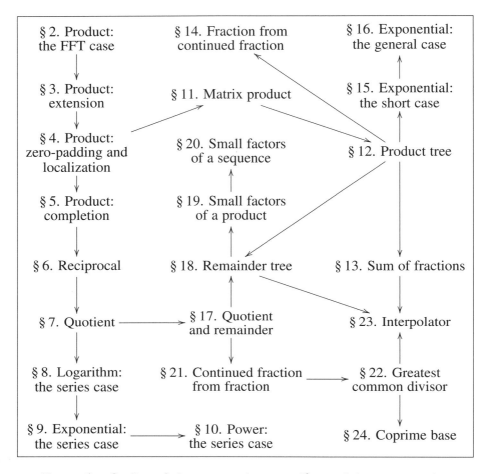

**Figure 1.** Outline of the paper. A vertex "§ N. F" here means that Section N describes an algorithm to compute the function F. Arrows here indicate prerequisite algorithms.

- History: How were these ideas developed?
- Improvements: The algorithm was chosen to be reasonably simple (subject to the $n^{1+o(1)}$ bound) at the expense of speed; how can the same function be computed even more quickly?

Sections 2 through 5 describe fast multiplication algorithms for various rings. The remaining sections describe various applications of fast multiplication. Here is a simplified summary of the functions being computed:

- § 6. Reciprocal. $f \mapsto 1/f$ approximation.
- § 7. Quotient. $f, h \mapsto h/f$ approximation.
- § 8. Logarithm. $f \mapsto \log f$ approximation.

- § 9. Exponential. $f \mapsto \exp f$ approximation. Also § 15, § 16.
- § 10. Power. $f, e \mapsto f^e$ approximation.
- § 11. Matrix product. $f, g \mapsto fg$ for $2 \times 2$ matrices.
- § 12. Product tree. $f_1, f_2, f_3, \ldots \mapsto$ tree of products including $f_1 f_2 f_3 \cdots$.
- § 13. Sum of fractions. $f_1, g_1, f_2, g_2, \ldots \mapsto f_1/g_1 + f_2/g_2 + \cdots$.
- § 14. Fraction from continued fraction. $q_1, q_2, \ldots \mapsto q_1 + 1/(q_2 + 1/(\cdots))$.
- § 17. Quotient and remainder. $f, h \mapsto \lfloor h/f \rfloor, h \bmod f$.
- § 18. Remainder tree. $h, f_1, f_2, \ldots \mapsto h \bmod f_1, h \bmod f_2, \ldots$.
- § 19. Small factors of a product. $S, h_1, h_2, \ldots \mapsto S(h_1 h_2 \cdots)$ where $S$ is a set of primes and $S(h)$ is the subset of $S$ dividing $h$.
- § 20. Small factors of a sequence. $S, h_1, h_2, \ldots \mapsto S(h_1), S(h_2), \ldots$.
- § 21. Continued fraction from fraction. $f_1, f_2 \mapsto q_1, q_2, q_3, \ldots$ with $f_1/f_2 = q_1 + 1/(q_2 + 1/(q_3 + 1/(\cdots)))$.
- § 22. Greatest common divisor. $f_1, f_2 \mapsto \gcd\{f_1, f_2\}$.
- § 23. Interpolator. $f_1, g_1, f_2, g_2, \ldots \mapsto h$ with $h \equiv f_j \pmod{g_j}$.
- § 24. Coprime base. $f_1, f_2, \ldots \mapsto$ coprime set $S$ with $f_1, f_2, \ldots \in \langle S \rangle$.

**Acknowledgments.** Thanks to Alice Silverberg, Paul Zimmermann, and the referee for their comments.

## 2. Product: the FFT case

**2.1. Input.** Let $n \geq 1$ be a power of 2. Let $c$ be a nonzero element of $\mathbf{C}$. The algorithm described in this section is given two elements $f, g$ of the ring $\mathbf{C}[x]/(x^n - c)$.

An element of $\mathbf{C}[x]/(x^n - c)$ is, by convention, represented as a sequence of $n$ elements of $\mathbf{C}$: the sequence $(f_0, f_1, \ldots, f_{n-1})$ represents $f_0 + f_1 x + \cdots + f_{n-1} x^{n-1}$.

**2.2. Output.** This algorithm computes the product $fg \in \mathbf{C}[x]/(x^n - c)$, represented in the same way. If the input is $f_0, f_1, \ldots, f_{n-1}, g_0, g_1, \ldots, g_{n-1}$ then the output is $h_0, h_1, \ldots, h_{n-1}$, where $h_i = \sum_{0 \leq j \leq i} f_j g_{i-j} + c \sum_{i+1 \leq j < n} f_j g_{i+n-j}$.

For example, for $n = 4$, the output is $f_0 g_0 + c f_1 g_3 + c f_2 g_2 + c f_3 g_1, f_0 g_1 + f_1 g_0 + c f_2 g_3 + c f_3 g_2, f_0 g_2 + f_1 g_1 + f_2 g_0 + c f_3 g_3, f_0 g_3 + f_1 g_2 + f_2 g_1 + f_3 g_0$.

**2.3. Model of computation.** Let $A$ be a commutative ring. An **operation in** $A$ is, by definition, one binary addition $a, b \mapsto a + b$, one binary subtraction $a, b \mapsto a - b$, or one binary multiplication $a, b \mapsto ab$. Here $a$ is an input, a constant, or a result of a previous operation; same for $b$.

For example, given $a, b \in \mathbf{C}$, one can compute $10a + 11b, 9a + 10b$ with four operations in $\mathbf{C}$: add $a$ and $b$ to obtain $a + b$; multiply by 10 to obtain $10a + 10b$; add $b$ to obtain $10a + 11b$; subtract $a$ from $10a + 10b$ to obtain $9a + 10b$.

Starting in Section 19 of this paper, the definition of **operation in** $A$ is expanded to allow equality tests. Starting in Section 21, the ring $A$ is assumed to be a field, and the definition of **operation in** $A$ is expanded to allow divisions (when the denominators are nonzero). Algorithms built out of additions, subtractions, multiplications, divisions, and equality tests are called **algebraic algorithms**. See [Bürgisser et al. 1997, Chapter 4] for a precise definition of this model of computation.

Warning: It is tempting to think of an algebraic algorithm (e.g., "add $a$ to $b$; multiply by $b$") as simply a chain of intermediate results (e.g., "$a+b$; $ab+b^2$"). Some authors *define* algebraic algorithms as chains of computable results; see, e.g., the definition of addition chains in [Knuth and Papadimitriou 1981]. But this simplification poses problems. Standard measurements of algebraic complexity, such as the number of multiplications, are generally not determined by the chain of intermediate results. (How many multiplications are in $2a$, $a^2$, $2a^2$?) An algebraic algorithm, properly defined, is not a chain of *computable results* but a chain of *computations*.

**2.4. Speed.** The algorithm in this section uses $O(n \lg n)$ operations — more precisely, $(9/2)n \lg n + 2n$ additions, subtractions, and multiplications — in **C**. Here $\lg = \log_2$.

**2.5. How it works.** If $n = 1$ then the algorithm simply multiplies $f_0$ by $g_0$ to obtain the output $f_0 g_0$.

The strategy for larger $n$ is to split an $n$-coefficient problem into two $(n/2)$-coefficient problems, which are handled by the same method recursively. One needs $\lg n$ levels of recursion to split the original problem into $n$ easy single-coefficient problems; each level of recursion involves $9/2$ operations per coefficient.

Consider, for any $n$, the functions $\varphi : \mathbf{C}[x]/(x^{2n} - c^2) \to \mathbf{C}[x]/(x^n - c)$ and $\varphi' : \mathbf{C}[x]/(x^{2n} - c^2) \to \mathbf{C}[x]/(x^n + c)$ that take $f_0 + \cdots + f_{2n-1} x^{2n-1}$ to $(f_0 + c f_n) + \cdots + (f_{n-1} + c f_{2n-1}) x^{2n-1}$ and $(f_0 - c f_n) + \cdots + (f_{n-1} - c f_{2n-1}) x^{2n-1}$ respectively. Given $f$, one can compute $\varphi(f)$, $\varphi'(f)$ with $n$ additions, $n$ subtractions, and $n$ multiplications by the constant $c$.

These functions $\varphi, \varphi'$ are $\mathbf{C}[x]$-algebra morphisms. In particular, they preserve multiplication: $\varphi(fg) = \varphi(f)\varphi(g)$ and $\varphi'(fg) = \varphi'(f)\varphi'(g)$. Furthermore, $\varphi \times \varphi'$ is injective: one can recover $f$ from $\varphi(f)$ and $\varphi'(f)$. It is simpler to recover $2f$: this takes $n$ additions, $n$ subtractions, and $n$ multiplications by the constant $1/c$.

Here, then, is how the algorithm computes $2nfg$, given $f, g \in \mathbf{C}[x]/(x^{2n} - c^2)$:

- Compute $\varphi(f), \varphi(g), \varphi'(f), \varphi'(g)$ with $2n$ additions, $2n$ subtractions, and $2n$ multiplications by $c$.

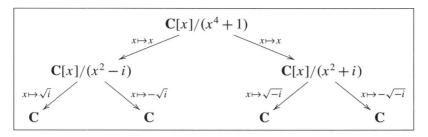

**Figure 2.** Splitting product in $\mathbf{C}[x]/(x^4+1)$ into products in $\mathbf{C}$.

- Recursively compute $n\varphi(f)\varphi(g) = \varphi(nfg)$ in $\mathbf{C}[x]/(x^n - c)$, and recursively compute $n\varphi'(f)\varphi'(g) = \varphi'(nfg)$ in $\mathbf{C}[x]/(x^n + c)$.
- Compute $2nfg$ from $\varphi(nfg)$, $\varphi'(nfg)$ with $n$ additions, $n$ subtractions, and $n$ multiplications by $1/c$.

For example, given $f = f_0 + f_1 x + f_2 x^2 + f_3 x^3$ and $g = g_0 + g_1 x + g_2 x^2 + g_3 x^3$, the algorithm computes $4fg$ in $\mathbf{C}[x]/(x^4+1) = \mathbf{C}[x]/(x^4 - i^2)$ as follows:

- Compute $\varphi(f) = (f_0 + i f_2) + (f_1 + i f_3)x$ and $\varphi'(f) = (f_0 - i f_2) + (f_1 - i f_3)x$, and similarly compute $\varphi(g)$ and $\varphi'(g)$.
- Recursively compute $2\varphi(f)\varphi(g)$ in $\mathbf{C}[x]/(x^2 - i)$, and recursively compute $2\varphi'(f)\varphi'(g)$ in $\mathbf{C}[x]/(x^2 + i)$.
- Recover $4fg$.

See Figure 2.

A straightforward induction shows that the total work to compute the product $nfg$, given $f, g \in \mathbf{C}[x]/(x^n - c)$, is $(3/2)n \lg n$ additions, $(3/2)n \lg n$ subtractions, $(3/2)n \lg n$ multiplications by various constants, and $n$ more multiplications. The algorithm then computes $fg$ with an additional $n$ multiplications by the constant $1/n$.

**2.6. Generalization.** More generally, let $A$ be a commutative ring in which 2 is invertible, let $n \geq 2$ be a power of 2, let $c$ be an invertible element of $A$, and let $\zeta$ be an $(n/2)$nd root of $-1$ in $A$.

By exactly the same method as above, one can multiply two elements of the ring $A[x]/(x^n - c^n)$ with $(9/2)n \lg n + 2n$ operations in $A$: specifically, $(3/2)n \lg n$ additions, $(3/2)n \lg n$ subtractions, $(3/2)n \lg n + n$ multiplications by constants, and $n$ more multiplications. The constants are $1/n$ and products of powers of $c$ and $\zeta$.

The assumption that $A$ has a primitive $n$th root of 1 is a heavy restriction on $A$. If $\mathbf{Z}/t$ has a primitive $n$th root of 1, for example, then every prime divisor of $t$ is in $1 + n\mathbf{Z}$. (This fact is a special case of Pocklington's primality test.) Section 3 explains how to handle more general rings $A$.

**2.7. Variant: radix 3.** Similarly, let $A$ be a commutative ring in which 3 is invertible, let $n \geq 3$ be a power of 3, let $c$ be an invertible element of $A$, and let $\zeta$ be an element of $A$ satisfying $1 + \zeta^{n/3} + \zeta^{2n/3} = 0$. Then one can multiply two elements of the ring $A[x]/(x^n - c^n)$ with $O(n \lg n)$ operations in $A$.

**2.8. History.** Gauss [1866, pages 265–327] was the first to point out that one can quickly compute a ring isomorphism from $\mathbf{R}[x]/(x^{2n} - 1)$ to $\mathbf{R}^2 \times \mathbf{C}^{n-1}$ when $n$ has no large prime factors. For example, Gauss [1866, pages 308–310] (in completely different language) mapped $\mathbf{R}[x]/(x^{12} - 1)$ to $\mathbf{R}[x]/(x^3 - 1) \times \mathbf{R}[x]/(x^3 + 1) \times \mathbf{C}[x]/(x^3 + i)$, then mapped $\mathbf{R}[x]/(x^3 - 1)$ to $\mathbf{R} \times \mathbf{C}$, mapped $\mathbf{R}[x]/(x^3 + 1)$ to $\mathbf{R} \times \mathbf{C}$, and mapped $\mathbf{C}[x]/(x^3 + i)$ to $\mathbf{C} \times \mathbf{C} \times \mathbf{C}$.

The **discrete Fourier transform**—this isomorphism from $\mathbf{R}[x]/(x^{2n} - 1)$ to $\mathbf{R}^2 \times \mathbf{C}^{n-1}$, or the analogous isomorphism from $\mathbf{C}[x]/(x^n - 1)$ to $\mathbf{C}^n$—was applied to many areas of scientific computation over the next hundred years. Gauss's method was reinvented several times, as discussed in [Heideman et al. 1985], and finally became widely known after it was reinvented and published by Cooley and Tukey [1965]. Gauss's method is now called the **fast Fourier transform** or simply the **FFT**.

Shortly after the Cooley–Tukey paper, Sande and Stockham pointed out that one can quickly multiply in $\mathbf{C}[x]/(x^n - 1)$ by applying the FFT, multiplying in $\mathbf{C}^n$, and applying the inverse FFT. See [Stockham 1966, page 229] and [Gentleman and Sande 1966, page 573].

Fiduccia [1972] was the first to point out that *each step* of the FFT is an algebra isomorphism. This fact is still not widely known, despite its tremendous expository value; most expositions of the FFT use only the *module* structure of each step. I have taken Fiduccia's idea much further in this paper and in [Bernstein 2001], identifying the ring morphisms behind all known multiplication methods.

**2.9. Improvements.** The algorithm explained above takes $15n \lg n + 8n$ operations in $\mathbf{R}$ to multiply in $\mathbf{C}[x]/(x^n - 1)$, if $n \geq 2$ is a power of 2 and $\mathbf{C}$ is represented as $\mathbf{R}[i]/(i^2 + 1)$:

- $5n \lg n$ to transform the first input from $\mathbf{C}[x]/(x^n - 1)$ to $\mathbf{C}^n$. The FFT takes $n \lg n$ additions and subtractions in $\mathbf{C}$, totalling $2n \lg n$ operations in $\mathbf{R}$, and $(1/2)n \lg n$ multiplications by various roots of 1 in $\mathbf{C}$, totalling $3n \lg n$ operations in $\mathbf{R}$.
- $5n \lg n$ to transform the second input from $\mathbf{C}[x]/(x^n - 1)$ to $\mathbf{C}^n$.
- $2n$ to scale one of the transforms, i.e., to multiply by $1/n$. One can eliminate most of these multiplications by absorbing $1/n$ into other constants.
- $6n$ to multiply the two transformed inputs in $\mathbf{C}^n$.
- $5n \lg n$ to transform the product from $\mathbf{C}^n$ back to $\mathbf{C}[x]/(x^n - 1)$.

One can reduce each $5n \lg n$ to $5n \lg n - 10n + 16$ for $n \geq 4$ by recognizing roots of 1 that allow easy multiplications: multiplications by 1 can be skipped, multiplications by $-1$ and $\pm i$ can be absorbed into subsequent computations, and multiplications by $\pm\sqrt{\pm i}$ are slightly easier than general multiplications.

Gentleman and Sande [1966] pointed out another algorithm, which I call the **twisted FFT**, to map $\mathbf{C}[x]/(x^n - 1)$ to $\mathbf{C}^n$ using $5n \lg n - 10n + 16$ operations. The twisted FFT maps $\mathbf{C}[x]/(x^{2n} - 1)$ to $\mathbf{C}[x]/(x^n - 1) \times \mathbf{C}[x]/(x^n + 1)$, twists $\mathbf{C}[x]/(x^n + 1)$ into $\mathbf{C}[x]/(x^n - 1)$ by mapping $x \to \zeta x$, and handles each $\mathbf{C}[x]/(x^n - 1)$ recursively.

The **split-radix FFT** is faster: it uses only $4n \lg n - 6n + 8$ operations for $n \geq 2$. The split-radix FFT is a mixture of Gauss's FFT with the Gentleman–Sande twisted FFT: it maps $\mathbf{C}[x]/(x^{4n} - 1)$ to $\mathbf{C}[x]/(x^{2n} - 1) \times \mathbf{C}[x]/(x^{2n} + 1)$, maps $\mathbf{C}[x]/(x^{2n} + 1)$ to $\mathbf{C}[x]/(x^n - i) \times \mathbf{C}[x]/(x^n + i)$, twists each $\mathbf{C}[x]/(x^n \pm i)$ into $\mathbf{C}[x]/(x^n - 1)$ by mapping $x \to \zeta x$, and recursively handles both $\mathbf{C}[x]/(x^{2n} - 1)$ and $\mathbf{C}[x]/(x^n - 1)$.

Another method is the **real-factor FFT**: map $\mathbf{C}[x]/(x^{4n} - (2\cos 2\alpha)x^{2n} + 1)$ to $\mathbf{C}[x]/(x^{2n} - (2\cos\alpha)x^n + 1) \times \mathbf{C}[x]/(x^{2n} + (2\cos\alpha)x^n + 1)$, and handle each factor recursively. If one represents elements of $\mathbf{C}[x]/(x^{2n} \pm \cdots)$ using the basis $(1, x, \ldots, x^{n-1}, x^{-n}, x^{1-n}, \ldots, x^{-1})$ then the real-factor FFT uses $4n \lg n + O(n)$ operations.

It is difficult to assign credit for the bound $4n \lg n + O(n)$. Yavne [1968, page 117] announced the bound $4n \lg n - 6n + 8$ (specifically, $3n \lg n - 3n + 4$ additions and subtractions and $n \lg n - 3n + 4$ multiplications), and apparently had in mind a method achieving that bound; but nobody, to my knowledge, has ever deciphered Yavne's explanation of the method. Ten years later, Bruun [1978] published the real-factor FFT. Several years after that, Duhamel and Hollmann [1984], Martens [1984], Vetterli and Nussbaumer [1984], and Stasinski (according to [Duhamel and Vetterli 1990, page 263]) independently discovered the split-radix FFT.

In 2004, Van Buskirk posted software demonstrating that the split-radix FFT is not optimal: the **tangent FFT** uses only $(34/9)n \lg n + O(n)$ operations, so multiplication in $\mathbf{C}[x]/(x^n - 1)$ uses only $(34/3)n \lg n + O(n)$ operations. The tangent FFT avoids the standard basis $1, x, \ldots, x^{n-1}$ of $\mathbf{C}[x]/(x^n - 1)$ and instead uses the basis $1/s_{n,0}, x/s_{n,1}, \ldots, x^{n-1}/s_{n,n-1}$ where

$$s_{n,k} = \prod_{\ell \geq 0} \max\left\{\left|\cos\frac{4^\ell 2\pi k}{n}\right|, \left|\sin\frac{4^\ell 2\pi k}{n}\right|\right\}.$$

Aside from this change, the tangent FFT maps $\mathbf{C}[x]/(x^{8n}-1)$ to $\mathbf{C}[x]/(x^{2n}-1) \times \mathbf{C}[x]/(x^{2n}-1) \times \mathbf{C}[x]/(x^{2n}-1) \times \mathbf{C}[x]/(x^n-1) \times \mathbf{C}[x]/(x^n-1)$ in essentially the same way as the split-radix FFT. See [Bernstein 2007] for further details and a cost analysis. See [Lundy and Van Buskirk 2007] and [Johnson and Frigo 2007] for two alternate explanations of the 34/9.

One can multiply in $\mathbf{R}[x]/(x^{2n}+1)$ with $(34/3)n \lg n + O(n)$ operations in $\mathbf{R}$, if $n$ is a power of 2: map $\mathbf{R}[x]/(x^{2n}+1)$ to $\mathbf{C}[x]/(x^n-i)$, twist $\mathbf{C}[x]/(x^n-i)$ into $\mathbf{C}[x]/(x^n-1)$, and apply the tangent FFT. This is approximately twice as fast as mapping $\mathbf{R}[x]/(x^{2n}+1)$ to $\mathbf{C}[x]/(x^{2n}+1)$.

One can also multiply in $\mathbf{R}[x]/(x^{2n}-1)$ with $(34/3)n \lg n + O(n)$ operations in $\mathbf{R}$, if $n$ is a power of 2: map $\mathbf{R}[x]/(x^{2n}-1)$ to $\mathbf{R}[x]/(x^n-1) \times \mathbf{R}[x]/(x^n+1)$; handle $\mathbf{R}[x]/(x^n-1)$ by the same method recursively; handle $\mathbf{R}[x]/(x^n+1)$ as above. This is approximately twice as fast as mapping $\mathbf{R}[x]/(x^{2n}-1)$ to $\mathbf{C}[x]/(x^{2n}-1)$. This speedup was announced by Bergland [1968], but it was already part of Gauss's FFT.

The general strategy of all of the above algorithms is to transform $f$, transform $g$, multiply the results, and then undo the transform to recover $fg$. There is some redundancy here if $f = g$: one can easily save a factor of $1.5 + o(1)$ by transforming $f$, squaring the result, and undoing the transform to recover $f^2$. (Of course, $f^2$ is much easier to compute if $2 = 0$ in $A$; this also saves time in Section 6.)

More generally, one can save the transform of each input, and reuse the transform in a subsequent multiplication if one knows (or observes) that the same input is showing up again. I call this technique **FFT caching**. FFT caching was announced in [Crandall and Fagin 1994, Section 9], but it was already widely known; see, e.g., [Montgomery 1992, Section 3.7].

Further savings are possible when one wants to compute a sum of products. Instead of undoing a transform to recover $ab$, undoing another transform to recover $cd$, and adding the results to obtain $ab + cd$, one can add first and then undo a single transform to recover $ab + cd$. I call this technique **FFT addition**.

There is much more to say about FFT performance, because there are much more sophisticated models of computation. Real computers have operation latency, memory latency, and instruction-decoding latency, for example; a serious analysis of constant factors takes these latencies into account.

## 3. Product: extension

**3.1. Input.** Let $A$ be a commutative ring in which 2 is invertible. Let $n \geq 1$ be a power of 2. The algorithm described in this section is given two elements $f, g$ of the ring $A[x]/(x^n+1)$.

An element of $A[x]/(x^n+1)$ is, by convention, represented as a sequence of $n$ elements of $A$: the sequence $(f_0, f_1, \ldots, f_{n-1})$ represents $f_0 + f_1 x + \cdots + f_{n-1} x^{n-1}$.

FAST MULTIPLICATION AND ITS APPLICATIONS    333

**3.2. Output.** This algorithm computes the product $fg \in A[x]/(x^n + 1)$.

**3.3. Speed.** This algorithm uses $O(n \lg n \lg \lg n)$ operations in $A$: more precisely, at most $(((9/2) \lg \lg(n/4) + 63/2) \lg(n/4) - 33/2)n$ operations in $A$ if $n \geq 8$.

**3.4. How it works.** For $n \leq 8$, use the definition of multiplication in $A[x]/(x^n + 1)$. This takes at most $n^2 + n(n-1) = (2n-1)n$ operations in $A$. If $n = 8$ then $\lg(n/4) = 1$ so $((9/2) \lg \lg(n/4) + 63/2) \lg(n/4) - 33/2 = 63/2 - 33/2 = 15 = 2n - 1$.

For $n \geq 16$, find the unique power $m$ of 2 such that $m^2 \in \{2n, 4n\}$. Notice that $8 \leq m < n$. Notice also that $\lg(n/4) - 1 \leq 2 \lg(m/4) \leq \lg(n/4)$, so $\lg \lg(m/4) \leq \lg \lg(n/4) - 1$ and $2 \lg(2n/m) \leq \lg(n/4) + 3$.

Define $B = A[x]/(x^m + 1)$. By induction, given $f, g \in B$, one can compute the product $fg$ with at most $(((9/2) \lg \lg(m/4) + 63/2) \lg(m/4) - 33/2)m$ operations in $A$. One can also compute any of the following with $m$ operations in $A$: the sum $f + g$; the difference $f - g$; the product $cf$, where $c$ is a constant element of $A$; the product $cf$, where $c$ is a constant power of $x$.

There is a $(2n/m)$th root of $-1$ in $B$, namely $x^{m^2/2n}$. Therefore one can use the algorithm explained in Section 2 to multiply quickly in $B[y]/(y^{2n/m} + 1)$ — and, consequently, to multiply in $A[x, y]/(y^{2n/m} + 1)$ if each input has $x$-degree smaller than $m/2$. This takes $(9/2)(2n/m) \lg(2n/m) + 2n/m$ easy operations in $B$ and $2n/m$ more multiplications in $B$.

Now, given $f, g \in A[x]/(x^n + 1)$, compute $fg$ as follows. Consider the $A[x]$-algebra morphism $\varphi : A[x, y]/(y^{2n/m} + 1) \to A[x]/(x^n + 1)$ that takes $y$ to $x^{m/2}$. Find $F \in A[x, y]/(y^{2n/m} + 1)$ such that $F$ has $x$-degree smaller than $m/2$ and $\varphi(F) = f$; explicitly, $F = \sum_j \sum_{0 \leq i < m/2} f_{i+(m/2)j} x^i y^j$ if $f = \sum f_i x^i$. Similarly construct $G$ from $g$. Compute $FG$ as explained above. Then compute $\varphi(FG) = fg$; this takes $n$ additional operations in $A$.

One multiplication in $A[x]/(x^n + 1)$ has thus been split into

- $2n/m$ multiplications in $B$, i.e., at most $((9 \lg \lg(m/4) + 63) \lg(m/4) - 33)n \leq (((9/2) \lg \lg(n/4) + 27) \lg(n/4) - 33)n$ operations in $A$;
- $(9n/m) \lg(2n/m) + 2n/m \leq ((9/2) \lg(n/4) + 31/2)n/m$ easy operations in $B$, i.e., at most $((9/2) \lg(n/4) + 31/2)n$ operations in $A$; and
- $n$ additional operations in $A$.

The total is at most $(((9/2) \lg \lg(n/4) + 63/2) \lg(n/4) - 33/2)n$ as claimed.

For example, given $f = f_0 + f_1 x + \cdots + f_7 x^7$ and $g = g_0 + g_1 x + \cdots + g_7 x^7$ in $A[x]/(x^8 + 1)$, define $F = (f_0 + f_1 x) + (f_2 + f_3 x)y + (f_4 + f_5 x)y^2 + (f_6 + f_7 x)y^3$ and $G = (g_0 + g_1 x) + (g_2 + g_3 x)y + (g_4 + g_5 x)y^2 + (g_6 + g_7 x)y^3$ in $A[x, y]/(y^4 + 1)$.

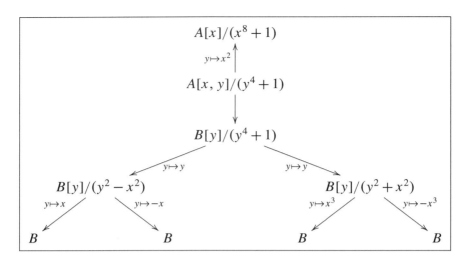

**Figure 3.** Splitting product in $A[x]/(x^8 + 1)$ into four products in $B = A[x]/(x^4 + 1)$, if 2 is invertible in $A$. Compare to Figure 2.

The product $FG$ has the form

$$(h_0 + h_1 x + h_2 x^2) + (h_3 + h_4 x + h_5 x^2)y$$
$$+ (h_6 + h_7 x + h_8 x^2)y^2 + (h_9 + h_{10} x + h_{11} x^2)y^3.$$

Compute this product in $A[x, y]/(x^4 + 1, y^4 + 1)$, and substitute $y = x^2$ to recover $fg = (h_0 - h_{11}) + h_1 x + (h_2 + h_3)x^2 + h_4 x^3 + (h_5 + h_6)x^4 + h_7 x^5 + (h_8 + h_9)x^6 + h_{10} x^7$. The multiplication in $A[x, y]/(x^4 + 1, y^4 + 1)$ splits into four multiplications in $A[x]/(x^4 + 1)$. See Figure 3.

**3.5. Variant: radix 3.** Similarly, let $A$ be a commutative ring in which 3 is invertible, and let $n \geq 3$ be a power of 3. One can multiply two elements of $A[x]/(x^{2n} + x^n + 1)$ with $O(n \lg n \lg \lg n)$ operations in $A$.

**3.6. The integer case; another model of computation.** Algorithms that multiply polynomials of high degree using very few coefficient operations are analogous to algorithms that multiply integers with many bits in very little time.

There are many popular definitions of time. In this paper, **time** means number of steps on a multitape Turing machine. See [Papadimitriou 1994, Section 2.3] for a precise definition of multitape Turing machines.

Let $n$ be a power of 2. There is an algorithm, analogous to the multiplication algorithm for $A[x]/(x^n + 1)$, that multiplies two elements of $\mathbf{Z}/(2^n + 1)$ in time $O(n \lg n \lg \lg n)$. Here an element of $\mathbf{Z}/(2^n + 1)$ is, by convention, represented as a sequence of $n + 1$ bits: the sequence $(f_0, f_1, \ldots, f_n)$ represents $f_0 + 2f_1 + \cdots + 2^n f_n$. Note that most numbers have two representations.

The multiplication algorithm for $\mathbf{Z}/(2^n+1)$ performs $2n/m$ multiplications in $\mathbf{Z}/(2^m+1)$, for $n \geq 16$, where $m^2 \in \{2n, 4n\}$. Splitting a $\mathbf{Z}/(2^n+1)$ multiplication into $\mathbf{Z}/(2^m+1)$ multiplications is analogous to, but slightly more complicated than, splitting an $A[x]/(x^n+1)$ multiplication into $A[x]/(x^m+1)$ multiplications. The complication is that a sum of $2n/m$ products of $(m/2)$-bit integers generally does not quite fit into $m$ bits. On the other hand, the sum does fit into $m+k$ bits for a small $k$, so it is determined by its images in $\mathbf{Z}/(2^m+1)$ and $\mathbf{Z}/2^k$. One multiplies in $\mathbf{Z}[y]/(y^{2n/m}+1)$ by multiplying recursively in $(\mathbf{Z}/(2^m+1))[y]/(y^{2n/m}+1)$ and multiplying straightforwardly in $(\mathbf{Z}/2^k)[y]/(y^{2n/m}+1)$.

**3.7. History.** The ideas in this section were developed first in the integer case. The crucial point is that one can multiply in $\mathbf{Z}[y]/(y^m \pm 1)$ by selecting $t$ so that $\mathbf{Z}/t$ has an appropriate root of 1, mapping $\mathbf{Z}[y]/(y^m \pm 1)$ to $(\mathbf{Z}/t)[y]/(y^m \pm 1)$, and applying an FFT over $\mathbf{Z}/t$. This multiplication method was suggested by Pollard [1971], independently by Nicholson [1971, page 532], and independently by Schönhage and Strassen [1971]. Schönhage and Strassen suggested the choice $t = 2^m + 1$ and proved the $O(n \lg n \lg \lg n)$ time bound.

An analogous method for polynomials was mentioned by Schönhage [1977] and presented in detail by Turk [1982, Section 2]. Schönhage also suggested using the radix-3 FFT to multiply polynomials over fields of characteristic 2.

Nussbaumer [1980] introduced a different polynomial-multiplication algorithm achieving the $O(n \lg n \lg \lg n)$ operation bound. Nussbaumer's algorithm starts in the same way, lifting (for example) $A[x]/(x^8+1)$ to $A[x, y]/(y^4+1)$ by $y \mapsto x^2$. It then maps $A[x, y]/(y^4+1)$ to $(A[y]/(y^4+1))[x]/(x^4-1)$ and applies an FFT over $A[y]/(y^4+1)$, instead of mapping $A[x, y]/(y^4+1)$ to $(A[x]/(x^4+1))[y]/(y^4+1)$ and applying an FFT over $A[x]/(x^4+1)$.

**3.8. Improvements.** Multiplication by a constant power of $x$ in $A[x]/(x^m+1)$ is easier than the above analysis indicates: multiplications by 1 in $A$ can be eliminated, and multiplications by $-1$ in $A$ can be absorbed into subsequent computations. The total operation count drops from $(9/2 + o(1))n \lg n \lg \lg n$ to $(3 + o(1))n \lg n \lg \lg n$.

The constant 3 here is the best known. There is much more to say about the $o(1)$. See [Bernstein 2001] for a survey of relevant techniques.

There is vastly more to say about integer multiplication, in part because Turing-machine time is a more complicated concept than algebraic complexity, and in part because real computers are more complicated than Turing machines. I will restrict my discussion of this area to one recent piece of news, namely that the Schönhage–Strassen time bound $O(n \lg n \lg \lg n)$ has been superseded: Fürer [2007] reduced $\lg \lg n$ to $2^{O(\lg^* n)}$. Here $\lg^* n = 0$ for $n = 1$; $\lg^* n = 1$ for

$2 \leq n < 4$; $\lg^* n = 2$ for $4 \leq n < 16$; $\lg^* n = 3$ for $16 \leq n < 65536$; $\lg^* n = 4$ for $65536 \leq n < 2^{65536}$; etc. All of the $\cdots \lg \lg n$ time bounds for integer operations later in this paper are therefore unnecessarily pessimistic.

Here is Fürer's idea in a nutshell. Recall that the split-radix FFT maps $\mathbf{C}[x]/(x^{2n}-c^2)$ to $\mathbf{C}[x]/(x^n-c) \times \mathbf{C}[x]/(x^n+c)$ when $c$ is an "easy" root of 1, specifically a power of $i$; otherwise it twists $\mathbf{C}[x]/(x^{2n}-c^2)$ into $\mathbf{C}[x]/(x^{2n}-1)$. Generalize from $\mathbf{C} = \mathbf{R}[i]/(i^2+1)$ to the ring $\mathbf{R}[i]/(i^{2^k}+1)$, with $\zeta$ chosen as an $(n/2^{k+1})$th root of $i$; for example, take

$$\zeta = \sum_{0 \leq d < 2^k} \frac{i^d}{2^{k-1}} \sum_{0 < j < 2^k;\ j\ \text{odd}} \cos\left(2\pi j \left(\frac{d}{2^{k+1}} - \frac{1}{n}\right)\right).$$

A split-radix FFT of size $n$ over $\mathbf{R}[i]/(i^{2^k}+1)$, with $b$ bits of precision in $\mathbf{R}$, then involves $\Theta(n \lg n)$ easy operations in $\mathbf{R}[i]/(i^{2^k}+1)$, each of which takes time $\Theta(2^k b)$, and only $\Theta((n \lg n)/k)$ hard multiplications in $\mathbf{R}[i]/(i^{2^k}+1)$, each of which can be expressed as an integer multiplication of size $\Theta(2^k b)$. Fürer takes both $b$ and $2^k$ on the scale of $\lg n$, reducing an integer multiplication of size $\Theta(n(\lg n)^2)$ to $\Theta((n \lg n)/\lg \lg n)$ integer multiplications of size $\Theta((\lg n)^2)$.

## 4. Product: zero-padding and localization

**4.1. Input.** Let $A$ be a commutative ring. Let $n$ be a positive integer. The algorithm in this section is given two elements $f, g$ of the polynomial ring $A[x]$ such that $\deg fg < n$: e.g., such that $n$ is the total number of coefficients in $f$ and $g$.

An element of $A[x]$ is, by convention, represented as a finite sequence of elements of $A$: the sequence $(f_0, f_1, \ldots, f_{d-1})$ represents $f_0 + f_1 x + \cdots + f_{d-1} x^{d-1}$.

**4.2. Output.** This algorithm computes the product $fg \in A[x]$.

**4.3. Speed.** The algorithm uses $O(n \lg n \lg \lg n)$ operations in $A$.

Equivalently: The algorithm uses at most $n\mu(n)$ operations in $A$, where $\mu : \mathbf{N} \to \mathbf{R}$ is a nondecreasing positive function with $\mu(n) \in O(\lg n \lg \lg n)$. The $\mu$ notation helps simplify the run-time analysis in subsequent sections of this paper.

**4.4. Special case: how it works if $A = \mathbf{C}$.** Given $f, g \in \mathbf{C}[x]$ such that $\deg fg < n$, one can compute $fg$ by using the algorithm of Section 2 to compute $fg \bmod (x^m - 1)$ in $\mathbf{C}[x]/(x^m - 1)$; here $m$ is the smallest power of 2 with $m \geq n$. This takes $O(m \lg m) = O(n \lg n)$ operations in $\mathbf{C}$.

For example, if $f = f_0 + f_1 x + f_2 x^2$ and $g = g_0 + g_1 x + g_2 x^2 + g_3 x^3$, use the algorithm of Section 2 to multiply the elements $f_0 + f_1 x + f_2 x^2 + 0 x^3 +$

$0x^4 + 0x^5 + 0x^6 + 0x^7$ and $g_0 + g_1 x + g_2 x^2 + g_3 x^3 + 0x^4 + 0x^5 + 0x^6 + 0x^7$ of $\mathbf{C}[x]/(x^8 - 1)$, obtaining $h_0 + h_1 x + h_2 x^2 + h_3 x^3 + h_4 x^4 + h_5 x^5 + 0x^6 + 0x^7$. Then $fg = h_0 + h_1 x + h_2 x^2 + h_3 x^3 + h_4 x^4 + h_5 x^5$. Appending zeros to an input — for example, converting $f_0, f_1, f_2$ to $f_0, f_1, f_2, 0, 0, 0, 0, 0$ — is called **zero-padding**.

In this special case $A = \mathbf{C}$, the aforementioned bound $\mu(n) \in O(\lg n \lg \lg n)$ is unnecessarily pessimistic: one can take $\mu(n) \in O(\lg n)$. Subsequent sections of this paper use the bound $\mu(n) \in O(\lg n \lg \lg n)$, and are correspondingly pessimistic.

Similar comments apply to other rings $A$ having appropriate roots of $-1$, and to nearby rings such as $\mathbf{R}$.

**4.5. Intermediate generality: how it works if 2 is invertible in $A$.** Let $A$ be any commutative ring in which 2 is invertible. Given $f, g \in A[x]$ with deg $fg < n$, one can compute $fg$ by using the algorithm of Section 3 to compute $fg$ mod $(x^m + 1)$ in $A[x]/(x^m + 1)$; here $m$ is the smallest power of 2 with $m \geq n$. This takes $O(m \lg m \lg \lg m) = O(n \lg n \lg \lg n)$ operations in $A$.

**4.6. Intermediate generality: how it works if 3 is invertible in $A$.** Let $A$ be any commutative ring in which 3 is invertible. The previous algorithm has a radix-3 variant that computes $fg$ using $O(n \lg n \lg \lg n)$ operations in $A$.

**4.7. Full generality: how it works for arbitrary rings.** What if neither 2 nor 3 is invertible? Answer: Map $A$ to the product of the localizations $2^{-N} A$ and $3^{-N} A$. This map is injective; 2 is invertible in $2^{-N} A$; and 3 is invertible in $3^{-N} A$.

In other words: Given polynomials $f, g$ over any commutative ring $A$, use the technique of Section 3 to compute $2^j fg$ for some $j$; use the radix-3 variant to compute $3^k fg$ for some $k$; and then compute $fg$ as a linear combination of $2^j fg$ and $3^k fg$. This takes $O(n \lg n \lg \lg n)$ operations in $A$ if deg $fg < n$.

Assume, for example, that deg $fg < 8$. Find $16 fg$ by computing $16 fg$ mod $(x^8 - 1)$, and find $9 fg$ by computing $9 fg$ mod $(x^{18} + x^9 + 1)$; then $fg = 4(16 fg) - 7(9 fg)$. The numbers 16 and 9 here are the denominators produced by the algorithm of Section 3.

**4.8. The integer case.** An analogous algorithm computes the product of two integers in time $O(n \lg n \lg \lg n)$, if the output size is known to be at most $n$ bits. (Given $f, g \in \mathbf{Z}$ with $|fg| < 2^n$, use the algorithm of Section 3 to compute $fg$ mod $(2^m + 1)$ in $\mathbf{Z}/(2^m + 1)$; here $m$ is the smallest power of 2 with $m \geq n+1$.)

Here an integer is, by convention, represented in **two's-complement notation**: a sequence of bits $(f_0, f_1, \ldots, f_{k-1}, f_k)$ represents $f_0 + 2 f_1 + \cdots + 2^{k-1} f_{k-1} - 2^k f_k$.

**4.9. History.** Karatsuba was the first to point out that integer multiplication can be done in subquadratic time; see [Karatsuba and Ofman 1963]. This result is often (e.g., in [Bürgisser et al. 1997, page 58]) incorrectly credited to Karatsuba and Ofman, but [Karatsuba and Ofman 1963, Theorem 2] explicitly credited the algorithm to Karatsuba alone.

Toom [1963] was the first to point out that integer multiplication can be done in essentially linear time: more precisely, time $n \exp(O(\sqrt{\log n}))$. Schönhage [1966] independently published the same observation a few years later. Cook [1966, page 53] commented that Toom's method could be used to quickly multiply polynomials over finite fields.

Stockham [1966, page 230] suggested zero-padding and FFT-based multiplication in $\mathbf{C}[x]/(x^n - 1)$ as a way to multiply in $\mathbf{C}[x]$.

The $O(n \lg n \lg \lg n)$ time bound for integers is usually credited to Schönhage and Strassen; see Section 3. Cantor and Kaltofen [1991] used $A \to 2^{-N}A \times 3^{-N}A$ to prove the $O(n \lg n \lg \lg n)$ operation bound for polynomials over any ring.

**4.10. Improvements.** The above algorithms take

- $(m/n)(9/2 + o(1))n \lg n$ operations in $\mathbf{C}$ to multiply in $\mathbf{C}[x]$; or
- $(m/n)(34/3 + o(1))n \lg n$ operations in $\mathbf{R}$ to multiply in $\mathbf{C}[x]$, using the tangent FFT; or
- $(m/n)(17/3 + o(1))n \lg n$ operations in $\mathbf{R}$ to multiply in $\mathbf{R}[x]$; or
- $(m/n)(3 + o(1))n \lg n \lg \lg n$ operations in any ring $A$ to multiply in $A[x]$, if 2 is invertible in $A$.

There are several ways to eliminate the $m/n$ factor here. One good way is to compute $fg$ modulo $x^m + 1$ for several powers $m$ of 2 with $\sum m \geq n$, then recover $fg$. For example, if $n = 80000$, one can recover $fg$ from $fg$ mod $(x^{65536} + 1)$ and $fg$ mod $(x^{16384} + 1)$. A special case of this technique was pointed out by Crandall and Fagin [1994, Section 7]. See [Bernstein 2001, Section 8] for an older technique.

One can save time at the beginning of the FFT when the input is known to be the result of zero-padding. For example, one does not need an operation to compute $f_0 + 0$. Similarly, one can save time at the end of the FFT when the output is known to have zeros: the zeros need not be recomputed.

In the context of FFT addition — for example, computing $ab + cd$ with only five transforms — the transform size does not need to be large enough for $ab$ and $cd$; it need only be large enough for $ab + cd$. This is useful in applications where $ab + cd$ is known to be small.

When $f$ has a substantially larger degree than $g$ (or vice versa), one can often save time by splitting $f$ into pieces of comparable size to $g$, and multiplying each piece by $g$. Similar comments apply in Section 7. In the polynomial case, this technique is most often called the "overlap-add method"; it was introduced by Stockham [1966, page 230] under the name "sectioning." The analogous technique for integers appears in [Knuth 1997, answer to Exercise 4.3.3–13] with credit to Schönhage.

See [Bernstein 2001] for a survey of further techniques.

## 5. Product: completion

**5.1. Input.** Let $A$ be a commutative ring. Let $n$ be a positive integer. The algorithm in this section is given the precision-$n$ representations of two elements $f, g$ of the power-series ring $A[\![x]\!]$.

The precision-$n$ representation of a power series $f \in A[\![x]\!]$ is, by definition, the polynomial $f \bmod x^n$. If $f = \sum_j f_j x^j$ then $f \bmod x^n = f_0 + f_1 x + \cdots + f_{n-1} x^{n-1}$. This polynomial is, in turn, represented in the usual way as its coefficient sequence $(f_0, f_1, \ldots, f_{n-1})$.

This representation does not carry complete information about $f$; it is only an approximation to $f$. It is nevertheless useful.

**5.2. Output.** This algorithm computes the precision-$n$ representation of the product $fg \in A[\![x]\!]$. If the input is $f_0, f_1, \ldots, f_{n-1}, g_0, g_1, \ldots, g_{n-1}$ then the output is $f_0 g_0, f_0 g_1 + f_1 g_0, f_0 g_2 + f_1 g_1 + f_2 g_0, \ldots, f_0 g_{n-1} + f_1 g_{n-2} + \cdots + f_{n-1} g_0$.

**5.3. Speed.** This algorithm uses $O(n \lg n \lg \lg n)$ operations in $A$: more precisely, at most $(2n-1)\mu(2n-1)$ operations in $A$.

**5.4. How it works.** Given $f \bmod x^n$ and $g \bmod x^n$, compute the polynomial product $(f \bmod x^n)(g \bmod x^n)$ by the algorithm of Section 4. Throw away the coefficients of $x^n, x^{n+1}, \ldots$ to obtain $(f \bmod x^n)(g \bmod x^n) \bmod x^n = fg \bmod x^n$.

For example, given the precision-3 representation $f_0, f_1, f_2$ of the series $f = f_0 + f_1 x + f_2 x^2 + \cdots$, and given the precision-3 representation $g_0, g_1, g_2$ of the series $g = g_0 + g_1 x + g_2 x^2 + \cdots$, first multiply $f_0 + f_1 x + f_2 x^2$ by $g_0 + g_1 x + g_2 x^2$ to obtain $f_0 g_0 + (f_0 g_1 + f_1 g_0) x + (f_0 g_2 + f_1 g_1 + f_2 g_0) x^2 + (f_1 g_2 + f_2 g_1) x^3 + f_2 g_2 x^4$; then throw away the coefficients of $x^3$ and $x^4$ to obtain $f_0 g_0, f_0 g_1 + f_1 g_0, f_0 g_2 + f_1 g_1 + f_2 g_0$.

**5.5. The integer case, easy completion: $\mathbf{Q} \to \mathbf{Q}_2$.** Consider the ring $\mathbf{Z}_2$ of 2-adic integers. The precision-$n$ representation of $f \in \mathbf{Z}_2$ is, by definition, the integer $f \bmod 2^n \in \mathbf{Z}$. This representation of elements of $\mathbf{Z}_2$ as nearby elements of $\mathbf{Z}$ is analogous in many ways to the representation of elements of $A[\![x]\!]$ as nearby elements of $A[x]$. In particular, there is an analogous multiplication algorithm: given $f$ mod $2^n$ and $g$ mod $2^n$, one can compute $fg$ mod $2^n$ in time $O(n \lg n \lg \lg n)$.

**5.6. The integer case, hard completion: Q → R.** Each real number $f \in \mathbf{R}$ is, by convention, represented as a nearby element of the localization $2^{-N}\mathbf{Z}$: an integer divided by a power of 2. If $|f| < 1$, for example, then there are one or two integers $d$ with $|d| \leq 2^n$ such that $|d/2^n - f| < 1/2^n$.

If another real number $g$ with $|g| < 1$ is similarly represented by an integer $e$ then $fg$ is *almost* represented by the integer $\lfloor de/2^n \rfloor$, which can be computed in time $O(n \lg n \lg \lg n)$. However, the distance from $fg$ to $\lfloor de/2^n \rfloor /2^n$ may be somewhat larger than $1/2^n$. This effect is called **roundoff error**: the output is known to slightly less precision than the input.

**5.7. History.** See [Knuth 1997, Section 4.1] for the history of positional notation.

**5.8. Improvements.** The coefficients of $x^n, x^{n+1}, \ldots$ in $fg$ are thrown away, so operations involved in multiplying $f$ mod $x^n$ by $g$ mod $x^n$ can be skipped if they are used only to compute those coefficients. The number of operations skipped depends on the multiplication method; optimizing $u, v \mapsto uv$ does not necessarily optimize $u, v \mapsto uv$ mod $x^n$. Similar comments apply to the integer case.

# 6. Reciprocal

**6.1. Input.** Let $A$ be a commutative ring. Let $n$ be a positive integer. The algorithm in this section is given the precision-$n$ representation of a power series $f \in A[\![x]\!]$ with $f(0) = 1$.

**6.2. Output.** This algorithm computes the precision-$n$ representation of the reciprocal $1/f = 1 + (1 - f) + (1 - f)^2 + (1 - f)^3 + \cdots \in A[\![x]\!]$. If the input is $1, f_1, f_2, f_3, \ldots, f_{n-1}$ then the output is $1, -f_1, f_1^2 - f_2, 2f_1 f_2 - f_1^3 - f_3, \ldots, \cdots - f_{n-1}$.

**6.3. Speed.** This algorithm uses $O(n \lg n \lg \lg n)$ operations in $A$: more precisely, at most $(8n + 2k - 8)\mu(2n - 1) + (2n + 2k - 2)$ operations in $A$ if $n \leq 2^k$.

**6.4. How it works.** If $n = 1$ then $(1/f)$ mod $x^n = 1$. There are 0 operations here; and $(8n + 2k - 8)\mu(2n - 1) + (2n + 2k - 2) = 2k\mu(1) + 2k \geq 0$ since $k \geq \lg n = 0$.

Otherwise define $m = \lceil n/2 \rceil$. Recursively compute $g_0 = (1/f)$ mod $x^m$; note that $m < n$. Then compute $(1/f)$ mod $x^n$ as $(g_0 - (fg_0 - 1)g_0)$ mod $x^n$, using the algorithm of Section 5 for the multiplications by $g_0$. This works because the difference $1/f - (g_0 - (fg_0 - 1)g_0)$ is exactly $f(1/f - g_0)^2$, which is a multiple of $x^{2m}$, hence of $x^n$.

For example, given the precision-4 representation $1 + f_1 x + f_2 x^2 + f_3 x^3$ of $f$, recursively compute $g_0 = (1/f) \bmod x^2 = 1 - f_1 x$. Multiply $f$ by $g_0$ modulo $x^4$ to obtain $1 + (f_2 - f_1^2)x^2 + (f_3 - f_1 f_2)x^3$. Subtract 1 and multiply by $g_0$ modulo $x^4$ to obtain $(f_2 - f_1^2)x^2 + (f_3 + f_1^3 - 2f_1 f_2)x^3$. Subtract from $g_0$ to obtain $1 - f_1 x + (f_1^2 - f_2)x^2 + (2f_1 f_2 - f_1^3 - f_3)x^3$. This is the precision-4 representation of $1/f$.

The proof of speed is straightforward. By induction, the recursive computation uses at most $(8m + 2(k-1) - 8)\mu(2m - 1) + (2m + 2(k-1) - 2)$ operations in $A$, since $m \leq 2^{k-1}$. The subtraction from $g_0$ and the subtraction of 1 use at most $n + 1$ operations in $A$. The two multiplications by $g_0$ use at most $2(2n - 1)\mu(2n - 1)$ operations in $A$. Apply the inequalities $m \leq (n+1)/2$ and $\mu(2m - 1) \leq \mu(2n - 1)$ to see that the total is at most $(8n + 2k - 8)\mu(2n - 1) + (2n + 2k - 2)$ as claimed.

### 6.5. The integer case, easy completion: $\mathbf{Q} \to \mathbf{Q}_2$. Let $f \in \mathbf{Z}_2$ be an odd 2-adic integer. Then $f$ has a reciprocal $1/f = 1 + (1 - f) + (1 - f)^2 + \cdots \in \mathbf{Z}_2$.

One can compute $(1/f) \bmod 2^n$, given $f \bmod 2^n$, by applying the same formula as in the power-series case: first recursively compute $g_0 = (1/f) \bmod 2^{\lceil n/2 \rceil}$; then compute $(1/f) \bmod 2^n$ as $(g_0 + (1 - fg_0)g_0) \bmod 2^n$. This takes time $O(n \lg n \lg \lg n)$.

### 6.6. The integer case, hard completion: $\mathbf{Q} \to \mathbf{R}$. Let $f \in \mathbf{R}$ be a real number between 0.25 and 1. Then $f$ has a reciprocal $g = 1 + (1 - f) + (1 - f)^2 + \cdots \in \mathbf{R}$. If $g_0$ is a close approximation to $1/f$, then $g_0 + (1 - fg_0)g_0$ is an approximation to $1/f$ with *nearly* twice the precision. Consequently one can compute a precision-$n$ representation of $1/f$, given a slightly higher-precision representation of $f$, in time $O(n \lg n \lg \lg n)$.

The details are, thanks to roundoff error, more complicated than in the power-series case, and are not included in this paper. See [Knuth 1997, Algorithm 4.3.3–R] or [Bernstein 1998, Section 8] for a complete algorithm.

### 6.7. History. Simpson [1740, page 81] presented the iteration $g \mapsto g - (fg - 1)g$ for reciprocals. Simpson also commented that one can carry out the second-to-last iteration at about 1/2 the desired precision, the third-to-last iteration at about 1/4 the desired precision, etc., so that the total time is comparable to the time for the last iteration. I have not been able to locate earlier use of this iteration.

Simpson considered, more generally, the iteration $g \mapsto g - p(g)/p'(g)$ for roots of a function $p$. The iteration $g \mapsto g - (fg - 1)g$ is the case $p(g) = g^{-1} - f$. The general case is usually called "Newton's method," but I see no evidence that Newton deserves credit for it. Newton used the iteration for polynomials

$p$, but so did previous mathematicians. Newton's descriptions never mentioned derivatives and were not amenable to generalization. See [Kollerstrom 1992] and [Ypma 1995] for further discussion.

Cook [1966, pages 81–86] published details of a variable-precision reciprocal algorithm for $\mathbf{R}$ taking essentially linear time, using the iteration $g \mapsto g - (fg - 1)g$ with Toom's multiplication algorithm. Sieveking [1972], apparently unaware of Cook's result, published details of an analogous reciprocal algorithm for $A[\![x]\!]$. The analogy was pointed out by Kung [1974].

## 6.8. Improvements.
Computing a reciprocal by the above algorithm takes $4 + o(1)$ times as many operations as computing a product. There are several ways that this constant 4 can be reduced. The following discussion focuses on $\mathbf{C}[\![x]\!]$ and assumes that $n$ is a power of 2. Analogous comments apply to other values of $n$; to $A[\![x]\!]$ for other rings $A$; to $\mathbf{Z}_2$; and to $\mathbf{R}$.

One can achieve $3 + o(1)$ by skipping some multiplications by low zeros. The point is that that $fg_0 - 1$ is a multiple of $x^m$. Write $u = ((fg_0 - 1) \bmod x^n)/x^m$ and $v = g_0 \bmod x^{n-m}$; then $u$ and $v$ are polynomials of degree below $n - m$, and $((fg_0 - 1)g_0) \bmod x^n = x^m uv \bmod x^n$. One can compute $uv \bmod x^n - 1$, extract the bottom $n - m$ coefficients of the product, and insert $m$ zeros, to obtain $((fg_0 - 1)g_0) \bmod x^n$.

One can achieve $2 + o(1)$ by skipping some multiplications by high zeros and by not recomputing a stretch of known coefficients. To compute $fg_0 \bmod x^n$, one multiplies $f \bmod x^n$ by $g_0$ and extracts the bottom $n$ coefficients. The point is that $(f \bmod x^n)g_0 - 1$ is a multiple of $x^m$, and has degree at most $m + n$, so it is easily computed from its remainder modulo $x^n - 1$: it has $m$ zeros, then the top $n - m$ coefficients of the remainder, then the bottom $n - m$ coefficients of the remainder.

One can achieve $5/3 + o(1)$ by applying FFT caching. There is a multiplication of $f \bmod x^n$ by $g_0$ modulo $x^n - 1$, and a multiplication of $(fg_0 - 1) \bmod x^n$ by $g_0$ modulo $x^n - 1$; the transform of $g_0$ can be reused rather than recomputed.

One can achieve $3/2 + o(1)$ by evaluating a cubic rather than two quadratics. The polynomial $((f \bmod x^n)g_0 - 1)g_0$ is a multiple of $x^m$ and has degree below $n + 2m$, so it is easily computed from its remainders modulo $x^n + 1$ and $x^m - 1$. One transforms $f \bmod x^n$, transforms $g_0$, multiplies the first transform by the square of the second, subtracts the second, and untransforms the result.

Brent [1976c] published $3 + o(1)$. Schönhage, Grotefeld, and Vetter [1994, page 256] announced $2 + o(1)$ without giving details. I published $28/15 + o(1)$ in 1998, and $3/2 + o(1)$ in 2000, with a rather messy algorithm; see [Bernstein 2004c]. Schönhage [2000] independently achieved $3/2 + o(1)$ with the simpler algorithm shown above.

## 7. Quotient

**7.1. Input.** Let $A$ be a commutative ring. Let $n$ be a positive integer. The algorithm in this section is given the precision-$n$ representations of power series $f, h \in A[\![x]\!]$ such that $f(0) = 1$.

**7.2. Output.** This algorithm computes the precision-$n$ representation of $h/f \in A[\![x]\!]$. If the input is $1, f_1, f_2, \ldots, f_{n-1}, h_0, h_1, h_2, \ldots, h_{n-1}$ then the output is

$$h_0,$$
$$h_1 - f_1 h_0,$$
$$h_2 - f_1 h_1 + (f_1^2 - f_2) h_0,$$
$$\vdots$$
$$h_{n-1} - \cdots + (\cdots - f_{n-1}) h_0.$$

**7.3. Speed.** This algorithm uses $O(n \lg n \lg \lg n)$ operations in $A$: more precisely, at most $(10n + 2k - 9)\mu(2n - 1) + (2n + 2k - 2)$ operations in $A$ if $n \le 2^k$.

**7.4. How it works.** First compute a precision-$n$ approximation to $1/f$ as explained in Section 6. Then multiply by $h$ as explained in Section 5.

**7.5. The integer case, easy completion: Q → $\mathbf{Q}_2$.** Let $h$ and $f$ be elements of $\mathbf{Z}_2$ with $f$ odd. Given $f \bmod 2^n$ and $h \bmod 2^n$, one can compute $(h/f) \bmod 2^n$ in time $O(n \lg n \lg \lg n)$ by the same method.

**7.6. The integer case, hard completion: Q → R.** Let $h$ and $f$ be elements of $\mathbf{R}$ with $0.5 \le f \le 1$. One can compute a precision-$n$ representation of $h/f$, given slightly higher-precision representations of $f$ and $h$, in time $O(n \lg n \lg \lg n)$. As usual, roundoff error complicates the algorithm.

**7.7. Improvements.** One can improve the number of operations for a reciprocal to $3/2 + o(1)$ times the number of operations for a product, as discussed in Section 6, so one can improve the number of operations for a quotient to $5/2 + o(1)$ times the number of operations for a product.

The reader may be wondering at this point why quotient deserves to be discussed separately from reciprocal. Answer: Further improvements are possible. Karp and Markstein [1997] pointed out that a quotient computation could profitably avoid some of the work in a reciprocal computation. I achieved a gap of $2.6/3 + o(1)$ in 1998 and $2/3 + o(1)$ in 2000, combining the Karp–Markstein idea with some FFT reuse; see [Bernstein 2004c]. In 2004, Hanrot and Zimmermann announced a gap of $1.75/3 + o(1)$: i.e., the number of operations for a quotient is $6.25/3 + o(1)$ times the number of operations for a product. More recently, Joris van der Hoeven [2006] announced $5/3 + o(1)$.

## 8. Logarithm: the series case

**8.1. Input.** Let $A$ be a commutative ring containing $\mathbf{Q}$. Let $n$ be a positive integer. The algorithm in this section is given the precision-$n$ representation of a power series $f \in A[\![x]\!]$ with $f(0) = 1$.

**8.2. Output.** This algorithm computes the precision-$n$ representation of the series $\log f = -(1-f) - (1-f)^2/2 - (1-f)^3/3 - \cdots \in A[\![x]\!]$. If the input is $1, f_1, f_2, f_3, \ldots$ then the output is $0, f_1, f_2 - f_1^2/2, f_3 - f_1 f_2 + f_1^3/3, \ldots$.

Define $D(\sum a_j x^j) = \sum j a_j x^j$. The reader may enjoy checking the following properties of log and $D$:

- $D(fg) = gD(f) + fD(g)$;
- $D(g^n) = ng^{n-1}D(g)$;
- if $f(0) = 1$ then $D(\log f) = D(f)/f$;
- if $f(0) = 1$ and $\log f = 0$ then $f = 1$;
- if $f(0) = 1$ and $g(0) = 1$ then $\log fg = \log f + \log g$;
- log is injective: i.e., if $f(0) = 1$ and $g(0) = 1$ and $\log f = \log g$ then $f = g$.

**8.3. Speed.** This algorithm uses $O(n \lg n \lg \lg n)$ operations in $A$: more precisely, at most $(10n + 2k - 9)\mu(2n - 1) + (4n + 2k - 4)$ operations in $A$ if $n \leq 2^k$.

**8.4. How it works.** Given $f \bmod x^n$, compute $D(f) \bmod x^n$ from the definition of $D$; compute $(D(f)/f) \bmod x^n$ as explained in Section 7; and recover $(\log f) \bmod x^n$ from the formula $D((\log f) \bmod x^n) = (D(f)/f) \bmod x^n$.

**8.5. The integer case.** This $A[\![x]\!]$ algorithm does not have a useful analogue for $\mathbf{Z}_2$ or $\mathbf{R}$, because $\mathbf{Z}_2$ and $\mathbf{R}$ do not have adequate replacements for the differential operator $D$. See, however, Section 16.

**8.6. History.** This algorithm was published by Brent [1976c, Section 13].

**8.7. Improvements.** See Section 7 for improved quotient algorithms. I do not know any way to compute $\log f$ more quickly than computing a generic quotient.

## 9. Exponential: the series case

**9.1. Input.** Let $A$ be a commutative ring containing $\mathbf{Q}$. Let $n$ be a positive integer. The algorithm in this section is given the precision-$n$ representation of a power series $f \in A[\![x]\!]$ with $f(0) = 0$.

**9.2. Output.** This algorithm computes the precision-$n$ representation of the series $\exp f = 1 + f + f^2/2! + f^3/3! + \cdots \in A[\![x]\!]$. If the input is $0, f_1, f_2, f_3, \ldots$ then the output is $1, f_1, f_2 + f_1^2/2, f_3 + f_1 f_2 + f_1^3/6, \ldots$.

The reader may enjoy checking the following properties of exp:

- if $f(0) = 0$ then $D(\exp f) = D(f) \exp f$;
- if $f(0) = 0$ then $\log \exp f = f$;
- if $g(0) = 1$ then $\exp \log g = g$;
- if $f(0) = 0$ and $g(0) = 0$ then $\exp(f + g) = (\exp f) \exp g$.

**9.3. Speed.** This algorithm uses $O(n \lg n \lg \lg n)$ operations in $A$: more precisely, at most $(24n + k^2 + 3k - 24)\mu(2n-1) + (12n + k^2 + 3k - 12)$ operations in $A$ if $n \leq 2^k$.

**9.4. How it works.** If $n = 1$ then $(\exp f) \bmod x^n = 1$. Otherwise define $m = \lceil n/2 \rceil$. Recursively compute $g_0 = (\exp f) \bmod x^m$. Compute $(\log g_0) \bmod x^n$ as explained in Section 8. Then compute $(\exp f) \bmod x^n$ as $(g_0 + (f - \log g_0)g_0) \bmod x^n$. This works because $\exp(f - \log g_0) - 1 - (f - \log g_0)$ is a multiple of $(f - \log g_0)^2$, hence of $x^{2m}$, hence of $x^n$.

The recursive step uses at most $(24(n+1)/2 + (k-1)^2 + 3(k-1) - 24)\mu(2n-1) + (12(n+1)/2 + (k-1)^2 + 3(k-1) - 12)$ operations by induction. The computation of $(\log g_0) \bmod x^n$ uses at most $(10n + 2k - 9)\mu(2n-1) + (4n + 2k - 4)$ operations. The subtraction from $f$ and the addition of $g_0$ use at most $2n$ operations. The multiplication by $g_0$ uses at most $(2n-1)\mu(2n-1)$ operations. The total is at most $(24n + k^2 + 3k - 24)\mu(2n-1) + (12n + k^2 + 3k - 12)$ as claimed.

**9.5. The integer case.** See Section 16.

**9.6. History.** The iteration $g \mapsto g + (f - \log g)g$ is an example of "Newton's method," i.e., Simpson's method.

Brent [1976c, Section 13] pointed out that this is a particularly efficient way to compute exp for $\mathbf{R}[\![x]\!]$, since log is so easy to compute for $\mathbf{R}[\![x]\!]$.

**9.7. Improvements.** Brent [1976c, Section 13] stated that the number of operations for an exponential in $\mathbf{R}[\![x]\!]$ could be improved to $22/3 + o(1)$ times the number of operations for a product. In fact, one can achieve $8.5/3 + o(1)$; see [Bernstein 2004c]. More recently, Joris van der Hoeven [2006] announced $7/3 + o(1)$.

## 10. Power: the series case

**10.1. Input.** Let $A$ be a commutative ring containing $\mathbf{Q}$. Let $n$ be a positive integer. The algorithm in this section is given the precision-$n$ representations of power series $f, e \in A[\![x]\!]$ such that $f(0) = 1$.

**10.2. Output.** This algorithm computes the precision-$n$ representation of the series $f^e = \exp(e \log f) \in A[\![x]\!]$. If the input is $1, f_1, f_2, \ldots, e_0, e_1, \ldots$ then the output is $1, e_0 f_1, e_1 f_1 + e_0 f_2 + e_0(e_0 - 1) f_1^2/2, \ldots$.

The reader may enjoy checking the following properties of $f, e \mapsto f^e$:

- $f^0 = 1$;
- $f^1 = f$;
- $f^{d+e} = f^d \cdot f^e$, so the notation $f^e$ for $\exp(e \log f)$ is, for positive integers $e$, consistent with the usual notation $f^e$ for $\prod_{1 \le j \le e} f$;
- $f^{-1} = 1/f$;
- $(f^d)^e = f^{de}$;
- $(fg)^e = f^e g^e$;
- $D(f^e) = D(e) f^e \log f + D(f) e f^{e-1}$;
- $f^e = 1 + e(f - 1) + (e(e-1)/2)(f - 1)^2 + \cdots$.

**10.3. Speed.** This algorithm uses $O(n \lg n \lg \lg n)$ operations in $A$: more precisely, at most $(36n + k^2 + 5k - 34)\mu(2n - 1) + (16n + k^2 + 5k - 16)$ operations in $A$ if $n \le 2^k$.

**10.4. How it works.** Given $f \bmod x^n$, compute $(\log f) \bmod x^n$ as explained in Section 8; compute $(e \log f) \bmod x^n$ as explained in Section 5; compute $(\exp(e \log f)) \bmod x^n$ as explained in Section 9.

**10.5. The integer case.** See Section 16.

**10.6. History.** According to various sources, Napier introduced the functions exp and log for $\mathbf{R}$, along with the idea of using exp and log to compute products in $\mathbf{R}$. I do not know the history of exp and log for $\mathbf{Z}_2$ and $A[\![x]\!]$.

**10.7. Improvements.** As in Sections 6, 7, and 9, one can remove some redundancy from the above algorithm. See [Bernstein 2004c].

Brauer [1939] pointed out that, if $e$ is a positive integer, one can compute $f^e$ with about $\lg e$ squarings and at most about $(\lg e)/\lg \lg e$ other multiplications. This is faster than the exp-log algorithm if $e$ is small. See [Bernstein 2002b] for further discussion of square-and-multiply exponentiation algorithms.

One can compute $f^e$ for any rational number $e$ with a generalization of the algorithm of Section 6. This takes essentially linear time for fixed $e$, as pointed out by Cook [1966, page 86]; it is faster than the exp-log algorithm if the height of $e$ is small, i.e., the numerator and denominator of $e$ are small. The special case $e = 1/2$ — i.e., square roots — is discussed in detail in [Bernstein 2004c].

## 11. Matrix product

**11.1. Input.** Let $A$ be a commutative ring. The algorithm in this section is given two $2 \times 2$ matrices $F = \begin{pmatrix} F_{11} & F_{12} \\ F_{21} & F_{22} \end{pmatrix}$ and $G = \begin{pmatrix} G_{11} & G_{12} \\ G_{21} & G_{22} \end{pmatrix}$ with entries in the polynomial ring $A[x]$.

**11.2. Output.** This algorithm computes the $2 \times 2$ matrix product $FG$.

**11.3. Speed.** This algorithm uses $O(n \lg n \lg \lg n)$ operations in $A$, where $n$ is the total number of input coefficients. More precisely, the algorithm uses at most $n(2\mu(n) + 2)$ operations in $A$. This bound is pessimistic.

Here, and elsewhere in this paper, **number of coefficients** means the number of elements of $A$ provided as input. Reader beware: the number of coefficients of an input polynomial is not determined by the polynomial; it depends on how the polynomial is represented. For example, the sequence $(5, 7, 0)$, with 3 coefficients, represents the same polynomial as the sequence $(5, 7)$, with 2 coefficients.

**11.4. How it works.** Multiply $F_{11}$ by $G_{11}$, multiply $F_{12}$ by $G_{21}$, add, etc., to obtain $FG = \begin{pmatrix} F_{11}G_{11} + F_{12}G_{21} & F_{11}G_{12} + F_{12}G_{22} \\ F_{21}G_{11} + F_{22}G_{21} & F_{21}G_{12} + F_{22}G_{22} \end{pmatrix}$.

**11.5. The integer case.** An analogous algorithm computes the product of two $2 \times 2$ matrices with entries in $\mathbf{Z}$ in time $O(n \lg n \lg \lg n)$, where $n$ is the number of input bits.

**11.6. History.** The matrix concept is generally credited to Sylvester and Cayley.

**11.7. Improvements.** The above algorithm involves 24 transforms. FFT caching — transforming each of the input polynomials $F_{11}, F_{12}, F_{21}, F_{22}, G_{11}, G_{12}, G_{21}, G_{22}$ just once — saves 8 transforms. FFT addition — untransforming $F_{11}G_{11} + F_{12}G_{21}$, for example, rather than separately untransforming $F_{11}G_{11}$ and $F_{12}G_{21}$ — saves 4 more transforms.

Strassen [1969] published a method to multiply $2 \times 2$ matrices using just 7 multiplications of entries and 18 additions or subtractions of entries, rather than 8 multiplications and 4 additions. Winograd observed that 18 could be replaced by 15; see, e.g., [Knuth 1997, page 500].

Many applications involve matrices of particular shapes: for example, matrices $F$ in which $F_{12} = 0$. One can often save time accordingly.

**11.8. Generalization: larger matrices.** Strassen [1969] published a general method to multiply $d \times d$ matrices using $O(d^\alpha)$ multiplications, additions, and subtractions of entries; here $\alpha = \log_2 7 = 2.807\ldots$. Subsequent work by Pan, Bini, Capovani, Lotti, Romani, Schönhage, Coppersmith, and Winograd showed that there is an algorithm to multiply $d \times d$ matrices using $d^{\beta+o(1)}$ multiplications and additions of entries, for a certain number $\beta < 2.38$. See [Bürgisser et al. 1997, Chapter 15] for a detailed exposition and further references.

It is not known whether matrix multiplication can be carried out in essentially linear time, when the matrix size is a variable.

## 12. Product tree

**12.1. Input.** Let $A$ be a commutative ring. Let $t$ be a nonnegative integer. The algorithm in this section is given $2 \times 2$ matrices $M_1, M_2, \ldots, M_t$ with entries in $A[x]$.

**12.2. Output.** This algorithm computes the **product tree** of $M_1, M_2, \ldots, M_t$, which is defined as follows. The root of the tree is the $2 \times 2$ matrix $M_1 M_2 \cdots M_t$. If $t \leq 1$ then that's the complete tree. If $t \geq 2$ then the left subtree is the product tree of $M_1, M_2, \ldots, M_s$, and the right subtree is the product tree of $M_{s+1}, M_{s+2}, \ldots, M_t$, where $s = \lceil t/2 \rceil$. For example, here is the product tree of $M_1, M_2, M_3, M_4, M_5, M_6$:

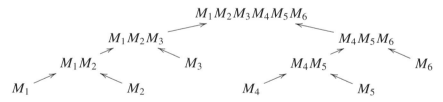

Most applications use only the root $M_1 M_2 \cdots M_t$ of the product tree. This root is often described in the language of linear recurrences as follows. Define $X_i = M_1 M_2 \cdots M_i$; then $X_i = X_{i-1} M_i$, i.e., $X_{i,j,k} = X_{i-1,j,0} M_{i,0,k} + X_{i-1,j,1} M_{i,1,k}$. The algorithm computes $X_{t,0,0}, X_{t,0,1}, X_{t,1,0}, X_{t,1,1}$, given the coefficients $M_{i,j,k}$ of the linear recurrence $X_{i,j,k} = X_{i-1,j,0} M_{i,0,k} + X_{i-1,j,1} M_{i,1,k}$, with the starting condition $(X_{0,0,0}, X_{0,0,1}, X_{0,1,0}, X_{0,1,1}) = (1, 0, 0, 1)$.

**12.3. Speed.** This algorithm uses $O(n(\lg n)^2 \lg \lg n)$ operations in $A$, where $n$ is the total number of coefficients in $M_1, M_2, \ldots, M_t$: more precisely, at most $nk(2\mu(n) + 2)$ operations in $A$ if $k$ is a nonnegative integer and $t \leq 2^k$.

**12.4. How it works.** If $t = 0$ then the answer is the identity matrix. If $t = 1$ then the answer is $M_1$. Otherwise recursively compute the product tree of $M_1, M_2, \ldots, M_s$ and the product tree of $M_{s+1}, M_{s+2}, \ldots, M_t$, where $s = \lceil t/2 \rceil$.

Multiply the roots $M_1 M_2 \cdots M_s$ and $M_{s+1} \cdots M_t$, as discussed in Section 11, to obtain $M_1 M_2 \cdots M_t$.

Time analysis: Define $m$ as the total number of coefficients in $M_1, \ldots, M_s$. By induction, the computation of the product tree of $M_1, \ldots, M_s$ uses at most $m(k-1)(2\mu(n)+2)$ operations, and the computation of the product tree of $M_{s+1}, \ldots, M_t$ uses at most $(n-m)(k-1)(2\mu(n)+2)$ operations. The final multiplication uses at most $n(2\mu(n)+2)$ operations. Add: $m(k-1) + (n-m)(k-1) + n = nk$.

**12.5. The integer case.** An analogous algorithm takes time $O(n(\lg n)^2 \lg \lg n)$ to compute the product tree of a sequence $M_1, M_2, \ldots, M_t$ of $2 \times 2$ matrices with entries in $\mathbf{Z}$. Here $n$ is the total number of input bits.

**12.6. Generalization: larger matrices.** One can use the same method to compute a product of several $d \times d$ matrices — in other words, to compute terms in linear recurrences of any order. It is not known whether this can be done in essentially linear time for variable $d$; see Section 11 for further comments.

**12.7. History.** Product trees are so simple and so widely applicable that they have been reinvented many times. They are not only a basic tool in the context of fast multiplication but also a basic tool for building low-depth parallel algorithms.

Unfortunately, most authors state product trees for particular applications, with no hint of the generality of the technique. They define ad-hoc product operations, and prove associativity of their product operations from scratch, never realizing that these operations are special cases of matrix product.

Weinberger and Smith [1958] published the "carry-lookahead adder," a low-depth parallel circuit for computing the sum of two nonnegative integers, with the inputs and output represented in the usual way as bit strings. This circuit computes (in different language) a product

$$\begin{pmatrix} a_1 & 0 \\ b_1 & 1 \end{pmatrix} \begin{pmatrix} a_2 & 0 \\ b_2 & 1 \end{pmatrix} \cdots \begin{pmatrix} a_t & 0 \\ b_t & 1 \end{pmatrix} = \begin{pmatrix} a_1 a_2 \cdots a_t & 0 \\ b_t + b_{t-1} a_t + b_{t-2} a_{t-1} a_t + \cdots + b_1 a_2 \cdots a_t & 1 \end{pmatrix}$$

of matrices over the Boole algebra $\{0, 1\}$ by multiplying pairs of matrices in parallel, then multiplying pairs of pairs in parallel, and so on for approximately $\lg t$ steps.

Estrin [1960] published a low-depth parallel algorithm for evaluating a one-variable polynomial. Estrin's algorithm computes (in different language) a product

$$\begin{pmatrix} a & 0 \\ b_1 & 1 \end{pmatrix} \begin{pmatrix} a & 0 \\ b_2 & 1 \end{pmatrix} \cdots \begin{pmatrix} a & 0 \\ b_t & 1 \end{pmatrix} = \begin{pmatrix} a^t & 0 \\ b_t + b_{t-1} a + b_{t-2} a^2 + \cdots + b_1 a^{t-1} & 1 \end{pmatrix}$$

by multiplying pairs, pairs of pairs, etc.

Schönhage, as reported in [Knuth 1971a, Exercise 4.4–13], pointed out that one can convert integers from base 10 to base 2 in essentially linear time. Schönhage's algorithm computes (in different language) a product

$$\begin{pmatrix} 10 & 0 \\ b_1 & 1 \end{pmatrix} \begin{pmatrix} 10 & 0 \\ b_2 & 1 \end{pmatrix} \cdots \begin{pmatrix} 10 & 0 \\ b_t & 1 \end{pmatrix} = \begin{pmatrix} 10^t & 0 \\ b_t + 10b_{t-1} + 100b_{t-2} + \cdots + 10^{t-1}b_1 & 1 \end{pmatrix}$$

by multiplying pairs of matrices, then pairs of pairs, etc.

Knuth [1971b, Theorem 1] published an algorithm to convert a continued fraction to a fraction in essentially linear time. Knuth's algorithm is (in different language) another example of the product-tree algorithm; see Section 14 for details.

Moenck and Borodin [1972, page 91] pointed out that one can compute the product tree — and thus the product — of a sequence of polynomials or a sequence of integers in essentially linear time; see also [Borodin and Moenck 1974, page 372]. Beware that the Moenck–Borodin "theorems" assume that all of the inputs are "single precision"; it is unclear what this is supposed to mean for integers.

Moenck and Borodin also pointed out an algorithm to add fractions in essentially linear time. This algorithm is (in different language) yet another example of the product-tree algorithm. See Section 13 for details and further historical notes.

Meanwhile, in the context of parallel algorithms, Stone and Kogge published the product-tree algorithm in a reasonable level of generality, with polynomial evaluation and continued-fraction-to-fraction conversion ("tridiagonal-linear-system solution") as examples. See [Stone 1973], [Kogge and Stone 1973], and [Kogge 1974]. Stone commented that linear recurrences of any order could be phrased as matrix products — see [Stone 1973, page 34] and [Stone 1973, page 37] — but, unfortunately, made little use of matrices elsewhere in his presentation.

Kogge and Stone [1973, page 792] credited Robert Downs, Harvard Lomax, and H. R. G. Trout for independent discoveries of general product-tree algorithms. They also stated that special cases of the algorithm were "known to J. J. Sylvester as early as 1853"; but I see no evidence that Sylvester ever formed a product tree in that context or any other context. Sylvester [1853] (cited in [Knuth 1971b] and [Stone 1973]) simply pointed out the associativity of continued fractions.

Brent [1976a, Section 6] pointed out that the numerator and denominator of $1 + 1/2 + 1/3! + \cdots + 1/t! \approx \exp 1$ could be computed quickly. Brent's algorithm formed (in different language) a product tree for

$$\begin{pmatrix} 1 & 0 \\ 1 & 1 \end{pmatrix} \begin{pmatrix} 2 & 0 \\ 1 & 1 \end{pmatrix} \cdots \begin{pmatrix} t & 0 \\ 1 & 1 \end{pmatrix} = \begin{pmatrix} t! & 0 \\ t! + t!/2 + \cdots + t(t-1) + t + 1 & 1 \end{pmatrix}.$$

Brent also addressed exp for more general inputs, as discussed in Section 15; and $\pi$, via arctan. Brent described his method as a mixed-radix adaptation of Schönhage's base-conversion algorithm. Evidently he had in mind the product

$$\begin{pmatrix} a_1 & 0 \\ b_1 & c \end{pmatrix} \begin{pmatrix} a_2 & 0 \\ b_2 & c \end{pmatrix} \cdots \begin{pmatrix} a_t & 0 \\ b_t & c \end{pmatrix} = \begin{pmatrix} a_1 a_2 \cdots a_t & 0 \\ c^{t-1} b_t + c^{t-2} b_{t-1} a_t + \cdots + b_1 a_2 \cdots a_t & c^t \end{pmatrix}$$

corresponding to the sum $\sum_{1 \le k \le t} c^{k-1} b_k / a_1 \cdots a_k$. Brent and McMillan [1980, page 308] mentioned that the sum $\sum_{1 \le k \le t} n^k (-1)^{k-1} / k! k$ could be handled similarly.

I gave a reasonably general statement of the product-tree algorithm in [Bernstein 1987], with a few series and continued fractions as examples. I pointed out that computing $M_1 M_2 \cdots M_t$ takes time $O(t (\lg t)^3 \lg \lg t)$ in the common case that the entries of $M_j$ are bounded by polynomials in $j$.

Gosper [1990] presented a wide range of illustrative examples of matrix products, emphasizing their "notational, analytic, and computational virtues." Gosper then [1990, page 263] gave a brief statement of the product-tree algorithm and credited it to Rich Schroeppel.

Chudnovsky and Chudnovsky [1990, pages 115–118] stated the product-tree algorithm for matrices $M_j$ whose entries depend rationally on $j$. They gave a broad class of series as examples in [Chudnovsky and Chudnovsky 1990, pages 123–134]. They called the algorithm "a well-known method to accelerate the (numerical) solution of linear recurrences."

Karatsuba used product trees (in different language) to evaluate various sums in several papers starting in 1991 and culminating in [Karatsuba 1999].

See [Haible and Papanikolaou 1997], [van der Hoeven 1999], [Borwein et al. 2000, Section 7], and [van der Hoeven 2001] for further uses of product trees to evaluate sums.

**12.8. Improvements.** One can change $s$ in the definition of a product tree and in the product-tree algorithm. The choice $s = \lceil t/2 \rceil$, balancing $s$ against $t - s$, is not necessarily the fastest way to compute $M_1 M_2 \cdots M_t$: when $M_1, M_2, \ldots, M_t$ have widely varying degrees, it is much better to balance $\deg M_1 + \deg M_2 + \cdots + \deg M_s$ against $\deg M_{s+1} + \deg M_{s+2} + \cdots + \deg M_t$. Strassen [1983, Theorem 3.2] proved that a slightly more complicated strategy is within a constant factor of optimal.

In some applications, $M_1, M_2, \ldots, M_t$ are known to commute. One can often permute $M_1, M_2, \ldots, M_t$ for slightly higher speed. Strassen [1983, Theorem

2.2] pointed out a particularly fast, and pleasantly simple, algorithm: find the two matrices of smallest degree, replace them by their product, and repeat. See [Bürgisser et al. 1997, Section 2.3] for an exposition.

Robert Kramer has recently pointed out another product-tree speedup. Suppose, as an illustration, that $M_1, M_2, M_3, M_4$ each have degree close to $n$. To multiply $M_1$ by $M_2$, one applies a size-$2n$ transform to each of $M_1$ and $M_2$, multiplies the transforms, and untransforms the result. To multiply $M_1 M_2$ by $M_3 M_4$, one starts by applying a size-$4n$ transform to $M_1 M_2$. Kramer's idea, which I call **FFT doubling**, is that the first half of the size-$4n$ transform of $M_1 M_2$ is exactly the size-$2n$ transform of $M_1 M_2$, which is already known. This idea saves two halves of every three transforms in a large balanced product-tree computation.

## 13. Sum of fractions

**13.1. Input.** Let $A$ be a commutative ring. Let $t$ be a positive integer. The algorithm in this section is given $2t$ polynomials $f_1, g_1, f_2, g_2, \ldots, f_t, g_t \in A[x]$.

**13.2. Output.** This algorithm computes $h = f_1 g_2 \cdots g_t + g_1 f_2 \cdots g_t + \cdots + g_1 g_2 \cdots f_t$, along with $g_1 g_2 \cdots g_t$.

The reader may think of this output as follows: the algorithm computes the sum $h/g_1 g_2 \cdots g_t$ of the fractions $f_1/g_1, f_2/g_2, \ldots, f_t/g_t$. The equation $h/g_1 g_2 \cdots g_t = f_1/g_1 + f_2/g_2 + \cdots + f_t/g_t$ holds in any $A[x]$-algebra where $g_1, g_2, \ldots, g_t$ are invertible: in particular, in the localization $g_1^{-N} \cdots g_t^{-N} A[x]$.

**13.3. Speed.** This algorithm uses $O(n(\lg n)^2 \lg \lg n)$ operations in $A$, where $n$ is the total number of coefficients in the input polynomials.

**13.4. How it works.** The matrix product $\begin{pmatrix} g_1 & f_1 \\ 0 & g_1 \end{pmatrix} \begin{pmatrix} g_2 & f_2 \\ 0 & g_2 \end{pmatrix} \cdots \begin{pmatrix} g_t & f_t \\ 0 & g_t \end{pmatrix}$ is exactly $\begin{pmatrix} g_1 g_2 \cdots g_t & h \\ 0 & g_1 g_2 \cdots g_t \end{pmatrix}$. Compute this product as described in Section 12.

The point is that adding fractions $a/b$ and $c/d$ to obtain $(ad+bc)/bd$ is the same as multiplying matrices $\begin{pmatrix} b & a \\ 0 & b \end{pmatrix}$ and $\begin{pmatrix} d & c \\ 0 & d \end{pmatrix}$ to obtain $\begin{pmatrix} bd & ad+bc \\ 0 & bd \end{pmatrix}$.

Another proof, using the language of recurrences: the quantities $p_j = g_1 \cdots g_j$ and $q_j = (f_1/g_1 + \cdots + f_j/g_j) p_j$ satisfy the recurrences $p_j = p_{j-1} g_j$ and $q_j = q_{j-1} g_j + p_{j-1} f_j$, i.e., $\begin{pmatrix} p_j & q_j \\ 0 & p_j \end{pmatrix} = \begin{pmatrix} p_{j-1} & q_{j-1} \\ 0 & p_{j-1} \end{pmatrix} \begin{pmatrix} g_j & f_j \\ 0 & g_j \end{pmatrix}$.

The reader may prefer to describe this algorithm without matrices: for $t \geq 2$, recursively compute $f_1/g_1 + \cdots + f_s/g_s$ and $f_{s+1}/g_{s+1} + \cdots + f_t/g_t$, and then add to obtain $f_1/g_1 + \cdots + f_t/g_t$. Here $s = \lceil t/2 \rceil$.

### 13.5. The integer case.
An analogous algorithm takes time $O(n(\lg n)^2 \lg \lg n)$ to compute $f_1 g_2 \cdots g_t + g_1 f_2 \cdots g_t + \cdots + g_1 g_2 \cdots f_t$ and $g_1 g_2 \cdots g_t$, given integers $f_1, g_1, f_2, g_2, \ldots, f_t, g_t$. Here $n$ is the total number of input bits.

### 13.6. History.
Horowitz [1972] published an algorithm to compute the polynomial

$$\left(\frac{b_1}{x-a_1} + \frac{b_2}{x-a_2} + \cdots + \frac{b_t}{x-a_t}\right)(x-a_1)(x-a_2)\cdots(x-a_t)$$

within a $\lg t$ factor of the time for polynomial multiplication. Horowitz's algorithm is essentially the algorithm stated above, except that it splits $t$ into $t/2, t/4, t/8, \ldots$ rather than $t/2, t/2$.

Borodin and Moenck [1974, Section 7] published a more general algorithm to add fractions, in both the polynomial case and the integer case, for the application described in Section 23.

### 13.7. Improvements.
See Section 12 for improved product-tree algorithms.

## 14. Fraction from continued fraction

### 14.1. Input.
Let $A$ be a commutative ring. Let $t$ be a nonnegative integer. The algorithm in this section is given $t$ polynomials $q_1, q_2, \ldots, q_t \in A[x]$ such that, for each $i$, at least one of $q_i, q_{i+1}$ is nonzero.

### 14.2. Output.
This algorithm computes the polynomials $F(q_1, q_2, \ldots, q_t) \in A[x]$ and $G(q_1, q_2, \ldots, q_t) \in A[x]$ defined recursively by $F() = 1$, $G() = 0$,

$$F(q_1, q_2, \ldots, q_t) = q_1 F(q_2, \ldots, q_t) + G(q_2, \ldots, q_t) \text{ for } t \geq 1, \text{ and}$$
$$G(q_1, q_2, \ldots, q_t) = F(q_2, \ldots, q_t) \text{ for } t \geq 1.$$

For example, $F(q_1, q_2, q_3, q_4) = q_1 q_2 q_3 q_4 + q_1 q_2 + q_1 q_4 + q_3 q_4 + 1$. In general, $F(q_1, q_2, \ldots, q_t)$ is the sum of all products of subsequences of $(q_1, q_2, \ldots, q_t)$ obtained by deleting any number of non-overlapping adjacent pairs.

The reader may think of this output as the numerator and denominator of a continued fraction:

$$\frac{F(q_1, q_2, \ldots, q_t)}{G(q_1, q_2, \ldots, q_t)} = q_1 + \cfrac{1}{\cfrac{F(q_2, \ldots, q_t)}{G(q_2, \ldots, q_t)}} = q_1 + \cfrac{1}{q_2 + \cfrac{1}{\ddots + \cfrac{1}{q_t}}}.$$

As in Section 13, these equations hold in any $A[x]$-algebra where all the divisions make sense.

**14.3. Speed.** This algorithm uses $O(n(\lg n)^2 \lg \lg n)$ operations in $A$, where $n$ is the total number of coefficients in the input polynomials.

**14.4. How it works.** The product $\begin{pmatrix} q_1 & 1 \\ 1 & 0 \end{pmatrix} \begin{pmatrix} q_2 & 1 \\ 1 & 0 \end{pmatrix} \begin{pmatrix} q_3 & 1 \\ 1 & 0 \end{pmatrix} \cdots \begin{pmatrix} q_t & 1 \\ 1 & 0 \end{pmatrix} \begin{pmatrix} 1 \\ 0 \end{pmatrix}$ is exactly $\begin{pmatrix} F(q_1, q_2, \ldots, q_t) \\ G(q_1, q_2, \ldots, q_t) \end{pmatrix}$ by definition of $F$ and $G$. Compute this product as described in Section 12.

The assumption that no two consecutive $q$'s are 0 ensures that the total number of coefficients in these matrices is in $O(n)$.

**14.5. The integer case.** An analogous algorithm, given integers $q_1, q_2, \ldots, q_t$, computes $F(q_1, q_2, \ldots, q_t)$ and $G(q_1, q_2, \ldots, q_t)$. This algorithm takes time $O(n(\lg n)^2 \lg \lg n)$, where $n$ is the total number of input bits.

**14.6. History.** See Section 12.

**14.7. Improvements.** See Section 12 for improved product-tree algorithms.

## 15. Exponential: the short case

**15.1. Input.** Let $A$ be a commutative ring containing $\mathbf{Q}$. Let $m$ and $n$ be positive integers. The algorithm in this section is given a polynomial $f \in A[x]$ with $\deg f < 2m$ and $f \bmod x^m = 0$. For example, if $m = 2$, the input is a polynomial of the form $f_2 x^2 + f_3 x^3$.

**15.2. Output.** This algorithm computes the precision-$n$ representation of the series $\exp f \in A[\![x]\!]$ defined in Section 9.

**15.3. Speed.** This algorithm uses $O(n(\lg n)^2 \lg \lg n)$ operations in $A$. It is usually slower than the algorithm of Section 9; its main virtue is that the same idea also works for $\mathbf{Z}_2$ and $\mathbf{R}$.

**15.4. How it works.** Define $k = \lceil n/m - 1 \rceil$. Compute the matrix product $\begin{pmatrix} u & v \\ 0 & w \end{pmatrix} = \begin{pmatrix} 1 & 1 \\ 0 & 1 \end{pmatrix} \begin{pmatrix} f & f \\ 0 & 1 \end{pmatrix} \begin{pmatrix} f & f \\ 0 & 2 \end{pmatrix} \cdots \begin{pmatrix} f & f \\ 0 & k \end{pmatrix}$ as described in Section 12. Then $(\exp f) \bmod x^n = (v/w) \bmod x^n$. Note that $w$ is simply the integer $k!$, so the division by $w$ is a multiplication by the constant $1/k!$.

The point is that $(u, v, w) = (f^k, k!(1 + f + f^2/2 + \cdots + f^k/k!), k!)$ by induction, so $(\exp f) - v/w = f^{k+1}/(k+1)! + f^{k+2}/(k+2)! + \cdots$; but $k$ was chosen so that $f^{k+1}$ is divisible by $x^n$.

**15.5. The integer case, easy completion: $\mathbf{Q} \to \mathbf{Q}_2$.** One can use the same method to compute a precision-$n$ representation of $\exp f \in \mathbf{Z}_2$, given an integer $f \in \{0, 2^m, (2)2^m, \ldots, (2^m - 1)2^m\}$, in time $O(n(\lg n)^2 \lg \lg n)$, for $m \geq 2$. Note that $k$ must be chosen somewhat larger in this case, because the final division of $v$ by $w = k!$ loses approximately $k$ bits of precision.

**15.6. The integer case, hard completion: Q → R.** One can compute a precision-$n$ representation of $\exp f$, given a real number $f$ such that $|f| < 2^{-m}$ and $f$ is a multiple of $2^{-2m}$, in time $O(n(\lg n)^2 \lg \lg n)$. As usual, roundoff error complicates the algorithm.

**15.7. History.** See Section 12.

**15.8. Improvements.** See Section 16.

# 16. Exponential: the general case

**16.1. Input.** Let $A$ be a commutative ring containing **Q**. Let $n$ be a positive integer. The algorithm in this section is given the precision-$n$ representation of a power series $f \in A[\![x]\!]$ with $f(0) = 0$.

**16.2. Output.** This algorithm computes the precision-$n$ representation of the series $\exp f \in A[\![x]\!]$ defined in Section 9.

**16.3. Speed.** This algorithm uses $O(n(\lg n)^3 \lg \lg n)$ operations in $A$. It is usually much slower than the algorithm of Section 9; its main virtue is that the same idea also works for $\mathbf{Z}_2$ and **R**.

**16.4. How it works.** Write $f$ as a sum $f_1 + f_2 + f_4 + f_8 + \cdots$ where $f_m$ mod $x^m = 0$ and $\deg f_m < 2m$. In other words, put the coefficient of $x^1$ into $f_1$; the coefficients of $x^2$ and $x^3$ into $f_2$; the coefficients of $x^4$ through $x^7$ into $f_4$; and so on.

Compute precision-$n$ approximations to $\exp f_1, \exp f_2, \exp f_4, \exp f_8, \ldots$ as described in Section 15. Multiply to obtain $\exp f$.

**16.5. The integer case.** Similar algorithms work for $\mathbf{Z}_2$ and **R**.

**16.6. History.** This method of computing exp is due to Brent [1976a, Theorem 6.2]. Brent also pointed out that, starting from a fast algorithm for exp, one can use "Newton's method" — i.e., Simpson's method — to quickly compute log and various other functions. Note the reversal of roles from Section 9, where exp was obtained by inverting log.

**16.7. Improvements.** Salamin, and independently Brent, observed that one could use the "arithmetic-geometric mean" to compute log and exp for **R** in time only $O(n(\lg n)^2 \lg \lg n)$. See [Beeler et al. 1972, Item 143], [Salamin 1976], [Brent 1976b], and [Brent 1976c, Section 9] for the basic idea; [Borwein and Borwein 1987] for much more information about the arithmetic-geometric mean; and my self-contained paper [Bernstein 2003] for constant-factor improvements.

## 17. Quotient and remainder

**17.1. Input.** Let $A$ be a commutative ring. Let $d$ and $e$ be nonnegative integers. The algorithm in this section is given two elements $f, h$ of the polynomial ring $A[x]$ such that $f$ is monic, $\deg f = d$, and $\deg h < e$.

**17.2. Output.** This algorithm computes $q, r \in A[x]$ such that $h = qf + r$ and $\deg r < d$. In other words, this algorithm computes $r = h \bmod f$ and $q = (h-r)/f$.

For example, say $d = 2$ and $e = 5$. Given $f = f_0 + f_1 x + x^2$ and $h = h_0 + h_1 x + h_2 x^2 + h_3 x^3 + h_4 x^4$, this algorithm computes

$$q = (h_2 - h_3 f_1 + h_4(f_1^2 - f_0)) + (h_3 - h_4 f_1)x + h_4 x^2$$

and

$$\begin{aligned}r = h \bmod f &= h - qf \\ &= (h_0 - h_2 f_0 + h_3 f_1 f_0 + h_4(f_1^2 - f_0)f_0) \\ &\quad + (h_1 - h_2 f_1 + h_3(f_1^2 - f_0) + h_4((f_1^2 - f_0)f_1 + f_1 f_0))x.\end{aligned}$$

**17.3. Speed.** This algorithm uses $O(e \lg e \lg \lg e)$ operations in $A$.

More precisely, the algorithm uses at most $(10(e-d) + 2k - 9)\mu(2(e-d) - 1) + (2(e-d) + 2k - 2) + e\mu(e) + e$ operations in $A$ if $1 \le e - d \le 2^k$. The algorithm uses no operations if $e \le d$.

For simplicity, subsequent sections of this paper use the relatively crude upper bound $12(e+1)(\mu(2e) + 1)$.

**17.4. How it works:** $A(x) \to A((x^{-1}))$. The point is that polynomial division in $A[x]$ is division in $A((x^{-1}))$; $A((x^{-1}))$, in turn, is isomorphic to $A((x))$.

If $e \le d$, the answer is $q = 0$ and $r = h$. Assume from now on that $e > d$.

Reverse the coefficient order in $f = \sum_j f_j x^j$ to obtain $F = \sum_j f_{d-j} x^j \in A[x]$; in other words, define $F = x^d f(x^{-1})$. Then $\deg F \le d$ and $F(0) = 1$. For example, if $d = 2$ and $f = f_0 + f_1 x + x^2$, then $F = 1 + f_1 x + f_0 x^2$.

Similarly, reverse $h = \sum_j h_j x^j$ to obtain $H = \sum_j h_{e-1-j} x^j \in A[x]$; in other words, define $H = x^{e-1} h(x^{-1})$. Then $\deg H < e$. For example, if $e = 5$ and $h = h_0 + h_1 x + h_2 x^2 + h_3 x^3 + h_4 x^4$, then $H = h_4 + h_3 x + h_2 x^2 + h_1 x^3 + h_0 x^4$.

Now compute $Q = (H/F) \bmod x^{e-d}$ as explained in Section 7. Then $\deg Q < e - d$. Reverse $Q = \sum_j q_{e-d-1-j} x^j$ to obtain $q = \sum_j q_j x^j \in A[x]$; in other words, define $q = x^{e-d-1} Q(x^{-1})$.

Compute $r = h - qf \in A[x]$ as explained in Section 4. Then $\deg r < d$. Indeed, $x^{e-1} r(x^{-1}) = H - QF$ is a multiple of $x^{e-d}$ by construction of $Q$.

**17.5. The $x$-adic case: $A(x) \to A((x))$.** Omit the reversal of coefficients in the above algorithm. The resulting algorithm, given two polynomials $f, h$ with $f(0) = 1$, $\deg f \leq d$, and $\deg h < e$, computes polynomials $q, r$ such that $h = qf + x^{\max\{e-d,0\}} r$ and $\deg r < d$.

**17.6. The integer case, easy completion: $\mathbf{Q} \to \mathbf{Q}_2$.** An analogous algorithm, given integers $f, h$ with $f$ odd, $|f| \leq 2^d$, and $|h| < 2^e$, computes integers $q, r$ such that $h = qf + 2^{\max\{e-d,0\}} r$ and $|r| < 2^d$. The algorithm takes time $O(n \lg n \lg \lg n)$, where $n$ is the total number of input bits.

**17.7. The integer case, hard completion: $\mathbf{Q} \to \mathbf{R}$.** An analogous algorithm, given integers $f, h$ with $f \neq 0$, computes integers $q, r$ such that $h = qf + r$ and $0 \leq r < |f|$. The algorithm takes time $O(n \lg n \lg \lg n)$, where $n$ is the total number of input bits.

It is often convenient to change the sign of $r$ when $f$ is negative; in other words, to replace $0 \leq r < |f|$ with $0 \leq r/f < 1$; in other words, to take $q = \lfloor h/f \rfloor$. The time remains $O(n \lg n \lg \lg n)$.

**17.8. History.** See Section 7 for historical notes on fast division in $A[\![x]\!]$ and $\mathbf{R}$.

The use of $x \mapsto x^{-1}$ for computing quotients dates back to at least 1973: Strassen [1973, page 240] (translated) commented that "the division of two formal power series can easily be used for the division of two polynomials with remainder." I have not attempted to trace the earlier history of the $x^{-1}$ valuation.

**17.9. Improvements.** One can often save some time, particularly in the integer case, by changing the problem, allowing a wider range of remainders. Most applications do not need the smallest possible remainder of $h$ modulo $f$; any reasonably small remainder is adequate.

A different way to divide $h$ by $f$ is to recursively divide the top half of $h$ by the top half of $f$, then recursively divide what's left. Moenck and Borodin [1972] published this algorithm (in the polynomial case), and observed that it takes time $O(n(\lg n)^2 \lg \lg n)$. Borodin and Moenck later [1974, Section 6] summarized the idea in two sentences and then dismissed it in favor of multiply-by-reciprocal. The Moenck–Borodin idea was reinvented many years later (in the integer case) by Jebelean [1997], by Daniel Ford (according to email I received from John Cannon in June 1998), and by Burnikel and Ziegler [1998]. The idea is claimed in [Burnikel and Ziegler 1998, Section 4] to be faster than multiply-by-reciprocal for fairly large values of $n$; on the other hand, the algorithm in [Burnikel and Ziegler 1998, Section 4.2] is certainly not the state of the art in reciprocal computation. Further investigation is needed.

Many applications of division in $\mathbf{R}$ can work equally well with division in $\mathbf{Z}_2$. This fact—widely known to number theorists since Hensel's introduction of $\mathbf{Z}_2$ (and more general completions) in the early 1900s—has frequently been applied to computations; replacing $\mathbf{R}$ with $\mathbf{Z}_2$ usually saves a little time and a considerable amount of effort. See, e.g., [Krishnamurthy 1977], [Hehner and Horspool 1979], [Gregory 1980], [Dixon 1982] (using $\mathbf{Z}_p$ where, for simplicity, $p$ is chosen to not divide an input), [Montgomery 1985], and [Jebelean 1993]. Often $\mathbf{Z}_2$ division is called "Montgomery reduction," but this gives too much credit to [Montgomery 1985].

In some applications, one knows in advance that a division will be exact, i.e., that the remainder will be zero. Schönhage and Vetter [1994] suggested computing the top half of the quotient with division in $\mathbf{R}$, and the bottom half of the quotient with division in $\mathbf{Z}_2$. These two half-size computations are faster than one full-size computation, because computation speed is not exactly linear. Similarly, for polynomials, one can combine $x$-adic division with the usual division.

Another exact-division method, for $h, f \in \mathbf{C}[x]$, is **deconvolution**: one solves $h = qf$ by transforming $h$, transforming $f$, dividing to obtain the transform of $q$, and untransforming the result to obtain $q$. Extra work is required if the transform of $f$ is noninvertible.

## 18. Remainder tree

**18.1. Input.** Let $A$ be a commutative ring. Let $t$ be a nonnegative integer. The algorithm in this section is given a polynomial $h \in A[x]$ and monic polynomials $f_1, f_2, \ldots, f_t \in A[x]$.

**18.2. Output.** This algorithm computes $h \bmod f_1, h \bmod f_2, \ldots, h \bmod f_t$. Actually, the algorithm computes more: the **remainder tree** of $h, f_1, f_2, \ldots, f_t$.

The remainder tree is defined as follows: for each vertex $v$ in the product tree of $f_1, f_2, \ldots, f_t$, there is a corresponding vertex $h \bmod v$ in the remainder tree of $h, f_1, f_2, \ldots, f_t$. In particular, the leaves of the product tree are $f_1, f_2, \ldots, f_t$, so the leaves of the remainder tree are $h \bmod f_1, h \bmod f_2, \ldots, h \bmod f_t$.

In other words: The root of the remainder tree is $h \bmod f_1 f_2 \cdots f_t$. If $t \leq 1$ then that's the complete tree. If $t \geq 2$ then the left subtree is the remainder tree of $h, f_1, \ldots, f_s$, and the right subtree is the remainder tree of $h, f_{s+1}, \ldots, f_t$, where $s = \lceil t/2 \rceil$.

For example, here is the remainder tree of $h, f_1, f_2, f_3, f_4, f_5, f_6$:

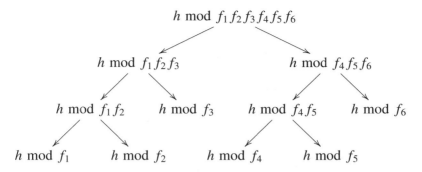

**18.3. Speed.** This algorithm uses $O(n(\lg n)^2 \lg \lg n)$ operations in $A$, where $n$ is the total number of coefficients in $h, f_1, f_2, \ldots, f_t$.

More precisely: Assume that $d, m, k$ are nonnegative integers, that $\deg h < m$, that $f_1, f_2, \ldots, f_t$ together have at most $d$ coefficients, and that $t \leq 2^k$. Then the algorithm uses at most $(12m + 26dk + 24 \cdot 2^k - 12)(\mu(2\max\{d, m\}) + 1)$ operations in $A$.

**18.4. How it works:** $A(x) \to A((x^{-1}))$. Here is a recursive algorithm that, given $h$ and the product tree $P$ of $f_1, \ldots, f_t$, computes the remainder tree $R$ of $h, f_1, f_2, \ldots, f_t$. This algorithm uses at most $12(m + 2dk + 2^{k+1} - 1)(\mu(2\max\{d, m\}) + 1)$ operations in $A$; add $2dk(\mu(2d) + 1)$ operations in $A$ to compute $P$ in the first place as explained in Section 12.

The root of $P$ is $f_1 f_2 \cdots f_t$. Compute $g = h \bmod f_1 f_2 \cdots f_t$ as explained in Section 17; this is the root of $R$. This uses at most $12(m + 1)(\mu(2m) + 1) \leq 12(m + 1)\mu(2\max\{d, m\}) + 1$ operations in $A$.

The strategy is to compute each remaining vertex in $R$ by reducing its parent vertex modulo the corresponding vertex in $P$. For example, the algorithm computes $h \bmod f_1 f_2 \cdots f_s$ as $(h \bmod f_1 f_2 \cdots f_t) \bmod f_1 f_2 \cdots f_s$.

If $t \leq 1$, stop. There are no operations here; and $k \geq 0$ so $12(m + 1) \leq 12(m + 2dk + 2^{k+1} - 1)$.

Otherwise define $s = \lceil t/2 \rceil$. Apply this algorithm recursively to $g$ and the left subtree of $P$ to compute the remainder tree of $g, f_1, f_2, \ldots, f_s$, which is exactly the left subtree of $R$. Apply this algorithm recursively to $g$ and the right subtree of $P$ to compute the remainder tree of $g, f_{s+1}, f_{s+2}, \ldots, f_t$, which is exactly the right subtree of $R$.

Time analysis: Define $c$ as the number of coefficients of $f_1, \ldots, f_s$. By induction, the recursion for the left subtree uses at most $12(d + 2c(k - 1) + 2^k - 1)$ times $\mu(2d) + 1 \leq \mu(2\max\{d, m\}) + 1$ operations, since $\deg g < d$. The recursion for the right subtree uses at most $12(d + 2(d - c)(k - 1) + 2^k - 1)$ times

$\mu(2\max\{d,m\})+1$ operations. Add: $m+1+d+2c(k-1)+2^k-1+d+2(d-c)(k-1)+2^k-1 = m+2dk+2^{k+1}-1$.

**18.5. The $x$-adic case: $A(x) \to A((x))$.** One obtains a simpler algorithm by omitting the reversals described in Section 17. The simpler algorithm, given polynomials $h, f_1, f_2, \ldots, f_t$ where $f_1(0) = f_2(0) = \cdots = f_t(0) = 1$, computes small polynomials $r_1, r_2, \ldots, r_t$ such that $h$ is congruent modulo $f_j$ to a certain power of $x$ times $r_j$. The algorithm uses $O(n(\lg n)^2 \lg \lg n)$ operations in $A$.

**18.6. The integer case, easy completion: $\mathbf{Q} \to \mathbf{Q}_2$.** An analogous algorithm, given an integer $h$ and odd integers $f_1, f_2, \ldots, f_t$, computes small integers $r_1, r_2, \ldots, r_t$ such that $h$ is congruent modulo $f_j$ to a certain power of 2 times $r_j$. The algorithm takes time $O(n(\lg n)^2 \lg \lg n)$, where $n$ is the total number of input bits.

**18.7. The integer case, hard completion: $\mathbf{Q} \to \mathbf{R}$.** An analogous algorithm, given an integer $h$ and nonzero integers $f_1, f_2, \ldots, f_t$, computes integers $r_1, r_2, \ldots, r_t$, with $0 \le r_j < |f_j|$, such that $h$ is congruent modulo $f_j$ to $r_j$. The algorithm takes time $O(n(\lg n)^2 \lg \lg n)$, where $n$ is the total number of input bits.

**18.8. History.** This algorithm was published by Moenck and Borodin for "single-precision" moduli $f_1, f_2, \ldots, f_t$. See [Moenck and Borodin 1972] and [Borodin and Moenck 1974, Sections 4–6].

**18.9. Improvements.** Montgomery [1992, Section 3.7] pointed out several opportunities to remove redundancy — for example, by FFT caching — within and across levels of computation of a remainder tree.

One can replace the remainder tree with the **scaled remainder tree** to save another constant factor in the computation of $h \bmod f_1, h \bmod f_2, \ldots, h \bmod f_t$. Where the remainder tree has integers such as $h \bmod f_1 f_2$, the scaled remainder tree has real numbers such as $(h \bmod f_1 f_2)/f_1 f_2$, represented in the usual way as nearby floating-point numbers. Here's the point: moving from $h \bmod f_1 f_2$ to $h \bmod f_1$ means dividing by $f_1$; moving from $(h \bmod f_1 f_2)/f_1 f_2$ to $(h \bmod f_1)/f_1$ means multiplying by $f_2$, which is faster. This speedup was achieved in the polynomial case by Bostan, Lecerf, and Schost, using a more complicated approach that does not work for integers; see [Bostan et al. 2003] and [Bostan et al. 2004, Section 3.1]. I found the scaled-remainder-tree structure, achieved the same speedup for integers, and then pointed out some redundancies that could be removed, saving even more time. See [Bernstein 2004d].

## 19. Small factors of a product

**19.1. Input.** Let $A$ be a commutative ring. Let $s$ be a positive integer, and let $t$ be a nonnegative integer. The algorithm in this section is given polynomials $h_1, h_2, \ldots, h_s \in A[x]$ and monic polynomials $f_1, f_2, \ldots, f_t \in A[x]$.

**19.2. Output.** This algorithm figures out which $f_i$'s divide $h_1 h_2 \cdots h_s$: it computes the subsequence $g_1, g_2, \ldots$ of $f_1, \ldots, f_t$ consisting of each $f_i$ that divides $h_1 h_2 \cdots h_s$.

The name "small factors" comes from the following important special case. Let $A$ be a finite field, and let $f_1, f_2, \ldots$ be all the small primes in $A[x]$, i.e., all the low-degree monic irreducible polynomials in $A[x]$. Then this algorithm computes the small factors of $h_1 h_2 \cdots h_s$.

For example, say $A = \mathbf{Z}/2$, $s = 4$, $t = 5$, $h_1 = 101111 = 1 + x^2 + x^3 + x^4 + x^5$, $h_2 = 1101011$, $h_3 = 00001011$, $h_4 = 0001111$, $f_1 = 01$, $f_2 = 11$, $f_3 = 111$, $f_4 = 1101$, and $f_5 = 1011$. This algorithm finds all the factors of $h_1 h_2 h_3 h_4$ among $f_1, f_2, f_3, f_4, f_5$. Its output is (01, 11, 111, 1011): the product $h_1 h_2 h_3 h_4$ is divisible by 01, 11, 111, and 1011, but not by 1101.

**19.3. Speed.** This algorithm uses $O(n (\lg n)^2 \lg \lg n)$ operations in $A$, where $n$ is the total number of coefficients in $h_1, h_2, \ldots, h_s, f_1, f_2, \ldots, f_t$.

More precisely: Assume that $d, m, j, k$ are nonnegative integers, that $s \leq 2^j$, that $t \leq 2^k$, that $h_1, \ldots, h_s$ together have at most $m$ coefficients, and that $f_1, \ldots, f_t$ together have at most $d$ coefficients. Then the algorithm uses at most $(2mj + 12m + 26dk + 24 \cdot 2^k - 12)(\mu(2 \max\{d, m\}) + 1) + d$ operations in $A$.

**19.4. How it works.** Compute $h = h_1 h_2 \cdots h_s$ as explained in Section 12. This uses at most $2mj(\mu(m) + 1) \leq 2mj(\mu(2 \max\{d, m\}) + 1)$ operations in $A$.

Compute $h \bmod f_1, \ldots, h \bmod f_t$ as explained in Section 18. This uses at most $(12m + 26dk + 24 \cdot 2^k - 12)(\mu(2 \max\{d, m\}) + 1)$ operations in $A$.

Check whether $h \bmod f_1 = 0$, $h \bmod f_2 = 0, \ldots, h \bmod f_t = 0$. This uses at most $d$ equality tests in $A$.

**19.5. The integer case.** An analogous algorithm, given integers $h_1, h_2, \ldots, h_s$ and nonzero integers $f_1, f_2, \ldots, f_t$, figures out which $f_i$'s divide $h_1 h_2 \cdots h_s$. The algorithm takes time $O(n (\lg n)^2 \lg \lg n)$, where $n$ is the total number of input bits.

**19.6. History.** See Section 20. This algorithm is a stepping-stone to the algorithm of Section 20.

**19.7. Improvements.** See Section 12 for improved product-tree algorithms, and Section 18 for improved remainder-tree algorithms.

As discussed in Section 17, Schönhage and Vetter combined $\mathbf{Z}_2$ and $\mathbf{R}$ to compute a quotient more quickly when the remainder was known in advance to be 0. A similar technique can be used to check more quickly whether a remainder is 0.

## 20. Small factors of a sequence

**20.1. Input.** Let $A$ be a commutative ring. Let $s$ be a positive integer, and let $t$ be a nonnegative integer. The algorithm in this section is given nonzero polynomials $h_1, h_2, \ldots, h_s \in A[x]$ and monic coprime polynomials $f_1, f_2, \ldots, f_t \in A[x]$ with $\deg f_i \geq 1$ for each $i$.

Here **coprime** means that $A[x] = f_i A[x] + f_j A[x]$ for every $i, j$ with $i \neq j$; in other words, there exist $u, v \in A[x]$ with $f_i u + f_j v = 1$. (Warning: Some authors instead say **pairwise coprime**, reserving "coprime" for the notion that $A[x] = f_1 A[x] + f_2 A[x] + \cdots + f_t A[x]$.)

The importance of coprimality is the **Chinese remainder theorem**: the $A[x]$-algebra morphism from $A[x]/f_1 f_2 \cdots f_t$ to $A[x]/f_1 \times A[x]/f_2 \times \cdots \times A[x]/f_t$ is an isomorphism. In particular, if each of $f_1, f_2, \ldots, f_t$ divides a polynomial $h$, then the product $f_1 f_2 \cdots f_t$ divides $h$. This is crucial for the speed of the algorithm.

**20.2. Output.** This algorithm figures out which $f_i$'s divide $h_1$, which $f_i$'s divide $h_2$, which $f_i$'s divide $h_3$, etc.

As in Section 19, the name "small factors" comes from the important special case that $A$ is a finite field and $f_1, f_2, \ldots$ are all of the small primes in $A[x]$. Then this algorithm computes the small factors of $h_1$, the small factors of $h_2$, etc.

For example, say $A = \mathbf{Z}/2$, $s = 4$, $t = 5$, $h_1 = 101111 = 1 + x^2 + x^3 + x^4 + x^5$, $h_2 = 1101011$, $h_3 = 00001011$, $h_4 = 0001111$, $f_1 = 01$, $f_2 = 11$, $f_3 = 111$, $f_4 = 1101$, and $f_5 = 1011$. This algorithm finds all the factors of $h_1, h_2, h_3, h_4$ among $f_1, f_2, f_3, f_4, f_5$. Its output is $(), (111), (01, 1011), (01, 11)$.

**20.3. Speed.** This algorithm uses $O(n(\lg n)^3 \lg \lg n)$ operations in $A$, where $n$ is the total number of coefficients in $h_1, h_2, \ldots, h_s, f_1, f_2, \ldots, f_t$.

More precisely: Assume, as in Section 19, that $d, m, j, k$ are nonnegative integers, that $h_1, \ldots, h_s$ together have at most $m$ coefficients, that $f_1, \ldots, f_t$ together have at most $d$ coefficients, that $s \leq 2^j$, and that $t \leq 2^k$. Then the algorithm uses at most $((104jk + j^2 + 109j + 12)m + 26dk + 24 \cdot 2^k)(\mu(2\max\{d, m\}) + 1) + d + 4mj$ operations in $A$.

**20.4. How it works.** Figure out which of $f_1, \ldots, f_t$ divide $h_1 \cdots h_s$, as explained in Section 19; write $(g_1, g_2, \ldots)$ for this subsequence of $(f_1, \ldots, f_t)$. This uses at most $(2mj + 12m + 26dk + 24 \cdot 2^k)(\mu(2\max\{d, m\}) + 1) + d$ operations in $A$, leaving $(104jk + j^2 + 107j)m(\mu(2\max\{d, m\}) + 1) + 4mj$ operations for the remaining steps in the algorithm.

If $s = 1$, the answer is $(g_1, g_2, \ldots)$. There are no further operations in this case, and $(104jk + j^2 + 107j)m(\mu(2\max\{d, m\}) + 1) + 4mj$ is nonnegative.

Otherwise apply the algorithm recursively to $h_1, h_2, \ldots, h_r$ and $g_1, g_2, \ldots$, and then apply the algorithm recursively to $h_{r+1}, h_{r+2}, \ldots, h_s$ and $g_1, g_2, \ldots$, where $r = \lceil s/2 \rceil$. This works because any $f$'s that divide $h_i$ also divide the product $h_1 h_2 \cdots h_s$ and are therefore included among the $g$'s.

The central point in the time analysis is that $\deg(g_1 g_2 \cdots) < m$. Indeed, $g_1, g_2, \ldots$ are coprime divisors of $h_1 \cdots h_s$, so their product is a divisor of $h_1 \cdots h_s$; but $h_1 \cdots h_s$ is a nonzero polynomial of degree smaller than $m$. Thus there are at most $\min\{m, t\}$ polynomials in $g_1, g_2, \ldots$, and the total number of coefficients in $g_1, g_2, \ldots$ is at most $\min\{2m, d\}$.

Define $\ell$ as the number of coefficients in $h_1, h_2, \ldots, h_r$, and define $e$ as the smallest nonnegative integer with $\min\{m, t\} \leq 2^e$. Then $\ell \leq m$; $e \leq k$; and $2^e \leq 2m$, since $m \geq 1$. The recursive computation for $h_1, \ldots, h_r, g_1, g_2, \ldots$ uses at most

$$((104(j-1)e + (j-1)^2 + 109(j-1) + 12)\ell + 26\min\{2m, d\}e + 24 \cdot 2^e)$$
$$(\mu(2\max\{\min\{2m, d\}, \ell\}) + 1) + \min\{2m, d\} + 4\ell(j-1)$$
$$\leq ((104(j-1)k + (j-1)^2 + 109(j-1) + 12)\ell + 52mk + 48m)$$
$$(\mu(2\max\{d, m\}) + 1) + 2m + 4\ell(j-1)$$

operations in $A$ by induction. Similarly, the recursive computation for $h_{r+1}, \ldots, h_s, g_1, g_2, \ldots$ uses at most

$$((104(j-1)k + (j-1)^2 + 109(j-1) + 12)(m - \ell) + 52mk + 48m)$$
$$(\mu(2\max\{d, m\}) + 1) + 2m + 4(m-\ell)(j-1)$$

operations in $A$. The total is exactly $(104jk + j^2 + 107j)m(\mu(2\max\{d, m\}) + 1) + 4mj$ as desired.

Here is an example of how the algorithm works. To factor $h_1 = 101111$, $h_2 = 1101011$, $h_3 = 00001011$, and $h_4 = 0001111$ over $A = \mathbf{Z}/2$ using the primes $01, 11, 111, 1101, 1011$, the algorithm first finds the factors $01, 11, 111, 1011$ of $h_1 h_2 h_3 h_4$ as explained in Section 19. It then recursively factors $h_1, h_2$ using $01, 11, 111, 1011$, and recursively factors $h_3, h_4$ using $01, 11, 111, 1011$.

At the first level of recursion in the same example: To factor $h_3, h_4$ using $01, 11, 111, 1011$, the algorithm finds the factors $01, 11, 1011$ of $h_3 h_4$ as explained in Section 19. It then recursively factors $h_3$ using $01, 11, 1011$, and recursively factors $h_4$ using $01, 11, 1011$.

**20.5. The integer case.** An analogous algorithm, given nonzero integers $h_1, h_2, \ldots, h_s$ and coprime integers $f_1, f_2, \ldots, f_t \geq 2$, figures out which $f_j$'s divide $h_1$, which $f_j$'s divide $h_2$, which $f_j$'s divide $h_3$, etc. The algorithm takes time $O(n(\lg n)^3 \lg \lg n)$, where $n$ is the total number of input bits.

An important special case is that $f_1, f_2, \ldots, f_t$ are the first $t$ prime numbers. Then this algorithm computes the small factors of $h_1$, the small factors of $h_2$, etc.

**20.6. History.** I introduced this algorithm in [Bernstein 2005, Section 21] and [Bernstein 2002a]. The version in [Bernstein 2005] is slower but more general: it relies solely on multiplication, exact division, and greatest common divisors.

**20.7. Improvements.** There are many previous algorithms to find small factors: for example, Legendre's root-finding method (when the inputs are polynomials over a finite field), sieving (when the inputs are successive values of a polynomial), Pollard's $p - 1$ method, and Lenstra's elliptic-curve method. These algorithms, and their applications, are recurring topics in this volume. See [Bernstein 2002a] for a comparison of speeds and for further pointers to the literature. Combinations and optimizations of these algorithms are an active area of research.

Many applications discard inputs that are not **smooth**, i.e., that do not factor completely over the primes $f_1, f_2, \ldots, f_t$. One can identify and discard those inputs without factoring them. Franke, Kleinjung, Morain, and Wirth [2004, Section 4] published a smoothness-detection algorithm that typically takes time $O(n(\lg n)^2 \lg \lg n)$. My paper [Bernstein 2004b] explains a slight variant that always takes time $O(n(\lg n)^2 \lg \lg n)$: first compute $f = f_1 f_2 \cdots f_t$; then compute $f \bmod h_1$, $f \bmod h_2$, $\ldots$; then, for each $i$, compute $f^{2^{b_i}} \bmod h_i$ by repeated squaring, for a sensible choice of $b_i$. The result is 0 if and only if $h_i$ is smooth.

# 21. Continued fraction from fraction

**21.1. Input.** Let $A$ be a field. Let $d$ be a nonnegative integer. The algorithm in this section is given polynomials $f_1, f_2 \in A[x]$, not both zero.

Both $f_1$ and $f_2$ are assumed to be represented without leading zero coefficients, so the algorithm can see the degrees of $f_1$ and $f_2$ without any equality tests in $A$.

**21.2. Output.** This algorithm computes polynomials $M_{11}, M_{12}, M_{21}, M_{22}$ in $A[x]$ such that $\deg M_{11}, \deg M_{12}, \deg M_{21}, \deg M_{22}$ are all at most $d$; $M_{11}M_{22} - M_{12}M_{21}$ is in $\{-1, 1\}$; and $\deg(M_{21}f_1 + M_{22}f_2) < \max\{\deg f_1, \deg f_2\} - d$. In particular, $M_{21}f_1 + M_{22}f_2 = 0$ in the important case $d = \max\{\deg f_1, \deg f_2\}$.

This algorithm also computes a factorization of the matrix $M = \begin{pmatrix} M_{11} & M_{12} \\ M_{21} & M_{22} \end{pmatrix}$: polynomials $q_1, q_2, \ldots, q_t \in A[x]$ with $M = \begin{pmatrix} 0 & 1 \\ 1 & -q_t \end{pmatrix} \cdots \begin{pmatrix} 0 & 1 \\ 1 & -q_2 \end{pmatrix} \begin{pmatrix} 0 & 1 \\ 1 & -q_1 \end{pmatrix}$.

In particular, $\begin{pmatrix} f_1 \\ f_2 \end{pmatrix} = \begin{pmatrix} q_1 & 1 \\ 1 & 0 \end{pmatrix} \begin{pmatrix} q_2 & 1 \\ 1 & 0 \end{pmatrix} \cdots \begin{pmatrix} q_t & 1 \\ 1 & 0 \end{pmatrix} M \begin{pmatrix} f_1 \\ f_2 \end{pmatrix}$. The reader may think of this equation as expanding $f_1/f_2$ into a continued fraction

$$q_1 + \cfrac{1}{q_2 + \cfrac{1}{\ddots + \cfrac{1}{q_t + \cfrac{g_2}{g_1}}}}$$

with $g_1 = M_{11} f_1 + M_{12} f_2$ and $g_2 = M_{21} f_1 + M_{22} f_2$; see Section 14.

Note for future reference that if $\deg f_1 \geq \deg f_2$ then $\deg g_1 = \deg f_1 - \deg M_{22} \geq \deg f_1 - d > \deg g_2$; if also $M_{21} \neq 0$ then $\deg g_1 = \deg f_2 - \deg M_{21}$. (Proof: $\deg M_{12} g_2 < d + \deg f_1 - d = \deg f_1$, and $M_{22} g_1 = M_{12} g_2 \pm f_1$, so $\deg M_{22} g_1 = \deg f_1$. If $M_{21} \neq 0$ then $\deg M_{21} f_1 \geq \deg f_1 \geq \deg f_1 - d > \deg g_2$, and $M_{22} f_2 = g_2 - M_{21} f_1$, so $\deg M_{22} f_2 = \deg M_{21} f_1$.)

**21.3. Speed.** This algorithm uses $O(d (\lg d)^2 \lg \lg d)$ operations in $A$.

More precisely: Assume that $2d \leq 2^k$ where $k$ is a nonnegative integer. Then this algorithm uses at most $(46dk + 44(2^{k+1} - 1))(\mu(4d + 8) + 1)$ operations in $A$. This bound is pessimistic.

**21.4. How it works.** The desired matrix $M$ is computed in nine steps shown below. The desired factorization of $M$ is visible from the construction of $M$, as is the (consequent) fact that $\det M \in \{-1, 1\}$.

There are several recursive calls in this algorithm. Most of the recursive calls reduce $d$; the other recursive calls preserve $d$ and reduce $\deg f_2$. The time analysis inducts on $(d, \deg f_2)$ in lexicographic order.

**Step 1: fix the input order.** If $\deg f_1 < \deg f_2$: Apply the algorithm recursively to $d, f_2, f_1$ to find a matrix $C$, of degree at most $d$, such that $\deg(C_{21} f_2 + C_{22} f_1) < \deg f_2 - d$. Compute $M = C \begin{pmatrix} 0 & 1 \\ 1 & 0 \end{pmatrix}$. The answer is $M$.

Time analysis: By induction, the recursive computation of $C$ uses at most $(46dk + 44(2^{k+1} - 1))(\mu(4d + 8) + 1)$ operations. The computation of $M$ uses no operations: $M$ is simply a reshuffling of $C$.

**Step 2: check for large** $d$. If $\deg f_1 < d$: Apply the algorithm recursively to $\deg f_1, f_1, f_2$ to find a matrix $M$, of degree at most $\deg f_1 < d$, such that $M_{21} f_1 + M_{22} f_2 = 0$. The answer is $M$.

Time analysis: By induction, the recursive computation of $M$ uses at most $(46(\deg f_1)k + 44(2^{k+1} - 1))(\mu(4(\deg f_1) + 8) + 1) \leq (46dk + 44(2^{k+1} - 1))(\mu(4d + 8) + 1)$ operations.

**Step 3: if no quotients are needed, stop.** If $\deg f_2 < \deg f_1 - d$: The answer is $\begin{pmatrix} 1 & 0 \\ 0 & 1 \end{pmatrix}$.

Time analysis: This computation uses $0 \leq (46dk + 44(2^{k+1} - 1))(\mu(4d + 8) + 1)$ operations.

**Step 4: focus on the top** $2d$ **coefficients.** Define $i = \deg f_1 - 2d$. If $i > 0$: Apply the algorithm recursively to $d$, $\lfloor f_1/x^i \rfloor$, $\lfloor f_2/x^i \rfloor$ to find a matrix $M$, of degree at most $d$, such that $\deg(M_{21}\lfloor f_1/x^i \rfloor + M_{22}\lfloor f_2/x^i \rfloor) < \deg(\lfloor f_1/x^i \rfloor) - d = \deg f_1 - i - d$.

The answer is $M$. Indeed, $x^i(M_{21}\lfloor f_1/x^i \rfloor + M_{22}\lfloor f_2/x^i \rfloor)$ has degree below $\deg f_1 - d$; $M_{21}(f_1 \bmod x^i)$ and $M_{22}(f_2 \bmod x^i)$ have degree below $d + i = \deg f_1 - d$; add to see that $M_{21} f_1 + M_{22} f_2$ has degree below $\deg f_1 - d$.

Time analysis: By induction, the recursive computation of $M$ uses at most $(46dk + 44(2^{k+1} - 1))(\mu(4d + 8) + 1)$ operations as claimed.

From now on, $0 \leq \deg f_1 - d \leq \deg f_2 \leq \deg f_1 \leq 2d$.

**Step 5: handle degree** 0. If $d = 0$ then $\deg f_1 = \deg f_2 = 0$; the answer is $\begin{pmatrix} 0 & 1 \\ 1 & -f_1/f_2 \end{pmatrix}$.

Time analysis: This computation uses $2 \leq (46dk + 44(2^{k+1} - 1))(\mu(4d + 8) + 1)$ operations as claimed.

From now on, $d \geq 1$, so $k \geq 1$.

**Step 6: compute the first half of the continued fraction.** Find a matrix $C$, of degree at most $\lfloor d/2 \rfloor$, with $\deg(C_{21} f_1 + C_{22} f_2) < \deg f_1 - \lfloor d/2 \rfloor$, by applying this algorithm to $\lfloor d/2 \rfloor$, $f_1$, $f_2$. Compute $g_2 = C_{21} f_1 + C_{22} f_2$. Compute $\deg g_2$.

Time analysis: Note that $2\lfloor d/2 \rfloor \leq d \leq 2^{k-1}$. The recursive computation of $C$ uses at most $(46(d/2)(k-1) + 44(2^k - 1))(\mu(4(d/2) + 8) + 1)$ operations by induction. The computation of $C_{21} f_1$ uses at most $3d\mu(3d)$ operations, since $\deg C_{21} \leq d$ and $\deg f_1 \leq 2d$. The computation of $C_{22} f_2$ uses at most $3d\mu(3d)$ operations. The computation of $g_2$ uses at most $2d$ additions, since $\deg g_2 < \deg f_1 \leq 2d$. The computation of $\deg g_2$ uses at most $2d$ equality tests.

**Step 7: if no more quotients are needed, stop.** If $\deg g_2 < \deg f_1 - d$: The answer is $C$. Time analysis: There are no operations here; for comparison, more than 0 operations are counted below.

From now on, $\deg f_1 - d \le \deg g_2 < \deg f_1 - \lfloor d/2 \rfloor$.

**Step 8: compute one more quotient.** Compute $g_1 = C_{11} f_1 + C_{12} f_2$. Compute polynomials $q, r \in A[x]$ such that $g_1 = q g_2 + r$ and $\deg r < \deg g_2$, as explained in Section 17. Compute $\deg r$.

Problem: Section 17 considers division by monic polynomials; $g_2$ is usually not monic. Solution: Divide $g_2$ by its leading coefficient, and adjust $q$ accordingly.

Observe that the matrix $\begin{pmatrix} 0 & 1 \\ 1 & -q \end{pmatrix} C = \begin{pmatrix} C_{21} & C_{22} \\ C_{11} - q C_{21} & C_{12} - q C_{22} \end{pmatrix}$ has degree at most $\deg f_1 - \deg g_2$. (Proof: Recall that $\deg C_{22} = \deg f_1 - \deg g_1$; so $\deg q C_{22} = \deg q + \deg C_{22} = (\deg g_1 - \deg g_2) + (\deg f_1 - \deg g_1) = \deg f_1 - \deg g_2$. Similarly, recall that $\deg C_{21} \le \deg f_2 - \deg g_1 \le \deg f_1 - \deg g_1$; so $\deg q C_{21} \le \deg f_1 - \deg g_2$. Finally, all of $C_{11}, C_{12}, C_{21}, C_{22}$ have degree at most $\lfloor d/2 \rfloor < \deg f_1 - \deg g_2$.)

Time analysis: The computation of $g_1$ uses at most $6d\mu(3d) + 2d + 1$ operations, since $\deg g_1 \le \deg f_1 \le 2d$. The division of $g_2$ by its leading coefficient uses at most $2d$ operations. The division of $g_1$ by the result uses at most $12(2d+2)(\mu(4d+2)+1)$ operations. The division of the quotient by the leading coefficient of $g_2$ uses at most $d+1$ operations since $\deg q \le \deg f_1 - \deg g_2 \le d$. The computation of $\deg r$ uses at most $2d$ equality tests.

**Step 9: compute the second half of the continued fraction.** Find a matrix $D$, of degree at most $\deg g_2 - (\deg f_1 - d)$, such that $\deg(D_{21} g_2 + D_{22} r) < \deg f_1 - d$, by applying the algorithm recursively to $\deg g_2 - (\deg f_1 - d), g_2, r$.

Compute $M = D \begin{pmatrix} 0 & 1 \\ 1 & -q \end{pmatrix} C$. Observe that $M_{21} f_1 + M_{22} f_2 = (0\ 1) M \begin{pmatrix} f_1 \\ f_2 \end{pmatrix} =$
$(0\ 1) D \begin{pmatrix} 0 & 1 \\ 1 & -q \end{pmatrix} C \begin{pmatrix} f_1 \\ f_2 \end{pmatrix} = (D_{21}\ D_{22}) \begin{pmatrix} 0 & 1 \\ 1 & -q \end{pmatrix} \begin{pmatrix} g_1 \\ g_2 \end{pmatrix} = (D_{21}\ D_{22}) \begin{pmatrix} g_2 \\ r \end{pmatrix} = D_{21} g_2 + D_{22} r$.

The answer is $M$. Indeed, the degree of $D$ is at most $\deg g_2 - \deg f_1 + d$, and the degree of $\begin{pmatrix} 0 & 1 \\ 1 & -q \end{pmatrix} C$ is at most $\deg f_1 - \deg g_2$, so the degree of $M$ is at most $d$; and $\deg(M_{21} f_1 + M_{22} f_2) = \deg(D_{21} g_2 + D_{22} r) < \deg f_1 - d$.

Time analysis: Note that $\deg g_2 - (\deg f_1 - d) \le \deg f_1 - \lfloor d/2 \rfloor - 1 - (\deg f_1 - d) \le \lfloor d/2 \rfloor$. By induction, the recursive computation of $D$ uses at most

$$(46(d/2)(k-1) + 44(2^k - 1))(\mu(4(d/2) + 8) + 1)$$

operations. The computation of $qC_{21}$ uses at most $(d+1)\mu(d+1)$ operations since $\deg qC_{21} \leq \deg f_1 - \deg g_2 \leq d$. The computation of $qC_{22}$ uses at most $(d+1)\mu(d+1)$ operations since $\deg qC_{22} = \deg f_1 - \deg g_2 \leq d$. The computation of $C_{11} - qC_{21}$ uses at most $d+1$ operations. The computation of $C_{12} - qC_{22}$ uses at most $d+1$ operations. The multiplication of $D$ by $\begin{pmatrix} 0 & 1 \\ 1 & -q \end{pmatrix} C$ uses at most $(8d+16)(\mu(4d+8)+1)$ operations.

Totals: at most $(46d(k-1) + 44(2^{k+1} - 2))(\mu(4d+8)+1)$ operations for the recursive computations of $C$ and $D$, and at most $(46d+42)\mu(4d+8) + 45d + 44 \leq (46d+44)(\mu(4d+8)+1)$ operations for everything else. The grand total is at most $(46dk + 44(2^{k+1} - 1))(\mu(4d+8)+1)$ operations as claimed.

**21.5. The integer case.** A more complicated algorithm, given a nonnegative integer $d$ and given integers $f_1$, $f_2$ not both zero, computes a (factored) $2 \times 2$ integer matrix $M$ with entries not much larger than $2^d$ in absolute value, with determinant in $\{-1, 1\}$, and with $|M_{21} f_1 + M_{22} f_2| < \max\{|f_1|, |f_2|\}/2^d$. This algorithm takes time $O(n(\lg n)^2 \lg \lg n)$, where $n$ is the total number of input bits.

The main complication here is that the answer for the top $2d$ bits of $f_1$ and $f_2$ is, in general, not exactly the answer for $f_1$ and $f_2$. One has to check whether $|M_{21} f_1 + M_{22} f_2|$ is too large, and divide a few more times if it is.

**21.6. History.** Almost all of the ideas in this algorithm were published by Lehmer [1938] in the integer case. Lehmer made the crucial observation that the top $2d$ bits of $f_1$ and $f_2$ determined approximately $d$ bits of the continued fraction for $f_1/f_2$. Lehmer suggested computing the continued fraction for $f_1/f_2$ by computing a small part of the continued fraction, computing another quotient, and then computing the rest of the continued fraction.

Shortly after fast multiplication was widely understood, Knuth [1971b] suggested replacing "a small part" with "half" in Lehmer's algorithm. Knuth proved that the continued fraction for $f_1/f_2$ could be computed within a $O((\lg n)^4)$ factor of multiplication time. Schönhage [1971] streamlined the Lehmer–Knuth algorithm and proved that the continued fraction for $f_1/f_2$ could be computed within a $O(\lg n)$ factor of multiplication time.

The $(\lg n)^3$ disparity between [Knuth 1971b] and [Schönhage 1971] arose as follows. Knuth lost one $\lg n$ factor from continually re-multiplying matrices $\begin{pmatrix} 0 & 1 \\ 1 & -q \end{pmatrix}$ instead of reusing their products $M$; another $\lg n$ factor from doing a binary search, again with no reuse of partial results, to determine how much of the continued fraction of $f_1/f_2$ matched the continued fraction of $\lfloor f_1/2^i \rfloor / \lfloor f_2/2^i \rfloor$; and one more $\lg n$ factor from an unnecessarily crude analysis.

Moenck [1973] claimed to have a simplified algorithm covering both the integer case and the polynomial case. In fact, Moenck's algorithm does not work in the integer case, does not work in the "nonnormal" situation of a quotient having degree different from 1, and does not work except when deg $f_1$ is a power of 2. The errors in [Moenck 1973] begin on page 143, where the "degree" function mapping an integer $A$ to $\lfloor \lg |A| \rfloor$ is claimed to be a homomorphism.

Brent, Gustavson, and Yun [1980, Section 3] outlined a simplified algorithm for the polynomial case. Strassen [1983, page 16] stated the remarkably clean algorithm shown above (under the assumption deg $f_1 \geq$ deg $f_2$), with one omission: Strassen's algorithm recurses forever when $d =$ deg $f_1 =$ deg $f_2 = 0$.

**21.7. Improvements.** There are many opportunities for FFT caching and FFT addition in this algorithm.

One can replace $\lfloor d/2 \rfloor$ in this algorithm by any integer between 0 and $d-1$. The optimal choice depends heavily on the exact speed of multiplication.

It is often helpful (for applications, and for recursion inside this algorithm) to compute $M_{21}f_1 + M_{22}f_2$, and sometimes $M_{11}f_1 + M_{12}f_2$, along with $M$. One can often save time by incorporating these computations into the recursion. For example, when $M$ is constructed from $D, q, C$, one can compute $M_{21}f_1 + M_{22}f_2$ as $D_{21}g_2 + D_{22}r$, and one can compute $M_{11}f_1 + M_{12}f_2$ as $D_{11}g_2 + D_{12}r$.

One can often save time by skipping $M_{11}$ and $M_{21}$, and working solely with $M_{12}, M_{22}, M_{11}f_1 + M_{12}f_2, M_{21}f_1 + M_{22}f_2$. Applications that need $M_{11}$ and $M_{21}$ can use formulas such as $M_{11} = ((M_{11}f_1 + M_{12}f_2) - M_{12}f_2)/f_1$. In [Knuth 1997, page 343] this observation is credited to Gordon H. Bradley.

Some applications need solely $M_{11}f_1 + M_{12}f_2$ and $M_{21}f_1 + M_{22}f_2$. The algorithm, and some of its recursive calls, can be sped up accordingly.

Often $f_1$ and $f_2$ have an easily detected common factor, such as $x$ or $x - 1$. Dividing out this factor speeds up the algorithm, perhaps enough to justify the cost of checking for the factor in the first place.

## 22. Greatest common divisor

**22.1. Input.** Let $A$ be a field. The algorithm in this section is given polynomials $f_1, f_2 \in A[x]$.

**22.2. Output.** This algorithm computes gcd$\{f_1, f_2\}$: in other words, a polynomial $g$ such that $gA[x] = f_1A[x] + f_2A[x]$ and such that $g$ is monic if it is nonzero.

This algorithm also computes polynomials $h_1, h_2 \in A[x]$, each of degree at most max{deg $f_1$, deg $f_2$}, such that $g = f_1h_1 + f_2h_2$.

In particular, if $f_1$ and $f_2$ are coprime, then $g = 1$; $h_1$ is a reciprocal of $f_1$ modulo $f_2$; and $h_2$ is a reciprocal of $f_2$ modulo $f_1$.

## 22.3. Speed.
This algorithm uses $O(n(\lg n)^2 \lg \lg n)$ operations in $A$, where $n$ is the total number of coefficients in $f_1, f_2$.

## 22.4. How it works.
If $f_1 = f_2 = 0$ then the answer is $0, 0, 0$. Assume from now on that at least one of $f_1$ and $f_2$ is nonzero.

Define $d = \max\{\deg f_1, \deg f_2\}$. Apply the algorithm of Section 21 to compute $M_{11}, M_{12}, M_{21}, M_{22}$ in $A[x]$, of degree at most $d$, with $M_{11}M_{22} - M_{12}M_{21} = \pm 1$ and $M_{21}f_1 + M_{22}f_2 = 0$.

Compute $u = M_{11}f_1 + M_{12}f_2$. Note that $\pm f_1 = M_{22}u$ and $\mp f_2 = M_{21}u$, so $uA[x] = f_1 A[x] + f_2 A[x]$. In particular, $u \neq 0$.

Compute $g = u/c$, $h = M_{11}/c$, and $h_2 = M_{12}/c$, where $c$ is the leading coefficient of $u$. Then $g$ is monic, and $gA[x] = f_1 A[x] + f_2 A[x]$, so $g = \gcd\{f_1, f_2\}$. The answer is $g, h_1, h_2$.

## 22.5. The integer case.
An analogous algorithm, given integers $f_1$ and $f_2$, computes $\gcd\{f_1, f_2\}$ and reasonably small integers $h_1, h_2$ with $\gcd\{f_1, f_2\} = f_1 h_1 + f_2 h_2$. This algorithm takes time $O(n(\lg n)^2 \lg \lg n)$, where $n$ is the total number of input bits.

## 22.6. History.
See Section 21. This application has always been one of the primary motivations for studying the problem of Section 21.

## 22.7. Improvements.
See Section 21.

The reader may have noticed that Section 21 and this section use division in $A((x^{-1}))$ and division in **R**. What about $\mathbf{Q}_2$? Answer: There are several "binary" algorithms to compute greatest common divisors of integers. See, e.g., [Sorenson 1994] and [Knuth 1997, pages 338–341; Exercises 4.5.2–38, 4.5.2–39, 4.5.2–40]. Stehlé and Zimmermann recently [2004] introduced a particularly clean "binary" gcd algorithm and proved that it takes time $O(n(\lg n)^2 \lg \lg n)$. The Stehlé–Zimmermann algorithm, given an odd integer $f_1$ and an even integer $f_2$, expands $f_1/f_2$ into what one might call a **simple 2-adic continued fraction**: a continued fraction with all quotients chosen from the set

$$\{\pm 1/2, \pm 1/4, \pm 3/4, \pm 1/8, \pm 3/8, \pm 5/8, \pm 7/8, \dots\}.$$

My initial impression is that this algorithm supersedes all previous work on gcd computation.

# 23. Interpolator

## 23.1. Input.
Let $A$ be a field. Let $t$ be a nonnegative integer. The algorithm in this section is given polynomials $f_1, f_2, \dots, f_t \in A[x]$ and nonzero coprime polynomials $g_1, g_2, \dots, g_t \in A[x]$.

**23.2. Output.** This algorithm computes $h \in A[x]$, with $\deg h < \deg g_1 g_2 \cdots g_t$, such that $h \equiv f_1 \pmod{g_1}$, $h \equiv f_2 \pmod{g_2}$, ..., $h \equiv f_t \pmod{g_t}$.

In particular, consider the special case that each $g_j$ is a monic linear polynomial $x - c_j$. The answer $h$ is a polynomial of degree below $t$ such that $h(c_1) = f_1(c_1)$, $h(c_2) = f_2(c_2)$, ..., $h(c_t) = f_t(c_t)$. Finding $h$ is usually called **interpolation** in this case, and I suggest using the same name for the general case. Another common name is **Chinese remaindering**.

**23.3. Speed.** This algorithm uses $O(n(\lg n)^2 \lg \lg n)$ operations in $A$, where $n$ is the total number of input coefficients.

**23.4. How it works.** For $t = 0$: The answer is 0.

Compute $G = g_1 \cdots g_t$ as explained in Section 12.

Compute $G \bmod g_1^2, \ldots, G \bmod g_t^2$ as explained in Section 18.

For each $j$ divide $G \bmod g_j^2$ by $g_j$, as explained in Section 17, to obtain $(G/g_j) \bmod g_j$. Note that $G/g_j$ and $g_j$ are coprime; thus $((G/g_j) \bmod g_j)$ and $g_j$ are coprime.

Compute a (reasonably small) reciprocal $p_j$ of $((G/g_j) \bmod g_j)$ modulo $g_j$, as explained in Section 22. Compute $q_j = f_j p_j \bmod g_j$ as explained in Section 17.

Now compute $h = (q_1/g_1 + \cdots + q_t/g_t)G$ as explained in Section 13. (Proof that, modulo $g_j$, this works: $h \equiv q_j(G/g_j) \equiv f_j p_j (G/g_j) \equiv f_j p_j (G/g_j \bmod g_j) \equiv f_j$.)

**23.5. The integer case.** An analogous algorithm, given integers $f_1, f_2, \ldots, f_t$ and given nonzero coprime integers $g_1, g_2, \ldots, g_t$, computes a reasonably small integer $h$ such that $h \equiv f_1 \pmod{g_1}$, $h \equiv f_2 \pmod{g_2}$, ..., $h \equiv f_t \pmod{g_t}$. The algorithm takes time $O(n(\lg n)^2 \lg \lg n)$, where $n$ is the total number of input bits.

**23.6. History.** Horowitz [1972] published most of the above algorithm, in the special case that each $g_j$ is a monic linear polynomial. Horowitz did not have a fast method (for large $t$) to compute $(G/g_1) \bmod g_1, \ldots, (G/g_t) \bmod g_t$ from $G$. Moenck and Borodin [1972, page 95] suggested the above solution in the ("single-precision") integer case; see also [Borodin and Moenck 1974, page 381].

The special case $t = 2$ was published first by Heindel and Horowitz [1971], along with a different essentially-linear-time interpolation algorithm for general $t$. The Heindel–Horowitz algorithm is summarized below; it takes time $O(n(\lg n)^3 \lg \lg n)$.

**23.7. Improvements.** When $g_j$ is a linear polynomial, $(G/g_j) \bmod g_j$ has degree 0, so $p_j$ is simply $1/((G/g_j) \bmod g_j)$, and $q_j$ is $(f_j \bmod g_j)/((G/g_j) \bmod g_j)$. More generally, whenever $g_j$ is very small, the algorithm of this section provides very small inputs to the modular-reciprocal algorithm of Section 22.

When $g_j$ is a monic linear polynomial, $(G/g_j) \bmod g_j$ is the same as $G' \bmod g_j$, where $G'$ is the derivative of $G$. Borodin and Moenck [1974, Sections 8–9] suggested computing $G' \bmod g_1, \ldots, G' \bmod g_t$ as explained in Section 18, instead of computing $G \bmod g_1^2, \ldots, G \bmod g_t^2$.

More generally, if $g_j$ and its derivative $g'_j$ are coprime, then $(G/g_j) \bmod g_j$ is the same as $(g'_j)^{-1} G' \bmod g_j$. One can compute $G' \bmod g_1, \ldots, G' \bmod g_t$; compute each reciprocal $(G')^{-1} \bmod g_j$; and compute $q_j = f_j g'_j (G')^{-1} \bmod g_j$.

Another way to compute $(G/g_j) \bmod g_j$, published by Bürgisser, Clausen, and Shokrollahi [1997, pages 77–78], is to first compute $G/g_1 + \cdots + G/g_t$ as explained in Section 13, then compute $(G/g_1 + \cdots + G/g_t) \bmod g_j = (G/g_j) \bmod g_j$ for all $j$ as explained in Section 18.

When $t = 2$, one can use the algorithm of Section 22 to simultaneously compute a reciprocal $p_2$ of $g_1 = G/g_2$ modulo $g_2$ and a reciprocal $p_1$ of $g_2 = G/g_1$ modulo $g_1$. The answer is then $(f_1 p_1 \bmod g_1) g_2 + (f_2 p_2 \bmod g_2) g_1$. It might be faster to compute $(f_1 p_1 g_2 + f_2 p_2 g_1) \bmod g_1 g_2$.

One can skip the computation of $p_1$ when $f_1 = 0$. One can reduce the general case to this case: interpolate $0, f_2 - f_1, f_3 - f_1, \ldots, f_t - f_1$ and then add $f_1$ to the result. In particular, for $t = 2$, the answer is $f_1 + ((f_2 - f_1) p_2 \bmod g_2) g_1$, if $f_1$ is small enough for that answer to be in the right range.

The Heindel–Horowitz algorithm interpolates pairs, then pairs of pairs, etc. This may be better than the Horowitz–Borodin–Moenck algorithm for small $t$.

One can cache the reciprocals $p_j$ for subsequent interpolations involving the same $g_1, \ldots, g_t$.

## 24. Coprime base

**24.1. Input.** Let $A$ be a field. Let $t$ be a nonnegative integer. The algorithm in this section is given monic polynomials $f_1, f_2, \ldots, f_t \in A[x]$.

**24.2. Output.** This algorithm computes a **coprime base** for $\{f_1, f_2, \ldots, f_t\}$: coprime monic polynomials $g_1, g_2, \ldots \in A[x]$ such that each $f_j$ can be factored as a product of powers of $g_1, g_2, \ldots$. In fact, the algorithm computes the **natural coprime base** for $\{f_1, f_2, \ldots, f_t\}$: the unique coprime base that does not contain 1 and that can be obtained from $f_1, f_2, \ldots, f_t$ by multiplication, exact division, and greatest common divisors.

Sample application: Given a polynomial $f_1 \in (\mathbf{Z}/2)[x]$, compute a $(\mathbf{Z}/2)$-basis $(f_2, \ldots, f_t)$ for the vector space $\{h \in (\mathbf{Z}/2)[x] : (f_1 h)' = h^2\}$. Then the natural coprime base for $f_1, \ldots, f_t$ contains all irreducible divisors of $f_1$. See [Göttfert 1994].

Many applications also want **factorization into coprimes**: the factorization of each $f_j$ over the coprime base $g_1, g_2, \ldots$. These factorizations can be computed quickly by an extension of the algorithm of Section 20.

**24.3. Speed.** This algorithm uses $O(n(\lg n)^7 \lg \lg n)$ operations in $A$, where $n$ is the total number of input coefficients.

**24.4. How it works.** Exercise for the reader! Three hints: (1) There is no need for any subroutines other than multiplication, exact division, and greatest common divisors. (2) The natural coprime base for $\{f^a, f^b\}$ is $\{f^{\gcd\{a,b\}}\} - \{1\}$; one can use a left-shift gcd algorithm. (3) Given coprime $g_1, g_2, \ldots$ one can quickly construct a *very small* set having natural coprime base $\{g_1, g_2, \ldots\} - \{1\}$.

**24.5. The integer case.** An analogous algorithm computes the natural coprime base of a set of positive integers $f_1, f_2, \ldots, f_t$. This algorithm takes time $O(n(\lg n)^7 \lg \lg n)$, where $n$ is the total number of input bits.

Sample application: If $f_1$ has at most $n$ bits then one can, in time $n(\lg n)^{O(1)}$, find the maximum integer $k$ such that $f_1$ is a $k$th power. The idea is as follows: compute good approximations $f_2, f_3, \ldots$ to $f_1^{1/2}, f_1^{1/3}, \ldots$; factor $f_1, f_2, f_3, \ldots$ into coprimes; and compute the greatest common divisor of the exponents of the factorization of $f_1$. See [Bernstein et al. 2007] for details.

**24.6. History.** I published this algorithm in [Bernstein 2005], after a decade of increasingly detailed outlines. No previous essentially-linear-time algorithms were known, even in the case $t = 2$. My newer paper [Bernstein 2004a] outlines an improvement from $(\lg n)^7$ to $(\lg n)^4$.

## References

[AFIPS 17] — (no editor), *AFIPS conference proceedings, volume 17: 1960 Western Joint Computer Conference*. ISSN 0095–6880. See [Estrin 1960].

[AFIPS 28] — (no editor), *AFIPS conference proceedings, volume 28: 1966 Spring Joint Computer Conference*, Spartan Books, Washington. See [Stockham 1966].

[AFIPS 29] — (no editor), *AFIPS conference proceedings, volume 29: 1966 Fall Joint Computer Conference*, Spartan Books, Washington. See [Gentleman and Sande 1966].

[AFIPS 33] — (no editor), *AFIPS conference proceedings, volume 33, part one: 1968 Fall Joint Computer Conference, December 9–11, 1968, San Francisco, California*, Thompson Book Company, Washington. See [Yavne 1968].

[ICM 1971.3] — (no editor), *Actes du congrès international des mathématiciens, tome 3*, Gauthier-Villars Éditeur, Paris. MR 54:5. See [Knuth 1971b].

[Aho 1973] Alfred V. Aho (chairman), *Proceedings of fifth annual ACM symposium on theory of computing: Austin, Texas, April 30–May 2, 1973*, Association for Computing Machinery, New York. See [Moenck 1973].

[Anderssen and Brent 1976] Robert S. Anderssen and Richard P. Brent (editors), *The complexity of computational problem solving*, University of Queensland Press, Brisbane. ISBN 0–7022–1213–X. URL: http://web.comlab.ox.ac.uk/oucl/work/richard.brent/pub/pub031.html. See [Brent 1976a].

[Beeler et al. 1972] Michael Beeler, R. William Gosper, and Richard Schroeppel, *HAKMEM*, Artificial Intelligence Memo No. 239, Massachusetts Institute of Technology. URL: http://www.inwap.com/pdp10/hbaker/hakmem/hakmem.html. Citations in this document: §16.7.

[Bergland 1968] Glenn D. Bergland, "A fast Fourier transform algorithm for real-valued series", *Communications of the ACM* **11**, 703–710. ISSN 0001–0782. URL: http://cr.yp.to/bib/entries.html#1968/bergland-real. Citations in this document: §2.9.

[Bernstein 1987] Daniel J. Bernstein, "New fast algorithms for $\pi$ and $e$", paper for the Westinghouse competition, distributed widely at the Ramanujan Centenary Conference. URL: http://cr.yp.to/papers.html#westinghouse. Citations in this document: §12.7.

[Bernstein 1998] Daniel J. Bernstein, "Detecting perfect powers in essentially linear time", *Mathematics of Computation* **67**, 1253–1283. ISSN 0025–5718. MR 98j:11121. URL: http://cr.yp.to/papers.html#powers. Citations in this document: §6.6.

[Bernstein 2001] Daniel J. Bernstein, "Multidigit multiplication for mathematicians". URL: http://cr.yp.to/papers.html#m3. Citations in this document: §2.8, §3.8, §4.10, §4.10.

[Bernstein 2002a] Daniel J. Bernstein, "How to find small factors of integers". URL: http://cr.yp.to/papers.html#sf. Citations in this document: §20.6, §20.7.

[Bernstein 2002b] Daniel J. Bernstein, "Pippenger's exponentiation algorithm". URL: http://cr.yp.to/papers.html#pippenger. Citations in this document: §10.7.

[Bernstein 2003] Daniel J. Bernstein, "Computing logarithm intervals with the arithmetic-geometric-mean iteration". URL: http://cr.yp.to/papers.html#logagm. ID 8f92 b1e3ec7918d37b28b9efcee5e97f. Citations in this document: §16.7.

[Bernstein 2004a] Daniel J. Bernstein, "Research announcement: Faster factorization into coprimes". URL: http://cr.yp.to/papers.html#dcba2. ID 53a2e278e21bcbb7287 b81c563995925. Citations in this document: §24.6.

[Bernstein 2004b] Daniel J. Bernstein, "How to find smooth parts of integers". URL: http://cr.yp.to/papers.html#smoothparts. ID 201a045d5bb24f43f0bd0d97fcf5355a. Citations in this document: §20.7.

[Bernstein 2004c] Daniel J. Bernstein, "Removing redundancy in high-precision Newton iteration". URL: http://cr.yp.to/papers.html#fastnewton. ID def7f1e35fb654671 c6f767b16b93d50. Citations in this document: §6.8, §7.7, §9.7, §10.7, §10.7.

[Bernstein 2004d] Daniel J. Bernstein, "Scaled remainder trees". URL: http://cr.yp.to/papers.html#scaledmod. ID e2b8da026cf72d01d97e20cf2874f278. Citations in this document: §18.9.

[Bernstein 2005] Daniel J. Bernstein, "Factoring into coprimes in essentially linear time", *Journal of Algorithms* **54**, 1–30. ISSN 0196–6774. URL: http://cr.yp.to/papers.html#dcba. ID f32943f0bb67a9317d4021513f9eee5a. Citations in this document: §20.6, §20.6, §24.6.

[Bernstein 2007] Daniel J. Bernstein, "The tangent FFT", pp. 291–300 in [Boztas and Lu 2007]. URL: http://cr.yp.to/papers.html#tangentfft. ID a9a77cef9a7b77f9b8b305 e276d5fe25. Citations in this document: §2.9.

[Bernstein et al. 2007] Daniel J. Bernstein, Hendrik W. Lenstra, Jr., and Jonathan Pila, "Detecting perfect powers by factoring into coprimes", *Mathematics of Computation* **76**, 385–388. Citations in this document: §24.5.

[Borodin and Moenck 1974] Allan Borodin and Robert T. Moenck, "Fast modular transforms", *Journal of Computer and System Sciences* **8**, 366–386; older version, not a subset, in [Moenck and Borodin 1972]. ISSN 0022–0000. MR 51:7365. URL: http://cr.yp.to/bib/entries.html#1974/borodin. Citations in this document: §12.7, §13.6, §17.9, §18.8, §23.6, §23.7.

[Borwein and Borwein 1987] Jonathan M. Borwein and Peter B. Borwein, *Pi and the AGM*, Wiley, New York. ISBN 0–471–83138–7. MR 89a:11134. Citations in this document: §16.7.

[Borwein et al. 2000] Jonathan M. Borwein, David M. Bradley, and Richard E. Crandall, "Computational strategies for the Riemann zeta function", *Journal of Computational and Applied Mathematics* **121**, 247–296. ISSN 0377–0427. MR 2001h:11110. URL: http://www.sciencedirect.com/science/article/B6TYH-4118GDF-F/1/64371ba75fa0 e923ba6b231779fb0673. Citations in this document: §12.7.

[Bostan et al. 2003] Alin Bostan, Grégoire Lecerf, and Éric Schost, "Tellegen's principle into practice", pp. 37–44 in [Hong 2003]. URL: http://cr.yp.to/bib/entries.html# 2003/bostan. Citations in this document: §18.9.

[Bostan et al. 2004] Alin Bostan, Grégoire Lecerf, Bruno Salvy, Éric Schost, and Bernd Wiebelt, "Complexity issues in bivariate polynomial factorization", pp. 42–49 in [Gutierrez 2004]. URL: http://cr.yp.to/bib/entries.html#2004/bostan. Citations in this document: §18.9.

[Boztas and Lu 2007] Serdar Boztas and Hsiao-Feng Lu (editors), *Applied algebra, algebraic algorithms and error-correcting codes: 17th international symposium, AAECC-17, Bangalore, India, December 2007, proceedings*, Lecture Notes in Computer Science **4851**, Springer-Verlag, Berlin. See [Bernstein 2007].

[Brauer 1939] Alfred Brauer, "On addition chains", *Bulletin of the American Mathematical Society* **45**, 736–739. ISSN 0273–0979. MR 1,40a. URL: http://cr.yp.to/bib/entries.html#1939/brauer. Citations in this document: §10.7.

[Brent 1976a] Richard P. Brent, "The complexity of multiple-precision arithmetic", pp. 126–165 in [Anderssen and Brent 1976]. URL: http://web.comlab.ox.ac.uk/oucl/work/richard.brent/pub/pub032.html. Citations in this document: §12.7, §16.6.

[Brent 1976b] Richard P. Brent, "Fast multiple-precision evaluation of elementary functions", *Journal of the ACM* **23**, 242–251. ISSN 0004–5411. MR 52:16111. URL: http://web.comlab.ox.ac.uk/oucl/work/richard.brent/pub/pub034.html. Citations in this document: §16.7.

[Brent 1976c] Richard P. Brent, "Multiple-precision zero-finding methods and the complexity of elementary function evaluation", pp. 151–176 in [Traub 1976]. MR 54: 11843. URL: http://web.comlab.ox.ac.uk/oucl/work/richard.brent/pub/pub028.html. Citations in this document: §6.8, §8.6, §9.6, §9.7, §16.7.

[Brent and McMillan 1980] Richard P. Brent and Edwin M. McMillan, "Some new algorithms for high-precision computation of Euler's constant", *Mathematics of Computation* **34**, 305–312. ISSN 0025–5718. MR 82g:10002. URL: http://web.comlab.ox.ac.uk/oucl/work/richard.brent/pub/pub049.html. Citations in this document: §12.7.

[Brent et al. 1980] Richard P. Brent, Fred G. Gustavson, and David Y. Y. Yun, "Fast solution of Toeplitz systems of equations and computation of Padé approximants", *Journal of Algorithms* **1**, 259–295. ISSN 0196–6774. MR 82d:65033. URL: http://web.comlab.ox.ac.uk/oucl/work/richard.brent/pub/pub059.html. Citations in this document: §21.6.

[Bruun 1978] Georg Bruun, "*z*-transform DFT filters and FFTs", *IEEE Transactions on Acoustics, Speech, and Signal Processing* **26**, 56–63. ISSN 0096–3518. URL: http://cr.yp.to/bib/entries.html#1978/bruun. Citations in this document: §2.9.

[Buell 2004] Duncan A. Buell (editor), *Algorithmic number theory: 6th international symposium, ANTS-VI, Burlington, VT, USA, June 2004, proceedings*, Lecture Notes in Computer Science **3076**, Springer-Verlag, Berlin. ISBN 3–540–22156–5. See [Franke et al. 2004], [Stehlé and Zimmermann 2004].

[Bürgisser et al. 1997] Peter Bürgisser, Michael Clausen, and Mohammed Amin Shokrollahi, *Algebraic complexity theory*, Springer-Verlag, Berlin. ISBN 3–540–60582–7. MR 99c:68002. Citations in this document: §2.3, §4.9, §11.8, §12.8, §23.7.

[Buhler 1998] Joe P. Buhler (editor), *Algorithmic number theory: ANTS-III*, Lecture Notes in Computer Science **1423**, Springer-Verlag, Berlin. ISBN 3–540–64657–4. MR 2000g:11002. See [Haible and Papanikolaou 1998].

[Burnikel and Ziegler 1998] Christoph Burnikel and Joachim Ziegler, *Fast recursive division*, MPI research report I-98-1-022. URL: http://data.mpi-sb.mpg.de/internet/reports.nsf/NumberView/1998-1-022. Citations in this document: §17.9, §17.9, §17.9.

[Cantor and Kaltofen 1991] David G. Cantor and Erich Kaltofen, "On fast multiplication of polynomials over arbitrary algebras", *Acta Informatica* **28**, 693–701. ISSN 0001–5903. MR 92i:68068. URL: http://www.math.ncsu.edu/~kaltofen/bibliography/. Citations in this document: §4.9.

[Chudnovsky and Chudnovsky 1990] David V. Chudnovsky and Gregory V. Chudnovsky, "Computer algebra in the service of mathematical physics and number theory", pp. 109–232 in [Chudnovsky and Jenks 1990]. MR 92g:11122. Citations in this document: §12.7, §12.7.

[Chudnovsky and Jenks 1990] David V. Chudnovsky and Richard D. Jenks (editors), *Computers in mathematics*, Lecture Notes in Pure and Applied Mathematics **125**, Marcel Dekker, New York. ISBN 0–8247–8341–7. MR 91e:00020. See [Chudnovsky and Chudnovsky 1990], [Gosper 1990].

[Cook 1966] Stephen A. Cook, *On the minimum computation time of functions*, Ph.D. thesis, Department of Mathematics, Harvard University. URL: http://cr.yp.to/bib/entries.html#1966/cook. Citations in this document: §4.9, §6.7, §10.7.

[Cooley and Tukey 1965] James W. Cooley and John W. Tukey, "An algorithm for the machine calculation of complex Fourier series", *Mathematics of Computation* **19**, 297–301. ISSN 0025–5718. MR 31:2843. URL: http://cr.yp.to/bib/entries.html#1965/cooley. Citations in this document: §2.8.

[Crandall and Fagin 1994] Richard Crandall and Barry Fagin, "Discrete weighted transforms and large-integer arithmetic", *Mathematics of Computation* **62**, 305–324. ISSN 0025–5718. MR 94c:11123. URL: http://cr.yp.to/bib/entries.html#1994/crandall. Citations in this document: §2.9, §4.10.

[Dixon 1982] John D. Dixon, "Exact solution of linear equations using $p$-adic expansions", *Numerische Mathematik* **40**, 137–141. ISSN 0029–599X. MR 83m:65025. URL: http://cr.yp.to/bib/entries.html#1982/dixon. Citations in this document: §17.9.

[Duhamel and Hollmann 1984] Pierre Duhamel and H. Hollmann, "Split-radix FFT algorithm", *Electronics Letters* **20**, 14–16. ISSN 0013–5194. URL: http://cr.yp.to/bib/entries.html#1984/duhamel. Citations in this document: §2.9.

[Duhamel and Vetterli 1990] Pierre Duhamel and Martin Vetterli, "Fast Fourier transforms: a tutorial review and a state of the art", *Signal Processing* **19**, 259–299. ISSN 0165–1684. MR 91a:94004. URL: http://cr.yp.to/bib/entries.html#1990/duhamel. Citations in this document: §2.9.

[Estrin 1960] Gerald Estrin, "Organization of computer systems — the fixed plus variable structure computer", pp. 33–40 in [AFIPS 17]. URL: http://cr.yp.to/bib/entries.html#1960/estrin. Citations in this document: §12.7.

[Fiduccia 1972] Charles M. Fiduccia, "Polynomial evaluation via the division algorithm: the fast Fourier transform revisited", pp. 88–93 in [Rosenberg 1972]. URL: http://cr.yp.to/bib/entries.html#1972/fiduccia-fft. Citations in this document: §2.8.

[Franke et al. 2004] Jens Franke, Thorsten Kleinjung, François Morain, and T. Wirth, "Proving the primality of very large numbers with fastECPP", pp. 194–207 in [Buell 2004]. URL: http://www.lix.polytechnique.fr/Labo/Francois.Morain/Articles/large.ps.gz. Citations in this document: §20.7.

[Fürer 2007] Martin Fürer, "Faster integer multiplication", pp. 57–66 in [Johnson and Feige 2007]. URL: http://www.cse.psu.edu/~furer/. Citations in this document: §3.8.

[Gauss 1866] Carl F. Gauss, *Werke, Band 3*, Königlichen Gesellschaft der Wissenschaften, Göttingen. URL: http://134.76.163.65/agora_docs/41929TABLE_OF_CONTENTS.html. Citations in this document: §2.8, §2.8.

[Gentleman and Sande 1966] W. Morven Gentleman and Gordon Sande, "Fast Fourier transforms — for fun and profit", pp. 563–578 in [AFIPS 29]. URL: http://cr.yp.to/bib/entries.html#1966/gentleman. Citations in this document: §2.8, §2.9.

[Göttfert 1994] Rainer Göttfert, "An acceleration of the Niederreiter factorization algorithm in characteristic 2", *Mathematics of Computation* **62**, 831–839. ISSN 0025-5718. MR 94g:11110. Citations in this document: §24.2.

[Gosper 1990] William Gosper, "Strip mining in the abandoned orefields of nineteenth century mathematics", pp. 261–284 in [Chudnovsky and Jenks 1990]. MR 91h:11154. URL: http://cr.yp.to/bib/entries.html#1990/gosper. Citations in this document: §12.7, §12.7.

[Gregory 1980] Robert T. Gregory, *Error-free computation: why it is needed and methods for doing it*, Robert E. Krieger Publishing Company, New York. ISBN 0-89874-240-4. MR 83f:65061. Citations in this document: §17.9.

[Gutierrez 2004] Jamie Gutierrez (editor), *Proceedings of the 2004 international symposium on symbolic and algebraic computation*, Association for Computing Machinery, New York. ISBN 1-58113-827-X. See [Bostan et al. 2004].

[Haible and Papanikolaou 1997] Bruno Haible and Thomas Papanikolaou, *Fast multiprecision evaluation of series of rational numbers*, Technical Report TI-7/97, Darmstadt University of Technology; see also newer version [Haible and Papanikolaou 1998]. URL: http://www.informatik.tu-darmstadt.de/TI/Mitarbeiter/papanik/Welcome.html. Citations in this document: §12.7.

[Haible and Papanikolaou 1998] Bruno Haible and Thomas Papanikolaou, "Fast multiprecision evaluation of series of rational numbers", pp. 338–350 in [Buhler 1998]; see also older version [Haible and Papanikolaou 1997]. MR 2000i:11197. URL: http://cr.yp.to/bib/entries.html#1998/haible.

[Hehner and Horspool 1979] Eric C. R. Hehner and R. Nigel Horspool, "A new representation of the rational numbers for fast easy arithmetic", *SIAM Journal on Computing* **8**, 124–134. ISSN 0097-5397. MR 80h:68027. URL: http://cr.yp.to/bib/entries.html#1979/hehner. Citations in this document: §17.9.

[Heideman et al. 1985] Michael T. Heideman, Don H. Johnson, and C. Sidney Burrus, "Gauss and the history of the fast Fourier transform", *Archive for History of Exact Sciences* **34**, 265–277. ISSN 0003-9519. MR 87f:01018. URL: http://cr.yp.to/bib/entries.html#1985/heideman. Citations in this document: §2.8.

[Heindel and Horowitz 1971] Lee E. Heindel and Ellis Horowitz, "On decreasing the computing time for modular arithmetic", pp. 126–128 in [Hennie 1971]. URL: http://cr.yp.to/bib/entries.html#1971/heindel. Citations in this document: §23.6.

[Hennie 1971] Fred C. Hennie (chairman), *12th annual symposium on switching and automata theory*, IEEE Computer Society, Northridge. See [Heindel and Horowitz 1971].

[Hong 2003] Hoon Hong (editor), *Proceedings of the 2003 international symposium on symbolic and algebraic computation*, Association for Computing Machinery, New York. ISBN 1–58113–641–2. See [Bostan et al. 2003].

[Horowitz 1972] Ellis Horowitz, "A fast method for interpolation using preconditioning", *Information Processing Letters* **1**, 157–163. ISSN 0020–0190. MR 47:4413. URL: http://cr.yp.to/bib/entries.html#1972/horowitz. Citations in this document: §13.6, §23.6.

[Jebelean 1993] Tudor Jebelean, "An algorithm for exact division", *Journal of Symbolic Computation* **15**, 169–180. ISSN 0747–7171. MR 93m:68092. URL: http://cr.yp.to/bib/entries.html#1993/jebelean. Citations in this document: §17.9.

[Jebelean 1997] Tudor Jebelean, "Practical integer division with Karatsuba complexity", pp. 339–341 in [Kuechlin 1997]. Citations in this document: §17.9.

[Johnson and Feige 2007] David S. Johnson and Uriel Feige (editors), *Proceedings of the 39th annual ACM symposium on theory of computing, San Diego, California, USA, June 11–13, 2007*, Association for Computing Machinery, New York. ISBN 978–1–59593–631–8. See [Fürer 2007].

[Johnson and Frigo 2007] Steven G. Johnson and Matteo Frigo, "A modified split-radix FFT with fewer arithmetic operations", *IEEE Transactions on Signal Processing* **55**, 111–119. Citations in this document: §2.9.

[Karatsuba 1999] Ekatharine A. Karatsuba, "Fast evaluation of hypergeometric functions by FEE", pp. 303–314 in [Papamichael et al. 1999]. MR 2000e:65030. URL: http://cr.yp.to/bib/entries.html#1999/karatsuba. Citations in this document: §12.7.

[Karatsuba and Ofman 1963] Anatoly A. Karatsuba and Y. Ofman, "Multiplication of multidigit numbers on automata", *Soviet Physics Doklady* **7**, 595–596. ISSN 0038–5689. URL: http://cr.yp.to/bib/entries.html#1963/karatsuba. Citations in this document: §4.9, §4.9.

[Karp 1972] Richard M. Karp (chairman), *13th annual symposium on switching and automata theory*, IEEE Computer Society, Northridge. See [Moenck and Borodin 1972].

[Karp and Markstein 1994] Alan H. Karp and Peter Markstein, *High-precision division and square root*, Technical Report HPL-93-42(R.1); see also newer version [Karp and Markstein 1997]. URL: http://www.hpl.hp.com/techreports/93/HPL-93-42.html.

[Karp and Markstein 1997] Alan H. Karp and Peter Markstein, "High-precision division and square root", *ACM Transactions on Mathematical Software* **23**, 561–589; see also older version [Karp and Markstein 1994]. ISSN 0098–3500. MR 1 671 702. URL: http://www.hpl.hp.com/personal/Alan_Karp/publications/publications.html. Citations in this document: §7.7.

[Knuth 1969] Donald E. Knuth, *The art of computer programming, volume 2: seminumerical algorithms*, 1st edition, 1st printing, Addison-Wesley, Reading; see also newer version [Knuth 1971a]. MR 44:3531.

[Knuth 1971a] Donald E. Knuth, *The art of computer programming, volume 2: seminumerical algorithms*, 1st edition, 2nd printing, Addison-Wesley, Reading; see also older version [Knuth 1969]; see also newer version [Knuth 1981]. MR 44:3531. Citations in this document: §12.7.

[Knuth 1971b] Donald E. Knuth, "The analysis of algorithms", pp. 269–274 in [ICM 1971.3]. MR 54:11839. URL: http://cr.yp.to/bib/entries.html#1971/knuth-gcd. Citations in this document: §12.7, §12.7, §21.6, §21.6.

[Knuth 1981] Donald E. Knuth, *The art of computer programming, volume 2: seminumerical algorithms*, 2nd edition, Addison-Wesley, Reading; see also older version [Knuth 1971a]; see also newer version [Knuth 1997]. ISBN 0–201–03822–6. MR 83i:68003.

[Knuth 1997] Donald E. Knuth, *The art of computer programming, volume 2: seminumerical algorithms*, 3rd edition, Addison-Wesley, Reading; see also older version [Knuth 1981]. ISBN 0–201–89684–2. Citations in this document: §4.10, §5.7, §6.6, §11.7, §21.7, §22.7.

[Knuth 2000] Donald E. Knuth (editor), *Selected papers on analysis of algorithms*, CSLI Publications, Stanford. ISBN 1–57586–212–3. MR 2001c:68066. See [Knuth and Papadimitriou 1981].

[Knuth and Papadimitriou 1981] Donald E. Knuth and Christos H. Papadimitriou, "Duality in addition chains", *Bulletin of the European Association for Theoretical Computer Science* **13**, 2–4; reprinted in [Knuth 2000]. ISSN 0252–9742. Citations in this document: §2.3.

[Kogge 1974] Peter M. Kogge, "Parallel solution of recurrence problems", *IBM Journal of Research and Development* **18**, 138–148. ISSN 0018–8646. MR 49:6552. URL: http://cr.yp.to/bib/entries.html#1974/kogge. Citations in this document: §12.7.

[Kogge and Stone 1973] Peter M. Kogge and Harold S. Stone, "A parallel algorithm for the efficient solution of a general class of recurrence equations", *IEEE Transactions on Computers* **22**, 786–793. ISSN 0018–9340. URL: http://cr.yp.to/bib/entries.html#1973/kogge. Citations in this document: §12.7, §12.7.

[Kollerstrom 1992] Nick Kollerstrom, "Thomas Simpson and 'Newton's method of approximation': an enduring myth", *British Journal for the History of Science* **1992**, 347–354. URL: http://www.ucl.ac.uk/sts/nk/newtonapprox.htm. Citations in this document: §6.7.

[Krishnamurthy 1977] E. V. Krishnamurthy, "Matrix processors using $p$-adic arithmetic for exact linear computations", *IEEE Transactions on Computers* **26**, 633–639. ISSN 0018–9340. MR 57:7963. URL: http://cr.yp.to/bib/entries.html#1977/krishnamurthy. Citations in this document: §17.9.

[Kuechlin 1997] Wolfgang Kuechlin (editor), *Symbolic and algebraic computation: ISSAC '97*, Association for Computing Machinery, New York. ISBN 0–89791–875–4. See [Jebelean 1997].

[Kung 1974] H. T. Kung, "On computing reciprocals of power series", *Numerische Mathematik* **22**, 341–348. ISSN 0029–599X. MR 50:3536. URL: http://cr.yp.to/bib/entries.html#1974/kung. Citations in this document: §6.7.

[Lehmer 1938] Derrick H. Lehmer, "Euclid's algorithm for large numbers", *American Mathematical Monthly* **45**, 227–233. ISSN 0002-9890. URL: http://links.jstor.org/sici?sici=0002-9890(193804)45:4<227:EAFLN>2.0.CO;2-Y. Citations in this document: §21.6.

[Lenstra and Tijdeman 1982] Hendrik W. Lenstra, Jr. and Robert Tijdeman (editors), *Computational methods in number theory I*, Mathematical Centre Tracts **154**, Mathematisch Centrum, Amsterdam. ISBN 90-6196-248-X. MR 84c:10002. See [Turk 1982].

[Lundy and Van Buskirk 2007] Thomas J. Lundy and James Van Buskirk, "A new matrix approach to real FFTs and convolutions of length $2^k$", *Computing* **80**, 23–45. ISSN 0010-485X. Citations in this document: §2.9.

[Martens 1984] Jean-Bernard Martens, "Recursive cyclotomic factorization — a new algorithm for calculating the discrete Fourier transform", *IEEE Transactions on Acoustics, Speech, and Signal Processing* **32**, 750–761. ISSN 0096-3518. MR 86b:94004. URL: http://cr.yp.to/bib/entries.html#1984/martens. Citations in this document: §2.9.

[Moenck 1973] Robert T. Moenck, "Fast computation of GCDs", pp. 142–151 in [Aho 1973]. URL: http://cr.yp.to/bib/entries.html#1973/moenck. Citations in this document: §21.6, §21.6.

[Moenck and Borodin 1972] Robert T. Moenck and Allan Borodin, "Fast modular transforms via division", pp. 90–96 in [Karp 1972]; newer version, not a superset, in [Borodin and Moenck 1974]. URL: http://cr.yp.to/bib/entries.html#1972/moenck. Citations in this document: §12.7, §17.9, §18.8, §23.6.

[Montgomery 1985] Peter L. Montgomery, "Modular multiplication without trial division", *Mathematics of Computation* **44**, 519–521. ISSN 0025-5718. MR 86e:11121. Citations in this document: §17.9, §17.9.

[Montgomery 1992] Peter L. Montgomery, *An FFT extension of the elliptic curve method of factorization*, Ph.D. thesis, University of California at Los Angeles. URL: http://cr.yp.to/bib/entries.html#1992/montgomery. Citations in this document: §2.9, §18.9.

[Nicholson 1971] Peter J. Nicholson, "Algebraic theory of finite Fourier transforms", *Journal of Computer and System Sciences* **5**, 524–547. ISSN 0022-0000. MR 44:4112. Citations in this document: §3.7.

[Nussbaumer 1980] Henri J. Nussbaumer, "Fast polynomial transform algorithms for digital convolution", *IEEE Transactions on Acoustics, Speech, and Signal Processing* **28**, 205–215. ISSN 0096-3518. MR 80m:94004. URL: http://cr.yp.to/bib/entries.html#1980/nussbaumer. Citations in this document: §3.7.

[Papadimitriou 1994] Christos M. Papadimitriou, *Computational complexity*, Addison-Wesley, Reading, Massachusetts. ISBN 0201530821. MR 95f:68082. Citations in this document: §3.6.

[Papamichael et al. 1999] Nicolas Papamichael, Stephan Ruscheweyh, and Edward B. Saff (editors), *Computational methods and function theory 1997: proceed-*

*ings of the third CMFT conference, 13–17 October 1997, Nicosia, Cyprus*, Series in Approximations and Decompositions **11**, World Scientific, Singapore. ISBN 9810236263. MR 2000c:00029. See [Karatsuba 1999].

[Pollard 1971] John M. Pollard, "The fast Fourier transform in a finite field", *Mathematics of Computation* **25**, 365–374. ISSN 0025-5718. MR 46:1120. URL: http://cr.yp.to/bib/entries.html#1971/pollard. Citations in this document: §3.7.

[Rosenberg 1972] Arnold L. Rosenberg (chairman), *Fourth annual ACM symposium on theory of computing*, Association for Computing Machinery, New York. MR 50:1553. See [Fiduccia 1972].

[Salamin 1976] Eugene Salamin, "Computation of $\pi$ using arithmetic-geometric mean", *Mathematics of Computation* **30**, 565–570. ISSN 0025-5718. MR 53:7928. Citations in this document: §16.7.

[Schönhage 1966] Arnold Schönhage, "Multiplikation großer Zahlen", *Computing* **1**, 182–196. ISSN 0010-485X. MR 34:8676. URL: http://cr.yp.to/bib/entries.html#1966/schoenhage. Citations in this document: §4.9.

[Schönhage 1971] Arnold Schönhage, "Schnelle Berechnung von Kettenbruchentwicklungen", *Acta Informatica* **1**, 139–144. ISSN 0001-5903. URL: http://cr.yp.to/bib/entries.html#1971/schoenhage-gcd. Citations in this document: §21.6, §21.6.

[Schönhage 1977] Arnold Schönhage, "Schnelle Multiplikation von Polynomen über Körpern der Charakteristik 2", *Acta Informatica* **7**, 395–398. ISSN 0001-5903. MR 55:9604. URL: http://cr.yp.to/bib/entries.html#1977/schoenhage. Citations in this document: §3.7.

[Schönhage 2000] Arnold Schönhage, "Variations on computing reciprocals of power series", *Information Processing Letters* **74**, 41–46. ISSN 0020-0190. MR 2001c:68069. Citations in this document: §6.8.

[Schönhage and Strassen 1971] Arnold Schönhage and Volker Strassen, "Schnelle Multiplikation großer Zahlen", *Computing* **7**, 281–292. ISSN 0010-485X. MR 45:1431. URL: http://cr.yp.to/bib/entries.html#1971/schoenhage-mult. Citations in this document: §3.7.

[Schönhage and Vetter 1994] Arnold Schönhage and Ekkehart Vetter, "A new approach to resultant computations and other algorithms with exact division", pp. 448–459 in [van Leeuwen 1994]. MR 96d:68109. URL: http://cr.yp.to/bib/entries.html#1994/schoenhage-exact. Citations in this document: §17.9.

[Schönhage et al. 1994] Arnold Schönhage, Andreas F. W. Grotefeld, and Ekkehart Vetter, *Fast algorithms: a multitape Turing machine implementation*, Bibliographisches Institut, Mannheim. ISBN 3-411-16891-9. MR 96c:68043. Citations in this document: §6.8.

[Sieveking 1972] Malte Sieveking, "An algorithm for division of powerseries", *Computing* **10**, 153–156. ISSN 0010-485X. MR 47:1257. URL: http://cr.yp.to/bib/entries.html#1972/sieveking. Citations in this document: §6.7.

[Simpson 1740] Thomas Simpson, *Essays on several curious and useful subjects in speculative and mix'd mathematics, illustrated by a variety of examples.* URL: http://cr.yp.to/bib/entries.html#1740/simpson. Citations in this document: §6.7.

[Sorenson 1994] Jonathan Sorenson, "Two fast GCD algorithms", *Journal of Algorithms* **16**, 110–144. ISSN 0196–6774. MR 94k:11135. Citations in this document: §22.7.

[Stehlé and Zimmermann 2004] Damien Stehlé and Paul Zimmermann, "A binary recursive gcd algorithm", pp. 411–425 in [Buell 2004]. Citations in this document: §22.7.

[Stockham 1966] Thomas G. Stockham, Jr., "High-speed convolution and correlation", pp. 229–233 in [AFIPS 28]. URL: http://cr.yp.to/bib/entries.html#1966/stockham. Citations in this document: §2.8, §4.9, §4.10.

[Stone 1973] Harold S. Stone, "An efficient parallel algorithm for the solution of a tridiagonal linear system of equations", *Journal of the ACM* **20**, 27–38. ISSN 0004–5411. MR 48:12792. URL: http://cr.yp.to/bib/entries.html#1973/stone. Citations in this document: §12.7, §12.7, §12.7, §12.7.

[Strassen 1969] Volker Strassen, "Gaussian elimination is not optimal", *Numerische Mathematik* **13**, 354–356. ISSN 0029–599X. MR 40:2223. URL: http://cr.yp.to/bib/entries.html#1969/strassen. Citations in this document: §11.7, §11.8.

[Strassen 1973] Volker Strassen, "Die Berechnungskomplexität von elementarsymmetrischen Funktionen und von Interpolationskoeffizienten", *Numerische Mathematik* **20**, 238–251. ISSN 0029–599X. MR 48:3296. Citations in this document: §17.8.

[Strassen 1981] Volker Strassen, "The computational complexity of continued fractions", pp. 51–67 in [Wang 1981]; see also newer version [Strassen 1983]. URL: http://cr.yp.to/bib/entries.html#1981/strassen.

[Strassen 1983] Volker Strassen, "The computational complexity of continued fractions", *SIAM Journal on Computing* **12**, 1–27; see also older version [Strassen 1981]. ISSN 0097–5397. MR 84b:12004. URL: http://cr.yp.to/bib/entries.html#1983/strassen. Citations in this document: §12.8, §12.8, §21.6.

[Sylvester 1853] James J. Sylvester, "On a fundamental rule in the algorithm of continued fractions", *Philosophical Magazine* **6**, 297–299. URL: http://cr.yp.to/bib/entries.html#1853/sylvester. Citations in this document: §12.7.

[Toom 1963] Andrei L. Toom, "The complexity of a scheme of functional elements realizing the multiplication of integers", *Soviet Mathematics Doklady* **3**, 714–716. ISSN 0197–6788. Citations in this document: §4.9.

[Traub 1976] Joseph F. Traub, *Analytic computational complexity*, Academic Press, New York. MR 52:15938. See [Brent 1976c].

[Turk 1982] Johannes W. M. Turk, "Fast arithmetic operations on numbers and polynomials", pp. 43–54 in [Lenstra and Tijdeman 1982]. MR 84f:10006. URL: http://cr.yp.to/bib/entries.html#1982/turk. Citations in this document: §3.7.

[van der Hoeven 1999] Joris van der Hoeven, "Fast evaluation of holonomic functions", *Theoretical Computer Science* **210**, 199–215. ISSN 0304–3975. MR 99h:65046. URL: http://www.math.u-psud.fr/~vdhoeven/. Citations in this document: §12.7.

[van der Hoeven 2001] Joris van der Hoeven, "Fast evaluation of holonomic functions near and in regular singularities", *Journal of Symbolic Computation* **31**, 717–743. ISSN 0747–7171. MR 2002j:30037. URL: http://www.math.u-psud.fr/~vdhoeven/. Citations in this document: §12.7.

[van der Hoeven 2006] Joris van der Hoeven, "Newton's method and FFT trading". URL: http://www.math.u-psud.fr/~vdhoeven/. Citations in this document: §7.7, §9.7.

[van Leeuwen 1994] Jan van Leeuwen (editor), *Algorithms — ESA '94: second annual European symposium, Utrecht, The Netherlands, September 26–28, 1994, proceedings*, Lecture Notes in Computer Science **855**, Springer-Verlag, Berlin. ISBN 3–540–58434–X. MR 96c:68002. See [Schönhage and Vetter 1994].

[Vetterli and Nussbaumer 1984] Martin Vetterli and Henri J. Nussbaumer, "Simple FFT and DCT algorithms with reduced number of operations", *Signal Processing* **6**, 262–278. ISSN 0165–1684. MR 85m:65128. URL: http://cr.yp.to/bib/entries.html#1984/vetterli. Citations in this document: §2.9.

[Wang 1981] Paul S. Wang (editor), *SYM-SAC '81: proceedings of the 1981 ACM Symposium on Symbolic and Algebraic Computation, Snowbird, Utah, August 5–7, 1981*, Association for Computing Machinery, New York. ISBN 0–89791–047–8. See [Strassen 1981].

[Weinberger and Smith 1958] Arnold Weinberger and J. L. Smith, "A logic for high-speed addition", *National Bureau of Standards Circular* **591**, 3–12. ISSN 0096–9648. URL: http://cr.yp.to/bib/entries.html#1958/weinberger. Citations in this document: §12.7.

[Yavne 1968] R. Yavne, "An economical method for calculating the discrete Fourier transform", pp. 115–125 in [AFIPS 33]. URL: http://cr.yp.to/bib/entries.html#1968/yavne. Citations in this document: §2.9.

[Ypma 1995] Tjalling J. Ypma, "Historical development of the Newton-Raphson method", *SIAM Review* **37**, 531–551. ISSN 1095–7200. MR 97b:01003. Citations in this document: §6.7.

DANIEL J. BERNSTEIN
DEPARTMENT OF MATHEMATICS, STATISTICS, AND COMPUTER SCIENCE
M/C 249
THE UNIVERSITY OF ILLINOIS AT CHICAGO
CHICAGO, IL 60607–7045
   djb@cr.yp.to

# Elementary thoughts on discrete logarithms

CARL POMERANCE

ABSTRACT. We give an introduction to the discrete logarithm problem in cyclic groups and treat the most important methods for solving them. These include the index calculus method, the rho and lambda methods, and the baby steps, giant steps method.

Given a cyclic group $G$ with generator $g$, and given an element $t$ in $G$, the discrete logarithm problem is that of computing an integer $l$ with $g^l = t$. The problem of computing discrete logarithms is fundamental in computational algebra, and of great importance in cryptography. In this lecture we shall examine how sometimes the problem may be reduced to the computation of discrete logarithms in smaller groups (though this reduction may not always lead to an easier problem). We give an example of how the reduction may be used profitably in taking "square roots" in cyclic groups of even order. We shall look at several exponential-time algorithms that work in a quite general setting, and we shall discuss the index calculus algorithm for taking discrete logarithms in the multiplicative group of integers modulo a prime.

## 1. "The" cyclic group of order $n$

I should begin by saying that the discrete logarithm (dl) problem is not always hard. Obviously it is easy if the target element $t$ is the group identity, or in general, some small power of $g$. Less obviously, there are entire families of cyclic groups for which the dl problem is easy. Take, for example, the additive group $G = \mathbf{Z}/n\mathbf{Z}$. If we use the generator $g = 1$, the problem of computing discrete logarithms is absolutely trivial. Here, and in the sequel, we identify elements of $\mathbf{Z}/n\mathbf{Z}$ with their least non-negative residue. As we shall see later in connection with the index calculus algorithm, the fact that in some groups we may naturally represent group elements as integers can be quite useful. If we change to another generator, it is still trivial. In fact, if $g$ is a generator, then it is is coprime to $n$. Finding the multiplicative inverse of $g \pmod{n}$, via Euclid's extended algorithm, as in [Buhler and Wagon 2008], suffices for finding the discrete logarithm of 1, and so we quickly get everything else. Let us take $n = 100$, $g = 11$, $t = 17$ by way of example. The multiplicative inverse of 11 modulo 100 is 91, so the discrete logarithm of 17 is $91 \times 17$, that is, 47.

Now let us take another cyclic group of order 100, namely $(\mathbf{Z}/101\mathbf{Z})^*$, the multiplicative group of reduced residues modulo 101. Coincidentally, 11 is still a generator. But finding an integer $l$ with $11^l \equiv 17 \pmod{101}$ is no longer immediate. Of course, in this small example we might simply try all possible values $l = 1, 2, \ldots, 100$. But if we replace 101 by larger primes this soon becomes *very* slow. Thus, if you ever hear someone talk of "*the* cyclic group of order $n$," beware. He is not talking about anything computational. The way the cyclic group is presented to you makes all the difference.

## 2. Reductions

We first embark on a tour of some fairly straightforward ways to reduce a dl problem in a cyclic group $G$ to dl problems in various subgroups. To begin, it is important to describe some ground rules. It is assumed that we know how to multiply and take inverses in $G$. In some situations it may be difficult to see if two elements in a group are equal, e.g., happen if the group is presented as a quotient structure, or perhaps as a group of binary quadratic forms, but we will always assume that the cost of determining whether two elements of $G$ are equal is of the same magnitude as performing a group operation. Finally, we shall assume that it is possible to assign symbols to the group elements so that they may be sorted.

For our first reduction, assume the order $n$ of the cyclic group $G$ may be nontrivially factored as $n = uv$, where $u$ and $v$ are coprime, i.e., $\gcd(u, v) = 1$. Then we may reduce the problem of solving for a discrete logarithm in $G$ to solving for discrete logarithms in the subgroups of $G$ of order $u$ and $v$. In particular, if $G = \langle g \rangle$, then $g^u$ generates the subgroup of $u$-th powers in $G$, which has order $v$, and similarly $g^v$ generates the subgroup of $v$-th powers, which has order $u$. Say we solve the for the discrete logs $l_u, l_v$ where

$$(g^u)^{l_u} = t^u, \quad (g^v)^{l_v} = t^v.$$

The powers $g^u, g^v, t^u, t^v$ are easy to find via repeated squaring, as in [Buhler and Wagon 2008]. Say we also find integers $a, b$ with $au + bv = 1$, using the extended Euclidean algorithm. Then

$$t = t^{au+bv} = (t^u)^a (t^v)^b = g^{ul_u a} g^{vl_v b} = g^{aul_u + bvl_v},$$

so that the discrete logarithm of $t$ is $aul_u + bvl_v$, and we are through.

The next reduction considers the case when the order of $G$ is a prime power, say $p^a$, where $a > 1$. The argument does not use the primality of $p$, but it may as well be assumed because of the first reduction. We will see that a dl problem in this group can be reduced to $a$ dl problems in the cyclic subgroup of $G$ of order $p$. Say, as usual, we are trying to find $l$ such that $g^l = t$. If $l$ is the least nonnegative value that works, and we write $l$ in the base $p$, we have

$$l = b_0 + b_1 p + \ldots + b_{a-1} p^{a-1},$$

with each $b_j$ an integer in $[0, p-1]$. We shall sequentially find $b_0, b_1, \ldots$ as follows. First note that

$$t^{p^{a-1}} = (g^l)^{p^{a-1}} = g^{l p^{a-1}} = g^{b_0 p^{a-1}} = (g^{p^{a-1}})^{b_0},$$

that is, $b_0$ is the solution of a dl problem in the cyclic subgroup of $p^{a-1}$-powers generated by $g^{p^{a-1}}$. Suppose that $b_0, \ldots, b_{j-1}$ have been computed. Consider $t_j = t g^{-b_0 - b_1 p - \ldots - b_{j-1} p^{j-1}}$. We have that $t_j$ is a $p^j$-power, so that $t_j^{p^{a-j-1}}$ is in the subgroup of $p^{a-1}$-powers. Solve the dl problem

$$t_j^{p^{a-j-1}} = (g^{p^{a-1}})^{b_j}$$

for $b_j$. This is the next base $p$ digit of $l$ that we are searching for.

As an illustration of these reductions, let's return to the example of $g = 11, t = 17$ in $(\mathbf{Z}/101\mathbf{Z})^*$. Since the order of the group is 100, it suffices to solve two dl problems each in groups of order 2 and 5. First, $17^{25} \equiv -1 \pmod{101}$, so in our subgroup of order 4, the element $17^{25}$ must have discrete logarithm 2, that is, $l_{25} = 2$. (We have solved now both dl problems in the group of order 2: the first is 0, the second 1, so $l_{25} = 0 + 1 \cdot 2 = 2$.) Next, we must find $l_4$ where $11^{l_4} \equiv 17^4 \equiv 95 \pmod{101}$, which is solving a dl problem in a cyclic group of order 25. This is reduced to two dl computations in a group of order 5. We begin by computing $17^{20} \equiv 95^5 \equiv 1 \pmod{101}$, so that $l_4$ is a multiple of 5. Also, $11^{20} \equiv 87 \pmod{101}$. We have to find a power of 87 that is congruent to 95 modulo 101, and we know the answer is 0, 1, 2, 3, or 4. Evidently it is not 0 or 1, and checking 2, we find that it works. Thus $l_4 = 0 + 2 \cdot 5 = 10$. Now $(-6) \cdot 4 + 1 \cdot 25 = 1$, so the discrete logarithm of 17 is $(-6) \cdot 4 \cdot 10 + 1 \cdot 25 \cdot 2 = -190$. The least nonnegative discrete logarithm is 10.

So, if we have the group order in our possession (which is not always the case), and since solving interesting dl problems is usually harder in practice than factoring, we might first run a factorization algorithm on the order, and reduce the problem to smaller cases as above. Smaller cases tend to be simpler, although we will see that for some methods, such an those in the final section of this paper involving smooth numbers, working in a subgroup can be as hard as working in the full group.

## 3. An application

Before continuing, I will describe a nice application to the second dl reduction, namely reducing a dl problem in a cyclic group of order $p^a$ to $a$ dl problems in a cyclic group of order $p$. The application is to taking square roots in a cyclic group. Say $G$ is a cyclic group of order $n$. If $n$ is odd, then every element is a square, and square roots are simple: the square root of an element $h$ is $h^{(n+1)/2}$. If $n$ is even, exactly half of the elements of $G$ are squares. There is a very simple test for squareness: $h$ is a square if and only if $h^{n/2} = 1$, where I am writing the group identity as 1. Suppose $n/2$ is odd. Then again, it is very easy to find a square root. If $h$ is a square, then $h^{(n/2+1)/2}$ is a square root. How can one find the other square root? This is easy if you can find a group element that is not a square. If $g$ is such a group element, then $x = g^{n/2}$ has order 2 and is not 1. Thus, if $y$ is a square root of $h$, then $xy$ is the other square root.

This last idea works in general, even when $n$ is divisible by a high power of 2. If $g$ is a nonsquare in $G$, then $g^{n/2}$ is an element of order 2 and can be used as $x$ in the above.

But how would one find even one square root of a square $h$ if $n$ is divisible by a power of 2 higher than the first power? We again will make use of a nonsquare $g$. Say $n = 2^u v$, where $v$ is odd. Then the element $g^v$ has order $2^u$. The element $h^v$ is in $\langle g^v \rangle$. Solve for the discrete log. As we have seen, this is very simple, since the order of the group is $2^u$. Say $h^v = (g^v)^l$. Of necessity, since $h$ is a square and $g$ is a nonsquare, $l$ must be even. Then a square root of $h$, as is easily checked, is $h^{(v+1)/2}(g^v)^{-l/2}$, and we are done.

This polynomial time algorithm has one small flaw. It is the production of a nonsquare $g$. Of course, if you are given a cyclic generator of $G$, then you may use this generator as a nonsquare. But what if you are not given this? For example, say $G = (\mathbf{Z}/p\mathbf{Z})^*$, where $p$ is a large prime. It may be hard to find a generator (a primitive root), especially if we don't know the prime factorization of $p - 1$. But surely, finding a nonsquare shouldn't be hard, since half of all elements in the group are nonsquares, and the test for one is simple. So, we have a random algorithm that will work very nicely. Choose elements from $G$ at random, and test for nonsquareness. The expected number of trials is 2. This method begs the question of how one is supposed to choose elements from a group at random. This is not so hard for $(\mathbf{Z}/p\mathbf{Z})^*$, but is conceivably a problem in general. So, modulo this problem of finding *some* nonsquare, taking square roots is easy.

As you might notice, this idea generalizes to taking $p$-th roots for all primes $p$. Further, if $p$ is small, the various dl problems that arise may all be handled quickly.

## 4. Baby steps, giant steps

As before, take a group $G = \langle g \rangle$ of order $n$, with $t \in G$. Our task is to find an integer $l$ with $0 \leq l \leq n-1$ and $g^l = t$. Suppose $m = \lceil \sqrt{n} \rceil$ and we write our discrete logarithm $l$ in the base $m$. Then $l = b_0 + b_1 m$ where $0 \leq b_0, b_1 \leq m-1$. It is evident that to find $l$, it is sufficient to find $b_0, b_1$.

We prepare two lists: $1, g, \ldots, g^{m-1}$ by taking 'baby steps', and $t, tg^{-m}, \ldots, tg^{-(m-1)m}$ by taking 'giant steps'. Since $g^{b_0} = tg^{-b_1 m}$, there must be a common element in the two lists. Moreover, any common element immediately allows us to find the discrete logarithm of $t$. We're done.

You might wonder how to find the element in common in the two lists. Maybe one should just sequentially make up to $m^2$ comparisons to see if an element from the first list matches with an element from the second list, expecting about $m^2/2$ comparisons on average. That is *one* way to do it, but there is a better way. As one of our ground rules, we assumed that we can label group elements in such a way that they can be sorted. Sort the elements in the first list. Then sequentially run through the second list to check for membership in the first list. The sorting can be done in $O(m \log m)$ comparisons, and each membership check, via a binary search, can be done in $O(\log m)$ comparisons. (A binary search involves identifying the midpoint of the sorted list, deciding if the searched-for element is in the first half or the second half, and then iterating in the appropriate half.) So in total we do about $O(m \log m) = O(\sqrt{n} \log n)$ comparisons after the lists are computed.

It is clear the first list, $1, g, \ldots, g^{m-1}$, can be computed in $m-2$ group operations. After these baby steps, the giant step $g^{-m}$ can be computed in two more operations, and then we can sequentially get the terms of the giant step sequence $t, tg^{-m}, tg^{-2m}, \ldots$ with one group operation per step. Note that after each giant step is taken, the look-up can be done in the baby step list, and we may stop as soon as the match is found. Thus, on average, we expect to traverse only about half of the giant steps before completion. Note too that if we wish to find the discrete logarithm of another group element $t'$, the same baby step sequence may be reused.

It may seem that one needs the group order $n$ to use baby steps, giant steps. However, all that is needed in the above is that $m \geq \sqrt{n}$. So start with a small choice for $m$, try the algorithm out, and if it fails try again with $2m$, etc. Eventually it will work, and when this happy event occurs, the total time spent is of the shape $O(\sqrt{n} \log n)$, even though we may still not know what $n$ is.

The baby steps, giant steps method was originally invented by Dan Shanks as a means of computing the order of an abelian group $G$ that is not necessarily cyclic. He was interested in particular in the class group of an imaginary quadratic number field. Here's how it works. By other means he gets a rough estimate of the order of the group: say it is in the interval $[x, x+y]$, where $y < x$. (In fact, using the Extended Riemann Hypothesis, he is able to get such an interval for the group order with $y < x^{1-c}$ for some positive $c$.) He then chooses a random element $h_1$ in the group, and via baby steps, giant steps, he finds an integer $n_1 \in [x, x+y]$ such that $h_1^{n_1} = 1$. By factoring $n_1$ into primes, it is then possible to compute the actual order $m_1$ of $h_1$, and if it is $n_1$, we are done; this must be the order of the group. If $m_1 < n_1$ there is more work to be done. Choose another random element $h_2$, use baby steps, giant steps to compute the order $m_2$ of the subgroup $\langle h_1, h_2 \rangle$, and so on. When finally a subgroup order $m_k$ is found in $[x, x+y]$, we have $n = m_k$ and we are done. In the case at hand of class groups, it is also possible to use the ERH to find a fairly small set of group elements known to generate the group, so that randomness is not needed at all.

## 5. The $\rho$-method

The baby steps, giant steps method, while a rigorous method of "square-root complexity," suffers from a high memory load, also about the square root of the group order. In contrast, John Pollard's $\rho$ method, which also runs in about the square root of the group order, has negligible space requirements. The down sides are that the $\rho$ method requires the group order, and it does not (yet) have a rigorous analysis. As with many other algorithms which produce a readily checkable answer, the fact that the method of achieving the answer is heuristic is not a practical concern, only a mathematical one.

The $\rho$ method is based heuristically on the birthday paradox. If you throw balls randomly into $n$ urns, where each urn is equally likely to receive a ball even if it already has one, how many balls should you expect to throw before some urn has two balls in it? The answer is surprisingly small, it is of magnitude $\sqrt{n}$. In particular, if $c\sqrt{n}$ balls are thrown, the probability that some urn has at least two balls is about $1 - e^{-c^2/2}$. So, in a room of 23 people, it is better than even odds that two of them have the same birthday.

Suppose $G = \langle g \rangle$ has order $n$. If $x_1, x_2, \ldots, x_k$ is a random sequence of elements of $G$, we would expect to see some $x_i = x_j$ when $k$ is of order $\sqrt{n}$. However, we do not wish to use a truly random sequence, even if we had the means of generating it. We specifically wish to use a pseudorandom sequence, in fact one where the next term $x_{i+1}$ depends in a specific manner on the current term $x_i$. As you will see, this conscious choice of avoiding randomness is important. It is also what keeps us at present from rigorously analyzing the algorithm.

So, we would like to define a function $f : G \to G$ which is both easy to compute, and seemingly random. Say we have some straightforward method of labeling the elements of $G$ with the integers $1, 2, \ldots, n$. For example, in the case $G = (\mathbf{Z}/p\mathbf{Z})^*$ we may label, as we have been doing, the elements of $G$ with the integers $1, 2, \ldots, p - 1$. It now makes sense to talk of elements in the first third of $G$, the second third, and the third third. Call these type I, type II, and type III elements, respectively.

Suppose we are trying to find the discrete logarithm of a group element $t$. For $x \in G$, let

$$f(x) = \begin{cases} tx & \text{if } x \text{ is type I,} \\ x^2 & \text{if } x \text{ is type II,} \\ gx & \text{if } x \text{ is type III.} \end{cases}$$

We shall take for our pseudorandom sequence, $g, f(g), f(f(g)), \ldots$. Let $x_i$ be the $i$-th term of this sequence, $i = 0, 1, 2, \ldots$. We also wish to keep track of the different types of elements we see as we traverse the sequence. Consider the sequence $(a_i, b_i)$ of pairs of residues modulo $n$ with initial term $(0, 1)$ and the rule

$$(a_{i+1}, b_{i+1}) = \begin{cases} (a_i + 1, b_i) & \text{if } x_i \text{ is type I,} \\ (2a_i, 2b_i) & \text{if } x_i \text{ is type II,} \\ (a_i, b_i + 1) & \text{if } x_i \text{ is type III.} \end{cases}$$

Then, as is easily checked, $x_i = t^{a_i} g^{b_i}$. And if it is discovered that $x_i = x_j$, then $g^{b_i - b_j} = t^{a_j - a_i}$. If $a_j - a_i$ is coprime to $n$, the discrete logarithm of $t$ is the inverse of $a_j - a_i$ modulo $n$ multiplied by $b_i - b_j$.

The condition that $a_j - a_i$ is coprime to $n$ is not a strenuous one. As we saw above, it is possible to reduce the dl problem to the case of prime group orders. So, it might be assumed that $n$ is prime. And so the only way for the equation $x_i = x_j$ not to lead to the discrete logarithm of $t$ is if the pair $(a_i, b_i)$ is identical to the pair $(a_j, b_j)$. In practice this event does not frequently occur. If it did occur, one could try for a new sequence where now one is searching for the discrete logarithm of $gt$, for example. Or, one could let the initial seed be $g^r$ for some random choice of $r$.

What remains to be discussed is an efficient way of searching for a pair $i, j$ with $x_i = x_j$. We certainly don't want to write down all of the terms and exhaustively check all pairs. Not only would this kill the square root running time, it would consume too much space. There is a very neat method for finding a repeat in the sequence, known as the Floyd cycle-finding algorithm. The idea is to compute the sequences $x_i, (a_i, b_i)$ twice, once at single speed, and once at double speed. That is, if you have $x_i, (a_i, b_i)$ and $x_{2i}, (a_{2i}, b_{2i})$, use the rules to compute $x_{i+1}, (a_{i+1}, b_{i+1})$, $x_{2i+1}, (a_{2i+1}, b_{2i+1})$, $x_{2i+2}, (a_{2i+2}, b_{2i+2})$, so that at each stage you have at hand $x_i, x_{2i}$. Check only these pairs for equality.

At first glance it would appear that a great deal of generality is lost if we insist that $j = 2i$ for our equation $x_i = x_j$. But here is where we use that our sequence is *not* random, but rather an orbit for our function $f$. Note that if $x_{i_0} = x_{j_0}$, where $i_0 < j_0$ is the first occurrence of equality, then so too do we have $x_{i_0+1} = x_{j_0+1}, x_{i_0+2} = x_{j_0+2}$, etc. Thus, if $\mu = j_0 - i_0$, then $x_u = x_v$ whenever $u, v \geq i_0$ and $u \equiv v \pmod{\mu}$. That is, the sequence becomes purely periodic starting at the $i_0$-th term. If we now take $u$ as the first multiple of $\mu$ that is at least $i_0$, namely, $u = \mu \lceil i_0/\mu \rceil$, then $x_u = x_{2u}$. Note that $u$ satisfies $i_0 \leq u < j_0$. That is, it is hardly a restriction at all to search for an equality of the form $x_u = x_{2u}$. We have so transformed a potentially quadratic search into a linear one. There are negligible memory requirements, since one only needs the current candidate $x_i, x_{2i}$ to find the next one.

The $\rho$-method can be used as a factoring algorithm, also an idea of Pollard. To factor $n$, use the function $f(x) = x^2 + a \bmod n$, where $a$ is not $0, -2$. Instead of checking for an equality $x_i = x_{2i}$, check for a nontrivial value of $\gcd(x_i - x_{2i}, n)$.

The $\rho$-method gets its name from the suggestive shape of the letter $\rho$, which can be thought of as the diagram for a sequence with a non-periodic beginning that eventually becomes periodic.

## 6. The $\lambda$-method

Pollard also suggested a version of the $\rho$-method that lends itself fairly easily to being parallelized, i.e., to many computers sharing the job of computing one discrete logarithm. The key part of the shape of the letter $\rho$ where the actual success is found is the point where the round part intersects the straight part. Focusing then on the convergence of two streams, the key Greek letter is $\lambda$.

Suppose we have $k$ computers each following its own random sequence in the group $G$. If the order of $G$ is $n$, then when the length of the sequences is about $\sqrt{n}/k$, we will begin to expect that some term in one computer's sequence will have a match with some term in another computer's sequence. Of course, we will not want to make every possible comparison. So we introduce the idea of a "distinguished point" and use a pseudorandom iteration that has the property that once there is a match between two streams, they stay identical from then on.

To be specific, suppose we use the same iteration as in the $\rho$-method, but we have computer $m$ initialize its pseudorandom sequence at $g^{r_m}$, where $r_m$ is a random number. We also make use of a perfect hash of group elements, an easily computable mapping of group elements to integers, which is 1:1. (This can be the same labeling as in the $\rho$-method.) We call a group element distinguished if its hash is divisible by $2^{20}$, say. Then each computer goes merrily along down its sequence, but whenever it arrives at a distinguished point, which occurs about every millionth iterate, it reports the event to a central computer. The central computer then sorts the incoming hashes of distinguished points, looking for a match. When one occurs, the data involved, if actually representing some $x_i = x_j$, can then be used to compute the desired discrete logarithm. So on average, our first match with distinguished points occurs only about a half-million iterations after the first match in any pair of streams. Note that it is possible for the match to occur between two reports of the same computer, namely for some reason, that computer had extraordinary luck with the $\rho$ method. That's fine, the $\lambda$ method will take it.

One can take other pseudorandom functions $f$. One way of choosing $f$ is to have a small set of integers $S$, and pre-compute $g^s$ for $s \in S$. Then, the function $f$ would send $x$ to some $g^s x$, which one depending on some property of the hash of $x$. An initial seed is $t^r$ for some random exponent $r$. This version is sometimes referred to as the kangaroo method, where the various values of $g^s$ are considered as "hops."

There are numerous ideas for fine-tuning and speeding up both the $\rho$ and $\lambda$ methods. For this I refer you to a survey paper by one of the most sophisticated practitioners, Edlyn Teske [2001]. As of this writing the champion calculations for groups of prime order involve primes near $2^{109}$ (one group arises from an elliptic curve over a prime finite field, the other for an elliptic curve over $\mathbf{F}_{2^{109}}$).

## 7. The index calculus and the search for smoothness

The methods we have described so far have an exponential run time. For some cyclic groups there are subexponential algorithms that involve smooth numbers, as introduced in [Pomerance 2008]. The idea here is to recognize the cyclic group as the unit group in a homomorphic image of the ring of integers $\mathbf{Z}$ or of another ring in which it makes sense to speak of smooth elements.

The prime example is the cyclic group $G = (\mathbf{Z}/p\mathbf{Z})^*$. For $p$ prime, $G$ is cyclic of order $p - 1$, and there is an obvious way to realize $G$ as the group of units of a homomorphic image of $\mathbf{Z}$. Whenever we represent elements of $G$ by integers, we are tacitly thinking in terms of this homomorphism. In particular, a multiplicative relation among integers leads to a multiplicative relation among group elements.

The idea with the index calculus method is to look at powers of a generator $g$ of $G$. Again, staying with our example, if $g^r$ is represented by an integer in the range $[1, p-1]$ and we happen to have the prime factorization of this integer, say as $p_1^{a_1} \cdots p_k^{a_k}$, then we have the index, or discrete logarithm, relation

$$r \equiv \log_g g^r \equiv a_1 \log_g p_1 + \ldots + a_k \log_g p_k \pmod{p-1}. \tag{1}$$

However, even though we know $r$ and $a_1, \ldots, a_k$, we don't know the discrete logarithms $\log_g p_1, \ldots, \log_g p_k$. Thus, we may view (1) as sort of an equation in $k$ unknowns.

Say we choose a smoothness bound $B$, and suppose that $p_1, \ldots, p_k$ are the primes up to $B$. We continually take random values of $r$, find the least positive integer representing $g^r$, and see if it is $B$-smooth. If it is, we get a relation as in (1). Continuing, suppose we assemble more than $k$ of these relations. Then presumably, linear algebra will allow us to solve for the unknowns, namely the discrete logarithms $\log_g p_1, \ldots, \log_g p_k$.

So, this would allow us to find the discrete logarithms of the small primes. However, suppose you are interested in the discrete logarithm of $t$, and the integer representing $t$ is not $B$-smooth. Then continue to choose random exponents $r$ until one is found with the integer representing $g^r t$ being $B$-smooth, say it is $p_1^{b_1} \cdots p_k^{b_k}$. Then

$$r + \log_g t \equiv b_1 \log_g p_1 + \ldots + b_k \log_g p_k \pmod{p-1}.$$

But now, the only thing unknown in our relation is $\log_g t$, and this is then found instantly.

Choosing an optimal value of $B$ and using Lenstra's elliptic curve factoring method discussed in [Poonen 2008] to recognize $B$-smooth integers, the expected complexity of the outlined method is $\exp((\sqrt{2}+o(1))\sqrt{\log p \log \log p})$. Moreover, once the initial work is done to find the discrete logarithms of the small primes, the additional time to find the discrete logarithms of a given group element is much smaller, only about $\exp((1/\sqrt{2}+o(1))\sqrt{\log p \log \log p})$. If a larger value of $B$ is used, the precomputation takes longer, but once it is done, the individual dl computations are even speedier.

Even though the elliptic curve factoring method has not been rigorously analyzed *in toto*, it can be shown that it recognizes sufficiently many smooth numbers in subexponential time, that it may be used as a subroutine in some rigorously analyzed algorithms. This somewhat paradoxical, but happy occurrence pertains to the index calculus method in $(\mathbf{Z}/p\mathbf{Z})^*$.

The key aspect of the index calculus algorithm is the use of smooth numbers. Can this always be done? That is an important question, and we really do not have a complete answer for various groups of interest. But for some groups, we can. For example, say we look at the multiplicative group of a finite field, $\mathbf{F}_q^*$, where $q = p^a$ is a prime power. We've just looked at the case $a = 1$. Let us look at the other extreme, $p = 2$. Does it make sense to speak of elements of $\mathbf{F}_{2^a}^*$ being smooth?

To answer this question, we think how the finite field $\mathbf{F}_{2^a}$ is constructed. One way is to view it as $\mathbf{F}_2[x]/(f(x))$, where $f(x)$ is an irreducible polynomial over $\mathbf{F}_2$ of degree $a$. We may then view our group as the group of units in a homomorphic image of the polynomial ring $\mathbf{F}_2[x]$. This ring is a unique factorization domain, where the prime elements are irreducible polynomials, and the degree of a polynomial gives us a measure of size. That is, we can say a polynomial is $B$-smooth if all of its irreducible factors have degrees at most $B$. There is a completely analogous development of the study of the distribution of smooth polynomials as with the study of smooth integers, and yes we can obtain a rigorous, subexponential discrete logarithm algorithm for $\mathbf{F}^*_{2^a}$. In fact this works more generally for $\mathbf{F}^*_{p^a}$, and I showed with Renet Lovorn Bender that it is subexponential in $p^a$ as long as $a \to \infty$ arbitrarily slowly.

What about the case $a > 1$, $a$ fixed? Then we can represent $\mathbf{F}_{p^a}$ as the quotient ring of $p$ in the ring of integers of an algebraic number field of degree $a$ in which $p$ is inert. And we may define smoothness in a number ring: an element is smooth if its norm to the rational integers is smooth. This heuristically gives a subexponential dl algorithm for all the cases of $a$ fixed or slowly growing, and it does so rigorously in the case $a = 2$, a result of Lovorn Bender.

What makes elliptic curve groups of prime order so attractive for cryptography at present, is that we know no way of introducing smooth numbers to solve dl's in them. We seem to be condemned to use the earlier exponential methods of this paper.

There are cryptosystems such as XTR that are based on the dl problem in large subgroups of very large cases of $\mathbf{F}^*_q$. What about index calculus? Yes, it can be used, but only in the parent group, which is very large. So, as a function of the size of the subgroup, the complexity is prohibitive, even though it is a subexponential function of $q$. So, another unsolved problem is to find a way of introducing smooth numbers directly into the subgroup.

The basic ideas of the index calculus can be taken much further, with tremendous gains in efficiency. In particular, the number field sieve for factoring integers may be adapted to the dl problem for the multiplicative group of a finite field, see [Schirokauer 2008].

For further reading, connections to cryptography, and references to original papers and other surveys, see [Crandall and Pomerance 2005; Odlyzko 2000; Schirokauer et al. 1996].

# References

[Buhler and Wagon 2008] J. P. Buhler and S. Wagon, "Basic algorithms in number theory", pp. 25–68 in *Surveys in algorithmic number theory*, edited by J. P. Buhler and P. Stevenhagen, Math. Sci. Res. Inst. Publ. **44**, Cambridge University Press, New York, 2008.

[Crandall and Pomerance 2005] R. Crandall and C. Pomerance, *Prime numbers: a computational perspective*, 2nd ed., Springer, New York, 2005.

[Odlyzko 2000] A. Odlyzko, "Discrete logarithms: the past and the future", *Des. Codes Cryptogr.* **19**:2-3 (2000), 129–145. Towards a quarter-century of public key cryptography.

[Pomerance 2008] C. Pomerance, "Smooth numbers and the quadratic sieve", pp. 69–81 in *Surveys in algorithmic number theory*, edited by J. P. Buhler and P. Stevenhagen, Math. Sci. Res. Inst. Publ. **44**, Cambridge University Press, New York, 2008.

[Poonen 2008] B. Poonen, "Elliptic curves", pp. 183–207 in *Surveys in algorithmic number theory*, edited by J. P. Buhler and P. Stevenhagen, Math. Sci. Res. Inst. Publ. **44**, Cambridge University Press, New York, 2008.

[Schirokauer 2008] O. Schirokauer, "The impact of the number field sieve on the discrete logarithm problem in finite fields", pp. 397–420 in *Surveys in algorithmic number theory*, edited by J. P. Buhler and P. Stevenhagen, Math. Sci. Res. Inst. Publ. **44**, Cambridge University Press, New York, 2008.

[Schirokauer et al. 1996] O. Schirokauer, D. Weber, and T. Denny, "Discrete logarithms: the effectiveness of the index calculus method", pp. 337–361 in *Algorithmic number theory* (ANTS II) (Talence, 1996), edited by H. Cohen, Lecture Notes in Comput. Sci. **1122**, Springer, Berlin, 1996.

[Teske 2001] E. Teske, "Square-root algorithms for the discrete logarithm problem (a survey)", pp. 283–301 in *Public-key cryptography and computational number theory* (Warsaw, 2000), edited by K. Alster et al., de Gruyter, Berlin, 2001.

CARL POMERANCE
DEPARTMENT OF MATHEMATICS
DARTMOUTH COLLEGE
HANOVER, NH 03755-3551
(603) 646-2415
carl.pomerance@dartmouth.edu

# The impact of the number field sieve on the discrete logarithm problem in finite fields

OLIVER SCHIROKAUER

## 1. Introduction

Let $p$ be a prime number and $n$ a positive integer, and let $q = p^n$. Let $\mathbb{F}_q$ be the field of $q$ elements and denote by $\mathbb{F}_q^*$ the multiplicative subgroup of $\mathbb{F}_q$. Assume $t$ and $u$ are elements in $\mathbb{F}_q^*$ with the property that $u$ is in the subgroup generated by $t$. The discrete logarithm of $u$ with respect to the base $t$, written $\log_t u$, is the least non-negative integer $x$ such that $t^x = u$.

In this paper we describe two methods to compute discrete logarithms, both of which derive from the number field sieve (NFS) factoring algorithm described in [Stevenhagen 2008] and [Lenstra and Lenstra 1993]. When factoring an integer $N$ with the NFS, we first choose a number ring $R$ as in [Stevenhagen 2008] for which there is a ring homomorphism $\phi : R \to \mathbb{Z}/N\mathbb{Z}$. Then we combine smooth elements in $R$ and in $\mathbb{Z}$ to obtain squares $\alpha_1 = a_1^2 \in R$ and $\alpha_2 = a_2^2 \in \mathbb{Z} \subseteq R$, such that $\phi(\alpha_1) = \phi(\alpha_2)$. If $\phi(a_1) \ne \pm\phi(a_2)$, then the gcd of $N$ and $a_1' - a_2$, where $a_1'$ is a representative in $\mathbb{Z}$ for $\phi(a_1)$, is a non-trivial factor of $N$. When using the strategy to compute discrete logarithms in $\mathbb{F}_q$, we choose either two number rings or two polynomial rings, call them $R_1$ and $R_2$, such that there are ring homomorphisms $\phi_1 : R_1 \to \mathbb{F}_q$ and $\phi_2 : R_2 \to \mathbb{F}_q$. We then construct two $(q-1)$-st powers $\alpha_1 \in R_1$ and $\alpha_2 \in R_2$ such that $\phi_1(\alpha_1) = t^x u \cdot \phi_2(\alpha_2)$ for some $x$. It follows that $x \equiv -\log_t u \bmod (q-1)$. When the chosen rings are number rings, the algorithm retains the title of number field sieve. In this case, if $q$ is prime, one of the rings can be taken to be $\mathbb{Z}$ as is done for factoring. However, it is often advantageous to use two non-trivial extensions of $\mathbb{Z}$ instead (see Section 2.2). When the chosen rings are polynomial rings, the algorithm is known as the function field sieve (FFS).

One difficulty encountered in the NFS and FFS discrete logarithm algorithms is that in order to combine smooth elements in $R_1$ into a $(q-1)$-st power which

397

is mapped by $\phi_1$ to a multiple of $t^x u$, it is necessary to find pre-images of $t$ and $u$ under $\phi_1$ which are smooth. This problem is the subject of Section 4 of the paper. Sections 2 and 3 are devoted to descriptions of the NFS and FFS, respectively, in the case that $t$ and $u$ come with smooth pre-images under $\phi_1$.

The importance of the number field and function field sieves lies in the fact that they are faster than other algorithms for computing discrete logarithms, both asymptotically and in practice. The expected running time of the NFS of Section 2 is conjectured to be

$$L_q[1/3; (64/9)^{1/3} + o(1)],$$

where

$$L_q[s; c] = \exp(c(\log q)^s (\log \log q)^{1-s})$$

and the $o(1)$ is for $q \to \infty$ subject to the constraint that $n$ does not grow too fast (see Section 2.8 for a precise formulation). The conjectured expected running time of the FFS of Section 3 is

$$L_q[1/3; (32/9)^{1/3} + o(1)], \qquad (1.1)$$

where again the $o(1)$ is for $q \to \infty$, this time with the restriction that $p$ does not grow too fast (see Section 3.4). Taken together, the algorithms have a conjectured running time of $L_q[1/3; O(1)]$ for all finite fields. By contrast, no other discrete logarithm algorithm has a proven or conjectural running time faster than $L_q[1/2; O(1)]$. Coppersmith's method for fields of characteristic two [Coppersmith 1984], which predates the FFS by a decade and which also runs in time (1.1), is considered here as a special case of the FFS (see Section 3.8).

In practice, the current record for computing logarithms in a prime field is held by Kleinjung [Kleinjung 2007], who used the NFS as described in [Joux and Lercier 2003] to compute logarithms in a field whose cardinality is a prime of 160 digits. In characteristic two, Joux and Lercier's computation of logarithms in the field of size $2^{613} \approx 3.399 \cdot 10^{184}$ using the FFS [Joux and Lercier 2005a] is the present record. Interest in computations of discrete logarithms in fields that are neither prime nor of characteristic two is a relatively recent phenomenon, spurred on in part by cryptographic applications. In particular, the fact that the discrete logarithm problem on an elliptic curve over a prime field $\mathbb{F}_p$ can be transported to the logarithm problem in an extension of $\mathbb{F}_p$ [Menezes et al. 1993], [Frey et al. 1999], has focused attention on these fields. Joux and Lercier's use of the FFS to compute logarithms in the field of size $370801^{30} \approx 1.186 \cdot 10^{167}$ [Joux and Lercier 2005b] is an indication that at current levels, the difficulty of computing logarithms in these "intermediate" fields is comparable to that of doing so in the prime and characteristic two cases.

It often occurs in practice that more than one logarithm in a given field is sought. When this is the case, the NFS and FFS can be split into a precomputation stage in which the logarithms of the "small" elements in the field are computed, and a fast reduction stage in which the desired logarithm is expressed in terms of the precomputed values. We refer the reader to [Schirokauer 2005] and [Joux and Lercier 2002] for descriptions of such versions of the NFS and FFS and note that they use the the same basic structure and techniques and have the same conjectural running times as the methods described in the present paper.

## 2. The number field sieve (NFS)

**2.1.** Let $p$ be an odd prime number. We adopt as a model for the finite field $\mathbb{F}_p$ the set of non-negative integers less than $p$, with addition and multiplication taken modulo $p$. Let $B$ be some positive real number, and recall that an integer is said to be $B$-smooth if each of its prime factors is at most $B$. We say that an element in $\mathbb{F}_p$ is $B$-smooth if it is $B$-smooth as an integer. Let $t$ and $u$ be elements in $\mathbb{F}_p^*$ which are $B$-smooth and for which $u \in \langle t \rangle$. In this section, we describe how to use the NFS to compute, not the residue of $\log_t u$ modulo $(p-1)$ as suggested in the introduction, but the residue of $\log_t u$ modulo an odd prime divisor $l$ of $p-1$. The reason for the restriction is given in Step 3 below. The algorithm we present can be modified to compute the residue of $\log_t u$ modulo any prime power divisor of $p-1$ [Schirokauer 1993]. Once these residues are known, $\log_t u$ is easily determined by means of the Chinese remainder theorem. We note that to compute the residue of $\log_t u$ modulo a power of a small prime, the exponential-time methods described in [Pomerance 2008] are preferable to the NFS. In what follows, therefore, we think of $l$ as being large.

**2.2. The NFS for prime fields.** In addition to a prime $p > 5$, an odd prime divisor $l$ of $p - 1$, smoothness bound $B$, and elements $t$ and $u$ as described above, the algorithm takes as input a parameter $C \geq 1$ and an integral parameter $d$ satisfying $\log_2 p > d \geq 1$. It outputs an integer $x$ which is likely to be congruent to $\log_t u$ modulo $l$.

**Step 1. Constructing the number rings.** The idea of the NFS is to produce a relation in $\mathbb{F}_p$ involving $t$ and $u$ by choosing two number rings which come with maps to $\mathbb{F}_p$ and then building $l$-th powers in these rings in such a way that they have the same image in $\mathbb{F}_p$. The challenge is to find rings in which the construction of suitable $l$-th powers requires as little work as possible.

One approach, presented in [Joux and Lercier 2003], is to choose an irreducible polynomial $f_1$ of degree $d$ with small, integral coefficients and a root modulo $p$, call it $m$. The set of vectors $(a_0, \ldots, a_{d-1}) \in \mathbb{Z}^d$ having the property that the polynomial $\sum a_i X^i$ has $m$ as a root mod $p$ is a lattice. Lattice reduction techniques can, therefore, be used to obtain a polynomial $f_2$ of degree $d-1$, having integral coefficients of size approximately $p^{1/d}$ and $m$ as a root mod $p$. We now proceed with the rings $R_1 = \mathbb{Z}[\alpha_1]$ and $R_2 = \mathbb{Z}[\alpha_2]$, where $\alpha_1$ and $\alpha_2$ are roots in $\mathbb{C}$ of $f_1$ and $f_2$ respectively. Note that for $j = 1, 2$ the map $\phi_j : R_j \to \mathbb{F}_p$ that sends the element $\sum b_i \alpha_j{}^i$, with $b_i \in \mathbb{Z}$, to the element in $\{0, \ldots, p-1\}$ congruent to $\sum b_i m^i$ mod $p$ is a ring homomorphism.

A second method to choose rings for the NFS is to define $m$ to be $\lfloor p^{1/d} \rfloor$ and let $f = \sum a_i X^i$ where the coefficients $a_0, \ldots, a_d$ are obtained by writing $p$ in the base $m$. In other words, $0 \leq a_i < m$ and

$$p = \sum_{i=0}^{d} a_i m^i.$$

Then $f$ is irreducible [Brillhart et al. 1981] and we choose as our two rings $\mathbb{Z}$ and $R = \mathbb{Z}[\alpha]$, where $\alpha \in \mathbb{C}$ satisfies $f(\alpha) = 0$. The required maps to $\mathbb{F}_p$ are the canonical projection from $\mathbb{Z}$ and its extension $\phi : R \to \mathbb{F}_p$ which sends $\sum b_i \alpha^i$, with $b_i \in \mathbb{Z}$, to the element in $\{0, \ldots, p-1\}$ congruent to $\sum b_i m^i$ mod $p$.

Despite the fact that of these two methods, the former is frequently the better one in practice, we continue with the latter approach for the remainder of the section. Not only does the resulting presentation more closely parallel the usual formulation of the NFS algorithm for factoring, but the exposition is made less cumbersome by having only one non-trivial extension of $\mathbb{Z}$ to handle. All the techniques described for this one ring can be applied to the two non-trivial extensions produced by the first method.

Proceeding, therefore, with $f$ and $R$ as above, we observe that

(i) the coefficients of $f$ are bounded by $p^{1/d}$;
(ii) $f$ has a root mod $p$ of size at most $p^{1/d}$;
(iii) $f$ is monic.

A fourth property, which $f$ is not guaranteed to satisfy but which we assume holds, is

(iv) $l$ does not divide the discriminant of $f$.

The bounds in (i) and (ii) are critical to the running time analysis of the algorithm as they determine a bound for the numbers tested for smoothness in the next step. Properties (iii) and (iv) are not necessary to the algorithm but simplify our exposition. See [Buhler et al. 1993] and [Schirokauer et al. 1996] for a version of the algorithm that does not require that (iii) hold and [Schirokauer 1993] for the modifications necessary if (iv) does not hold.

**Step 2. Sieving.** Let $\mathcal{O}$ be the ring of integers of the number field $\mathbb{Q}(\alpha)$ and $N : \mathbb{Q}(\alpha) \to \mathbb{Q}$ the norm map. An element $\gamma \in \mathcal{O}$ is said to be $B$-smooth if $N(\gamma)$ is $B$-smooth in $\mathbb{Z}$, or equivalently, if each prime ideal dividing the ideal generated by $\gamma$ lies over a rational prime $\leq B$. In this step, we use sieving techniques to find the set $S$ of elements $(a, b) \in \mathbb{Z} \times \mathbb{Z}$ such that $|a|, |b| \leq C$ and both $a - b\alpha$ and $a - bm$ are $B$-smooth. This stage is exactly like the sieving step in the NFS for factoring. We refer the reader to [Buhler et al. 1993] and [Lenstra et al. 1993b] for details.

For $a, b \in \mathbb{Z}$ satisfying $|a|, |b| \leq C$, we have

$$|a - bm| \leq C(p^{1/d} + 1),$$
$$|N(a - b\alpha)(a - bm)| = |b^d f(a/b)(a - bm)| \leq 2^d (d + 1) C^d p^{1/d}. \quad (2.3)$$

The product of the bounds in (2.3) then is a bound on the size of the integer that must be $B$-smooth for a pair $(a, b)$ to be in $S$. When the values $C$ and $d$ are chosen optimally, this product is significantly smaller than the size of the candidates for smoothness in other index calculus algorithms, such as those described in [McCurley 1990] and [Schirokauer et al. 1996]. It is for this reason that the NFS is faster than these methods.

**Step 3. Computing exponent vectors.** Let $\pi(B)$ be the number of rational primes $\leq B$ and let $q_1, \ldots, q_{\pi(B)}$ be a list of these primes. Similarly, let $\Pi(B)$ be the number of prime ideals of $\mathcal{O}$ which are either of norm $\leq B$ and of degree 1 or lie over a rational prime $\leq B$ dividing $[\mathcal{O} : R]$, and let $\mathfrak{q}_1, \ldots, \mathfrak{q}_{\Pi(B)}$ be a list of these ideals. As explained in [Buhler et al. 1993], for each $(a, b) \in S$, the prime ideal factors of $(a - b\alpha)$ are contained in this list. For each rational prime $q$ and integer $g$, let $v_q(g)$ be the exponent to which $q$ divides $g$. Similarly, for each prime ideal $\mathfrak{q} \subseteq \mathcal{O}$ and element $\gamma \in \mathcal{O}$, let $v_\mathfrak{q}(\gamma)$ be the exponent to which $\mathfrak{q}$ divides the ideal generated by $\gamma$.

For $(a, b) \in S$, compute the vector $V_{a,b}$ of length $\pi(B) + \Pi(B) + d$ whose first $\pi(B)$ entries are

$$v_{q_1}(a - bm), \ldots, v_{q_{\pi(B)}}(a - bm),$$

whose next $\Pi(B)$ entries are

$$v_{\mathfrak{q}_1}(a - b\alpha), \ldots, v_{\mathfrak{q}_{\Pi(B)}}(a - b\alpha),$$

and whose last $d$ entries are the images of $a - b\alpha$ under the character maps defined in the next paragraph. The values $v_{q_i}(a - bm)$can be read off of the prime factorization of $a - bm$ and are easily obtained from the sieve in Step 2. The values $v_{\mathfrak{q}_i}(a - b\alpha)$ can be read off of the prime factorization of $N(a - b\alpha)$ for all $\mathfrak{q}_i$ for which the localization of $R$ at $R \cap \mathfrak{q}_i$ is integrally closed and hence is a discrete valuation ring. In this case, the sieve in Step 2 again produces the needed entries. For the remaining prime ideals, each of which lies over a prime dividing $[\mathcal{O} : R]$, the desired values can be efficiently computed using the method sketched in [Lenstra 1992] and described in detail in [Cohen 1993]. See [Stevenhagen 2008, formula (7.4)] and [Buhler et al. 1993] for a discussion of the relationship between the factorization of $(a - b\alpha)$ and $N(a - b\alpha)$.

Let
$$\Gamma = \{\gamma \in \mathcal{O} \mid N(\gamma) \not\equiv 0 \bmod l\}.$$

Let $\varepsilon$ be the least common multiple of the orders of the multiplicative groups $(\mathcal{O}/\ell)^*$, where $\ell$ ranges over the prime ideals lying above $l$. Since $l$ does not divide the discriminant of $f$ and is therefore unramified in $\mathbb{Q}(\alpha)$, we have for all $\gamma \in \Gamma$,

$$\gamma^\varepsilon \equiv 1 \bmod l. \tag{2.4}$$

Let $\lambda : \Gamma \to l\mathcal{O}/l^2\mathcal{O}$ be the map sending $\gamma$ to $(\gamma^\varepsilon - 1) + l^2\mathcal{O}$. We obtain $d$ maps

$$\lambda_j : \Gamma \to \mathbb{Z}/l\mathbb{Z}$$

by fixing a module basis $\{b_j l + l^2 \mathcal{O}\}_{j=1,\ldots,d}$ for $l\mathcal{O}/l^2\mathcal{O}$ over $\mathbb{Z}/l\mathbb{Z}$ and projecting $\lambda$ onto each coordinate. In other words, the $\lambda_j$ are given by the congruence

$$\gamma^\varepsilon - 1 \equiv \sum_{j=1}^{d} \lambda_j(\gamma) b_j l \bmod l^2.$$

Since $\lambda(\gamma\gamma') = \lambda(\gamma) + \lambda(\gamma')$ and $\lambda_j(\gamma\gamma') = \lambda_j(\gamma) + \lambda_j(\gamma')$, the maps $\lambda$ and $\lambda_j$ are homomorphisms on $\mathcal{O}^*$. We include in the vector $V_{a,b}$ the values $\lambda_1(a - b\alpha)$, $\ldots$, $\lambda_d(a - b\alpha)$.

The role of the maps $\lambda_j$ is to enable us to construct elements in $\mathcal{O}$ which are $l$-th powers. In the next step, we produce an element $\gamma \in \Gamma$ such that $v_\mathfrak{q}(\gamma) \equiv 0 \bmod l$ for all prime ideals $\mathfrak{q} \in \mathcal{O}$ and such that $\lambda(\gamma) = 0$. If the class number of $K$ is prime to $l$, as we expect for large $l$, then $\gamma$ is certain to generate the $l$-th power of a principal ideal and hence is the product of an $l$-th power and a unit $\omega$. Since any $l$-th power is mapped to 0 by $\lambda$, we find that $\omega$ is mapped to 0 as well. We claim that it is likely in this case that $\omega$, and in turn $\gamma$, is an $l$-th power. For more on the $\lambda_j$, including a precise formulation of the above claim and a heuristic argument supporting it, see [Schirokauer 1993]. Finally, note that it is in order to calculate the $\lambda_j$ that we need to work with a single prime divisor of $p - 1$. If analogous, logarithmic maps with values in $\mathbb{Z}/(p-1)\mathbb{Z}$ could be computed without factoring $p - 1$, then $\log_t u$ could be obtained without knowing the factorization of $p - 1$.

**Step 4. Linear algebra.** Let $V_t$ be the vector of length $\pi(B) + \Pi(B) + d$ whose first $\pi(B)$ entries are $v_{q_1}(t), \ldots, v_{q_{\pi(B)}}(t)$ and whose last $\Pi(B) + d$ coordinates are all 0. Define $V_u$ similarly, with $u$ in place of $t$. Let $A$ be the matrix whose first column is $V_t$ and remaining columns are the vectors $V_{a,b}$. Now solve the congruence

$$AX \equiv -V_u \bmod l. \tag{2.5}$$

If $V_u$ is not in the column space of $A$, increase the parameter $C$ in order to enlarge the set $S$ and, one expects, the rank of $A$.

When the parameters $B, C$, and $d$ are chosen in order to minimize the time required for the sieving in Step 2, subject to the constraint that $C$ is large enough that (2.5) can be solved, one finds that the time required for Step 2 is equal to $r^{2+o(1)}$, where $r$ is the column length of $A$ and the $o(1)$ is for $p \to \infty$. Since one would like to be able to solve (2.5) within the same amount of time, Gaussian elimination, which requires $O(r^3)$ steps, is not a good choice. Moreover, Gaussian elimination is not practical for matrices of the size arising in current implementations. Instead, it is best to use a method which takes advantage of the fact that almost all the entries of $A$ are 0. The most useful of these from a theoretical standpoint is the coordinate recurrence method which is described in [Wiedemann 1986] and which can be shown to solve (2.5) in time $r^{2+o(1)}$, as desired. Combinations and adaptations of three other methods, the conjugate gradient method, the Lanczos algorithm, and structured Gaussian elimination, have had success in practice (see [Odlyzko 2000] for discussion and references). Nevertheless, the linear algebra continues to be a significant practical concern and accounts in part for the fact that computing discrete logarithms with the NFS is more difficult than factoring, in which case the linear algebra is done modulo 2.

The entries in a solution to (2.5), excluding the first which corresponds to the element $t$, can be indexed by the pairs $(a, b) \in S$. Let $(x, \ldots, x_{a,b}, \ldots)$ be one such solution. Let

$$\delta = t^x u \prod (a - bm)^{x_{a,b}} \quad \text{and} \quad \gamma = \prod (a - b\alpha)^{x_{a,b}},$$

where $t$ and $u$ are thought of as integers. Then $v_q(\delta) \equiv 0 \bmod l$ for all primes $q$, and therefore $\delta$ is an $l$-th power. Similarly $v_\mathfrak{q}(\gamma) \equiv 0 \bmod l$ for all prime ideals $\mathfrak{q} \subseteq \mathcal{O}$. In addition, $\lambda_i(\gamma) = 0$ for $i = 1, \ldots, d$. As remarked earlier, it is likely that $\gamma$ is then an $l$-th power in $\mathcal{O}$. We assume that this is the case; see [Schirokauer 1993] for comments on how to weaken this assumption. Clearly, $\delta$ is the $l$-th power of an element in $R$. The same may not be true of $\gamma$. However, since $f'(\alpha)\mathcal{O} \subseteq R$, we see that both $f'(\alpha)^l \delta$ and $f'(\alpha)^l \gamma$ are $l$-th powers of an element in $R$. Recall that $\phi : R \to \mathbb{F}_p$ is the ring homomorphism that satisfies $\phi(\alpha) = \phi(m)$. Since $\phi(f'(\alpha)^l \delta)$ and $\phi(f'(\alpha)^l \gamma)$ are $l$-th powers and $\phi(f'(\alpha)^l \delta) = t^x u \phi(f'(\alpha)^l \gamma)$, we find that $t^x u$ is an $l$-th power in $\mathbb{F}_p^*$ and conclude that $x \equiv -\log_t u \bmod l$.

We observe that the smoothness of $t$ and $u$ is used in the algorithm to ensure that these two elements appear in the relations constructed in Step 4. However, it suffices to know two elements $\tau$ and $\upsilon$ in $R$ such that $\phi(\tau) = t$ and $\phi(\upsilon) = u$ and such that the ideals $(\tau)$ and $(\upsilon)$ in $\mathbb{O}$ factor over the set $q_1, \ldots, q_{\Pi(B)}$ introduced in Step 3. In this case, the vectors $V_t$ and $V_u$ are replaced by vectors $V_\tau$ and $V_\upsilon$ containing the exponents appearing in the prime ideal factorizations of $(\tau)$ and $(\upsilon)$ in $\mathbb{O}$, as well as the values $\lambda_j(\tau)$ and $\lambda_j(\upsilon)$, and the linear algebra yields products

$$\prod (a - bm)^{x_{a,b}} \quad \text{and} \quad \tau^x \upsilon \prod (a - b\alpha)^{x_{a,b}},$$

which are expected to be $l$-th powers.

EXAMPLE 2.6. Let $p$ be the Mersenne prime $2^{127} - 1$ discovered by Lucas in 1876. Since $p$ is of the form $r^e - s$ with $r$ and $s$ small in absolute value, we can use the techniques of the special number field sieve described in [Lenstra et al. 1993b]. The analysis given there reveals that the optimal value of $d$ in the present example is 3. We proceed to construct a cubic extension of $\mathbb{Q}$ by noting the $p$ divides $2^{129} - 4 = (2^{43})^3 - 4$. Hence the polynomial $f = X^3 - 4$ has a root $m \bmod p$ which is close to $p^{1/3}$ and has extremely small coefficients. It is, therefore, ideally suited to our purpose, and we let $R = \mathbb{Z}[\sqrt[3]{4}]$ and $\phi : R \to \mathbb{F}_p$ be the map which sends $\sqrt[3]{4}$ to $2^{43}$. In Step 2, we look for pairs $a, b$ such that $a - 2^{43}b$ and $N(a - b\sqrt[3]{4}) = a^3 - 4b^3$ are both smooth. In Step 3, we encounter a potential complication due to the fact that $R$ is not the full ring of integers $\mathbb{O}$ of $\mathbb{Q}(\sqrt[3]{4})$, as demonstrated by the fact that $(\sqrt[3]{4})^2/2$ is equal to $\sqrt[3]{2}$ and so is integral. This difficulty is readily handled, however, by observing that $\mathbb{O} = \mathbb{Z}[\sqrt[3]{2}]$ and that the only prime dividing $[\mathbb{O} : R]$ is 2; see [Marcus 1977, Chapter 2]. Since $2 = (\sqrt[3]{2})^3$, there is only one prime ideal of $\mathbb{O}$ lying above 2 and its residue degree is one. The exponent to which this ideal divides $(a - b\sqrt[3]{4})$ is therefore equal to $v_2(N(a - b\sqrt[3]{4}))$. Thus all the entries in the exponent vectors $V_{a,b}$ in Step 3 can be obtained from the factorizations of $a - 2^{43}b$ and $N(a - b\sqrt[3]{4})$.

We consider briefly the computation of the residue of $\log_t \upsilon$ modulo the largest prime divisor of $p - 1$, a prime of eleven digits which we denote by $l$ and which divides $p - 1$ only once. Since $x^3 - 2$ splits completely over $\mathbb{F}_l$, the value of $\varepsilon$ in (2.4) is $l - 1$. We note that $\mathbb{O}$ is a principal ideal domain, as is easily seen by computing the Minkowski constant, and that $\lambda$ induces an injective $\mathbb{F}_l$-linear map from $\mathbb{O}^*/(\mathbb{O}^*)^l$ to $l\mathbb{O}/l^2\mathbb{O}$, a fact which can be proved by showing that $(\sqrt[3]{2} - 1)^{-1}$ is the fundamental unit of $\mathbb{O}$ (see [Marcus 1977, Exercise 5.36]) and checking that $\lambda(\sqrt[3]{2} - 1) \neq 0$. Thus any element in the kernel of $\lambda$ is an

$l$-th power. We conclude that the product $\prod(a-b\sqrt[3]{4})^{x_{a,b}}$ obtained in Step 4 is itself an $l$-th power and that the algorithm produces the desired residue.

One consequence of the special nature of $f$ in this example is that we could have computed generators for the prime ideals and unit group of $\mathcal{O}$ and explicitly factored the smooth elements found in Step 2 into a product of powers of these generators. In this case, we would not have needed the additive characters, but would have constructed $(p-1)$-st powers by finding a dependency modulo $p-1$ among the vectors containing the exponents appearing in these factorizations. In particular, we could have avoided factoring $p-1$. For more on this approach, which may be particularly attractive for number fields of small discriminant, see [Lenstra et al. 1993b; 1993a].

**2.7. The NFS for general fields.** Let $p$ be a prime and $n > 1$, and let $q = p^n$. In this section, we briefly describe two methods for computing logarithms in $\mathbb{F}_q$. In the first, we build the NFS on top of a number field $F$ of small discriminant having $\mathbb{F}_q$ as a residue field. To find a suitable field, we look for a prime $r$ having the property that $\mathbb{Q}(\zeta)$, where $\zeta$ is a primitive $r$th root of unity, has a subfield of degree $n$ over $\mathbb{Q}$ in which $p$ is inert. This subfield serves as $F$. If the Extended Riemann Hypothesis holds, then we are guaranteed to find such a prime $r$ satisfying $r < (\log q)^{O(1)}$ [Shoup 1992]. One attractive feature of our choice of $F$ is that its ring of integers $\mathcal{O}_F$ has an integral basis consisting of the conjugates of the trace of $\zeta$ in $F$. See [Schirokauer 2000] for details. With $F$ in hand, we construct an extension of $F$ of degree $d$ by adjoining to it a root $\alpha$ of the polynomial obtained by expanding $p$ in base $m$, where $m = \lfloor p^{1/d} \rfloor$. Next, we look for pairs $a, b \in \mathcal{O}_F \times \mathcal{O}_F$ such that $a - b\alpha$ and $a - bm$ are both $B$-smooth. As before, we consider approximately $C^2$ many pairs, chosen so that for any candidate, the coefficients appearing in the expressions $a = \sum a_j t_j$ and $b = \sum b_j t_j$, where $\{t_1, \ldots, t_n\}$ is the integral basis specified above, are at most $C^{1/n}$ in absolute value. The $B$-smooth pairs can be identified using sieving techniques or the elliptic curve factoring method [Lenstra 1987]. For each such pair, as well as for $t$ and $u$, we construct exponent vectors, including this time the values of the additive characters for both $F$ and $F(\alpha)$. Finally, we perform linear algebra mod $l$ to obtain an integer which is likely to be the residue of $\log_t u \mod l$.

In the second approach, proposed in [Joux et al. 2006], two number rings with maps to $\mathbb{F}_q$ are obtained by letting $f_1$ be any polynomial of degree $n$ with small coefficients which is irreducible mod $p$ and letting $f_2 = f_1 + p$. Then for $i = 1, 2$, we set $R_i = \mathbb{Z}[\alpha_i]$, where $\alpha_i \in \mathbb{C}$ is a root of $f_i$. Because $f_1$ is irreducible mod $p$, there exists $v \in \mathbb{F}_q$ such that $\mathbb{F}_q = \mathbb{F}_p(v)$ and $v$ is a root of the polynomial in $\mathbb{F}_p[X]$ obtained by reducing the coefficients of $f_1$ mod $p$. For $i = 1, 2$, we let $\phi_i : R_i \to \mathbb{F}_q$ be the map that sends $\alpha_i$ to $v$. The natural next step in the NFS is to look for pairs $a, b \in \mathbb{Z} \times \mathbb{Z}$ such that $a - b\alpha_1$ and $a - b\alpha_2$ are $B$-smooth. We might expect this search to go more quickly than the corresponding step in the previous approach, since $\alpha_1$ and $\alpha_2$ are of degree $n$ over $\mathbb{Q}$, whereas $\alpha$ and $m$ above are of degrees $dn$ and $n$ respectively. However, this advantage is countered by the fact that instead of having $2n$ coefficients $a_1, \ldots, a_n, b_1, \ldots, b_n$ available as was the case before, there are only two parameters $a$ and $b$ to vary. In order to increase the pool of smoothness candidates, it is necessary to expand the search to include higher degree polynomials in $\alpha_1$ and $\alpha_2$. A bound on this degree enters as a new parameter in the analysis of the method. Once enough pairs are found, linear algebra is performed on the associated exponent vectors in order to obtain the desired residue of $\log_t u \bmod l$.

**2.8. Running time.** We consider first the running time of the sieving stage in the case that the field in which we compute is prime. As we have seen, the candidates for smoothness in this case are bounded by

$$C(p^{1/d} + 1) \cdot (d+1)C^d p^{1/d} \leq 2dC^{d+1} p^{2/d}. \tag{2.9}$$

Let $x$ denote the right hand side of (2.9). In order to be able to solve (2.5), we expect that the number of double-smooth pairs which are needed, call it $N$, is slightly larger than the length of the vectors $V_{a,b}$ appearing in Step 3. This length is equal to $\pi(B) + \Pi(B) + d$, which is bounded by $(d+1)B + d$. Thus, if $d = B^{o(1)}$ for $p \to \infty$, we have

$$N = B^{1+o(1)}. \tag{2.10}$$

Let $\psi(x, B)$ be the number of positive integers $\leq x$ which are $B$-smooth. We adopt the critical assumption that the integers which are tested for smoothness in Step 2 behave like random numbers with regard to the property of being $B$-smooth. Then we can interpret the quotient $xN/\psi(x, B)$ as the number of pairs that need to be tested. The following theorem, which is copied verbatim from [Buhler et al. 1993], tells us the optimal value for this quantity in the case that (2.10) holds.

THEOREM 2.11. *Suppose $g$ is a function defined for all $y \geq 2$ that satisfies $g(y) \geq 1$ and $g(y) = y^{1+o(1)}$ for $y \to \infty$. Then as $x \to \infty$,*

$$xg(y)/\psi(x, y) \geq L_x[1/2; \sqrt{2} + o(1)]$$

*uniformly for all $y \geq 2$. In addition,*

$$xg(y)/\psi(x, y) = L_x[1/2; \sqrt{2} + o(1)]$$

*for $x \to \infty$ if and only if $y = L_x[1/2; \sqrt{2}/2 + o(1)]$ for $x \to \infty$.*

Theorem 2.11 reveals that if (2.10) is valid, then

$$C^2 \geq L_x[1/2; \sqrt{2}+o(1)]. \tag{2.12}$$

In the best case that equality holds in (2.12), the values of $C$ and $d$ which minimize $C$ satisfy

$$C = L_p[1/3; (8/9)^{1/3} + o(1)],$$
$$d = ((3+o(1)) \log p / \log \log p)^{1/3}, \tag{2.13}$$

where the limit implicit in the $o(1)$ is for $p \to \infty$. We thus arrive at the conjecture that the number of pairs that need to be tested in Step 2 for the algorithm to succeed is equal to

$$L_p[1/3; (64/9)^{1/3} + o(1)]. \tag{2.14}$$

In [Buhler et al. 1993] and [Schirokauer 2000], the reader will find a much more careful analysis in support of this claim. A quick calculation shows that when (2.13) holds, $x = L_p[2/3; (64/3)^{1/3} + o(1)]$, and a second calculation using Theorem 2.11 reveals that the optimal value of $B$ satisfies

$$B = L_p[1/3; (8/9)^{1/3} + o(1)]. \tag{2.15}$$

Finally, investigation of the entire method in the case that (2.13) and (2.15) hold, leads to the conjecture that (2.14) not only represents the sieving time in Step 2 but is also a bound for the running time of the other steps of the algorithm. We note that in the case of the special number field sieve exhibited in Example 2.6, the numbers being tested for smoothness are bounded in absolute value by $2dC^{d+1}p^{1/d}$. The fact that the power of $p$ is smaller in this expression than in (2.9) results in a reduced running time of $L_p[1/3; (32/9)^{1/3} + o(1)]$. The gain from $(64/9)^{1/3}$ to $(32/9)^{1/3}$ means that for large $p$, logarithms can be computed in $\mathbb{F}_p$ when $p$ is special in nearly the same amount of time as is needed to handle a general prime field of size $\sqrt{p}$.

We now turn to the complexity of the methods described for non-prime fields. To analyze the first approach, in which the NFS is built on top of a field $F$, we define the height of $\gamma \in \mathcal{O}_F$ by the formula $h(\gamma) = \max\{|\sigma\gamma|\}$, where $\sigma$ ranges over the embeddings of $F$ into $\mathbb{R}$. Then the norm of $\gamma$ is at most $h(\gamma)^n$ in absolute value. The elements in $\mathcal{O}_F$ that are tested for smoothness in this version of the NFS are of the form $b^d f(a/b)(a-bm)$ and are of height at most

$$2dC^{(d+1)/n}r^{d+1}p^{2/d}.$$

Hence, the integers being tested for $B$-smoothness are bounded in absolute value by

$$(2d)^n C^{d+1} q^{2/d} r^{(d+1)n} \leq d^n C^{d+1} q^{2/d} (\log q)^{O(dn)}.$$

We see immediately that the way in which $n$ grows as $q \to \infty$ has an effect on the running time of the algorithm. Indeed, if

$$n \leq o\left(\frac{\log q}{\log \log q}\right)^{1/3}, \tag{2.16}$$

then, when $d = (\log q / \log \log q)^{1/3}$, the factor $d^n (\log q)^{O(dn)}$ is at most $L_q[2/3; o(1)]$. In this case, it is conjectured that the algorithm runs in time $L_q[1/3; (64/9)^{1/3} + o(1)]$ as in the prime case. Moreover, if the little-oh in (2.16) is replaced by a big-oh, the secondary constant $(64/9)^{1/3}$ is lost but the primary constant $1/3$ is retained in the running time. We leave it as an exercise to show more generally that if $q \to \infty$ subject to the restriction that $n \leq O(\log q / \log \log q)^{e_n}$ for some constant $e_n$, then the running time is conjecturally equal to $L_q[\max\{1/3, (1+e_n)/4\}; O(1)]$.

To analyze the second approach given, in which $R_1$ is generated by a polynomial $f_1$ with small coefficients and $R_2$ by the polynomial $f_2 = f_1 + p$, we again seek a bound on the integers being tested for smoothness. Considering $R_1$ first, we note that the norm of an element of the form $\sum_{i=0}^{e} a_i \alpha_1{}^i$ is equal to the resultant of $f_1$ and the polynomial $\sum a_i X^i$. This resultant, in turn, is equal to the determinant of the associated Sylvester matrix and is therefore bounded in absolute value by $(n+e)^{n+e} M^n D_1^e$, where $M$ is a bound on the absolute value of the coefficients $a_i$ and $D_1$ is a bound, assumed to be small, on the absolute value of the coefficients of $f_1$. Combining this quantity with the corresponding value for $f_2$, we obtain a bound on the smoothness candidates of

$$(n+e)^{2(n+e)} M^{2n} (pD_1 + D_1^2)^e. \tag{2.17}$$

In order to compare this bound with the one arising in the prime field case, we assume that the number of candidates tested is $C^2$. It follows that $M < C^{2/e}$, and we can replace (2.17) with

$$x = (n+e)^{2(n+e)} C^{4n/e} (pD_1 + D_1^2)^e. \tag{2.18}$$

The constraint on $C$ is then given, as before, by (2.12). It is now straightforward to verify that (2.12) and (2.18) can be simultaneously satisfied with $C = L_q[1/3; O(1)]$ if and only if $n$ is bounded by $O(\log q / \log \log q)^{2/3}$ but not bounded by $O(\log q / \log \log q)^k$ for any $k < 1/3$. Indeed, for a given $n = (\log q / \log \log q)^{e_n}$ with $1/3 \leq e_n \leq 2/3$, the bound $e$ on the degree of the polynomials in $\alpha_1$ and $\alpha_2$ that are tested for smoothness should be set equal to $\lceil (\log q / \log \log q)^{e_n - 1/3} \rceil$. We refer the reader to [Joux et al. 2006] for more details and conclude by remarking that the two NFS methods for general fields which we have presented, taken in conjunction, conjecturally run in time $L[1/3; O(1)]$ so long as $q \to \infty$ with $n \leq O(\log q / \log \log q)^{2/3}$.

## 3. The function field sieve (FFS)

Adleman [1994] describes a function field analogue of the number field sieve which he calls the function field sieve. In order to compute logarithms in a finite field of characteristic $p$, the algorithm makes use of an algebraic extension of $\mathbb{F}_p(X)$ in the same way that the number field sieve makes use of an algebraic extension of $\mathbb{Q}$. Like the number field sieve, it is conjectured to run in time $L_q[1/3; O(1)]$, provided that the cardinality $q$ of the finite field tends to $\infty$ in a restricted fashion. In this section, we give a sketch of the FFS and a conjecture as to its complexity. We define a notion of smoothness for elements in $\mathbb{F}_q$ and as we did in the preceding section, restrict ourselves to the special case that the logarithm base and the element whose logarithm we seek are smooth. The algorithm we describe below is not the one found in [Adleman 1994] but instead is a modification of the simpler and improved version which is presented in [Adleman and Huang 1999] and which incorporates some of the techniques found in the algorithm of Coppersmith for fields of characteristic two [Coppersmith 1984]. The relationship between the FFS and Coppersmith's method is explored in Section 3.8.

**3.1. The FFS.** Let $p$ be a prime and $q = p^n$. We begin by choosing a model for the finite field $\mathbb{F}_q$. Let $g$ be a polynomial of minimal degree such that $X^n + g$ is irreducible and at least one root of $g$ in some algebraic closure of $\mathbb{F}_p$ is of multiplicity one. Let $f = X^n + g$ and fix as a model for $\mathbb{F}_q$ the set of polynomials in $\mathbb{F}_p[X]$ of degree $< n$, with addition and multiplication taken modulo $f$.

Recall that a polynomial in $\mathbb{F}_p[X]$ is said to be $B$-smooth if it factors into irreducibles all of which are of degree at most $B$. Thinking of the elements of $\mathbb{F}_q$ as polynomials, we can apply the notion of smoothness to elements in $\mathbb{F}_q$. The version of the FFS we now describe takes as input a smoothness bound $B \geq 1$, two $B$-smooth elements $t, u \in \mathbb{F}_q^*$ satisfying $u \in \langle t \rangle$, and two integral parameters $C \geq 0$ and $d \geq 1$. It outputs $\log_t u$.

**Step 1. Constructing an extension field.** Let $F = \mathbb{F}_p(X)$. We build an extension of $F$ by adjoining to it a root of a polynomial of degree $d$. The special form of $f$ allows us to find a suitable polynomial with particularly small coefficients. Let $k$ be the smallest multiple of $d$ greater than or equal to $n$ and let $H = Y^d + X^{k-n}g \in \mathbb{F}_p[X, Y]$. The fact that $g$ has a root of multiplicity prime to $d$ implies by Eisenstein's criterion that $H$ is absolutely irreducible. Let $\overline{F}$ be an algebraic closure of $F$ and let $\alpha \in \overline{F}$ be a root of $H$, considered as a polynomial in $Y$ over $\mathbb{F}_p[X]$. Let $R = \mathbb{F}_p[X][\alpha]$ and denote by $K$ the field of fractions of $R$. Let $\phi : R \to \mathbb{F}_q$ be the map which sends an element $h(X, \alpha)$ to the polynomial of degree $<n$ congruent to $h(X, X^{k/d})$ mod $f$. Since $f$ divides $(X^{k/d})^d + X^{k-n}g$, we see that $\phi$ is a ring homomorphism.

**Step 2. Sieving.** Let $N : K \to F$ be the norm map. We say that an element in $R$ is $B$-smooth if its image under $N$ is $B$-smooth. Let $S$ be the set of pairs $(a, b) \in \mathbb{F}_p[X] \times \mathbb{F}_p[X]$ such that $\deg(a)$ and $\deg(b)$ are at most $C$ and both $a - b\alpha$ and $a - bm$ are $B$-smooth. In this step, we use a sieve to identify the elements in $S$. We refer the reader to [Gao and Howell 1999; Gordon 1993b; Joux and Lercier 2002; Thomé 2001] for details and note that testing candidates for smoothness is of no consequence to the complexity analysis of the algorithm as it is possible to factor polynomials in polynomial time using the algorithm of [Berlekamp 1970].

**Step 3. Computing valuation vectors.** Let $M_F$ be the set of discrete valuations of $F$ of degree $\leq B$ (see [Stichtenoth 1993] for background information on discrete valuations and places in function fields). Let $M_K$ be the set of discrete valuations of $K$ which are extensions of the valuations in $M_F$. For each $(a, b) \in S$, we construct a vector $V_{a,b}$ containing the values $v(a - bm)$ for all $v \in M_F$, and $v(a - b\alpha)$ for all $v \in M_K$.

The valuations in $M_F$, excluding the valuation at $\infty$ which we denote by $v_\infty$, are in one-to-one correspondence with the irreducible polynomials in $F$. Indeed, for each such polynomial $h$, we obtain a valuation $v_h$ whose value at an element $\gamma$ is the exponent to which $h$ divides $\gamma$. Thus, to determine $v_h(a - bm)$ for some pair $(a, b) \in S$, it suffices to factor $(a - bm)$ into irreducibles, a task already accomplished in Step 2. Since $v_h(a - bm) = 0$ for all $h$ of degree $> B$, and since for all $\gamma \in F$,

$$\sum_{h \text{ irreducible}} \deg(h) v_h(\gamma) = -v_\infty(\gamma),$$

we see that $v_\infty(a - bm)$ is obtained immediately.

Let $v$ be an element in $M_K$ which does not lie over $v_\infty$. Let $Q = \{\gamma \in K \mid v(\gamma) > 0\}$ be the associated place. As was the case with the NFS, if the localization of $R$ at $R \cap Q$ is a discrete valuation ring, then $v(a - b\alpha)$ can be read off the factorization of $N(a - b\alpha)$. If not, $v(a - b\alpha)$ can be computed by a method that uses Newton polygons to construct a fractional power series containing the information needed to evaluate $v$. This technique is discussed in detail in [Adleman and Huang 1999].

In the case that $v$ is an extension of $v_\infty$, we proceed by homogenizing $H$ with respect to a new variable $Z$ and dehomogenizing with respect to either $X$ or $Y$. Doing so yields a new polynomial $H'$. Renaming variables, we obtain a domain $\mathbb{F}_p[U, V]/H'(U, V)$ whose field of fractions $K'$ is isomorphic to $K$ under a map which sends the places associated to the extensions of $v_\infty$ to finite places in $K'$. The associated valuations can now be computed with the Newton polygon method alluded to above.

**Step 4. Linear algebra.** Let $V_t$ be the vector of length $|M_F|+|M_K|$ whose first $|M_F|$ entries are the values $v(t)$ for $v \in M_F$ and whose remaining coordinates are 0's. Define $V_u$ in the same way but with $t$ replaced by $u$. Let $A$ be the matrix whose first column is $V_t$ and remaining columns are the vectors $V_{a,b}$, and solve the congruence

$$AX \equiv -V_u \mod (q-1)/(p-1). \tag{3.2}$$

Unlike what we encountered with the NFS, the linear algebra this time is done modulo a number which may be composite. See [McCurley 1990; Schirokauer 1993] for some comments on this situation. We continue under the assumption that we are able to produce a solution $(x, \ldots, x_{a,b}, \ldots)_{(a,b) \in S}$ to (3.2). Let

$$\delta = t^x u \prod (a-bm)^{x_{a,b}} \quad \text{and} \quad \gamma = \prod (a-b\alpha)^{x_{a,b}},$$

where $t$ and $u$ are thought of here as elements in $\mathbb{F}_p[X]$. Since

$$v(\delta) \equiv 0 \mod (q-1)/(p-1) \quad \text{for all } v \in M_F,$$

we know that $\delta$ is a product of an element in $\mathbb{F}_p^*$ and a $(q-1)/(p-1)$-st power. Since any such power in $\mathbb{F}_q^*$ is in $\mathbb{F}_p^*$, we find that $\phi(\delta) \in \mathbb{F}_p^*$. Let $h$ be the gcd of $(q-1)/(p-1)$ and the class number of $K$. If $h = 1$, then the fact that $v(\gamma) \equiv 0 \mod (q-1)/(p-1)$ for all $v \in M_K$ implies that $\gamma$ is the product of an an element in $\mathbb{F}_p^*$ and a $(q-1)/(p-1)$-st power. Hence, $\phi(\gamma) \in \mathbb{F}_p^*$. In this case, $t^x u = \phi(\delta)\phi(\gamma)^{-1} = \mu$ for some $\mu \in \mathbb{F}_p^*$, and we have

$$x \equiv -\log_t u \mod (q-1)/(p-1).$$

All that remains is the computation of a logarithm in $\mathbb{F}_p^*$, namely $\log_{t'} \mu$ where $t' = t^{(q-1)/(p-1)}$. This can be accomplished with the NFS or one of the methods described in [Pomerance 2008]. If $h > 1$, we adopt the modification presented in [Schirokauer 2002].

EXAMPLE 3.3. Let $q = 2^{127}$ and assume we are trying to compute logarithms in $\mathbb{F}_q$. We adopt as our model for $\mathbb{F}_q$ the set of polynomials of degree at most 126, with addition and multiplication done modulo $f = X^{127} + X + 1$. The same optimization that was used in our NFS example indicates that we should let $d = 3$. Taking advantage of the special form of $f$, we let

$$H = Y^3 + X^2(X+1).$$

The map $\phi$ in this case sends $\alpha$ to $X^{43}$ and the pairs of polynomials tested for smoothness in Step 2 are of the form $a^3 + b^3 X^2(X+1)$ and $a + bX^{43}$.

We consider the problem of computing the valuation vectors described in Step 3. We do not give citations for the statements made below but refer the reader to [Stichtenoth 1993, Chapter III] for the theorems on which they are based.

Let $F = \mathbb{F}_2(X)$. Let $P_X$ and $P_{X+1}$ be the places of $F$ corresponding to the irreducible polynomials $X$ and $X+1$ respectively, and let $P_\infty$ be the place of $F$ at infinity. Let $Q$ be a place of $K$ lying over a place $P$ of $F$ other than $P_X$, $P_{X+1}$ or $P_\infty$, and let $v_Q$ be the associated valuation. Since $\alpha^3 = X^2(X+1)$ in $K$, we see that $v_Q(\alpha) = 0$. It follows that $R \cap Q$ is integrally closed, and therefore, the values $v_Q(a+b\alpha)$ can be read off of the factorization of $N(a+b\alpha)$. Since $K$ is a pure cubic extension of $F$ and $v_X(X^2(X+1))$ and $v_{X+1}(X^2(X+1))$ are both prime to $[K:F]$, we find that $P_X$ and $P_{X+1}$ are totally ramified in $K$. It follows that there is only one place, with residue degree one, lying above each of these places. We conclude, by the same reasoning given for the prime lying above 2 in Example 2.6, that $v_Q(a+b\alpha) = v_X(N(a+b\alpha))$, where $Q$ is the lone place above $P_X$ and $v_{Q'}(a+b\alpha) = v_{X+1}(N(a+b\alpha))$, where $Q'$ lies over $P_{X+1}$. We note that in the latter case, $R \cap Q'$ is a discrete valuation ring. Indeed, $H$ is non-singular at the point $(1,0)$. By contrast $H$ is singular at $(0,0)$ and the local ring $R \cap Q$ is not integrally closed, as is made explicit by the fact that $\alpha^2/X$ is a cube root of $X(X^2+1)$.

Finally, we consider $P_\infty$. Homogenizing the curve $Y^3 + X^2(X+1)$ with respect to a third variable $Z$ and dehomogenizing with respect to $X$ yields the curve $Y^3 + Z + 1$. Let $\beta$ be a root of $V^3 + U + 1$ in an algebraic closure of $\mathbb{F}_2(U)$. Let $E = \mathbb{F}_2(U)(\beta)$ and note that the map $\psi : K \to E$ given by

$$\psi(X) = 1/U \quad \text{and} \quad \psi(\alpha) = \beta/U$$

is an isomorphism. The place $P_\infty \subset F$ is mapped by $\psi$ to the place $P_U$ of $\mathbb{F}_2(U)$ corresponding to the polynomial $U$. To determine the splitting behavior of this place in $E$, we look at the splitting behavior of the image of the polynomial $V^3 + U + 1$ in the residue field of $P_U$. This image is $V^3 + 1$, which is the product of a linear factor and an irreducible quadratic factor over $\mathbb{F}_2$. We conclude that $P_U$ splits into a place of degree 1 and a place of degree 2 in $E$. Let $v_1$ and $v_2$ be the corresponding valuations and observe that any element of the form $a + b\alpha \in K$, where $a$ and $b$ are polynomials in $X$, is mapped to an element in $E$ of the form $U^\varepsilon(a' + b'\beta)$ for some $\varepsilon$, where $a'$ and $b'$ are polynomials in $U$. We can now use the fact that $v_1(a' + b'\beta)$ is the exponent to which $U$ appears in the norm of $(a' + b'\beta)$ in $\mathbb{F}_2[U]$ and that $v_2(a' + b'\beta) = 0$ to complete the computation of the valuation vector for the pair $(a, b)$.

**3.4. Running time.** In the analysis of the FFS, we adopt the same assumption as we did in the the case of the NFS, namely that the elements being tested for smoothness behave as random elements. In this case, the elements are polynomials in $\mathbb{F}_p[X]$ and so to make use of our assumption, we need results concerning the probability that a polynomial of degree $M$ picked at random is $B$-smooth. Such results can be found in [Bender and Pomerance 1998] and reveal that the probability of interest is, roughly speaking, equal to the probability that an integer of size at most $p^M$ is $p^B$-smooth. It follows that the size of the factor base and number of pairs $(a,b)$ tested in Step 2 should be asymptotically equal to the analogous quantities in the special number field sieve. Note that we use the special NFS here because it and the FFS both make use of a small field extension obtained by taking advantage of a special representation of $\mathbb{F}_q$. Since the linear algebra problem is the same in both the special NFS and FFS, we arrive at the conjecture that the running time of the FFS, like that of the special NFS, is

$$L_q[1/3; (32/9)^{1/3} + o(1)]. \tag{3.5}$$

In [Adleman and Huang 1999], the authors provide a heuristic argument in support of this conjecture which analyzes the FFS directly and does not proceed by analogy with the NFS.

The only obstruction to our conjecture is the requirement that the smoothness bound be at least one. As a consequence, we find that the factor base in the algorithm is at least size $p$ and the linear algebra in Step 4 requires at least $p^2$ many steps. As $q \to \infty$, it may be the case that $p^2$ is greater than (3.5) and that (3.5) is therefore not valid. The reader can easily check that this does not happen if

$$p \leq n^{o(\sqrt{n})} \tag{3.6}$$

as $q \to \infty$. It may, however, happen if (3.6) is relaxed to

$$p \leq n^{O(\sqrt{n})}. \tag{3.7}$$

In this case the consequences are not so dramatic as the primary constant of 1/3 in (3.5) is retained in the running time of the algorithm. Outside the range defined by (3.7), however, the FFS no longer runs in time $L_q[1/3; O(1)]$. For these fields, the time required by the algorithm is $p^{2+o(1)}$; see [Schirokauer 2002].

Since $p = L_q[s; c]$ if and only if $n = c^{-1}(\log q / \log \log q)^{1-s}$, we see from the above discussion that for $q \to \infty$ such that $n \geq (\log q / \log \log q)^e$, the FFS runs conjecturally in time $L_q[\max\{1/3, 1-e\}; O(1)]$. Combining this result with that for the NFS presented in Section 2.8, we conjecture that the algorithm which chooses for a given $q$ the faster of the NFS and FFS runs in time $L_q[1/3; O(1)]$ for all fields.

**3.8. Coppersmith's algorithm.** Coppersmith [1984] presents a method for computing logarithms in fields of characteristic two which has a conjectural expected running time of $L_q[1/3; c + o(1)]$, where $q$ is the cardinality of the field, the $o(1)$ is for $q \to \infty$, and $c$ is a constant which is equal to $(32/9)^{1/3}$ in the case that a certain quantity appearing in the algorithm is close to a power of two, and which is slightly larger otherwise. Coppersmith includes in his article a description of how to use his method to compute logarithms in the finite field considered in Example 3.3. He uses the same model for $\mathbb{F}_q$ as we give and observes that for any pair $(a, b) \in \mathbb{F}_2[X] \times \mathbb{F}_2[X]$,

$$(a + bX^{33})^4 \equiv a^4 + b^4 X(X+1) \bmod (X^{127} + X + 1). \qquad (3.9)$$

Thus, for each $(a, b)$ for which both sides of (3.9) are smooth, we obtain a relation in $\mathbb{F}_q$ involving only polynomials of degree at most the smoothness bound. Each such relation yields a linear relation among the logarithms of the elements appearing in it. If $t$ and $u$ are smooth, then once enough relations are found, $\log_t u$ can be determined using linear algebra mod $(q-1)$.

Assume now that we decide to compute logarithms in $\mathbb{F}_q$ with the FFS with $d = 4$, instead of the optimal value of 3. Then $H = Y^4 + X(X+1)$ and $m = X^{33}$. Let $\alpha, R, K$ and $\phi$ be as given in the description of the FFS in Section 3.1, and observe that $K$ is a purely inseparable extension of $\mathbb{F}_2(X)$. The norm map $N : K \to \mathbb{F}_2(X)$ in this case sends $\gamma \in K$ to $\gamma^4$ and is additive as well as multiplicative. Since the norm of any element in the kernel of $\phi$ is in $(X^{127} + X + 1)$, we see that for any pair $\delta, \gamma \in R$ such that $\phi(\delta) = \phi(\gamma)$,

$$N(\delta) \equiv N(\gamma) \bmod (X^{127} + X + 1).$$

In the case that $\delta = a + bX^{33}$ and $\gamma = a + b\alpha$, we obtain congruence (3.9). Thus, finding $a, b$ such that the norms of $a + bX^{33}$ and $a + b\alpha$ are smooth, as required by the FFS, is equivalent to finding $a, b$ such that both sides of (3.9) are smooth. In this way, we see that Coppersmith's algorithm is a special case of the FFS in which the extension field disappears and the relations can be realized in the base ring $\mathbb{F}_2[X]$.

## 4. General discrete logarithms

Let $p$ be a prime and $q = p^n$, and let $R_1$ and $R_2$ be rings, together with maps $\phi_1 : R_1 \to \mathbb{F}_q$ and $\phi_2 : R_2 \to \mathbb{F}_q$, chosen for use in the NFS or FFS discrete logarithm algorithm. As we have seen, these methods proceed in two stages. First, sieving techniques are used to find pairs of smooth elements $(\delta_1, \delta_2) \in R_1 \times R_2$ satisfying $\phi_1(\delta_1) = \phi_2(\delta_2)$, and then a linear algebra computation produces the desired logarithm. We note that the particular logarithm problem being tackled comes into play in the second stage but has no bearing on the search in the first stage. In fact, once sufficiently many smooth pairs are collected, the linear algebra can be tailored to produce $\log_t u$ for any $t$ and $u$ so long as they are the images in $\mathbb{F}_q$ of smooth elements in $R_1$ or $R_2$. In this section we discuss how to proceed if either $t$ or $u$ is an element for which we do not have a smooth pre-image under $\phi_1$ or $\phi_2$.

We begin with the case that $u$ fails to come with a smooth pre-image. One approach to this difficulty is to replace $R_1$ or $R_2$ by a ring $S$ that does contain a known smooth pre-image of $u$. The challenge is to do so while retaining all the desired features of the replaced ring. In the original adaptation of the number field sieve to the discrete logarithm problem, Gordon [1993a] provides a suitable method in the case that the field is prime and $u$ is represented by a moderately sized integer. To make use of the strategy, a reduction step is performed first in which a value $z$ is found such that $t^z u$ is represented by an integer with moderately sized factors. The algorithm runs in time $L_p[1/3; (64/9)^{13} + o(1)]$ for $p \to \infty$, but is not used in practice due to the fact that the entire NFS must be run multiple times in order to compute a single logarithm.

A second approach, which has its origins in Coppersmith's paper [1984] on computing discrete logarithms in characteristic two and is the method currently employed in practice, uses as its building block the following technique. Let $\gamma$ be an element in $R_1$ and assume $(\delta_1, \delta_2) \in R_1 \times R_2$ satisfies

(i) $\gamma | \delta_1$
(ii) $\delta_1/\gamma$ and $\delta_2$ are smooth.
(iii) $\phi_1(\delta_1) = \phi_2(\delta_2)$.

Then $\phi_1(\gamma) = \phi_2(\delta_2)/\phi_1(\delta_1/\gamma)$ in $\mathbb{F}_q$, and since $\delta_2$ and $\delta_1/\gamma$ are smooth, the logarithms of their images can be computed. Hence, the logarithm of $\phi_1(\gamma)$ can be determined.

At first glance, it might appear that this strategy is sufficient for computing $\log_t u$ for arbitrary $u$. Indeed, given the fact that only one pair of smooth elements is needed, the algorithm should be faster than those described earlier. The problem is that if we set the smoothness bound equal to the one used in the NFS or FFS, then for most $u$, any pre-image of $u$ in $R_1$ will be so large that, because of condition (i), the time needed to find $\delta_1$ and $\delta_2$ will greatly exceed the time need to compute the logarithms of $\phi_2(\delta_2)$ and $\phi_1(\delta_1/u)$ in $\mathbb{F}_q$.

The solution is to repeat the process multiple times, with more moderate expectations each time. In particular, we adopt a sequence of smoothness bounds $B_1 > \cdots > B_k = B$, where $B$ is the smoothness bound used in the NFS or FFS. Initially, we follow the above approach to reduce the problem of computing $\log_t u$ to the problem of computing the logarithms of a collection of images of elements in $R_1$ and $R_2$ of norm size at most $B_1$. Here, size refers to absolute value in the case of the NFS and degree in the case of the FFS. We then implement the procedure for each of these elements, using the bound $B_2$, thereby reducing the problem to the computation of logarithms of images of elements with norm size at most $B_2$. After $k$ rounds of descent, the original problem is reduced to the one which can be solved by the NFS or FFS. In fact, if we work back up the tree making all the necessary substitutions, we obtain $B$-smooth elements

$\sigma_1, \sigma_1' \in R_1$ and $\sigma_2, \sigma_2' \in R_2$ such that $u = \phi_1(\sigma_1)\phi_1(\sigma_1')^{-1}\phi_2(\sigma_2)\phi_2(\sigma_2')^{-1}$. Given this representation of $u$, the valuation vector $V_u$ is easily produced and $\log_t u$ can be computed with one running of the NFS or FFS.

One difficulty that arises in the analysis of this descent technique is that, because smoothness bounds are used that are larger than those in the NFS or FFS, the time required to sieve with all the primes of size up to such a bound dominates the entire computation. It thus becomes necessary to factor smoothness candidates individually. When using the reduction in conjunction with the NFS, for instance, these factorizations must be done with the elliptic curve factoring method and are costly. Nevertheless, Commeine and Semaev [2006] have shown that in the case of a prime field $\mathbb{F}_p$ the descent runs conjecturally in time $L_p[1/3; 3^{1/3} + o(1)]$ for $p \to \infty$. This is faster than the NFS of Section 2. On the FFS side, where polynomials can be factored quickly using polynomial gcd's, Joux and Lercier [2005a] provide an analysis showing that the running time of the reduction is bounded by that of the FFS of Section 3 so long as $p \leq n^{O(\sqrt{n})}$. Recall that this is the entire range in which the FFS runs in time $L_q[1/3; O(1)]$. Though it does not seem unlikely that the descent approach will lead to an $L[1/3; O(1)]$ reduction for all fields, the details in cases other than those cited have yet to be worked out.

In practice, the descent method appears to be very fast, as a look at any of the implementation announcements cited in the introduction will reveal. This may seem surprising, particularly in the NFS case, given our comments concerning the large smoothness bounds that are used and the apparent need to test smoothness candidates individually. However, an initial reduction step [Joux and Lercier 2003], which we do not describe here, and the use of sieving techniques despite the large bounds, speed the method up greatly. In this context, we note that condition (i) forces the search for smooth elements to take place in a lattice when working over $\mathbb{Z}$, or the analogous structure when working over $\mathbb{F}_p[X]$. For example, when the reduction is applied to $\gamma \in R_1$ in the case that $R_1 = \mathbb{Z}[\alpha]$ for some algebraic integer $\alpha$ and the elements tested for smoothness are linear in $\alpha$, the search is confined to the lattice

$$\{(a, b) \in \mathbb{Z}^2 \mid a - b\alpha \equiv 0 \bmod \gamma\}.$$

Sieving over a lattice structure is regularly done in implementations of the NFS and FFS and does not pose a difficulty. See [Pollard 1993] for an introduction to the subject.

We conclude this section by addressing the question of what to do if we know of no smooth element in $R_1$ or $R_2$ which maps to the base $t$. One way to proceed is to pick elements in $\mathbb{F}_q^*$ until a primitive element $t'$ with a known smooth preimage in $R_1$ or $R_2$ is found. The desired logarithm can then be determined by means of the identity

$$\log_t u \equiv \frac{\log_{t'} u}{\log_{t'} t} \mod (q-1).$$

If $q = p$ is prime and the Extended Riemann Hypothesis (ERH) holds, then we are assured of finding such a $t'$ by testing the elements in $\mathbb{F}_p^*$ represented by the integers $\leq (\log p)^{O(1)}$ [Shoup 1992]. More generally, and again under the assumption that the ERH is true, a primitive element in $\mathbb{F}_q^*$ having a smooth pre-image in the ring $\mathcal{O}_F$ introduced in Section 2.7 can be obtained in time $(\log q)^{O(n)}$ [Buchmann and Shoup 1996]. If one represents $\mathbb{F}_q$ as a quotient of $\mathbb{F}_p[X]$ and searches for $t'$ among the elements in $\mathbb{F}_q^*$ represented by polynomials in $\mathbb{F}_p[X]$ of small degree, one is certain of succeeding within $(np)^{O(1)}$ trials [Shoup 1992]. Thus, we see that so long as $n \leq o(\log q/\log\log q)^{1/3}$ or $p \leq n^{o(\sqrt{n})}$, finding a primitive element with smooth pre-image can be accomplished without affecting the running time of the NFS or FFS.

In the event that we are unable to switch primitive elements, there is a second option. We can apply to $t$ the descent reduction just described for handling $u$. Doing so produces a representation of $t$ as the product of the images of smooth elements, or their inverses, from $R_1$ and $R_2$. Such a representation can then be used to obtain a valuation vector $V_t$, which in turn can be incorporated into the linear algebra computation. This approach highlights the fact that, although $t$ and $u$ have different positions in the logarithm problem, they are incorporated in the same way into the NFS and FFS methods described in this paper.

## References

[Adleman 1994] L. M. Adleman, "The function field sieve", pp. 108–121 in *Algorithmic number theory* (Ithaca, NY, 1994), edited by L. M. Adleman and M.-D. Huang, Lecture Notes in Comput. Sci. **877**, Springer, Berlin, 1994.

[Adleman and Huang 1999] L. M. Adleman and M.-D. A. Huang, "Function field sieve method for discrete logarithms over finite fields", *Inform. and Comput.* **151**:1-2 (1999), 5–16.

[Bender and Pomerance 1998] R. L. Bender and C. Pomerance, "Rigorous discrete logarithm computations in finite fields via smooth polynomials", pp. 221–232 in *Computational perspectives on number theory* (Chicago, 1995), edited by D. A. Buell and J. T. Teitelbaum, AMS/IP Stud. Adv. Math. **7**, Amer. Math. Soc., Providence, RI, 1998.

[Berlekamp 1970] E. R. Berlekamp, "Factoring polynomials over large finite fields", *Math. Comp.* **24** (1970), 713–735.

[Brillhart et al. 1981] J. Brillhart, M. Filaseta, and A. Odlyzko, "On an irreducibility theorem of A. Cohn", *Canad. J. Math.* **33**:5 (1981), 1055–1059.

[Buchmann and Shoup 1996] J. Buchmann and V. Shoup, "Constructing nonresidues in finite fields and the extended Riemann hypothesis", *Math. Comp.* **65**:215 (1996), 1311–1326.

[Buhler et al. 1993] J. P. Buhler, H. W. Lenstra, Jr., and C. Pomerance, "Factoring integers with the number field sieve", pp. 50–94 in *The development of the number field sieve*, edited by A. K. Lenstra and H. W. Lenstra, Jr., Lecture Notes in Math. **1554**, Springer, Berlin, 1993.

[Cohen 1993] H. Cohen, *A course in computational algebraic number theory*, Graduate Texts in Mathematics **138**, Springer, Berlin, 1993.

[Commeine and Semaev 2006] A. Commeine and I. Semaev, "An algorithm to solve the discrete logarithm problem with the number field sieve", pp. 174–190 in *Public key cryptography*, edited by M. Yung et al., Lecture Notes in Comput. Sci. **3958**, 2006.

[Coppersmith 1984] D. Coppersmith, "Fast evaluation of logarithms in fields of characteristic two", *IEEE Trans. Inform. Theory* **30**:4 (1984), 587–594.

[Frey et al. 1999] G. Frey, M. Müller, and H.-G. Rück, "The Tate pairing and the discrete logarithm applied to elliptic curve cryptosystems", *IEEE Trans. Inform. Theory* **45**:5 (1999), 1717–1719.

[Gao and Howell 1999] S. Gao and J. Howell, "A general polynomial sieve: Designs and codes — a memorial tribute to Ed Assmus", *Des. Codes Cryptogr.* **18**:1-3 (1999), 149–157.

[Gordon 1993a] D. M. Gordon, "Discrete logarithms in $GF(p)$ using the number field sieve", *SIAM J. Discrete Math.* **6**:1 (1993), 124–138.

[Gordon 1993b] K. Gordon, D. andMcCurley, "Massively parallel computation of discrete logarithms", pp. 312–324 in *Advances in Cryptology — Crypto '92*, edited by E. F. Brickell, Lecture Notes in Comput. Sci. **740**, Springer, Berlin, 1993.

[Joux and Lercier 2002] A. Joux and R. Lercier, "The function field sieve is quite special", pp. 431–445 in *Algorithmic number theory* (Sydney, 2002), Lecture Notes in Comput. Sci. **2369**, Springer, Berlin, 2002.

[Joux and Lercier 2003] A. Joux and R. Lercier, "Improvements to the general number field sieve for discrete logarithms in prime fields. A comparison with the Gaussian integer method", *Math. Comp.* **72**:242 (2003), 953–967.

[Joux and Lercier 2005a] A. Joux and R. Lercier, "Discrete logarithms in $GF(2^{607})$ and $GF(2^{613})$", email to the NMBRTHRY mailing list, 23 September 2005.

[Joux and Lercier 2005b] A. Joux and R. Lercier, "Discrete logarithms in $GF(370801^{30})$ — 168 digits — 556 bits", email to the NMBRTHRY mailing list, 9 November 2005.

[Joux et al. 2006] A. Joux, R. Lercier, N. P. Smart, and F. Vercauteren, "The function field sieve in the medium prime case", pp. 326–344 in *Advances in Cryptology – CRYPTO 2006*, edited by C. Dwork, Lecture Notes in Comput. Sci. **4117**, 2006.

[Kleinjung 2007] T. Kleinjung, "Discrete logarithms in $GF(p)$ — 160 digits", email to the NMBRTHRY mailing list, 5 February 2007.

[Lenstra 1987] H. W. Lenstra, Jr., "Factoring integers with elliptic curves", *Ann. of Math.* (2) **126**:3 (1987), 649–673.

[Lenstra 1992] H. W. Lenstra, Jr., "Algorithms in algebraic number theory", *Bull. Amer. Math. Soc.* (N.S.) **26**:2 (1992), 211–244.

[Lenstra and Lenstra 1993] A. K. Lenstra and H. W. Lenstra, Jr. (editors), *The development of the number field sieve*, Lecture Notes in Mathematics **1554**, Springer, Berlin, 1993.

[Lenstra et al. 1993a] A. K. Lenstra, H. W. Lenstra, Jr., M. S. Manasse, and J. M. Pollard, "The factorization of the ninth Fermat number", *Math. Comp.* **61**:203 (1993), 319–349.

[Lenstra et al. 1993b] A. K. Lenstra, H. W. Lenstra, Jr., M. S. Manasse, and J. M. Pollard, "The number field sieve", pp. 11–42 in *The development of the number field sieve*, edited by A. K. Lenstra and H. W. Lenstra, Jr., Lecture Notes in Math. **1554**, Springer, Berlin, 1993.

[Marcus 1977] D. A. Marcus, *Number fields*, Springer, New York, 1977.

[McCurley 1990] K. S. McCurley, "The discrete logarithm problem", pp. 49–74 in *Cryptology and computational number theory* (Boulder, CO, 1989), edited by C. Pomerance, Proc. Sympos. Appl. Math. **42**, Amer. Math. Soc., Providence, RI, 1990.

[Menezes et al. 1993] A. J. Menezes, T. Okamoto, and S. A. Vanstone, "Reducing elliptic curve logarithms to logarithms in a finite field", *IEEE Trans. Inform. Theory* **39**:5 (1993), 1639–1646.

[Odlyzko 2000] A. Odlyzko, "Discrete logarithms: the past and the future", *Des. Codes Cryptogr.* **19**:2-3 (2000), 129–145.

[Pollard 1993] J. M. Pollard, "The lattice sieve", pp. 43–49 in *The development of the number field sieve*, edited by A. K. Lenstra and H. W. Lenstra, Jr., Lecture Notes in Math. **1554**, Springer, Berlin, 1993.

[Pomerance 2008] C. Pomerance, "Elementary thoughts on discrete logarithms", pp. 385–396 in *Surveys in algorithmic number theory*, edited by J. P. Buhler and P. Stevenhagen, Math. Sci. Res. Inst. Publ. **44**, Cambridge University Press, New York, 2008.

[Schirokauer 1993] O. Schirokauer, "Discrete logarithms and local units", *Philos. Trans. Roy. Soc. London Ser. A* **345**:1676 (1993), 409–423.

[Schirokauer 2000] O. Schirokauer, "Using number fields to compute logarithms in finite fields", *Math. Comp.* **69**:231 (2000), 1267–1283.

[Schirokauer 2002] O. Schirokauer, "The special function field sieve", *SIAM J. Discrete Math.* **16**:1 (2002), 81–98.

[Schirokauer 2005] O. Schirokauer, "Virtual logarithms", *J. Algorithms* **57**:2 (2005), 140–147.

[Schirokauer et al. 1996] O. Schirokauer, D. Weber, and T. Denny, "Discrete logarithms: the effectiveness of the index calculus method", pp. 337–361 in *Algorithmic number theory* (Talence, 1996), edited by H. Cohen, Lecture Notes in Comput. Sci. **1122**, Springer, Berlin, 1996.

[Shoup 1992] V. Shoup, "Searching for primitive roots in finite fields", *Math. Comp.* **58**:197 (1992), 369–380.

[Stevenhagen 2008] P. Stevenhagen, "The number field sieve", pp. 83–100 in *Surveys in algorithmic number theory*, edited by J. P. Buhler and P. Stevenhagen, Math. Sci. Res. Inst. Publ. **44**, Cambridge University Press, New York, 2008.

[Stichtenoth 1993] H. Stichtenoth, *Algebraic function fields and codes*, Springer, Berlin, 1993.

[Thomé 2001] E. Thomé, "Computation of discrete logarithms in $\mathbb{F}_{2^{607}}$", pp. 107–124 in *Advances in cryptology — ASIACRYPT 2001* (Gold Coast), edited by C. Boyd, Lecture Notes in Comput. Sci. **2248**, Springer, Berlin, 2001.

[Wiedemann 1986] D. H. Wiedemann, "Solving sparse linear equations over finite fields", *IEEE Trans. Inform. Theory* **32**:1 (1986), 54–62.

OLIVER SCHIROKAUER
DEPARTMENT OF MATHEMATICS
OBERLIN COLLEGE
10 NORTH PROFESSOR STREET
OBERLIN, OH 44074-1019
UNITED STATES
oliver.schirokauer@oberlin.edu

# Reducing lattice bases to find small-height values of univariate polynomials

DANIEL J. BERNSTEIN

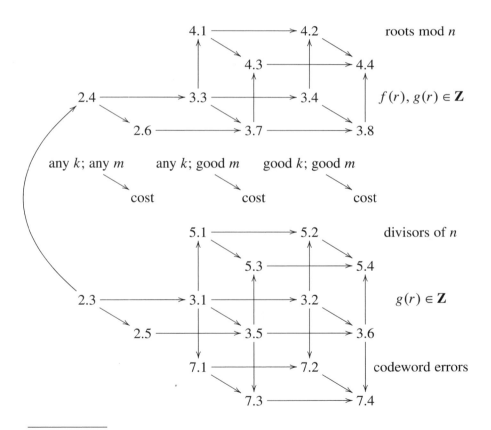

*Mathematics Subject Classification:* Primary 11Y16. Secondary 94B35.

Permanent ID of this document: 82f82c041b7e2bdce94a5e1f94511773. Date: 2008.05.02. The author was supported by the National Science Foundation under grant DMS–0140542, and by the Alfred P. Sloan Foundation.

| Find | $f(r)$ | $k$ | $m$ | History |
|---|---|---|---|---|
| divisors, in $u + v\mathbf{Z}$, of $n$ | $(r+uw)/n$ where $wv \in 1 + n\mathbf{Z}$ | 1 | 3 | Lenstra [1984], for proving primality |
| divisors, in an interval, of $n$ | $(r+w)/n$ for one $w$ | 1 | 3 | Rivest Shamir [1986], for breaking cryptosystems; independent of [Lenstra 1984] |
| roots of $p(x)$ mod $n$ | $p(r)/n$ | 1 | $d+1$ | Håstad [1988, Section 3]; first use of nonlinear $f$; independently: Vallée Girault Toffin [1989] (using dual lattice; more difficult) |
| roots of $p(x)$ mod $n$ | $p(r)/n$ | big | big | Coppersmith [1996a] (using dual), for breaking cryptosystems; first use of big $m, k$; simplified: Howgrave-Graham [1997] (explicitly avoiding dual) |
| divisors, in an interval, of $n$ | $(r+w)/n$ | big | big | Coppersmith [1996b] (in a more complicated way); simplified: Howgrave-Graham [1997] |
| divisors, in $1 + v\mathbf{Z}$, of $n$ | $(r+w)/n$ | 2 | 5 | Konyagin Pomerance [1997, Algorithm 3.2]; independent of [Coppersmith 1996a] |
| divisors, in $u + v\mathbf{Z}$, of $n$ | $(r+uw)/n$ | big | big | Coppersmith Howgrave-Graham Nagaraj [Howgrave-Graham 1998, Section 5.5] |
| large values of $\gcd\{x + w, n\}$ | $(r+w)/n$ | 1 | big | Goldreich Ron Sudan [1999] (using dual), for error correction; previous function-field version: Sudan [1997]; independent of [Coppersmith 1996a] |
| high-power divisors, in an interval, of $n$ | $(r+w)^d/n$ | big | big | Boneh Durfee Howgrave-Graham [1999] |
| large values of $\gcd\{x + w, n\}$ | $(r+w)/n$ | big | big | Boneh [2000], for error correction; independently: Howgrave-Graham [2001, Section 3]; previous function-field version: Guruswami Sudan [1999] |
| large values of $\gcd\{p(x), n\}$ | $p(r)/n$ | big | big | Boneh [2000, Section 4] |

ABSTRACT. This paper illustrates, improves, and unifies a variety of previous results on finding divisors in residue classes (Lenstra, Konyagin, Pomerance, Coppersmith, Howgrave-Graham, Nagaraj), divisors in short intervals (Rivest, Shamir, Coppersmith, Howgrave-Graham), modular roots (Håstad, Vallée, Girault, Toffin, Coppersmith, Howgrave-Graham), high-power divisors (Boneh, Durfee, Howgrave-Graham), and codeword errors beyond half distance (Sudan, Guruswami, Goldreich, Ron, Boneh).

# 1. Introduction

Consider the fraction $(r^3 - s)/n$, where $n$ is a large integer with no known factors. For typical integers $r, s$ there is no cancellation between the numerator $r^3 - s$ and the denominator $n$. In other words, the height of $(r^3 - s)/n$ is usually $\max\{|r^3 - s|, n\}$. Here the **height** of a rational number $m/n$ is, by definition, $\max\{|m|, |n|\}/\gcd\{m, n\}$.

However, if $r$ is a cube root of $s$ modulo $n$, then one can remove $n$ from both the numerator and denominator. In other words, the height of $(r^3 - s)/n$ is only $\max\{|(r^3 - s)/n|, 1\}$. The problem of finding a cube root of $s$ modulo $n$ can thus be viewed as the problem of finding small-height values of the polynomial $(x^3 - s)/n$.

Many other useful properties of numbers $r$ can be recast in the form "$f(r)$ has small height" for various polynomials $f$. For example, the problem of factoring $n$ can be viewed as the problem of finding all $r$ such that $r/n$ has small height.

There is a surprisingly fast method, using lattice-basis reduction, to find all numbers $r$ such that *both* $r$ and $f(r)$ have small height. This paper presents a very general statement of the method (see Theorem 2.3); asymptotically optimal parameters (see Section 3); and an exposition of various applications of the method (see Sections 4, 5, and 7). The theorems and algorithms can easily be switched from $\mathbf{Q}$ to the rational function field $\mathbf{F}_q(t)$ over a finite field $\mathbf{F}_q$, although better algorithms are often available in the function-field case.

I have made no attempt to cover analogous methods for higher-degree global fields or for polynomials in more variables. There are several papers on small-height values of bivariate polynomials, but each application seems to pose a new optimization problem. I will leave it to future writers to unify the literature on this topic.

**History.** The table on the second page of this paper fits previous results into the framework of Theorem 2.3. Notation: $f$ is the polynomial with useful small-height values; $d$ is the degree of $f$; $m$ is the lattice rank; $k$ is the highest $f$ exponent used in defining the lattice. Results improve primarily as $m$ increases, secondarily as $k$ increases.

It was recognized by Howgrave-Graham [1997] and Boneh, Durfee, and Howgrave-Graham [1999] that "$r + w$ divides $n$" and "$(r + w)^d$ divides $n$" could be handled by the same technique as "$p(r)$ is divisible by $n$." Meanwhile, "$\gcd\{r + w, n\}$ is large" was published independently by Goldreich, Ron, and Sudan [1999]. A unified "$\gcd\{p(r), n\}$ is large" algorithm was finally published, with insufficient fanfare, by Boneh [2000, Section 4].

**Index of theorems in this paper.** Algorithms in this paper are expressed in two ways: as theorems stating that the algorithms produce the desired results, and as "cost" theorems stating that there exist low-cost algorithms (in a particular cost measure) producing the desired results. Readers who want to understand what the algorithms achieve, without worrying at first about how the algorithms work, should start with the cost theorems, such as Theorem 4.4.

The chart on the first page of this paper has three rows for algorithms aimed at specific applications: "roots mod $n$," "divisors of $n$," and "codeword errors." It also has two rows for more general algorithms that can be used for other applications: an "$f(r), g(r) \in \mathbf{Z}$" row generalizing "roots mod $n$," and a "$g(r) \in \mathbf{Z}$" row generalizing all of these applications.

Algorithms in the "any $k$; any $m$" column of the chart have two parameters $(k, m)$ affecting their speed and output; the user can tune these parameters for the application at hand. Algorithms in the "good $k$; good $m$" column fix choices of $(k, m)$ that work reasonably well for a wide variety of applications, although they are often not exactly optimal. Readers who find themselves overwhelmed by the flexibility of $k$ and $m$ should start with the algorithms in the "good $k$; good $m$" column.

## 2. The general method

This section explains how to find all rational numbers $r$ such that $f(r)$ and $g(r)$ simultaneously have small height. Here $f, g \in \mathbf{Q}[x]$ are polynomials, each of positive degree, each with positive leading coefficient. Write $d = \deg f$, and assume for simplicity that $\deg g = 1$.

Theorem 2.2 below gives a more precise definition of "small height." The height bound depends on two integer parameters $k \geq 1$ and $m \geq dk + 1$. A typical special case is $k = 1$ and $m = 2d$. See Section 3 for further comments on the choice of $k$ and $m$.

**The lattice.** Define $L \subset \mathbf{Q}[x]$ as the **Z**-module

$$\begin{array}{llllll}
\mathbf{Z} & + \mathbf{Z}g & + \mathbf{Z}g^2 & + & \cdots & + \mathbf{Z}g^{d-1} \\
+ \mathbf{Z}f & + \mathbf{Z}gf & + \mathbf{Z}g^2 f & + & \cdots & + \mathbf{Z}g^{d-1} f \\
+ \mathbf{Z}f^2 & + \mathbf{Z}gf^2 & + \mathbf{Z}g^2 f^2 & + & \cdots & + \mathbf{Z}g^{d-1} f^2 \\
+ \cdots \\
+ \mathbf{Z}f^{k-1} & + \mathbf{Z}gf^{k-1} & + \mathbf{Z}g^2 f^{k-1} & + & \cdots & + \mathbf{Z}g^{d-1} f^{k-1} \\
+ \mathbf{Z}f^k & + \cdots & + \mathbf{Z}g^{m-dk-1} f^k.
\end{array}$$

For example, if $k=1$ and $m=d+1$, then $L=\mathbf{Z}+\mathbf{Z}g+\mathbf{Z}g^2+\cdots+\mathbf{Z}g^{d-1}+\mathbf{Z}f$; if $k=1$ and $m=2d$, then $L=\mathbf{Z}+\mathbf{Z}g+\mathbf{Z}g^2+\cdots+\mathbf{Z}g^{d-1}+\mathbf{Z}f+\cdots+\mathbf{Z}g^{d-1}f$.

The specified basis elements $1, g, \ldots, g^{d-1}, f, \ldots, g^{m-dk-1} f^k$ have degrees $0, 1, 2, \ldots, m-1$ respectively. Thus $L$ is a lattice of rank $m$ under the usual coefficient-vector metric on $\mathbf{Q}[x]$, namely $\varphi \mapsto |\varphi| = \sqrt{\varphi_0^2 + \varphi_1^2 + \varphi_2^2 + \cdots}$, where $\varphi = \varphi_0 + \varphi_1 x + \varphi_2 x^2 + \cdots$.

The basis elements have leading coefficients $1, g_1, g_1^2, \ldots, g_1^{m-dk-1} f_d^k$, where $g_1$ is the leading coefficient of $g$ and $f_d$ is the leading coefficient of $f$. Thus

$$\det L = g_1^{kd(d-1)/2+(m-dk)(m-dk-1)/2} f_d^{dk(k-1)/2+k(m-dk)}$$
$$= g_1^{m(m-1)/2} (g_1^d / f_d)^{dk(k+1)/2 - mk}.$$

For example, if $k=1$ and $m=2d$, then $\det L = g_1^{d(d-1)} f_d^d = g_1^{d(2d-1)} (g_1^d / f_d)^{-d}$.

**Theorem 2.1.** *Let $d, k, m$ be positive integers with $m \geq dk+1$. Let $f \in \mathbf{Q}[x]$ be a polynomial of degree $d$ with leading coefficient $f_d > 0$. Let $g \in \mathbf{Q}[x]$ be a polynomial of degree $1$ with leading coefficient $g_1 > 0$. Define $L$ as above. If $\varphi \in L$, $r \in \mathbf{Q}$, and $\gcd\{1, f(r)\}^k \gcd\{1, g(r)\}^{\max\{d-1, m-dk-1\}} > |(1, r, \ldots, r^{m-1})| \, |\varphi|$, then $\varphi(r) = 0$.*

Example: if $k=1$, $m=2d$, $\varphi \in L$, $r \in \mathbf{Q}$, and $\gcd\{1, f(r)\} \gcd\{1, g(r)\}^{d-1} > |(1, r, \ldots, r^{2d-1})| \, |\varphi|$, then $\varphi(r) = 0$.

The reader should interpret $\gcd\{1, f(r)\} > \cdots$ in Theorem 2.1 as "$f(r)$ has small denominator"; $\gcd\{1, g(r)\} > \cdots$ as "$g(r)$ has small denominator"; and $|(1, r, \ldots, r^{m-1})| < \cdots$ as "$f(r)$ and $g(r)$ have small numerators." Theorem 2.1 can thus be summarized as "$\varphi(r) = 0$ if $f(r)$ and $g(r)$ both have small height."

*Proof.* $\varphi \in \mathbf{Z} + \mathbf{Z}g + \cdots + \mathbf{Z}g^{d-1} f^{k-1} + \mathbf{Z}f^k + \cdots + \mathbf{Z}g^{m-dk-1} f^k$, so $\varphi(r) \in \mathbf{Z} + \mathbf{Z}g(r) + \cdots + \mathbf{Z}g(r)^{d-1} f(r)^{k-1} + \mathbf{Z}f(r)^k + \cdots + \mathbf{Z}g(r)^{m-dk-1} f(r)^k \subseteq \mathbf{Z} \gcd\{1, f(r)\}^k \gcd\{1, g(r)\}^{\max\{d-1, m-dk-1\}}$. But $|\varphi(r)| \leq |(1, \ldots, r^{m-1})| \, |\varphi| < \gcd\{1, f(r)\}^k \gcd\{1, g(r)\}^{\max\{d-1, m-dk-1\}}$. Thus $\varphi(r) = 0$. □

**Theorem 2.2.** *Let $d, k, m$ be positive integers with $m \geq dk + 1$. Let $f \in \mathbf{Q}[x]$ be a polynomial of degree $d$ with leading coefficient $f_d > 0$. Let $g \in \mathbf{Q}[x]$ be a polynomial of degree 1 with leading coefficient $g_1 > 0$. Define $L$ as above. Let $\varphi \in L$ be a nonzero vector such that $|\varphi| \leq 2^{(m-1)/2}(\det L)^{1/m}$. If $r \in \mathbf{Q}$ and*

$$\gcd\{1, f(r)\}^k \gcd\{1, g(r)\}^{\max\{d-1, m-dk-1\}}$$
$$> \left|(1, r, \ldots, r^{m-1})\right| (2g_1)^{(m-1)/2} (g_1^d/f_d)^{dk(k+1)/2m-k}$$

*then $\varphi(r) = 0$.*

*Proof.* $(\det L)^{1/m} = g_1^{(m-1)/2}(g_1^d/f_d)^{dk(k+1)/2m-k}$. Apply Theorem 2.1. □

For example, if $k = 1$ and $m = 2d$, then $\varphi(r) = 0$ for every $r \in \mathbf{Q}$ such that $\gcd\{1, f(r)\} \gcd\{1, g(r)\}^{d-1} > \left|(1, r, \ldots, r^{2d-1})\right| (2g_1)^{d-1/2}(g_1^d/f_d)^{-1/2}$.

**Theorem 2.3.** *Let $d, k, m$ be positive integers with $m \geq dk + 1$. Let $f \in \mathbf{Q}[x]$ be a polynomial of degree $d$ with leading coefficient $f_d > 0$. Let $g \in \mathbf{Q}[x]$ be a polynomial of degree 1 with leading coefficient $g_1 > 0$. Define $L$ as above. Let $\varphi \in L$ be a nonzero vector such that $|\varphi| \leq 2^{(m-1)/2}(\det L)^{1/m}$. Define $\gamma = m^{1/2k}(2g_1)^{(m-1)/2k}(g_1^d/f_d)^{d(k+1)/2m-1}$. If $r \in \mathbf{Q}$, $|r| \leq 1$, $\gcd\{1, f(r)\} > \gamma$, and $g(r) \in \mathbf{Z}$, then $\varphi(r) = 0$.*

For example, if $k = 1$ and $m = 2d$, then $\varphi(r) = 0$ for every $r \in \mathbf{Q}$ such that $|r| \leq 1$, $g(r) \in \mathbf{Z}$, and $\gcd\{1, f(r)\} > \gamma$, where $\gamma = (2d)^{1/2}(2g_1)^{d-1/2}(g_1^d/f_d)^{-1/2}$.

*Proof.* $\gamma^k = m^{1/2}(2g_1)^{(m-1)/2}(g_1^d/f_d)^{dk(k+1)/2m-k}$; $\left|(1, r, \ldots, r^{m-1})\right| \leq m^{1/2}$; and $\gcd\{1, g(r)\} = 1$. Apply Theorem 2.2. □

**Theorem 2.4.** *Let $d, k, m$ be positive integers with $m \geq dk + 1$. Let $f \in \mathbf{Q}[x]$ be a polynomial of degree $d$ with leading coefficient $f_d > 0$. Let $g \in \mathbf{Q}[x]$ be a polynomial of degree 1 with leading coefficient $g_1 > 0$. Define $L$ as above. Let $\varphi \in L$ be a nonzero vector such that $|\varphi| \leq 2^{(m-1)/2}(\det L)^{1/m}$. Assume that $g_1 < (g_1^d/f_d)^{2k/(m-1)-dk(k+1)/m(m-1)}/2m^{1/(m-1)}$. If $r \in \mathbf{Q}$, $|r| \leq 1$, $f(r) \in \mathbf{Z}$, and $g(r) \in \mathbf{Z}$, then $\varphi(r) = 0$.*

For example, if $k = 1$ and $m = 2d$, then $\varphi(r) = 0$ for every $r \in \mathbf{Q}$ such that $|r| \leq 1$, $f(r) \in \mathbf{Z}$, and $g(r) \in \mathbf{Z}$, provided that $2g_1 < (g_1^d/2df_d)^{1/(2d-1)}$.

*Proof.* By assumption $m^{1/2k}(2g_1)^{(m-1)/2k}(g_1^d/f_d)^{d(k+1)/2m-1} < 1 = \gcd\{1, f(r)\}$. Apply Theorem 2.3. □

**Computation.** It is easy to compute all of the rational numbers $r$ identified in Theorems 2.2, 2.3, and 2.4:

- Feed the basis vectors $1, g, \ldots, g^{d-1}, f, \ldots, g^{m-dk-1}f^k$ of $L$ to a lattice-basis-reduction algorithm, such as the Lenstra–Lenstra–Lovasz algorithm, to obtain a nonzero vector $\varphi \in L$ such that $|\varphi| \leq 2^{(m-1)/2}(\det L)^{1/m}$. See [Lenstra

et al. 1982] or [Lenstra 2008]. The theorems now state that all of the desired numbers $r$ are roots of $\varphi$.
- Compute the rational roots of $\varphi$, by approximating the real (or 2-adic) roots of $\varphi$ to high precision. See, e.g., [Loos 1983]. By construction $\varphi$ has degree at most $m-1$, so it has at most $m-1$ roots.
- Check each root $r$ to see whether it satisfies the stated conditions.

Each step is reasonably fast if $f$, $g$, $k$, and $m$ are reasonably small.

One way to measure the complexity of this algorithm is to measure its output size, i.e., to count the number of qualifying $r$'s. Theorems 2.5 and 2.6 state bounds on this measure of algorithm complexity. I will leave it to the reader to formulate theorems regarding other measures.

**Theorem 2.5.** *Let $d, k, m$ be positive integers with $m \geq dk+1$. Let $f \in \mathbf{Q}[x]$ be a polynomial of degree $d$ with leading coefficient $f_d > 0$. Let $g \in \mathbf{Q}[x]$ be a polynomial of degree 1 with leading coefficient $g_1 > 0$. Then there are at most $m-1$ values $r \in \mathbf{Q}$ such that $g(r) \in \mathbf{Z}$, $|r| \leq 1$, and $\gcd\{1, f(r)\} > m^{1/2k}(2g_1)^{(m-1)/2k}(g_1^d/f_d)^{d(k+1)/2m-1}$.*

Take, for example, $k=1$ and $m=2d$: there are at most $2d-1$ values $r \in \mathbf{Q}$ such that $|r| \leq 1$, $\gcd\{1, f(r)\} > (2d)^{1/2}(2g_1)^{d-1/2}(g_1^d/f_d)^{-1/2}$, and $g(r) \in \mathbf{Z}$.

*Proof.* Apply lattice-basis reduction to Theorem 2.3.

In more detail: Define $\gamma = m^{1/2k}(2g_1)^{(m-1)/2k}(g_1^d/f_d)^{d(k+1)/2m-1}$, and define $L$ as above. There is a nonzero vector $\varphi \in L$ such that $|\varphi| \leq 2^{(m-1)/2}(\det L)^{1/m}$. By Theorem 2.3, each qualifying value $r \in \mathbf{Q}$ is a root of $\varphi$. The degree of $\varphi$ is at most $m-1$ by construction of $L$, so there are at most $m-1$ roots of $\varphi$. □

**Theorem 2.6.** *Let $d, k, m$ be positive integers with $m \geq dk+1$. Let $f \in \mathbf{Q}[x]$ be a polynomial of degree $d$ with leading coefficient $f_d > 0$. Let $g \in \mathbf{Q}[x]$ be a polynomial of degree 1 with leading coefficient $g_1 > 0$. Assume that $g_1 < (g_1^d/f_d)^{2k/(m-1)-dk(k+1)/m(m-1)}/2m^{1/(m-1)}$. Then there are at most $m-1$ values $r \in \mathbf{Q}$ such that $|r| \leq 1$, $f(r) \in \mathbf{Z}$, and $g(r) \in \mathbf{Z}$.*

Take, for example, $k=1$ and $m=2d$: if $2g_1 < (g_1^d/2df_d)^{1/(2d-1)}$ then there are at most $2d-1$ values $r \in \mathbf{Q}$ such that $|r| \leq 1$, $f(r) \in \mathbf{Z}$, and $g(r) \in \mathbf{Z}$.

*Proof.* Apply lattice-basis reduction to Theorem 2.4. □

## 3. Parameter choice and other optimizations

This section discusses the choice of $k$ and $m$ in Section 2, and other ways to speed up the computation of the desired numbers $r$.

The history of this subject shows each application progressing from simple choices of $k$ and $m$ to near-optimal choices of $k$ and $m$; see the second page of this paper. It turns out to be possible to unify all of these application-specific optimizations into a few straightforward formulas: Theorem 3.2 states near-optimal choices of $k$ and $m$ for Theorem 2.3, and Theorem 3.4 states near-optimal choices of $k$ and $m$ for Theorem 2.4. Future applications should be able to reuse these unified theorems, rather than wasting time redoing the same optimizations from scratch.

**Parameter choice for Theorem** 2.3. Theorem 2.3 assumes that $\gcd\{1, f(r)\} > \gamma$, where $\gamma = m^{1/2k}(2g_1)^{(m-1)/2k}(g_1^d/f_d)^{d(k+1)/2m-1}$. How small can one make this lower bound $\gamma$ by varying $m$ and $k$?

Assume that $g_1$ and $1/f_d$ exceed 1. Theorem 3.1 then says that $\gamma$ is smaller than $\beta = m^{1/2k}(2g_1)^{\alpha d(1+1/2k)} f_d/g_1^d$, where $\alpha = \sqrt{1 + (\lg(1/f_d))/\lg((2g_1)^d)}$, if $m$ is chosen as $\lceil \alpha d(k+1) \rceil$. This choice of $m$ approximately balances the factors $(2g_1)^{(m-1)/2k}$ and $(g_1^d/f_d)^{d(k+1)/2m}$ in Theorem 2.3. Note that $\alpha \geq 1$, so $m \geq dk + d$. Note also that $m$ is not difficult to compute: comparing $\alpha d(k+1)$ to an integer boils down to comparing integer powers of $f_d$ and $2g_1$.

As $k$ increases (slowing down the computation of a short nonzero vector $\varphi$ in $L$), $\beta$ converges to $(2g_1)^{\alpha d} f_d/g_1^d$, which is very close to a lower bound on $\gamma$. The quantity $(2g_1)^{\alpha d}$ is the doubly-geometric average of $(2g_1)^d$ and $(2g_1)^d/f_d$. Theorem 3.2 considers the special case $k = \lceil \alpha d \lceil \lg 2g_1 \rceil /2 \rceil$, which balances the desire for a small $\beta$ against the desire for small lattice ranks.

For comparison: If $k = 1$, the optimal choice of $m$ is approximately $\sqrt{2}\alpha d$ for large $\alpha d$, with $\gamma \approx (2g_1)^{\sqrt{2}\alpha d} f_d/g_1^d$. Allowing larger $k$ thus changes the exponent of $2g_1$ by a factor of approximately $\sqrt{2}$.

**Theorem 3.1.** *Let $d$ be a positive integer. Let $f \in \mathbf{Q}[x]$ be a polynomial of degree $d$ with leading coefficient $f_d \in (0, 1]$. Let $g \in \mathbf{Q}[x]$ be a polynomial of degree 1 with leading coefficient $g_1 \geq 1$. Let $k$ be a positive integer. Define $\alpha = \sqrt{1 + (\lg(1/f_d))/\lg((2g_1)^d)}$, $m = \lceil \alpha d(k+1) \rceil$, $\beta = m^{1/2k}(2g_1)^{\alpha d(1+1/2k)} f_d/g_1^d$, and $L$ as above. Let $\varphi \in L$ be a nonzero vector such that $|\varphi| \leq 2^{(m-1)/2}(\det L)^{1/m}$. If $r \in \mathbf{Q}$, $|r| \leq 1$, $\gcd\{1, f(r)\} \geq \beta$, and $g(r) \in \mathbf{Z}$, then $\varphi(r) = 0$.*

*Proof.* First $m - 1 \leq \alpha d(k+1)$ so $(2g_1)^{(m-1)/2k} \leq (2g_1)^{\alpha d(k+1)/2k}$. Second $1/m \leq 1/\alpha d(k+1)$ so $(g_1^d/f_d)^{d(k+1)/2m} \leq (g_1^d/f_d)^{1/2\alpha} < ((2g_1)^d/f_d)^{1/2\alpha} = (2g_1)^{\alpha d/2}$ by choice of $\alpha$. Thus $m^{1/2k}(2g_1)^{(m-1)/2k}(g_1^d/f_d)^{d(k+1)/2m} f_d/g_1^d < \beta$. Now apply Theorem 2.3. □

**Theorem 3.2.** *Let $d$ be a positive integer. Let $f \in \mathbf{Q}[x]$ be a polynomial of degree $d$ with leading coefficient $f_d \in (0, 1]$. Let $g \in \mathbf{Q}[x]$ be a polynomial of degree 1 with leading coefficient $g_1 \geq 1$. Define $\alpha = \sqrt{1 + (\lg(1/f_d))/\lg((2g_1)^d)}$, $k = \lceil \alpha d \lceil \lg 2g_1 \rceil /2 \rceil$, $m = \lceil \alpha d(k+1) \rceil$, and $L$ as above. Let $\varphi \in L$ be a nonzero vector such that $|\varphi| \leq 2^{(m-1)/2}(\det L)^{1/m}$. If $r \in \mathbf{Q}$, $|r| \leq 1$, $\gcd\{1, f(r)\} \geq 2m^{1/2k}(2g_1)^{\alpha d} f_d/g_1^d$, and $g(r) \in \mathbf{Z}$, then $\varphi(r) = 0$.*

*Proof.* $k \geq \alpha d(\lg 2g_1)/2$, so $1 \geq (\lg 2g_1)\alpha d/2k$, so $2 \geq (2g_1)^{\alpha d/2k}$. Therefore $\gcd\{1, f(r)\} \geq \beta$ where $\beta = m^{1/2k}(2g_1)^{\alpha d(1+1/2k)} f_d/g_1^d$. Apply Theorem 3.1. □

**Parameter choice for Theorem 2.4.** Theorem 2.4 assumes that $g_1$ is smaller than $(g_1^d/f_d)^{2k/(m-1)-dk(k+1)/m(m-1)}/2m^{1/(m-1)}$. How large can one make this exponent $2k/(m-1) - dk(k+1)/m(m-1)$ by varying $m$ and $k$?

Theorem 3.3 chooses $m = dk+d$, achieving exponent $k/(dk+d-1)$, which is reasonably close to optimal. As $k$ increases (slowing down the computation of $\varphi$), the exponent converges to $1/d$. Theorem 3.4 considers the special case $k = \lceil \lceil \lg(g_1^d/2^d f_d) \rceil /d \rceil$, which balances the desire for a large exponent against the desire for small lattice ranks.

**Theorem 3.3.** *Let $f \in \mathbf{Q}[x]$ be a polynomial of positive degree $d$ with leading coefficient $f_d > 0$. Let $g \in \mathbf{Q}[x]$ be a polynomial of degree 1 with leading coefficient $g_1 > 0$. Let $k$ be a positive integer. Define $m = dk+d$ and $L = \mathbf{Z} + \mathbf{Z}g + \cdots + \mathbf{Z}g^{d-1} + \mathbf{Z}f + \mathbf{Z}gf + \cdots + \mathbf{Z}g^{d-1}f + \cdots + \mathbf{Z}f^k + \mathbf{Z}gf^k + \cdots + \mathbf{Z}g^{d-1}f^k$. Let $\varphi \in L$ be a nonzero vector such that $|\varphi| \leq 2^{(m-1)/2}(\det L)^{1/m}$. Assume that $g_1 < (g_1^d/f_d)^{k/(m-1)}/2m^{1/(m-1)}$. If $r \in \mathbf{Q}$, $|r| \leq 1$, $f(r) \in \mathbf{Z}$, and $g(r) \in \mathbf{Z}$, then $\varphi(r) = 0$.*

*Proof.* $d(k+1)/m = 1$ so $2k/(m-1) - dk(k+1)/m(m-1) = 2k/(m-1) - k/(m-1) = k/(m-1)$. Apply Theorem 2.4. □

**Theorem 3.4.** *Let $f \in \mathbf{Q}[x]$ be a polynomial of positive degree $d$ with leading coefficient $f_d \in (0, 1/8^d)$. Let $g \in \mathbf{Q}[x]$ be a polynomial of degree 1 with leading coefficient $g_1 \geq 1/4$. Define $k = \lceil \lceil \lg(g_1^d/2^d f_d) \rceil /d \rceil$, $m = dk+d$, and $L = \mathbf{Z} + \mathbf{Z}g + \cdots + \mathbf{Z}g^{d-1} + \mathbf{Z}f + \mathbf{Z}gf + \cdots + \mathbf{Z}g^{d-1}f + \cdots + \mathbf{Z}f^k + \mathbf{Z}gf^k + \cdots + \mathbf{Z}g^{d-1}f^k$. Let $\varphi \in L$ be a nonzero vector such that $|\varphi| \leq 2^{(m-1)/2}(\det L)^{1/m}$. If $r \in \mathbf{Q}$, $|r| \leq 1$, $f(r) \in \mathbf{Z}$, and $g(r) \in \mathbf{Z}$, then $\varphi(r) = 0$.*

*Proof.* First $g_1^d/2^d f_d > (1/4)^d/(2/8)^d = 1$, so $k$ is a positive integer.

Next $m = d(k+1) \geq 2$, so $\lg m \leq m-1$, so $1 \leq 2/m^{1/(m-1)}$.

Next $m - 1 = dk+d-1 \geq \lg(g_1^d/2^d f_d) + d - 1 \geq ((d-1)/d)\lg(g_1^d/2^d f_d) + d - 1 = ((m-1-dk)/d)\lg(g_1^d/f_d)$ so $d(m-1) \geq (m-1-dk)\lg(g_1^d/f_d)$ so $1 \geq (1/d - k/(m-1))\lg(g_1^d/f_d)$; i.e., $(g_1^d/f_d)^{1/d-k/(m-1)} \leq 2$.

Put everything together: $g_1 = (g_1^d/f_d)^{1/d-k/(m-1)}(g_1^d/f_d)^{k/(m-1)} f_d^{1/d}(1) < (2)(g_1^d/f_d)^{k/(m-1)}(1/8)(2/m^{1/(m-1)}) = (g_1^d/f_d)^{k/(m-1)}/2m^{1/(m-1)}$. Finally apply Theorem 3.3. □

**Computation.** Theorems 3.1, 3.2, 3.3, and 3.4, like Theorems 2.3 and 2.4, can easily be converted into algorithms to compute the set of $r$'s. Theorems 3.5, 3.6, 3.7, and 3.8, like Theorems 2.5 and 2.6, measure the complexity of these algorithms by stating bounds on the output size.

**Theorem 3.5.** *Let $d$ be a positive integer. Let $f \in \mathbf{Q}[x]$ be a polynomial of degree $d$ with leading coefficient $f_d \in (0, 1]$. Let $g \in \mathbf{Q}[x]$ be a polynomial of degree 1 with leading coefficient $g_1 \geq 1$. Let $k$ be a positive integer. Define $\alpha = \sqrt{1 + (\lg(1/f_d))/\lg((2g_1)^d)}$ and $m = \lceil \alpha d(k+1) \rceil$. Then there are at most $m - 1$ values $r \in \mathbf{Q}$ such that $|r| \leq 1$, $\gcd\{1, f(r)\} \geq m^{1/2k}(2g_1)^{\alpha d(1+1/2k)} f_d/g_1^d$, and $g(r) \in \mathbf{Z}$.*

*Proof.* Apply lattice-basis reduction to Theorem 3.1. □

**Theorem 3.6.** *Let $d$ be a positive integer. Let $f \in \mathbf{Q}[x]$ be a polynomial of degree $d$ with leading coefficient $f_d \in (0, 1]$. Let $g \in \mathbf{Q}[x]$ be a polynomial of degree 1 with leading coefficient $g_1 \geq 1$. Define $\alpha = \sqrt{1 + (\lg(1/f_d))/\lg((2g_1)^d)}$, $k = \lceil \alpha d \lceil \lg 2g_1 \rceil /2 \rceil$, and $m = \lceil \alpha d(k+1) \rceil$. Then there are at most $m - 1$ values $r \in \mathbf{Q}$ such that $|r| \leq 1$, $\gcd\{1, f(r)\} \geq 2m^{1/2k}(2g_1)^{\alpha d} f_d/g_1^d$, and $g(r) \in \mathbf{Z}$.*

The bound $m - 1$ is approximately $(\lg((2g_1)^d) + \lg(1/f_d))d/2$. The limit on $\gcd\{1, f(r)\}$ is approximately $f_d/g_1^d$ times the doubly-geometric average of $(2g_1)^d$ and $(2g_1)^d/f_d$.

*Proof.* Apply lattice-basis reduction to Theorem 3.2. □

**Theorem 3.7.** *Let $f \in \mathbf{Q}[x]$ be a polynomial of positive degree $d$ with leading coefficient $f_d > 0$. Let $g \in \mathbf{Q}[x]$ be a polynomial of degree 1 with leading coefficient $g_1 > 0$. Let $k$ be a positive integer. Define $m = dk + d$. Assume that $g_1 < (g_1^d/f_d)^{k/(m-1)}/2m^{1/(m-1)}$. Then there are at most $m - 1$ values $r \in \mathbf{Q}$ such that $|r| \leq 1$, $f(r) \in \mathbf{Z}$, and $g(r) \in \mathbf{Z}$.*

*Proof.* Apply lattice-basis reduction to Theorem 3.3. □

**Theorem 3.8.** *Let $f \in \mathbf{Q}[x]$ be a polynomial of positive degree $d$ with leading coefficient $f_d \in (0, 1/8^d)$. Let $g \in \mathbf{Q}[x]$ be a polynomial of degree 1 with leading coefficient $g_1 \geq 1/4$. Then there are fewer than $\lg(g_1^d/f_d) + d - 1$ values $r \in \mathbf{Q}$ such that $|r| \leq 1$, $f(r) \in \mathbf{Z}$, and $g(r) \in \mathbf{Z}$.*

*Proof.* Apply lattice-basis reduction to Theorem 3.4, using $m < \lg(g_1^d/f_d) + d$. □

**Combining Theorem 3.3 with brute force.** Theorem 3.3, applied to $f$ and $g$, finds all rational numbers $r \in [-1, 1]$ with $f(r), g(r) \in \mathbf{Z}$. The same theorem, applied to $f(x + 2)$ and $g(x + 2)$, finds all rational numbers $r \in [1, 3]$ with $f(r), g(r) \in \mathbf{Z}$. With $c$ such computations, involving $c$ lattices of rank $m = dk + d$, one can cover an $r$ interval of length $2c$.

One can view Theorem 3.3 as searching the rationals $r$ with $g(r) \in \mathbf{Z}$, to see which ones have $f(r) \in \mathbf{Z}$. An interval of length $2c$ has approximately $2cg_1 < c(g_1^d/f_d)^{k/(dk+d-1)}$ rationals $r$ with $g(r) \in \mathbf{Z}$, so the number of $r$'s searched per unit time is approximately $(g_1^d/f_d)^{k/(dk+d-1)}$ divided by the time to handle a lattice of rank $dk + d$. Given $f$ and $g$, one can choose $k$ to (approximately) maximize this ratio. This idea appears in [Coppersmith 1996a].

**Smaller improvements.** Another way to expand the number of $r$'s searched is to perform several rational-root calculations per lattice, searching for roots of shifts of $\varphi$. Example: The roots of $\varphi - 2, \varphi - 1, \varphi, \varphi + 1, \varphi + 2$ include all $r \in \mathbf{Q}$ such that $|r| \leq 1$, $f(r) \in \mathbf{Z}$, and $g(r) \in \mathbf{Z}$, provided that $g_1 < 3(g_1^d/f_d)^{k/(m-1)}/2m^{1/(m-1)}$; note the 3 here. I learned this idea from Lenstra.

The choice of $m$ in Theorem 3.1 is not exactly optimal. It is better to have the computer run through all pairs $(k, m)$, in increasing order of the $r$ computation time, until finding a pair $(k, m)$ where the bound in Theorem 2.3 is satisfactory. Similar comments apply to Theorem 3.3.

I quoted lattice-basis reduction in Section 2 as producing nonzero vectors $\varphi \in L$ such that $|\varphi|$ is at most $2^{(m-1)/2}(\det L)^{1/m}$. Slower reduction algorithms can shrink the factor $2^{(m-1)/2}$; even without this extra work, lattice-basis reduction often produces a vector $\varphi$ with $|\varphi| < (\det L)^{1/m}$. Bounds that depend on $\varphi$, as in Theorem 2.1, are slightly better than bounds that depend solely on $\det L$.

In Theorems 2.3, 2.4, 3.1, and 3.3, the lattice $L$ can be replaced by a slightly smaller lattice, namely $\mathbf{Z} + \mathbf{Z}g + \mathbf{Z}g(g-1)/2 + \mathbf{Z}g(g-1)(g-2)/6 + \cdots$. The point is that $g(r)(g(r)-1)/2$ etc. are integers if $g(r)$ is an integer. This idea was published in [Coppersmith 2001], with credit to Howgrave-Graham and Lenstra independently.

A few years earlier, Howgrave-Graham [1998, Section 4.5.2] had made the similar observation that $f$ could be replaced by $f/d!$ in many situations, after suitable tweaking of the coefficients of $f$.

Yet another slight improvement is to change the metric used to define the lattice, replacing $1, x, x^2, \ldots, x^{m-1}$ with Chebyshev polynomials. This idea was published in [Coppersmith 2001, page 24], with partial credit (of unclear scope) to Boneh.

## 4. Example: roots mod $n$ given their high bits

This section explains how to search an interval $[-H, H]$ for integer roots of an integer polynomial $p$ modulo $n$, if $H$ is not too large. For example, this section explains how to search the interval $[t - H, t + H]$ for cube roots of $s$ modulo $n$, if $H$ is not too large; here $p = (x + t)^3 - s$.

As in previous sections, the search method is parametrized by an exponent $k$. Theorem 4.2 uses a particular $k$ that works well for most applications; Theorem 4.1 is more general and allows $k$ to be tuned for the reader's application. The other theorems in this section measure the cost of the resulting computations.

The choice of $k$ in Theorem 4.2 allows $H$ up to about $n^{1/d}$. For example, one can find cube roots of $s$ modulo $n$ in any interval of length about $n^{1/3}$. This generalizes the obvious fact that one can quickly compute $r$ from $r^3 \bmod n$ if $0 \le r < n^{1/3}$. For comparison, the simpler choice $k = 1$ allows $H$ up to only about $n^{2/d(d+1)}$; for example, about $n^{1/6}$ for $d = 3$.

**Theorem 4.1.** *Let $n$ be a positive integer. Let $p \in \mathbf{Z}[x]$ be a monic polynomial of positive degree $d$. Let $k$ be a positive integer. Define $m = dk + d$. Let $H$ be a positive integer smaller than $n^{k/(m-1)}/2m^{1/(m-1)}$. Define $f = p(Hx)/n \in \mathbf{Q}[x]$, $g = Hx \in \mathbf{Q}[x]$, and $L = \mathbf{Z} + \mathbf{Z}g + \cdots + \mathbf{Z}g^{d-1} + \mathbf{Z}f + \mathbf{Z}gf + \cdots + \mathbf{Z}g^{d-1}f + \cdots + \mathbf{Z}f^k + \mathbf{Z}gf^k + \cdots + \mathbf{Z}g^{d-1}f^k$. Let $\varphi \in L$ be a nonzero vector such that $|\varphi| \le 2^{(m-1)/2}(\det L)^{1/m}$. If $s \in \mathbf{Z}$, $p(s) \in n\mathbf{Z}$, and $|s| \le H$, then $\varphi(s/H) = 0$.*

*Proof.* Define $r = s/H$. By hypothesis $r \in \mathbf{Q}$, $|r| \le 1$, $f(r) = p(s)/n \in \mathbf{Z}$, $g(r) = s \in \mathbf{Z}$, and $g_1 = H < n^{k/(m-1)}/2m^{1/(m-1)} = (g_1^d/f_d)^{k/(m-1)}/2m^{1/(m-1)}$. Apply Theorem 3.3. □

**Theorem 4.2.** *Let $n$ be a positive integer. Let $p \in \mathbf{Z}[x]$ be a monic polynomial of positive degree $d$. Let $H$ be a positive integer smaller than $n^{1/d}/8$. Define $k = \lceil (\lg n)/d \rceil - 1$ and $m = dk + d$. Define $f = p(Hx)/n \in \mathbf{Q}[x]$, $g = Hx \in \mathbf{Q}[x]$, and $L = \mathbf{Z} + \mathbf{Z}g + \cdots + \mathbf{Z}g^{d-1} + \mathbf{Z}f + \mathbf{Z}gf + \cdots + \mathbf{Z}g^{d-1}f + \cdots + \mathbf{Z}f^k + \mathbf{Z}gf^k + \cdots + \mathbf{Z}g^{d-1}f^k$. Let $\varphi \in L$ be a nonzero vector such that $|\varphi| \le 2^{(m-1)/2}(\det L)^{1/m}$. If $s \in \mathbf{Z}$, $p(s) \in n\mathbf{Z}$, and $|s| \le H$, then $\varphi(s/H) = 0$.*

*Proof.* The leading coefficient $f_d$ of $f$ is $H^d/n \in (0, 1/8^d)$, and the leading coefficient $g_1$ of $g$ is $H > 1/4$. The quotient $g_1^d/2^d f_d$ is $H^d/2^d(H^d/n) = n/2^d$. Consequently $k = \lceil (\lg n - d)/d \rceil = \lceil \lceil \lg n - d \rceil /d \rceil = \lceil \lceil \lg(g_1^d/2^d f_d) \rceil /d \rceil$.

Define $r = s/H$. By hypothesis $r \in \mathbf{Q}$, $|r| \le 1$, $f(r) = p(s)/n \in \mathbf{Z}$, and $g(r) = s \in \mathbf{Z}$. Apply Theorem 3.4. □

**Theorem 4.3.** *Let $n$ be a positive integer. Let $p \in \mathbf{Z}[x]$ be a monic polynomial of positive degree $d$. Let $k$ be a positive integer. Define $m = dk + d$. Let $H$ be a positive integer smaller than $n^{k/(m-1)}/2m^{1/(m-1)}$. Then there are at most $m - 1$ integers $s \in \{-H, \ldots, -1, 0, 1, \ldots, H - 1, H\}$ such that $p(s) \in n\mathbf{Z}$.*

*Proof.* Apply lattice-basis reduction to Theorem 4.1. □

**Theorem 4.4.** *Let $n$ be a positive integer. Let $p \in \mathbf{Z}[x]$ be a monic polynomial of positive degree $d$. Let $H$ be a positive integer smaller than $n^{1/d}/8$. Then there are fewer than $\lg n + d - 1$ integers $s \in \{-H, \ldots, -1, 0, 1, \ldots, H - 1, H\}$ such that $p(s) \in n\mathbf{Z}$.*

*Proof.* Apply lattice-basis reduction to Theorem 4.2, using $m < \lg n + d$. □

**History.** The $n^{2/d(d+1)}$ result was first published by Håstad, and the $n^{1/d}$ result was first published by Coppersmith. Both authors used their results to break various naive forms of the RSA cryptosystem.

These results also have a positive application to cryptography: compressing RSA (or Rabin) signatures. Instead of transmitting a cube root (or square root) of $s$ modulo $n$, one can transmit the top $2/3$ (or $1/2$) of the bits of the root. But this application is now obsolete, because Bleichenbacher [2004] proposed a different compression mechanism allowing substantially faster decompression and verification: compress the cube root to an integer $v$ such that the remainder $v^3 s \bmod n$ is a cube in $\mathbf{Z}$.

**Numerical example.** Define $n = 28448470441146665947699244512\allowbreak 63$. How do we find, near the integer 124918005771231374100000000000, a square root of 19825184643242306916705771650\allowbreak 29 modulo $n$? In other words: How do we find a small root of $p = (x + 124918005771231374100000000000)^2 - 19825184643242306916705771650\allowbreak 29$ modulo $n$?

Choose $k=2$ and $H=10^{12}/2$. Define $d=\deg p=2$ and $m=dk+d=6$. Then $m(2H)^{m-1} = 6 \cdot 10^{60} < n^2$ so $H < n^{k/(m-1)}/2m^{1/(m-1)}$. Define $f = p(Hx)/n$, $g = Hx$, and $L = \mathbf{Z} + \mathbf{Z}g + \mathbf{Z}f + \mathbf{Z}gf + \mathbf{Z}f^2 + \mathbf{Z}gf^2$.

Reduce the basis $1, g, f, gf, f^2, gf^2$ to find a nonzero vector in $L$ of length at most $2^{(m-1)/2}(\det L)^{1/m} = 2^{5/2}H^{5/2}/n \approx 0.352$. I did this and found a vector $\varphi$ of length approximately 0.019, namely

$$3gf^2$$
$$- 1499016069254776489264474669541 4 f^2$$
$$+ 1645555060488421911465440990695 3 gf$$
$$- 7073107916020226404213966825942253\allowbreak 63949260 f$$
$$+ 45130857618317564931536880639086452818402141\allowbreak 35915989672783463 g$$
$$+ (\cdots)1$$
$$=$$
$$(93750000000000000000000000000000000000000000000000000000000/n^2)x^5$$
$$- (40296668463375000000000000000000000000000000000000000000000/n^2)x^4$$
$$- (85214077087705062073168750000000000000000000000000000000000/n^2)x^3$$
$$+ (75495591489572741344321511190096580000000000000000000000000/n^2)x^2$$
$$+ (85256085569824577108175046901012420957509825119500000000000/n^2)x$$
$$- (73391645786690147620682490399407175727933183364776412308271/n^2)1.$$

The only rational root of $\varphi$ is $372834385559/H$. Check that $p(372834385559)$ is a multiple of $n$, i.e., that 124918005771231374372834385559 is a square root of 19825184643242306916705771650\allowbreak 29 modulo $n$.

Theorem 4.1 guaranteed that this procedure would find every integer root of $p$ modulo $n$ in the interval $[-H, H]$. (Theorem 2.1 guaranteed an even wider interval after $|\varphi|$ turned out to be noticeably smaller than $2^{(m-1)/2}(\det L)^{1/m}$.) This is much faster than separately checking each of the $10^{12} + 1$ integers in this interval.

## 5. Example: constrained divisors of $n$

This section explains how to search for small integers $s$ such that

- $u + s$ divides $n$; or, more generally,
- $u + vs$ divides $n$, where $v$ is coprime to $n$; or, more generally,
- $(u + vs)^d$ divides $n$, where $v$ is coprime to $n$.

For example, by choosing $d = 1$ and choosing $v$ as a large power of 2, one can search for divisors of $n$ having specified low bits.

As in previous sections, the search method has a parameter $k$. Theorem 5.2 uses a particular $k$ that works well for most applications; Theorem 5.1 is more general and allows $k$ to be tuned for the reader's application. Theorems 5.3 and 5.4 measure the cost of the resulting computations.

Section 6 combines this search method with brute force to search a somewhat wider range of $s$. Conclusion in a nutshell: if $v \geq n^{1/4}$, and $v$ is coprime to $n$, then one can quickly find all divisors of $n$ in $(u + v\mathbf{Z}) \cap [1, n^{1/2}]$.

**Theorem 5.1.** *Let $d, n, u, v, w, H$ be positive integers such that $vw - 1 \in n\mathbf{Z}$ and $n \geq H^d$. Let $k$ be a positive integer. Define $\alpha = \sqrt{(\lg 2^d n)/\lg 2^d H^d}$, $m = \lceil \alpha d(k+1) \rceil$, $\lambda = m^{1/2kd}(2H)^{\alpha(1+1/2k)}$, $f = (uw + Hx)^d/n \in \mathbf{Q}[x]$, $g = Hx \in \mathbf{Q}[x]$, and $L$ as above. Let $\varphi \in L$ be a nonzero vector such that $|\varphi| \leq 2^{(m-1)/2}(\det L)^{1/m}$. If $s \in \mathbf{Z}$, $|s| \leq H$, $u + vs \geq \lambda$, and $n \in (u + vs)^d \mathbf{Z}$, then $\varphi(s/H) = 0$.*

The polynomial $(uw + Hx)^d/n$ used here is better than $(u + vHx)^d/n$ when $v > 1$: it has a smaller leading coefficient, so it produces a smaller lattice $L$.

*Proof.* By hypothesis $u + vs \geq \lambda > 0$. Note that $u + vs$ divides $uw + s$. Indeed, $u + vs$ divides $(u + vs)w = uw + s + (vw - 1)s$; but $u + vs$ also divides $(u + vs)^d$, hence $n$, hence $vw - 1$.

Define $r = s/H$. Then $f(r) = (uw + s)^d/n$. The numerator $(uw + s)^d$ and the denominator $n$ are both divisible by $(u + vs)^d$, so $\gcd\{1, f(r)\} \geq (u + vs)^d/n \geq \lambda^d/n = m^{1/2k}(2H)^{\alpha d(1+1/2k)}/n$.

By hypothesis $g_1 = H \geq 1$; $1/f_d = n/H^d \geq 1$; $\alpha = \sqrt{1 + \lg(1/f_d)/\lg((2g_1)^d)}$; $r \in \mathbf{Q}$; $|r| = |s|/H \leq 1$; $\gcd\{1, f(r)\} \geq m^{1/2k}(2g_1)^{\alpha d(1+1/2k)} f_d/g_1^d$; and $g(r) = s \in \mathbf{Z}$. Apply Theorem 3.1. □

**Theorem 5.2.** Let $d, n, u, v, w, H$ be positive integers such that $vw - 1 \in n\mathbf{Z}$ and $n \geq H^d$. Define $\alpha = \sqrt{(\lg 2^d n)/\lg 2^d H^d}$, $k = \lceil \alpha d \lceil \lg 2H \rceil / 2 \rceil$, $m = \lceil \alpha d(k+1) \rceil$, $f = (uw + Hx)^d/n \in \mathbf{Q}[x]$, $g = Hx \in \mathbf{Q}[x]$, and $L$ as above. Let $\varphi \in L$ be a nonzero vector such that $|\varphi| \leq 2^{(m-1)/2}(\det L)^{1/m}$. If $s \in \mathbf{Z}$, $|s| \leq H$, $u + vs \geq 2^{1/d} m^{1/2kd}(2H)^\alpha$, and $n \in (u+vs)^d \mathbf{Z}$, then $\varphi(s/H) = 0$.

The lattice rank $m$ here is larger than $(d/2) \lg 2^d n$. It is only slightly larger for typical values of $d, n, H$.

*Proof.* By hypothesis $2H \geq 2$ so $\lg 2H \geq 1$; hence $k$ is a positive integer. Also $2k \geq \alpha d \lg 2H$ so $2^{1/d} \geq (2H)^{\alpha/2k}$ so $u + vs \geq \lambda$ where $\lambda = m^{1/2kd}(2H)^{\alpha(1+1/2k)}$. Apply Theorem 5.1. □

**Theorem 5.3.** Let $d, n, u, v, H$ be positive integers such that $\gcd\{v, n\} = 1$ and $n \geq H^d$. Let $k$ be a positive integer. Define $\alpha = \sqrt{(\lg 2^d n)/\lg 2^d H^d}$ and $m = \lceil \alpha d(k+1) \rceil$. Then there are at most $m - 1$ integers $s \in \{-H, \ldots, -1, 0, 1, \ldots, H-1, H\}$ such that $u + vs \geq m^{1/2kd}(2H)^{\alpha(1+1/2k)}$ and $n \in (u+vs)^d \mathbf{Z}$.

*Proof.* Find a positive integer $w$ with $vw - 1 \in n\mathbf{Z}$. Apply lattice-basis reduction to Theorem 5.1. □

**Theorem 5.4.** Let $d, n, u, v, H$ be positive integers such that $\gcd\{v, n\} = 1$ and $n \geq H^d$. Define $\alpha = \sqrt{(\lg 2^d n)/\lg 2^d H^d}$, $k = \lceil \alpha d \lceil \lg 2H \rceil / 2 \rceil$, and $m = \lceil \alpha d(k+1) \rceil$. Then there are at most $m - 1$ integers $s \in \{-H, \ldots, -1, 0, 1, \ldots, H-1, H\}$ such that $u + vs \geq 2^{1/d} m^{1/2kd}(2H)^\alpha$ and $n \in (u+vs)^d \mathbf{Z}$.

*Proof.* Find a positive integer $w$ with $vw - 1 \in n\mathbf{Z}$. Apply lattice-basis reduction to Theorem 5.2. □

**History.** Results of this type were developed in two contexts independently. The first context is proving primality of $n$: the Adleman–Pomerance–Rumely method [1983] exhibits some arithmetic progressions and proves, using factors of unit groups of extensions of $\mathbf{Z}/n$, that every divisor of $n$ is in one of those progressions. The second context is factoring an RSA public key $n$ given part of the secret key: for example, finding a divisor of $n$ given the low bits of the divisor.

In the first context, Lenstra [1984] showed how to find all divisors of $n$ in an arithmetic progression $u + v\mathbf{Z}$ with $\lg v > (1/3) \lg n$. Konyagin and Pomerance [1997, Algorithm 3.2] improved $(1/3) \lg n$ to $0.3 \lg n$, in the special case $u = 1$. This $0.3 \lg n$ result, for any $u$, follows from Theorem 2.3 with $m = 5$ and $k = 2$; I have not checked whether the resulting algorithm is equivalent to the Konyagin–Pomerance algorithm.

In the second context, Rivest and Shamir [1986] gave a heuristic outline of a method to find a divisor of $n$ given about $(1/3) \lg n$ high bits of the divisor.

Coppersmith [1996b] proved that a much more complicated bivariate algorithm would find a divisor of $n$ given $(0.25+\varepsilon)\lg n$ high bits of the divisor. Howgrave-Graham [1997] achieved $(0.25+\varepsilon)\lg n$ with the simpler algorithm shown here. Each of these authors commented that the method also applied to low bits, but they did not generalize to other arithmetic progressions.

These two threads in the literature were eventually combined: Coppersmith, Howgrave-Graham, and Nagaraj improved the Konyagin–Pomerance $0.3\lg n$ to $(0.25+\varepsilon)\lg n$. See [Howgrave-Graham 1998, Section 5.5] and [Coppersmith et al. 2004]. Lenstra subsequently pointed out that the $\varepsilon$ could be eliminated; see Section 6 for further discussion.

Boneh, Durfee, and Howgrave-Graham [1999] pointed out, at least for $v=1$, the further generalization from divisors $u+vs$ to divisors $(u+vs)^d$. As $d$ increases, the allowable range of $H$ shrinks, but the range of interesting divisors shrinks more quickly. At an extreme, for $d$ larger than about $\sqrt{\lg n}$, this method finds $d$-power divisors of $n$ more quickly than the elliptic-curve method.

**Numerical example.** Consider the problem of finding $p \approx 1814430925000000$ such that $p^2$ divides $n = 376737519824311248322897466745610595514463036\overline{7}$.

Define $d=2$, $u=1814430925000000$, $v=1$, $w=1$, $k=2$, and $H=10^6$. Define $\alpha = \sqrt{(\lg 4n)/\lg 4H^2} \approx 1.91424$ and $m = \lceil \alpha d(k+1)\rceil = 12$. Then $u - H \geq \lambda$ where $\lambda = m^{1/2kd}(2H)^{\alpha(1+1/2k)}$. Define $f = (uw + Hx)^d/n = (u+Hx)^2/n$, $g = Hx$, and $L = \mathbf{Z} + \mathbf{Z}g + \mathbf{Z}f + \mathbf{Z}gf + \mathbf{Z}f^2 + \mathbf{Z}gf^2 + \mathbf{Z}g^2f^2 + \mathbf{Z}g^3f^2 + \mathbf{Z}g^4f^2 + \mathbf{Z}g^5f^2 + \mathbf{Z}g^6f^2 + \mathbf{Z}g^7f^2$.

Find a nonzero vector in $L$ of length at most $2^{(m-1)/2}(\det L)^{1/m}$ by reducing the basis $1, g, f, gf, f^2, gf^2, g^2f^2, g^3f^2, g^4f^2, g^5f^2, g^6f^2, g^7f^2$. I did this and found the vector

$$8654285929051698536731156579739732909254403370124466963870118306516 f^2$$
$$- 6050109444904732893967670609502978242326457349320354 f$$
$$- 27255412018787295847722163555072174417628911011368059 gf^2$$
$$- 13217375993392331719811049580402472\overline{8}4$$
$$- 6668878229472208312826600694772455332 gf$$
$$+ 7510732878996292723404180926729165467 g^2 f^2$$
$$- 832523980748052892274 g$$
$$- 16557770862327878578539 g^3 f^2$$
$$+ 22814 g^4 f^2$$

of length approximately $2.3 \cdot 10^{-38}$. The only rational root of this polynomial is $339897/H$. Check that $1814430925339897^2$ is a divisor of $n$.

Theorem 5.1 guaranteed that this procedure would find all divisors $(u+s)^2$ of $n$ with $-H \leq s \leq H$. In fact, Theorem 2.3 guaranteed that $k=2$ and $m=7$ would have done the same job, and that $k=1$ and $m=5$ would have worked for the smaller interval $-450000 \leq s \leq 450000$.

## 6. Partitioning an arithmetic progression

Consider the problem of finding all divisors of $n$ in $(u+v\mathbf{Z}) \cap [1, n^{1/2}]$. Here $u, v, n$ are positive integers with $v \geq n^{1/4}$ and $\gcd\{v, n\} = 1$.

One can use Theorem 5.2 to find all divisors of $n$ in the arithmetic progression $u - vH, u - v(H-1), \ldots, u + v(H-1), u + vH$. But there is a limitation here: the smallest entry $u - vH$ must exceed $2m^{1/2k}(2H)^\alpha$, approximately the doubly-geometric average of $n$ and $H^d$. Another way to view the lower bound on $u - vH$ is as follows: if the smallest entry $u - vH$ is approximately $n^{1/\alpha}$ then the number of entries is limited to approximately $n^{1/\alpha^2}$. In particular, if this method is searching for divisors around $n^{1/2}$, then it will search at most about $n^{1/4}$ entries in a specified arithmetic progression.

This might not sound like a serious limitation: by hypothesis $v \geq n^{1/4}$, so there are at most $n^{1/4}$ elements of $(u + v\mathbf{Z}) \cap [1, n^{1/2}]$. But one cannot search $n^{1/4}$ elements unless the *smallest* element searched is close to $n^{1/2}$.

The point of this section is that one can cover $(u+v\mathbf{Z}) \cap [1, n^{1/2}]$ with $O((\lg n)^{1/2})$ arithmetic progressions and $O((\lg n)^{1/2})$ extra integers, where each progression meets the conditions of Theorem 5.2. Therefore one can quickly find all the divisors of $n$ in $(u+v\mathbf{Z}) \cap [1, n^{1/2}]$. See Theorem 6.4 for a bound on the cost of this computation.

My bounds here are completely explicit. Various constants can be improved; my goal in selecting constants was not to obtain optimal cost bounds, but to simplify the statements and the proofs as far as possible while still achieving $O((\lg n)^{1/2})$.

**Theorem 6.1.** *Let $n$ be an integer with $n \geq 2^{24}$. Let $v$ be a positive integer with $\gcd\{v, n\} = 1$. Let $H$ be an integer with $2 \leq H \leq n$. Define $\alpha = \sqrt{(\lg 2n)/\lg 2H}$. Let $z$ be an integer with $z \geq 4(2H)^\alpha$. Then there are at most $2\lg 2n + \sqrt{\lg 2n}$ divisors of $n$ in $\{z, z+v, z+2v, \ldots, z+2vH\}$.*

*Proof.* The difference $2^{r\sqrt{2}} - 4r^2 - 2r$ is positive for all real numbers $r \geq 5$: its value at $r=5$ is $2^{5\sqrt{2}} - 100 - 10 > 2^7 - 110 > 0$; its derivative at $r=5$ is $2^{5\sqrt{2}} \sqrt{2} \log 2 - 40 - 2 > 0$; and its second derivative is $2^{r\sqrt{2}}(\sqrt{2}\log 2)^2 - 8 > 0$ for $r \geq 5$. In particular, $\sqrt{\lg 2n} \geq \sqrt{25} = 5$, so $2^{\sqrt{2\lg 2n}} \geq 4\lg 2n + 2\sqrt{\lg 2n}$.

Define $k = \lceil \alpha \lceil \lg 2H \rceil / 2 \rceil$. By hypothesis $H \geq 2$ so $\lg 2H \geq \lg 4 = 2$ so $2k \geq \alpha \lg 2H = \sqrt{(\lg 2n)\lg 2H} \geq \sqrt{2\lg 2n}$. Furthermore $H \leq n$ so $\alpha \geq 1$ so

$\alpha \lceil \lg 2H \rceil /2 \geq 1$ so $k \leq 2\alpha \lceil \lg 2H \rceil /2 = \alpha \lceil \lg 2H \rceil \leq 2\alpha \lg 2H$ so $\alpha(k+1) \leq 2\alpha^2 \lg 2H + \alpha = 2 \lg 2n + \alpha \leq 2 \lg 2n + \sqrt{\lg 2n}$.

Define $m = \lceil \alpha(k+1) \rceil$. Again $\alpha \geq 1$ so $\alpha(k+1) \geq 1$; thus $m \leq 2\alpha(k+1) \leq 4 \lg 2n + 2\sqrt{\lg 2n} \leq 2\sqrt{2 \lg 2n} \leq 2^{2k}$. Consequently $z \geq 2m^{1/2k}(2H)^\alpha$.

Define $d = 1$ and $u = z + vH$. By Theorem 5.4, $n$ has at most $m - 1$ divisors in $\{u - vH, \ldots, u - v, u, u + v, \ldots, u + vH\} \cap [2m^{1/2k}(2H)^\alpha, \infty) = \{z, z + v, z + 2v, \ldots, z + 2vH\}$. Finally $m - 1 \leq \alpha(k+1) \leq 2 \lg 2n + \sqrt{\lg 2n}$. □

**Theorem 6.2.** *Let $n, u, v$ be integers with $v \geq n^{1/4} \geq 2^{64}$ and $\gcd\{v, n\} = 1$. Let $i$ be an integer with $8 \leq i \leq \sqrt{\lg n}/2$. Then there are at most $2 \lg 2n + \sqrt{\lg 2n}$ divisors of $n$ in $(u + v\mathbf{Z}) \cap [n^{1/2 - 2/i}, n^{1/2 - 2/(i+1)}]$.*

*Proof.* Define $H = \lfloor n^{1/4 - 2/(i+1)}/2 \rfloor$. Note that $n^{1/4 - 2/(i+1)} \geq n^{1/4 - 2/9} = n^{1/36} \geq 4$, so $H \geq 2$; and $H \leq n^{1/4 - 2/(i+1)} \leq n$. Define $\alpha = \sqrt{(\lg 2n)/\lg 2H}$. Define $z$ as the smallest element of $(u + v\mathbf{Z}) \cap [n^{1/2 - 2/i}, \infty)$. Note that $z \geq n^{1/2 - 2/i} \geq n^{1/2 - 2/8} = n^{1/4} \geq 2^{64}$.

I claim that $z + 2Hv + v > n^{1/2 - 2/(i+1)}$. Proof: $H + 1 > n^{1/4 - 2/(i+1)}/2$, so $z + 2Hv + v \geq (1 + 2H + 1)n^{1/4} > n^{1/4 - 2/(i+1)} n^{1/4} = n^{1/2 - 2/(i+1)}$.

I also claim that $z \geq 4(2H)^\alpha$. Proof: $i^2 \geq (i+1)(i-1)$; so $2/(i+1) \geq 2(i-1)/i^2$; so $2/(i+1) - 2(i-2)/i^2 \geq 2/i^2 \geq 2/(\sqrt{\lg n}/2)^2 = 8/\lg n$; so

$$\left(\left(\frac{1}{2} - \frac{2}{i}\right)\lg n - 2\right)^2 - \left(\frac{1}{4} - \frac{2}{i+1}\right)(\lg 2n)\lg n$$
$$= \left(\frac{2}{i+1} - \frac{2(i-2)}{i^2}\right)(\lg n)^2 - \left(4\left(\frac{1}{2} - \frac{2}{i}\right) + \left(\frac{1}{4} - \frac{2}{i+1}\right)\right)\lg n + 4$$
$$\geq \frac{8}{\lg n}(\lg n)^2 - \left(4\left(\frac{1}{2}\right) + \left(\frac{1}{4}\right)\right)\lg n + 4 = \frac{23}{4}\lg n + 4 \geq 0;$$

so

$$\alpha^2(\lg 2H)^2 = \lg 2n \lg 2H$$
$$\leq (\lg 2n)\left(\frac{1}{4} - \frac{2}{i+1}\right)\lg n \leq \left(\left(\frac{1}{2} - \frac{2}{i}\right)\lg n - 2\right)^2;$$

so $\alpha \lg 2H \leq |(1/2 - 2/i)\lg n - 2| = (1/2 - 2/i)\lg n - 2 \leq \lg z - 2$.

Now apply Theorem 6.1 to see that there are at most $2 \lg 2n + \sqrt{\lg 2n}$ divisors of $n$ in $\{z, z + v, \ldots, z + 2vH\}$. Finally $(u + v\mathbf{Z}) \cap [n^{1/2 - 2/i}, n^{1/2 - 2/(i+1)}] \subseteq \{z, z + v, \ldots, z + 2vH\}$. □

**Theorem 6.3.** *Let $n, u, v$ be integers with $v \geq n^{1/4} \geq 2^{75}$ and $\gcd\{v, n\} = 1$. Let $i$ be an integer with $1 \leq i \leq \lceil 16\sqrt{\lg n} \rceil$. Then there are at most $2 \lg 2n + \sqrt{\lg 2n} + 1$ divisors of $n$ in $(u + v\mathbf{Z}) \cap [n^{1/2}/2^{i/4}, n^{1/2}/2^{(i-1)/4}]$.*

*Proof.* Define $H = \lfloor n^{1/4}/2^{(i+13)/4} \rfloor$. Note that $(\sqrt{\lg n} - 8)^2 \geq (\sqrt{4 \cdot 75} - 8)^2 \geq 82$, so $\lg n - 16\sqrt{\lg n} \geq 82 - 8^2 = 18$, so $\lg n - i \geq 17$, so $n^{1/4}/2^{(i+13)/4} = 2^{(\lg n - i - 13)/4} \geq 2^{4/4} = 2$, so $H \geq 2$; and $H \leq n^{1/4} \leq n$. Define $\alpha = \sqrt{(\lg 2n)/\lg 2H}$.

Define $z$ as the smallest element of $(u+v\mathbf{Z}) \cap [n^{1/2}/2^{i/4}, \infty)$.

I claim that $z + 2Hv + 2v > n^{1/2}/2^{(i-1)/4}$. Proof: $H+1 > n^{1/4}/2^{(i+13)/4}$, and $1 + 2^{-9/4} \geq 2^{1/4}$, so $z + 2(H+1)v > n^{1/2}/2^{i/4} + 2^{-9/4}n^{1/2}/2^{i/4} \geq n^{1/2}/2^{(i-1)/4}$.

I claim that $z \geq 4(2H)^\alpha$. Proof:

$$\left(\frac{1}{2}\lg n - \frac{i}{4} - 2\right)^2 - \frac{(\lg 2n)(\lg n - i - 9)}{4} = \left(\frac{i}{4} + 2\right)^2 + \frac{i+9}{4} \geq 0;$$

so $\alpha^2(\lg 2H)^2 = (\lg 2n)\lg 2H \leq (\lg 2n)(\lg n - i - 9)/4 \leq ((1/2)\lg n - i/4 - 2)^2$; and $(1/2)\lg n - i/4 - 2 \geq (\lg n - i)/4 - 2 \geq 17/4 - 2 \geq 0$, so $\alpha \lg 2H \leq |(1/2)\lg n - i/4 - 2| = (1/2)\lg n - i/4 - 2 \leq \lg z - 2$.

Now apply Theorem 6.1 to see that the set $\{z, z+v, \ldots, z+2vH\}$ has at most $2\lg 2n + \sqrt{\lg 2n}$ divisors of $n$; so $\{z, z+v, \ldots, z+2vH+v\}$ has at most $2\lg 2n + \sqrt{\lg 2n} + 1$ divisors of $n$. Finally $(u+v\mathbf{Z}) \cap [n^{1/2}/2^{i/4}, n^{1/2}/2^{(i-1)/4}] \subseteq \{z, z+v, \ldots, z+2Hv+v\}$. □

**Theorem 6.4.** *Let $n, u, v$ be integers with $v \geq n^{1/4} \geq 2^{75}$ and $\gcd\{v, n\} = 1$. Define $\ell = \lg 2n$. Then there are at most $33\ell^{1.5} + 4.5\ell + 10\ell^{0.5} + 2$ divisors of $n$ in $(u+v\mathbf{Z}) \cap [1, n^{1/2}]$.*

*Proof.* There is at most one divisor of $n$ in $(u+v\mathbf{Z}) \cap [1, n^{1/4}]$, since $v \geq n^{1/4}$.

Write $s = \lfloor \sqrt{\lg n}/2 \rfloor$. Then $s \geq 8$. Also $s+1 > \sqrt{\lg n}/2$, so $n^{1/2-2/(s+1)} > n^{1/2-4/\sqrt{\lg n}}$. Apply Theorem 6.2 for each $i \in \{8, 9, \ldots, s\}$ to cover the intervals $[n^{1/2-2/8}, n^{1/2-2/9}], [n^{1/2-2/9}, n^{1/2-2/10}], \ldots, [n^{1/2-2/s}, n^{1/2-2/(s+1)}]$: there are at most $(s-7)(2\ell + \ell^{0.5})$ divisors of $n$ in $(u+v\mathbf{Z}) \cap [n^{1/2-2/8}, n^{1/2-2/(s+1)}] \supseteq (u+v\mathbf{Z}) \cap [n^{1/4}, n^{1/2-4/\sqrt{\lg n}}]$.

Write $t = \lceil 16\sqrt{\lg n} \rceil$. Then $t/4 \geq 4\sqrt{\lg n} = (4/\sqrt{\lg n})\lg n$, so $n^{1/2}/2^{t/4} \leq n^{1/2-4/\sqrt{\lg n}}$. Apply Theorem 6.3 for each $i \in \{1, 2, \ldots, t\}$ to cover the intervals $[n^{1/2}/2^{1/4}, n^{1/2}/2^{0/4}], [n^{1/2}/2^{2/4}, n^{1/2}/2^{1/4}], \ldots, [n^{1/2}/2^{t/4}, n^{1/2}/2^{(t-1)/4}]$: there are at most $t(2\ell + \ell^{0.5} + 1)$ divisors of $n$ in $(u+v\mathbf{Z}) \cap [n^{1/2}/2^{t/4}, n^{1/2}/2^{0/4}] \supseteq (u+v\mathbf{Z}) \cap [n^{1/2-4/\sqrt{\lg n}}, n^{1/2}]$.

Finally add: there are at most $1 + (s-7)(2\ell + \ell^{0.5}) + t(2\ell + \ell^{0.5} + 1) \leq 1 + (\ell^{0.5}/2 - 7)(2\ell + \ell^{0.5}) + (1 + 16\ell^{0.5})(2\ell + \ell^{0.5} + 1) = 33\ell^{1.5} + 4.5\ell + 10\ell^{0.5} + 2$ divisors of $n$ in $(u+v\mathbf{Z}) \cap [1, n^{1/2}]$. □

**History.** Coppersmith, Howgrave-Graham, and Nagaraj constructed lattices of total rank $O(\varepsilon^{-3/2})$ that would handle all $v \geq n^{1/4+\varepsilon}$ for all sufficiently large $n$. See [Howgrave-Graham 1998, Section 5.5] and [Coppersmith et al. 2004]. It is not clear whether one can take $\varepsilon \approx 1/\lg n$ here: Coppersmith,

Howgrave-Graham, and Nagaraj did not give simple formulas for their partition of $[1/4, 1/2]$ as a function of $\varepsilon$, and did not quantify "sufficiently large" as a function of $\varepsilon$.

Lenstra constructed lattices of total rank $O((\lg n)^2)$ handling all $v \geq n^{1/4}$, and asked whether one could achieve $O((\lg n)^{3/2})$. I constructed $O((\lg n)^{1/2})$ lattices of total rank $O((\lg n)^{3/2})$ handling all $v \geq n^{1/4}$; see Theorem 6.4.

The essential difference between these proofs is in the analysis of how much progress is made by a $(2H+1)$-entry arithmetic progression starting at $z$. The Coppersmith–Howgrave-Graham–Nagaraj proof has an advantage in handling small divisors: it chooses $H$ much larger than $z/v$, producing a large lower bound on $\lg 2Hv$ and thus on $\lg(z+2Hv)$, as in Theorem 6.2 here. Lenstra's proof has an advantage in handling large divisors: it allows $H$ to be as small as, e.g., $0.1z/v$, and then observes that $\lg(z+2Hv) \geq \lg 1.2z > \lg z + 0.25$, as in Theorem 6.3 here. My proof combines these advantages, and does some extra work to make all the bounds explicit.

Coppersmith, Howgrave-Graham, and Nagaraj tuned their choices of $(k, m)$ more tightly than I have done, and they computed particularly good partitions (at least for the number-of-outputs cost measure) for several specific values of $\varepsilon$. As usual, I am leaving this level of optimization to the reader.

## 7. Example: codeword errors past half the minimum distance

Fix a positive integer $H$. Fix finitely many distinct primes $p_1, p_2, p_3, \ldots$. Assume that the product $n = p_1 p_2 \cdots$ is much larger than $H$. The **residue representation** of an integer $s \in [-H, H]$ is, by definition, the vector ($s \bmod p_1$, $s \bmod p_2$, $s \bmod p_3$, ...).

There must be many differences between the residue representations of $s$ and $s'$ if $s' \neq s$. Specifically: Define the **distance** between $(v_1, v_2, \ldots)$ and $(v_1', v_2', \ldots)$ as the sum of $\lg p_i$ for all $i$ such that $v_i \neq v_i'$, and define the **distance** between integers $s$ and $s'$ as the distance between the residue representations of $s$ and $s'$. Then the distance between $s$ and $s'$ is exactly $\lg n - \lg \gcd\{s'-s, n\}$, which is at least $\lg n - \lg 2H$ since $\gcd\{s'-s, n\} \leq 2H$.

Thus the residue representation can tolerate some errors. For any vector $v$, there is at most one $s$ whose representation has distance $< (\lg n - \lg 2H)/2$ from $v$.

This section explains how to efficiently recover $s$ from a vector at any distance up to about $\lg n - \sqrt{(\lg 2n) \lg 2H}$. One first interpolates the vector into an integer $u \in \{0, 1, \ldots, n-1\}$, and then finds $s$ such that $\gcd\{u-s, n\}$ is large. For distances above $(\lg n - \lg 2H)/2$, there might be several possibilities for $s$; this section explains how to find them all.

As in previous sections, the user can choose a parameter $k$. Theorem 7.2 uses a particular parameter $k$ that works well for most applications; Theorem 7.4 measures the cost of the resulting computation. Theorem 7.1 is more general and allows $k$ to be tuned for the reader's application; Theorem 7.3 measures the cost of the resulting computation. The simplest case $k = 1$, $m = 2$ of Theorem 7.1 finds all $s$ with $\gcd\{u - s, n\} > (4Hn)^{1/2}$, i.e., with distance smaller than $(\lg n - \lg 4H)/2$; there is at most one such $s$.

**Theorem 7.1.** *Let $n, u, H, k$ be positive integers such that $n \geq H$. Define $\alpha = \sqrt{(\lg 2n)/\lg 2H}$, $m = \lceil \alpha(k+1) \rceil$, $\lambda = m^{1/2k}(2H)^{\alpha(1+1/2k)}$, $f = (Hx - u)/n \in \mathbf{Q}[x]$, $g = Hx \in \mathbf{Q}[x]$, $d = 1$, and $L$ as above. Let $\varphi \in L$ be a nonzero vector such that $|\varphi| \leq 2^{(m-1)/2}(\det L)^{1/m}$. If $s \in \mathbf{Z}$, $|s| \leq H$, and $\gcd\{u - s, n\} \geq \lambda$, then $\varphi(s/H) = 0$.*

Compare to the case $v = 1$, $w = 1$, $d = 1$ of Theorem 5.1.

*Proof.* Define $r = s/H$. By hypothesis $g_1 = H \geq 1$; $1/f_d = n/H \geq 1$; $\alpha = \sqrt{1 + \lg(1/f_d)/\lg(2g_1)}$; $r \in \mathbf{Q}$; $|r| = |s|/H \leq 1$; $g(r) = s \in \mathbf{Z}$; and $f(r) = (s - u)/n$, so $\gcd\{1, f(r)\} \geq \lambda/n = m^{1/2k}(2g_1)^{\alpha(1+1/2k)} f_d/g_1$. Apply Theorem 3.1. □

**Theorem 7.2.** *Let $n, u, H$ be positive integers such that $n \geq H$. Define $\alpha = \sqrt{(\lg 2n)/\lg 2H}$, $k = \lceil \alpha \lceil \lg 2H \rceil /2 \rceil$, $m = \lceil \alpha(k+1) \rceil$, $f = (Hx - u)/n \in \mathbf{Q}[x]$, $g = Hx \in \mathbf{Q}[x]$, $d = 1$, and $L$ as above. Let $\varphi \in L$ be a nonzero vector such that $|\varphi| \leq 2^{(m-1)/2}(\det L)^{1/m}$. If $s \in \mathbf{Z}$, $|s| \leq H$, and $\gcd\{u - s, n\} \geq 2m^{1/2k}(2H)^\alpha$, then $\varphi(s/H) = 0$.*

*Proof.* By hypothesis $2H \geq 2$ so $\lg 2H \geq 1$; hence $k$ is a positive integer. Also $2k \geq \alpha \lg 2H$ so $2 \geq (2H)^{\alpha/2k}$ so $\gcd\{u - s, n\} \geq \lambda$ where $\lambda = m^{1/2k}(2H)^{\alpha(1+1/2k)}$. Apply Theorem 7.1. □

**Theorem 7.3.** *Let $n, u, H, k$ be positive integers such that $n \geq H$. Define $\alpha = \sqrt{(\lg 2n)/\lg 2H}$ and $m = \lceil \alpha(k+1) \rceil$. Then there are at most $m - 1$ integers $s \in \{-H, \ldots, 0, 1, \ldots, H\}$ such that $\gcd\{u - s, n\} \geq m^{1/2k}(2H)^{\alpha(1+1/2k)}$.*

*Proof.* Apply lattice-basis reduction to Theorem 7.1. □

**Theorem 7.4.** *Let $n, u, H$ be positive integers such that $n \geq H$. Define $\alpha = \sqrt{(\lg 2n)/\lg 2H}$, $k = \lceil \alpha \lceil \lg 2H \rceil /2 \rceil$, and $m = \lceil \alpha(k+1) \rceil$. Then there are at most $m - 1$ integers $s \in \{-H, \ldots, 0, 1, \ldots, H - 1, H\}$ such that $\gcd\{u - s, n\} \geq 2m^{1/2k}(2H)^\alpha$.*

*Proof.* Apply lattice-basis reduction to Theorem 7.2. □

**History.** The rational-function-field version of the simple case $k = 1$, $m = 2$ is the "Berlekamp–Massey algorithm" for decoding "Reed–Solomon codes."

The fact that one can efficiently correct larger errors was pointed out first in the function-field case by Sudan [1997], and then in the number-field case by Goldreich, Ron, and Sudan [1999]. These results are tantamount to optimizing $m$ in Theorem 2.3 with $k = 1$. Increasing $k$ produces an asymptotic $\sqrt{2}$ exponent improvement, as discussed in Section 3; this $\sqrt{2}$ improvement was pointed out in the function-field case by Guruswami and Sudan [1999], and in the number-field case by Boneh [2000].

Algorithms that may produce several values of $s$ are called "list decoding" algorithms. Of course, the resulting list is most useful when it has just one value of $s$.

**Numerical example.** Define $H = 1000000$, $n = 101 \cdot 103 \cdot 107 \cdot 109 \cdot 113 \cdot 127 \cdot 131 \cdot 137 \cdot 139 \cdot 149 \cdot 151 \cdot 157 \cdot 163 \cdot 167 \cdot 173 \cdot 179 \cdot 181 \cdot 191 \cdot 193 \cdot 197 \cdot 199$, and $u = 4765345845193600442153574482968114946568\allowbreak 48207$. The goal here is to find every $s \in [-H, H]$ with residue representation close to $(u \bmod 101, u \bmod 103, \ldots, u \bmod 199)$, i.e., close to $(94, 43, 17, 71, 103, 77, 64, 25, 114, 9, 106, 16, 62, 134, 75, 13, 155, 26, 138, 21, 105)$.

Choose $k = 3$. Define $\alpha = \sqrt{(\lg 2n)/\lg 2H} \approx 2.697$ and $m = \lceil \alpha(k+1) \rceil = 11$. Define $f = (Hx - u)/n$, $g = Hx$, and $L = \mathbf{Z} + \mathbf{Z}f + \mathbf{Z}f^2 + \mathbf{Z}f^3 + \mathbf{Z}gf^3 + \mathbf{Z}g^2 f^3 + \mathbf{Z}g^3 f^3 + \mathbf{Z}g^4 f^3 + \mathbf{Z}g^5 f^3 + \mathbf{Z}g^6 f^3 + \mathbf{Z}g^7 f^3$.

Reduce the basis $1, f, f^2, f^3, gf^3, g^2f^3, g^3f^3, g^4f^3, g^5f^3, g^6f^3, g^7f^3$ to find a nonzero vector in $L$ of length at most $2^{(m-1)/2}(\det L)^{1/m}$: for example, the vector

$(25587000000000000000000000000000000000000000000000/n^3)x^8$
$- (11721491973000000000000000000000000000000000000000/n^3)x^7$
$- (16962979598492916000000000000000000000000000000000/n^3)x^6$
$- (90804950564045088121550000000000000000000000000000/n^3)x^5$
$- (16268872584260106363071226779000000000000000000000/n^3)x^4$
$- (47860927326254884030215835975433610000000000000000/n^3)x^3$
$- (68525665600669610580610714527465995863869000000000/n^3)x^2$
$- (48664703743008291510964005462444491801551604010000/n^3)x$ 
$+ (19654220351564720341671319570621613333140807708304\allowbreak07/n^3)1.$

The only rational root of this polynomial is $s/H$ where $s = 476511$. The residue representation of $s$ is $(94, 33, 40, 72, 103, 7, 64, 25, 19, 9, 106, 16, 62, 60, 69, 13, 119, 157, 187, 165, 105)$; the distance from $s$ to $u$ is approximately 79.41.

Theorem 7.1 guaranteed that this procedure would find every $s \in [-H, H]$ within distance $\lg n - \lg \lambda \approx 84.8$ of $u$; here $\lambda = m^{1/2k}(2H)^{\alpha(1+1/2k)}$. Even better,

Theorem 2.3 guaranteed that this procedure would find every $s$ within distance $-\lg \gamma \approx 88.28$ of $u$; here $\gamma = m^{1/2k}(2H)^{(m-1)/2k}n^{(k+1)/2m-1}$. Both bounds are far above $(\lg n - \lg 2H)/2 \approx 65.16$.

# References

[STOC 1999] — (no editor), *Annual ACM symposium on theory of computing: proceedings of the 31st symposium (STOC '99) held in Atlanta, GA, May 1–4, 1999*, Association for Computing Machinery, New York. ISBN 1-58113-067-8. MR 2001f:68004. See [Goldreich et al. 1999].

[STOC 2000] — (no editor), *Proceedings of the 32nd annual ACM symposium on theory of computing*, Association for Computing Machinery, New York. ISBN 1-58113-184-4. See [Boneh 2000].

[Adleman et al. 1983] Leonard M. Adleman, Carl Pomerance, and Robert S. Rumely, "On distinguishing prime numbers from composite numbers", *Annals of Mathematics* **117**, 173–206. ISSN 0003-486X. MR 84e:10008. Citations in this document: §5.

[Bleichenbacher 2004] Daniel Bleichenbacher, "Compressing Rabin signatures", pp. 126–128 in [Okamoto 2004]. Citations in this document: §4.

[Boneh 2000] Dan Boneh, "Finding smooth integers in short intervals using CRT decoding", pp. 265–272 in [STOC 2000]; see also newer version [Boneh 2002]. Citations in this document: §0, §0, §1, §7.

[Boneh 2002] Dan Boneh, "Finding smooth integers in short intervals using CRT decoding", *Journal of Computer and System Sciences* **64**, 768–784; see also older version [Boneh 2000]. ISSN 0022-0000. MR 1 912 302. URL: http://crypto.stanford.edu/~dabo/abstracts/CRTdecode.html.

[Boneh et al. 1999] Dan Boneh, Glenn Durfee, and Nick Howgrave-Graham, "Factoring $N = p^r q$ for large $r$", pp. 326–337 in [Wiener 1999]. URL: http://crypto.stanford.edu/~dabo/abstracts/prq.html. Citations in this document: §0, §1, §5.

[Buhler and Stevenhagen 2008] Joe P. Buhler and Peter Stevenhagen (editors), *Surveys in algorithmic number theory*, Mathematical Sciences Research Institute Publications **44**, Cambridge University Press, New York; this book. See [Lenstra 2008].

[Coppersmith 1996a] Don Coppersmith, "Finding a small root of a univariate modular equation", pp. 155–165 in [Maurer 1996]; see also newer version [Coppersmith 1997]. MR 97h:94008. Citations in this document: §0, §0, §0, §3.

[Coppersmith 1996b] Don Coppersmith, "Finding a small root of a bivariate integer equation; factoring with high bits known", pp. 178–189 in [Maurer 1996]; see also newer version [Coppersmith 1997]. MR 97h:94009. Citations in this document: §0, §5.

[Coppersmith 1997] Don Coppersmith, "Small solutions to polynomial equations, and low exponent RSA vulnerabilities", *Journal of Cryptology* **10**, 233–260; see also older version [Coppersmith 1996a] and [Coppersmith 1996b]. ISSN 0933–2790. MR 99b:94027.

[Coppersmith 2001] Don Coppersmith, "Finding small solutions to small degree polynomials", pp. 20–31 in [Silverman 2001]. MR 2003f:11034. URL: http://cr.yp.to/bib/entries.html#2001/coppersmith. Citations in this document: §3, §3.

[Coppersmith et al. 2004] Don Coppersmith, Nick Howgrave-Graham, and S. V. Nagaraj, "Divisors in residue classes, constructively". URL: http://eprint.iacr.org/2004/339. Citations in this document: §5, §6.

[Darnell 1997] Michael Darnell (editor), *Cryptography and coding: proceedings of the 6th IMA International Conference held at the Royal Agricultural College, Cirencester, December 17–19, 1997*, Lecture Notes in Computer Science **1355**, Springer-Verlag. ISBN 3–540–63927–6. MR 99g:94019. See [Howgrave-Graham 1997].

[Goldreich et al. 1999] Oded Goldreich, Dana Ron, and Madhu Sudan, "Chinese remaindering with errors", pp. 225–234 in [STOC 1999]; see also newer version [Goldreich et al. 2000]. MR 2001i:68050. URL: http://theory.lcs.mit.edu/~madhu/papers.html. Citations in this document: §0, §1, §7.

[Goldreich et al. 2000] Oded Goldreich, Dana Ron, and Madhu Sudan, "Chinese remaindering with errors", *IEEE Transactions on Information Theory* **46**, 1330–1338; see also older version [Goldreich et al. 1999]. ISSN 0018–9448. MR 2001k:11005. URL: http://theory.lcs.mit.edu/~madhu/papers.html.

[Graham and Nešetřil 1997] Ronald L. Graham and Jaroslav Nešetřil (editors), *The mathematics of Paul Erdős. I*, Algorithms and Combinatorics **13**, Springer-Verlag, Berlin. ISBN 3–540–61032–4. MR 97f:00032. See [Konyagin and Pomerance 1997].

[Guruswami and Sudan 1999] Venkatesan Guruswami and Madhu Sudan, "Improved decoding of Reed-Solomon and algebraic-geometry codes", *IEEE Transactions on Information Theory* **45**, 1757–1767. ISSN 0018–9448. MR 2000j:94033. URL: http://theory.lcs.mit.edu/~madhu/bib.html. Citations in this document: §0, §7.

[Håstad 1988] Johan Håstad, "Solving simultaneous modular equations of low degree", *SIAM Journal on Computing* **17**, 336–341. ISSN 0097–5397. MR 89e:68049. URL: http://www.nada.kth.se/~johanh/papers.html. Citations in this document: §0.

[Howgrave-Graham 1997] Nicholas Howgrave-Graham, "Finding small roots of univariate modular equations revisited", pp. 131–142 in [Darnell 1997]. MR 99j:94049. Citations in this document: §0, §0, §1, §5.

[Howgrave-Graham 1998] Nicholas Howgrave-Graham, *Computational mathematics inspired by RSA*, Ph.D. thesis. URL: http://cr.yp.to/bib/entries.html#1998/howgrave-graham. Citations in this document: §0, §3, §5, §6.

[Howgrave-Graham 2001] Nicholas Howgrave-Graham, "Approximate integer common divisors", pp. 51–66 in [Silverman 2001]. MR 2003h:11160. URL: http://cr.yp.to/bib/entries.html#2001/howgrave-graham. Citations in this document: §0.

[Konyagin and Pomerance 1997] Sergei Konyagin and Carl Pomerance, "On primes recognizable in deterministic polynomial time", pp. 176–198 in [Graham and Nešetřil 1997]. MR 98a:11184. URL: http://cr.yp.to/bib/entries.html#1997/konyagin. Citations in this document: §0, §5.

[Lenstra et al. 1982] Arjen K. Lenstra, Hendrik W. Lenstra, Jr., and László Lovász, "Factoring polynomials with rational coefficients", *Mathematische Annalen* **261**, 515–534. ISSN 0025–5831. MR 84a:12002. URL: http://cr.yp.to/bib/entries.html#1982/lenstra-lll. Citations in this document: §2.

[Lenstra 1984] Hendrik W. Lenstra, Jr., "Divisors in residue classes", *Mathematics of Computation* **42**, 331–340. ISSN 0025–5718. MR 85b:11118. URL: http://www.jstor.org/sici?sici=0025-5718(198401)42:165<331:DIRC>2.0.CO;2-6. Citations in this document: §0, §0, §5.

[Lenstra 2008] Hendrik W. Lenstra, Jr., "Lattices", pp. 127–181 in [Buhler and Stevenhagen 2008]. Citations in this document: §2.

[Loos 1983] Rüdiger Loos, "Computing rational zeros of integral polynomials by $p$-adic expansion", *SIAM Journal on Computing* **12**, 286–293. ISSN 0097–5397. MR 85b:11123. Citations in this document: §2.

[Maurer 1996] Ueli M. Maurer (editor), *Advances in cryptology—EUROCRYPT '96: Proceedings of the Fifteenth International Conference on the Theory and Application of Cryptographic Techniques held in Saragossa, May 12–16, 1996*, Lecture Notes in Computer Science **1070**, Springer-Verlag, Berlin. ISBN 3–540–61186–X. MR 97g:94002. See [Coppersmith 1996a], [Coppersmith 1996b].

[Mora 1989] Teo Mora (editor), *Applied algebra, algebraic algorithms and error-correcting codes: proceedings of the sixth international conference (AAECC-6) held in Rome, July 4–8, 1988*, Lecture Notes in Computer Science **357**, Springer-Verlag, Berlin. ISBN 3–540–51083–4. MR 90d:94002. See [Vallée et al. 1989].

[Okamoto 2004] Tatsuaki Okamoto (editor), *Topics in cryptology—CT-RSA 2004: the cryptographers' track at the RSA Conference 2004, San Francisco, CA, USA, February 23–27, 2004, proceedings*, Lecture Notes in Computer Science, Springer, Berlin. ISBN 3–540–20996–4. MR 2005d:94157. See [Bleichenbacher 2004].

[Pichler 1986] Franz Pichler (editor), *Advances in cryptology—EUROCRYPT '85: proceedings of a workshop on the theory and application of cryptographic techniques (EUROCRYPT '85) held in Linz, April 1985*, Lecture Notes in Computer Science **219**, Springer-Verlag. ISBN 3–540–16468–5. MR 87d:94003. See [Rivest and Shamir 1986].

[Rivest and Shamir 1986] Ronald L. Rivest and Adi Shamir, "Efficient factoring based on partial information", pp. 31–34 in [Pichler 1986]. MR 851 581. Citations in this document: §0, §5.

[Silverman 2001] Joseph H. Silverman (editor), *Cryptography and lattices: proceedings of the 1st International Conference (CaLC 2001) held in Providence, RI, March 29–30, 2001*, Lecture Notes in Computer Science **2146**, Springer-Verlag, Berlin. ISBN 3–540–42488–1. MR 2002m:11002. See [Coppersmith 2001], [Howgrave-Graham 2001].

[Sudan 1997] Madhu Sudan, "Decoding of Reed Solomon codes beyond the error-correction bound", *Journal of Complexity* **13**, 180–193. ISSN 0885–064X. MR 98f:94024. URL: http://theory.lcs.mit.edu/~madhu/bib.html. Citations in this document: §0, §7.

[Vallée et al. 1989] Brigitte Vallée, Marc Girault, and Philippe Toffin, "How to guess $\ell$th roots modulo $n$ by reducing lattice bases", pp. 427–442 in [Mora 1989]. MR 90k:11168. URL: http://cr.yp.to/bib/entries.html#1989/vallee. Citations in this document: §0.

[Wiener 1999] Michael Wiener (editor), *Advances in cryptology—CRYPTO '99*, Lecture Notes in Computer Science **1666**, Springer-Verlag, Berlin. ISBN 3–5540–66347–9. MR 2000h:94003. See [Boneh et al. 1999].

DANIEL J. BERNSTEIN
DEPARTMENT OF MATHEMATICS, STATISTICS, AND COMPUTER SCIENCE
M/C 249
THE UNIVERSITY OF ILLINOIS AT CHICAGO
CHICAGO, IL 60607–7045
UNITED STATES
   djb@cr.yp.to

# Computing Arakelov class groups

RENÉ SCHOOF

ABSTRACT. Shanks's infrastructure algorithm and Buchmann's algorithm for computing class groups and unit groups of rings of integers of algebraic number fields are most naturally viewed as computations inside Arakelov class groups. In this paper we discuss the basic properties of Arakelov class groups and of the set of reduced Arakelov divisors. As an application we describe Buchmann's algorithm in this context.

CONTENTS

| | |
|---|---|
| 1. Introduction | 447 |
| 2. The Arakelov class group | 449 |
| 3. Étale $\mathbb{R}$-algebras | 451 |
| 4. Hermitian line bundles and ideal lattices | 452 |
| 5. The oriented Arakelov class group | 456 |
| 6. Metrics on Arakelov class groups | 459 |
| 7. Reduced Arakelov divisors | 464 |
| 8. Quadratic fields | 472 |
| 9. Reduced Arakelov divisors; examples and counterexamples | 474 |
| 10. Computations with reduced Arakelov divisors | 479 |
| 11. A deterministic algorithm | 484 |
| 12. Buchmann's algorithm | 489 |
| Acknowledgements | 493 |
| References | 493 |

## 1. Introduction

Daniel Shanks [1972] observed that the forms in the principal cycle of reduced binary quadratic forms of positive discriminant exhibit a group-like behavior. This was a surprising phenomenon, because the principal cycle itself constitutes the trivial class of the class group. Shanks called this group-like structure 'inside' the neutral element of the class group the *infrastructure*. He exploited it by designing an efficient algorithm to compute the regulator of a real quadratic number field. Later, H. W. Lenstra [1982] (see also [Schoof 1982]) made Shanks' observations more precise, introducing a certain topological group and providing a satisfactory framework for Shanks's algorithm. Both Shanks [1976, Section 1; 1979, 4.4], and Lenstra [1982, section 15] indicated that the infrastructure ideas could be generalized to arbitrary number fields. This was done first by H. Williams and his students [Williams et al. 1983] for complex cubic fields, then by J. Buchmann [1987a; 1987b; 1987c] and by Buchmann and Williams [1989]. Finally Buchmann [1990; 1991] described an algorithm for computing the class group and regulator of an arbitrary number field that, under reasonable assumptions, has a subexponential running time. It has been implemented in the computer algebra packages LiDIA, MAGMA and PARI.

In these expository notes we present a natural setting for the infrastructure phenomenon and for Buchmann's algorithm. It is provided by Arakelov theory [Szpiro 1985, 1987; Van der Geer and Schoof 2000]. We show that Buchmann's algorithm for computing the class number and regulator of a number field $F$ has a natural description in terms of the *Arakelov class group* $\text{Pic}_F^0$ of $F$ and the set $\text{Red}_F$ of *reduced Arakelov divisors*. We show that Lenstra's topological group is essentially equal to the Arakelov class group of a real quadratic field. We also introduce the *oriented Arakelov class group* $\widetilde{\text{Pic}}_F^0$. This is a natural generalization of $\text{Pic}_F^0$, useful for analyzing Buchmann's algorithm and for computing the units of the ring of integers $O_F$ themselves rather than just the regulator.

The main result of this paper is formulated in Theorems 7.4 and 7.7. It says that the finite set $\text{Red}_F$ of reduced Arakelov divisors is, in a precise sense, regularly distributed in the compact Arakelov class groups $\text{Pic}_F^0$ and $\widetilde{\text{Pic}}_F^0$.

In Section 2 we introduce the Arakelov class group of a number field $F$. In Section 3 we study the étale $\mathbb{R}$-algebra $F \otimes_\mathbb{Q} \mathbb{R}$. In Section 4 we discuss the relations between Arakelov divisors, Hermitian line bundles and ideal lattices. In Section 5 we define the oriented Arakelov class group and in Section 6 we give both Arakelov class groups a natural translation invariant Riemannian structure. The rest of the notes is devoted to computational issues. Section 7 contains the main results. Here we introduce *reduced* Arakelov divisors and describe their basic properties. In Section 8, we work out the details for quadratic number fields. In Section 9 we present explicit examples illustrating various properties of reduced divisors. In Section 10 we discuss the computational aspects of reduced Arakelov divisors. In Section 11 we present a *deterministic* algorithm to compute the Arakelov class group. Finally, in Section 12 we present Buchmann's algorithm from the point of view of Arakelov theory. See [Marcus 1977] for the basic properties of algebraic number fields.

## 2. The Arakelov class group

In this section we introduce the Arakelov class group of a number field $F$. This group is analogous to the degree zero subgroup of the Picard group of a complete algebraic curve. In order to have a good analogy with the geometric situation, we formally 'complete' the spectrum of the ring of integers $O_F$ by adjoining primes at infinity. An *infinite* prime of $F$ is a field homomorphism $\sigma : F \longrightarrow \mathbb{C}$, considered up to complex conjugation. An infinite prime $\sigma$ is called *real* when $\sigma(F) \subset \mathbb{R}$ and *complex* otherwise. We let $r_1$ and $r_2$ denote the number of real and complex infinite primes, respectively. We have $r_1 + 2r_2 = n$ where $n = [F : \mathbb{Q}]$.

An *Arakelov divisor* is a formal finite sum $D = \sum_{\mathfrak{p}} n_{\mathfrak{p}}\mathfrak{p} + \sum_{\sigma} x_{\sigma}\sigma$, where $\mathfrak{p}$ runs over the nonzero prime ideals of $O_F$ and $\sigma$ runs over the infinite primes of $F$. The coefficients $n_{\mathfrak{p}}$ are in $\mathbb{Z}$ but the $x_{\sigma}$ can be any number in $\mathbb{R}$. The Arakelov divisors form an additive group, the Arakelov divisor group $\mathrm{Div}_F$. It is isomorphic to $\bigoplus_{\mathfrak{p}} \mathbb{Z} \times \bigoplus_{\sigma} \mathbb{R}$. The *principal* Arakelov divisor associated to an element $f \in F^*$ is the divisor $(f) = \sum_{\mathfrak{p}} n_{\mathfrak{p}}\mathfrak{p} + \sum_{\sigma} x_{\sigma}\sigma$ with $n_{\mathfrak{p}} = \mathrm{ord}_{\mathfrak{p}}(f)$ and $x_{\sigma}(f) = -\log|\sigma(f)|$. The principal Arakelov divisors form a subgroup of $\mathrm{Div}_F$.

Since it is analogous to the Picard group of an algebraic curve, the quotient of $\mathrm{Div}_F$ by its subgroup of principal Arakelov divisors is denoted by $\mathrm{Pic}_F$. A principal Arakelov divisor $(f)$ is trivial if and only if $f$ is a unit of $O_F$ all of whose conjugates have absolute value equal to 1. It follows that $(f)$ is trivial if and only if $f$ is contained in the group of roots of unity $\mu_F$. Therefore there is an exact sequence

$$0 \longrightarrow \mu_F \longrightarrow F^* \longrightarrow \mathrm{Div}_F \longrightarrow \mathrm{Pic}_F \longrightarrow 0.$$

We call $I = \prod_{\mathfrak{p}} \mathfrak{p}^{-n_{\mathfrak{p}}}$ the *ideal associated* to an Arakelov divisor $D = \sum_{\mathfrak{p}} n_{\mathfrak{p}}\mathfrak{p} + \sum_{\sigma} x_{\sigma}\sigma$. The ideal associated to the zero Arakelov divisor is the ring of integers $O_F$. The ideal associated to a principal Arakelov divisor $(f)$ is the principal ideal $f^{-1}O_F$. Here and in the rest of the paper we often call fractional ideals simply 'ideals'. If we want to emphasize that an ideal is integral, we call it an $O_F$-ideal.

The map that sends a divisor $D$ to its associated ideal $I$ is a homomorphism from $\mathrm{Div}_F$ to the group of fractional ideals $\mathrm{Id}_F$ of $F$. Its kernel is the group $\bigoplus_{\sigma} \mathbb{R}$ of divisors supported in the infinite primes. We have the commutative diagram at the top of the next page, the rows and columns of which are exact. In the diagram, $\mathrm{Pid}_F$ denotes the group of principal ideals of $F$. The map $F^*/\mu_F \longrightarrow \mathrm{Div}_F$ induces a homomorphism from $O_F^*/\mu_F$ to $\bigoplus_{\sigma} \mathbb{R}$. This homomorphism is given by $\varepsilon \mapsto (-\log|\sigma(\varepsilon)|)_{\sigma}$ and its cokernel is denoted by $T$.

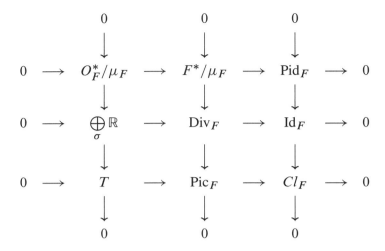

The *norm* $N(\mathfrak{p})$ of a nonzero prime ideal $\mathfrak{p}$ of $O_F$ is the order of its residue field $O_F/\mathfrak{p}$. The *degree* $\deg(\mathfrak{p})$ of $\mathfrak{p}$ is defined as $\log N(\mathfrak{p})$. The degree of an infinite prime $\sigma$ is equal to 1 or 2 depending on whether $\sigma$ is real or complex. The degree extends by linearity to a surjective homomorphism $\deg : \mathrm{Div}_F \longrightarrow \mathbb{R}$. The *norm* $N(D)$ of a divisor $D$ is defined as $N(D) = e^{\deg(D)}$. The divisors of degree 0 form a subgroup $\mathrm{Div}_F^0$ of $\mathrm{Div}_F$. By the product formula, $\mathrm{Div}_F^0$ contains the principal Arakelov divisors.

DEFINITION 2.1. Let $F$ be a number field. The *Arakelov class group* $\mathrm{Pic}_F^0$ of $F$ is the quotient of $\mathrm{Div}_F^0$ by its subgroup of principal divisors.

The degree map $\deg : \mathrm{Div}_F \longrightarrow \mathbb{R}$ factors through $\mathrm{Pic}_F$ and the Arakelov class group is the kernel of the induced homomorphism $\deg : \mathrm{Pic}_F \longrightarrow \mathbb{R}$. We let $(\bigoplus_\sigma \mathbb{R})^0$ denote the subgroup of divisors in $\bigoplus_\sigma \mathbb{R}$ that have degree zero and $T^0$ the cokernel of the homomorphism $O_F^* \longrightarrow (\bigoplus_\sigma \mathbb{R})^0$. In other words, $T^0$ is the quotient of the vector space $\{(v_\sigma)_\sigma \in \bigoplus_\sigma \mathbb{R} : \sum_\sigma \deg(\sigma) v_\sigma = 0\}$ by the group of vectors $\{(\log|\sigma(\varepsilon)|)_\sigma : \varepsilon \in O_F^*\}$. By Dirichlet's unit theorem, $T^0$ is a compact real torus.

PROPOSITION 2.2. *There is a natural exact sequence*
$$0 \longrightarrow T^0 \longrightarrow \mathrm{Pic}_F^0 \longrightarrow Cl_F \longrightarrow 0.$$

PROOF. Since $F$ has at least one infinite prime, the composite map $\mathrm{Div}_F^0 \hookrightarrow \mathrm{Div}_F \longrightarrow \mathrm{Id}_F$ is still surjective. The result now follows by replacing the groups $\mathrm{Div}_F$, $\mathrm{Pic}_F$, $T$ and $\bigoplus_\sigma \mathbb{R}$ in the diagram above by their degree 0 subgroups. □

The group $T^0$ is the connected component of the identity of the topological group $\mathrm{Pic}_F^0$. It follows that $\mathrm{Pic}_F^0$, being an extension of the finite class group by $T^0$, is a compact real Lie group of dimension $r_1 + r_2 - 1$.

DEFINITION 2.3. The natural homomorphism $\mathrm{Div}_F^0 \longrightarrow \mathrm{Id}_F$ admits a section

$$d : \mathrm{Id}_F \longrightarrow \mathrm{Div}_F^0.$$

It is given by $d(I) = D$ where $D = \sum_{\mathfrak{p}} n_{\mathfrak{p}} \mathfrak{p} + \sum_{\sigma} x_{\sigma} \sigma$ is the Arakelov divisor for which $I = \prod_{\mathfrak{p}} \mathfrak{p}^{-n_{\mathfrak{p}}}$ and $x_{\sigma} = (1/n) \log N(I)$ for every infinite prime $\sigma$.

PROPOSITION 2.4. *Let* $\bar{d} : \mathrm{Id}_F \longrightarrow \mathrm{Pic}_F^0$ *denote the homomorphism that maps $I$ to the class of the divisor $d(I)$. Then the sequence*

$$0 \longrightarrow \{f \in F^* : \text{all } |\sigma(f)| \text{ are equal}\}/\mu_F \longrightarrow \mathrm{Id}_F \xrightarrow{\bar{d}} \mathrm{Pic}_F^0$$

*is exact. Moreover, the image of $\bar{d}$ is dense in* $\mathrm{Pic}_F^0$.

This proposition is not used in the rest of the paper. It can be proved along the lines of (and in fact follows immediately from) Proposition 6.4 below. The kernel of $\bar{d}$ is not a very convenient group to work with, and this is one of the reasons for introducing *oriented* Arakelov divisors below.

Finally we remark that there is a natural surjective continuous homomorphism $\mathbf{A}_F^* \longrightarrow \mathrm{Div}_F$ from the idèle group $\mathbf{A}_F^*$ to the Arakelov divisor group. It follows that $\mathrm{Pic}_F$ is a quotient of the idèle class group. We do not make any use of this fact in the rest of the paper.

## 3. Étale $\mathbb{R}$-algebras

Let $F$ be a number field of degree $n$. In this section we study the $\mathbb{R}$-algebra $F_{\mathbb{R}} = F \otimes_{\mathbb{Q}} \mathbb{R}$.

For any infinite prime $\sigma$ of $F$, we write $F_\sigma$ for $\mathbb{R}$ or $\mathbb{C}$ depending on whether $\sigma$ is real or complex. The natural map $F \longrightarrow \prod_\sigma F_\sigma$ sending $f \in F$ to the vector $(\sigma(f))_\sigma$ induces an isomorphism $F_{\mathbb{R}} = F \otimes_{\mathbb{Q}} \mathbb{R} \cong \prod_\sigma F_\sigma$ of $\mathbb{R}$-algebras. Let $u \mapsto \bar{u}$ denote the canonical conjugation of the étale algebra $F_{\mathbb{R}}$. In terms of the isomorphism $F_{\mathbb{R}} \cong \prod_\sigma F_\sigma$, it is simply the morphism that maps a vector $u = (u_\sigma)_\sigma$ to $\bar{u} = (\bar{u}_\sigma)_\sigma$. In these terms it is also easy to describe the set of invariants of the canonical conjugation. It is the subalgebra $\prod_\sigma \mathbb{R}$ of $\prod_\sigma F_\sigma$.

For any $u \in F_{\mathbb{R}}$, we define the norm $N(u)$ and trace $\mathrm{Tr}(u)$ of $u$ as the determinant and trace of an $n \times n$-matrix (with respect to any $\mathbb{R}$-basis) of the $\mathbb{R}$-linear map $F_{\mathbb{R}} \longrightarrow F_{\mathbb{R}}$ given by multiplication by $u$. In terms of coordinates, we have for $u = (u_\sigma)_\sigma \in \prod_\sigma F_\sigma$ that $\mathrm{Tr}(u) = \sum_\sigma \deg(\sigma)\mathrm{Re}(u_\sigma)$ while $N(u) = \prod_{\sigma \text{ real}} u_\sigma \prod_{\sigma \text{ complex}} u_\sigma \bar{u}_\sigma$.

Being an étale $\mathbb{R}$-algebra, $F_{\mathbb{R}}$ admits a canonical Euclidean structure; see for instance [Groenewegen 2001]. It is given by the scalar product

$$\langle u, v \rangle = \mathrm{Tr}(u\bar{v}) \quad \text{for } u, v \in F_{\mathbb{R}}.$$

This scalar product has the 'Hermitian' property $\langle \lambda u, v\rangle = \langle u, \bar{\lambda} v\rangle$ for $u, v, \lambda \in F_{\mathbb{R}}$. In terms of coordinates, we have for $u = (u_\sigma)_\sigma$ and $v = (v_\sigma)_\sigma$ in $F_{\mathbb{R}} \cong \prod_\sigma F_\sigma$ that

$$\langle u, v\rangle = \sum_\sigma \deg(\sigma)\mathrm{Re}(u_\sigma \bar{v}_\sigma).$$

We write $\|u\| = \langle u, u\rangle^{1/2}$ for the *length* of $u \in F_{\mathbb{R}}$. For the element $1 \in F \subset F_{\mathbb{R}}$ we have $\|1\| = \sqrt{n}$. For every $u \in F_{\mathbb{R}}$, all coordinates of the product $u\bar{u} \in \prod_\sigma F_\sigma$ are nonnegative real numbers. We define $|u|$ to be the vector

$$|u| = (|u_\sigma|)_\sigma$$

in the group $\prod_\sigma \mathbb{R}_+^* \subset F_{\mathbb{R}}^*$. Here we let $\mathbb{R}_+^* = \{x \in \mathbb{R}^* : x > 0\}$. We have $|u|^2 = u\bar{u}$. The map $u \mapsto |u|$ is a homomorphism. It is a section of the inclusion map $\prod_\sigma \mathbb{R}_+^* \subset F_{\mathbb{R}}^*$.

PROPOSITION 3.1. *Let $F$ be a number field of degree $n$. For every $u \in F_{\mathbb{R}}$,*

(i) $\qquad\qquad\qquad N(u\bar{u})^{1/n} \leq \frac{1}{n}\mathrm{Tr}(u\bar{u});$

(ii) $\qquad\qquad\qquad |N(u)| \leq n^{-n/2}\|u\|^n.$

*In either case, equality holds if and only if $u$ is contained in the subalgebra $\mathbb{R}$ of $F_{\mathbb{R}}$.*

PROOF. Since all coordinates of $u\bar{u}$ are nonnegative, (i) is just the arithmetic-geometric mean inequality. The second inequality follows from (i) and the fact that $N(\bar{u}) = N(u)$. $\square$

## 4. Hermitian line bundles and ideal lattices

In this section we introduce the Hermitian line bundles and ideal lattices associated to Arakelov divisors and study some of their properties.

Let $F$ be a number field of degree $n$ and let $D = \sum_{\mathfrak{p}} n_{\mathfrak{p}}\mathfrak{p} + \sum_\sigma x_\sigma \sigma$ be an Arakelov divisor. By $I = \prod_{\mathfrak{p}} \mathfrak{p}^{-n_{\mathfrak{p}}}$ we denote the ideal associated to $D$ in Section 2 and by $u$ the unit $(\exp(-x_\sigma))_\sigma \in \prod_\sigma \mathbb{R}_+^* \subset F_{\mathbb{R}}^*$. This leads to the following definition.

DEFINITION 4.1. Let $F$ be a number field. A *Hermitian line bundle* is a pair $(I, u)$ where $I$ is a fractional $F$-ideal and $u$ a unit of the algebra $F_{\mathbb{R}} \cong \prod_\sigma F_\sigma$ all of whose coordinates are positive real numbers.

As we explained above, to every Arakelov divisor $D$ there corresponds a Hermitian line bundle $(I, u)$. This correspondence is bijective and we often identify the two notions. The zero Arakelov divisor corresponds to the trivial bundle $(O_F, 1)$. A principal Arakelov divisor $(f)$ corresponds to the Hermitian line bundle $(f^{-1}O_F, |f|)$ and the divisor $d(I)$ associated to a fractional ideal $I$ at the end of Section 2, corresponds to the pair $(I, N(I)^{-1/n})$. Note that $N(I)^{-1/n}$ is contained in the 'diagonal' subgroup $\mathbb{R}_+^*$ of $\prod_\sigma \mathbb{R}_+^*$. It follows from the formulas for $N(u)$ given in the previous section that the degree of an Arakelov divisor $D = (I, u)$ is equal to $-\log(|N(u)|N(I))$.

DEFINITION 4.2. *Let $F$ be a number field. An* ideal lattice *of $F$ is a projective $O_F$-module $L$ of rank 1 equipped with a real-valued positive definite scalar product on $L \otimes_\mathbb{Z} \mathbb{R}$ satisfying $\langle \lambda x, y \rangle = \langle x, \bar\lambda y \rangle$ for $x, y \in L \otimes_\mathbb{Z} \mathbb{R}$ and $\lambda \in F_\mathbb{R}$. Two ideal lattices $L, L'$ are called* isometric *if there is an $O_F$-isomorphism $L \cong L'$ that is compatible with the scalar products on $L \otimes_\mathbb{Z} \mathbb{R}$ and $L' \otimes_\mathbb{Z} \mathbb{R}$.*

Here $\lambda \mapsto \bar\lambda$ is the canonical algebra involution of the étale $\mathbb{R}$-algebra $F_\mathbb{R}$ introduced in Section 3. Note that it need not preserve $F$. Note also that $L \otimes_\mathbb{Z} \mathbb{R}$ has the structure of an $F_\mathbb{R}$-module. See [Bayer-Fluckiger 1999; Groenewegen 2001] for more on ideal lattices. There is a natural way to associate an ideal lattice to an Arakelov divisor $D$. It is most naturally expressed in terms of the Hermitian line bundle $(I, u)$ associated to $D$. The $O_F$-module $I$ is projective and of rank 1. Multiplication by $u$ gives an $O_F$-isomorphism with $uI = \{ux : x \in I\} \subset F_\mathbb{R}$. The canonical scalar product on $F_\mathbb{R}$ introduced in Section 3 gives $uI$ the structure of an ideal lattice. Alternatively, putting

$$\|f\|_D = \|uf\|, \quad \text{for } f \in I,$$

we obtain a scalar product on $I$ itself that we extend by linearity to $I \otimes_\mathbb{Z} \mathbb{R}$. In additive notation, if $f \in I$ and $u \in F_\mathbb{R}^*$ is equal to $\exp((-x_\sigma)_\sigma)$, then $uf$ is equal to the vector $(\sigma(f)e^{-x_\sigma})_\sigma \in F_\mathbb{R}$ and we have $\|f\|_D^2 = \|uf\|^2 = \sum_\sigma \deg(\sigma)|\sigma(f)e^{-x_\sigma}|^2$ for $f \in I$.

The ideal lattice corresponding to the zero Arakelov divisor, i.e. to the trivial bundle $(O_F, 1)$, is the ring of integers $O_F$ viewed as a subset of $F \subset F_\mathbb{R}$ equipped with its canonical Euclidean structure. The covolume of this lattice is equal to $\sqrt{|\Delta_F|}$, where $\Delta_F$ denotes the discriminant of the number field $F$. The covolume of the lattice associated to an arbitrary divisor $D = (I, u)$ is equal to

$$\operatorname{covol}(D) = \sqrt{|\Delta_F|}\,N(I)|N(u)| = \sqrt{|\Delta_F|}/N(D) = \sqrt{|\Delta_F|}\,e^{-\deg(D)}.$$

For any ideal $I$, the lattice associated to the Arakelov divisor

$$d(I) = (I, N(I)^{-1/n})$$

can be thought of as the lattice $I \subset F \subset F_\mathbb{R}$ equipped with the canonical scalar product of $F_\mathbb{R}$, but *scaled* with a factor $N(I)^{-1/n}$ so that its covolume is equal to $\sqrt{|\Delta_F|}$.

PROPOSITION 4.3. *Let $F$ be a number field of discriminant $\Delta_F$.*

(i) *The map that associates the ideal lattice $uI$ to an Arakelov divisor $D = (I, u)$, induces a bijection between the group $\text{Pic}_F$ and the set of isometry classes of ideal lattices.*

(ii) *The same map induces a bijection between the group $\text{Pic}_F^0$ and the set of isometry classes of ideal lattices of covolume $\sqrt{|\Delta_F|}$.*

PROOF. Let $D = (I, u)$ be an Arakelov divisor and let $D' = D + (g)$ for some $g \in F^*$. Then $D' = (g^{-1}I, u|g|)$ and multiplication by $g$ induces an isomorphism $g^{-1}I \cong I$ of $O_F$-modules. This map is also an isometry between the associated lattices since

$$\|g^{-1}f\|_{D'} = \|u|g|g^{-1}f\| = \|uf\| = \|f\|_D$$

for all $f \in I \otimes_{\mathbb{Z}} \mathbb{R}$. Here we have used the fact that $v = |g|g^{-1}$ satisfies $v\bar{v} = 1$ and that therefore $\|vh\| = \text{Tr}(vh\bar{v}\bar{h}) = \text{Tr}(h\bar{h}) = \|h\|$ for all $h \in I \otimes_{\mathbb{Z}} \mathbb{R}$. We conclude that the map that sends an Arakelov divisor to its associated ideal lattice induces a well defined map from $\text{Pic}_F$ to the set of isometry classes of ideal lattices. This map is *injective*. Indeed, if $D = (I, u)$ and $D' = (I', u')$ give rise to isometric lattices, then there exists $g \in F^*$ so that $I' = gI$ and $\|gf\|_{D'} = \|f\|_D$ for all $f \in I \otimes_{\mathbb{Z}} \mathbb{R}$. This means that $\|u'gf\| = \|uf\|$ for all $f \in I \otimes_{\mathbb{Z}} \mathbb{R} = F_{\mathbb{R}}$. For any infinite prime $\sigma$, we let $e_\sigma \in F_{\mathbb{R}}$ be the idempotent for which $\sigma(e_\sigma) = 1$ while $\sigma'(e_\sigma) = 0$ for all $\sigma' \neq \sigma$. Substituting $f = e_\sigma$, we find that $|\sigma(g)u'_\sigma| = |u_\sigma|$ for every $\sigma$. It follows that $|g| = u/u'$, implying that $D' = D + (g)$ as required.

To see that the map is *surjective*, consider an ideal lattice $L$ with Hermitian scalar product $\langle\langle -, - \rangle\rangle$ on $L \otimes_{\mathbb{Z}} \mathbb{R} = F_{\mathbb{R}}$. We may assume that $L$ is actually an $O_F$-ideal. The idempotent elements $e_\sigma$ in $F_{\mathbb{R}} \cong \prod_\sigma F_\sigma$ are invariant under the canonical involution. This implies that the $e_\sigma$ are pairwise orthogonal because $\langle\langle e_\sigma, e_{\sigma'} \rangle\rangle = \langle\langle e_\sigma^2, e_{\sigma'} \rangle\rangle = \langle\langle e_\sigma, e_\sigma e_{\sigma'} \rangle\rangle = 0$. Therefore the real numbers $u_\sigma = \langle\langle e_\sigma, e_\sigma \rangle\rangle^{1/2}$ determine the metric on $I \otimes_{\mathbb{Z}} \mathbb{R}$. The Arakelov divisor $(L, u)$ with $u = (u_\sigma)_\sigma \in \prod_\sigma \mathbb{R}_+^*$ is then mapped to the isometry class of $L$.

This proves (i). Part (ii) follows immediately from this. □

The following proposition deals with the lengths of the shortest nonzero vectors in the lattices associated to Arakelov divisors.

PROPOSITION 4.4. *Let $F$ be a number field of degree $n$ and let $D = (I, u)$ be an Arakelov divisor.*

(i) *For every nonzero $f$ in $I$ we have*

$$\|f\|_D \geq \sqrt{n}e^{-\frac{1}{n}\deg D}.$$

Moreover, equality holds if and only if $D = (fO_F, \lambda |f|^{-1})$ for some $\lambda > 0$. In other words, if and only if $D$ is equal to the principal Arakelov divisor $-(f)$, scaled by a positive factor $\lambda$.

(ii) There exists a nonzero $f \in I$ such that $|u_\sigma \sigma(f)| < (2/\pi)^{r_2/n} \operatorname{covol}(D)^{1/n}$ for every $\sigma$ and hence

$$\|f\|_D \leq \sqrt{n} \cdot (2/\pi)^{r_2/n} \operatorname{covol}(D)^{1/n}.$$

Here $r_2$ is the number of complex primes of $F$.

PROOF. (i) Take $f \in I$. By Proposition 3.1 we have

$$\|f\|_D^2 = \|uf\|^2 \geq n|N(uf)|^{2/n}.$$

Since $|N(f)| \geq N(I)$ we find that

$$\|f\|_D^2 \geq n|N(u)N(I)|^{2/n} = ne^{-(2/n)\deg(D)}.$$

The last inequality follows from the fact that $\deg(D) = -\log|N(u)N(I)|$. This proves the first statement. By Proposition 3.1, equality holds if and only if all $|u_\sigma \sigma(f)|$ are equal to some $\lambda > 0$ and if $I$ is the principal ideal generated by $f$. This implies that $D$ is of the form $(f^{-1}O_F, |f|^{-1}\lambda)$ as required.

(ii) Consider the set $V = \{(y_\sigma)_\sigma \in F_{\mathbb{R}} : |y_\sigma| \leq (2/\pi)^{r_2/n} \operatorname{covol}(D)^{1/n}$ for all $\sigma\}$. This is a bounded symmetric convex set of volume

$$2^{r_1}(2\pi)^{r_2}(2/\pi)^{r_2}\operatorname{covol}(D) = 2^n \operatorname{covol}(D).$$

By Minkowski's Convex Body Theorem there exists a nonzero element $f \in I$ for which $(u_\sigma \sigma(f))_\sigma \in uI \subset F_{\mathbb{R}}$ is in $V$. This implies (ii). □

We mention the following special case of the proposition.

COROLLARY 4.5. *Let $D = (I, u)$ be an Arakelov divisor of degree 0. Then any nonzero $f \in I$ has the property that $\|f\|_D \geq \sqrt{n}$, with equality if and only if $D = -(f)$. On the other hand, there exists a nonzero $f \in I$ with*

$$\|f\|_D \leq \sqrt{n}\,(2/\pi)^{r_2/n}\sqrt{|\Delta_F|}^{1/n}.$$

Proposition 4.4(i) says that the lattices $uI$ associated to Arakelov divisors $D = (I, u)$ are rather 'nice'. They are not very skew in the sense that they do not contain any nonzero vectors that are extremely short with respect to $\operatorname{covol}(D)^{1/n}$. This property can be expressed by means of the *Hermite constant* $\gamma(D) = \gamma(uI)$. The latter is defined as the square of the length of the shortest nonzero vector in the lattice $uI$ associated to $D$ divided by $\operatorname{covol}(D)^{2/n}$. The more skew the lattice, the smaller is its Hermite constant. The constant $\gamma(D)$ only depends on the class of $D$ in $\operatorname{Pic}_F$.

COROLLARY 4.6. *Let $F$ be a number field of degree $n$ and let $D = (I, u)$ be an Arakelov divisor. Then*

$$\frac{n}{|\Delta_F|^{1/n}} \leq \gamma(D) \leq n\left(\frac{2}{\pi}\right)^{2r_2/n}.$$

*The lower bound is attained if and only if $D$ is a principal divisor scaled by some $\lambda > 0$ as in Proposition 4.4(i).*

The function $h^0(D) = \log(\sum_{f \in I} \exp(-\pi \|f\|_D^2))$ introduced in [Van der Geer and Schoof 2000] and briefly discussed in Section 10 is related to the Hermite constant $\gamma(D)$. Indeed, for most Arakelov divisors $D = (I, u)$ the shortest nonzero vectors in the associated lattice are equal to products of a root of unity by one fixed shortest vector. Moreover, for most $D$ the contributions of the zero vector and these vectors contribute the bulk to the infinite sum $\sum_{f \in I} \exp(-\pi \|f\|_D^2)$. Therefore, for most Arakelov divisors $D$ the quantity $(h^0(D) - 1)/w_F$ is close to $\exp(-\pi \gamma(D) \operatorname{covol}(D)^{2/n})$. Here $w_F$ denotes the number of roots of unity in the field $F$.

## 5. The oriented Arakelov class group

In this section we introduce the oriented Arakelov divisor group $\widetilde{\operatorname{Pic}}_F$ associated to a number field $F$.

In Section 4 we have associated to an Arakelov divisor $D$ a Hermitian line bundle $(I, u)$. Here $I$ is an ideal and $u$ is a unit in the *subgroup* $F_{\mathbb{R},+}^* = \prod_\sigma \mathbb{R}_+^*$ of $F_\mathbb{R}^*$. An *oriented* Hermitian line bundle is a pair $(I, u)$ where $I$ is an ideal and $u$ is an *arbitrary* unit in $F_\mathbb{R}^* \cong \prod_\sigma F_\sigma^*$. The corresponding *oriented Arakelov divisors* are formal sums $\sum_\mathfrak{p} n_\mathfrak{p} \mathfrak{p} + \sum_\sigma x_\sigma \sigma$ with $n_\mathfrak{p} \in \mathbb{Z}$ and $x_\sigma \in F_\sigma^*$. They form a group $\widetilde{\operatorname{Div}}_F$ and we have

$$\widetilde{\operatorname{Div}}_F \cong \operatorname{Id}_F \times F_\mathbb{R}^* \cong \bigoplus_\mathfrak{p} \mathbb{Z} \times \prod_\sigma F_\sigma^*.$$

The *principal* oriented Arakelov divisor associated to $f \in F^*$ is simply the oriented divisor corresponding to the oriented Hermitian bundle $(f^{-1} O_F, f)$, where the second coordinate $f$ is viewed as an element of $F_\mathbb{R}^*$. The cokernel of the injective homomorphism $F^* \longrightarrow \widetilde{\operatorname{Div}}_F$ is denoted by $\widetilde{\operatorname{Pic}}_F$. The inclusion $\operatorname{Div}_F \subset \widetilde{\operatorname{Div}}_F$ admits the natural section $\widetilde{\operatorname{Div}}_F \longrightarrow \operatorname{Div}_F$ given by $(I, u) \mapsto (I, |u|)$. The degree $\deg(D)$ of an oriented Arakelov divisor $D = (I, u)$ is by definition the degree of the 'ordinary' Arakelov divisor $(I, |u|)$. In this way principal oriented Arakelov divisors have degree 0.

DEFINITION 5.1. The quotient of the group $\widetilde{\mathrm{Div}}_F^0$ of oriented Arakelov divisors of degree 0 by the subgroup of principal divisors is called the *oriented Arakelov class group*. It is denoted by $\widetilde{\mathrm{Pic}}_F^0$.

The commutative diagram below has exact rows and columns. The bottom row relates the groups $\widetilde{\mathrm{Pic}}_F^0$ and $\mathrm{Pic}_F^0$ to one another.

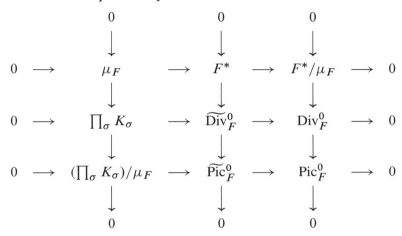

Here $K_\sigma$ denotes the maximal compact subgroup of $F_\sigma^*$. In other words $K_\sigma = \{1, -1\}$ if $\sigma$ is real, while $K_\sigma = \{z \in \mathbb{C}^* : |z| = 1\}$ if $\sigma$ is complex. Since $\mathrm{Pic}_F^0$ and the groups $K_\sigma$ are compact, it follows from the exactness of the bottom row of the diagram that $\widetilde{\mathrm{Pic}}_F^0$ is compact as well.

In order to see the topological structure of $\widetilde{\mathrm{Pic}}_F^0$ better, we construct a second exact sequence. Let $F_{\mathbb{R},\mathrm{conn}}^*$ denote the connected component of $1 \in F_{\mathbb{R}}^*$. It is isomorphic to a product of copies of $\mathbb{R}_+^*$ for the real primes and $F_\sigma^* = \mathbb{C}^*$ for the complex ones. It is precisely the kernel of the homomorphism

$$\widetilde{\mathrm{Div}}_F \longrightarrow \mathrm{Id}_F \times \prod_{\sigma \text{ real}} \{\pm 1\},$$

given by mapping $D = (I, u)$ to $(I, \mathrm{sign}(u))$. Here $\mathrm{sign}(u)$ denotes the vector $(\mathrm{sign}(u_\sigma))_{\sigma \text{ real}}$.

DEFINITION 5.2. By $\widetilde{T}$ we denote the quotient of the group $F_{\mathbb{R},\mathrm{conn}}^*$ by its subgroup $O_{F,+}^* = \{\varepsilon \in O_F^* : \sigma(\varepsilon) > 0 \text{ for all real } \sigma\}$. Taking degree zero subgroups, we put

$$(F_{\mathbb{R},\mathrm{conn}}^*)^0 = \{u \in F_{\mathbb{R},\mathrm{conn}}^* : N(u) = 1\} \quad \text{and} \quad \widetilde{T}^0 = (F_{\mathbb{R},\mathrm{conn}}^*)^0 / O_{F,+}^*.$$

The map $\widetilde{T} \longrightarrow \widetilde{\mathrm{Pic}}_F$ given by $v \mapsto (O_F, v)$ is a well defined homomorphism. So is the map $\widetilde{\mathrm{Pic}}_F \longrightarrow Cl_{F,+}$ that sends the class of the divisor $(I, u)$ to the narrow ideal class of $gI$ where $g \in F^*$ is any element for which $\mathrm{sign}(g) =$

sign($u$). Here the *narrow* ideal class group $Cl_{F,+}$ is defined as the group of ideals modulo the principal ideals that are generated by $f \in F_+^* = \{f \in F^* : \sigma(f) > 0 \text{ for all real } \sigma\}$. It is a finite group.

The following proposition says that the groups $\widetilde{T}$ and $\widetilde{T}^0$ are the connected components of identity of $\widetilde{\mathrm{Pic}}_F$ and $\widetilde{\mathrm{Pic}}_F^0$ respectively. It provides an analogue to Proposition 2.2.

PROPOSITION 5.3. *Let $F$ be a number field of degree $n$.*

(i) *The natural sequences*

$$0 \longrightarrow \widetilde{T} \longrightarrow \widetilde{\mathrm{Pic}}_F \longrightarrow Cl_{F,+} \longrightarrow 0$$

*and*

$$0 \longrightarrow \widetilde{T}^0 \longrightarrow \widetilde{\mathrm{Pic}}_F^0 \longrightarrow Cl_{F,+} \longrightarrow 0$$

*are exact.*

(ii) *The groups $\widetilde{T}$ and $\widetilde{T}^0$ are the connected components of identity of $\widetilde{\mathrm{Pic}}_F$ and $\widetilde{\mathrm{Pic}}_F^0$ respectively. The group $\widetilde{T}$ has dimension $n$ while $\widetilde{T}^0$ is a compact torus of dimension $n - 1$.*

PROOF. (i) Let $\widetilde{\mathrm{Pid}}_F$ denote the image of the map

$$F^* \longrightarrow \widetilde{\mathrm{Div}}_F \longrightarrow \mathrm{Id}_F \times \prod_{\sigma \text{ real}} \{\pm 1\}.$$

This leads to a commutative diagram with exact rows:

$$\begin{array}{ccccccccc} 0 & \longrightarrow & O_{F,+}^* & \longrightarrow & F^* & \longrightarrow & \widetilde{\mathrm{Pid}}_F & \longrightarrow & 0 \\ & & \downarrow & & \downarrow & & \downarrow & & \\ 0 & \longrightarrow & F_{\mathbb{R},\mathrm{conn}}^* & \longrightarrow & \widetilde{\mathrm{Div}}_F & \longrightarrow & \mathrm{Id}_F \times \prod_{\sigma \text{ real}} \{\pm 1\} & \longrightarrow & 0, \end{array}$$

where the vertical maps are all injective. An application of the snake lemma shows the sequence of cokernels to be exact: this is the first exact sequence of (i). Indeed, the kernel of the surjective homomorphism $\mathrm{Id}_F \times \prod_{\sigma \text{ real}} \{\pm 1\} \to Cl_{F,+}$ given by mapping a pair $(I, s)$ to the narrow ideal class of $gI$, where $g \in F^*$ is any element for which $\mathrm{sign}(g) = \mathrm{sign}(s)$, is precisely equal to $\widetilde{\mathrm{Pid}}_F$. The second exact sequence is obtained by taking degree-zero parts.

(ii) Since $Cl_{F,+}$ is finite and both groups $\widetilde{T}$ and $\widetilde{T}^0$ are connected, the first statement is clear. Since the Lie group $F_{\mathbb{R},\mathrm{conn}}^*$ has dimension $n$, so do the groups $\widetilde{T}$ and $\widetilde{\mathrm{Pic}}_F$. It follows that the groups $\widetilde{T}^0$ and $\widetilde{\mathrm{Pic}}_F^0$ have dimension $n - 1$. □

The classes of two oriented Arakelov divisors $(I, u)$ and $(J, v)$ are on the same connected component of $\widetilde{\mathrm{Pic}}_F^0$ if and only if $J = gI$ for some $g \in F^*$ for which $u_\sigma v_\sigma \sigma(g) > 0$ for each real $\sigma$.

DEFINITION 5.4. An *embedded ideal lattice* is an ideal lattice $L$ together with an $O_F$-linear isometric embedding $L \hookrightarrow F_\mathbb{R}$. To every oriented Arakelov divisor $D = (I, u)$ we associate the ideal lattice $uI$ together with the embedding $uI \subset F_\mathbb{R}$. Two embedded ideal lattices are called isometric if there is an isometry of ideal lattices that commutes with the embeddings. We have the following analogue of Proposition 4.3.

PROPOSITION 5.5. *Let $F$ be a number field of discriminant $\Delta_F$.*

(i) *The map that associates to an oriented Arakelov divisor $D = (I, u)$ its associated embedded ideal lattice, induces a bijection between the oriented Arakelov class group $\widetilde{\mathrm{Pic}}_F$ and the set of isometry classes of embedded ideal lattices.*

(ii) *The same map induces a bijection between $\widetilde{\mathrm{Pic}}^0_F$ and the set of isometry classes of embedded ideal lattices of covolume $\sqrt{|\Delta_F|}$.*

PROOF. If two oriented Arakelov divisors $D = (I, u)$ and $D' = (I', u')$ differ by a principal divisor $(f^{-1}O_F, f)$, then multiplication by $f$ induces an isometry between the embedded lattices $uI$ and $u'I'$. Therefore the map in (i) is well defined. If the embedded lattices $uI$ and $u'I'$ are isometric, then this isometry is given by multiplication by some $x \in F_\mathbb{R}^*$. Then $f = u^{-1}xu'$ is contained in $F^*$ and we have $D - D' = (f^{-1}O_F, f)$. This shows that the map is injective. To see that the map is surjective, let $I$ be a fractional ideal and let $\iota : I \hookrightarrow F_\mathbb{R}$ be an $O_F$-linear embedding. Tensoring $I$ with $\mathbb{R}$, we obtain an $F_\mathbb{R}$-linear isomorphism $F_\mathbb{R} \cong I \otimes_\mathbb{Z} \mathbb{R} \longrightarrow F_\mathbb{R}$, which is necessarily multiplication by some $u \in F_\mathbb{R}^*$. Therefore $\iota(I) = uI$ and the oriented divisor $(I, u)$ maps to the embedded ideal lattice $\iota : I \hookrightarrow F_\mathbb{R}$. □

We will not use this in the rest of the paper, but note that there is a natural surjective continuous homomorphism from the idèle group $\mathbf{A}_F^*$ to the oriented Arakelov divisor group $\widetilde{\mathrm{Div}}_F$. It follows that the group $\widetilde{\mathrm{Pic}}_F$ is a quotient of the idèle class group.

## 6. Metrics on Arakelov class groups

Let $F$ be a number field. In this section we provide the Arakelov class groups $\mathrm{Pic}_F$ and $\widetilde{\mathrm{Pic}}_F$ with translation invariant Riemannian structures.

By the diagram in Section 2, the connected component $T$ of the group $\mathrm{Pic}_F$ is isomorphic to $\bigoplus_\sigma \mathbb{R}$ modulo the closed discrete subgroup $\Lambda = \{(\log |\sigma(\varepsilon)|)_\sigma : \varepsilon \in O_F^*\}$. Therefore the tangent space at 0 is isomorphic to $\bigoplus_\sigma \mathbb{R}$. Identifying this vector space with the subalgebra $\prod_\sigma \mathbb{R}$ of $F_\mathbb{R} = \prod_\sigma F_\sigma$, it inherits the canonical scalar product from $F_\mathbb{R}$. Since this $\mathbb{R}$-valued scalar product is positive definite, both groups $T$ and $\text{Pic}_F$ are in this way equipped with a translation invariant Riemannian structure.

For $u \in \prod_\sigma \mathbb{R}_+^* \subset F_\mathbb{R}^*$ we let $\log u$ denote the element $(\log \sigma(u))_\sigma \in \prod_\sigma \mathbb{R} \subset F_\mathbb{R}$. We have

$$\|\log u\|^2 = \sum_\sigma \deg(\sigma) |\log \sigma(u)|^2.$$

DEFINITION 6.1. For $u \in T$ we put

$$\|u\|_{\text{Pic}} = \min_{\substack{u' \in F_{\mathbb{R},+}^* \\ u' \equiv u \pmod{\Lambda}}} \|\log u'\| = \min_{\varepsilon \in O_F^*} \|\log(|\varepsilon|u)\|.$$

Every divisor class in $T$ is represented by a divisor of the form $D = (O_F, u)$ for some $u \in \bigoplus_\sigma \mathbb{R}_+^*$. Here $u$ is unique up to multiplication by units $\varepsilon \in O_F^*$. For such a divisor class in $T$ we define

$$\|D\|_{\text{Pic}} = \|u\|_{\text{Pic}}.$$

The function $\|u\|_{\text{Pic}}$ on $T$ satisfies the triangle inequality. It gives rise to a distance function that induces the natural topology of $\text{Pic}_F$. The distance is only defined for divisor classes $D$ and $D'$ that lie on the same connected component. By Proposition 2.2, the class of the difference $D - D'$ is then equal to $(O_F, u)$ for some unique $u \in T$ and we define the *distance* $\|D - D'\|_{\text{Pic}}$ between $D$ and $D'$ as $\|u\|_{\text{Pic}}$. The closed subgroups $T^0$ and $\text{Pic}_F^0$ inherit Riemannian structures from $\text{Pic}_F$.

The Euclidean structures of the ideal lattices corresponding to Arakelov divisors and the metric on $\text{Pic}_F$ are not unrelated. The following proposition says that the difference between the Euclidean structures of two Arakelov divisors $D, D'$ is bounded in terms of $\|D - D'\|_{\text{Pic}}$.

PROPOSITION 6.2. *Let $F$ be a number field and let $D = (I, u)$ and $D' = (I, u')$ be two Arakelov divisors. Then there exists a unit $\varepsilon \in O_F^*$ for which the divisor $D'' = (I, u'|\varepsilon|)$ satisfies*

$$e^{-\|D-D''\|_{\text{Pic}}} \leq \frac{\|x\|_D}{\|x\|_{D''}} \leq e^{\|D-D''\|_{\text{Pic}}}, \qquad \text{for every } x \in I.$$

*The classes of $D'$ and $D''$ in $\text{Pic}_F$ are the same, so $\|D - D'\|_{\text{Pic}} = \|D - D''\|_{\text{Pic}}$.*

PROOF. Let $\varepsilon \in O_F^*$ be such that the expression $\sum_\sigma \deg(\sigma) |\log(|\sigma(\varepsilon)|u'_\sigma/u_\sigma)|^2$ is minimal. Let $D'' = (I, |\varepsilon|u')$. Putting $v = u'|\varepsilon|/u$ we have as a consequence

$$\|D - D''\|_{\text{Pic}}^2 = \sum_\sigma \deg(\sigma) |\log v_\sigma|^2.$$

For any $x \in I$ we have

$$\|x\|_{D''}^2 = \|u'\varepsilon x\|^2 = \|vux\|^2 = \sum_\sigma \deg(\sigma)|u_\sigma v_\sigma \sigma(x)|^2$$

$$\leq \max_\sigma |v_\sigma|^2 \sum_\sigma \deg(\sigma)|u_\sigma \sigma(x)|^2 = \left(\max_\sigma |v_\sigma|\right)^2 \|x\|_D^2.$$

Since

$$\log \max_\sigma |v_\sigma| = \max_\sigma \log |v_\sigma| \leq \max_\sigma |\log |v_\sigma|| \leq \|D - D''\|_{\text{Pic}},$$

the first inequality follows. The second follows by symmetry. The last line of the proposition is clear. □

We now define a similar metric on the *oriented* Arakelov class group. By Proposition 5.3, the connected component of $\widetilde{\text{Pic}}_F$ is $\widetilde{T} = F_{\mathbb{R},\text{conn}}^* / O_{F,+}^*$. We recall that $F_{\mathbb{R},\text{conn}}^*$ is the connected component of identity of the group $F_{\mathbb{R}}^*$. It is isomorphic to a product of copies of $\mathbb{R}_+^*$, one for each real prime, and of $\mathbb{C}^*$, one for each complex prime. The group $O_{F,+}^*$ is the subgroup of $\varepsilon \in O_F^*$ for which $\sigma(\varepsilon) > 0$ for every real infinite prime $\sigma$.

The exponential homomorphism $\exp : F_{\mathbb{R}} \longrightarrow F_{\mathbb{R}}^*$ is defined in terms of the usual exponential function by $\exp(u) = (\exp(u_\sigma))_\sigma$ for $u = (u_\sigma) \in F_{\mathbb{R}} \cong \prod_\sigma F_\sigma$. The image of the exponential function is precisely the group $F_{\mathbb{R},\text{conn}}^*$. The preimage of $O_{F,+}^*$ is a discrete closed subgroup $\Lambda$ of $F_{\mathbb{R}}$. We have a natural isomorphism of Lie groups

$$\exp : F_{\mathbb{R}}/\Lambda \xrightarrow{\cong} F_{\mathbb{R},\text{conn}}^* / O_{F,+}^* = \widetilde{T}.$$

Therefore the tangent space of $\widetilde{T}$ at 0 is isomorphic to $F_{\mathbb{R}}$. The canonical scalar product on $F_{\mathbb{R}}$ provides both groups $\widetilde{T}$ and $\widetilde{\text{Pic}}$ with a translation invariant Riemannian structure.

DEFINITION 6.3. For $u \in \widetilde{T}$ we put

$$\|u\|_{\widetilde{\text{Pic}}} = \min_{\substack{y \in F_{\mathbb{R}} \\ \exp(y) \equiv u \pmod{O_{F,+}^*}}} \|y\|$$

Explicitly, for $u \in F_{\mathbb{R}}^* = \prod_\sigma F_\sigma^*$ we let $\log u$ denote the element $(\log(\sigma(u)))_\sigma \in \prod_\sigma F_\sigma \subset F_{\mathbb{R}}$. Here we use the principal branch of the complex logarithm. We have

$$\|u\|_{\widetilde{\text{Pic}}}^2 = \min_{\varepsilon \in O_{F,+}^*} \|\log(\varepsilon u)\|^2 = \min_{\varepsilon \in O_{F,+}^*} \sum_\sigma \deg(\sigma) \left|\log \sigma(\varepsilon u)\right|^2.$$

Every divisor class in $\widetilde{T}$ can be represented by a divisor of the form $D=(O_F, u)$ for some $u \in F^*_{\mathbb{R},\text{conn}}$. Here $u$ is unique up to multiplication by units $\varepsilon \in O^*_{F,+}$. For any divisor $D$ of the form $(O_F, u)$ with $u \in \widetilde{T}$ we define

$$\|D\|_{\widetilde{\text{Pic}}} = \|u\|_{\widetilde{\text{Pic}}}.$$

The function $\|u\|_{\widetilde{\text{Pic}}}$ on $\widetilde{T}$ satisfies the triangle inequality and this gives rise to a distance function that induces the natural topology on $\widetilde{\text{Pic}}_F$. The distance is only defined for divisor classes $D$ and $D'$ that lie on the same connected component. By Proposition 5.3, the class of the difference $D - D'$ is then equal to $(O_F, u)$ for some unique $u \in \widetilde{T}$ and we define the *distance* $\|D - D'\|_{\widetilde{\text{Pic}}}$ between $D$ and $D'$ as $\|u\|_{\widetilde{\text{Pic}}}$.

The closed subgroups $\widetilde{T}^0$ and $\widetilde{\text{Pic}}^0_F$ inherit Riemannian structures from $\text{Pic}_F$. We leave to the reader the task of proving an "oriented" version of Proposition 6.2.

The morphism $d : \text{Id}_F \longrightarrow \widetilde{\text{Div}}^0_F$ given by $d(I) = (I, N(I)^{-1/n})$ is a section of the natural map $\widetilde{\text{Div}}^0_F \longrightarrow \text{Id}_F$. The embedded ideal lattice associated to $d(I)$ is the ideal lattice $I \subset F_{\mathbb{R}}$ scaled by a factor $N(I)^{-1/n}$. This lattice has covolume $\sqrt{|\Delta_F|}$.

Next we prove an oriented version of Proposition 2.4. It says that the classes of the divisors of the form $d(I)$ are dense in $\widetilde{\text{Pic}}^0_F$ and it implies Proposition 2.4. The exactness of the first sequence of [Lenstra 1982, Section 9] is a special case.

PROPOSITION 6.4. *Let $F$ be a number field of degree $n$. Let $\bar{d} : \text{Id}_F \longrightarrow \widetilde{\text{Pic}}^0_F$ be the map that sends $I$ to the class of the oriented Arakelov divisor $d(I)$ in $\widetilde{\text{Pic}}^0_F$. Then the sequence*

$$0 \longrightarrow \text{Id}_{\mathbb{Q}} \longrightarrow \text{Id}_F \xrightarrow{\bar{d}} \widetilde{\text{Pic}}^0_F$$

*is exact. The image of the map $\bar{d}$ is dense in $\widetilde{\text{Pic}}^0_F$.*

PROOF. Every ideal in $\text{Id}_{\mathbb{Q}}$ is generated by some $f \in \mathbb{Q}^*_{>0}$. Let $f \in \mathbb{Q}^*_{>0}$. Then $\bar{d}$ maps the $F$-ideal $fO_F$ to the class of the oriented Arakelov divisor $(fO_F, |N(f)|^{-1/n})$. Since $|N(f)| = |f|^n$, this divisor is equal to $(fO_F, f^{-1})$. Therefore its image in $\widetilde{\text{Pic}}^0_F$ is trivial.

Conversely, suppose that a fractional ideal $I$ has the property that the class of $(I, N(I)^{-1/n})$ is trivial in $\widetilde{\text{Pic}}^0_F$. This means that $I = fO_F$ for some $f \in F^*$ and that $f = N(I)^{1/n}$. In other words, $\sigma(f) = N(I)^{1/n}$ for all infinite primes $\sigma$. Thus all conjugates of $f$ are equal, so that $f \in \mathbb{Q}^*$. This shows that the sequence is exact.

To show that the image of $\bar{d}$ is dense, we let $0 < \varepsilon < 1$ and pick $D = (I, u) \in \widetilde{\mathrm{Div}}^0_F$. Note that $N(I)|N(u)| = 1$. Consider the set

$$B = \{(v_\sigma)_\sigma \in F_\mathbb{R} : |v_\sigma - u_\sigma| < \varepsilon |u_\sigma| \text{ for all } \sigma\}.$$

Then $B$ is a an open subset of $F_\mathbb{R}^*$ and all $v \in B$ have the same signature as $u$. Since $F$ is dense in $F_\mathbb{R}$, there is an element $f \in B \cap F$.

The difference between $d(fI)$ and the divisor $D$ is equal to

$$(fO_F, N(fI)^{-1/n} u^{-1}).$$

Since $N(u)N(I) = 1$, this is equivalent to the Arakelov divisor $(O_F, v)$, where

$$v = N(f/u)^{-1/n} u^{-1} f \in F_\mathbb{R}^*.$$

Therefore the distance between $D$ and $\bar{d}(f)$ is $\|v\|_{\widetilde{\mathrm{Pic}}}$. Since

$$\left| \frac{\sigma(f)}{u_\sigma} - 1 \right| < \varepsilon,$$

it follows from the Taylor series expansion of the principal branch of the logarithm that

$$\left| \log \frac{\sigma(f)}{u_\sigma} \right| < \frac{\varepsilon}{1-\varepsilon}$$

for all $\sigma$ and hence

$$\frac{1}{n} \left| \log \frac{N(f)}{N(u)} \right| < \frac{\varepsilon}{1-\varepsilon}.$$

It follows that

$$\|v\|_{\widetilde{\mathrm{Pic}}} \leq \sqrt{n} \max_\sigma |\log(N(f/u)^{-1/n} u_\sigma^{-1} \sigma(f))|$$

$$\leq \sqrt{n} \left( \frac{1}{n} \left| \log \frac{N(f)}{N(u)} \right| + \max_\sigma \left| \log \frac{\sigma(f)}{u_\sigma} \right| \right) < \frac{2\varepsilon \sqrt{n}}{1-\varepsilon}.$$

This implies that the image of $\bar{d}$ is dense, as required. $\square$

Finally we compute the volumes of the compact Riemannian manifolds $\mathrm{Pic}^0_F$ and $\widetilde{\mathrm{Pic}}^0_F$.

PROPOSITION 6.5. *Let $F$ be a number field of degree $n$ and discriminant $\Delta_F$. Then:*

(i) $$\mathrm{vol}(\mathrm{Pic}^0_F) = \frac{w_F \sqrt{n}}{2^{r_1} (2\pi \sqrt{2})^{r_2}} \cdot |\Delta_F|^{1/2} \cdot \operatorname*{Res}_{s=1} \zeta_F(s).$$

(ii) $$\mathrm{vol}(\widetilde{\mathrm{Pic}}^0_F) = \sqrt{n} \cdot |\Delta_F|^{1/2} \cdot \operatorname*{Res}_{s=1} \zeta_F(s).$$

Here $r_1$ is the number of real primes and $r_2$ is the number of complex primes of $F$. By $w_F$ we denote the number of roots of unity and by $\zeta_F(s)$ the Dedekind zeta function of $F$.

PROOF. (i) The subspace $(\bigoplus_\sigma \mathbb{R})^0$ of divisors of degree 0 is the orthogonal complement of 1 in the subalgebra $\prod_\sigma \mathbb{R}$ of $F_\mathbb{R}$. Using the fact that $\|1\| = \sqrt{n}$, one checks that the volume of $\text{Pic}_F^0$ is equal to $\sqrt{n}\, 2^{-r_2/2} R_F$ where $R_F$ is the regulator of $F$. It follows from the exact sequence of Proposition 2.2 that the compact group $\text{Pic}_F^0$ has volume $\sqrt{n}\, 2^{-r_2/2} h_F R_F$ where $h_F = \#Cl_F$ is the class number of $F$. The formula [Marcus 1977] for the residue of the zeta function at $s=1$ now easily implies (i).

(ii) Since the natural volume of the group $K_\sigma$ is 2 or $2\pi\sqrt{2}$ depending on whether $\sigma$ is real or complex, it follows from the commutative diagram following Definition 5.1 that $\text{vol}(\widetilde{\text{Pic}}_F^0)$ is equal to $2^{r_1}(2\pi\sqrt{2})^{r_2}/w_F$ times the volume of $\text{Pic}_F^0$. This implies (ii). □

## 7. Reduced Arakelov divisors

Let $F$ be a number field of degree $n$. In this section we introduce *reduced* Arakelov divisors associated to $F$. These form a finite subset of $\text{Div}_F^0$. The main result of this section is that the image of this set in the groups $\text{Pic}_F^0$ and $\widetilde{\text{Pic}}_F^0$ is in a certain precise sense regularly distributed.

The results of this section extend work by Lenstra [1982] and Buchmann and Williams [1988] and make certain statements by Buchmann [1987b; 1990; 1991] more precise. In particular, Theorems 7.4 and 7.6 and Corollary 7.9 extend [Buchmann 1987b, Section 2; 1988, Proposition 2.7; 1990, Section 3.3]. Note that in deducing the Corollaries below we did not make any particular effort to obtain the best possible estimates. They can most certainly be improved upon.

Let $I$ be a fractional ideal. A nonzero element $f \in I$ is called *minimal* if it is nonzero and if the only element $g \in I$ for which $|\sigma(g)| < |\sigma(f)|$ for all infinite primes $\sigma$, is $g = 0$. If $f \in I$ is minimal, then for every $h \in F^*$, the element $hf$ is minimal in the ideal $hI$. In particular, if $h \in O_F^*$, the element $hf$ is minimal in the same ideal $I$. Therefore there are, in general, infinitely many minimal elements in $I$.

If $D = (I, u)$ is an Arakelov divisor, the minimal elements $f \in I$ are precisely the ones for which the open boxes $\{(y_\sigma)_\sigma \in F_\mathbb{R} : |y_\sigma| < |u_\sigma \sigma(f)|\text{ for all }\sigma\}$ contain only the point 0 of the lattice $uI$. Note, however, that the notion of minimality depends only on $I$ and is *independent* of the metric induced by the element $u$. *Shortest* elements $f \in I$ are the elements for which $\|f\|_D = \min\{\|g\|_D : g \in I - \{0\}\}$. This notion depends on the divisor $D = (I, u)$ and hence on the lattice $uI$. It does not merely depend on $I$. Since $\|g\|_D = \|ug\|$ for each $g \in I$, the vector $uf$ for any shortest $f \in I$ is a shortest nonzero vector of the lattice $uI$ associated to $D$. The number of shortest elements in $I$ is always finite. Shortest vectors are clearly minimal, but the converse is not true. It may even happen that a minimal element $f \in I$ is not a shortest element of the lattice $D = (I, u)$ for *any* choice of $u$. See Section 9 for an explicit example.

DEFINITION. An Arakelov divisor or oriented Arakelov divisor $D$ in $\text{Div}_F$ is called *reduced* if it is of the form $D = d(I) = (I, N(I)^{-1/n})$ for some fractional ideal $I$, and if 1 is a minimal element of $I$. The set of reduced Arakelov divisors is denoted by $\text{Red}_F$.

Since reduced Arakelov divisors have degree zero, the covolume of the lattices associated to reduced Arakelov divisors is $\sqrt{|\Delta_F|}$. With respect to the natural metric, $1 \in O_F$ is a shortest and hence minimal element. Therefore the trivial Arakelov divisor $(O_F, 1)$ is reduced. In general, if $D = d(I)$ is reduced, the element $1 \in I$ is merely minimal and need not be a shortest element. However, the next proposition shows that it is not too far away from being so.

PROPOSITION 7.1. *Let $F$ be a number field of degree $n$ and let $D = d(I) = (I, N(I)^{-1/n})$ be a reduced Arakelov divisor. Then*

$$\|1\|_D \leq \sqrt{n}\|x\|_D \quad \text{for all nonzero } x \in I.$$

*In particular, the element $1 \in I$ is at most $\sqrt{n}$ times as long as the shortest element in $I$.*

PROOF. We have $\|1\|_D = \sqrt{n}N(I)^{-1/n}$. Since $1 \in I$ is minimal, every nonzero $x \in I$ satisfies $|\sigma(x)| \geq 1$ for some embedding $\sigma : F \longrightarrow \mathbb{C}$. Therefore $\|x\|_D \geq N(I)^{-1/n}|\sigma(x)| \geq N(I)^{-1/n}$. □

If $D = (I, u)$ is an Arakelov divisor and $f \in I$ is minimal, then $1 \in f^{-1}I$ is again minimal and the divisor $d(f^{-1}I) = (f^{-1}I, N(fI^{-1})^{1/n})$ is reduced. In particular, if $f \in I$ is a shortest element, the divisor $d(f^{-1}I)$ is reduced. However, even though the element $1 \in f^{-1}I$ is minimal, it *need not be* a shortest element. Indeed, even if 1 *is* a shortest vector of the lattice associated to $(f^{-1}I, |f|^{-1}u)$, it may not be a shortest vector of the lattice $d(f^{-1}I) = (f^{-1}I, N(fu^{-1})^{1/n})$, which has a different metric. In Section 9 we present an example of this phenomenon.

It is not so easy to say in terms of the associated ideal lattice $uI$ precisely what it means for a divisor $D = (I, u)$ to be reduced. We make the following imprecise observation. When $1 \in I$ is not merely minimal, but happens to be a shortest element in $I$, then all roots of unity in $F$ are also shortest elements in $I$. Usually, these are the *only* shortest elements in $I$. In that case the arithmetic-geometric mean inequality implies that $\gamma(D) = \gamma(uI)$, viewed as a function on $\text{Pic}_F^0$, attains a local minimum at $D = (I, N(I)^{-1/n})$. So, the lattice corresponding to a reduced divisor is the "skewest" $O_F$-lattice in a small neighborhood in $\text{Pic}^0_F$. But this is just a rule of thumb; it is not always true.

DEFINITION. Let $F$ be a number field. Let $\Delta_F$ denote its discriminant and $r_2$ its number of complex infinite primes. Then we put

$$\partial_F = \left(\frac{2}{\pi}\right)^{r_2} \sqrt{|\Delta_F|}.$$

PROPOSITION 7.2. *Let $F$ be a number field of degree $n$.*

(i) *Let $I$ be a fractional ideal. If $d(I) = (I, N(I)^{-1/n})$ is a reduced Arakelov divisor, the inverse $I^{-1}$ of $I$ is an $O_F$-ideal of norm at most $\partial_F$.*
(ii) *The set $\text{Red}_F$ of reduced Arakelov divisors is finite.*
(iii) *The natural map $\text{Red}_F \longrightarrow \widetilde{\text{Pic}}^0_F$ is injective.*

PROOF. Since $1 \in I$, the ideal $I^{-1}$ is contained in $O_F$. By Proposition 4.4(ii) there exists a nonzero $f \in I$ for which $|N(I)^{-1/n}\sigma(f)| < \partial_F^{1/n}$ for each $\sigma$. Therefore, if $N(I^{-1}) > \partial_F$, we have $|\sigma(f)| < 1$ for each $\sigma$, contradicting the minimality of $1 \in I$. This proves (i). Part (ii) follows at once from (i) and the fact that there are only finitely many $O_F$-ideals of bounded norm.

To prove (iii), suppose that the reduced Arakelov divisors $D = d(I)$ and $D' = d(I')$ have the same image in $\widetilde{\text{Pic}}^0_F$. Then there exists $f \in F^*$ so that $I' = fI$ and $N(I')^{1/n} = N(I)^{1/n} f$. As in the proof of Proposition 6.4, it follows that all conjugates of $f$ are equal and hence that $f \in \mathbb{Q}^*$. Since both $I$ and $I'$ contain 1 as a minimal vector, this implies that $f = \pm 1$. Since $f = N(I'I^{-1})^{1/n} > 0$, we have $f = 1$ and hence $D = D'$ as required. □

Part (iii) of Proposition 7.2 does not hold when we replace $\widetilde{\text{Pic}}^0_F$ by $\text{Pic}^0_F$. See Example 9.3 for an example. Incidentally, Theorem 7.7 below strengthens the statement considerably.

Before we begin our discussion of the distribution of the reduced divisors in the Arakelov class groups, we characterize them 'geometrically'. This characterization plays no role in the sequel. For every fractional ideal $I$ with $1 \in I$ consider the following set of divisors of degree zero:

$$\Sigma_I = \{(I, v) \in \text{Div}^0_F : \log v_\sigma \leq (1/n) \log \partial_F \text{ for all } \sigma\}.$$

The set $\Sigma_I$ is not empty if and only if $N(I^{-1}) \leq \partial_F$. Indeed, under this condition $\Sigma_I$ contains the divisor $(I, N(I)^{-1/n})$ and its elements have the form $(I, N(I)^{-1/n}) + (O_F, w)$ with $w$ running over the exponentials of the vectors $y \in \left(\bigoplus_\sigma \mathbb{R}\right)^0$ satisfying

$$y_\sigma \leq \frac{1}{n}(\log \partial_F + \log N(I)) \quad \text{for every } \sigma.$$

Since $\sum_\sigma \deg(\sigma) y_\sigma = 0$, the set $\Sigma_I$ is a bounded simplex.

The following proposition says what it means for a divisor

$$d(I) = (I, N(I)^{-1/n})$$

to be reduced in terms of its simplex $\Sigma_I$.

PROPOSITION 7.3. *Let $I$ be a fractional ideal with $1 \in I$. The Arakelov divisor $D = d(I) = (I, N(I)^{-1/n})$ is reduced if and only if it has the property that for every fractional ideal $I'$ with $1 \in I'$ for which $\Sigma_I \subset \Sigma_{I'} + (f)$ for some $f \in F^*$, we necessarily have $fI = I'$ and $|\sigma(f)| = 1$ for all $\sigma$.*

PROOF. Suppose that $D = (I, N(I)^{-1/n})$ is reduced. Let $I'$ be a fractional ideal with $1 \in I'$ and $f \in F^*$. Suppose that for some $f \in F^*$ the simplex $\Sigma_I$ is contained in the translated simplex $\Sigma_{I'} + (f) = \{(I', v) + (f) : (I', v) \in \Sigma_{I'}\}$. This implies $I' = fI$. In addition, we have $\log(v_\sigma/|\sigma(f)|) \leq (1/n) \log \partial_F$ whenever $\log v_\sigma \leq (1/n) \log \partial_F$. It follows that $|\sigma(f)| \geq 1$ for all $\sigma$. Since $1 \in I$ is minimal, so is $f \in I'$. Since $1 \in I'$, this implies $|\sigma(f)| = 1$ for every $\sigma$.

Conversely, suppose that $D = (I, N(I)^{-1/n})$ has the property described in the proposition. We want to show that $1 \in I$ is minimal. Let therefore $g \in I$ such that $|\sigma(g)| \leq 1$ for all $\sigma$. Consider the $O_F$-ideal $I' = g^{-1}I$. Then we have $1 \in I'$ and $\Sigma_I \subset \Sigma_{I'} + (g^{-1})$. Indeed, if $(I, v) \in \Sigma_I$, then $\log v_\sigma \leq (1/n) \log \partial_F$ and hence $\log(v_\sigma |\sigma(g)|) \leq (1/n) \log \partial_F$. This means precisely that $(I, v)$ is contained in $\Sigma_{I'} + (g^{-1})$. We conclude that $|\sigma(g)| = 1$ for every $\sigma$. It follows that $1 \in I$ is minimal, as required. □

When $F$ is totally real, we necessarily have $f = \pm 1$ and hence $I = I'$ in Proposition 7.3. The proposition says therefore that, in a certain sense, the image in $\mathrm{Pic}_F^0$ of $\Sigma_I$ is not contained in the image of any other simplex. When $F$ is not totally real, this is still true for most $I$.

In the rest of this section we study the distribution of the image of the set $\mathrm{Red}_F$ in the compact groups $\mathrm{Pic}_F^0$ and $\widetilde{\mathrm{Pic}}_F^0$ and estimate its size. First we look at the image of the set $\mathrm{Red}_F$ in $\mathrm{Pic}_F^0$. Theorem 7.4 says that $\mathrm{Red}_F$ is rather *dense* in $\mathrm{Pic}_F^0$.

THEOREM 7.4. *Let $F$ be a number field of degree $n$ admitting $r_2$ complex infinite primes.*

(i) *For any Arakelov divisor $D = (I, u)$ of degree 0 there is a reduced divisor $D'$ and an element $f \in F^*$ so that*

$$D - D' = (f) + (O_F, v)$$

*with*

$$\log |v_\sigma| \leq \frac{1}{n} \log \partial_F \quad \text{for each } \sigma.$$

*In particular,*

$$\|D - D'\|_{\text{Pic}} \leq \log \partial_F.$$

(ii) *The natural map*

$$\bigcup_D \Sigma_I \longrightarrow \text{Pic}_F^0$$

*is surjective. Here the union runs over the reduced Arakelov divisors $D = (I, N(I)^{-1/n})$.*

PROOF. By Minkowski's Theorem (Proposition 4.4(ii)), there is a nonzero element $f \in I$ satisfying

$$|u_\sigma \sigma(f)| \leq \partial_F^{1/n} \quad \text{for every } \sigma.$$

Then there is also a shortest and hence a minimal such element $f$. The divisor $D' = d(f^{-1}I)$ is then *reduced*. It lies on the same component of $\text{Pic}_F^0$ as $D$. We have

$$D - D' + (f) = (O_F, v),$$

where $v$ is the vector $(v_\sigma)_\sigma \in \prod_\sigma \mathbb{R}_+^*$ with $v_\sigma = u_\sigma |\sigma(f)| N(f^{-1}I)^{1/n}$ and hence $\log |v_\sigma| = \log |u_\sigma \sigma(f)| + (1/n) \log(N(f^{-1}I))$ for every $\sigma$. Because $N(f^{-1}I) \leq 1$, this implies that $\log |v_\sigma| \leq \log |u_\sigma \sigma(f)|$ which by assumption is at most $\frac{1}{n} \log \partial_F$, as required.

Since $\sum_\sigma \deg(\sigma) \log v_\sigma = 0$, Lemma 7.5 below implies that

$$\|D - D'\|_{\text{Pic}}^2 = \|v\|_{\text{Pic}}^2 \leq n(n-1)\left(\frac{1}{n} \log \partial_F\right)^2.$$

This proves (i). Part (ii) is merely a reformulation of part (i). □

LEMMA 7.5. *Let $x_i \in \mathbb{R}$ for $i = 1, \ldots, n$. Suppose that $\sum_{i=1}^n x_i = 0$ and that $x \in \mathbb{R}$ has the property that $x_i \leq x$ for all $i = 1, \ldots, n$. Then $\sum_{i=1}^n x_i^2 \leq n(n-1)x^2$.*

We leave the proof of the lemma to the reader. The theorem says that $\text{Pic}_F^0$ can be covered with simplices $\Sigma_I$ centered in the reduced divisors $D$. We use the theorem to estimate the volume of the Arakelov class group $\text{Pic}_F^0$ in terms of the number of reduced divisors.

COROLLARY 7.6. *Let $F$ be a number field of degree $n$ with $r_1$ real and $r_2$ complex infinite primes. We have*

$$\text{vol}(\text{Pic}_F^0) \leq \frac{2^{-r_2/2} n^{-1/2}}{(r_1 + r_2 - 1)!} (\log \partial_F)^{r_1 + r_2} \#\text{Red}_F$$

$$\leq (\log |\Delta_F|)^n \#\text{Red}_F.$$

PROOF. Let $D = d(I) = (I, N(I)^{-1/n})$ be a reduced divisor. The set $\Sigma_I$ is given by

$$\left\{(I, N(I)^{-1/n}) + (O_F, v) : \log v_\sigma \leq \frac{1}{n}\big(\log \partial_F + \log N(I)\big)\right\}.$$

By Proposition 7.2(i) we have $N(I^{-1}) \leq \partial_F$. This implies that the set $\Sigma_I$ is a nonempty simplex of volume $\left(\frac{1}{n}\log(\partial_F N(I))\right)^{r_1+r_2}$ times the volume of the standard simplex

$$\{(y_\sigma) \in \bigoplus_\sigma \mathbb{R} : \textstyle\sum_\sigma y_\sigma = 0 \text{ and } y_\sigma \leq 1 \text{ for each } \sigma\},$$

which one checks to be equal to $2^{-r_2/2}n^{r_1+r_2-1/2}/(r_1+r_2-1)!$. This leads to the inequality

$$\mathrm{vol}(\widetilde{\mathrm{Pic}}_F^0) \leq \frac{2^{-r_2/2}n^{r_1+r_2-1/2}}{(r_1+r_2-1)!} \sum_D \left(\frac{1}{n}\log(\partial_F N(I))\right)^{r_1+r_2}.$$

Here the sum runs over the reduced divisors $D = (I, N(I)^{-1/n})$ of $F$.

Since $N(I) \leq 1$, the first estimate follows. The second inequality follows by a rather crude estimate from the first one. □

Next we prove a kind of converse to Theorem 7.4. The following theorem and its corollary say that the image of the set $\mathrm{Red}_F$ is rather *sparse* in the group $\widetilde{\mathrm{Pic}}_F^0$. Recall that $F_+^* = \{x \in F^* : \sigma(x) > 0 \text{ for all real } \sigma\}$.

THEOREM 7.7. *Let $F$ be a number field.*

(i) *Let $D$ and $D'$ be two reduced divisors in $\widetilde{\mathrm{Div}}_F^0$. If there exists an element $f \in F_+^*$ for which*

$$D - D' + (f) = (O_F, v),$$

*with $|\log v_\sigma| < \log \frac{4}{3}$ for each $\sigma$, then $D = D'$ in $\widetilde{\mathrm{Div}}_F^0$. Similarly, if $\|v\|_{\widetilde{\mathrm{Pic}}} < \log \frac{4}{3}$, we have $D = D'$ in $\widetilde{\mathrm{Div}}_F^0$.*

(ii) *The natural map from*

$$\bigcup_{D' \in \mathrm{Red}_F} \{D' + (O_F, v) : v \in (F_{\mathbb{R},\,\mathrm{conn}}^*)^0 \text{ and } |\log v_\sigma| < \tfrac{1}{2}\log \tfrac{4}{3} \text{ for each } \sigma\}$$

*to $\widetilde{\mathrm{Pic}}_F^0$ is injective.*

PROOF. Suppose that $D = d(I)$ and $D' = d(I')$ are two reduced divisors with the property that $D - D' + (f) = (O_F, v)$ with $f \in F^*$ for which $\sigma(f) > 0$ for all real $\sigma$. By Proposition 5.3, the images of $D$ and $D'$ in $\widetilde{\mathrm{Pic}}_F^0$ lie on the same

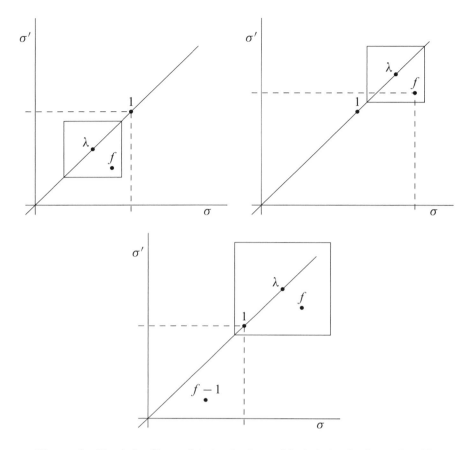

**Figure 1.** Top left: Since $f$ is in the box of $\lambda$, it is in the box of $1$. Top right: Since $f$ is in the box of $\lambda$, $1$ is in the box of $f$. Bottom: Since $f$ is in the box of $\lambda$, $f-1$ is in the box of $1$.

connected component of $\widetilde{\mathrm{Pic}}_F^0$. We put $\lambda = N(I/I')^{1/n}$. Then $\sigma(f)/\lambda = v_\sigma$. Since $|\log v_\sigma| < \log \frac{4}{3}$, we have

$$\left|\frac{\sigma(f)}{\lambda} - 1\right| = |v_\sigma - 1|$$
$$= |\exp(\log v_\sigma) - 1| \le \exp|\log(v_\sigma)| - 1 < \exp(\log \tfrac{4}{3}) - 1 = \tfrac{1}{3},$$

and hence

$$|\sigma(f) - \lambda| < \tfrac{1}{3}\lambda \quad \text{for every } \sigma.$$

Since $D$ and $D'$ are reduced, the element $1$ is minimal in both $I$ and $I'$. Therefore both $1$ and $f$ are minimal in $fI' = I$.

If $\lambda$ is small, i.e., if $0 < \lambda < \frac{1}{2}$, we have $|\sigma(f)| \leq |\sigma(f) - \lambda| + |\lambda| < \frac{1}{3}\lambda + \lambda < \frac{4}{3} \cdot \frac{1}{2} < 1$ for each $\sigma$. In other words, $|\sigma(f)| < |\sigma(1)|$ for all $\sigma$, contradicting the fact that $1 \in I$ is minimal. If $\lambda$ is large, i.e., if $\lambda > \frac{3}{2}$, we have that $|\sigma(f)| \geq |\lambda| - |\sigma(f) - \lambda| \geq \lambda - \frac{1}{3}\lambda > \frac{2}{3} \cdot \frac{3}{2} = 1$ for each $\sigma$. In other words, $|\sigma(1)| < |\sigma(f)|$ for all $\sigma$, contradicting the fact that $f \in I$ is a minimal vector. Therefore $\frac{1}{2} \leq \lambda \leq \frac{3}{2}$. This implies that the element $f - 1 \in I$ satisfies

$$|\sigma(f-1)| \leq |\sigma(f) - \lambda| + |\lambda - 1| < \frac{1}{3}\lambda + |\lambda - 1| \leq \frac{1}{3} \cdot \frac{3}{2} + \frac{1}{2} = 1 = |\sigma(1)|$$

for all $\sigma$. Since $1 \in I$ is a minimal vector, this implies that $f - 1 = 0$. Therefore $I = I'$ and hence $D = D'$. This proves the first statement.

If that $\|v\|_{\widetilde{\text{Pic}}} < \log \frac{4}{3}$, there is a totally positive unit $\varepsilon$ with $|\log(\sigma(\varepsilon)v_\sigma)| < \log \frac{4}{3}$ for each $\sigma$. Replacing $f$ by $\varepsilon f$ if necessary, we may then assume that $|\log(v_\sigma)| < \log \frac{4}{3}$ for each $\sigma$ and we are back in the earlier situation. This proves (i).

Part (ii) follows, because (i) implies that the sets

$$\{D' + (O_F, v) : v \in (F^*_{\mathbb{R},\text{conn}})^0 \text{ and } |\log(v_\sigma)| < \tfrac{1}{2}\log \tfrac{4}{3} \text{ for each } \sigma\}$$

map injectively to $\widetilde{\text{Pic}}^0_F$ and that their images are mutually disjoint. This proves the theorem. $\square$

COROLLARY 7.8. *There is a constant $c > 0$, so that for every number field $F$ of degree $n$, the number of reduced divisors contained in a ball of radius 1 in $\text{Pic}^0_F$ is at most $(cn)^{n/2}$.*

PROOF. The reduced divisors whose images in $\text{Pic}^0_F$ are contained in a ball of radius 1 lie in a subset $S$ of $\widetilde{\text{Pic}}^0_F$ of volume $2^{r_1}(2\pi\sqrt{2})^{r_2}/w_F$ times the volume of a unit ball in $\text{Pic}^0_F$. By Theorem 7.7, the balls of radius $\frac{1}{2}\log(\frac{4}{3})$ centered at reduced divisors are mutually disjoint in $\widetilde{\text{Pic}}^0_F$. Comparing the volume of the union of the disjoint balls with the volume of $S$ leads to the estimate. $\square$

COROLLARY 7.9. *Let $F$ be a number field of degree $n$. Then*

$$\#\text{Red}_F \leq \text{vol}(\widetilde{\text{Pic}}^0_F) \cdot 6^n.$$

PROOF. Theorem 7.7(ii) implies that the volume of $\widetilde{\text{Pic}}^0_F$ is at least $\#\text{Red}_F$ times the volume of the simplex $\{v \in ((F^*_{\mathbb{R},\text{conn}})^0 : |\log(v_\sigma)| < \frac{1}{2}\log\frac{4}{3}$ for each $\sigma\}$, which is equal to

$$\frac{2^{-r_2/2}n^{r_1+r_2-1/2}}{(r_1+r_2-1)!}\left(\tfrac{1}{2}\log\tfrac{4}{3}\right)^{r_1+r_2}.$$

Since this is at least $6^{-n}$, the result follows. $\square$

COROLLARY 7.10. *Let F be a number field. Then*

$$(\log|\Delta_F|)^{-n} \leq \frac{\#\mathrm{Red}_F}{\mathrm{vol}(\mathrm{Pic}_F^0)} \leq 18^n.$$

PROOF. The volume of $\widetilde{\mathrm{Pic}}_F^0$ is $2^{r_1}(2\pi\sqrt{2})^{r_2}/w_F$ times the volume of $\mathrm{Pic}_F^0$. Since $2^{r_1}(2\pi\sqrt{2})^{r_2}/w_F \leq (2\pi\sqrt{2})^{n/2}$, the inequalities follow from Corollaries 7.6 and 7.9 respectively. □

We recall the following estimates for the volume of $\mathrm{Pic}_F^0$. They say that in a sense the volume of $\mathrm{Pic}_F^0$ is approximately equal to $\sqrt{|\Delta_F|}$.

PROPOSITION 7.11. *Let $n \geq 1$. For every number field F of degree n we have:*

(i) $\qquad \mathrm{vol}(\mathrm{Pic}_F^0) \leq \sqrt{|\Delta_F|} (\log|\Delta_F|)^{n-1}$ ;

(ii) (GRH) *there exists a constant $c > 0$ only depending on the degree n so that*

$$\mathrm{vol}(\mathrm{Pic}_F^0) \geq c\sqrt{|\Delta_F|}/\log\log|\Delta_F|.$$

PROOF. Part (i) follows from Corollary 7.7, the fact that for every reduced divisor $d(I)$ the ideal $I^{-1}$ is integral and has norm at most $(2/\pi)^{r_2}\sqrt{|\Delta_F|}$ and the estimate for the number of $O_F$-ideals of bounded norm provided in [Lenstra 1992, Theorem 6.5]. Under assumption of the generalized Riemann Hypothesis (GRH) for the zeta function of the normal closure of $F$, Buchmann and Williams [1989, (3.2)] obtained the estimate in (ii). □

## 8. Quadratic fields

Since the class group of $\mathbb{Q}$ is trivial and $\mathbb{Z}^* = \{\pm 1\}$, the group $\mathrm{Pic}_{\mathbb{Q}}^0$ is trivial and the degree map induces an isomorphism $\mathrm{Pic}_{\mathbb{Q}} \cong \mathbb{R}$. The narrow class group of $\mathbb{Q}$ is also trivial and it follows from Definition 5.1 that $\widetilde{\mathrm{Pic}}_{\mathbb{Q}}^0 = 0$ and that $\widetilde{\mathrm{Pic}}_{\mathbb{Q}}$ is isomorphic to $\mathbb{R}_+^*$.

This is the whole story as far as $\mathbb{Q}$ is concerned. We now briefly work out the theory of the previous sections for quadratic number fields. For these fields the language of binary quadratic forms is often used [Lenstra 1982; Shanks 1972].

EXAMPLE 8.1. For complex quadratic fields $F$, the torus $T^0$ of Section 2 is trivial, so that the group $\mathrm{Pic}_F^0$ is canonically isomorphic to the class group $Cl_F$ of $F$. The group $\widetilde{\mathrm{Pic}}_F^0$ is an extension of $Cl_F$ by a circle group of length $2\pi\sqrt{2}/w_F$. Here $w_F = 2$ except when $F = \mathbb{Q}(i)$ or $\mathbb{Q}((1+\sqrt{-3})/2)$, in which case $w_F = 4$ or 6 respectively.

We describe the reduced Arakelov divisors of $F$. Let $D = (I, N(I)^{-1/2})$ be reduced. The fact that 1 is a minimal element of $I$ simply means that it is a shortest vector in the corresponding lattice in $F_{\mathbb{R}} \cong \mathbb{C}$. We write $I = \mathbb{Z} + f\mathbb{Z}$

for some $f$ in the upper half-plane $\{z \in \mathbb{C} : \text{Im}(z) > 0\}$. Since $O_F \cdot I \subset I$, we have $f = (b + \sqrt{\Delta_F})/(2a)$ for certain $a, b \in \mathbb{Z}$, $a > 0$ and $b^2 - 4ac = \Delta_F$ for some $c \in \mathbb{Z}$ with $\gcd(a, b, c) = 1$. The $O_F$-ideal $I^{-1}$ is generated by $a$ and $\frac{1}{2}(b - \sqrt{\Delta_F})$ and has norm $a$. For complex quadratic fields the simplices $\Sigma_I$ introduced in Section 6 are simply points.

Since $f$ is unique up to addition of an integer, the $\text{SL}_2(\mathbb{Z})$-equivalence class of the binary quadratic form $N(X + fY)/N(I) = aX^2 + bXY + cY^2$ is well defined. The form has discriminant $\Delta_F$. If we choose $f$ to lie in the usual fundamental domain for the action of $\text{SL}_2(\mathbb{Z})$ on the upper half-plane, the corresponding quadratic form is reduced in the sense of Gauss. There is a slight ambiguity here. If $|f| = 1$, the reduced Arakelov divisors $d(\mathbb{Z} + f\mathbb{Z})$ and $d(\mathbb{Z} + \bar{f}\mathbb{Z})$ give rise to the quadratic forms $aX^2 + bXY + aY^2$ and $aX^2 - bXY + aY^2$ respectively. If $f$ is not a root of unity, the Arakelov divisors are distinct, but the two quadratic forms are $\text{SL}_2(\mathbb{Z})$-equivalent and only one of them is reduced. Apart from this ambiguity, the map that associates to a reduced Arakelov divisor its associated reduced quadratic form, is a bijection.

EXAMPLE 8.2. Any real quadratic field $F$ can be written as $\mathbb{Q}(\sqrt{\Delta_F})$, where $\Delta_F$ denotes its discriminant. The group $\text{Pic}_F^0$ is an extension of the class group by a circle group and the group $\widetilde{\text{Pic}}_F^0$ is an extension of the *narrow* class group by a circle group. We describe the reduced Arakelov divisors of $F$. Let $\sigma$ and $\sigma'$ denote the two infinite primes of $F$. To be definite, we let $\sigma$ denote the embedding that maps $\sqrt{\Delta_F}$ to the positive square root of $\Delta_F$ in $\mathbb{R}$. Let $D = d(I) = (I, N(I)^{-1/2})$ be reduced. The fact that $1 \in I$ is minimal implies that we can write $I = \mathbb{Z} + f\mathbb{Z}$ for a unique $f$ satisfying $\sigma(f) > 1$ and $-1 < \sigma'(f) < 0$. The fact that $O_F \cdot I \subset I$ implies that $f = (b + \sqrt{\Delta_F})/(2a)$ where $\Delta_F = b^2 - 4ac$ for some $c \in \mathbb{Z}$ with $\gcd(a, b, c) = 1$. The conditions on $\sigma(f)$ and $\sigma'(f)$ say that $a > 0$ and $|\sqrt{\Delta_F} - 2a| < b < \sqrt{\Delta_F}$. The $O_F$-ideal $I^{-1}$ is generated by $a$ and $\frac{1}{2}(b - \sqrt{\Delta_F})$. Its norm is $a$. The simplex $\Sigma_I$ of Section 6 is an interval of length $\sqrt{2} \log(\sqrt{\Delta_F}/a)$ centered in $D$.

The map that associates the quadratic form $aX^2 + bXY + cY^2$ to the reduced divisor $D = (I, N(I)^{-1/2})$, is a bijection between the set of reduced Arakelov divisors of $F$ and the set of reduced binary quadratic forms of discriminant $\Delta_F$ with $a > 0$.

The element $1 \in I$ is a shortest vector precisely when both $\|f\|$ and $\|f - 1\|$ are at least $\|1\| = \sqrt{2}$. This condition is not always satisfied. Drawing a picture, one sees that it is if $\sigma(f) - \sigma'(f) \geq 2$, or equivalently if $a < \frac{1}{2}\sqrt{\Delta_F}$, but this is not a necessary condition.

When $D = d(I)$ and $I = \mathbb{Z} + f\mathbb{Z}$ as above, the vector $f$ is a minimal element of $I$. Therefore $D' = d(f^{-1}I)$ is a reduced Arakelov divisor. We have $D = D' + (f) + (O_F, v)$, where $v \in F_{\mathbb{R}}^* \cong \mathbb{R}^* \times \mathbb{R}^*$ is the vector

$(|\sigma'(f)/\sigma(f)|^{1/2}, -|\sigma(f)/\sigma'(f)|^{1/2})$. The distance between the images of $D$ and $D'$ in $\text{Pic}_F^0$ is equal to $\|v\|_{\text{Pic}}$. Since $f = (b + \sqrt{\Delta})/(2a)$, we have $\|v\|_{\text{Pic}} = 2^{-3/2} \log |(b+\sqrt{\Delta_F})/(b-\sqrt{\Delta_F})|$. In this way we recover Lenstra's distance formula [Lenstra 1982, (11.1)]. The divisor $D'$ is the 'successor' of $D$ in its component, in the sense that there are no reduced divisors on the circle between $D$ and $D'$. In order to obtain $D$'s 'predecessor', take $g$ the shortest minimum such that $|\sigma(g)| < |\sigma'(g)|$. Then the Arakelov divisor $d(g^{-1}I)$ is the predecessor of $D$.

Lenstra's group $\mathcal{F}$, or rather its topological completion $\overline{\mathcal{F}}$, is closely related to the oriented Arakelov class group of the real quadratic field $F$, and several of the results in [Lenstra 1982] are special cases of the ones in this paper. The group $\overline{\mathcal{F}}$ is not quite *equal* to $\widetilde{\text{Pic}}_F^0$ but it admits a degree 2 cover onto it. More generally, for a number field $F$ we let $\text{Pic}_F^+$ denote the group $\widetilde{\text{Div}}_F^0$ modulo its subgroup $\pm F_+^*$. When $F$ is totally complex, i.e., when $r_1 = 0$, this is simply $\widetilde{\text{Pic}}_F^0$. When $r_1 > 0$ however, there is an exact sequence

$$0 \longrightarrow \{\pm 1\}^{r_1}/\{\pm 1\} \longrightarrow \text{Pic}_F^+ \longrightarrow \widetilde{\text{Pic}}_F^0 \longrightarrow 0.$$

Let $(F_{\mathbb{R}}^*)^0 = \{u \in F_{\mathbb{R}} : |N(u)| = 1\}$. The topological structure of $\text{Pic}_F^+$ can be seen from the exact sequence

$$0 \longrightarrow (F_{\mathbb{R}}^*)^0/\pm O_{F,+}^* \longrightarrow \text{Pic}_F^+ \longrightarrow Cl_{F,+} \longrightarrow 0,$$

realizing $\text{Pic}_F^+$ as an extension of the narrow class group $Cl_{F,+}$ by a $2^{r_1-1}$-component Lie group. When $F$ is real quadratic, the group $\text{Pic}_F^+$ is equal to Lenstra's group $\overline{\mathcal{F}}$.

## 9. Reduced Arakelov divisors; examples and counterexamples

Let $F$ be a number field of degree $n$ and discriminant $\Delta_F$. Theorems 7.4 and 7.7 say that the image of the set $\text{Red}_F$ of reduced Arakelov divisors is, in a precise sense, rather regularly distributed in the groups $\text{Pic}_F^0$ and $\widetilde{\text{Pic}}_F^0$. In this section we discuss these results and we consider variations in the definition of the set of reduced divisors.

Theorem 7.4 says that the image of $\text{Red}_F$ is rather 'dense' in $\text{Pic}_F^0$. After a first draft of this paper was written, a similar result was obtained for the larger group $\widetilde{\text{Pic}}_F^0$.

PROPOSITION 9.1 (BUHLER ET AL.[1]). *Let $L \subset \mathbb{R}^n$ be a lattice and suppose that all nonzero vectors of $L$ have all their coordinates different from zero. Then there exists a minimal vector $(x_i) \in L$ with $x_i > 0$ for all $i$.*

---

[1] Personal communication from Joe Buhler describing discussions between Buhler, Randy Dougherty, Chris Freiling, Dan Mauldin, Nghi Nguyen, Peter Ostapenko, and Ken Zeger.

PROOF. Let $x \in \mathbb{R}^n$ be an *extreme* point of the convex hull of the set

$$S = \{(x_i) \in L : x_i > 0 \text{ for all } i\}.$$

This means that no open line segment inside the convex hull of $S$ contains $x$. It follows that $x$ is contained in $L$. If $x$ were not a minimal vector of $L$, there would be a non-zero vector $y = (y_i)$ in $L$ for which $-x_i < y_i < x_i$ for all $i$, so that both vectors $x - y$ and $x + y$ are in $S$. Since $x$ is contained in the line segment connecting $x - y$ and $x + y$, this contradicts the fact that $x$ is an extreme point. It follows that $x$ is minimal. □

This result easily implies that all components of the oriented Arakelov class group $\widetilde{\text{Pic}}_F^0$ contain reduced Arakelov divisors. In addition, Joe Buhler and his collaborators obtain a bound for the length of the shortest vector $(x_i) \in L$ with $x_i > 0$ for all $i$. It leads to an analogue of Theorem 7.4 for $\widetilde{\text{Pic}}_F^0$. The dependence on $n$ is a little worse. I do not know how to compute these vectors efficiently.

In the other direction, Theorem 7.7 implies that the image of $\text{Red}_F$ in $\widetilde{\text{Pic}}_F^0$ is rather 'sparse'. When we replace $\widetilde{\text{Pic}}_F^0$ by $\text{Pic}_F^0$, the theorem is no longer true. First of all the map $\text{Red}_F \longrightarrow \text{Pic}_F$ is in general not injective. In addition, it may happen that distinct reduced divisors have images in $\text{Pic}_F^0$ that are much closer to one another than the bound $\log(\frac{4}{3})$ of Theorem 7.7. However, by Corollary 7.9, the *number* of reduced divisors in a ball in $\text{Pic}_F^0$ of radius 1 is bounded by a constant depending only on the degree of $F$.

LEMMA 9.2. *Let $F$ be a number field of degree $n$, let $D = (I, u)$ be an Arakelov divisor and suppose $f \in I$.*

(i) $d(f^{-1}I) = d(I)$ in $\text{Div}_F^0$ *if and only if $f$ is a unit of $O_F$.*
(ii) *The classes of $d(f^{-1}I)$ and $d(I)$ in $\text{Pic}_F^0$ are equal if and only if $f$ is the product of a unit and an element $g \in F^*$ all of whose absolute values $|\sigma(g)|$ are equal.*
(iii) $\|d(I) - d(f^{-1}I)\|_{\text{Pic}} < 2\sqrt{n} \max_\sigma |\log |\sigma(f)||.$

PROOF. Part (i) follows from the fact that $I = f^{-1}I$ if and only if $f \in O_F^*$. Since

$$d(f^{-1}I) - d(I) = (f^{-1}O_F, |N(f)|^{1/n}),$$

the class of this divisor is trivial in $\text{Pic}_F^0$ if and only if there is $g \in F^*$ for which $f = \varepsilon g$ for some unit $\varepsilon \in O_F^*$ and $|\sigma(g)|^{-1} = |N(f)|^{-1/n}$ for all $\sigma$. Since $|N(g)| = |N(f)|$, the second relation is equivalent to the fact that the $|\sigma(g)|$ are all equal. This proves (ii).

To prove (iii) we note that

$$\|d(I) - d(f^{-1}I)\|_{\text{Pic}} \leq \sqrt{n} \max_\sigma |\log |\sigma(f)/N(f)^{1/n}||,$$

which is at most $\sqrt{n}$ times $\max_\sigma |\log|\sigma(f)|| + \frac{1}{n} \sum_\sigma \deg(\sigma) \log |\sigma(f)|$. The estimate follows easily.

This implies that $f$ is a root of unity. □

Proposition 7.2(iii) says that the natural map from the set of reduced divisors $\text{Red}_F$ to the oriented Arakelov class group $\widetilde{\text{Pic}}^0_F$ is injective. The following example shows that, in general, the map $\text{Red}_F \longrightarrow \text{Pic}^0_F$ is *not*.

EXAMPLE 9.3. Let $a > b \geq 1$ and put $\Delta = b^2 - 4a^2$. Suppose that $\Delta$ is squarefree and let $F$ denote the complex quadratic number field $\mathbb{Q}(\sqrt{\Delta})$. Let $I$ denote the fractional $O_F$-ideal $\mathbb{Z} + f\mathbb{Z}$, where $f = (b + \sqrt{\Delta})/(2a)$. Then $1 \in I$ is minimal. Let $\sigma : F \longrightarrow \mathbb{C}$ denote the unique infinite prime. Since $\sigma(f)$ has absolute value 1, the element $f$ is also minimal. Since $f$ is not a unit of $O_F$, Lemma 9.2 implies that the reduced divisors $d(I)$ and $d(f^{-1}I)$ are distinct, but that their classes in $\text{Pic}^0_F$ are equal.

Theorem 7.7 says that the distance between the images of the reduced divisors in $\widetilde{\text{Pic}}^0_F$ is bounded from below by an absolute constant. The following example shows that this is false for the Arakelov class group $\text{Pic}^0_F$.

EXAMPLE 9.4. Let $n$ be a large even integer such that $\Delta = n^2 + 1$ is squarefree and consider the field $F = \mathbb{Q}(\sqrt{\Delta})$. Let $f = (1 + \sqrt{\Delta})/n \in F$. Then 1 is a minimal element in $I = \mathbb{Z} + f\mathbb{Z}$. The conjugates $\sigma(f)$ are close to 1 and $-1$ respectively. Indeed, we have $|\log|\sigma(f)|| \approx \Delta^{-1/2}$ for each infinite prime $\sigma$. It follows from Lemma 9.2(iii) that the classes of the reduced divisors $d(I)$ and $d(f^{-1}I)$ are at distance at most $2\sqrt{2}\,\Delta^{-1/2}$ in $\text{Pic}^0_F$.

The definition of the set $\text{Red}_F$ is rather delicate, as we'll see now by considering slight variations of it. We let $\text{Red}'_F$ denote the set of divisors $d(I)$ for which $1 \in I$ is a *shortest* rather than a *minimal* vector and write $\text{Red}''_F$ for the set of divisors $d(I)$ for which we have $N(I^{-1}) \leq \partial_F = (2/\pi)^{r_2}\sqrt{|\Delta_F|}$ and for which $1 \in I$ is merely *primitive*, i.e., not divisible by an integer $d \geq 2$. Since *shortest* implies *minimal* and *minimal* implies *primitive*, we have the inclusions

$$\text{Red}'_F \subset \text{Red}_F \subset \text{Red}''_F$$

of finite sets. Theorem 7.4 says that the set $\text{Red}_F$ is rather 'dense' in the Arakelov divisor class group. It is not clear whether the set $\text{Red}'_F$ has the same property. The proof of Theorem 7.4, showing that every $D = (I, u)$ of degree 0 is close to a reduced divisor $D' \in \text{Red}_F$, does not work for $\text{Red}'_F$. Indeed, tracing the steps of the proof of Theorem 7.4, we see that if $f \in I$ is a shortest vector, it is also minimal and hence the element $1 \in f^{-1}I$ is *minimal*. It follows that the divisor $d(f^{-1}I)$ is in $\text{Red}_F$. However, 1 need not be a *shortest* vector in $f^{-1}I$ so that $d(f^{-1}I)$ may not be contained in $\text{Red}'_F$.

The following example shows that this phenomenon actually occurs. It shows that the set $\text{Red}'_F$ is, at least in this sense, too small.

EXAMPLE 9.5. We present examples of reduced Arakelov divisors $D = d(I)$ with the property that the element $1 \in I$ is *not* a shortest vector of the lattice $I$ associated to $(I, u)$ for *any* $u \in F_{\mathbb{R}}^*$. This implies that $D$ is not equal to $d(f^{-1}J)$ for any divisor $D' = (J, v)$ and a shortest element $f \in J$. Indeed, if that were the case, 1 would be shortest vector in the lattice associated to the Arakelov divisor $(I, f^{-1}v)$.

Let $F$ be a real quadratic number field of discriminant $\Delta$. Then $F = \mathbb{Q}(\sqrt{\Delta})$. Suppose that $d(I)$ is a reduced Arakelov divisor. We write $I = \mathbb{Z} + f\mathbb{Z}$ where $f > 0$ and $-1 < \bar{f} < 0$. Here we identify $F$ with its image in $\mathbb{R}$ through one of its embeddings and we write $f \mapsto \bar{f}$ for the other embedding.

CLAIM. *If* $N(f - \frac{1}{2}) > -\frac{3}{4}$, *then* 1 *is not a shortest element of* $I$ *for any Arakelov divisor* $(I, u)$ *of degree zero.*

PROOF. Suppose that $D = (I, u)$ has degree 0. Then we have

$$u = \left(\frac{v}{\sqrt{N(I)}}, \frac{v^{-1}}{\sqrt{N(I)}}\right)$$

for some $v \in \mathbb{R}_{>0}^*$. Suppose that $1 \in I$ is a shortest vector in the lattice associated to $D$. This implies in particular that $\|1\|_D \leq \|f\|_D$ and $\|1\|_D \leq \|f - 1\|_D$. This means that $v^{-2} + v^2 \leq v^{-2}f^2 + v^2 \bar{f}^2$ and that $v^{-2} + v^2 \leq v^{-2}(f - 1)^2 + v^2(\bar{f} - 1)^2$. In other words we have that $v^4 \leq (f^2 - 1)/(1 - \bar{f}^2)$ and $v^4 \geq (2f - f^2)/(\bar{f}^2 - 2\bar{f})$ respectively. Therefore, if the upper bound for $v^4$ is smaller than the lower bound, there cannot exist such an $v$. This happens precisely when $(f - \bar{f})(2f\bar{f} - f - \bar{f} + 2) > 0$. Since $f - \bar{f}$ is positive, this means that $2f\bar{f} - f - \bar{f} + 2 > 0$ which is equivalent to $N(f - \frac{1}{2}) > -\frac{3}{4}$. This proves the claim. □

When $f = (b + \sqrt{\Delta})/(2a)$ as in Section 8, a sufficient condition for the inequality of the claim to hold is that $a \geq \sqrt{\Delta/3}$. As an explicit example, take the field $\mathbb{Q}(\sqrt{21})$ and the reduced divisor $d(I)$ associated to $I = \mathbb{Z} + f\mathbb{Z}$, with $f = (3 + \sqrt{21})/6$.

In the other direction, it may happen that the image of $\text{Red}''_F$ is very dense in $\widetilde{\text{Pic}}_F^0$, so that an analogue of Theorem 7.7 does not hold for this set. We present two examples, due to H. W. Lenstra, showing that for some number fields certain small open balls in $\widetilde{\text{Pic}}_F^0$ contain the images of very many $D \in \text{Red}''_F$. Both examples exploit the existence of certain 'very small' elements in $F$. In the first example these are contained in a proper subfield, but this is not the case in the second example.

EXAMPLE 9.6. Let $F$ be a number field of degree $n$ containing $\mathbb{Q}(i)$. Let $m, m' \in \mathbb{Z}$ satisfy $\frac{1}{2}|\Delta_F|^{1/2n} < m, m' < |\Delta_F|^{1/2n} - 1$. Let $I$ and $I'$ denote the inverses of the $O_F$-ideals generated by $m - i$ and $m' - i$ respectively. Then 1 is primitive in both $I$ and $I'$ and the norms of $I^{-1}$ and $I'^{-1}$ do not exceed $\partial_F = (2/\pi)^{r_2} |\Delta_F|^{1/2}$. It follows that $d(I)$ and $d(I')$ are in $\mathrm{Red}''_F$. If the images of $d(I)$ and $d(I')$ in $\widetilde{\mathrm{Pic}}^0_F$ are equal, Proposition 6.4 implies that $I = mI'$ for some $m \in \mathbb{Q}^*$. Since 1 is primitive in both $I$ and $I'$, it follows that $m = \pm 1$. This implies that $I = I'$ and hence that $N(I) = m^2 + 1$ is equal to $N(I') = m'^2 + 1$, so that $m = m'$. Therefore $d(I)$ and $d(I')$ are distinct in $\widetilde{\mathrm{Pic}}^0_F$, whenever $m$ and $m'$ are.

Assume in addition that $|m - m'| < |\Delta_F|^{1/3n}$ and that $|\Delta_F| > 4^{6n}$. Then the distance between $m$ and $m'$ is much smaller than $m$ and $m'$ themselves. The distance between the Arakelov divisors $d(I)$ and $d(I')$ in $\widetilde{\mathrm{Pic}}^0_F$ is at most $\sqrt{n} |\log((m-i)/(m'-i))|$. This does not exceed

$$\sqrt{n}|m - m'|/(|m-i| - |m-m'|) \leq \sqrt{n}|\Delta_F|^{1/3n}/\left(\tfrac{1}{2}|\Delta_F|^{1/2n} - |\Delta_F|^{1/3n}\right)$$
$$\leq 4\sqrt{n}|\Delta_F|^{-1/6n}.$$

In this way we obtain $|\Delta_F|^{1/3n}$ elements of $\mathrm{Red}''_F$ whose images in $\widetilde{\mathrm{Pic}}^0_F$ are distinct, but are as close as $4\sqrt{n}|\Delta_F|^{-1/6n}$ to one another. By varying $F$ over degree $n/2$ extensions of $\mathbb{Q}(i)$, we can make $|\Delta_F|$ as large as we like. One may replace $\mathbb{Q}(i)$ by any number field and proceed similarly.

EXAMPLE 9.7. Let $n \geq 4$ and $a \in \mathbb{Z}$ be such that the polynomial $X^n - a$ is irreducible over $\mathbb{Q}$. Let $\alpha$ denote a zero and put $F = \mathbb{Q}(\alpha)$. Suppose that the ring of integers of $F$ is equal to $\mathbb{Z}[\alpha]$. There are infinitely many such integers $a$. Then $|\Delta_F| = n^n |a|^{n-1}$ and $|\sigma(\alpha)| = |a|^{1/n}$ for every infinite prime $\sigma$. Let $m, m' \in \mathbb{Z}$ satisfy $\frac{1}{2}|a|^{1/2-1/2n} + |a|^{1/n} < m, m' < |a|^{1/2-1/2n}$ and $|m - m'| \leq |a|^{1/4}$. Consider two Arakelov divisors $d(I)$ and $d(J)$ given by $I^{-1} = (m - \alpha)O_F$ and $J^{-1} = (m' - \alpha)O_F$. The norms of $I^{-1}$ and $J^{-1}$ are at most $\partial_F$. Since both $I$ and $J$ contain 1 as a primitive element, we have $d(I), d(J) \in \mathrm{Red}''_F$. The argument used in Example 9.6 shows that the images of $d(I)$ and $d(J)$ in $\widetilde{\mathrm{Pic}}^0_F$ are distinct when $m \neq m'$. The difference between $d(I)$ and $d(J)$ is equal to $(IJ^{-1}, N(IJ^{-1})^{1/n})$, which is equivalent to $(O_F, v)$, where

$$v = \frac{m - \sigma(\alpha)}{m' - \sigma(\alpha)} \left| N\left(\frac{m' - \alpha}{m - \alpha}\right) \right|^{1/n}.$$

It follows that $\|d(I) - d(J)\|_{\widetilde{\mathrm{Pic}}}$ is at most

$$2\sqrt{n} \max_\sigma \left| \log\left(\frac{m - \sigma(\alpha)}{m' - \sigma(\alpha)}\right) \right|.$$

Since $(m - \sigma(\alpha))/(m' - \sigma(\alpha)) = 1 + (m - m')/(m' - \sigma(\alpha))$ and since $|m' - \sigma(\alpha)| \geq m' - |\sigma(\alpha)| \geq \frac{1}{2}|a|^{1/2-1/2n}$, the absolute value of the logarithm of $(m-\sigma(\alpha))/(m'-\sigma(\alpha))$ is at most $4|m-m'|/|a|^{1/2-1/2n}$ for each $\sigma$. It follows that $\|d(I) - d(J)\|_{\widetilde{\mathrm{Pic}}}$ is at most $4\sqrt{n}|a|^{-1/4+1/2n}$, which becomes arbitrarily small as $|a|$ grows.

## 10. Computations with reduced Arakelov divisors

In this section we discuss the set of reduced Arakelov divisors from a computational point of view. Our presentation is informal; in particular, we do not say much about the accuracy of the approximations required to perform the computations with the real and complex numbers involved. (See [Thiel 1995] for a more rigorous approach.) Since Arakelov divisors can be represented as lattices in the Euclidean space $F_\mathbb{R}$, lattice reduction algorithms play an important role. When the degree $n$ of the number field is large, the celebrated Lenstra–Lenstra–Lovász (LLL) reduction algorithm [Lenstra et al. 1982; Lenstra 2008] is an important tool.

We suppose that the number field $F$ is given as $\mathbb{Q}(\alpha)$, where $\alpha$ is the zero of some irreducible monic polynomial $\varphi(X) \in \mathbb{Z}[X]$. We assume that we have already computed an LLL-reduced basis $\{\omega_1, \ldots, \omega_n\}$ for the ring of integers $O_F$ embedded in $F_\mathbb{R}$. In other words, we have an explicit lattice

$$O_F = \omega_1 \mathbb{Z} + \ldots + \omega_n \mathbb{Z} \subset F_\mathbb{R},$$

with, say, an LLL-reduced basis $\{\omega_1, \ldots, \omega_n\}$. Such a basis can be computed as explained in [Lenstra 1992, Section 4] or [Cohen 1993, Section 6.1] combined with a basis reduction algorithm. We have also computed a multiplication table i.e., coefficients $\lambda_{ijk} \in \mathbb{Z}$ for which $\omega_i \omega_j = \sum_k \lambda_{ijk} \omega_k$. The discriminant $\Delta_F$ of $F$ is the integer given by $\Delta_F = \det(\mathrm{Tr}(\omega_i \omega_j))$. By [Lenstra 1992, Section 2.10] we have $\lambda_{ijk} = |\Delta_F|^{O(n)}$. We view the degree $n$ of $F$ as fixed and estimate the running times of the algorithms in terms of $|\Delta_F|$.

An Arakelov divisor or oriented Arakelov divisor $D = (I, u)$ is determined by its associated ideal $I$ and the vector $u \in F_\mathbb{R}^* \cong \prod_\sigma F_\sigma^*$. It can be represented by an $n \times n$ matrix $\lambda_{ij}$ having the property that the vectors $\sum_{ij} \lambda_{ij} \omega_j$ form an LLL-reduced basis for the lattice $I \subset F_\mathbb{R}$, together with a sufficiently accurate approximation of the vector $u = (u_\sigma)_\sigma$. We have $\lambda_{ij} = O(N(I))$ (see [Thiel 1995]). In practice, one might want to take logarithms and work with the vectors $(\log u_\sigma)_\sigma$. There are efficient algorithms to multiply ideals, to compute inverses and to test for equality. See [Cohen 1993, sections 4.6–4.8]. Using these one can compute efficiently in the group $\mathrm{Div}_F$. The algorithms have been implemented in LiDIA, MAGMA and PARI.

Rather than the Arakelov divisor group, we are interested in computing in the *Arakelov class group* $\text{Pic}_F^0$. We do calculations in this group by means of the set $\text{Red}_F$ of *reduced* divisors in $\text{Div}_F^0$. By Theorems 7.4 and 7.7, the image of the finite set $\text{Red}_F$ is in a certain sense regularly distributed in the compact groups $\text{Pic}_F^0$ and $\widetilde{\text{Pic}}_F^0$. Reduced divisors have one further property that is important for our application: a reduced divisor $D$ is of the form $D = d(I) = (I, N(I)^{-1/n})$ where $I^{-1}$ is and integral ideal of norm at most $\partial_F = (2/\pi)^{r_2}|\Delta_F|^{1/2}$. Therefore $D$ can be represented using only $(\log|\Delta_F|)^{O(n)}$ bits.

Before describing the algorithms, we formulate a lemma concerning the LLL algorithm.

LEMMA 10.1. *Let* $\mathbf{b}_1, \ldots, \mathbf{b}_n$ *be an LLL-reduced basis of a real vector space* $V$. *Then for every vector* $\mathbf{x} = \sum_{i=1}^{n} m_i \mathbf{b}_i$ *of* $V$ *we have*

$$|m_i| \|\mathbf{b}_i^*\| \leq \left(\frac{3}{\sqrt{2}}\right)^{n-i} \|\mathbf{x}\| \quad \text{for } 1 \leq i \leq n.$$

Here $\mathbf{b}_1^*, \ldots, \mathbf{b}_n^*$ *denotes the Gram–Schmidt orthogonalization of the basis* $\mathbf{b}_1, \ldots, \mathbf{b}_n$.

For the proof, see [Lenstra 2008].

COROLLARY 10.2. *Let* $\mathbf{b}_1, \ldots, \mathbf{b}_n$ *be an LLL-reduced basis of a real vector space* $V$. *Then we have for any vector* $\mathbf{x} = \sum_{i=1}^{n} m_i \mathbf{b}_i$ *in* $V$ *that*

$$|m_i| \leq 2^{(i-1)/2} \left(\frac{3}{\sqrt{2}}\right)^{n-i} \frac{\|\mathbf{x}\|}{\|\mathbf{b}_1\|} \quad \text{for } 1 \leq i \leq n.$$

PROOF. The LLL conditions imply that $\|\mathbf{b}_1^*\| \leq 2^{(i-1)/2} \|\mathbf{b}_i^*\|$ for every $i = 1, 2, \ldots, n$. Since $\mathbf{b}_1 = \mathbf{b}_1^*$, the result follows from Lemma 10.1. □

We have the following basic algorithms at our disposal. For number fields of fixed degree $n$, each runs in time polynomial in $\log|\Delta_F|$.

ALGORITHM 10.3 (REDUCTION ALGORITHM). Given an Arakelov divisor $D = (I, u) \in \text{Div}_F^0$,

– check whether it is reduced or not;
– compute a reduced divisor $D'$ that is close to $D$ in $\text{Pic}_F^0$.

*Description.* We compute an LLL-reduced basis $\mathbf{b}_1, \ldots, \mathbf{b}_n$ of the lattice $uI \subset F_\mathbb{R}$. Then we compute a shortest vector $\mathbf{x}$ in $uI$ as follows. Any shortest vector $\mathbf{x} = \sum_{i=1}^{n} m_i \mathbf{b}_i$ in the lattice satisfies $\|\mathbf{x}\|/\|\mathbf{b}_1\| \leq 1$. Therefore Corollary 10.2 implies that the coordinates $m_i \in \mathbb{Z}$ are bounded independent of the discriminant of $F$. To compute a shortest vector in the lattice in time polynomial in $\log|\Delta_F|$, we may therefore just try all possible $m_i$.

To find a reduced divisor $D'$ that is close to $D$ in $\text{Pic}_F^0$, we compute a shortest vector $f$ in the lattice $I$ associated to $D$. The divisor $D' = d(f^{-1}I)$ is then reduced. Moreover, by Theorem 7.4 or rather its proof, the divisor $D'$ has the property that $\|D - D'\|_{\text{Pic}} \leq \log \partial_F$, so that $D'$ is close to $D$.

In a similar way one can check that a given divisor $D = (I, u)$ is reduced. First of all we must have that $u = N(I)^{-1/n}$. Then we check that 1 is contained in $I$. To see whether 1 is a *minimal* element of $I$, we need to make sure that the box
$$B = \{(y_\sigma) \in F_\mathbb{R} : |y_\sigma| < 1 \text{ for all } \sigma.\}.$$
contains no nonzero points of the lattice $I \subset F_\mathbb{R}$. The box $B$ contains all vectors of length at most 1. On the other hand, every vector in $B$ has length at most $\sqrt{n}$.

If the first vector $\mathbf{b}_1$ of the LLL-reduced basis has length less than 1, it is contained in $B$ and the element $1 \in I$ is *not* minimal. In this case we are done. Suppose therefore that we have $\|\mathbf{b}_1\| \geq 1$. It suffices now to compute all vectors $\mathbf{x}$ in the lattice that have length less than $\sqrt{n}$ and see whether they are in the box $B$ or not. By Corollary 10.2, the vectors $\mathbf{x} = \sum_{i=1}^n m_i \mathbf{b}_i$ of length at most $\sqrt{n}$ have the property that
$$|m_i| \leq 2^{(i-1)/2} \left(\frac{3}{\sqrt{2}}\right)^{n-i} \frac{\|\mathbf{x}\|}{\|\mathbf{b}_1\|} \leq 2^{(n-1)/2} \left(\frac{3}{2}\right)^{n-i} \sqrt{n}.$$
So, the number of vectors to be checked is bounded independently of the discriminant of $F$. This completes the description of the algorithm. Both algorithms run in time polynomial in $\log |\Delta_F|$, $\log \|u\|$ and the logarithmic height of $N(I)$.

ALGORITHM 10.4 (COMPOSITION ALGORITHM). Given two reduced Arakelov divisors $D = d(I)$ and $D' = d(J)$, compute a reduced divisor that is close to the sum $D + D'$ in $\text{Pic}_F^0$.

*Description.* One first adds $D$ and $D'$ as divisors. Since $N(I^{-1}), N(J^{-1}) \leq \partial_F$, the result $(IJ, N(IJ)^{-1/n})$ can be computed in time polynomial in $\log |\Delta_F|$. Then one reduces the result by means of Algorithm 10.3. The resulting reduced divisor is then close to $D + D'$. Since $N(IJ)^{-1} \leq \partial_F^2$, the running time of this second step is also polynomial in $\log |\Delta_F|$.

ALGORITHM 10.5 (INVERSION ALGORITHM). Given a reduced Arakelov divisor $D = d(I)$, compute a reduced divisor that is close to $-D$ in $\text{Pic}_F^0$.

*Description.* One just computes the inverse ideal $I^{-1}$ and reduces the divisor $d(I^{-1})$ by means of Algorithm 10.3. Since $N(I^{-1}) \leq \partial_F$, the running time of this algorithm is also polynomial in $\log |\Delta_F|$.

I owe the next algorithm to Hendrik Lenstra. See [Buchmann 1987a; 1987c; Thiel 1995] for a different approach. We first prove a lemma.

LEMMA 10.6. *Let $D = (I, u)$ be an Arakelov divisor of degree 0 and let $\varepsilon > 0$. Then every reduced divisor at distance at most $\varepsilon$ from $D$ is of the form $d(I\mu^{-1})$ where $\mu$ is a minimal element of $I$ satisfying*

$$\|\mu\|_D < \sqrt{n}\, e^{2\varepsilon} \|y\|_D \quad \text{for all nonzero } y \in I.$$

*In particular, the inequality holds for a nonzero $y \in I$ that is shortest with respect to the metric of $D$.*

PROOF. Let $D'$ be a reduced divisor for which we have $\|D - D'\|_{\text{Pic}} < \varepsilon$. Then we have $D' = d(I\mu^{-1})$ for some minimal element $\mu \in I$. By Proposition 6.2 there is a unit $\eta$ so that for $D'' = D + (\mu) + (O_F, |\eta|)$ we have

$$e^{-\|D' - D''\|_{\text{Pic}}} \leq \frac{\|x\|_{D'}}{\|x\|_{D''}} \leq e^{\|D' - D''\|_{\text{Pic}}} \quad \text{for every } x \in I\mu^{-1}.$$

We multiply $\mu$ by $\eta$. Then $\mu$ remains a minimal element of $I$ and the divisor $D'$ does not change. But now $D''$ is equal to $D + (\mu)$. Since $\|D - D'\|_{\text{Pic}} = \|D' - D''\|_{\text{Pic}}$, the inequality above and Proposition 7.1 imply that

$$\|\mu\|_D = \|1\|_{D+(\mu)} \leq e^{\varepsilon} \|1\|_{D'}$$
$$\leq e^{\varepsilon} \sqrt{n} \|x\|_{D'} \leq e^{2\varepsilon} \sqrt{n} \|x\|_{D+(\mu)} = e^{2\varepsilon} \sqrt{n} \|x\mu\|_D,$$

for any nonzero $x \in I\mu^{-1}$. Hence $\|\mu\|_D \leq \sqrt{n} e^{2\varepsilon} \|y\|_D$ for all non-zero $y \in I$. $\square$

ALGORITHM 10.7 (SCAN ALGORITHM). Let $D = (I, u)$ be an Arakelov divisor of degree 0. Compute all reduced Arakelov divisors in a ball in the Arakelov class group $\text{Pic}_F^0$ of radius 1 and center $D$ in time polynomial in $\log |\Delta_F|$.

*Description.* Choose $\varepsilon, \varepsilon' \in \mathbb{R}$ such that $0 < \varepsilon' < \varepsilon < 1$. Inside the open ball of divisors in $\text{Pic}_F^0$ having distance at most $1 + \varepsilon$ from $D$, we compute a *web* of regularly distributed points. The points $P$ in the web are at most $\varepsilon$ and at least $\varepsilon'$ apart. By Theorem 7.4 every $P$ is the class of a divisor of the form $D' + (O_F, v)$ for some reduced divisor $D' = d(J)$ and a totally positive $v \in F_{\mathbb{R}}^*$ satisfying $\|v\|_{\text{Pic}} < \log \partial_F$. Therefore an LLL-reduced basis for the lattice associated to each $P$ can be computed in time polynomial in $\log |\Delta_F|$.

By Lemma 10.6, the reduced divisors we are looking for are among the divisors of the form $d(J\mu^{-1})$ where $D' = d(J)$ is reduced, $P = D' + (O_F, v)$ is in the web and $\mu \in J$ is a *minimal* element for which $\|\mu\|_P$ is at most $e^{2\varepsilon}\sqrt{n}$ times the length of a nonzero element $y \in J$ that is shortest with respect to the metric induced by $P$. So, it suffices to compute the elements $\mu$ for all $P$ in the web. For a given $P$, Corollary 10.2 says that the number of vectors $\mu \in J$ of length at most $e^{2\varepsilon}\sqrt{n}$ times the length of the shortest nonzero vector, is bounded independently of $P$ and even of the discriminant of $F$. They can be computed in time polynomial in $\log |\Delta_F|$. Minimality of the elements $\mu$ can be tested by means of Algorithm 10.3. Finally, since the divisors $P$ are at least $\varepsilon'$ apart, the number of points in the web is proportional to the volume of the ball.

ALGORITHM 10.8 (JUMP ALGORITHM). Given a divisor

$$D = \sum_{\mathfrak{p}} n_{\mathfrak{p}} \mathfrak{p} + \sum_{\sigma} x_{\sigma} \sigma$$

of degree 0, compute a reduced Arakelov divisor whose image in $\mathrm{Pic}_F^0$ has distance less than $\log \partial_F$ from $D$.

*Description.* We assume that at most $O(\log |\Delta_F|)$ coefficients of $D$ are non-zero, that the coefficients themselves have size $|\Delta_F|^{O(1)}$ and that $N(\mathfrak{p})$ has size $|\Delta_F|^{O(1)}$ for the prime ideals $\mathfrak{p}$ with $n_{\mathfrak{p}} \neq 0$. Directly applying the reduction algorithm to $D = (I, v)$ with $I = \prod_{\mathfrak{p}} \mathfrak{p}^{n_{\mathfrak{p}}}$ and $v = \exp(x_\sigma)_\sigma$, is not a very good idea, since the LLL-algorithm and therefore the reduction algorithm run in time polynomial in the coefficients $|x_\sigma|$, which is *exponential* in terms of $\log |\Delta_F|$. Therefore we proceed differently.

We have $D = (I, 1) + (O_F, v)$. We compute reduced divisors close to $(I, 1)$ and to $(O_F, v)$. Adding these as in Algorithm 10.4, we may then compute a reduced divisor close to $D$, in the sense that its distance to $D$ is at most $\log \partial_F$. For each prime ideal $\mathfrak{p}$ with $n_{\mathfrak{p}} \neq 0$ we use Algorithm 10.3 to compute a reduced divisor $D_{\mathfrak{p}}$ close to $(\mathfrak{p}, 1)$. We compute a reduced divisor close to $\sum_{\mathfrak{p}} n_{\mathfrak{p}} D_{\mathfrak{p}}$ by composing and reducing as in Algorithm 10.4. The result is a reduced divisor close to $(I, 1)$.

Next we explain how to compute efficiently a reduced divisor close to the divisor $(O_F, v)$. Let $t \geq 0$ be the smallest integer for which the vector $y = (y_\sigma)_\sigma$ given by $y_\sigma = 2^{-t} x_\sigma$ satisfies $n|y_\sigma| < \log \partial_F$ for all $\sigma$. Then $t \in O(\log |\Delta_F|)$. Put $w = \exp(y_\sigma)_\sigma$. We have $w^{2^t} = v$. We inductively compute reduced Arakelov divisors $D_i = d(I_i)$ for which

$$\|D_i - (O_F, w^{2^i})\|_{\mathrm{Pic}} < \log \partial_F, \qquad \text{for } i = 0, 1, \ldots, t,$$

as follows. We put $D_0 = (O_F, 1)$. We compute $D_{i+1}$ from $D_i$ by doubling. More precisely, by induction we have $D_i = (O_F, w^{2^i}) + (O_F, w_i)$ in $\mathrm{Pic}_F^0$ for some $w_i \in F_{\mathbb{R}}^*$ with $\|w_i\|_{\mathrm{Pic}} < \log \partial_F$. Let $D_{i+1}$ be a reduced divisor whose distance to $2D_i - (O_F, w_i^2)$ is at most $\log \partial_F$. Then we have

$$D_{i+1} - (O_F, w^{2^{i+1}}) = (O_F, w_{i+1})$$

in $\mathrm{Pic}_F^0$ for some $w_{i+1} \in F_{\mathbb{R}}$ satisfying $\|w_{i+1}\|_{\mathrm{Pic}} < \log \partial_F$. Using Algorithm 10.3, we see that $D_{i+1}$ is of the form $d(I_{i+1})$ where $I_{i+1} = I_i^2/(x)$ for some element $x \in I_i^2$ that has the property that $w_i^{-2} x$ is a shortest vector in the lattice $w_i^{-2} I_i^2 \subset F_\mathbb{R}$. Since $\|w_i\|_{\text{Pic}} < \log \partial_F$ and since $D_i = d(I_i)$ is reduced, the computation of $D_{i+1}$ can be performed in time polynomial in $\log |\Delta_F|$. This completes the description of the algorithm.

Mutatis mutandis, we have the same algorithms for the group $\widetilde{\text{Div}}_F^0$ of *oriented* divisors and for the oriented Arakelov class group $\widetilde{\text{Pic}}_F^0$. The only difference is that the unit $u$ of an oriented Arakelov divisor $D = (I, u)$ is a complex rather than a positive real number. The image of the set of reduced Arakelov divisors in this group is probably also reasonably dense in $\widetilde{\text{Pic}}_F^0$ and that's all we need for the Jump algorithm to work. See Proposition 9.1.

APPLICATION 10.9. We present an algorithm to compute the function $h^0(D)$, introduced in [Van der Geer and Schoof 2000]. For an Arakelov divisor $D = (I, u)$, the number $h^0(D)$ should be viewed as the arithmetic analogue of the dimension of the space of global sections of a divisor $D$ on an algebraic curve. The number $h^0(D)$ depends only on the class of $D$ in $\text{Pic}_F^0$ and is defined as

$$h^0(D) = \log \sum_{f \in I} \exp(-\pi \|f\|_D^2).$$

See Section 4 for the close relation between the function $h^0(D)$ and the Hermite constant $\gamma(D)$ of the ideal lattice associated to $D$. Since the short vectors $f \in I$ contribute the most to this exponentially quickly converging sum, the function $h^0(D)$ can be evaluated most efficiently when we know a good, i.e., a reasonably orthogonal basis for $I$. As we explained above, a direct application of a lattice reduction algorithm to $D$ may be very time consuming. Therefore we apply the *Jump algorithm*. We jump to a reduced divisor $D' = d(J)$ close to $D$ in $\text{Pic}_F^0$. Then $D$ is equivalent to $D' + (O_F, v)$ for some short $v \in F_\mathbb{R}^*$ and

$$h^0(D) = h^0(D' + (O_F, v)) = \log \sum_{f \in J} \exp\left(-\pi \|f\|_{D'+(O_F,v)}^2\right)$$

$$= \log \sum_{f \in J} \exp\left(-\pi N(J)^{-2/n} \sum_\sigma \deg(\sigma)|\sigma(f)|^2 v_\sigma^2\right).$$

Since $D'$ is reduced and the vector $v = (v_\sigma)_\sigma$ is short, an LLL reduced basis for the lattice associated to $D' + (O_F, v)$ can be computed efficiently. This is because $J^{-1}$ is an integral ideal of norm at most $|\Delta_F|^{1/2}$. This completes the description of the algorithm to compute $h^0(D)$.

## 11. A deterministic algorithm

In this section we describe a *deterministic* algorithm to compute the Arakelov class group of a number field $F$ of degree $n$ and discriminant $\Delta_F$. It runs in time proportional to $\sqrt{|\Delta_F|}$ times a power of $\log |\Delta_F|$.

LEMMA 11.1. *Let $B > 0$. Then any ideal $J \subset O_F$ with $N(J) < B$ is of the form $J = xI^{-1}$, where the Arakelov divisor $D = (I, N(I)^{-1/n})$ is reduced, the element $u = N(x)^{1/n}/|x|$ of $F_\mathbb{R}^*$ satisfies $\|u\|_{\text{Pic}} < \log \partial_F$, and the element $x$ is contained in $I$ and satisfies $\|x\|_{D+(O_F,u)} < \sqrt{n}B^{1/n}$.*

PROOF. Suppose that $J \subset O_F$ satisfies $N(J) < B$. By Minkowski's Theorem there exists $y \in J^{-1}$, a shortest vector in $J^{-1} \subset F_\mathbb{R}$, satisfying $|\sigma(y)| < N(J)^{-1/n}\partial_F^{1/n}$ for every $\sigma$. We pick such an element $y$, put $x = 1/y$ and $I = xJ^{-1}$. Then the Arakelov divisor $D = (I, N(I)^{-1/n})$ is reduced. Moreover, since $xI^{-1} = J \subset O_F$, we have $x \in I$.

Writing $u = N(x)^{1/n}/|x|$, all coordinates of the vector $N(I)^{-1/n}ux$ have absolute value $N(I)^{-1/n}N(x)^{1/n} = N(J)^{1/n}$, so

$$\|x\|_{D+(O_F,u)} = \sqrt{n}N(J)^{1/n} < \sqrt{n}B^{1/n}.$$

Finally, we estimate $\|u\|_{\text{Pic}}$. Since $N(I) \leq 1$, we have

$$|u_\sigma| = \frac{|N(x)|^{1/n}}{|\sigma(x)|} = |\sigma(y)||N(x)|^{1/n} \leq N(J)^{-1/n}\partial_F^{1/n}|N(x)|^{1/n}$$
$$= N(I)^{1/n}\partial_F^{1/n} \leq \partial_F^{1/n}.$$

Lemma 7.5 then implies $\|u\|_{\text{Pic}} \leq (1 - 1/n)^{1/2}\log \partial_F \leq \log \partial_F$, as required. □

It is not hard to see that the lemma's converse also holds: any ideal $J \subset O_F$ for which the three conditions are satisfied automatically has norm at most $B$.

ALGORITHM 11.2. *Suppose we have computed all reduced divisors in a connected component of the Arakelov class group $\text{Pic}_F^0$. In that component, detect all divisors that are of the form $(J^{-1}, N(J)^{1/n})$ with $J \subset O_F$ and $N(J) < \partial_F$.*

*Description.* Let $\varepsilon, \varepsilon' \in \mathbb{R}$ be such that $0 < \varepsilon' < \varepsilon$. For each reduced divisor $D = (I, N(I)^{-1/n})$ in the given connected component, we make a web in the ball of center $D = (I, N(I)^{-1/n})$ and radius $\log \partial_F$, whose members $P = D + (O_F, v) = (I, N(I)^{-1/n}v)$ are at most $\varepsilon$ and at least $\varepsilon'$ apart. For each divisor $P = D + (O_F, v)$ in the web, we compute the vectors $x$ for which we have $\|x\|_P \leq \sqrt{n}e^{2\varepsilon}\partial_F^{1/n}$. This is done as follows. First we compute an LLL-reduced basis $\mathbf{b}_1, \ldots, \mathbf{b}_n$ for the lattice associated to the Arakelov divisor $P$. Let $\mathbf{b}_1^*, \ldots, \mathbf{b}_n^*$ denote its Gram–Schmidt orthogonalization. By Lemma 10.1 we have for any vector $x = \sum_{i=1}^n m_i \mathbf{b}_i$ in the lattice for which $\|x\|_P$ is at most $\sqrt{n}e^{2\varepsilon}\partial_F^{1/n}$, that

$$|m_i|\|\mathbf{b}_i^*\| \leq \left(\frac{3}{\sqrt{2}}\right)^{n-1}\sqrt{n}e^{2\varepsilon}\partial_F^{1/n}.$$

We simply try *all* coefficients $m_i$ satisfying this inequality.

For each such element $x$ we then compute the corresponding ideals $J = I^{-1}x$. The ideals $J$ that we compute in this way are contained in $O_F$. Moreover, every ideal $J \subset O_F$ of norm at most $\partial_F$ and for which the Arakelov divisor $(J^{-1}, N(J)^{1/n})$ lies on the given component, is obtained in this way. Indeed, if we have $N(J) < \partial_F$, Lemma 11.2 with $B = \partial_F$ implies that $J = xI^{-1}$ for some reduced divisor $d(I) = (I, N(I)^{-1/n})$ and some $x \in I$. Moreover, we have $\|x\|_{D+(O_F,u)} < \sqrt{n}\partial_F^{1/n}$ for some $u$ satisfying $\|u\|_{\text{Pic}} < \log \partial_F$. This means that the divisor $D + (O_F, u)$ is contained in the ball of center $D = (I, N(I)^{-1/n})$ and radius $\log \partial_F$. Therefore there is a member $P = D + (O_F, v)$ of the web at distance at most $\varepsilon$ from $D + (O_F, u)$. Proposition 6.2 implies then that

$$\|x\|_P \leq e^{2\varepsilon} \sqrt{n}\partial_F^{1/n},$$

as required.

This shows that we encounter all ideals $J$ that we are after. But we'll find many more and we'll find each ideal many times. Indeed, the vectors $x = \sum_{i=1}^n m_i \mathbf{b}_i$ that we consider in the computation above satisfy

$$|m_i|\|\mathbf{b}_i^*\| \leq \left(\frac{3}{\sqrt{2}}\right)^{n-1} \sqrt{n} e^{2\varepsilon}\partial_F^{1/n}$$

for each $i$ and hence

$$\|x\|_P \leq n \left(\frac{3}{\sqrt{2}}\right)^{n-1} e^{2\varepsilon}\partial_F^{1/n}.$$

It follows from the arithmetic geometric mean inequality that for the ideal $J = xI^{-1}$ we have

$$N(J) = N(xI^{-1}) \leq n^{n/2} e^{2\varepsilon n} \left(\frac{3}{\sqrt{2}}\right)^{n(n-1)} \partial_F.$$

In order to estimate the running time of this algorithm, we estimate the number of ideals $J$ that we compute, *and in addition* we estimate for how many divisors $P$ in the web and how many vectors $x$, we obtain each ideal $J$. By [Lenstra 1992, Theorem 6.5], the number of ideals $J$ is bounded by $\sqrt{|\Delta_F|}$ times a power of $\log |\Delta_F|$ times a constant that depends only on the degree $n$. Next we bound the number of times we find each ideal $J$.

First, suppose that for some divisor $P = (I, N(I)^{-1/n}) + (O_F, v)$ in the web, there are two elements $x, x' \in I^{-1}$ satisfying

$$\max(\|x\|_P, \|x'\|_P) \leq \sqrt{n}e^{2\varepsilon}\partial_F^{1/n},$$

for which the ideals $xI^{-1}$ and $x'I^{-1}$ are the *same*. Then we have

$$|\sigma(x)N(I)^{-1/n}v_\sigma| \leq \sqrt{n}e^{2\varepsilon}\partial_F^{1/n}$$

for each $\sigma$. Since we have $|N(v)| = 1$, the product over $\sigma$ satisfies

$$\prod_\sigma |\sigma(x) N(I)^{-1/n} v_\sigma|^{\deg(\sigma)} = N(xI^{-1}) \geq 1.$$

Therefore

$$-(n-1)\log(\sqrt{n}e^{2\varepsilon}\partial_F^{1/n}) \leq \log|\sigma(x)N(I)^{-1/n}v_\sigma| \leq \log(\sqrt{n}e^{2\varepsilon}\partial_F^{1/n})$$

for every $\sigma$. We have the same inequalities for $x'$. It follows that the unit $\eta = x'/x$ satisfies

$$-\log\partial_F - n\log(\sqrt{n}e^{2\varepsilon}) \leq \log|\sigma(\eta)| = \log\left|\frac{\sigma(x')}{\sigma(x)}\right|$$
$$\leq \log\partial_F + n\log(\sqrt{n}e^{2\varepsilon}),$$

for every $\sigma$ and hence we have

$$\|\log|\eta|\| \leq \sqrt{n}\log\partial_F + n^{3/2}\log(\sqrt{n}e^{2\varepsilon}).$$

By [Dobrowolski 1979], there exists an absolute constant $c > 0$ such that any unit $\eta \in O_F^*$ that is not a root of unity satisfies $\|\log|\eta|\| > cn^{-3/2}$. Since the number of roots of unity in $F$ is $O(n\log n)$, the number of units satisfying the bounds above is bounded by some power of $\log\partial_F$. It follows that the number of distinct elements $x \in I$ for which the ideals $xI^{-1}$ are equal to the same ideal $J \subset O_F$ is also bounded by some power of $\log\partial_F$.

Next, suppose that an ideal $J \subset O_F$ of norm at most $\partial_F$ is of the form $xI^{-1}$ where $D = (I, N(I)^{-1/n})$ is a reduced divisor and $x \in I$ satisfies $\|x\|_P \leq e^{2\varepsilon}\sqrt{n}\partial_F^{1/n}$ for some divisor $P = D + (O_F, v)$ in the web constructed. In particular, $v$ satisfies $\|v\|_{\text{Pic}} < \log\partial_F$. This implies that

$$\left|\frac{\sigma(x)}{N(x)^{1/n}}v_\sigma^{-1}\right| = |\sigma(x)N(I)^{1/n}v_\sigma^{-1}|\frac{1}{N(J)^{1/n}} < \frac{\sqrt{n}e^{2\varepsilon}\partial_F^{1/n}}{N(J)^{1/n}} \leq \sqrt{n}e^{2\varepsilon}\partial_F^{1/n}.$$

It follows that the Arakelov divisors $P$ and $(J^{-1}, N(J)^{1/n})$ are rather close to one another in $\text{Pic}_F^0$. Indeed, we have

$$\|P - (J^{-1}, N(J)^{1/n})\|_{\text{Pic}} = \|(O_F, |x|N(x)^{-1/n}v_\sigma^{-1})\|_{\text{Pic}}.$$

Since we have $\log|\sigma(x)N(x)^{-1/n}v_\sigma^{-1}| < \log(\sqrt{n}e^{2\varepsilon}\partial_F^{1/n})$ for every infinite prime $\sigma$, it follows from Lemma 7.5 that we have

$$\|P - (J, N(J)^{1/n})\|_{\text{Pic}} < \log(n^{2/n}e^{2\varepsilon/n}\partial_F).$$

By Corollary 7.9, the number of reduced divisors in a ball is bounded by some constant, depending only on the degree of the number field, times its volume.

Therefore the number of web members $P$ for which we encounter a given ideal $J \subset O_F$, is bounded by a polynomial expression in $\log \partial_F$.

This completes the description and our analysis of the algorithm.

**A deterministic algorithm.** Finally we explain the deterministic algorithm to compute the Arakelov class group of a number field $F$. This algorithm seems to have been known to the experts. It was explained to me by Hendrik Lenstra. We start at the neutral element $(O_F, 1)$ of the Arakelov class group. We use Algorithm 10.3 to determine all reduced Arakelov divisors in the ball of radius $2 \log \partial_F$ and center $(O_F, 1)$. Then we do the same with the reduced divisors $D$ we found: determine all reduced Arakelov divisors in the ball of radius $2 \log \partial_F$ and center $D$. Proceeding in a systematic way that is somewhat complicated to write down, we find in this way *all* reduced divisors in the connected component of identity. Keeping track of their positions in terms of the coordinates in $\prod_\sigma F_\sigma$ one computes in this way the absolute values of a set of generators of the unit group $O_F^*$. The running time is proportional to the volume of the connected component of identity and is polynomial in $\log |\Delta_F|$.

Next we use Algorithm 11.2 and make a list $\mathfrak{L}$ of all integral ideals $J \subset O_F$ of norm at most $\partial_F$, for which $(J^{-1}, N(J)^{1/n})$ is on the connected component of identity. The amount of work is again proportional to the volume of the connected component of identity and polynomial in $\log |\Delta_F|$. By Minkowski's Theorem, the prime ideals of norm at most $\partial_F$ generate the ideal class group of $F$. Therefore we check whether all prime ideals of norm at most $\partial_F$ are in the list. This involves computing gcd's of the polynomial that defines the number field $F$ with the polynomials $X^{p^i} - X$ for $i = 1, 2, \ldots, n$ for prime numbers $p$ that are smaller than the Minkowski bound $\partial_F$. One reads off the degrees of the prime ideals over $p$ and hence the number of primes of norm $p^i$ for $i = 1, 2, \ldots$. The amount of work is linear in the length of the list and polynomial time in $\log p$ for each prime $p$. If all prime ideals of norm at most $\partial_F$ are in the list $\mathfrak{L}$, then we are done. The class number is 1 and the Arakelov class group is connected.

However, if we do encounter a prime number $p$, for which a prime ideal $\mathfrak{p}$ of norm $p^i < \partial_F$ is missing, then we compute it. This involves factoring a polynomial of degree $n$ modulo $p$. When we do this with a simple minded trial division algorithm, the amount of work is at most $p^i < \partial_F$ times a power of $\log |\Delta_F|$. By successive multiplications and reductions, we compute for $j = 1, 2, \ldots$ reduced divisor $D_j$ in the connected components of the Arakelov class groups that contain divisors of the form $(\mathfrak{p}^j, u)$ for some $u$. Each time we check whether $D_j$ is already in the list $\mathfrak{L}$. If it is, we stop computing divisors $D_j$.

Then we repeat the algorithm, but this time we work with the connected components of the divisors $D_j$ rather than $(O_F, 1)$: we use Algorithm 10.3 to determine all reduced Arakelov divisors in the balls of radius $2 \log \partial_F$ and center $D_j$. Then we do the same with the reduced divisors we found, and so on. Once we have computed all reduced divisors on the connected components of $D_j$, we use Algorithm 11.2 to compute all integral ideals $J \subset O_F$ of norm at most $\partial_F$, for which $(J^{-1}, N(J)^{1/n})$ is on the connected components of the divisors $D_j$ and we add these to the list $\mathfrak{L}$.

When we are done with this, the list $\mathfrak{L}$ contains all integral ideals $J \subset O_F$ of norm at most $\partial_F$, whose classes are in the group generated by the ideal class of $\mathfrak{p}$. We check again whether all prime ideals of norm at most $\partial_F$ are in the list. If this turns out to be the case, we are done. The ideal class group is cyclic, generated by the class of $\mathfrak{p}$. If, on the other hand, we do encounter a second prime number $q$, for which a prime ideal $\mathfrak{q}$ of norm $q^i < \partial_F$ is missing, then we compute it. We compute reduced divisors that are in the components of the powers of $\mathfrak{q}$ ... etc.

For each new prime that we find is *not* in the list $\mathfrak{L}$, we factor a polynomial and the amount of work to do this is at most $\partial_F$. However, since the ideal class group has order at most $\sqrt{|\Delta_F|}$ times power of $\log|\Delta_F|$, we need to do this at most $\log|\Delta_F|$ times. As a result this algorithm takes time at most $\sqrt{|\Delta_F|}$ times power of $\log|\Delta_F|$.

## 12. Buchmann's algorithm

In this section we briefly sketch Buchmann's algorithm [1990; 1991] for computing the Arakelov divisor class group and, as a corollary, the class group and regulator of a number field $F$. This algorithm combines the infrastructure idea with an algorithm for complex quadratic number fields presented by J. Hafner and K. McCurley [1989]. When we fix the degree of $F$, the algorithm is under reasonable assumptions subexponential in the discriminant of the number field $F$. A practical approach is described in [Cohen 1993, Section 6.5]. The algorithm has been implemented in LiDIA, MAGMA and PARI. See also [Thiel 1995].

Let $F$ be a number field of degree $n$. The structure of Buchmann's algorithm is very simple. Our first description involves the Arakelov class group $\mathrm{Pic}_F^0$ rather than the oriented group $\widetilde{\mathrm{Pic}}_F^0$.

**Step 1: Estimate the volume of $\mathrm{Pic}_F^0$.** By Proposition 6.5 the volume of the compact Lie group $\mathrm{Pic}_F^0$ is given by

$$\mathrm{vol}(\mathrm{Pic}_F^0) = \frac{w_F \sqrt{n}}{2^{r_1}(2\pi\sqrt{2})^{r_2}} \cdot |\Delta_F|^{1/2} \cdot \mathrm{Res}_{s=1} \zeta_F(s).$$

The computation of $r_1, r_2$ and $w_F = \#\mu_F$ is easy. The discriminant is computed as a byproduct of the calculation of the ring of integers $O_F$. Approximating the residue of the zeta function

$$\zeta_F(s) = \prod_{\mathfrak{p}} \left(1 - \frac{1}{N(\mathfrak{p})^s}\right)^{-1}$$

at $s = 1$ is done by dividing $\zeta_F(s)$ by the zeta function of $\mathbb{Q}$ and by directly evaluating a truncated Euler product

$$\prod_{p \leq X} \frac{1 - 1/p}{\prod_{\mathfrak{p}|p}(1 - 1/N(\mathfrak{p}))}.$$

This involves factoring the ideals $pO_F$ for all prime numbers $p < X$; for efficient methods to do this, see [Cohen 1993]. The Euler product converges rather slowly. Under assumption of the Generalized Riemann Hypothesis for the zeta function of $F$, using the primes $p < X$, the relative error is $O(X^{-1/2} \log |\Delta_F X|)$. Here the O-symbol only depends on the degree of the number field $F$. See [Buchmann and Williams 1989; Schoof 1982]. Therefore, there is a constant $c$ only depending on the degree of $F$, so that if we truncate the Euler product at $X = c \log^2 |\Delta_F|$, the relative error in the approximation of $\text{vol}(\text{Pic}_F^0)$ is at most $1/2$.

**Step 2: Compute a factor basis.** We compute a factor base $\mathcal{B}$, that is, a list of prime ideals $\mathfrak{p}$ of $O_F$ of norm less than $Y$ for some $Y > 0$. Computing a factor basis involves factoring the ideals $pO_F$ for various prime numbers $p$. It is convenient to do this alongside the computation of the Euler factors in Step 1. We add the infinite primes to our factor basis. By normalizing, we obtain in this way a factor basis of Arakelov divisors of degree 0. The factor basis should be so large that the natural homomorphism

$$\left(\bigoplus_{\mathfrak{p} \in \mathcal{B}} \mathbb{Z} \times \bigoplus_{\sigma} \mathbb{R}\right)^0 \longrightarrow \text{Pic}_F^0$$

is surjective. By Proposition 2.2 this means that the classes of the primes in $\mathcal{B}$ must generate the ideal class group. Under assumption of the Generalized Riemann Hypothesis for the $L$-functions $L(s, \chi)$ associated to characters $\chi$ of the ideal class group $Cl_F$ of $F$, this is the case for $Y > c' \log^2 |\Delta_F|$ for some constant $c' > 0$ that only depends on the degree of $F$. Taking $\mathcal{B}$ this big, we have

$$\text{Pic}_F^0 = \left(\bigoplus_{\mathfrak{p} \in \mathcal{B}} \mathbb{Z} \times \bigoplus_{\sigma} \mathbb{R}\right)^0 \Big/ H,$$

where $H$ is the discrete subgroup of principal divisors of $\mathcal{B}$-units, i.e., the group of divisors $(f)$ where $f \in F^*$ are elements whose prime factorizations involve only prime ideals $\mathfrak{p} \in \mathcal{B}$.

**Step 3: Compute many elements in $H$.** An Arakelov divisor $D = (I, u)$ is called $\mathcal{B}$-smooth if $I$ is a product of powers of primes in $\mathcal{B}$. We need to find elements $f \in F^*$ for which $(f)$ is $\mathcal{B}$-smooth and hence $(f) \in H$. This is achieved by repeatedly doing the following. For at most $O(\log|\Delta_F|)$ prime ideals $\mathfrak{p} \in \mathcal{B}$ pick random exponents $m_\mathfrak{p} \in \mathbb{Z}$ of absolute value not larger than $|\Delta_F|$. In addition, pick random $x_\sigma \in \mathbb{R}$ of absolute value not larger than $|\Delta_F|$. Replacing $x_\sigma$ by $x_\sigma N(D)^{-1/n}$, scale the Arakelov divisor

$$D = \sum_\mathfrak{p} m_\mathfrak{p} \mathfrak{p} + \sum_\sigma x_\sigma \sigma$$

so that it acquires degree zero. Then the class of $D$ is a random element of $\mathrm{Pic}_F^0$. We use the Jump Algorithm described in Section 10 and "jump to $D$". The result is a reduced divisor $D' = (I, N(I)^{-1/n})$ whose image in $\mathrm{Pic}_F^0$ is not too far from the image of $D$. This means that

$$D = (f) + D' + (O_F, v)$$

for some $f \in F^*$ and $v = (v_\sigma) \in \left(\prod_\sigma \mathbb{R}_+^*\right)^0$ for which $\|v\|_{\mathrm{Pic}}$ is small, say at most $\log \partial_F$. There is no need to compute $f$, but when one applies the Jump Algorithm one should keep track of the infinite components and compute $v$ or its logarithm.

Since the divisor $D$ is random, it seems reasonable to think of the reduced divisor $D' = (I, N(I)^{-1/n})$ as being "random" as well. Next we attempt to factor the integral ideal $I^{-1}$ into a product of prime ideals $\mathfrak{p} \in \mathcal{B}$. Since $D'$ is random and since the norm of $I^{-1}$ is at most $\partial_F = (2/\pi)^{r_2}|\Delta_F|^{1/2}$ and hence relatively small, we have a fair chance to succeed. If we do, then we have $D' = \sum_{\mathfrak{p} \in \mathcal{B}} n_\mathfrak{p} \mathfrak{p} + \sum_\sigma y_\sigma \sigma$ and hence $(f) \in H$. This factorization leads to a relation of the form

$$(f) = D - D' - (O_F, v) = \sum_{\mathfrak{p} \in \mathcal{B}} (m_\mathfrak{p} - n_\mathfrak{p})\mathfrak{p} + \sum_\sigma (x_\sigma - y_\sigma + v_\sigma)\sigma.$$

In this way we have computed an explicit element in $H$.

Since we want to find many such relations, we need to be successful relatively often. In other words, the 'random' reduced divisors $D'$ that we obtain, should be $\mathcal{B}$-*smooth* relatively often. This is the weakest point of our analysis of the algorithm. In Section 9 the set $\mathrm{Red}_F''$ of Arakelov divisors $d(I)$ for which $1 \in I$ is primitive and $N(I^{-1}) \leq \sqrt{|\Delta_F|}$ was introduced. Under the assumption of the Generalized Riemann Hypothesis, Buchmann and Hollinger [1996] showed that when $Y \approx \exp(\sqrt{\log|\Delta_F|})$, the proportion of $\mathcal{B}$-smooth ideals $J$ with $d(J^{-1}) \in \mathrm{Red}_F''$ is at least $\exp(-\sqrt{\log|\Delta_F|\log\log|\Delta_F|})$. Here the Riemann Hypothesis for the zeta-function of the normal closure of $F$ is used to guarantee the existence of sufficiently many prime ideals of norm at most $\sqrt{|\Delta_F|}$ and degree 1. It is likely, but at present not known whether the proportion of $\mathcal{B}$-smooth ideals $I$ for which $d(I)$ is contained in the subset $\text{Red}_F$ rather than $\text{Red}''_F$, is *also* at least $\exp(-\sqrt{\log |\Delta_F|} \log \log |\Delta_F|)$. Even if this were the case, there is the problem that the divisor $D'$ that comes out of the reduction algorithm is not a 'random' reduced divisor. Indeed, Example 9.5 provides examples of reduced divisors that are not the reduction of *any* Arakelov divisor. These reduced divisors will never show up in our calculations, since everything we compute is a result of the reduction algorithm. It would be of interest to know how many such reduced divisors there may be.

For the next step we need to have computed approximately as many elements in $H$ as the size of the factor base $\mathcal{B}$. This implies that we expect to have to repeat the computation explained above about $\exp(\sqrt{\log |\Delta_F|} \log \log |\Delta_F|)$ times. When the discriminant $|\Delta_F|$ is large, this is more work than we need to do in Steps 1, 2 and 4. Step 3 is in practice the dominating part of the algorithm. It follows that the algorithm is subexponential and runs in time

$$O\bigl(\exp(\sqrt{\log |\Delta_F|} \log \log |\Delta_F|)\bigr).$$

**Step 4: Verify that the elements computed in Step 3 actually generate $H$.**
Let $H'$ denote the subgroup of $H$ generated by the divisors

$$(f) = \sum_{\mathfrak{p} \in \mathcal{B}} k_{\mathfrak{p}} \mathfrak{p} + \sum_{\sigma} y_{\sigma} \sigma$$

computed in Step 3. The quotient group $\bigl(\bigoplus_{\mathfrak{p} \in \mathcal{B}} \mathbb{Z} \times \bigoplus_{\sigma} \mathbb{R}\bigr)^0 / H'$ admits a natural map onto $\text{Pic}_F^0$. Its volume is equal to the determinant of a square matrix of size $\#\mathcal{B}$ whose rows are the coefficients of a set of $\#\mathcal{B}$ independent principal divisors that generate $H'$. If the quotient of the volume by the estimate of $\text{vol}(\text{Pic}_F^0)$ computed in Step 1 is less than $1/2$, then $H' = H$ and $\bigl(\bigoplus_{\mathfrak{p} \in \mathcal{B}} \mathbb{Z} \times \bigoplus_{\sigma} \mathbb{R}\bigr)/H'$ is actually *isomorphic* to $\text{Pic}_F^0$ and we are done.

In practice this means that once we have computed somewhat more divisors $(f)$ in $H$ than $\#\mathcal{B}$, we "reduce" the coefficient matrix. From the "reduced" matrix we can read off the structure of the ideal class group as well approximations to the logarithms of the absolute values of a set of units $\varepsilon$ that generate the unit group $O_F^*$. This enables us to compute the regulator $R_F$.

This completes our description of Buchmann's algorithm. It seems difficult to compute approximations to the numbers $\sigma(\varepsilon)$ themselves from approximations to their absolute values $|\sigma(\varepsilon)|$. If one wants to obtain such approximations, one should apply the algorithm above to the *oriented* Arakelov class group. The computations are the same, but rather than real, one carries complex coordinates $x_\sigma$ along. More precisely,

$$\widetilde{\text{Pic}}_F^0 = \Bigl(\bigoplus_{\mathfrak{p} \in \mathcal{B}} \mathbb{Z} \times \bigoplus_{\sigma} F_\sigma^*\Bigr)^0 \Big/ \widetilde{H}$$

for the discrete subgroup $\widetilde{H}$ that consists of elements $f \in F^*$ whose prime factorizations involve only prime ideals $\mathfrak{p} \in \mathcal{B}$. In this way one obtains approximations to $\sigma(\varepsilon_i)$ for a basis $\varepsilon_i$ of the unit group $O_F^*$. In principle, once one has such approximations one may solve the linear system $\sigma(\varepsilon_i) = \sum_j \lambda_{ij} \sigma(\omega_j)$ and compute $\lambda_{ij} \in \mathbb{Z}$ so that $\varepsilon_i = \sum_j \lambda_{ij} \omega_j$ for $1 \le i \le r_1 + r_2 - 1$. However, it is well known that the size of the coefficients $\lambda_{ij}$ may grow doubly exponentially quickly in $\log |\Delta_F|$ and it is therefore not reasonable to ask for an efficient algorithm that computes a set of generators of the unit group as linear combination of the basis $\omega_k$ of the additive group $O_F$.

What can be done efficiently, is to compute a *compact representation* of a set of generators of the unit group $O_F^*$. Briefly, this works as follows. Using the notation used in the description of the Jump Algorithm of Section 10, one finds for each fundamental unit $\varepsilon_j$ integers $m_{ij}$ such that $\prod_i v_i^{m_{ij}}$ is close to $\varepsilon_j$. The Arakelov divisors $(O_F, v_i)$ are equivalent to reduced divisors $d(f_i^{-1})$. While jumping towards the fundamental unit, one keeps track of the principal ideals that are encountered on the way. For instance, if in the process one computes the sum of the divisors $(O_F, v_i)$ and $(O_F, v_j)$ and reduces the result by means of a shortest vector $f$, then the result is equivalent to the reduced divisor $d((f f_i f_j)^{-1})$. The size of the elements $f_i$, $f_j$ and $f$ ... etc. is bounded by $(\log |\Delta_F|)^{O(1)}$. With a good strategy one can jump reasonably close to the unit. The number of jumps we need to reach this point is also bounded by $(\log |\Delta_F|)^{O(1)}$. Using the approximations to the fundamental units and to the vectors $f_i$, $f_j$, $f$ ... etc, we can approximate a small element $g \in F^*$, so that the difference between the divisor we jumped to and the fundamental unit is equivalent to a divisor of the form $(O_F, g)$. Since $g$ is small, we can compute it in time bounded by $\log |\Delta_F|^{O(1)}$ from its the approximations of the various $\sigma(g)$. From this we easily obtain the fundamental unit $\varepsilon_j$.

## Acknowledgements

I thank the Clay Foundation for financial support during my stay at MSRI in the fall of 2000, Burcu Baran, Hendrik Lenstra, Sean Hallgren and Takao Watanabe for several useful remarks, Silvio Levy for the production of Figure 1 and YoungJu Choie for inviting me to lecture on 'infrastructure' at KIAS in June 2001.

## References

[Bayer-Fluckiger 1999] E. Bayer-Fluckiger, "Lattices and number fields", pp. 69–84 in *Algebraic geometry: Hirzebruch 70* (Warsaw, 1998), edited by P. Pragacz et al., Contemp. Math. **241**, Amer. Math. Soc., Providence, RI, 1999.

[Buchmann 1987a] J. Buchmann, "On the computation of units and class numbers by a generalization of Lagrange's algorithm", *J. Number Theory* **26**:1 (1987), 8–30.

[Buchmann 1987b] J. Buchmann, "On the period length of the generalized Lagrange algorithm", *J. Number Theory* **26**:1 (1987), 31–37.

[Buchmann 1987c] J. Buchmann, *Zur Komplexität der Berechnung von Einheiten und Klassenzahlen algebraischer Zahlkörper*, Habilitationsschrift, Univ. Düsseldorf, 1987.

[Buchmann 1990] J. Buchmann, "A subexponential algorithm for the determination of class groups and regulators of algebraic number fields", pp. 27–41 in *Séminaire de Théorie des Nombres* (Paris, 1988–1989), edited by C. Goldstein, Progr. Math. **91**, Birkhäuser, Boston, 1990.

[Buchmann and Düllmann 1991] J. Buchmann and S. Düllmann, "A probabilistic class group and regulator algorithm and its implementation", pp. 53–72 in *Computational number theory* (Debrecen, 1989), edited by A. Pethö et al., de Gruyter, Berlin, 1991.

[Buchmann and Hollinger 1996] J. A. Buchmann and C. S. Hollinger, "On smooth ideals in number fields", *J. Number Theory* **59**:1 (1996), 82–87.

[Buchmann and Williams 1988] J. Buchmann and H. C. Williams, "On the infrastructure of the principal ideal class of an algebraic number field of unit rank one", *Math. Comp.* **50**:182 (1988), 569–579.

[Buchmann and Williams 1989] J. Buchmann and H. C. Williams, "On the computation of the class number of an algebraic number field", *Math. Comp.* **53**:188 (1989), 679–688.

[Cohen 1993] H. Cohen, *A course in computational algebraic number theory*, Graduate Texts in Mathematics **138**, Springer, Berlin, 1993.

[Dobrowolski 1979] E. Dobrowolski, "On a question of Lehmer and the number of irreducible factors of a polynomial", *Acta Arith.* **34**:4 (1979), 391–401.

[Van der Geer and Schoof 2000] G. Van der Geer and R. Schoof, "Effectivity of Arakelov divisors and the theta divisor of a number field", *Selecta Math.* (*N.S.*) **6**:4 (2000), 377–398.

[Groenewegen 2001] R. P. Groenewegen, "An arithmetic analogue of Clifford's theorem", *J. Théor. Nombres Bordeaux* **13**:1 (2001), 143–156.

[Hafner and McCurley 1989] J. L. Hafner and K. S. McCurley, "A rigorous subexponential algorithm for computation of class groups", *J. Amer. Math. Soc.* **2**:4 (1989), 837–850.

[Lenstra 1982] H. W. Lenstra, Jr., "On the calculation of regulators and class numbers of quadratic fields", pp. 123–150 in *Journées Arithmétiques* (Exeter, 1980), edited by J. V. Armitage, London Math. Soc. Lecture Note Ser. **56**, Cambridge Univ. Press, Cambridge, 1982.

[Lenstra 1992] H. W. Lenstra, Jr., "Algorithms in algebraic number theory", *Bull. Amer. Math. Soc.* (*N.S.*) **26**:2 (1992), 211–244.

[Lenstra 2008] H. W. Lenstra, Jr., "Lattices", pp. 127–181 in *Surveys in algorithmic number theory*, edited by J. P. Buhler and P. Stevenhagen, Math. Sci. Res. Inst. Publ. **44**, Cambridge University Press, New York, 2008.

[Lenstra et al. 1982] A. K. Lenstra, H. W. Lenstra, Jr., and L. Lovász, "Factoring polynomials with rational coefficients", *Math. Ann.* **261**:4 (1982), 515–534.

[Marcus 1977] D. A. Marcus, *Number fields*, Springer, New York, 1977.

[Schoof 1982] R. J. Schoof, "Quadratic fields and factorization", pp. 235–286 in *Computational methods in number theory, Part II* (Amsterdam, 1982), edited by H. W. Lenstra, Jr. and R. Tijdeman, Math. Centre Tracts **155**, Math. Centrum, Amsterdam, 1982.

[Shanks 1972] D. Shanks, "The infrastructure of a real quadratic field and its applications", pp. 217–224 in *Proceedings of the Number Theory Conference* (Boulder, CO, 1972), Univ. Colorado, Boulder, 1972.

[Shanks 1976] D. Shanks, "A survey of quadratic, cubic and quartic algebraic number fields (from a computational point of view)", pp. 15–40. Congressus Numerantium, No. XVII in *Proceedings of the Seventh Southeastern Conference on Combinatorics, Graph Theory, and Computing* (Baton Rouge, LA, 1976), edited by F. Hoffman et al., Utilitas Math., Winnipeg, Man., 1976.

[Szpiro 1985] L. Szpiro, "Degrés, intersections, hauteurs", pp. 11–28 in *Séminaire sur les pinceaux arithmétiques: La conjecture de Mordell*, Astérisque **127**, Soc. math. de France, Paris, 1985.

[Szpiro 1987] L. Szpiro, "Présentation de la théorie d'Arakélov", pp. 279–293 in *Current trends in arithmetical algebraic geometry* (Arcata, CA, 1985), edited by K. Ribet, Contemp. Math. **67**, Amer. Math. Soc., Providence, RI, 1987.

[Thiel 1995] C. Thiel, *On the complexity of some problems in algorithmic algebraic number theory*, Ph.D. thesis, Universität des Saarlandes, Saarbrücken, 1995.

[Williams and Shanks 1979] H. C. Williams and D. Shanks, "A note on class-number one in pure cubic fields", *Math. Comp.* **33**:148 (1979), 1317–1320.

[Williams et al. 1983] H. C. Williams, G. W. Dueck, and B. K. Schmid, "A rapid method of evaluating the regulator and class number of a pure cubic field", *Math. Comp.* **41**:163 (1983), 235–286.

RENÉ SCHOOF
DIPARTIMENTO DI MATEMATICA
UNIVERSITÀ DI ROMA 2 "TOR VERGATA"
VIA DELLA RICERCA SCIENTIFICA
I-00133 ROMA
ITALY
schoof@mat.uniroma2.it, schoof@science.uva.nl

# Computational class field theory

HENRI COHEN AND PETER STEVENHAGEN

ABSTRACT. Class field theory furnishes an intrinsic description of the abelian extensions of a number field which is in many cases not of an immediate algorithmic nature. We outline the algorithms available for the explicit computation of such extensions.

## CONTENTS

| | |
|---|---|
| 1. Introduction | 497 |
| 2. Class field theory | 499 |
| 3. Local aspects: ideles | 503 |
| 4. Computing class fields: preparations | 508 |
| 5. Class fields as Kummer extensions | 509 |
| 6. Class fields arising from complex multiplication | 515 |
| 7. Class fields from modular functions | 522 |
| 8. Class invariants | 529 |
| Acknowledgements | 532 |
| References | 533 |

## 1. Introduction

Class field theory is a twentieth century theory describing the set of finite *abelian* extensions $L$ of certain base fields $K$ of arithmetic type. It provides a canonical description of the Galois groups $\mathrm{Gal}(L/K)$ in terms of objects defined 'inside $K$', and gives rise to an explicit determination of the maximal abelian quotient $G_K^{\mathrm{ab}}$ of the absolute Galois group $G_K$ of $K$. In the classical examples, $K$ is either a *global field*, that is, a number field or a function field in one variable over a finite field, or a *local field* obtained by completing a global field at one of its primes. In this paper, which takes an algorithmic approach, we restrict to the fundamental case in which the base field $K$ is a number field. By doing so, we avoid the complications arising for $p$-extensions in characteristic $p > 0$.

Class field theory describes $G_K^{\text{ab}}$ for a number field $K$ in a way that can be seen as a first step towards a complete description of the full group $G_K \subset G_{\mathbb{Q}}$. At the moment, such a description is still far away, and it is not even clear what kind of description one might hope to achieve. Grothendieck's anabelian Galois theory and his theory of *dessins d'enfant* [Schneps 1994] constitute one direction of progress, and the largely conjectural *Langlands program* [Bump et al. 2003] provides an other approach. Despite all efforts and partial results [Völklein 1996], a concrete question such as the *inverse problem of Galois theory* — which asks whether, for a number field $K$, all finite groups $G$ occur as the Galois group of some finite extension $L/K$ — remains unanswered for all $K$.

A standard method for gaining insight into the structure of $G_K$, and for realizing certain types of Galois groups over $K$ as quotients of $G_K$, consists of studying the action of $G_K$ on 'arithmetical objects' related to $K$, such as the division points in $\overline{\mathbb{Q}}$ of various algebraic groups defined over $K$. A good example is the *Galois representation* arising from the group $E[m](\overline{\mathbb{Q}})$ of $m$-torsion points of an elliptic curve $E$ that is defined over $K$. The action of $G_K$ on $E[m](\overline{\mathbb{Q}})$ factors via a finite quotient $T_m \subset \text{GL}_2(\mathbb{Z}/m\mathbb{Z})$ of $G_K$, and much is known [Serre 1989] about the groups $T_m$. Elliptic curves with *complex multiplication* by an order in an imaginary quadratic field $K$ give rise to *abelian* extensions of $K$ and yield a particularly explicit instance of class field theory.

For the much simpler example of the multiplicative group $\mathbf{G}_m$, the division points of $\mathbf{G}_m(\overline{\mathbb{Q}})$ are the *roots of unity* in $\overline{\mathbb{Q}}$. The extensions of $K$ they generate are the *cyclotomic extensions* of $K$. Because the Galois group of the extension $K \subset K(\zeta_m)$ obtained by adjoining a primitive $m$-th root of unity $\zeta_m$ to $K$ naturally embeds into $(\mathbb{Z}/m\mathbb{Z})^*$, all cyclotomic extensions are abelian. For $K = \mathbb{Q}$, Kronecker discovered in 1853 that *all* abelian extensions are accounted for in this way.

THEOREM 1.1 (KRONECKER–WEBER). *Every finite abelian extension $\mathbb{Q} \subset L$ is contained in some cyclotomic extension $\mathbb{Q} \subset \mathbb{Q}(\zeta_m)$.*

Over number fields $K \neq \mathbb{Q}$, there are more abelian extensions than just cyclotomic ones, and the analogue of Theorem 1.1 is what class field theory provides: every abelian extension $K \subset L$ is contained in some *ray class field extension* $K \subset H_{\mathfrak{m}}$. Unfortunately, the theory does not provide a 'natural' system of generators for the fields $H_{\mathfrak{m}}$ that plays the role of the roots of unity in Theorem 1.1. Finding such a system for all $K$ is one of the Hilbert problems from 1900 that is still open. Notwithstanding this problem, class field theory is in principle constructive, and, once one finds in some way a possible generator of $H_{\mathfrak{m}}$ over $K$, it is not difficult to verify that it does generate $H_{\mathfrak{m}}$. The information we have on $H_{\mathfrak{m}}$ is essentially an intrinsic description, in terms of the splitting and ramification of the primes in the extension $K \subset H_{\mathfrak{m}}$, of the Galois group

Gal($H_\mathfrak{m}/K$) as a *ray class group* Cl$_\mathfrak{m}$. This group replaces the group $(\mathbb{Z}/m\mathbb{Z})^*$ that occurs implicitly in Theorem 1.1 as the underlying Galois group:

$$(\mathbb{Z}/m\mathbb{Z})^* \xrightarrow{\sim} \mathrm{Gal}(\mathbb{Q}(\zeta_m)/\mathbb{Q}), \quad (a \bmod m) \longmapsto (\sigma_a : \zeta_m \mapsto \zeta_m^a). \quad (1\text{-}2)$$

We can in principle find generators for any specific class field by combining our knowledge of its ramification data with a classical method to generate arbitrary solvable field extensions, namely, the adjunction of *radicals*. More formally, we call an extension $L$ of an arbitrary field $K$ a *radical extension* if $L$ is contained in the splitting field over $K$ of a finite collection of polynomials of the form $X^n - a$, with $n \in \mathbb{Z}_{\geq 1}$ not divisible by char($K$) and $a \in K$. If the collection of polynomials can be chosen so that $K$ contains a primitive $n$-th root of unity for each polynomial $X^n - a$ in the collection, then the radical extension $K \subset L$ is said to be a *Kummer extension*. Galois theory tells us that every Kummer extension is abelian and, conversely, that an abelian extension $K \subset L$ of exponent $n$ is Kummer if $K$ contains a primitive $n$-th root of unity. Here the *exponent* of an abelian extension $K \subset L$ is the smallest positive integer $n$ that annihilates Gal($L/K$). Thus, for every finite abelian extension $K \subset L$ of a number field $K$, there exists a cyclotomic extension $K \subset K(\zeta)$ such that the 'base-changed' extension $K(\zeta) \subset L(\zeta)$ is Kummer.

In Section 5, we compute the class fields of $K$ as subfields of Kummer extensions of $K(\zeta)$ for suitable cyclotomic extensions $K(\zeta)$ of $K$. The practical problem of the method is that the auxiliary fields $K(\zeta)$ may be much larger than the base field $K$, and this limits its use to not-too-large examples.

If $K$ is imaginary quadratic, elliptic curves with complex multiplication solve the Hilbert problem for $K$, and this yields methods that are much faster than the Kummer extension constructions for general $K$. We describe these complex multiplication methods in some detail in our Sections 6 to 8. We do not discuss their extension to abelian varieties with complex multiplication [Shimura 1998]; nor do we discuss the analytic generation of class fields of totally real number fields $K$ using *Stark units* [Cohen 2000, Chapter 6].

## 2. Class field theory

Class field theory generalizes Theorem 1.1 by focusing on the Galois group $(\mathbb{Z}/m\mathbb{Z})^*$ of the cyclotomic extension $\mathbb{Q} \subset \mathbb{Q}(\zeta_m)$ rather than on the specific generator $\zeta_m$. The extension $\mathbb{Q} \subset \mathbb{Q}(\zeta_m)$ is unramified at all primes $p \nmid m$, and the splitting behavior of such $p$ only depends on the residue class $(p \bmod m) \in (\mathbb{Z}/m\mathbb{Z})^*$. More precisely, the residue class degree $f_p = [\mathbf{F}_p(\zeta_m) : \mathbf{F}_p]$ of the primes over $p \nmid m$ equals the order of the *Frobenius automorphism* $(\sigma_p : \zeta_m \mapsto \zeta_m^p) \in \mathrm{Gal}(\mathbb{Q}(\zeta_m)/\mathbb{Q})$, and this is the order of $(p \bmod m) \in (\mathbb{Z}/m\mathbb{Z})^*$ under the standard identification (1-2).

Now let $K \subset L$ be *any* abelian extension of number fields. Then for each prime $\mathfrak{p}$ of $K$ that is unramified in $L$, by [Stevenhagen 2008, Section 15] there is a unique element $\mathrm{Frob}_\mathfrak{p} \in \mathrm{Gal}(L/K)$ that induces the Frobenius automorphism $x \mapsto x^{\#k_\mathfrak{p}}$ on the residue class field extensions $k_\mathfrak{p} \subset k_\mathfrak{q}$ for the primes $\mathfrak{q}$ in $L$ extending $\mathfrak{p}$. The order of this *Frobenius automorphism* $\mathrm{Frob}_\mathfrak{p}$ of $\mathfrak{p}$ in $\mathrm{Gal}(L/K)$ equals the residue class degree $[k_\mathfrak{q} : k_\mathfrak{p}]$, and the subgroup $\langle \mathrm{Frob}_\mathfrak{p} \rangle \subset \mathrm{Gal}(L/K)$ is the decomposition group of $\mathfrak{p}$.

We define the *Artin map* for $L/K$ as the homomorphism

$$\psi_{L/K} : I_K(\Delta_{L/K}) \longrightarrow \mathrm{Gal}(L/K), \quad \mathfrak{p} \longmapsto \mathrm{Frob}_\mathfrak{p} \qquad (2\text{-}1)$$

on the group $I_K(\Delta_{L/K})$ of fractional $\mathbb{Z}_K$-ideals generated by the primes $\mathfrak{p}$ of $K$ that do not divide the discriminant $\Delta_{L/K}$ of the extension $K \subset L$. Such primes $\mathfrak{p}$ are known to be unramified in $L$ by [Stevenhagen 2008, Theorem 8.5]. For an ideal $\mathfrak{a} \in I_K(\Delta_{L/K})$, we call $\psi_{L/K}(\mathfrak{a})$ the *Artin symbol* of $\mathfrak{a}$ in $\mathrm{Gal}(L/K)$.

For $K = \mathbb{Q}$, we can rephrase Theorem 1.1 as follows.

THEOREM 2.2 (KRONECKER–WEBER). *If $\mathbb{Q} \subset L$ is an abelian extension, there exists an integer $m \in \mathbb{Z}_{>0}$ such that the kernel of the Artin map $\psi_{L/\mathbb{Q}}$ contains all $\mathbb{Z}$-ideals $x\mathbb{Z}$ with $x > 0$ and $x \equiv 1 \bmod m$.*

The equivalence of Theorems 1.1 and 2.2 follows from the analytic fact that an extension of number fields is trivial if all primes outside a density zero subset split completely in it. Thus, if all primes $p \equiv 1 \bmod m$ split completely in $\mathbb{Q} \subset L$, then all primes of degree one are split in $\mathbb{Q}(\zeta_m) \subset L(\zeta_m)$ and $L$ is contained in the cyclotomic field $\mathbb{Q}(\zeta_m)$.

The positivity condition on $x$ in Theorem 2.2 can be omitted if the primes $p \equiv -1 \bmod m$ also split completely in $L$, that is, if $L$ is totally real and contained in the maximal real subfield $\mathbb{Q}(\zeta_m + \zeta_m^{-1})$ of $\mathbb{Q}(\zeta_m)$. The allowed values of $m$ in Theorem 2.2 are the multiples of some minimal positive integer, the *conductor* of $\mathbb{Q} \subset L$. It is the smallest integer $m$ for which $\mathbb{Q}(\zeta_m)$ contains $L$. The prime divisors of the conductor are exactly the primes that ramify in $L$, and $p^2$ divides the conductor if and only if $p$ is *wildly* ramified in $L$.

For a quadratic field $L$ of discriminant $d$, the conductor equals $|d|$, and Theorem 2.2 says that the Legendre symbol $\left(\frac{d}{x}\right)$ only depends on $x$ modulo $|d|$. This is Euler's version of the quadratic reciprocity law. The main statement of class field theory is the analogue of Theorem 2.2 over arbitrary number fields $K$.

THEOREM 2.3 (ARTIN'S RECIPROCITY LAW). *If $K \subset L$ is an abelian extension, there exists a nonzero ideal $\mathfrak{m}_0 \subset \mathbb{Z}_K$ such that the kernel of the Artin map $\psi_{L/K}$ in (2-1) contains all principal $\mathbb{Z}_K$-ideals $x\mathbb{Z}_K$ with $x$ totally positive and $x \equiv 1 \bmod \mathfrak{m}_0$.*

This innocuous-looking statement is highly nontrivial. It shows there is a powerful global connection relating the splitting behavior in $L$ of *different* primes of $K$. Just as Theorem 2.2 implies the quadratic reciprocity law, Artin's reciprocity law implies the general *power reciprocity laws* from algebraic number theory; see [Artin and Tate 1990, Chapter 12, §4; Cassels and Fröhlich 1967, p. 353].

It is customary to treat the positivity conditions at the real primes of $K$ and the congruence modulo $\mathfrak{m}_0$ in Theorem 2.3 on equal footing. To this end, one formally defines a *modulus* $\mathfrak{m}$ of $K$ to be a nonzero $\mathbb{Z}_K$-ideal $\mathfrak{m}_0$ times a subset $\mathfrak{m}_\infty$ of the real primes of $K$. For a modulus $\mathfrak{m} = \mathfrak{m}_0 \mathfrak{m}_\infty$, we write

$$x \equiv 1 \bmod^* \mathfrak{m}$$

if $x$ satisfies $\text{ord}_\mathfrak{p}(x-1) \geq \text{ord}_\mathfrak{p}(\mathfrak{m}_0)$ at the primes $\mathfrak{p}$ dividing the *finite part* $\mathfrak{m}_0$ and if $x$ is positive at the real primes in the *infinite part* $\mathfrak{m}_\infty$ of $\mathfrak{m}$.

In the language of moduli, Theorem 2.3 asserts that there exists a modulus $\mathfrak{m}$ such that the kernel $\ker \psi_{L/K}$ of the Artin map contains the *ray group* $R_\mathfrak{m}$ of principal $\mathbb{Z}_K$-ideals $x\mathbb{Z}_K$ generated by elements $x \equiv 1 \bmod^* \mathfrak{m}$. As in the case of Theorem 2.2, the set of these *admissible* moduli for $K \subset L$ consists of the multiples $\mathfrak{m}$ of some minimal modulus $\mathfrak{f}_{L/K}$, the *conductor* of $K \subset L$. The primes occurring in $\mathfrak{f}_{L/K}$ are the primes of $K$, both finite and infinite, that ramify in $L$. An infinite prime of $K$ is said to ramify in $L$ if it is real but has complex extensions to $L$. As for $K = \mathbb{Q}$, a finite prime $\mathfrak{p}$ occurs with higher multiplicity in the conductor if and only if it is wildly ramified in $L$.

If $\mathfrak{m} = \mathfrak{m}_0 \mathfrak{m}_\infty$ is an admissible modulus for $K \subset L$ and $I_\mathfrak{m}$ denotes the group of fractional $\mathbb{Z}_K$-ideals generated by the primes $\mathfrak{p}$ coprime to $\mathfrak{m}_0$, then the Artin map induces a homomorphism

$$\psi_{L/K} : \text{Cl}_\mathfrak{m} = I_\mathfrak{m}/R_\mathfrak{m} \longrightarrow \text{Gal}(L/K), \quad [\mathfrak{p}] \longmapsto \text{Frob}_\mathfrak{p} \qquad (2\text{-}4)$$

on the *ray class group* $\text{Cl}_\mathfrak{m} = I_\mathfrak{m}/R_\mathfrak{m}$ modulo $\mathfrak{m}$. Our earlier remark on the triviality of extensions in which almost all primes split completely implies that it is *surjective*. By the Chebotarev density theorem [Stevenhagen and Lenstra 1996], even more is true: the Frobenius automorphisms $\text{Frob}_\mathfrak{p}$ for $\mathfrak{p} \in I_\mathfrak{m}$ are *equidistributed* over the Galois group $\text{Gal}(L/K)$. In particular, a modulus $\mathfrak{m}$ is admissible for an abelian extension $K \subset L$ if and only if (almost) all primes $\mathfrak{p} \in R_\mathfrak{m}$ of $K$ split completely in $L$.

Since the order of the Frobenius automorphism $\text{Frob}_\mathfrak{p} \in \text{Gal}(L/K)$ equals the residue class degree $f_\mathfrak{p}$ of the primes $\mathfrak{q}$ in $L$ lying over $\mathfrak{p}$, the norm $N_{L/K}(\mathfrak{q}) = \mathfrak{p}^{f_\mathfrak{p}}$ of every prime ideal $\mathfrak{q}$ in $\mathbb{Z}_L$ coprime to $\mathfrak{m}$ is contained in the kernel of the Artin map. A nontrivial index calculation shows that the norms of the $\mathbb{Z}_L$-ideals coprime to $\mathfrak{m}$ actually generate the kernel in (2-4). In other words, the *ideal group* $A_\mathfrak{m} \subset I_\mathfrak{m}$ that corresponds to $L$, in the sense that we have $\ker \psi_{L/K} =$

$A_\mathfrak{m}/R_\mathfrak{m}$, is equal to
$$A_\mathfrak{m} = N_{L/K}(I_{\mathfrak{m}\mathbb{Z}_L}) \cdot R_\mathfrak{m}. \qquad (2\text{-}5)$$

The *existence theorem* from class field theory states that for every modulus $\mathfrak{m}$ of $K$, there exists an extension $K \subset L = H_\mathfrak{m}$ for which the map $\psi_{L/K}$ in (2-4) is an isomorphism. Inside some fixed algebraic closure $\overline{K}$ of $K$, the extension $H_\mathfrak{m}$ is uniquely determined as the maximal abelian extension $L$ of $K$ in which all primes in the ray group $R_\mathfrak{m}$ split completely. It is the *ray class field* $H_\mathfrak{m}$ modulo $\mathfrak{m}$ mentioned in the introduction, for which the analogue of Theorem 1.1 holds over $K$. If $K \subset L$ is abelian, we have $L \subset H_\mathfrak{m}$ whenever $\mathfrak{m}$ is an admissible modulus for $L$. For $L = H_\mathfrak{m}$, we have $A_\mathfrak{m} = R_\mathfrak{m}$ in (2-5) and an Artin isomorphism $\mathrm{Cl}_\mathfrak{m} \xrightarrow{\sim} \mathrm{Gal}(H_\mathfrak{m}/K)$.

EXAMPLE 2.6.1. It will not come as a surprise that for $K = \mathbb{Q}$, the ray class field modulo $(m) \cdot \infty$ is the cyclotomic field $\mathbb{Q}(\zeta_m)$, and the ray class group $\mathrm{Cl}_{(m) \cdot \infty}$ is the familiar group $(\mathbb{Z}/m\mathbb{Z})^*$ acting on the $m$-th roots of unity. Leaving out the real prime $\infty$ of $\mathbb{Q}$, we find the ray class field modulo $(m)$ to be the maximal real subfield $\mathbb{Q}(\zeta_m + \zeta_m^{-1})$ of $\mathbb{Q}(\zeta_m)$. This is the maximal subfield in which the real prime $\infty$ is unramified.

EXAMPLE 2.6.2. The ray class field of conductor $\mathfrak{m} = (1)$ is the *Hilbert class field* $H = H_1$ of $K$. It is the largest abelian extension of $K$ that is unramified at all primes of $K$, both finite and infinite. Since $I_1$ and $R_1$ are the groups of all fractional and all principal fractional $\mathbb{Z}_K$-ideals, respectively, the Galois group $\mathrm{Gal}(H/K)$ is isomorphic to the ordinary class group $\mathrm{Cl}_K$ of $K$, and the primes of $K$ that split completely in $H$ are precisely the *principal* prime ideals of $K$. This peculiar fact makes it possible to derive information about the class group of $K$ from the existence of unramified extensions of $K$, and conversely.

The ray group $R_\mathfrak{m}$ is contained in the subgroup $P_\mathfrak{m} \subset I_\mathfrak{m}$ of principal ideals in $I_\mathfrak{m}$, and the quotient $I_\mathfrak{m}/P_\mathfrak{m}$ is the class group $\mathrm{Cl}_K$ of $K$ for all $\mathfrak{m}$. Thus, the ray class group $\mathrm{Cl}_\mathfrak{m} = I_\mathfrak{m}/R_\mathfrak{m}$ is an extension of $\mathrm{Cl}_K$ by a finite abelian group $P_\mathfrak{m}/R_\mathfrak{m}$ that generalizes the groups $(\mathbb{Z}/m\mathbb{Z})^*$ from (1-2). More precisely, we have a natural exact sequence

$$\mathbb{Z}_K^* \longrightarrow (\mathbb{Z}_K/\mathfrak{m})^* \longrightarrow \mathrm{Cl}_\mathfrak{m} \longrightarrow \mathrm{Cl}_K \longrightarrow 0 \qquad (2\text{-}7)$$

in which the residue class of $x \in \mathbb{Z}_K$ coprime to $\mathfrak{m}_0$ in the finite group

$$(\mathbb{Z}_K/\mathfrak{m})^* = (\mathbb{Z}_K/\mathfrak{m}_0)^* \times \prod_{\mathfrak{p}|\mathfrak{m}_\infty} \langle -1 \rangle$$

consists of its ordinary residue class modulo $\mathfrak{m}_0$ and the signs of its images under the real primes $\mathfrak{p}|\mathfrak{m}_\infty$. This group naturally maps onto $P_\mathfrak{m}/R_\mathfrak{m} \subset \mathrm{Cl}_\mathfrak{m}$, with a kernel reflecting the fact that generators of principal $\mathbb{Z}_K$-ideals are only unique up to multiplication by units in $\mathbb{Z}_K$.

Interpreting both class groups in (2-7) as Galois groups, we see that all ray class fields contain the Hilbert class field $H = H_1$ from Example 2.6.2, and that we have an Artin isomorphism

$$(\mathbb{Z}_K/\mathfrak{m})^*/\mathrm{im}[\mathbb{Z}_K^*] \xrightarrow{\sim} \mathrm{Gal}(H_\mathfrak{m}/H) \tag{2-8}$$

for their Galois groups over $H$. By Example 2.6.1, this is a generalization of the isomorphism (1-2).

In class field theoretic terms, we may specify an abelian extension $K \subset L$ by giving an admissible modulus $\mathfrak{m}$ for the extension together with the corresponding ideal group

$$A_\mathfrak{m} = \ker[I_\mathfrak{m} \to \mathrm{Gal}(L/K)] \tag{2-9}$$

arising as the kernel of the Artin map (2-4). In this way, we obtain a *canonical bijection* between abelian extensions of $K$ inside $\overline{K}$ and ideal groups $R_\mathfrak{m} \subset A_\mathfrak{m} \subset I_\mathfrak{m}$ of $K$, provided that one allows for the fact that the 'same' ideal group $A_\mathfrak{m}$ can be defined modulo different multiples $\mathfrak{m}$ of its *conductor*, that is, the conductor of the corresponding extension. More precisely, we call the ideal groups $A_{\mathfrak{m}_1}$ and $A_{\mathfrak{m}_2}$ *equivalent* if they satisfy $A_{\mathfrak{m}_1} \cap I_\mathfrak{m} = A_{\mathfrak{m}_2} \cap I_\mathfrak{m}$ for some common multiple $\mathfrak{m}$ of $\mathfrak{m}_1$ and $\mathfrak{m}_2$.

Both from a theoretical and an algorithmic point of view, (2-5) provides an immediate description of the ideal group corresponding to $L$ as the *norm group* $A_\mathfrak{m} = N_{L/K}(I_{\mathfrak{m}\mathbb{Z}_L}) \cdot R_\mathfrak{m}$ as soon as we are able to find an admissible modulus $\mathfrak{m}$ for $L$. In the reverse direction, finding the *class field $L$* corresponding to an ideal group $A_\mathfrak{m}$ is much harder. Exhibiting practical algorithms to do so is the principal task of computational class field theory, and the topic of this paper. Already in the case of the Hilbert class field $H$ of $K$ from Example 2.6.2, we know no 'canonical' generator of $H$, and the problem is nontrivial.

## 3. Local aspects: ideles

Over $K = \mathbb{Q}$, all abelian Galois groups are described as quotients of the groups $(\mathbb{Z}/m\mathbb{Z})^*$ for some modulus $m \in \mathbb{Z}_{\geq 1}$. One may avoid the ubiquitous choice of moduli that arises when dealing with abelian fields by combining the Artin isomorphisms (1-2) at all 'finite levels' $m$ into a single *profinite* Artin isomorphism

$$\varprojlim_m (\mathbb{Z}/m\mathbb{Z})^* = \widehat{\mathbb{Z}}^* \xrightarrow{\sim} \mathrm{Gal}(\mathbb{Q}_{\mathrm{ab}}/\mathbb{Q}) \tag{3-1}$$

between the unit group $\widehat{\mathbb{Z}}^*$ of the profinite completion $\widehat{\mathbb{Z}}$ of $\mathbb{Z}$ and the absolute abelian Galois group of $\mathbb{Q}$. The group $\widehat{\mathbb{Z}}^*$ splits as a product $\prod_p \mathbb{Z}_p^*$ by the Chinese remainder theorem, and $\mathbb{Q}_{\mathrm{ab}}$ is obtained correspondingly as a compositum of the fields $\mathbb{Q}(\zeta_{p^\infty})$ generated by the $p$-power roots of unity. The automorphism corresponding to $u = (u_p)_p \in \widehat{\mathbb{Z}}^*$ acts as $\zeta \mapsto \zeta^{u_p}$ on $p$-power roots of unity. Note that the component group $\mathbb{Z}_p^* \subset \widehat{\mathbb{Z}}^*$ maps to the inertia group at $p$ in any finite quotient $\mathrm{Gal}(L/\mathbb{Q})$ of $\mathrm{Gal}(\mathbb{Q}_{\mathrm{ab}}/\mathbb{Q})$.

For arbitrary number fields $K$, one can take the projective limit in (2-7) over all moduli and describe $\mathrm{Gal}(K_{\mathrm{ab}}/K)$ by an exact sequence

$$1 \longrightarrow \mathbb{Z}_K^* \longrightarrow \widehat{\mathbb{Z}}_K^* \times \prod_{\mathfrak{p}\ \mathrm{real}} \langle -1 \rangle \xrightarrow{\psi_K} \mathrm{Gal}(K_{\mathrm{ab}}/K) \longrightarrow \mathrm{Cl}_K \longrightarrow 1, \quad (3\text{-}2)$$

which treats somewhat asymmetrically the finite primes occurring in $\widehat{\mathbb{Z}}_K^* = \prod_{\mathfrak{p}\ \mathrm{finite}} U_\mathfrak{p}$ and the infinite primes. Here $\psi_K$ maps the element $-1$ at a real prime $\mathfrak{p}$ to the complex conjugation at the extensions of $\mathfrak{p}$. The image of $\psi_K$ is the Galois group $\mathrm{Gal}(K_{\mathrm{ab}}/H)$ over the Hilbert class field $H$, which is of finite index $h_K = \#\mathrm{Cl}_K$ in $\mathrm{Gal}(K_{\mathrm{ab}}/K)$. For an abelian extension $L$ of $K$ containing $H$, the image of the component group $U_\mathfrak{p} \subset \widehat{\mathbb{Z}}_K^*$ in $\mathrm{Gal}(L/H)$ is again the inertia group at $\mathfrak{p}$ in $\mathrm{Gal}(L/K)$. As $H$ is totally unramified over $K$, the same is true if $L$ does not contain $H$: the inertia groups for $\mathfrak{p}$ in $\mathrm{Gal}(LH/K)$ and $\mathrm{Gal}(L/K)$ are isomorphic under the restriction map.

A more elegant description of $\mathrm{Gal}(K_{\mathrm{ab}}/K)$ than that provided by the sequence (3-2) is obtained if one treats all primes of $K$ in a uniform way and redefines the Artin map $\psi_K$ — as we will do in (3-7) — using the *idele group*

$$\mathbf{A}_K^* = {\prod_\mathfrak{p}}' K_\mathfrak{p}^* = \{(x_\mathfrak{p})_\mathfrak{p} : x_\mathfrak{p} \in U_\mathfrak{p} \text{ for almost all } \mathfrak{p}\}$$

of $K$. This group [Stevenhagen 2008, Section 14], consists of those elements in the Cartesian product of the multiplicative groups $K_\mathfrak{p}^*$ at *all* completions $K_\mathfrak{p}^*$ of $K$ that have their $\mathfrak{p}$-component in the local unit group $U_\mathfrak{p}$ for almost all $\mathfrak{p}$. Here $U_\mathfrak{p}$ is, as before, the unit group of the valuation ring at $\mathfrak{p}$ if $\mathfrak{p}$ is a finite prime of $K$; for infinite primes $\mathfrak{p}$, the choice of $U_\mathfrak{p}$ is irrelevant as there are only finitely many such $\mathfrak{p}$. We take $U_\mathfrak{p} = K_\mathfrak{p}^*$, and write $U_\infty$ to denote $\prod_{\mathfrak{p}\ \mathrm{infinite}} K_\mathfrak{p}^* = K \otimes_\mathbb{Q} \mathbb{R}$. Note that we have $\prod_{\mathfrak{p}\ \mathrm{finite}} U_\mathfrak{p}^* = \widehat{\mathbb{Z}}_K^*$.

The topology on $\mathbf{A}_K^*$ is the *restricted* product topology: elements are close if they are $\mathfrak{p}$-adically close at finitely many $\mathfrak{p}$ *and* have a quotient in $U_\mathfrak{p}$ for all other $\mathfrak{p}$. With this topology, $K^*$ embeds diagonally into $\mathbf{A}_K^*$ as a discrete subgroup. As the notation suggests, $\mathbf{A}_K^*$ is the unit group of the *adele ring* $\mathbf{A}_K = {\prod_\mathfrak{p}}' K_\mathfrak{p}$, the subring of $\prod_\mathfrak{p} K_\mathfrak{p}$ consisting of elements having integral components for almost all $\mathfrak{p}$.

To any idele $x = (x_\mathfrak{p})_\mathfrak{p}$, we can associate an ideal $x\mathbb{Z}_K = \prod_{\mathfrak{p}\ \mathrm{finite}} \mathfrak{p}^{\mathrm{ord}_\mathfrak{p}(x_\mathfrak{p})}$, and this makes the group $I_K$ of fractional $\mathbb{Z}_K$-ideals into a quotient of $\mathbf{A}_K^*$. For a global element $x \in K^* \subset \mathbf{A}_K^*$, the ideal $x\mathbb{Z}_K$ is the principal $\mathbb{Z}_K$-ideal generated by $x$, and so we have an exact sequence

$$1 \longrightarrow \mathbb{Z}_K^* \longrightarrow \widehat{\mathbb{Z}}_K^* \times U_\infty \longrightarrow \mathbf{A}_K^*/K^* \longrightarrow \mathrm{Cl}_K \longrightarrow 1 \quad (3\text{-}3)$$

that describes the *idele class group* $\mathbf{A}_K^*/K^*$ of $K$ in a way reminiscent of (3-2).

To obtain $\mathrm{Gal}(K_{\mathrm{ab}}/K)$ as a quotient of $\mathbf{A}_K^*/K^*$, we show that the ray class groups $\mathrm{Cl}_{\mathfrak{m}}$ defined in the previous section are natural quotients of $\mathbf{A}_K^*/K^*$. To do so, we associate to a modulus $\mathfrak{m} = \mathfrak{m}_0 \mathfrak{m}_\infty$ of $K$ an open subgroup $W_{\mathfrak{m}} \subset \mathbf{A}_K^*$, as follows. Write $\mathfrak{m} = \prod_{\mathfrak{p}} \mathfrak{p}^{n(\mathfrak{p})}$ as a formal product, with $n(\mathfrak{p}) = \mathrm{ord}_{\mathfrak{p}}(\mathfrak{m}_0)$ for finite $\mathfrak{p}$, and $n(\mathfrak{p}) \in \{0, 1\}$ to indicate the infinite $\mathfrak{p}$ in $\mathfrak{m}_\infty$. Now put

$$W_{\mathfrak{m}} = \prod_{\mathfrak{p}} U_{\mathfrak{p}}^{(n(\mathfrak{p}))}$$

for subgroups $U_{\mathfrak{p}}^{(k)} \subset K_{\mathfrak{p}}^*$ that are defined by

$$U_{\mathfrak{p}}^{(k)} = \begin{cases} U_{\mathfrak{p}} & \text{if } k = 0; \\ 1 + \mathfrak{p}^k & \text{if } \mathfrak{p} \text{ is finite and } k > 0; \\ U_{\mathfrak{p}}^+ \subset U_{\mathfrak{p}} = \mathbb{R}^* & \text{if } \mathfrak{p} \text{ is real and } k = 1. \end{cases}$$

Here we write $U_{\mathfrak{p}}^+$ for real $\mathfrak{p}$ to denote the subgroup of positive elements in $U_{\mathfrak{p}}$. Because $\mathbb{C}^*$ and $\mathbb{R}_{>0}^*$ have no proper open subgroups, one sees from the definition of the restricted product topology on $\mathbf{A}_K^*$ that a subgroup $H \subset \mathbf{A}_K^*$ is open if and only if it contains $W_{\mathfrak{m}}$ for some modulus $\mathfrak{m}$.

LEMMA 3.4. *For every modulus* $\mathfrak{m} = \prod_{\mathfrak{p}} \mathfrak{p}^{n(\mathfrak{p})}$ *of* $K$, *there is an isomorphism*

$$\mathbf{A}_K^*/K^* W_{\mathfrak{m}} \xrightarrow{\sim} \mathrm{Cl}_{\mathfrak{m}}$$

*that maps* $(x_{\mathfrak{p}})_{\mathfrak{p}}$ *to the class of* $\prod_{\mathfrak{p} \text{ finite}} \mathfrak{p}^{\mathrm{ord}_{\mathfrak{p}}(y x_{\mathfrak{p}})}$. *Here* $y \in K^*$ *is a global element satisfying* $y x_{\mathfrak{p}} \in U_{\mathfrak{p}}^{n(\mathfrak{p})}$ *for all* $\mathfrak{p} | \mathfrak{m}$.

PROOF. Note first that the global element $y$ required in the definition exists by the approximation theorem. The precise choice of $y$ is irrelevant, since for any two elements $y$ and $y'$ satisfying the requirement, we have $y/y' \equiv 1 \bmod^* \mathfrak{m}$. We obtain a homomorphism $\mathbf{A}_K^* \to \mathrm{Cl}_{\mathfrak{m}}$ that is surjective since it maps a prime element $\pi_{\mathfrak{p}}$ at a finite prime $\mathfrak{p} \nmid \mathfrak{m}$ to the class of $\mathfrak{p}$. Its kernel consists of the ideles that can be multiplied into $W_{\mathfrak{m}}$ by a global element $y \in K^*$. □

If $\mathfrak{m}$ is an admissible modulus for the finite abelian extension $K \subset L$, we can compose the isomorphism in Lemma 3.4 with the Artin map (2-4) for $K \subset L$ to obtain an idelic Artin map

$$\widehat{\psi}_{L/K} : \mathbf{A}_K^*/K^* \longrightarrow \mathrm{Gal}(L/K) \qquad (3\text{-}5)$$

that no longer refers to the choice of a modulus $\mathfrak{m}$. This map, which exists as a corollary of Theorem 2.3, is a continuous surjection that maps the class of a prime element $\pi_{\mathfrak{p}} \in K_{\mathfrak{p}}^* \subset \mathbf{A}_K^*$ to the Frobenius automorphism $\mathrm{Frob}_{\mathfrak{p}} \in \mathrm{Gal}(L/K)$ whenever $\mathfrak{p}$ is finite and unramified in $K \subset L$.

For a finite extension $L$ of $K$, the adele ring $\mathbf{A}_L$ is obtained from $\mathbf{A}_K$ by a base change $K \subset L$, so we have a norm map $N_{L/K} : \mathbf{A}_L \to \mathbf{A}_K$ that maps $\mathbf{A}_L^*$ to $\mathbf{A}_K^*$ and restricts to the field norm on $L^* \subset \mathbf{A}_L^*$. Since it induces the ideal norm $I_L \to I_K$ on the quotient $I_L$ of $\mathbf{A}_K^*$, one deduces that the kernel of (3-5) equals $(K^* \cdot N_{L/K}[\mathbf{A}_L^*])$ mod $K^*$, and that we have isomorphisms

$$\mathbf{A}_K^* / K^* N_{L/K}[\mathbf{A}_L^*] \cong I_\mathfrak{m}/A_\mathfrak{m} \xrightarrow{\sim} \mathrm{Gal}(L/K), \tag{3-6}$$

with $A_\mathfrak{m}$ the ideal group modulo $\mathfrak{m}$ that corresponds to $L$ in the sense of (2-9). Taking the limit in (3-5) over all finite abelian extensions $K \subset L$ inside $\overline{K}$, one obtains the idelic Artin map

$$\psi_K : \mathbf{A}_K^*/K^* \longrightarrow G_K^{\mathrm{ab}} = \mathrm{Gal}(K_{\mathrm{ab}}/K). \tag{3-7}$$

This is a continuous surjection that is uniquely determined by the property that the $\psi_K$-image of the class of a prime element $\pi_\mathfrak{p} \in K_\mathfrak{p}^* \subset \mathbf{A}_K^*$ maps to the Frobenius automorphism $\mathrm{Frob}_\mathfrak{p} \in \mathrm{Gal}(L/K)$ for every finite abelian extension $K \subset L$ in which $\mathfrak{p}$ is unramified. It exhibits all abelian Galois groups over $K$ as a quotient of the idele class group $\mathbf{A}_K^*/K^*$ of $K$.

The kernel of the Artin map (3-7) is the connected component of the unit element in $\mathbf{A}_K^*/K^*$. In the idelic formulation, the finite abelian extensions of $K$ inside $\overline{K}$ correspond bijectively to the open subgroups of $\mathbf{A}_K^*/K^*$ under the map

$$L \longmapsto \psi_K^{-1}[\mathrm{Gal}(K_{\mathrm{ab}}/L)] = (K^* \cdot N_{L/K}[\mathbf{A}_L^*]) \bmod K^*.$$

In this formulation, computational class field theory amounts to generating, for any given open subgroup of $\mathbf{A}_K^*/K^*$, the abelian extension $K \subset L$ corresponding to it.

EXAMPLE 3.8. Before continuing, let us see what the idelic reformulation of (3-1) comes down to for $K = \mathbb{Q}$. Every idele $x = ((x_p)_p, x_\infty) \in \mathbf{A}_\mathbb{Q}^*$ can uniquely be written as the product of the rational number

$$\mathrm{sign}(x_\infty) \prod_p p^{\mathrm{ord}_p(x_p)} \in \mathbb{Q}^*$$

and a 'unit idele' $u_x \in \prod_p \mathbb{Z}_p^* \times \mathbb{R}_{>0} = \widehat{\mathbb{Z}}^* \times \mathbb{R}_{>0}$. In this way, the Artin map (3-7) becomes a continuous surjection

$$\psi_\mathbb{Q} : \mathbf{A}_\mathbb{Q}^*/\mathbb{Q}^* \cong \widehat{\mathbb{Z}}^* \times \mathbb{R}_{>0} \longrightarrow \mathrm{Gal}(\mathbb{Q}_{\mathrm{ab}}/\mathbb{Q}).$$

Its kernel is the connected component $\{1\} \times \mathbb{R}_{>0}$ of the unit element in $\mathbf{A}_\mathbb{Q}^*/\mathbb{Q}^*$. Comparison with (3-1) leads to a commutative diagram of isomorphisms

$$\begin{array}{ccc} \widehat{\mathbb{Z}}^* & \xrightarrow{-1} & \widehat{\mathbb{Z}}^* \\ \text{can}\Big\downarrow\sim & (3\text{-}1)\Big\downarrow\sim & \\ \mathbf{A}_{\mathbb{Q}}^*/(\mathbb{Q}^*\cdot\mathbb{R}_{>0}) & \xrightarrow{\sim} & \mathrm{Gal}(\mathbb{Q}_{\mathrm{ab}}/\mathbb{Q}) \end{array} \qquad (3\text{-}8)$$

in which the upper horizontal map is *not* the identity. To see this, note that the class of the prime element $\ell \in \mathbb{Q}_\ell^* \subset \mathbf{A}_\mathbb{Q}^*$ in $\mathbf{A}_\mathbb{Q}^*/(\mathbb{Q}^*\cdot\mathbb{R}_{>0})$ is represented by the idele $x = (x_p)_p \in \widehat{\mathbb{Z}}^*$ having components $x_p = \ell^{-1}$ for $p \neq \ell$ and $x_\ell = 1$. This idele maps to the Frobenius of $\ell$, which raises roots of unity of order coprime to $\ell$ to their $\ell$-th power. Since $x$ is in all $W_\mathfrak{m}$ for all conductors $\mathfrak{m} = \ell^k$, it fixes $\ell$-power roots of unity. Thus, the upper isomorphism $-1$ is *inversion* on $\widehat{\mathbb{Z}}^*$.

Even though the idelic and the ideal group quotients on the left hand side of the arrow in (3-6) are the 'same' finite group, it is the idelic quotient that neatly encodes information at the *ramifying* primes $\mathfrak{p}|\mathfrak{m}$, which seem 'absent' in the other group. More precisely, we have for all primes $\mathfrak{p}$ an injective map $K_\mathfrak{p}^* \to \mathbf{A}_K^*/K^*$ that can be composed with (3-7) to obtain a *local* Artin map $\psi_{K_\mathfrak{p}} : K_\mathfrak{p}^* \to \mathrm{Gal}(L/K)$ at every prime $\mathfrak{p}$ of $K$. If $\mathfrak{p}$ is finite and unramified in $K \subset L$, we have $U_\mathfrak{p} \subset \ker \psi_{K_\mathfrak{p}}$ and an induced isomorphism of finite cyclic groups

$$K_\mathfrak{p}^*/\langle\pi_\mathfrak{p}^{f_\mathfrak{p}}\rangle U_\mathfrak{p} = K_\mathfrak{p}^*/N_{L_\mathfrak{q}/K_\mathfrak{p}}[L_\mathfrak{q}^*] \xrightarrow{\sim} \langle\mathrm{Frob}_\mathfrak{p}\rangle = \mathrm{Gal}(L_\mathfrak{q}/K_\mathfrak{p}),$$

since $\mathrm{Frob}_\mathfrak{p}$ generates the decomposition group of $\mathfrak{p}$ in $\mathrm{Gal}(L/K)$, which may be identified with the Galois group of the local extension $K_\mathfrak{p} \subset L_\mathfrak{q}$ at a prime $\mathfrak{q}|\mathfrak{p}$ in $L$. It is a nontrivial fact that (3-5) induces for *all* primes $\mathfrak{p}$ of $K$, including the ramifying and the infinite primes, a *local Artin isomorphism*

$$\psi_{L_\mathfrak{q}/K_\mathfrak{p}} : K_\mathfrak{p}^*/N_{L_\mathfrak{q}/K_\mathfrak{p}}[L_\mathfrak{q}^*] \xrightarrow{\sim} \mathrm{Gal}(L_\mathfrak{q}/K_\mathfrak{p}). \qquad (3\text{-}10)$$

In view of our observation after (3-2), it maps $U_\mathfrak{p}/N_{L_\mathfrak{q}/K_\mathfrak{p}}[U_\mathfrak{q}]$ for finite $\mathfrak{p}$ isomorphically onto the inertia group of $\mathfrak{p}$.

We can use (3-10) to *locally* compute the exponent $n(\mathfrak{p})$ to which $\mathfrak{p}$ occurs in the conductor of $K \subset L$: it is the smallest nonnegative integer $k$ for which we have $U_\mathfrak{p}^{(k)} \subset N_{L_\mathfrak{q}/K_\mathfrak{p}}[L_\mathfrak{q}^*]$. For unramified primes $\mathfrak{p}$ we obtain $n(\mathfrak{p}) = 0$, as the local norm is then surjective on the unit groups. For tamely ramified primes we have $n(\mathfrak{p}) = 1$, and for wildly ramified primes $\mathfrak{p}$, the exponent $n(\mathfrak{p})$ may be found by a local computation. In many cases it is sufficient to use an upper bound coming from the fact that every $d$-th power in $K_\mathfrak{p}^*$ is a norm from $L_\mathfrak{q}$, with $d$ the degree of $K_\mathfrak{p} \subset L_\mathfrak{q}$ (or even $K \subset L$). Using Hensel's Lemma [Buhler and Wagon 2008], one then finds

$$n(\mathfrak{p}) \leq e(\mathfrak{p}/p)\Big(\frac{1}{p-1} + \mathrm{ord}_p(e_\mathfrak{p})\Big) + 1, \qquad (3\text{-}11)$$

where $e(\mathfrak{p}/p)$ is the absolute ramification index of $\mathfrak{p}$ over the underlying rational prime $p$ and $e_\mathfrak{p}$ is the ramification index of $\mathfrak{p}$ in $K \subset L$. Note that $e_\mathfrak{p}$ is independent of the choice of an extension prime as $K \subset L$ is Galois.

## 4. Computing class fields: preparations

Our fundamental problem is the computation of the class field $L$ that corresponds to a given ideal group $A_\mathfrak{m}$ of $K$ in the sense of (2-5). One may 'give' $A_\mathfrak{m}$ by specifying $\mathfrak{m}$ and a list of ideals for which the classes in the ray class group $\text{Cl}_\mathfrak{m}$ generate $A_\mathfrak{m}$. The first step in computing $L$ is the computation of the group $I_\mathfrak{m}/A_\mathfrak{m}$ that will give us control of the Artin isomorphism $I_\mathfrak{m}/A_\mathfrak{m} \xrightarrow{\sim} \text{Gal}(L/K)$. Because linear algebra over $\mathbb{Z}$ provides us with good algorithms [Cohen 2000, Section 4.1] to deal with finite or even finitely generated abelian groups, this step essentially reduces to computing the finite group $\text{Cl}_\mathfrak{m}$ of which $I_\mathfrak{m}/A_\mathfrak{m}$ is quotient.

For the computation of the ray class group $\text{Cl}_\mathfrak{m}$ modulo $\mathfrak{m} = \mathfrak{m}_0 \cdot \mathfrak{m}_\infty$, one computes, in line with [Schoof 2008], the three other groups in the exact sequence (2-7) in which it occurs, and the maps between them. The class group $\text{Cl}_K$ and the unit group $\mathbb{Z}_K^*$ in (2-7) can be computed using the algorithm described in [Stevenhagen 2008, Section 12], which factors *smooth* elements of $\mathbb{Z}_K$ over a *factor base*. As this takes exponential time as a function of the base field $K$, it can only be done for moderately sized $K$. For the group $(\mathbb{Z}_K/\mathfrak{m}_0)^*$, one uses the Chinese remainder theorem to decompose it into a product of local multiplicative groups the form $(\mathbb{Z}_K/\mathfrak{p}^k)^*$. Here we need to assume that we are able to factor $\mathfrak{m}_0$, but this is a safe assumption as we are unlikely to deal with extensions for which we cannot even factor the conductor. The group $(\mathbb{Z}_K/\mathfrak{p}^k)^*$ is a product of the cyclic group $k_\mathfrak{p}^* = (\mathbb{Z}_K/\mathfrak{p})^*$ and the subgroup $(1+\mathfrak{p})/(1+\mathfrak{p}^k)$, the structure of which can be found inductively using the standard isomorphisms $(1+\mathfrak{p}^a)/(1+\mathfrak{p}^{a+1}) \cong k_\mathfrak{p}$ and, more efficiently,

$$(1+\mathfrak{p}^a)/(1+\mathfrak{p}^{2a}) \xrightarrow{\sim} \mathfrak{p}^a/\mathfrak{p}^{2a}$$

between multiplicative and additive quotients. In many cases, the result can be obtained in one stroke using the p-adic logarithm [Cohen 2000, Section 4.2.2]. Finding $\text{Cl}_\mathfrak{m}$ from the other groups in (2-7) is now a standard application of linear algebra over $\mathbb{Z}$. The quotient $I_\mathfrak{m}/A_\mathfrak{m}$ gives us an explicit description of the Galois group $\text{Gal}(L/K)$ in terms of Artin symbols of $\mathbb{Z}_K$-ideals.

For the ideal group $A_\mathfrak{m}$, we next compute its conductor $\mathfrak{f}$, which may be a proper divisor of $\mathfrak{m}$. This comes down to checking whether we have $A_\mathfrak{m} \supset I_\mathfrak{m} \cap R_\mathfrak{n}$ for some modulus $\mathfrak{n}|\mathfrak{m}$. Even in the case $A_\mathfrak{m} = R_\mathfrak{m}$, the conductor can be smaller than $\mathfrak{m}$, as the trivial isomorphism $(\mathbb{Z}/6\mathbb{Z})^* \xrightarrow{\sim} (\mathbb{Z}/3\mathbb{Z})^*$ of ray class groups over $K = \mathbb{Q}$ shows. The conductor $\mathfrak{f}$ obtained, which is the same as the conductor $\mathfrak{f}_{L/K}$ of the corresponding extension, is exactly divisible by the primes that ramify in $K \subset L$. In particular, we know the signature of $L$ from the real primes dividing $\mathfrak{f}$. With some extra effort, one can even compute the discriminant $\Delta_{L/K}$ using Hasse's *Führerdiskriminantenproduktformel*

$$\Delta_{L/K} = \prod_{\chi: I_\mathfrak{m}/A_\mathfrak{m} \to \mathbb{C}^*} \mathfrak{f}(\chi)_0. \qquad (4\text{-}1)$$

Here $\chi$ ranges over the characters of the finite group $I_\mathfrak{m}/A_\mathfrak{m} \cong \mathrm{Gal}(L/K)$, and $\mathfrak{f}(\chi)_0$ denotes the finite part of the conductor $\mathfrak{f}(\chi)$ of the ideal group $A_\chi$ modulo $\mathfrak{m}$ satisfying $A_\chi/A_\mathfrak{m} = \ker \chi$. All these quantities can be computed by the standard algorithms for finite abelian groups.

EXAMPLE 4.2. If $K \subset L$ is cyclic of prime degree $\ell$, we have a trivial character of conductor (1) and $\ell - 1$ characters of conductor $\mathfrak{f}_{L/K}$, so (4-1) reduces to

$$\Delta_{L/K} = (\mathfrak{f}_{L/K})_0^{\ell-1}.$$

In particular, we see that the discriminant of a quadratic extension $K \subset L$ is not only for $K = \mathbb{Q}$, but generally equal to the finite part of the conductor of the extension.

Having at our disposal the Galois group $\mathrm{Gal}(L/K)$, the discriminant $\Delta_{L/K}$, and the Artin isomorphism $I_\mathfrak{m}/A_\mathfrak{m} \xrightarrow{\sim} \mathrm{Gal}(L/K)$ describing the splitting behavior of the primes in $K \subset L$, we proceed with the computation of a generator for $L$ over $K$, that is, an irreducible polynomial in $K[X]$ with the property that its roots in $\overline{K}$ generate $L$.

Because the computation of class fields is not an easy computation, it is often desirable to decompose $\mathrm{Gal}(L/K)$ as a product $\prod_i \mathrm{Gal}(L_i/K)$ of Galois groups $\mathrm{Gal}(L_i/K)$ and to realize $L$ as a compositum of extensions $L_i$ that are computed separately. This way one can work with extensions $L/K$ that are cyclic of prime power order, or at least of prime power exponent. The necessary reduction of the global class field theoretic data for $L/K$ to those for each of the $L_i$ is only a short computation involving finite abelian groups.

## 5. Class fields as Kummer extensions

Let $K$ be *any* field containing a primitive $n$-th root of unity $\zeta_n$, and let $K \subset L$ be an abelian extension of *exponent* dividing $n$. In this situation, Kummer theory [Lang 2002, Chapter VIII, §6–8] tells us that $L$ can be obtained by adjoining to $K$ the $n$-th roots of certain elements of $K$. More precisely, let $W_L = K^* \cap L^{*n}$ be the subgroup of $K^*$ of elements that have an $n$-th root in $L$. Then we have

$L = K(\sqrt[n]{W_L})$, and there is the canonical *Kummer pairing*

$$\text{Gal}(L/K) \times W_L/K^{*n} \longrightarrow \langle \zeta_n \rangle$$
$$(\sigma, w) \longmapsto \langle \sigma, w \rangle = (w^{1/n})^{\sigma-1} = \frac{\sigma(\sqrt[n]{w})}{\sqrt[n]{w}}. \quad (5\text{-}1)$$

By *canonical*, we mean that the natural action of an automorphism $\tau \in \text{Aut}(\overline{K})$ on the pairing for $K \subset L$ yields the Kummer pairing for $\tau K \subset \tau L$, that is,

$$\langle \tau\sigma\tau^{-1}, \tau w \rangle = \langle \sigma, w \rangle^{\tau}. \quad (5\text{-}2)$$

The Kummer pairing is *perfect*, that is, it induces an isomorphism

$$W_L/K^{*n} \xrightarrow{\sim} \text{Hom}(\text{Gal}(L/K), \mathbb{C}^*). \quad (5\text{-}3)$$

In the case where $\text{Gal}(L/K)$ is cyclic of order $n$, this means that $L = K(\sqrt[n]{\alpha})$ and $W_L = K^* \cap L^{*n} = \langle \alpha \rangle \cdot K^{*n}$ for some $\alpha \in K$. If $\sqrt[n]{\beta}$ also generates $L$ over $K$, then $\alpha$ and $\beta$ are powers of each other modulo $n$-th powers.

We will apply Kummer theory to generate the class fields of a number field $K$. Thus, let $L$ be the class field of $K$ from Section 4 that is to be computed. Suppose that we have computed a 'small' modulus $\mathfrak{f}$ for $L$ that is only divisible by the ramifying primes, such as the conductor $\mathfrak{f}_{L/K}$, and an ideal group $A_\mathfrak{f}$ for $L$ by the methods of Section 4. With this information, we control the Galois group of our extension via the Artin isomorphism $I_\mathfrak{f}/A_\mathfrak{f} \xrightarrow{\sim} \text{Gal}(L/K)$. Let $n$ be the exponent of $\text{Gal}(L/K)$. Then we can directly apply Kummer theory if $K$ contains the required $n$-th roots of unity; if not, we need to pass to a cyclotomic extension of $K$ first. This leads to a natural case distinction.

*Case 1*: $K$ contains a primitive $n$-th root of unity $\zeta_n$. Under the restrictive assumption that $K$ contains $\zeta_n$, the class field $L$ is a Kummer extension of $K$, and generating $L = K(\sqrt[n]{W_L})$ comes down to finding generators for $W_L/K^{*n}$. We first compute a *finite* group containing $W_L/K^{*n}$. This reduction is a familiar ingredient from the *proofs* of class field theory [Artin and Tate 1990; Cassels and Fröhlich 1967].

LEMMA 5.4. *Let $K \subset L$ be finite abelian of exponent $n$, and assume $\zeta_n \in K$. Suppose $S$ is a finite set of primes of $K$ containing the infinite primes such that*

(1) *$K \subset L$ is unramified outside $S$;*
(2) *$\text{Cl}_K/\text{Cl}_K^n$ is generated by the classes of the finite primes in $S$.*

*Then the image of the group $U_S$ of $S$-units in $K^*/K^{*n}$ is finite of order $n^{\#S}$, and it contains the group $W_L/K^{*n}$ from (5-1).*

The first condition in Lemma 5.4 means that $S$ contains all the primes that divide our small modulus $\mathfrak{f}$. The second condition is automatic if the class number of $K$ is prime to $n$, and it is implied by the first if the classes of the ramifying primes generate $\text{Cl}_K/\text{Cl}_K^n$. Any set of elements of $\text{Cl}_K$ generating $\text{Cl}_K/\text{Cl}_K^n$ actually generates the full '$n$-part' of the class group, that is, the product of the $p$-Sylow subgroups of $\text{Cl}_K$ at the primes $p|n$. In general, there is a lot of freedom in the choice of primes in $S$ outside $\mathfrak{f}$. One tries to have $S$ 'small' in order to minimize the size $n^{\#S}$ of the group $(U_S \cdot K^{*n})/K^{*n}$ containing $W_L/K^{*n}$.

PROOF OF LEMMA 5.4. By the Dirichlet unit theorem [Stevenhagen 2008, Theorem 10.9], the group $U_S$ of $S$-units of $K$ is isomorphic to $\mu_K \times \mathbb{Z}^{\#S-1}$. As $\mu_K$ contains $\zeta_n$, the image $(U_S \cdot K^{*n})/K^{*n} \cong U_S/U_S^n$ of $U_S$ in $K^*/K^{*n}$ is finite of order $n^{\#S}$.

To show that $(U_S \cdot K^{*n})/K^{*n}$ contains $W_L/K^{*n}$, pick any $\alpha \in W_L$. Since $K \subset K(\sqrt[n]{\alpha})$ is unramified outside $S$, we have $(\alpha) = \mathfrak{a}_S \mathfrak{b}^n$ for some product $\mathfrak{a}_S$ of prime ideals in $S$ and $\mathfrak{b}$ coprime to all finite primes in $S$. As the primes in $S$ generate the $n$-part of $\text{Cl}_K$, we can write $\mathfrak{b} = \mathfrak{b}_S \mathfrak{c}$ with $\mathfrak{b}_S$ a product of prime ideals in $S$ and $\mathfrak{c}$ an ideal of which the class in $\text{Cl}_K$ is of order $u$ coprime to $n$. Now $\alpha^u$ generates an ideal of the form $(\alpha^u) = \mathfrak{a}'_S(\gamma^n)$ with $\mathfrak{a}'_S$ a product of prime ideals in $S$ and $\gamma \in K^*$. It follows that $\alpha^u \gamma^{-n} \in K^*$ is an $S$-unit, and so $\alpha^u$ and therefore $\alpha$ is contained in $U_S \cdot K^{*n}$. □

In the situation of Lemma 5.4, we see that $K \subset L$ is a subextension of the Kummer extension $K \subset N = K(\sqrt[n]{U_S})$ of degree $n^{\#S}$. We have to find the subgroup of $U_S/U_S^n$ corresponding to $L$. This amounts to a computation in linear algebra using the Artin map and the Kummer pairing. For ease of exposition, we assume that the set $S$ we choose to satisfy Lemma 5.4 contains all primes dividing $n$. This implies that $N$ is the maximal abelian extension of exponent $n$ of $K$ that is unramified outside $S$.

As we compute $L$ as a subfield of the abelian extension $K \subset N$, we replace the modulus $\mathfrak{f}$ of $K \subset L$ by some multiple $\mathfrak{m}$ that is an admissible modulus for $K \subset N$. Clearly $\mathfrak{m}$ only needs to be divisible by the ramified primes in $K \subset N$, which are all in $S$. Wild ramification only occurs at primes $\mathfrak{p}$ dividing $n$, and for these primes we can take $\text{ord}_{\mathfrak{p}}(\mathfrak{m})$ equal to the bound given by (3-11). The ideal group modulo $\mathfrak{m}$ corresponding to $N$ is $I_{\mathfrak{m}}^n \cdot P_{\mathfrak{m}}$ because $N$ is the maximal exponent-$n$ extension of $K$ of conductor $\mathfrak{m}$; hence the Artin map for $K \subset N$ is

$$I_{\mathfrak{m}} \longrightarrow I_{\mathfrak{m}}/(I_{\mathfrak{m}}^n \cdot P_{\mathfrak{m}}) = \text{Cl}_{\mathfrak{m}}/\text{Cl}_{\mathfrak{m}}^n \xrightarrow{\sim} \text{Gal}(N/K). \qquad (5\text{-}5)$$

The induced map $I_{\mathfrak{m}} \to \text{Gal}(L/K)$ is the Artin map for $K \subset L$, which has the ideal group $A_{\mathfrak{m}}$ corresponding to $L$ as its kernel. Let $\Sigma_L \subset I_{\mathfrak{m}}$ be a finite set of ideals of which the classes generate the $\mathbb{Z}/n\mathbb{Z}$-module $A_{\mathfrak{m}}/(I_{\mathfrak{m}}^n \cdot P_{\mathfrak{m}}) \cong \text{Gal}(N/L)$. We then have to determine the subgroup $V_L \subset U_S$ consisting of those $S$-units $v \in U_S$ that have the property that $\sqrt[n]{v}$ is left invariant by the Artin symbols of all ideals in $\Sigma_L$, since the class field we are after is $L = K(\sqrt[n]{V_L})$.

We are here in a situation to apply linear algebra over $\mathbb{Z}/n\mathbb{Z}$, because the Kummer pairing (5-1) tells us that the action of the Artin symbols $\psi_{N/K}(\mathfrak{a})$ of the ideals $\mathfrak{a} \in I_\mathfrak{m}$ on the $n$-th roots of the $S$-units is described by the pairing of $\mathbb{Z}/n\mathbb{Z}$-modules given by

$$I_\mathfrak{m}/I_\mathfrak{m}^n \times U_S/U_S^n \longrightarrow \langle \zeta_n \rangle$$
$$(\mathfrak{a}, u) \longmapsto \langle \psi_{N/K}(\mathfrak{a}), u \rangle = (u^{1/n})^{\psi_{N/K}(\mathfrak{a})-1}. \quad (5\text{-}6)$$

Making this computationally explicit amounts to computing the pairing for some choice of basis elements of the three modules involved.

For $\langle \zeta_n \rangle$ we have the obvious $\mathbb{Z}/n\mathbb{Z}$-generator $\zeta_n$, and $I_\mathfrak{m}/I_\mathfrak{m}^n$ is a free $\mathbb{Z}/n\mathbb{Z}$-module generated by the primes $\mathfrak{p} \notin S$. If $K$ is of moderate degree, the general algorithm [Stevenhagen 2008, Section 12] for computing units and class groups can be used to compute generators for $U_S$, which then form a $\mathbb{Z}/n\mathbb{Z}$-basis for $U_S/U_S^n$. In fact, finding $s - 1 = \#S - 1$ independent units in $U_S$ that generate a subgroup of index coprime to $n$ is enough: together with a root of unity generating $\mu_K$, these will generate $U_S/U_S^n$. This is somewhat easier than finding actual generators for $U_S$, because maximality modulo $n$-th powers is not difficult to establish for a subgroup $U \subset U_S$ having the right rank $s = \#S$. Indeed, each reduction modulo a small prime $\mathfrak{p} \notin S$ provides a character $U \subset U_S \to k_\mathfrak{p}^*/(k_\mathfrak{p}^*)^n \cong \langle \zeta_n \rangle$, the $n$-th power residue symbol at $\mathfrak{p}$. By finding $s$ independent characters, one shows that the intersection of their kernels equals $U^n = U \cap U_S^n$.

For a prime $\mathfrak{p} \notin S$ and $u \in U_S$, the definitions of the Kummer pairing and the Frobenius automorphism yield

$$\langle \text{Frob}_\mathfrak{p}, u \rangle = (u^{1/n})^{\text{Frob}_\mathfrak{p} - 1} \equiv u^{(N\mathfrak{p}-1)/n} \in k_\mathfrak{p}^*,$$

where $N\mathfrak{p} = \#k_\mathfrak{p}$ is the absolute norm of $\mathfrak{p}$. Thus $\langle \text{Frob}_\mathfrak{p}, u \rangle$ is simply the power of $\zeta_n$ that is congruent to $u^{(N\mathfrak{p}-1)/n} \in k_\mathfrak{p}^*$. Even when $\mathfrak{p}$ is large, this is not an expensive discrete logarithm problem in $k_\mathfrak{p}^*$, since in practice the exponent $n \leq [L : K]$ is small: one can simply check all powers of $\bar\zeta_n \in k_\mathfrak{p}^*$. Since $\langle \zeta_n \rangle$ reduces injectively modulo primes $\mathfrak{p} \nmid n$, the $n$-root of unity $\langle \text{Frob}_\mathfrak{p}, u \rangle$ can be recovered from its value in $k_\mathfrak{p}^*$.

From the values $\langle \text{Frob}_\mathfrak{p}, u \rangle$, we compute all symbols $\langle \psi_{N/K}(\mathfrak{a}), u \rangle$ by linearity. It is now a standard computation in linear algebra to find generators for the subgroup $V_L/U_S^n \subset U_S/U_S^n$ that is annihilated by the ideals $\mathfrak{a} \in \Sigma_L$ under the pairing (5-6). This yields explicit generators for the Kummer extension $L = K(\sqrt[n]{V_L})$, and concludes the computation of $L$ in the case where $K$ contains $\zeta_n$, with $n$ the exponent of $\text{Gal}(L/K)$.

*Case 2*: $K$ does not contain $\zeta_n$. In this case $L$ is not a Kummer extension of $K$, but $L' = L(\zeta_n)$ is a Kummer extension of $K' = K(\zeta_n)$.

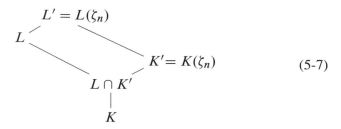

To find generators of $L'$ over $K'$ by the method of Case 1, we need to 'lift' the class field theoretic data from $K$ to $K'$ to describe $L'$ as a class field of $K'$. Lifting the modulus $\mathfrak{f} = \mathfrak{f}_0 \mathfrak{f}_\infty$ for $K \subset L$ is easy: as $K'$ is totally complex, $\mathfrak{f}' = \mathfrak{f}_0 \mathbb{Z}_{K'}$ is admissible for $K' \subset L'$. From the definition of the Frobenius automorphism, it is immediate that we have a commutative diagram

$$\begin{array}{ccc} I_{K',\mathfrak{f}'} & \xrightarrow{\text{Artin}} & \text{Gal}(L'/K') \\ \downarrow N_{K'/K} & & \downarrow \text{res} \\ I_{K,\mathfrak{f}} & \xrightarrow{\text{Artin}} & \text{Gal}(L/K) \end{array}$$

As the restriction map on the Galois groups is injective, we see that the inverse norm image $N_{K'/K}^{-1} A_{\mathfrak{f}} \subset I_{K',\mathfrak{f}'}$ is the ideal group of $K'$ corresponding to the extension $K' \subset L'$. Because $N_{K'/K}^{-1} A_{\mathfrak{f}}$ contains $P_{K',\mathfrak{f}'}$, computing this inverse image takes place inside the finite group $\text{Cl}_{K',\mathfrak{f}'}$, a ray class group for $K'$.

We perform the algorithm from Case 1 for the extension $K' \subset L'$ to find generators of $L'$ over $K'$. We are then working with (ray) class groups and $S$-units in $K'$ rather than in $K$, and $S$ has to satisfy Lemma 5.4 condition (2) for $\text{Cl}_{K'}$. All this is only feasible if $K'$ is of moderate degree, and this seriously restricts the values of $n$ one can handle in practice. Our earlier observation that we may decompose $I_{\mathfrak{f}}/A_{\mathfrak{f}} \cong \text{Gal}(L/K)$ into a product of cyclic groups of prime power order and generate $L$ accordingly as a compositum of cyclic extensions of $K$ is particularly relevant in this context, as it reduces our problem to a number of instances where $K \subset L$ is cyclic of prime power degree. Current implementations [Fieker 2001] deal with prime power values up to 20.

We further assume for simplicity that we are indeed in the case where $K \subset L$ is cyclic of prime power degree $n$, with $K' = K(\zeta_n) \neq K$. Suppose that, using the algorithm from Case 1, we have computed a Kummer generator $\theta \in L'$ for which we have $L' = K'(\theta) = K(\zeta_n, \theta)$ and $\theta^n = \alpha \in K'$. We then need to 'descend' $\theta$ efficiently to a generator $\eta$ of $L$ over $K$. If $n$ is prime, one has

$L = K(\eta)$ for the trace

$$\eta = \mathrm{Tr}_{L'/L}(\theta). \qquad (5\text{-}8)$$

For prime powers this does not work in all cases. One can however replace $\theta$ by $\theta + k\zeta_n$ for some small integer $k \in \mathbb{Z}$ to ensure that $\theta$ generates $L'$ over $K$, and then general field theory tells us that the coefficients of the irreducible polynomial

$$f_L^\theta = \prod_{\tau \in \mathrm{Gal}(L'/L)} (X - \tau(\theta)) \in L[X] \qquad (5\text{-}9)$$

of $\theta$ over $L$ generate $L$ over $K$. As we took $K \subset L$ to be cyclic of prime power degree, one of the coefficients is actually a generator, and in practice the trace works. In all cases, one needs an explicit description of the action of the Galois group $\mathrm{Gal}(L'/L)$ on $\theta$ and $\zeta_n$ in order to compute the trace (5-8), and possibly other coefficients of $f_L^\theta$ in (5-9). Finally, if we have $L = K(\eta)$, we need the action of $\mathrm{Gal}(L/K)$ on $\eta$ in order to write down the generating polynomial

$$f_K^\eta = \prod_{\sigma \in \mathrm{Gal}(L/K)} (X - \sigma(\eta)) \in K[X]$$

for $K \subset L$ that we are after.

As before, the Artin map gives us complete control over the action of the abelian Galois group $\mathrm{Gal}(L'/K)$ on $L' = K(\zeta_n, \theta)$, provided that we describe the elements of $\mathrm{Gal}(L'/K)$ as Artin symbols. We let $\mathfrak{m}$ be an admissible modulus for $K \subset L'$; the least common multiple of $\mathfrak{f}_{L/K}$ and $n \cdot \prod_{\mathfrak{p}\text{ real}} \mathfrak{p}$ is an obvious choice for $\mathfrak{m}$. All we need to know is the explicit action of the Frobenius automorphism $\mathrm{Frob}_\mathfrak{p} \in \mathrm{Gal}(L'/K)$ of a prime $\mathfrak{p} \nmid \mathfrak{m}$ of $K$ on the generators $\zeta_n$ and $\theta$ of $L'$ over $K$. Note that $\mathfrak{p}$ does not divide $n$, and that we may assume that $\alpha = \theta^n$ is a unit at $\mathfrak{p}$.

The cyclotomic action of $\mathrm{Frob}_\mathfrak{p} \in \mathrm{Gal}(L'/K)$ is given by $\mathrm{Frob}_\mathfrak{p}(\zeta_n) = \zeta_n^{N\mathfrak{p}}$, with $N\mathfrak{p} = \#k_\mathfrak{p}$ the absolute norm of $\mathfrak{p}$. This provides us with the Galois action on $K'$ and yields canonical isomorphisms

$$\mathrm{Gal}(K'/K) \cong \mathrm{im}[N_{K/\mathbb{Q}} : I_\mathfrak{m} \longrightarrow (\mathbb{Z}/n\mathbb{Z})^*],$$
$$\mathrm{Gal}(K'/(L \cap K')) \cong \mathrm{im}[N_{K/\mathbb{Q}} : A_\mathfrak{m} \longrightarrow (\mathbb{Z}/n\mathbb{Z})^*].$$

In order to understand the action of $\mathrm{Frob}_\mathfrak{p}$ on $\theta = \sqrt[n]{\alpha}$, we first observe that $K \subset L' = K'(\theta)$ can only be abelian if $\alpha$ is in the *cyclotomic* eigenspace of $(K')^*$ modulo $n$-th powers under the action of $\mathrm{Gal}(K'/K)$. More precisely, applying (5-2) for $K' \subset L'$ with $\tau = \mathrm{Frob}_\mathfrak{p}$, we have $\mathrm{Frob}_\mathfrak{p} \cdot \sigma \cdot \mathrm{Frob}_\mathfrak{p}^{-1} = \sigma$ since $\mathrm{Gal}(L'/K)$ is abelian, and therefore

$$\langle \sigma, \mathrm{Frob}_\mathfrak{p}(\alpha) \rangle = \langle \sigma, \alpha \rangle^{\mathrm{Frob}_\mathfrak{p}} = \langle \sigma, \alpha \rangle^{N\mathfrak{p}} = \langle \sigma, \alpha^{N\mathfrak{p}} \rangle$$

for all $\sigma \in \mathrm{Gal}(L'/K')$. By (5-3), we conclude that

$$\mathrm{Frob}_{\mathfrak{p}}(\alpha) = \alpha^{N\mathfrak{p}} \cdot \gamma_{\mathfrak{p}}^{n} \qquad (5\text{-}10)$$

for some element $\gamma_{\mathfrak{p}} \in K'$. Knowing how $\mathrm{Frob}_{\mathfrak{p}}$ acts on $\alpha \in K' = K(\zeta_n)$, we can compute $\gamma_{\mathfrak{p}}$ by extracting some $n$-th root of $\mathrm{Frob}_{\mathfrak{p}}(\alpha)\alpha^{-N\mathfrak{p}}$ in $K'$. The element $\gamma_{\mathfrak{p}}$ is only determined up to multiplication by $n$-th roots of unity by (5-10). Because we took $\alpha$ to be a unit at $\mathfrak{p}$, we have $\gamma_{\mathfrak{p}}^n \equiv 1 \bmod \mathfrak{p}$ by definition of the Frobenius automorphism, and so there is a unique element $\gamma_{\mathfrak{p}} \equiv 1 \bmod \mathfrak{p}$ satisfying (5-10). With this choice of $\gamma_{\mathfrak{p}}$, we have

$$\mathrm{Frob}_{\mathfrak{p}}(\theta) = \theta^{N\mathfrak{p}} \cdot \gamma_{\mathfrak{p}}$$

because the $n$-th powers of both quantities are the same by (5-10), and they are congruent modulo $\mathfrak{p}$. This provides us with the explicit Galois action of $\mathrm{Frob}_{\mathfrak{p}}$ on $\theta$ for unramified primes $\mathfrak{p}$.

The description of the Galois action on $\theta$ and $\zeta_n$ in terms of Frobenius symbols is all we need. The Galois group $\mathrm{Gal}(L'/L) \cong \mathrm{Gal}(K'/(L \cap K'))$, which we may identify with the subgroup $N_{K/\mathbb{Q}}(A_{\mathfrak{m}})$ of $(\mathbb{Z}/n\mathbb{Z})^*$, is either cyclic or, if $n$ is a power of 2, generated by 2 elements. Picking one or two primes $\mathfrak{p}$ in $A_{\mathfrak{m}}$ with norms in suitable residue classes modulo $n$ is all it takes to generate $\mathrm{Gal}(L'/L)$ by Frobenius automorphisms, and we can use these elements to descend $\theta$ to a generator $\eta$ for $L$ over $K$. We also control the Galois action of $\mathrm{Gal}(L/K) = I_{\mathfrak{m}}/A_{\mathfrak{m}}$ on $\eta$, and this makes it possible to compute the irreducible polynomial $f_K^{\eta}$ for the generator $\eta$ of $L$ over $K$.

## 6. Class fields arising from complex multiplication

As we observed in Example 2.6.1, the ray class fields over the rational number field $\mathbb{Q}$ are the cyclotomic fields. For these fields, we have explicit generators over $\mathbb{Q}$ that arise 'naturally' as the values of the analytic function $q : x \mapsto e^{2\pi i x}$ on the unique archimedean completion $\mathbb{R}$ of $\mathbb{Q}$. The function $q$ is periodic modulo the ring of integers $\mathbb{Z} \subset \mathbb{R}$ of $\mathbb{Q}$, and it induces an isomorphism

$$\begin{aligned}\mathbb{R}/\mathbb{Z} &\xrightarrow{\sim} T = \{z \in \mathbb{C} : z\bar{z} = 1\} \subset \mathbb{C} \\ x &\mapsto q(x) = e^{2\pi i x}\end{aligned} \qquad (6\text{-}1)$$

between the quotient group $\mathbb{R}/\mathbb{Z}$ and the 'circle group' $T$ of complex numbers of absolute value 1. The Kronecker–Weber theorem 1.1 states that the values of the analytic function $q$ at the points of the *torsion subgroup* $\mathbb{Q}/\mathbb{Z} \subset \mathbb{R}/\mathbb{Z}$ generate the maximal abelian extension $\mathbb{Q}_{\mathrm{ab}}$ of $\mathbb{Q}$. More precisely, the $q$-values at the $m$-torsion subgroup $\frac{1}{m}\mathbb{Z}/\mathbb{Z}$ of $\mathbb{R}/\mathbb{Z}$ generate the $m$-th cyclotomic field $\mathbb{Q}(\zeta_m)$. Under this parametrization of roots of unity by $\mathbb{Q}/\mathbb{Z}$, the Galois action on the $m$-torsion values comes from multiplications on $\frac{1}{m}\mathbb{Z}/\mathbb{Z}$ by integers $a \in \mathbb{Z}$ coprime to $m$, giving rise to the Galois group

$$\mathrm{Gal}(\mathbb{Q}(\zeta_m)/\mathbb{Q}) = \mathrm{Aut}(\tfrac{1}{m}\mathbb{Z}/\mathbb{Z}) = (\mathbb{Z}/m\mathbb{Z})^* \qquad (6\text{-}2)$$

from (1-2). Taking the projective limit over all $m$, one obtains the identification of $\mathrm{Gal}(\mathbb{Q}_{\mathrm{ab}}/\mathbb{Q})$ with $\mathrm{Aut}(\mathbb{Q}/\mathbb{Z}) = \widehat{\mathbb{Z}}^*$ from (3-1), and we saw in Example 3.8 that the relation with the Artin isomorphism is given by the commutative diagram (3-8). To stress the analogy with the complex multiplication case, we rewrite (3-8) as

$$\begin{array}{ccc} \widehat{\mathbb{Z}}^* & \xrightarrow{\;-1\;} & \widehat{\mathbb{Z}}^* = \mathbf{A}_{\mathbb{Q}}^*/(\mathbb{Q}^* \cdot \mathbb{R}_{>0}) \\ {\scriptstyle\mathrm{can}}\downarrow\wr & & \wr\downarrow{\scriptstyle\mathrm{Artin}} \\ \mathrm{Aut}(\mathbb{Q}/\mathbb{Z}) & \xrightarrow{\;\sim\;} & \mathrm{Gal}(\mathbb{Q}_{\mathrm{ab}}/\mathbb{Q}) \end{array} \qquad (6\text{-}3)$$

where $-1$ denotes inversion on $\widehat{\mathbb{Z}}^*$.

From now on, we take $K$ to be an imaginary quadratic field. Then $K$ has a single archimedean completion $K \to \mathbb{C}$, and much of what we said for the analytic function $q$ on $\mathbb{R}/\mathbb{Z}$ has an analogue for the quotient group $\mathbb{C}/\mathbb{Z}_K$. In complete analogy, we will define an analytic function $f_K : \mathbb{C}/\mathbb{Z}_K \to \mathbb{P}^1(\mathbb{C})$ in (6-14) with the property that its finite values at the $m$-torsion subgroup $\frac{1}{m}\mathbb{Z}_K/\mathbb{Z}_K$ of $\mathbb{C}/\mathbb{Z}_K$ generate the ray class field $H_m$ of $K$ of conductor $m\mathbb{Z}_K$. However, to define this *elliptic function* $f_K$ on the complex *elliptic curve* $\mathbb{C}/\mathbb{Z}_K$, we need an algebraic description of $\mathbb{C}/\mathbb{Z}_K$, which exists over an *extension* of $K$ that is usually larger than $K$ itself. The Hilbert class field $H = H_1$ of $K$ from Example 2.6.2 is the smallest extension of $K$ that one can use, and the torsion values of $f_K$ generate class fields over $H$. This makes the construction of $H$ itself into an important preliminary step that does not occur over $\mathbb{Q}$, as $\mathbb{Q}$ is its own Hilbert class field.

In this section, we give the classical algorithms for constructing the extensions $K \subset H$ and $H \subset H_m$. The next section provides some theoretical background and different views on complex multiplication. Our final Section 8 shows how such views lead to algorithmic improvements.

Complex multiplication starts with the fundamental observation [Silverman 1986, Chapter VI] that for every lattice $\Lambda \subset \mathbb{C}$, the complex torus $\mathbb{C}/\Lambda$ admits a meromorphic function, the Weierstrass $\wp$-function

$$\wp_\Lambda : z \mapsto z^{-2} + \sum_{\omega \in \Lambda \setminus \{0\}} [(z-\omega)^{-2} - \omega^{-2}],$$

that has period lattice $\Lambda$ and is holomorphic except for double poles at the points of $\Lambda$. The corresponding Weierstrass map

$$W : \mathbb{C}/\Lambda \longrightarrow E_\Lambda \subset \mathbb{P}^2(\mathbb{C}), \quad z \mapsto [\wp_\Lambda(z) : \wp'_\Lambda(z) : 1]$$

is a complex analytic isomorphism between the torus $\mathbb{C}/\Lambda$ and the complex elliptic curve $E_\Lambda \subset \mathbb{P}^2(\mathbb{C})$ defined by the affine Weierstrass equation

$$y^2 = 4x^3 - g_2(\Lambda)x - g_3(\Lambda).$$

The Weierstrass coefficients

$$g_2(\Lambda) = 60 \sum_{\omega \in \Lambda \setminus \{0\}} \omega^{-4} \quad \text{and} \quad g_3(\Lambda) = 140 \sum_{\omega \in \Lambda \setminus \{0\}} \omega^{-6} \qquad (6\text{-}4)$$

of $E_\Lambda$ are the *Eisenstein series* of weight 4 and 6 for the lattice $\Lambda$. The natural addition on $\mathbb{C}/\Lambda$ translates into an algebraic group structure on $E_\Lambda(\mathbb{C})$ sometimes referred to as 'chord and tangent addition'. On the Weierstrass model $E_\Lambda$, the point $O = [0 : 1 : 0] = W(0 \bmod \Lambda)$ at infinity is the zero point, and any line in $\mathbb{P}^2(\mathbb{C})$ intersects the curve $E_\Lambda$ in 3 points, counting multiplicities, that have sum $O$.

All complex analytic maps $\mathbb{C}/\Lambda_1 \to \mathbb{C}/\Lambda_2$ fixing the zero point are multiplications $z \mapsto \lambda z$ with $\lambda \in \mathbb{C}$ satisfying $\lambda \Lambda_1 \subset \Lambda_2$. These are clearly group homomorphisms, and in the commutative diagram

$$\begin{array}{ccc} \mathbb{C}/\Lambda_1 & \xrightarrow{\lambda} & \mathbb{C}/\Lambda_2 \\ W_1 \downarrow \sim & & W_1 \downarrow \sim \\ E_{\Lambda_1} & \xrightarrow{\phi_\lambda} & E_{\Lambda_2} \end{array} \qquad (6\text{-}5)$$

the corresponding maps $\phi_\lambda : E_{\Lambda_1} \longrightarrow E_{\Lambda_2}$ between algebraic curves are known as *isogenies*. For $\lambda \neq 0$, the isogeny $\phi_\lambda$ is a finite algebraic map of degree $[\Lambda_2 : \lambda \Lambda_1]$, and $E_{\Lambda_1}$ and $E_{\Lambda_2}$ are isomorphic as complex algebraic curves if and only if we have $\lambda \Lambda_1 = \Lambda_2$ for some $\lambda \in \mathbb{C}$. The isogenies $E_\Lambda \to E_\Lambda$ form the *endomorphism ring*

$$\text{End}(E_\Lambda) = \{\lambda \in \mathbb{C} : \lambda \Lambda \subset \Lambda\} \qquad (6\text{-}6)$$

of the curve $E_\Lambda$, which we can view as a discrete subring of $\mathbb{C}$. The $\lambda$-value of the analytically defined endomorphism 'multiplication by $\lambda \in \mathbb{C}$' is reflected algebraically as a true multiplication by $\lambda$ of the *invariant differential* $dx/y$ on $E_\Lambda$ coming from $dz = d(\wp_\Lambda)/\wp'_\Lambda$. If $\text{End}(E_\Lambda)$ is strictly larger than $\mathbb{Z}$, it is a complex quadratic order $\mathcal{O}$ and $E_\Lambda$ is said to have *complex multiplication* (CM) by $\mathcal{O}$.

To generate the class fields of our imaginary quadratic field $K$, we employ an elliptic curve $E_\Lambda$ having CM by $\mathbb{Z}_K$. Such a curve can be obtained by taking $\Lambda$ equal to $\mathbb{Z}_K$ or to a fractional $\mathbb{Z}_K$-ideal $\mathfrak{a}$, but the Weierstrass coefficients (6-4) for $\Lambda = \mathfrak{a}$ will not in general be algebraic.

In order to find an *algebraic* model for the complex curve $E_\Lambda$, we scale $\Lambda$ to a *homothetic* lattice $\lambda\Lambda$ to obtain a $\mathbb{C}$-isomorphic model

$$E_{\lambda\Lambda} : y^2 = 4x^3 - \lambda^{-4}g_2(\Lambda)x - \lambda^{-6}g_3(\Lambda)$$

under the Weierstrass map. The discriminant $\Delta = g_2^3 - 27g_3^2$ of the Weierstrass polynomial $4x^3 - g_2 x - g_3$ does not vanish, and the lattice function

$$\Delta(\Lambda) = g_2(\Lambda)^3 - 27g_3(\Lambda)^2$$

is of weight 12: it satisfies $\Delta(\lambda\Lambda) = \lambda^{-12}\Delta(\Lambda)$. Thus, the $j$-invariant

$$j(\Lambda) = 1728 \frac{g_2(\Lambda)^3}{\Delta(\Lambda)} = 1728 \frac{g_2(\Lambda)^3}{g_2(\Lambda)^3 - 27g_3(\Lambda)^2} \qquad (6\text{-}7)$$

is of weight zero and is an invariant of the homothety class of $\Lambda$ or, equivalently, the isomorphism class of the complex elliptic curve $\mathbb{C}/\Lambda$. It generates the minimal field of definition over which $\mathbb{C}/\Lambda$ admits a Weierstrass model.

If $E_\Lambda$ has CM by $\mathbb{Z}_K$, then $\Lambda$ is homothetic to some $\mathbb{Z}_K$-ideal $\mathfrak{a}$. It follows that, up to isomorphism, there are only finitely many complex elliptic curves $E_\Lambda$ having CM by $\mathbb{Z}_K$, one for each ideal class in $\text{Cl}_K$. Because any automorphism of $\mathbb{C}$ maps the algebraic curve $E_\Lambda$ to an elliptic curve with the same endomorphism ring, we find that the $j$-invariants of the ideal classes of $\mathbb{Z}_K$ form a set of $h_K = \#\text{Cl}_K$ distinct *algebraic* numbers permuted by the absolute Galois group $G_\mathbb{Q}$ of $\mathbb{Q}$. This allows us to define the *Hilbert class polynomial* of $K$ as

$$\text{Hil}_K(X) = \prod_{[\mathfrak{a}]\in\text{Cl}_K} (X - j(\mathfrak{a})) \in \mathbb{Q}[X]. \qquad (6\text{-}8)$$

Its importance stems from the following theorem, traditionally referred to as the *first main theorem of complex multiplication*.

THEOREM 6.9. *The Hilbert class field $H$ of $K$ is the splitting field of the polynomial $\text{Hil}_K(X)$ over $K$. This polynomial is irreducible in $K[X]$, and the Galois action of the Artin symbol $\sigma_\mathfrak{c} = \psi_{H/K}(\mathfrak{c})$ of the ideal class $[\mathfrak{c}] \in \text{Cl}_K \cong \text{Gal}(H/K)$ on the roots $j(\mathfrak{a})$ of $\text{Hil}_K(X)$ is given by $j(\mathfrak{a})^{\sigma_\mathfrak{c}} = j(\mathfrak{a}\mathfrak{c}^{-1})$.*

To compute $\text{Hil}_K(X)$ from its definition (6-8), one compiles a list of $\mathbb{Z}_K$-ideal classes in the style of Gauss, who did this in terms of binary quadratic forms. Every $\mathbb{Z}_K$-ideal class $[\mathfrak{a}]$ has a representative of the form $\mathbb{Z}\tau + \mathbb{Z}$, with $\tau \in K$ a root of some irreducible polynomial $aX^2 + bX + c \in \mathbb{Z}[X]$ of discriminant $b^2 - 4ac = \Delta_{K/\mathbb{Q}}$. If we take for $\tau$ the root in the complex upper half plane $\mathbf{H}$, the *orbit* of $\tau$ under the natural action

$$\begin{pmatrix} \alpha & \beta \\ \gamma & \delta \end{pmatrix}(z) = \frac{\alpha z + \beta}{\gamma z + \delta}$$

of the modular group $SL_2(\mathbb{Z})$ on **H** is uniquely determined by $[\mathfrak{a}] \in Cl_K$. In this orbit, there is a unique element

$$\tau_\mathfrak{a} = \frac{-b + \sqrt{b^2 - 4ac}}{2a} \in \mathbf{H}$$

that lies in the standard fundamental domain for the action of $SL_2(\mathbb{Z})$ on **H** consisting of those $z \in \mathbf{H}$ that satisfy the two inequalities $|\text{Re}(z)| \leq \frac{1}{2}$ and $z\bar{z} \geq 1$ and, in case we have equality in either of them, also $\text{Re}(z) \leq 0$. This yields a description

$$[\mathfrak{a}] = \left[\mathbb{Z} \cdot \frac{-b + \sqrt{b^2 - 4ac}}{2a} + \mathbb{Z}\right] \qquad \longleftrightarrow \qquad (a, b, c)$$

of the elements of $Cl_K$ as *reduced* integer triples $(a, b, c)$ whose discriminant is $b^2 - 4ac = \Delta_{K/\mathbb{Q}}$. As we have $\text{Re}(\tau_\mathfrak{a}) = -b/2a$ and $\tau_\mathfrak{a}\bar{\tau}_\mathfrak{a} = c/a$, the reduced integer triples $(a, b, c)$ corresponding to $\tau_\mathfrak{a}$ in the fundamental domain for $SL_2(\mathbb{Z})$ are those satisfying

$$|b| \leq a \leq c \qquad \text{and} \qquad b^2 - 4ac = \Delta_{K/\mathbb{Q}},$$

where $b$ is nonnegative if $|b| = a$ or $a = c$. For reduced forms, one sees from the inequality $\Delta_{K/\mathbb{Q}} = b^2 - 4ac \leq a^2 - 4a^2 = -3a^2$ that we have bounds $|b| \leq a \leq \sqrt{|\Delta_{K/\mathbb{Q}}|/3}$, so the list is indeed finite and can easily be generated [Cohen 1993, Algorithm 5.3.5]. See [Cox 1989] for the classical interpretation of the triples $(a, b, c)$ as positive definite integral binary quadratic forms $aX^2 + bXY + cY^2$ of discriminant $b^2 - 4ac = \Delta_{K/\mathbb{Q}}$.

If we put $j(\tau) = j(\mathbb{Z}\tau + \mathbb{Z})$, the $j$-function (6-7) becomes a holomorphic function $j : \mathbf{H} \to \mathbb{C}$ invariant under the action of $SL_2(\mathbb{Z})$. As it is in particular invariant under $\tau \mapsto \tau + 1$, it can be expressed in various ways in terms of the variable $q = e^{2\pi i \tau}$ from (6-1). Among them is the well-known *integral* Fourier expansion

$$j(\tau) = j(q) = q^{-1} + 744 + 196884q + \ldots \in q^{-1} + \mathbb{Z}[[q]] \qquad (6\text{-}10)$$

that explains the normalizing factor 1728 in the definition (6-7) of $j$. It implies [Lang 1987, Chapter 5, §2] that the roots of $\text{Hil}_K(X)$ in (6-8) are algebraic integers, and so $\text{Hil}_K(X)$ is a polynomial in $\mathbb{Z}[X]$ that can be computed *exactly* from complex approximations of its roots that are sufficiently accurate to yield the right hand side of (6-8) in $\mathbb{C}[X]$ to 'one-digit precision'. For numerical computations of $j(\tau)$, one uses approximate values of the Dedekind $\eta$-function

$$\eta(\tau) = q^{1/24} \prod_{n \geq 1}(1 - q^n) = q^{1/24} \sum_{n \in \mathbb{Z}}(-1)^n q^{n(3n-1)/2}, \qquad (6\text{-}11)$$

which has a lacunary Fourier expansion that is better suited for numerical purposes than (6-10). From $\eta$-values one computes $\mathfrak{f}_2(\tau) = \sqrt{2}\eta(2\tau)/\eta(\tau)$ and finally $j(\tau)$ as

$$j(\tau) = \frac{(\mathfrak{f}_2^{24}(\tau) + 16)^3}{\mathfrak{f}_2^{24}(\tau)}. \tag{6-12}$$

This finishes the description of the classical algorithm to compute the Hilbert class field $H$ of $K$.

Having computed the irreducible polynomial $\operatorname{Hil}_K(X)$ of $j_K = j(\mathbb{Z}_K)$, we can write down a Weierstrass model $E_K$ for $\mathbb{C}/\mathbb{Z}_K$ over $H = K(j_K)$ (or even over $\mathbb{Q}(j_K)$) and use it to generate the ray class field extensions $H \subset H_m$. Choosing $E_K$ is easy in the special cases $K = \mathbb{Q}(\zeta_3), \mathbb{Q}(i)$, when one of $g_2 = g_2(\mathbb{Z}_K)$ and $g_3 = g_3(\mathbb{Z}_K)$ vanishes and the other can be scaled to have any nonzero rational value. For $K \neq \mathbb{Q}(\zeta_3), \mathbb{Q}(i)$, the number $\lambda = \sqrt{g_3/g_2} \in \mathbb{C}^*$ is determined up to sign, and since we have

$$c_K = \lambda^{-4} g_2 = \lambda^{-6} g_3 = g_2^3 g_3^{-2} = 27 \frac{j_K}{j_K - 1728},$$

the model $y^2 = 4x^3 - c_K x - c_K$ for $\mathbb{C}/\lambda\mathbb{Z}_K$ is defined over $\mathbb{Q}(j_K) \subset H$. A more classical choice is $\lambda^2 = \Delta/(g_2 g_3)$, with $g_2, g_3$ and $\Delta$ associated to $\mathbb{Z}_K$, giving rise to the model

$$E_K : y^2 = w_K(x) = 4x^3 - \frac{c_K}{(c_K - 27)^2} x - \frac{c_K}{(c_K - 27)^3}. \tag{6-13}$$

Any scaled Weierstrass parametrization $W_K : \mathbb{C}/\mathbb{Z}_K \xrightarrow{\sim} E_K$ with $E_K$ defined over $H$ can serve as the imaginary quadratic analogue of the isomorphism $q : \mathbb{R}/\mathbb{Z} \xrightarrow{\sim} T$ in (6-1). For the model $E_K$ in (6-13), the $x$-coordinate $\wp_{\lambda \mathbb{Z}_K}(\lambda z) = \lambda^{-2} \wp_{\mathbb{Z}_K}(z)$ of $W_K(z)$ is given by the *Weber function*

$$f_K(z) = \frac{g_2(\mathbb{Z}_K) g_3(\mathbb{Z}_K)}{\Delta(\mathbb{Z}_K)} \wp_{\mathbb{Z}_K}(z). \tag{6-14}$$

It has 'weight 0' in the sense that the right side is invariant under simultaneous scaling $(\mathbb{Z}_K, z) \to (\lambda \mathbb{Z}_K, \lambda z)$ by $\lambda \in \mathbb{C}^*$ of the lattice $\mathbb{Z}_K$ and the argument $z$.

In the special cases $K = \mathbb{Q}(i)$ and $\mathbb{Q}(\zeta_3)$ that have $\mathbb{Z}_K^*$ of order 4 and 6, there are slightly different Weber functions $f_K$ that are not the $x$-coordinates on a Weierstrass model for $\mathbb{C}/\mathbb{Z}_K$ over $H$, but an appropriately scaled *square* and *cube* of such $x$-coordinates, respectively. In all cases, the analogue of the Kronecker–Weber theorem for $K$ is the following *second main theorem of complex multiplication*.

THEOREM 6.15. *The ray class field $H_m$ of conductor $m\mathbb{Z}_K$ of $K$ is generated over the Hilbert class field $H$ of $K$ by the values of the Weber function $f_K$ at the nonzero $m$-torsion points of $\mathbb{C}/\mathbb{Z}_K$.*

In the non-special cases, the values of $f_K$ at the $m$-torsion points of $\mathbb{C}/\mathbb{Z}_K$ are the $x$-coordinates of the nonzero $m$-torsion points of the elliptic curve $E_K$ in (6-13). For $K = \mathbb{Q}(i)$ and $\mathbb{Q}(\zeta_3)$, one uses squares and cubes of these coordinates. In all cases, generating $H_m$ over $H$ essentially amounts to computing *division polynomials* $T_m \in \mathbb{C}[X]$ that have these $x$-coordinates as their roots. We will define these polynomials as elements of $H[x]$, because the recursion formulas at the end of this section show that their coefficients are elements of the ring generated over $\mathbb{Z}$ by the coefficients of the Weierstrass model of $E_K$.

If $m$ is odd, the nonzero $m$-torsion points come in pairs $\{P, -P\}$ with the same $x$-coordinate $x_P = x_{-P}$, and we can define a polynomial $T_m(x) \in H[x]$ of degree $(m^2 - 1)/2$ up to sign by

$$T_m(x)^2 = m^2 \prod_{\substack{P \in E_K[m](\mathbb{C}), \\ P \neq O}} (x - x_P).$$

For even $m$, we adapt the definition by excluding the 2-torsion points satisfying $P = -P$ from the product, and define $T_m(X) \in H[x]$ of degree $(m^2 - 4)/2$ by

$$T_m(x)^2 = (m/2)^2 \prod_{\substack{P \in E_K[m](\mathbb{C}), \\ 2P \neq O}} (x - x_P).$$

The 'missing' $x$-coordinates of the nonzero 2-torsion points are the zeros of the cubic polynomial $w_K \in H[x]$ in (6-13). This is a square in the *function field* of the elliptic curve (6-13), and in many ways the natural object to consider is the 'division polynomial'

$$\psi_m(x, y) = \begin{cases} T_m(x) & \text{if } m \text{ is odd}; \\ 2y T_m(x) & \text{if } m \text{ is even.} \end{cases}$$

This is an element of the function field $\mathbb{C}(x, y)$ living in the quadratic extension $H[x, y]$ of the polynomial ring $H[x]$ defined by $y^2 = w_K(x)$. It is uniquely defined up to the sign choice we have for $T_m$. Most modern texts take the sign of the highest coefficient of $T_m$ equal to 1. Weber [1908, p. 197] takes it equal to $(-1)^{m-1}$, which amounts to a sign change $y \mapsto -y$ in $\psi_m(x, y)$.

By construction, the function $\psi_m$ has divisor $(1 - m^2)[O] + \sum_{P \neq O, mP = O}[P]$. The normalizing highest coefficients $m$ and $m/2$ in $T_m$ lead to neat recursive

formulas
$$\psi_{2m+1} = \psi_{m+2}\psi_m^3 - \psi_{m+1}^3\psi_{m-1},$$
$$\psi_{2m} = (2y)^{-1}\psi_m(\psi_{m+2}\psi_{m-1}^2 - \psi_{m+1}^2\psi_{m-2})$$

for $\psi_m$ that are valid for $m > 1$ and $m > 2$. These can be used to compute $\psi_m$ and $T_m$ recursively, using 'repeated doubling' of $m$. One needs the initial values $T_1 = T_2 = 1$ and

$$T_3 = 3X^4 + 6aX^2 + 12bX - a^2,$$
$$T_4 = 2X^6 + 10aX^4 + 40bX^3 - 10a^2X^2 - 8abX - 16b^2 - 2a^3,$$

where we have written the Weierstrass polynomial in (6-13) as $w_K = 4(x^3 + ax + b)$ to indicate the relation with the nowadays more common affine model $y^2 = x^3 + ax + b$ to which $E_K$ is isomorphic under $(x, y) \mapsto (x, y/2)$.

## 7. Class fields from modular functions

The algorithms in the previous section are based on the main Theorems 6.9 and 6.15 of complex multiplication, which can be found already in Weber's textbook [1908] and predate the class field theory for general number fields. The oldest proofs of 6.9 and 6.15 are of an analytic nature, and derive arithmetic information from congruence properties of Fourier expansions such as (6-10). Assuming general class field theory, one can shorten these proofs as it suffices, just as after Theorem 2.2, to show that, up to sets of primes of zero density, the 'right' primes split completely in the purported class fields. In particular, it is always possible to restrict attention to the primes of $K$ of residue degree one in such arguments. Deuring [1958] provides analytic proofs of both kinds in his survey monograph.

Later proofs [Lang 1987, Part 2] of Deuring and Shimura combine class field theory with the *reduction* of the endomorphisms in (6-6) modulo primes, which yields endomorphisms of elliptic curves over finite fields. These proofs are firmly rooted in the algebraic theory of elliptic curves [Silverman 1986; Silverman 1994]. Here one takes for $E_A$ in (6-6) an elliptic curve $E$ that has CM by $\mathbb{Z}_K$ and is given by a Weierstrass equation over the splitting field $H'$ over $K$ of the Hilbert class polynomial $\text{Hil}_K(X)$ in (6-8). In this case the Weierstrass equation can be considered modulo any prime q of $H'$, and for almost all primes, known as the primes *of good reduction*, this yields an elliptic curve $E_q = E \bmod q$ over the finite field $k_q = \mathbb{Z}_{H'}/q$. For such q, the choice of an extension of q to $\overline{\mathbb{Q}}$ yields a reduction homomorphism $E_K(\overline{\mathbb{Q}}) \to E_q(\bar{k}_q)$ on points that is *injective* on torsion points of order coprime to q. The endomorphisms of $E$ are given by rational functions with coefficients in $H'$, and for

primes $\mathfrak{q}$ of $H'$ of good reduction there is a natural reduction homomorphism
$$\mathrm{End}(E) \to \mathrm{End}(E_\mathfrak{q})$$
that is injective and preserves degrees. The 'complex multiplication' by an element $\alpha \in \mathrm{End}(E) = \mathbb{Z}_K$ multiplies the invariant differential $dx/y$ on $E$ by $\alpha$, so it becomes inseparable in $\mathrm{End}(E_\mathfrak{q})$ if and only if $\mathfrak{q}$ divides $\alpha$. The first main theorem of complex multiplication (Theorem 6.9), which states that $H'$ equals the Hilbert class field $H$ of $K$ and provides the Galois action on the roots of $\mathrm{Hil}_K(X)$, can now be derived as follows.

PROOF OF THEOREM 6.9. Let $\mathfrak{p}$ be a prime of degree one of $K$ that is coprime to the discriminant of $\mathrm{Hil}_K(X)$, and let $\alpha \in \mathbb{Z}_K$ be an element of order 1 at $\mathfrak{p}$, say $\alpha \mathfrak{p}^{-1} = \mathfrak{b}$ with $(\mathfrak{b}, \mathfrak{p}) = 1$. Let $\mathfrak{a}$ be a fractional $\mathbb{Z}_K$-ideal. Then the complex multiplication $\alpha : \mathbb{C}/\mathfrak{a} \to \mathbb{C}/\mathfrak{a}$ factors in terms of complex tori as
$$\mathbb{C}/\mathfrak{a} \xrightarrow{\mathrm{can}} \mathbb{C}/\mathfrak{ap}^{-1} \xrightarrow[\alpha]{\sim} \mathbb{C}/\mathfrak{ab} \xrightarrow{\mathrm{can}} \mathbb{C}/\mathfrak{a}.$$

If $E$ and $E'$ denote Weierstrass models over $H'$ for $\mathbb{C}/\mathfrak{a}$ and $\mathbb{C}/\mathfrak{ap}^{-1}$, we obtain isogenies $E \to E' \to E$ of degree $p = N\mathfrak{p}$ and $N\mathfrak{b}$ with composition $\alpha$. If we assume that $E$ and $E'$ have good reduction above $\mathfrak{p}$, we can reduce the isogeny $E \to E'$ at some prime $\mathfrak{q}|\mathfrak{p}$ to obtain an isogeny $E_\mathfrak{q} \to E'_\mathfrak{q}$ of degree $p$. This isogeny is inseparable as $\alpha$ lies in $\mathfrak{q}$, and therefore equal to the Frobenius morphism $E_\mathfrak{q} \to E_\mathfrak{q}^{(p)}$ followed by an isomorphism $E_\mathfrak{q}^{(p)} \xrightarrow{\sim} E'_\mathfrak{q}$. The result is an equality $j(E_\mathfrak{q}^{(p)}) = j(E'_\mathfrak{q})$ of $j$-invariants that amounts to $j(\mathfrak{a})^p = j(\mathfrak{ap}^{-1}) \bmod \mathfrak{q}$, and this implies the Frobenius automorphism $\sigma_\mathfrak{q} \in \mathrm{Gal}(H'/K)$ acts as $j(\mathfrak{a})^{\sigma_\mathfrak{q}} = j(\mathfrak{ap}^{-1})$, independent of the choice of the extension prime $\mathfrak{q}|\mathfrak{p}$. Because the $j$-function is an invariant for the homothety class of a lattice, we have $j(\mathfrak{a}) = j(\mathfrak{ap}^{-1})$ if and only if $\mathfrak{p}$ is principal. It follows that up to finitely many exceptions, the primes $\mathfrak{p}$ of degree one splitting completely in $K \subset H'$ are the principal primes, so $H'$ equals the Hilbert class field $H$ of $K$, and the splitting primes in $K \subset H$ are *exactly* the principal primes. Moreover, we have $j(\mathfrak{a})^{\sigma_\mathfrak{c}} = j(\mathfrak{ac}^{-1})$ for the action of the Artin symbol $\sigma_\mathfrak{c}$, and $\mathrm{Hil}_K$ is irreducible over $K$ as its roots are transitively permuted by $\mathrm{Gal}(H/K) = \mathrm{Cl}_K$. □

In a similar way, one can understand the content of the second main theorem of complex multiplication. If $E_K$ is a Weierstrass model for $\mathbb{C}/\mathbb{Z}_K$ defined over the Hilbert class field $H$ of $K$, then the torsion points in $E_K(\mathbb{C})$ have algebraic coordinates. As the group law is given by algebraic formulas over $H$, the absolute Galois group $G_H$ of $H$ acts by group automorphisms on
$$E_K^{\mathrm{tor}}(\mathbb{C}) \cong K/\mathbb{Z}_K.$$
Moreover, the action of $G_H$ commutes with the complex multiplication action of $\mathrm{End}(E_K) \cong \mathbb{Z}_K$, which is given by isogenies defined over $H$. It follows that

$G_H$ acts by $\mathbb{Z}_K$-module automorphisms on $E_K^{\text{tor}}(\mathbb{C})$. For the cyclic $\mathbb{Z}_K$-module $E_K[m](\mathbb{C}) \cong \frac{1}{m}\mathbb{Z}_K/\mathbb{Z}_K$ of $m$-torsion points, the resulting Galois representation

$$G_H \longrightarrow \text{Aut}_{\mathbb{Z}_K}(\tfrac{1}{m}\mathbb{Z}_K/\mathbb{Z}_K) \cong (\mathbb{Z}_K/m\mathbb{Z}_K)^*$$

of $G_H$ is therefore *abelian*. It shows that, just as in the cyclotomic case (6-2), the Galois action *over $H$* on the *$m$-division field* of $E_K$, which is the extension of $H$ generated by the $m$-torsion points of $E_K$, comes from *multiplications* on $\frac{1}{m}\mathbb{Z}_K/\mathbb{Z}_K$ by integers $\alpha \in \mathbb{Z}_K$ coprime to $m$. The content of Theorem 6.15 is that, in line with (2-8), we obtain the $m$-th ray class field of $K$ from this $m$-division field by taking invariants under the action of $\mathbb{Z}_K^* = \text{Aut}(E_K)$. In the 'generic case' where $\mathbb{Z}_K^* = \{\pm 1\}$ has order 2, adjoining $m$-torsion points 'up to inversion' amounts to the equality

$$H_m = H(\{x_P : P \in E_K[m](\mathbb{C}), P \neq O\}) \tag{7-1}$$

occurring in Theorem 6.15, since the $x$-coordinate $x_P$ determines $P$ up to multiplication by $\pm 1$. More generally, a root of unity $\zeta \in \mathbb{Z}_K^*$ acts as an automorphism of $E_K$ by $x_{[\zeta]P} = \zeta^{-2} x_P$, and so in the special cases where $K$ equals $\mathbb{Q}(i)$ or $\mathbb{Q}(\zeta_3)$ and $\mathbb{Z}_K^*$ has order $2k$ with $k = 2, 3$, one replaces $x_P$ by $x_P^k$ in (7-1). The classical Weber functions replacing (6-14) for $K = \mathbb{Q}(i)$ and $K = \mathbb{Q}(\zeta_3)$ are $f_K(z) = (g_2^2(\mathbb{Z}_K)/\Delta(\mathbb{Z}_K))\wp_{\mathbb{Z}_K}^2(z)$ and $f_K(z) = (g_3(\mathbb{Z}_K)/\Delta(\mathbb{Z}_K))\wp_{\mathbb{Z}_K}^3(z)$.

PROOF OF THEOREM 6.15. As in the case of Theorem 6.9, we show that the primes of degree one of $K$ that split completely in the extension $K \subset H'_m$ defined by adjoining to $H$ the $m$-torsion points of $E_K$ 'up to automorphisms' are, up to a zero density subset of primes, the primes in the ray group $R_m$. Primes $\mathfrak{p}$ of $K$ splitting in $H'_m$ are principal as they split in $H$. For each $\mathbb{Z}_K$-generator $\pi$ of $\mathfrak{p}$, which is uniquely determined up to multiplication by $\mathbb{Z}_K^*$, one obtains a complex multiplication by $\pi \in \mathbb{Z}_K \cong \text{End}(E_K)$ that fixes $E_K[m](\mathbb{C})$ 'up to automorphisms' if and only if $\mathfrak{p} \in R_m$.

Let $\mathfrak{p} = \pi \mathbb{Z}_K$ be a prime of degree 1 over $p$ for which $E_K$ has good reduction modulo $\mathfrak{p}$. Then the isogeny $\phi_\pi : E_K \to E_K$, which corresponds to multiplication by $\pi$ as in (6-5), reduces modulo a prime $\mathfrak{q}|\mathfrak{p}$ of the $m$-division field of $E_K$ to an endomorphism of degree $p$. Since $\pi$ is in $\mathfrak{q}$, this reduction is inseparable, and so it equals the Frobenius endomorphism of $E_{K,\mathfrak{q}}$ up to an automorphism. One shows [Lang 1987, p. 125] that this *local* automorphism of $E_{K,\mathfrak{q}}$ is induced by a global automorphism of $E_K$, that is, a complex multiplication by a unit in $\mathbb{Z}_K^*$, and concludes that $\phi_\pi$ induces a Frobenius automorphism above $\mathfrak{p}$ on $H'_m$. As the reduction modulo $\mathfrak{q}$ induces an isomorphism $E_K[m] \xrightarrow{\sim} E_{K,\mathfrak{q}}[m]$ on the $m$-torsion points, this Frobenius automorphism is trivial if and only if we have $\pi \equiv 1 \bmod^* m\mathbb{Z}_K$ for a suitable choice of $\pi$. Thus, $\mathfrak{p}$ splits completely in $H'_m$ if and only if $\mathfrak{p}$ is in $R_m$, and $H'_m$ is the ray class field $H_m$. $\square$

The argument just given shows that we have a concrete realization of the Artin isomorphism $(\mathbb{Z}_K/m\mathbb{Z}_K)^*/\mathrm{im}[\mathbb{Z}_K^*] \xrightarrow{\sim} \mathrm{Gal}(H_m/H)$ from (2-8) by complex multiplications. Passing to the projective limit, this yields the analogue

$$\widehat{\mathbb{Z}}_K^*/\mathbb{Z}_K^* \xrightarrow{\sim} \mathrm{Gal}(K_{\mathrm{ab}}/H) \subset G_K^{\mathrm{ab}}$$

of (3-1). For the analogue of (6-3), we note first that for imaginary quadratic $K$, the subgroup $U_\infty$ in (3-3), which equals $\mathbb{C}^*$, maps isomorphically to the connected component of the unit element in $\mathbf{A}_K^*/K^*$. Because it is the kernel of the Artin map $\psi_K$ in (3-7), we obtain a commutative diagram

$$\begin{array}{ccccc}
\widehat{\mathbb{Z}}_K^* & \xrightarrow{-1} & \mathbb{Z}_K^*/\mathbb{Z}_K^* & \subset & \mathbf{A}_K^*/(K^* \cdot \mathbb{C}^*) \\
{\scriptstyle \mathrm{can}}\downarrow\,\wr & & \downarrow\wr & & {\scriptstyle \mathrm{Artin}}\downarrow\,\wr \\
\mathrm{Aut}_{\mathbb{Z}_K}(K/\mathbb{Z}_K) & \longrightarrow & \mathrm{Gal}(K_{\mathrm{ab}}/H) & \subset & \mathrm{Gal}(K_{\mathrm{ab}}/K)
\end{array} \qquad (7\text{-}2)$$

in which the inversion map $-1$ arises just as in (3-8). A slight difference with the diagram (6-3) for $\mathbb{Q}$ is that the horizontal arrows now have a small finite kernel coming from the unit group $\mathbb{Z}_K^*$. Moreover, we have only accounted for the automorphisms of $K_{\mathrm{ab}}$ over $H$, not over $K$. Automorphisms of $H_m$ that are not the identity on $H$ arise as Artin maps $\sigma_{\mathfrak{c}}$ of nonprincipal ideals $\mathfrak{c}$ coprime to $m$, and the proof of Theorem 6.9 shows that for the isogeny $\phi_{\mathfrak{c}}$ in the commutative diagram

$$\begin{array}{ccc}
\mathbb{C}/\mathbb{Z}_K & \xrightarrow{\mathrm{can}} & \mathbb{C}/\mathfrak{c}^{-1} \\
w\downarrow\,\wr & & w'\downarrow\,\wr \\
E_K & \xrightarrow{\phi_{\mathfrak{c}}} & E_K',
\end{array} \qquad (7\text{-}3)$$

we have to compute the restriction $\phi_{\mathfrak{c}} : E_K[m] \to E_K'[m]$ to $m$-torsion points. To do so in an efficient way, we view the $j$-values and $x$-coordinates of torsion points involved as weight zero functions on complex lattices such as $\mathbb{Z}_K$ or $\mathfrak{c}$. As we may scale all lattices as we did for (6-10) to $\mathbb{Z}\tau + \mathbb{Z}$ with $\tau \in \mathbf{H}$, such functions are *modular functions* $\mathbf{H} \to \mathbb{C}$ as defined in [Lang 1987, Chapter 6].

The $j$-function itself is the primordial modular function: a holomorphic function on $\mathbf{H}$ that is invariant under the full modular group $\mathrm{SL}_2(\mathbb{Z})$. Every meromorphic function on $\mathbf{H}$ that is invariant under $\mathrm{SL}_2(\mathbb{Z})$ and, when viewed as a function of $q = e^{2\pi i \tau}$, meromorphic in $q = 0$, is in fact a rational function of $j$. The Weber function $f_K$ in (6-14) is a function

$$f_\tau(z) = \frac{g_2(\tau)g_3(\tau)}{\Delta(\tau)} \wp_{[\tau,1]}(z)$$

that depends on the lattice $\mathbb{Z}_K = \mathbb{Z}\tau + \mathbb{Z} = [\tau, 1]$, and fixing some *choice* of a generator $\tau$ of $\mathbb{Z}_K$ over $\mathbb{Z}$, we can label its $m$-torsion values used in generating

$H_m$ as

$$F_u(\tau) = f_\tau(u_1\tau + u_2) \quad \text{with } u = (u_1, u_2) \in \tfrac{1}{m}\mathbb{Z}^2/\mathbb{Z}^2 \setminus \{(0,0)\}. \tag{7-4}$$

For $m > 1$, the functions $F_u : \mathbf{H} \to \mathbb{C}$ in (7-4) are known as the *Fricke functions* of level $m$. These are holomorphic functions on $\mathbf{H}$ that are $x$-coordinates of $m$-torsion points on a 'generic elliptic curve' over $\mathbb{Q}(j)$ with $j$-invariant $j$. As they are zeroes of division polynomials in $\mathbb{Q}(j)[X]$, they are algebraic over $\mathbb{Q}(j)$ and generate a finite algebraic extension of $\mathbb{Q}(j)$, the $m$-th modular function field

$$\mathcal{F}_m = \mathbb{Q}(j, \{F_u\}_{u \in (\tfrac{1}{m}\mathbb{Z}/\mathbb{Z})^2 \setminus \{(0,0)\}}). \tag{7-5}$$

Note above that $\mathcal{F}_1 = \mathbb{Q}(j)$. We may now rephrase the main theorems of complex multiplication in the following way.

THEOREM 7.6. *Let $K$ be an imaginary quadratic field with ring of integers $\mathbb{Z}[\tau]$. Then the $m$-th ray class field extension $K \subset H_m$ is generated by the finite values $f(\tau)$ of the functions $f \in \mathcal{F}_m$.*

For generic $K$ this is directly clear from Theorems 6.9 and 6.15. In the special cases $K = \mathbb{Q}(i), \mathbb{Q}(\zeta_3)$, the functions $F_u(\tau)$ in (7-4) vanish at the generator $\tau$ of $\mathbb{Z}_K$, so an extra argument [Lang 1987, p. 128] involving modified Weber functions in $\mathcal{F}_m$ is needed.

It is not really necessary to take $\tau$ in Theorem 7.6 to be a generator of $\mathbb{Z}_K$; it suffices that the elliptic curve $\mathbb{C}/[\tau, 1]$ is an elliptic curve having CM by $\mathbb{Z}_K$.

In computations, it is essential to have the explicit action of $\mathrm{Gal}(K_{\mathrm{ab}}/K)$ on the values $f(\tau)$ from Theorem 7.6 for arbitrary functions $f$ in the modular function field $\mathcal{F} = \cup_{m \geq 1} \mathcal{F}_m$. As class field theory gives us the group $\mathrm{Gal}(K_{\mathrm{ab}}/K)$ in (7-2) as an explicit quotient of the idele class group $\mathbf{A}_K^*/K^*$ under the Artin map (3-7), this means that we need to find the natural action of $x \in \mathbf{A}_K^*$ on the values $f(\tau)$ in Theorem 7.6. We will do so by reinterpreting the action of the Artin symbol $\sigma_x \in \mathrm{Gal}(K_{\mathrm{ab}}/K)$ on the function value of $f$ at $\tau$ as the value of some *other* modular function $f^{g_\tau(x)}$ at $\tau$, that is,

$$(f(\tau))^{\sigma_x} = f^{g_\tau(x)}(\tau), \tag{7-7}$$

for some natural homomorphism $g_\tau : \mathbf{A}_K^* \to \mathrm{Aut}(\mathcal{F})$ induced by $\tau$.

To understand the automorphisms of $\mathcal{F}$, we note first that the natural left action of $\mathrm{SL}_2(\mathbb{Z})$ on $\mathbf{H}$ gives rise to a right action on $\mathcal{F}_m$ that is easily made explicit for the Fricke functions (7-4), using the 'weight 0' property of $f_K$. For $M = \begin{pmatrix} \alpha & \beta \\ \gamma & \delta \end{pmatrix} \in \mathrm{SL}_2(\mathbb{Z})$ we have

$$F_u(M\tau) = F_u\left(\frac{\alpha\tau + \beta}{\gamma\tau + \delta}\right) = \frac{g_2(\tau)g_3(\tau)}{\Delta(\tau)} \wp_{[\tau,1]}(u_1(\alpha\tau + \beta) + u_2(\gamma\tau + \delta))$$
$$= F_{uM}(\tau).$$

As $u = (u_1, u_2)$ is in $\frac{1}{m}\mathbb{Z}^2/\mathbb{Z}^2$, we only need to know $M$ modulo $m$, so the Fricke functions of level $m$ are invariant under the congruence subgroup

$$\Gamma(m) = \ker[\mathrm{SL}_2(\mathbb{Z}) \to \mathrm{SL}_2(\mathbb{Z}/m\mathbb{Z})]$$

of $\mathrm{SL}_2(\mathbb{Z})$, and they are permuted by $\mathrm{SL}_2(\mathbb{Z})$. As we have $F_{-u_1,-u_2} = F_{u_1,u_2}$, we obtain a natural right action of $\mathrm{SL}_2(\mathbb{Z}/m\mathbb{Z})/\{\pm 1\}$ on $\mathscr{F}_m$.

Besides this 'geometric action', there is a cyclotomic action of $(\mathbb{Z}/m\mathbb{Z})^*$ on the functions $f \in \mathscr{F}_m$ via their Fourier expansions, which lie in $\mathbb{Q}(\zeta_m)((q^{1/m}))$ since they involve rational expansions in

$$e^{2\pi i(a_1\tau + a_2)/m} = \zeta_m^{a_2} q^{a_1/m} \quad \text{for } a_1, a_2 \in \mathbb{Z}.$$

On the Fricke function $F_u = F_{(u_1,u_2)}$, the automorphism $\sigma_k : \zeta_m \mapsto \zeta_m^k$ clearly induces $\sigma_k : F_{(u_1,u_2)} \mapsto F_{(u_1,ku_2)}$. Thus, the two actions may be combined to give an action of $\mathrm{GL}_2(\mathbb{Z}/m\mathbb{Z})/\{\pm 1\}$ on $\mathscr{F}_m$, with $\mathrm{SL}_2(\mathbb{Z}/m\mathbb{Z})/\{\pm 1\}$ acting geometrically and $(\mathbb{Z}/m\mathbb{Z})^*$ acting as the subgroup $\{\pm \begin{pmatrix} 1 & 0 \\ 0 & k \end{pmatrix} : k \in (\mathbb{Z}/m\mathbb{Z})^*\}/\{\pm 1\}$. The invariant functions are $\mathrm{SL}_2(\mathbb{Z})$-invariant with rational $q$-expansion; so they lie in $\mathbb{Q}(j)$, and we have a natural isomorphism

$$\mathrm{GL}_2(\mathbb{Z}/m\mathbb{Z})/\{\pm 1\} \xrightarrow{\sim} \mathrm{Gal}(\mathscr{F}_m/\mathbb{Q}(j)), \tag{7-8}$$

or, if we take the union $\mathscr{F} = \bigcup_m \mathscr{F}_m$ on the left hand side and the corresponding projective limit on the right hand side,

$$\mathrm{GL}_2(\widehat{\mathbb{Z}})/\{\pm 1\} \xrightarrow{\sim} \mathrm{Gal}(\mathscr{F}/\mathbb{Q}(j)). \tag{7-9}$$

Note that $\mathscr{F}_m$ contains $\mathbb{Q}(\zeta_m)(j)$ as the invariant field of $\mathrm{SL}_2(\mathbb{Z}/m\mathbb{Z})/\{\pm 1\}$, and that the action of $\mathrm{GL}_2(\widehat{\mathbb{Z}})/\{\pm 1\}$ on the subextension $\mathbb{Q}(j) \subset \mathbb{Q}_{\mathrm{ab}}(j)$ with group $\mathrm{Gal}(\mathbb{Q}_{\mathrm{ab}}/\mathbb{Q}) \cong \widehat{\mathbb{Z}}^*$ is via the determinant map $\det : \mathrm{GL}_2(\widehat{\mathbb{Z}})/\{\pm 1\} \to \widehat{\mathbb{Z}}^*$.

To discover the explicit form of the homomorphism $g_\tau$ in (7-7), let $\mathfrak{p} = \pi \mathbb{Z}_K$ be a principal prime of $K$. Then the Artin symbol $\sigma_\mathfrak{p}$ is the identity on $H$, and the proof of Theorem 6.15 shows that its action on the $x$-coordinates of the $m$-torsion points of $E_K$ for $m$ not divisible by $\pi$ can be written as

$$F_u(\tau)^{\sigma_\mathfrak{p}} = F_u(\pi\tau) = F_{uM_\pi}(\tau),$$

where $M_\pi$ is the matrix in $\mathrm{GL}_2(\mathbb{Z}/m\mathbb{Z})$ that represents the multiplication by $\pi$ on $\frac{1}{m}\mathbb{Z}_K/\mathbb{Z}_K$ with respect to the basis $\{\tau, 1\}$. In explicit coordinates, this means that if $\tau \in \mathbf{H}$ is a zero of the polynomial $X^2 + BX + C$ of discriminant $B^2 - 4C = \Delta_K$ and $\pi = x_1\tau + x_2$ is the representation of $\pi$ on the $\mathbb{Z}$-basis $[\tau, 1]$ of $\mathbb{Z}_K$, then we have

$$M_\pi = \begin{pmatrix} -Bx_1 + x_2 & -Cx_1 \\ x_1 & x_2 \end{pmatrix} \in \mathrm{GL}_2(\mathbb{Z}/m\mathbb{Z}). \tag{7-10}$$

As the Fricke functions of level $m$ generate $\mathcal{F}_m$, we obtain in view of (7-8) the identity

$$f(\tau)^{\sigma_{\mathfrak{p}}} = f^{M_\pi}(\tau) \quad \text{for } f \in \mathcal{F}_m \text{ and } \mathfrak{p} \nmid m,$$

which is indeed of the form (7-7). We can rewrite this in the style of the diagram (7-2) by observing that the Artin symbol of $\pi \in K_\mathfrak{p}^* \subset \mathbf{A}_K^*$ acts as $\sigma_\mathfrak{p}$ on torsion points of order $m$ coprime to $\mathfrak{p}$, and trivially on $\pi$-power torsion points. Moreover, $(\pi \bmod K^*) \in \mathbf{A}_K/K^*$ is in the class of the idele $x \in \widehat{\mathbb{Z}}_K^*$ having component 1 at $\mathfrak{p}$ and $\pi^{-1}$ elsewhere. Thus, if we define

$$\begin{aligned} g_\tau : \widehat{\mathbb{Z}}_K^* &\longrightarrow \mathrm{GL}_2(\widehat{\mathbb{Z}}) \\ x &\longmapsto M_x^{-1} \end{aligned} \tag{7-11}$$

by sending $x = x_1 \tau + x_2 \in \widehat{\mathbb{Z}}_K$ to the *inverse* of the matrix $M_x$ describing multiplication by $x$ on $\widehat{\mathbb{Z}}_K$ with respect to the basis $[\tau, 1]$, then $M_x$ is given explicitly as in (7-10), and formula (7-7) holds for $f \in \mathcal{F}$ and $x \in \widehat{\mathbb{Z}}_K^*$ if we use the natural action of $\mathrm{GL}_2(\widehat{\mathbb{Z}})$ on $\mathcal{F}$ from (7-9).

To obtain complex multiplication by arbitrary ideles, we note that on the one hand, the idele class quotient $\mathbf{A}_K^*/(K^* \cdot \mathbb{C}^*)$ from (7-2), which is isomorphic to $\mathrm{Gal}(K_{\mathrm{ab}}/K)$ under the Artin map, is the quotient of the unit group $\widehat{K}^*$ of the *finite adele ring*

$$\widehat{K} = \widehat{\mathbb{Z}}_K \otimes_\mathbb{Z} \mathbb{Q} = {\prod_{\mathfrak{p} \text{ finite}}}' K_\mathfrak{p} \subset \mathbf{A}_K = \widehat{K} \times \mathbb{C}$$

by the subgroup $K^* \subset \widehat{K}^*$ of principal ideles. On the other hand, not all automorphisms of $\mathcal{F}$ come from $\mathrm{GL}_2(\widehat{\mathbb{Z}})$ as in (7-9): there is also an action of the projective linear group $\mathrm{PGL}_2(\mathbb{Q})^+ = \mathrm{GL}_2(\mathbb{Q})^+/\mathbb{Q}^*$ of rational matrices of positive determinant, which naturally act on $\mathbf{H}$ by linear fractional transformations. It does not fix $j$, as it maps the elliptic curve $\mathbb{C}/[\tau, 1]$ defined by $\tau \in \mathbf{H}$ not to an isomorphic, but to an isogenous curve. More precisely, if we pick $M = \begin{pmatrix} \alpha & \beta \\ \gamma & \delta \end{pmatrix} \in \mathrm{GL}_2(\mathbb{Q})^+$ in its residue class modulo $\mathbb{Q}^*$ such that $M^{-1}$ has integral coefficients, then the lattice

$$(\gamma\tau + \delta)^{-1}[\tau, 1] = \left[ \frac{\tau}{\gamma\tau + \delta}, \frac{1}{\gamma\tau + \delta} \right] = M^{-1}\left[ (\alpha\tau + \beta)/(\gamma\tau + \delta), 1 \right]$$

is a sublattice of finite index $\det M^{-1}$ in $[(\alpha\tau + \beta)/(\gamma\tau + \delta), 1]$, and putting $\mu = (\gamma\tau + \delta)^{-1}$, we have a commutative diagram

$$\begin{array}{ccc} \mathbb{C}/[\tau, 1] & \xrightarrow{\mu} & \mathbb{C}/[M\tau, 1] = \mathbb{C}/[(\alpha\tau + \beta)/(\gamma\tau + \delta), 1] \\ {\scriptstyle W} \downarrow \wr & & {\scriptstyle W'} \downarrow \wr \\ E_\tau & \xrightarrow{\phi_\mu} & E_{M\tau} \end{array}$$

as in (7-3). Moreover, the torsion point $u_1\tau + u_2$ having coordinates $u = (u_1, u_2)$ with respect to $[\tau, 1]$ is mapped to the torsion point with coordinates $uM^{-1}$ with respect to the basis $[M\tau, 1]$.

We let $\widehat{\mathbb{Q}} = \widehat{\mathbb{Z}} \otimes_{\mathbb{Z}} \mathbb{Q} = \prod_p' \mathbb{Q}_p$ be the ring of finite $\mathbb{Q}$-ideles. Then every element in the ring $\mathrm{GL}_2(\widehat{\mathbb{Q}})$ can be written as $UM$ with $U \in \mathrm{GL}_2(\widehat{\mathbb{Z}})$ and $M \in \mathrm{GL}_2(\mathbb{Q})^+$. This representation is not unique since $\mathrm{GL}_2(\widehat{\mathbb{Z}})$ and $\mathrm{GL}_2(\mathbb{Q})^+$ have nontrivial intersection $\mathrm{SL}_2(\mathbb{Z})$, but we obtain a well-defined action $\mathrm{GL}_2(\widehat{\mathbb{Q}}) \to \mathrm{Aut}(\mathcal{F})$ by putting $f^{UM}(\tau) = f^U(M\tau)$. We now extend, for the zero $\tau \in \mathbf{H}$ of a polynomial $X^2 + BX + C \in \mathbb{Q}[X]$, the map $g_\tau$ in (7-11) to

$$g_\tau : \widehat{K}^* = (\widehat{\mathbb{Q}}\tau + \widehat{\mathbb{Q}})^* \longrightarrow \mathrm{GL}_2(\widehat{\mathbb{Q}})$$
$$x = x_1\tau + x_2 \longmapsto M_x^{-1} = \begin{pmatrix} -Bx_1 + x_2 & -Cx_1 \\ x_1 & x_2 \end{pmatrix}^{-1} \quad (7\text{-}12)$$

to obtain the complete Galois action of $\mathrm{Gal}(K_{\mathrm{ab}}/K) \cong \widehat{K}^*/K^*$ on modular function values $f(\tau)$. The result is known as *Shimura's reciprocity law*:

THEOREM 7.13. *Let $\tau \in \mathbf{H}$ be imaginary quadratic, $f \in \mathcal{F}$ a modular function that is finite at $\tau$, and $x \in \widehat{K}^*/K^*$ a finite idele for $K = \mathbb{Q}(\tau)$. Then $f(\tau)$ is abelian over $K$, and the idele $x$ acts on it via its Artin symbol by*

$$f(\tau)^x = f^{g_\tau(x)}(\tau),$$

*where $g_\tau$ is defined as in (7-12).*

## 8. Class invariants

Much work has gone into algorithmic improvements of the classical algorithms in Section 6, with the aim of reducing the size of the class polynomials obtained. Clearly the *degree* of the polynomials involved cannot be lowered, as these are the degrees of the field extensions one wants to compute. There are however methods to reduce the size of their coefficients. These already go back to Weber, who made extensive use of 'smaller' functions than $j$ to compute class fields in his algebra textbook [Weber 1908]. The function $\mathfrak{f}_2$ that we used to compute $j$ in (6-12), and that carries Weber's name (as does the elliptic function in (6-14)) provides a good example. A small field such as $K = \mathbb{Q}(\sqrt{-71})$, for which the class group of order 7 is easily computed by hand, already has the sizable Hilbert class polynomial

$$\begin{aligned}\mathrm{Hil}_K(X) = {}& X^7 + 313645809715\,X^6 - 3091990138604570\,X^5 \\ & + 98394038810047812049302\,X^4 \\ & - 823534263439730779968091389\,X^3 \\ & + 5138800366453976780323726329446\,X^2 \\ & - 4253194739461396032746051511 87659\,X \\ & + 737707086760731113357714241006081263\,.\end{aligned}$$

However, the Weber function $\mathfrak{f}_2$, when evaluated at an appropriate generator of $\mathbb{Z}_K$ over $\mathbb{Z}$, also yields a generator for $H$ over $K$, with irreducible polynomial

$$X^7 + X^6 - X^5 - X^4 - X^3 + X^2 + 2X - 1.$$

As Weber showed, the function $\mathfrak{f}_2$ can be used to generate $H$ over $K$ when 2 splits and 3 does not ramify in $\mathbb{Q} \subset K$. The general situation illustrated by this example is that, despite the content of Theorem 7.6, it is sometimes possible to use a function $f$ of high level, like the Weber function $\mathfrak{f}_2 \in \mathcal{F}_{48}$ of level 48, to generate the Hilbert class field $H$ of conductor 1. The attractive feature of such high level functions $f$ is that they can be much smaller than the $j$-function itself. In the case of $\mathfrak{f}_2$, the extension $\mathbb{Q}(j) \subset \mathbb{Q}(\mathfrak{f}_2)$ is of degree 72 by (6-12), and this means that the size of the coefficients of class polynomials using $\mathfrak{f}_2$ is about a factor 72 smaller than the coefficients of $\mathrm{Hil}_K(X)$ itself. Even though this is only a constant factor, and complex multiplication is an intrinsically 'exponential' method, the computational improvement is considerable. For this reason, Weber's use of 'small' functions has gained renewed interest in present-day computational practice.

Shimura's reciprocity law Theorem 7.13 is a convenient tool to understand the occurrence of *class invariants*, that is, modular functions $f \in \mathcal{F}$ of higher level that generate the Hilbert class field of $K$ when evaluated at an appropriate generator $\tau$ of $\mathbb{Z}_K$. Classical examples of such functions used by Weber are $\gamma_2 = \sqrt[3]{j}$ and $\gamma_3 = \sqrt{j - 1728}$, which have level 3 and 2. As is clear from (6-12), the $j$-function can also be constructed out of even smaller building blocks involving the Dedekind $\eta$-function (6-11). Functions that are currently employed in actual computations are

$$\frac{\eta(pz)}{\eta(z)} \quad \text{and} \quad \frac{\eta(pz)\eta(qz)}{\eta(pqz)\eta(z)}, \tag{8-1}$$

which are of level $24p$ and $24pq$. These functions, or sometimes small powers of them, can be used to generate $H$, and the resulting minimal polynomials have much smaller coefficients than $\mathrm{Hil}_K(X)$. We refer to [Cohen 2000, Section 6.3] for the precise theorems, and indicate here how to use Theorem 7.13 to obtain such results for arbitrary modular functions $f \in \mathcal{F}$.

Let $f \in \mathcal{F}$ be any modular function of level $m$, and assume $\mathbb{Q}(f) \subset \mathcal{F}$ is Galois. Suppose we have an explicit Fourier expansion in $\mathbb{Q}(\zeta_m)((q^{1/m}))$ that we can use to approximate its values numerically. Suppose also that we know the explicit action of the generators $S : z \mapsto 1/z$ and $T : z \mapsto z+1$ on $f$. Then we can determine the Galois orbit of $f(\tau)$ for an element $\tau \in \mathbf{H}$ that generates $\mathbb{Z}_K$ in the following way. First, we determine elements $x = x_1\tau + x_2 \in \mathbb{Z}_K$ with the property that they generate $(\mathbb{Z}_K/m\mathbb{Z}_K)^*/\mathbb{Z}_K^*$. Then the Galois orbit of $f(\tau)$ over $H$ is determined using Theorem 7.13, and amounts to computing the (repeated) action of the matrices $g_\tau(x) \in \mathrm{GL}_2(\mathbb{Z}/m\mathbb{Z})$ (given by the right hand side of (7-10)) on $f$. This involves writing $g_\tau(x)$ as a product of powers of $S$ and $T$ and a matrix $\begin{pmatrix} 1 & 0 \\ 0 & k \end{pmatrix}$ acting on $f$ via its Fourier coefficients. Although $f$ may have a large $\mathrm{GL}_2(\mathbb{Z}/m\mathbb{Z})$-orbit over $\mathbb{Q}(j)$, the matrices $g_\tau(x)$ only generate a small subgroup of $\mathrm{GL}_2(\mathbb{Z}/m\mathbb{Z})$ isomorphic to $(\mathbb{Z}_K/m\mathbb{Z}_K)^*/\mathbb{Z}_K^*$, and one often finds that the orbit of $f$ under this subgroup is quite small. In many cases, one can slightly modify $f$, multiplying it by suitable roots of unity or raising it to small powers, to obtain an orbit of length one. This means that $f \in \mathcal{F}$ is invariant under $g_\tau[\widehat{\mathbb{Z}}_K^*] \subset \mathrm{GL}_2(\widehat{\mathbb{Z}})$. As we have the fundamental equivalence

$$f(\tau)^x = f(\tau) \quad \Longleftrightarrow \quad f^{g_\tau(x)} = f, \tag{8-2}$$

this is equivalent to finding that $f(\tau)$ is a class invariant for $K = \mathbb{Q}(\tau)$. The verification that $g_\tau[\widehat{\mathbb{Z}}_K^*]$ stabilizes $f$ takes place modulo the level $m$ of $f$, so it follows from (7-12) that if $f(\tau)$ is a class invariant for $K = \mathbb{Q}(\tau)$, then $f(\tau')$ is a class invariant for $K' = \mathbb{Q}(\tau')$ whenever $\tau' \in \mathbf{H}$ is a generator of $\mathbb{Z}_{K'}$ that has an irreducible polynomial congruent modulo $m$ to that of $\tau$. In particular, a function of level $m$ that yields class invariants does so for families of quadratic fields for which the discriminant is in certain congruence classes modulo $4m$.

If $f(\tau)$ is found to be a class invariant, we need to determine its conjugates over $K$ to determine its irreducible polynomial over $K$ as we did in (6-8) for $j(\tau)$. This amounts to computing $f(\tau)^{\sigma_\mathfrak{c}}$ as in Theorem 6.9, with $\mathfrak{c}$ ranging over the ideal classes of $\mathrm{Cl}_K$. If we list the ideal classes of $\mathrm{Cl}_K$ as in Section 6 as integer triples $(a, b, c)$ representing the reduced quadratic forms of discriminant $\Delta_K$, the Galois action of their Artin symbols in Theorem 6.9 may be given by

$$j(\tau)^{(a,-b,c)} = j\left(\frac{-b+\sqrt{b^2-4ac}}{2a}\right).$$

For a class invariant $f(\tau)$ a similar formula is provided by Shimura's reciprocity law. Let $\mathfrak{a} = \mathbb{Z} \cdot ((-b + \sqrt{b^2 - 4ac})/2) + \mathbb{Z} \cdot a$ be a $\mathbb{Z}_K$-ideal in the ideal class corresponding to the form $(a, b, c)$. Then the $\widehat{\mathbb{Z}}_K$-ideal $\mathfrak{a}\widehat{\mathbb{Z}}_K$ is principal since $\mathbb{Z}_K$-ideals are locally principal, and we let $x \in \widehat{\mathbb{Z}}_K$ be a generator. The element

$x$ is a finite idele in $\widehat{K}^*$, and the Artin symbol of $x^{-1}$ acts on $f(\tau)$ as the Artin symbol of the form $(a, -b, c)$. We have $U = g_\tau(x^{-1}) M^{-1} \in \mathrm{GL}_2(\widehat{\mathbb{Z}})$ for the matrix $M \in \mathrm{GL}_2(\mathbb{Q})^+$ defined by

$$[\tau, 1] M = \left[ \frac{b + \sqrt{b^2 - 4ac}}{2}, 2a \right],$$

since $U$ stabilizes the $\widehat{\mathbb{Z}}_K$-lattice spanned by the basis $[\tau, 1]$. Applying Theorem 7.13 for the idele $x^{-1}$ yields the desired formula

$$f(\tau)^{(a,-b,c)} = f^U \left( \frac{-b + \sqrt{b^2 - 4ac}}{2a} \right).$$

This somewhat abstract description may be phrased as a simple explicit recipe for the coefficients of $U \in \mathrm{GL}_2(\widehat{\mathbb{Z}})$, which we only need to know modulo $m$, see [Stevenhagen 2001].

There are limits to the improvements coming from intelligent choices of modular functions to generate class fields. For any nonconstant function $f \in \mathcal{F}$, there is a polynomial relation $\Psi(j, f) = 0$ between $j$ and $f$, with $\Psi \in \mathbb{C}[X, Y]$ some irreducible polynomial with algebraic coefficients. The *reduction* factor one obtains by using class invariants coming from $f$ (if these exist) instead of the classical $j$-values is defined as

$$r(f) = \frac{\deg_f(\Psi(f, j))}{\deg_j(\Psi(f, j))}.$$

By [Hindry and Silverman 2000, Proposition B.3.5], this is, asymptotically, the *inverse* of the factor

$$\lim_{h(j(\tau)) \to \infty} \frac{h(f(\tau))}{h(j(\tau))}.$$

Here $h$ is the absolute logarithmic height, and we take the limit over all CM-points $\mathrm{SL}_2(\mathbb{Z}) \cdot \tau \in \mathbf{H}$. It follows from gonality estimates for modular curves [Bröker and Stevenhagen 2008, Theorem 4.1] that $r(f)$ is bounded above by $1/(24\lambda_1)$, where $\lambda_1$ is 'Selberg's eigenvalue' as defined in [Sarnak 1995]. The currently proved bounds [Kim 2003, p. 176] on $\lambda_1$ yield $r(f) \leq 32768/325 \approx 100.8$, and conjectural bounds imply $r(f) \leq 96$. Thus Weber's function $\mathfrak{f}_2$, which has $r(f) = 72$ and yields class invariants for a positive density subset of all discriminants, is close to being optimal.

## Acknowledgements

Useful comments on earlier versions of this paper were provided by Reinier Bröker, René Schoof and Marco Streng. Bjorn Poonen provided us with the reference to [Kim 2003].

# References

[Artin and Tate 1990] E. Artin and J. Tate, *Class field theory*, 2nd ed., Advanced Book Classics, Addison-Wesley, Redwood City, CA, 1990.

[Bröker and Stevenhagen 2008] R. Bröker and P. Stevenhagen, "Constructing elliptic curves of prime order", pp. 17–28 in *Computational Arithmetic Geometry*, edited by K. E. Lauter and K. A. Ribet, Contemp. Math. **463**, 2008.

[Buhler and Wagon 2008] J. P. Buhler and S. Wagon, "Basic algorithms in number theory", pp. 25–68 in *Surveys in algorithmic number theory*, edited by J. P. Buhler and P. Stevenhagen, Math. Sci. Res. Inst. Publ. **44**, Cambridge University Press, New York, 2008.

[Bump et al. 2003] D. Bump, J. W. Cogdell, E. de Shalit, D. Gaitsgory, E. Kowalski, and S. S. Kudla, *An introduction to the Langlands program*, Birkhäuser Boston Inc., Boston, MA, 2003. Lectures presented at the Hebrew University of Jerusalem, Jerusalem, March 12–16, 2001, Edited by Joseph Bernstein and Stephen Gelbart.

[Cassels and Fröhlich 1967] J. W. S. Cassels and A. Fröhlich (editors), *Algebraic number theory*, Academic Press, London, 1967.

[Cohen 1993] H. Cohen, *A course in computational algebraic number theory*, Graduate Texts in Mathematics **138**, Springer, Berlin, 1993.

[Cohen 2000] H. Cohen, *Advanced topics in computational number theory*, Graduate Texts in Mathematics **193**, Springer, New York, 2000.

[Cox 1989] D. A. Cox, *Primes of the form $x^2 + ny^2$: Fermat, class field theory and complex multiplication*, John Wiley & Sons, New York, 1989.

[Deuring 1958] M. Deuring, *Die Klassenkörper der komplexen Multiplikation*, Enzyklopädie der mathematischen Wissenschaften, Band $I_2$, Heft 10, Teil II, Teubner, Stuttgart, 1958.

[Fieker 2001] C. Fieker, "Computing class fields via the Artin map", *Math. Comp.* **70**:235 (2001), 1293–1303.

[Hindry and Silverman 2000] M. Hindry and J. H. Silverman, *Diophantine geometry: an introduction*, Graduate Texts in Mathematics **201**, Springer, New York, 2000.

[Kim 2003] H. H. Kim, "Functoriality for the exterior square of $GL_4$ and the symmetric fourth of $GL_2$", *J. Amer. Math. Soc.* **16**:1 (2003), 139–183.

[Lang 1987] S. Lang, *Elliptic functions*, Second ed., Graduate Texts in Mathematics **112**, Springer, New York, 1987.

[Lang 2002] S. Lang, *Algebra*, Third ed., Graduate Texts in Mathematics **211**, Springer, New York, 2002.

[Sarnak 1995] P. Sarnak, "Selberg's eigenvalue conjecture", *Notices Amer. Math. Soc.* **42**:11 (1995), 1272–1277.

[Schneps 1994] L. Schneps (editor), *The Grothendieck theory of dessins d'enfants* (Luminy, 1993), London Math. Soc. Lecture Note Ser. **200**, Cambridge Univ. Press, Cambridge, 1994.

[Schoof 2008] R. J. Schoof, "Computing Arakelov class groups", pp. 447–495 in *Surveys in algorithmic number theory*, edited by J. P. Buhler and P. Stevenhagen, Math. Sci. Res. Inst. Publ. **44**, Cambridge University Press, New York, 2008.

[Serre 1989] J.-P. Serre, *Abelian l-adic representations and elliptic curves*, Second ed., Advanced Book Classics, Addison-Wesley, Redwood City, CA, 1989.

[Shimura 1998] G. Shimura, *Abelian varieties with complex multiplication and modular functions*, Princeton Mathematical Series **46**, Princeton University Press, Princeton, NJ, 1998.

[Silverman 1986] J. H. Silverman, *The arithmetic of elliptic curves*, Graduate Texts in Mathematics **106**, Springer, New York, 1986.

[Silverman 1994] J. H. Silverman, *Advanced topics in the arithmetic of elliptic curves*, Graduate Texts in Mathematics **151**, Springer, New York, 1994.

[Stevenhagen 2001] P. Stevenhagen, "Hilbert's 12th problem, complex multiplication and Shimura reciprocity", pp. 161–176 in *Class field theory — its centenary and prospect* (Tokyo, 1998), edited by K. Miyake, Adv. Stud. Pure Math. **30**, Math. Soc. Japan, Tokyo, 2001.

[Stevenhagen 2008] P. Stevenhagen, "The arithmetic of number rings", pp. 209–266 in *Surveys in algorithmic number theory*, edited by J. P. Buhler and P. Stevenhagen, Math. Sci. Res. Inst. Publ. **44**, Cambridge University Press, New York, 2008.

[Stevenhagen and Lenstra 1996] P. Stevenhagen and H. W. Lenstra, Jr., "Chebotarëv and his density theorem", *Math. Intelligencer* **18**:2 (1996), 26–37.

[Völklein 1996] H. Völklein, *Groups as Galois groups: an introduction*, Cambridge Studies in Advanced Mathematics **53**, Cambridge Univ. Press, Cambridge, 1996.

[Weber 1908] H. Weber, *Lehrbuch der Algebra*, F. Vieweg und Sohn, Braunschweig, 1908. Reprinted by Chelsea Pub., New York, 1961.

HENRI COHEN
LABORATOIRE A2X, U.M.R. 5465 DU C.N.R.S.
UNIVERSITÉ BORDEAUX I
351 COURS DE LA LIBÉRATION
33405 TALENCE CEDEX
FRANCE
  cohen@math.u-bordeaux1.fr

PETER STEVENHAGEN
MATHEMATISCH INSTITUUT,
UNIVERSITEIT LEIDEN, POSTBUS 9512
2300 RA LEIDEN
THE NETHERLANDS
  psh@math.leidenuniv.nl

# Protecting communications against forgery

DANIEL J. BERNSTEIN

ABSTRACT. This paper is an introduction to cryptography. It covers secret-key message-authentication codes, unpredictable random functions, public-key secret-sharing systems, and public-key signature systems.

## 1. Introduction

Cryptography protects communications against espionage: an eavesdropper who intercepts a message will be unable to decipher it. This is useful for many types of information: credit-card transactions, medical records, love letters.

Cryptography also protects communications against sabotage: a forger who fabricates or modifies a message will be unable to deceive the receiver. This is useful for *all* types of information. If the receiver does not care about the authenticity of a message, why is he listening to the message in the first place?

This paper explains how cryptography prevents forgery. Section 2 explains how to protect $n$ messages if the sender and receiver share $128(n+1)$ secret bits. Section 3 explains how the sender and receiver can generate many shared secret bits from a short shared secret. Section 4 explains how the sender and receiver can generate a short shared secret from a public conversation. Section 5 explains how the sender can protect a message sent to many receivers, without sharing any secrets.

*Mathematics Subject Classification:* 94A62.

Permanent ID of this document: `9774ae5a1749a7b256cc923a7ef9d4dc`. Date: 2008.05.01.

## 2. Unbreakable secret-key authenticators

Here is a protocol for transmitting a message when the sender and receiver both know certain secrets:

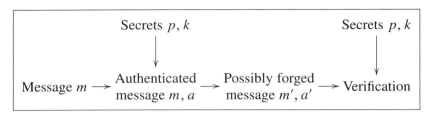

The message is a polynomial $m \in F[x]$ with $m(0) = 0$ and $\deg m \leq 1000000$. Here $F$ is the field $(\mathbf{Z}/2)[y]/(y^{128} + y^9 + y^7 + y^2 + 1)$ of size $2^{128}$. The secrets are two independent uniform random elements $p, k$ of $F$.

The sender transmits $(m, a)$ where $a = m(p) + k$. The forger replaces $(m, a)$ with some $(m', a')$; if the forger is inactive then $(m', a') = (m, a)$. The receiver discards $(m', a')$ unless $a' = m'(p) + k$.

The extra information $a$ is called an **authenticator**.

**Security.** I claim that the forger has chance smaller than $2^{-108}$ of fooling the receiver, i.e., of finding $(m', a')$ with $m' \neq m$ and $a' = m'(p) + k$. The proof is easy. Fix $(m, a)$ and $(m', a')$, and count pairs $(p, k)$:

- There are exactly $2^{128}$ pairs $(p, k)$ satisfying $a = m(p) + k$. Indeed, there is exactly one possible $k$ for each possible $p$.
- Fewer than $2^{20}$ of these pairs also satisfy $a' = m'(p) + k$, if $m'$ is different from $m$. Indeed, any qualifying $p$ would have to be a root of the nonzero polynomial $m - m' - a + a'$; this polynomial has degree at most 1000000, so it has at most $1000000 < 2^{20}$ roots.

Thus the conditional probability that $a' = m'(p) + k$, given that $a = m(p) + k$, is smaller than $2^{20}/2^{128} = 2^{-108}$.

In practice, the receiver will continue listening for messages after discarding a forgery, so the forger can try again and again. Consider a persistent, wealthy, long-lived forger who tries nearly $2^{75}$ forgeries by flooding the receiver with one billion messages per second for one million years. His chance of success — his chance of producing at least one $(m', a')$ with $a' = m'(p) + k$ and with $m'$ not transmitted by the sender — is still smaller than $2^{-108} 2^{75} = 2^{-33}$.

**Handling many messages.** One can use a single $p$ with many $k$'s to protect a series of messages:

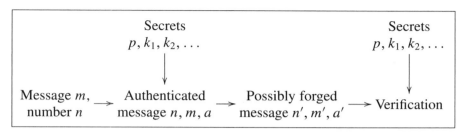

The sender and receiver share secrets $p, k_1, k_2, k_3, \ldots$; as in the single-message case, $(p, k_1, k_2, k_3, \ldots)$ is a uniform random sequence of elements of $F$. The sender transmits the $n$th message $m$ as $(n, m, a)$ where $a = m(p) + k_n$. The receiver discards $(n', m', a')$ unless $a' = m'(p) + k_{n'}$.

In this context $n$ is called a **nonce** and $a$ is again called an **authenticator**. The random function $(n, m) \mapsto m(p) + k_n$ is called a **message-authentication code** (MAC).

The forger's chance of success — his chance of producing at least one forgery $(n', m', a')$ with $a' = m'(p) + k_{n'}$ and with $m'$ different from all of the messages transmitted by the sender — is smaller than $2^{-108} D$, where $D$ is the number of forgery attempts. This is true even if the forger sees all the messages transmitted by the sender. It is true even if the forger can influence the choice of those messages, perhaps responding dynamically to previous authenticators. In fact, it is true even if the forger has complete control over each message!

Define an **attack** as an algorithm that chooses a message $m_1$, sees the sender's authenticator $m_1(p) + k_1$, chooses a message $m_2$, sees the sender's authenticator $m_2(p) + k_2$, etc., and finally chooses $(n', m', a')$. Define the attack as **successful** if $a' = m'(p) + k_{n'}$ and $m' \notin \{m_1, m_2, \ldots\}$. Then the attack is successful with probability smaller than $2^{-108}$. The proof is, as in the single-message case, a simple matter of counting.

Of course, if the forger actually has the power to choose a message $m_1$ for the sender to authenticate, then the forger does not need to modify messages in transit. Real senders restrict the messages $m_1, m_2, \ldots$ that they authenticate, and thus restrict the possible set of attacks. But the security guarantee does not rely on any such restrictions.

**History.** Gilbert, MacWilliams, and Sloane [1974, Section 9] introduced the first easy-to-compute unbreakable authenticator, using a long shared secret for a long message. Wegman and Carter [1981, Section 3] proposed the form $h(m) + k_n$ for an authenticator and pointed out that a short secret could handle a long message.

There is now a huge literature on unbreakable MACs. For two surveys see [Nevelsteen and Preneel 1999] and [Bernstein 2004, Sections 8–10]. For three state-of-the-art systems see [Black et al. 1999], [Bernstein 2005], and [Bernstein 2007].

## 3. Conjecturally unpredictable random functions

Here is a protocol that is *conjectured* to protect a series of messages:

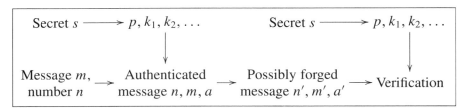

The sender and receiver share a secret uniform random 128-bit string $s$. The sender and receiver both compute $p = \text{SLASH}(0, s)$, $k_1 = \text{SLASH}(1, s)$, $k_2 = \text{SLASH}(2, s)$, etc. The sender transmits the $n$th message $m$ as $(n, m, a)$ where $a = m(p) + k_n$. The receiver discards $(n', m', a')$ unless $a' = m'(p) + k_{n'}$.

The function SLASH — see below for the definition — takes 512 bits of input. Message numbers $n$ are assumed to be at most $2^{128} - 1$; a pair $(n, s)$ is then encoded as a 512-bit input $(n_0, n_1, \ldots, n_{127}, 0, 0, \ldots, 0, s_0, s_1, \ldots, s_{127})$ where $n = n_0 + 2n_1 + \cdots + 2^{127} n_{127}$. SLASH produces 128 bits of output. The result has no apparent structure aside from its computability.

Note that the sender and receiver can compute $\text{SLASH}(n, s)$ when they need it, rather than storing the long string $(p, k_1, k_2, \ldots)$.

**Security.** A forger, given several authenticated messages, might try to solve for $s$. Presumably only one choice for $s$ is consistent with all the authenticators. However, the fastest *known* method of solving for $s$ is to search through all $2^{128}$ possibilities. This is far beyond the computer power available today.

Is there a faster attack? Perhaps. We believe that this protocol is unbreakable, but we have no proof. (The random string $(p, k_1, k_2, \ldots)$ is not uniform, so the proof in Section 2 does not apply.) On the other hand, this protocol has the advantage of using only 128 shared secret bits to handle any number of messages.

**Unpredictability.** Let $u$ be a uniform random function from $\{0, 1, 2, \ldots\}$ to $F$. Consider oracle algorithms $A$ that print 0 or 1. What is the difference between

- the probability that $A$ prints 1 using $n \mapsto \text{SLASH}(n, s)$ as an oracle and
- the probability that $A$ prints 1 using $u$ as an oracle?

The difference is conjectured to be smaller than $2^{-40}$ for every $A$ that finishes in at most $2^{80}$ steps. In other words, $n \mapsto \mathrm{SLASH}(n, s)$ is conjectured to be **unpredictable**.

If $n \mapsto \mathrm{SLASH}(n, s)$ is, in fact, unpredictable, then this multiple-message short-secret authentication protocol is unbreakable: a fast algorithm that makes $D$ forgery attempts cannot succeed with probability larger than $2^{-108} D + 2^{-40}$.

**The SLASH definition.** Say $x_0, x_1, \ldots, x_{15}$ are 32-bit strings. For $i \geq 16$ define $x_i = x_{i-16} + ((x_{i-1} + \delta_i) \oplus (x_{i-1} \lll 7))$. Then $\mathrm{SLASH}(x_0, x_1, \ldots, x_{15})$ is the 256-bit string $(x_0 \oplus x_{520}, x_1 \oplus x_{521}, \ldots, x_7 \oplus x_{527})$. In contexts where only 128 bits are required, the first 128 bits are used.

Notation: $(a_0, a_1, \ldots, a_{31}) + (b_0, b_1, \ldots, b_{31}) = (c_0, c_1, \ldots, c_{31})$ means that $a_0 + 2a_1 + \cdots + 2^{31} a_{31} + b_0 + 2b_1 + \cdots + 2^{31} b_{31} \equiv c_0 + 2c_1 + \cdots + 2^{31} c_{31} \pmod{2^{32}}$; $(a_0, a_1, \ldots, a_{31}) \oplus (b_0, b_1, \ldots, b_{31}) = (c_0, c_1, \ldots, c_{31})$ means that $a_i + b_i \equiv c_i \pmod{2}$ for each $i$; $(a_0, a_1, \ldots, a_{31}) \lll 7 = (c_0, c_1, \ldots, c_{31})$ means that $c_7 = a_0, c_8 = a_1, \ldots, c_{31} = a_{24}, c_0 = a_{25}, \ldots, c_6 = a_{31}$; and $\delta_i$ means the string $(c_0, c_1, \ldots, c_{31})$ such that $c_0 + 2c_1 + \cdots + 2^{31} c_{31} \equiv 2654435769 \lfloor i/16 \rfloor \pmod{2^{32}}$.

**History.** Turing [1950] introduced the concept of unpredictability: "Suppose we could be sure of finding [laws of behaviour] if they existed. Then given a discrete-state machine it should certainly be possible to discover by observation sufficient about it to predict its future behaviour, and this within a reasonable time, say a thousand years. But this does not seem to be the case. I have set up on the Manchester computer a small programme using only 1000 units of storage, whereby the machine supplied with one sixteen figure number replies with another within two seconds. I would defy anyone to learn from these replies sufficient about the programme to be able to predict any replies to untried values."

The literature is full of very quickly computable short random functions that seem difficult to predict. **Short** means that the random function is determined by a short uniform random string. See the surveys [Schneier 1996], [Menezes et al. 1996], [Nechvatal et al. 1999], and [Nechvatal et al. 2001] for many examples. A typical example is more complicated than SLASH but somewhat faster.

Beware that the literature is also full of definitions that distract attention from unpredictability. For example, a **block cipher** is a short random inverse pair of functions $(f, f^{-1})$. One hopes that $(f, f^{-1})$ is indistinguishable from a uniform random inverse pair of functions. This indistinguishability implies unpredictability of $f$ if the input size of $f$ is large enough, say 256 bits; but the extra constraint of invertibility is unnecessary for applications and excludes many good designs.

Blum, Blum, and Shub [1986] constructed a fast short random function with a small input, and proved that any fast algorithm to predict that function could be turned into a surprisingly fast algorithm to factor integers. Naor and Reingold [1997] constructed fast random functions with large inputs and with similar guarantees of unpredictability. These "provable" functions are never used in practice, because they are not nearly as fast as state-of-the-art block ciphers; but they show that unpredictability is not a silly concept.

Unpredictability has an interesting application to complexity theory: one can use it to convert fast probabilistic algorithms into reasonably fast deterministic algorithms. This was pointed out by Yao [1982]. It is now widely believed that the complexity classes BPP and P are identical, i.e., that everything decidable in polynomial time with the help of randomness is also decidable in polynomial time deterministically. One exposition of the topic is [Goldreich 1999, Section 3.4].

The name "unpredictable" has several aliases in the literature. See [Bernstein 1999, Section 2] for further discussion.

## 4. Public-key secret sharing

Here is a protocol for the sender and receiver to generate a 128-bit shared secret from a public conversation:

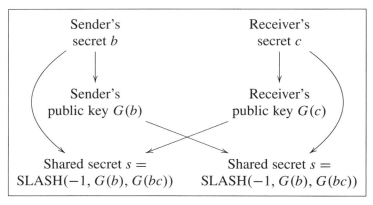

The sender starts from a secret uniform random $b \in 16\mathbf{Z}$ with $0 < b \leq 2^{225}$. The sender computes and announces a **public key** $G(b)$, namely the $x$-coordinate of the $b$th multiples of the points $(53(2^{224} - 1)/(2^8 - 1), \pm \ldots)$ on the elliptic curve $y^2 = x^3 + 7530x^2 + x$ over the field $\mathbf{Z}/(2^{226} - 5)$. It is not difficult to compute $G(b)$ from $b$; see, e.g., [Blake et al. 2000], [Hankerson et al. 2004], [Doche and Lange 2005], and the chapter [Poonen 2008] in this volume.

Similarly, the receiver starts from a secret uniform random $c \in 16\mathbf{Z}$ with $0 < c \leq 2^{225}$. The receiver computes and announces a public key $G(c)$.

The sender and receiver are assumed to receive correct copies of $G(b)$ and $G(c)$ from each other. Subsequent messages are protected against forgery, but the public keys themselves must be protected by something outside this protocol.

The sender now computes $G(bc)$; it is not difficult to compute $G(bc)$ from $b$ and $G(c)$, both of which are known to the sender. The receiver computes $G(bc)$ from $c$ and $G(b)$ in the same way. Finally, the sender and receiver both compute $s = \text{SLASH}(-1, G(b), G(bc))$. Here $(-1, G(b), G(bc))$ is encoded as the 512-bit string $(g_0, g_1, \ldots, g_{225}, 1, 1, 1, \ldots, 1, h_0, h_1, \ldots, h_{225})$ where $G(b) = g_0 + 2g_1 + \cdots + 2^{225}g_{225}$ and $G(bc) = h_0 + 2h_1 + \cdots + 2^{225}h_{225}$.

As in Section 3, the sender and receiver can use this shared secret $s$ to protect the authenticity of a series of messages:

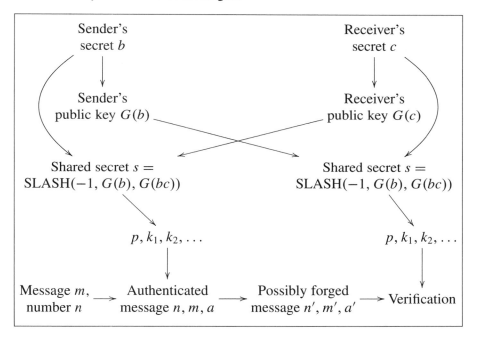

The sender can also reuse his secret $b$ with other receivers: given the public key $G(d)$ of another receiver, the sender computes the corresponding shared secret $\text{SLASH}(-1, G(b), G(bd))$ and continues exactly as above. Furthermore, the sender and receiver can reverse roles, using $\text{SLASH}(-1, G(c), G(bc))$ and $\text{SLASH}(-1, G(d), G(bd))$ for messages sent in the opposite direction.

**Security.** The complete definition of security here is more complicated than it was in Sections 2 and 3, because the forger has more power. In particular, the forger is given the public keys. The forger can also feed a number $G(c)$ to the sender (without necessarily knowing what $c$ is) and receive authenticators computed using $\text{SLASH}(-1, G(b), G(bc))$.

The fastest *known* attack is to start from the public key $G(b)$, perform about $2^{112}$ elliptic-curve operations, and deduce the secret $b$, after which the forger can compute $s = \text{SLASH}(-1, G(b), G(bc))$ in the same way as the sender. As in Section 3, this is beyond the computer power available today, but there may be faster attacks.

This attack does not depend on the details of SLASH. To formalize this notion, consider a **generic protocol** in which the sender and receiver use an oracle for any 128-bit function in place of SLASH; then there is a **generic attack** in which the forger, having access to the same oracle, succeeds in forgeries after about $2^{112}$ elliptic-curve operations.

A generic attack that succeeds for all 128-bit functions can be converted into an algorithm at comparable speed that, given $G(b)$ and $G(c)$, computes $G(bc)$. A generic attack that succeeds with probability $p$ on average over all 128-bit functions can be converted into an algorithm at comparable speed that, given $G(b)$ and $G(c)$, computes $G(bc)$ with probability comparable to $p$. The idea of the proof is that if the algorithm never feeds $G(bc)$ to the oracle then it has no information about the shared secret. Of course, the value of this proof is limited, for two reasons: first, there might be faster non-generic attacks that exploit the structure of SLASH; second, we have no proof that computing $G(bc)$ from $G(b)$ and $G(c)$ is difficult.

**History.** Diffie and Hellman [1976] introduced the general idea of sharing a secret through a public channel. They also introduced the specific approach of exchanging public keys $2^b \bmod \ell$ and $2^c \bmod \ell$ to share a secret $2^{bc} \bmod \ell$; here $\ell$ is a fixed prime. The problem of computing $2^{bc} \bmod \ell$ from $(2^b \bmod \ell, 2^c \bmod \ell)$ is called the **Diffie–Hellman problem**.

There are surprisingly fast techniques to compute $b$ from $2^b \bmod \ell$. See [Schirokauer 2008] in this volume. Consequently one must choose a rather large prime $\ell$ in the Diffie–Hellman system.

Miller [1986], and independently Koblitz [1987], suggested replacing the unit group $(\mathbf{Z}/\ell)^*$ with an elliptic curve over $\mathbf{Z}/\ell$. No surprisingly fast techniques are known for the "elliptic-curve Diffie–Hellman problem" for most curves with near-prime order, so we *believe* that a relatively small value of $\ell$, such as $\ell = 2^{226}-5$, is safe. My elliptic curve $y^2 = x^3 + 7530x^2 + x$ over the field $\mathbf{Z}/(2^{226}-5)$ has order $(2^{226}-5)+1-120004032613757866556879513972474 36$, which is 16 times a prime. See [Bernstein 2006] for discussion of a similar curve.

Elliptic-curve computations involve more effort than unit-group operations, but this increase is outweighed by the reduction in the size of $\ell$, so the Miller–Koblitz elliptic-curve variant is faster than the original Diffie–Hellman system. It also has shorter keys. The variant is becoming increasingly popular.

Fiat and Shamir [1987] proved that a generic attack on one protocol could be converted into an algorithm to solve an easy-to-state mathematical problem. Bellare and Rogaway [1993] expanded the idea to more protocols. Many such proofs have now been published. For an exposition see [Koblitz and Menezes 2007].

## 5. Public-key signatures

Here is a protocol — with no shared secrets — for the sender to protect many messages sent to many receivers:

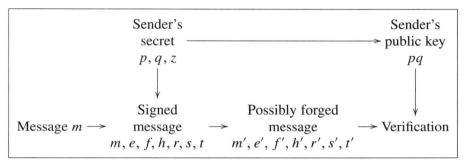

The sender starts from a secret uniform random 256-bit string $z$, and secret uniform random primes $p, q$ in the interval $[2^{768}, 2^{768} + 2^{766}]$ with $p \bmod 8 = 3$ and $q \bmod 8 = 7$; primality can be tested quickly, as explained in [Schoof 2008] in this volume. The sender computes and announces the product $pq$, which is assumed to be transmitted correctly to all receivers. Subsequent messages are protected against forgery as follows.

Given a message $m$, the sender computes

- $r = \text{SLASH}(-2, z, m) \bmod 16$;
- $h = H(r, m)$ where $H(r, m) = \text{SLASH}(-12, r, m) + 2^{128} \text{SLASH}(-13, r, m) + \cdots + 2^{1408} \text{SLASH}(-23, r, m) + 1$;
- $u = h^{(q+1)/4} \bmod q$;
- $e = 1$ if $u^2 \equiv h \pmod{q}$, else $e = -1$;
- $v = (eh)^{(p+1)/4} \bmod p$;
- $f = 1$ if $v^2 \equiv eh \pmod{p}$, else $f = 2$;
- $w = f^{(3q-5)/4} u \bmod q$;
- $x = f^{(3p-5)/4} v \bmod p$;
- $y = w + q(q^{p-2}(x - w) \bmod p)$;
- $s = \min\{y, pq - y\}$; and
- $t = (fs^2 - eh)/pq$.

The sender then transmits $(m, e, f, h, r, s, t)$.

At this point $(e, f, h, r, s, t)$ is a **signature** of $m$ under the public key $pq$. This means, by definition, that $e \in \{1, -1\}$; $f \in \{1, 2\}$; $r \in \{0, 1, \ldots, 15\}$; $s$ and $t$ are in $\{0, 1, \ldots, 2^{1536} - 1\}$; $h = H(r, m)$; and $fs^2 = tpq + eh$.

The receiver discards $(m', e', f', h', r', s', t')$ if $(e', f', h', r', s', t')$ is not a signature of $m'$. The receiver can save time here by checking the equation $f'(s')^2 = t'pq + e'h'$ modulo a secret 128-bit prime.

Observe that signatures are different from authenticators: a signature can be verified by anyone, while an authenticator can be verified only by people who could have created the authenticator. The receiver can convince third parties that the sender signed a message; the receiver cannot convince third parties that the sender authenticated a message. Signatures are appropriate for public communications; authenticators are appropriate for private communications.

**Security.** Like the protocols in Sections 3 and 4, this protocol *appears* to make forgeries difficult, even if the forger can inspect signatures on messages under his control. There are surprisingly fast techniques to factor $pq$ into $p, q$ — see [Pomerance 2008] and [Stevenhagen 2008] in this volume — but for large $pq$ these computations are beyond the computer power available today.

One can prove that any generic attack against this protocol can be converted into an algorithm at comparable speed to factor $pq$ with comparable success probability. However, as in Section 4, the value of this proof is limited: there might be faster non-generic attacks, and we have no proof that factorization is difficult.

**Message length.** The above description of signatures presumes that $(-2, z, m)$ and $(-12, r, m)$ and so on are encoded as 512-bit strings to be fed to SLASH. Thus messages $m$ must be very short.

One can handle longer messages by modifying SLASH to allow larger inputs. One can, for example, define SLASH$(x_0, x_1, x_2, x_3)$, where each $x_i$ is a 256-bit string, as SLASH(SLASH(SLASH(SLASH(0, 0, $x_0$), 1, $x_1$), 2, $x_2$), 3, $x_3$).

**History.** The concept of public-key signatures was introduced by Diffie and Hellman [1976]. Rivest, Shamir, and Adleman [1978] are often credited with the first useful example; but the original RSA system is obviously breakable.

(In the original RSA system, $s$ is a signature of $m$ under a public key $(n, e)$ if $s^e \equiv m \pmod{n}$. First obvious attack: the forger immediately computes the message $2^e \bmod n$ with signature 2. Second obvious attack: starting from $m$, the forger obtains from the sender a signature on the message $2^e m \bmod n$, and then divides the result by 2 modulo $n$.)

Rabin [1979] introduced the first useful signature system. Rabin's signature system, with various improvements by Williams [1980], Barwood, Wigley, and me, is the system described in this section. Recent results of Bleichenbacher,

Coppersmith, and Gentry show that signatures and public keys in this system can be compressed to a surprising extent. See [Bernstein 2008] for a survey and comparison of Rabin-type systems.

There are many "cryptographic hash functions" that can be used in place of $H$; see, e.g., the survey [Menezes et al. 1996, Sections 9.3–9.4]. On the other hand, some hash functions have been broken; for example, Wang et al. [2004] found collisions in the widely used "MD5" function. I offer $1000 to the first person to publish a SLASH input whose output is 128 all-zero bits, or two different 512-bit SLASH inputs with the same 256-bit output.

There are other signature systems. One interesting example is the ElGamal system [1985b], which uses Diffie–Hellman public keys. Keys and signatures in elliptic-curve variants of ElGamal's system are smaller than keys and signatures in Rabin-type systems; on the other hand, signature verification is slower. Rabin-type systems and ElGamal-type systems are both widely used.

# References

[Ashby 1993] Victoria Ashby (editor), *First ACM conference on computer and communications security*, Association for Computing Machinery, New York. See [Bellare and Rogaway 1993].

[Bellare and Rogaway 1993] Mihir Bellare and Phillip Rogaway, "Random oracles are practical: a paradigm for designing efficient protocols", pp. 62–73 in [Ashby 1993]. Citations in this document: §4.

[Bernstein 1999] Daniel J. Bernstein, "How to stretch random functions: the security of protected counter sums", *Journal of Cryptology* **12**, 185–192. ISSN 0933-2790. URL: http://cr.yp.to/papers.html#stretch. Citations in this document: §3.

[Bernstein 2004] Daniel J. Bernstein, "Floating-point arithmetic and message authentication". URL: http://cr.yp.to/papers.html#hash127. ID dabadd3095644704 c5cbe9690ea3738e. Citations in this document: §2.

[Bernstein 2005] Daniel J. Bernstein, "The Poly1305-AES message-authentication code", pp. 32–49 in [Gilbert and Handschuh 2005]. URL: http://cr.yp.to/papers.html#poly1305. ID 0018d9551b5546d97c340e0dd8cb5750. Citations in this document: §2.

[Bernstein 2006] Daniel J. Bernstein, "Curve25519: new Diffie-Hellman speed records", pp. 207–228 in [Yung et al. 2006]. URL: http://cr.yp.to/papers.html#curve25519. ID 4230efdfa673480fc079449d90f322c0. Citations in this document: §4.

[Bernstein 2007] Daniel J. Bernstein, "Polynomial evaluation and message authentication". URL: http://cr.yp.to/papers.html#pema. ID b1ef3f2d385a926123e 1517392e20f8c. Citations in this document: §2.

[Bernstein 2008] Daniel J. Bernstein, "RSA signatures and Rabin–Williams signatures: the state of the art". URL: http://cr.yp.to/papers.html#rwsota. ID 5e92b45abdf8abc4e55ea02607400599. Citations in this document: §5.

[Black et al. 1999] John Black, Shai Halevi, Hugo Krawczyk, Ted Krovetz, and Phillip Rogaway, "UMAC: fast and secure message authentication", pp. 216–233 in [Wiener 1999]. URL: http://www.cs.ucdavis.edu/~rogaway/umac/. Citations in this document: §2.

[Blake et al. 2000] Ian F. Blake, Gadiel Seroussi, and Nigel P. Smart, *Elliptic curves in cryptography*, Cambridge University Press, Cambridge. ISBN 0–521–65374–6. MR 1 771 549. Citations in this document: §4.

[Blakley and Chaum 1985] G. R. Blakley and David Chaum (editors), *Advances in cryptology: CRYPTO '84*, Lecture Notes in Computer Science **196**, Springer-Verlag, Berlin. ISBN 3–540–15658–5. MR 86j:94003. See [ElGamal 1985a].

[Blum et al. 1986] Lenore Blum, Manuel Blum, and Michael Shub, "A simple unpredictable pseudo-random number generator", *SIAM Journal on Computing* **15**, 364–383. ISSN 0097–5397. MR 87k:65007. URL: http://cr.yp.to/bib/entries.html#1986/blum. Citations in this document: §3.

[Buhler and Stevenhagen 2008] Joe P. Buhler and Peter Stevenhagen (editors), *Surveys in algorithmic number theory*, Mathematical Sciences Research Institute Publications **44**, Cambridge University Press, New York; this book. See [Pomerance 2008], [Poonen 2008], [Schirokauer 2008], [Schoof 2008], [Stevenhagen 2008].

[Cohen and Frey 2005] Henri Cohen and Gerhard Frey (editors), *Handbook of elliptic and hyperelliptic curve cryptography*, CRC Press. ISBN 1–58488–518–1. See [Doche and Lange 2005].

[Diffie and Hellman 1976] Whitfield Diffie and Martin Hellman, "New directions in cryptography", *IEEE Transactions on Information Theory* **22**, 644–654. ISSN 0018–9448. MR 55:10141. Citations in this document: §4, §5.

[Doche and Lange 2005] Christophe Doche and Tanja Lange, "Arithmetic of elliptic curves", pp. 267–302 in [Cohen and Frey 2005]. MR 2162729. Citations in this document: §4.

[ElGamal 1985a] Taher ElGamal, "A public key cryptosystem and a signature scheme based on discrete logarithms", pp. 10–18 in [Blakley and Chaum 1985]; see also newer version [ElGamal 1985b]. MR 87b:94037.

[ElGamal 1985b] Taher ElGamal, "A public key cryptosystem and a signature scheme based on discrete logarithms", *IEEE Transactions on Information Theory* **31**, 469–472; see also older version [ElGamal 1985a]. ISSN 0018–9448. MR 86j:94045. Citations in this document: §5.

[Fiat and Shamir 1987] Amos Fiat and Adi Shamir, "How to prove yourself: practical solutions to identification and signature problems", pp. 186–194 in [Odlyzko 1987]. MR 88m:94023. Citations in this document: §4.

[FOCS 1979] — (no editor), *20th annual symposium on foundations of computer science*, IEEE Computer Society, New York. MR 82a:68004. See [Wegman and Carter 1979].

[FOCS 1982] — (no editor), *23rd annual symposium on foundations of computer science*, IEEE Computer Society, New York. MR 85k:68007. See [Yao 1982].

[FOCS 1997] — (no editor), *38th annual symposium on foundations of computer science*, IEEE Computer Society Press, Los Alamitos. ISBN 0–8186–8197–7. See [Naor and Reingold 1997].

[Gilbert and Handschuh 2005] Henri Gilbert and Helena Handschuh (editors), *Fast software encryption: 12th international workshop, FSE 2005, Paris, France, February 21–23, 2005, revised selected papers*, Lecture Notes in Computer Science **3557**, Springer, Berlin. ISBN 3–540–26541–4. See [Bernstein 2005].

[Gilbert et al. 1974] Edgar N. Gilbert, F. Jessie MacWilliams, and Neil J. A. Sloane, "Codes which detect deception", *Bell System Technical Journal* **53**, 405–424. ISSN 0005–8580. MR 55:5306. Citations in this document: §2.

[Goldreich 1999] Oded Goldreich, *Modern cryptography, probabilistic proofs and pseudorandomness*, Springer-Verlag, Berlin. ISBN 3–540–64766–X. MR 2000f:94029. Citations in this document: §3.

[Hankerson et al. 2004] Darrel Hankerson, Alfred Menezes, and Scott Vanstone, *Guide to elliptic curve cryptography*, Springer, New York. ISBN 0–387–95273–X. MR 2054891. Citations in this document: §4.

[Koblitz 1987] Neal Koblitz, "Elliptic curve cryptosystems", *Mathematics of Computation* **48**, 203–209. ISSN 0025–5718. MR 88b:94017. Citations in this document: §4.

[Koblitz and Menezes 2005] Neal Koblitz and Alfred J. Menezes, "Another look at 'provable security' ", revised 4 May 2005; see also newer version [Koblitz and Menezes 2007]. URL: http://eprint.iacr.org/2004/152/.

[Koblitz and Menezes 2007] Neal Koblitz and Alfred J. Menezes, "Another look at 'provable security' ", *Journal of Cryptology* **20**, 3–37; see also older version [Koblitz and Menezes 2005]. ISSN 0933–2790. Citations in this document: §4.

[Menezes et al. 1996] Alfred J. Menezes, Paul C. van Oorschot, and Scott A. Vanstone, *Handbook of applied cryptography*, CRC Press, Boca Raton, Florida. ISBN 0–8493–8523–7. MR 99g:94015. URL: http://cacr.math.uwaterloo.ca/hac. Citations in this document: §3, §5.

[Miller 1986] Victor S. Miller, "Use of elliptic curves in cryptography", pp. 417–426 in [Williams 1986]. MR 88b:68040. Citations in this document: §4.

[Naor and Reingold 1997] Moni Naor and Omer Reingold, "Number-theoretic constructions of efficient pseudo-random functions", pp. 458–467 in [FOCS 1997]. URL: http://www.wisdom.weizmann.ac.il/~naor/onpub.html. Citations in this document: §3.

[Nechvatal et al. 1999] James Nechvatal, Elaine Barker, Donna Dodson, Morris Dworkin, James Foti, and Edward Roback, "Status report on the first round of the development of the Advanced Encryption Standard", *Journal of Research of the National Institute of Standards and Technology* **104**. URL: http://nvl.nist.gov/pub/nistpubs/jres/104/5/cnt104-5.htm. Citations in this document: §3.

[Nechvatal et al. 2001] James Nechvatal, Elaine Barker, Lawrence Bassham, William Burr, Morris Dworkin, James Foti, and Edward Roback, "Report on the development of the Advanced Encryption Standard (AES)", *Journal of Research of the National Institute of Standards and Technology* **106**. URL: http://nvl.nist.gov/pub/nistpubs/jres/106/3/cnt106-3.htm. Citations in this document: §3.

[Nevelsteen and Preneel 1999] Wim Nevelsteen and Bart Preneel, "Software performance of universal hash functions", pp. 24–41 in [Stern 1999]. Citations in this document: §2.

[Odlyzko 1987] Andrew M. Odlyzko (editor), *Advances in cryptology — CRYPTO '86: proceedings of the conference on the theory and applications of cryptographic techniques held at the University of California, Santa Barbara, Calif., August 11–15, 1986,* Lecture Notes in Computer Science **263**, Springer-Verlag, Berlin. ISBN 3-540-18047-8. MR 88h:94004. See [Fiat and Shamir 1987].

[Pomerance 2008] Carl Pomerance, "Smooth numbers and the quadratic sieve", pp. 69–81 in [Buhler and Stevenhagen 2008]. Citations in this document: §5.

[Poonen 2008] Bjorn Poonen, "Elliptic curves", pp. 183–207 in [Buhler and Stevenhagen 2008]. Citations in this document: §4.

[Rabin 1979] Michael O. Rabin, *Digitalized signatures and public-key functions as intractable as factorization*, Technical Report 212, MIT Laboratory for Computer Science. URL: http://ncstrl.mit.edu/Dienst/UI/2.0/Describe/ncstrl.mit_lcs/MIT/LCS/TR-212. Citations in this document: §5.

[Rivest et al. 1978] Ronald L. Rivest, Adi Shamir, and Leonard M. Adleman, "A method for obtaining digital signatures and public-key cryptosystems", *Communications of the ACM* **21**, 120–126. ISSN 0001-0782. URL: http://cr.yp.to/bib/entries.html#1978/rivest. Citations in this document: §5.

[Schirokauer 2008] Oliver Schirokauer, "The impact of the number field sieve on the discrete logarithm problem in finite fields", pp. 397–420 in [Buhler and Stevenhagen 2008]. Citations in this document: §4.

[Schneier 1996] Bruce Schneier, *Applied cryptography: protocols, algorithms, and source code in C*, 2nd edition, Wiley, New York. ISBN 0-471-12845-7. Citations in this document: §3.

[Schoof 2008] René Schoof, "Four primality testing algorithms", pp. 101–125 in [Buhler and Stevenhagen 2008]. Citations in this document: §5.

[Stern 1999] Jacques Stern (editor), *Advances in cryptology: EUROCRYPT '99*, Lecture Notes in Computer Science **1592**, Springer-Verlag, Berlin. ISBN 3-540-65889-0. MR 2000i:94001. See [Nevelsteen and Preneel 1999].

[Stevenhagen 2008] Peter Stevenhagen, "The number field sieve", pp. 83–100 in [Buhler and Stevenhagen 2008]. Citations in this document: §5.

[Turing 1950] Alan M. Turing, "Computing machinery and intelligence", *MIND* **59**, 433–460. ISSN 0026–4423. MR 12,208c. Citations in this document: §3.

[Wang et al. 2004] Xiaoyun Wang, Dengguo Feng, Xuejia Lai, and Hongbo Yu, "Collisions for hash functions MD4, MD5, HAVAL–128 and RIPEMD". URL: http://eprint.iacr.org/2004/199/. Citations in this document: §5.

[Wegman and Carter 1979] Mark N. Wegman and J. Lawrence Carter, "New classes and applications of hash functions", pp. 175–182 in [FOCS 1979]; see also newer version [Wegman and Carter 1981]. URL: http://cr.yp.to/bib/entries.html#1979/wegman.

[Wegman and Carter 1981] Mark N. Wegman and J. Lawrence Carter, "New hash functions and their use in authentication and set equality", *Journal of Computer and System Sciences* **22**, 265–279; see also older version [Wegman and Carter 1979]. ISSN 0022–0000. MR 82i:68017. URL: http://cr.yp.to/bib/entries.html#1981/wegman. Citations in this document: §2.

[Wiener 1999] Michael Wiener (editor), *Advances in cryptology — CRYPTO '99*, Lecture Notes in Computer Science **1666**, Springer-Verlag, Berlin. ISBN 3–5540–66347–9. MR 2000h:94003. See [Black et al. 1999].

[Williams 1980] Hugh C. Williams, "A modification of the RSA public-key encryption procedure", *IEEE Transactions on Information Theory* **26**, 726–729. ISSN 0018–9448. URL: http://cr.yp.to/bib/entries.html#1980/williams. Citations in this document: §5.

[Williams 1986] Hugh C. Williams (editor), *Advances in cryptology: CRYPTO '85*, Lecture Notes in Computer Science **218**, Springer, Berlin. ISBN 3–540–16463–4. See [Miller 1986].

[Yao 1982] Andrew C. Yao, "Theory and applications of trapdoor functions", pp. 80–91 in [FOCS 1982]. Citations in this document: §3.

[Yung et al. 2006] Moti Yung, Yevgeniy Dodis, Aggelos Kiayias, and Tal Malkin (editors), *9th international conference on theory and practice in public-key cryptography, New York, NY, USA, April 24–26, 2006, Proceedings*, Lecture Notes in Computer Science **3958**, Springer, Berlin. ISBN 978-3-540-33851-2. See [Bernstein 2006].

DANIEL J. BERNSTEIN
DEPARTMENT OF MATHEMATICS, STATISTICS, AND COMPUTER SCIENCE
M/C 249
THE UNIVERSITY OF ILLINOIS AT CHICAGO
CHICAGO, IL 60607–7045
UNITED STATES
   djb@cr.yp.to

# Algorithmic theory of zeta functions over finite fields

DAQING WAN

ABSTRACT. We give an introductory account of the general algorithmic theory of the zeta function of an algebraic set defined over a finite field.

## CONTENTS

| | |
|---|---|
| 1. Introduction | 551 |
| 2. Generalities on computing zeta functions | 552 |
| 3. Reduction to hypersurfaces | 559 |
| 4. Hypersurface examples | 563 |
| 5. Pure weight decomposition | 565 |
| 6. Pure slope decomposition | 570 |
| 7. Zeta functions modulo $p$ | 572 |
| Acknowledgment | 576 |
| References | 576 |

## 1. Introduction

Let $\mathbb{F}_q$ be a finite field of $q$ elements and $p$ its characteristic. Let $X$ be an algebraic set defined over $\mathbb{F}_q$. For each positive integer $k$, let $N_k$ denote the number of $\mathbb{F}_{q^k}$-rational points on $X$. The zeta function $Z(X)$ of $X$ is the generating function

$$Z(X) = Z(X, T) = \exp\left(\sum_{k=1}^{\infty} \frac{N_k}{k} T^k\right).$$

---

The author is partially supported by the NSF and the NSFC.

The zeta function contains important arithmetic and geometric information concerning $X$. It has been studied extensively in connection with the celebrated Weil conjectures [1949].

Both practical applications and theoretical investigations make a good understanding of the zeta function from an algorithmic point of view increasingly important. The aim of this paper is to present a brief introductory account of the various fundamental problems and results in the emerging algorithmic theory of zeta functions. We shall focus on general properties rather than on results that are restricted to special cases. In particular, in most of this paper we do not assume $X$ to be smooth and projective, although in that case one can often say more.

The contents are organized as follows. In Section 2 we review general properties of zeta functions from an algorithmic point of view. A naive effective algorithm for computing the zeta function is given. If the characteristic $p$ is small, one can use Dwork's $p$-adic method to obtain a polynomial time algorithm for computing the zeta function in the case that the numbers of variables and defining equations for $X$ are fixed.

In Section 3, we show that the general case of algebraic sets can be reduced in various ways to the case that $X$ is a hypersurface. A more detailed discussion of that crucial case is given in Section 4, with emphasis on the smooth projective case. In Section 5 we consider the complex pure weight decomposition. Using the LLL factorization algorithm and Deligne's main theorem, we show that, when the zeta function is given, one can compute in polynomial time how many zeros and poles with a given complex absolute value it has. In Section 6, which is devoted to the $p$-adic pure slope decomposition, we use the theory of Newton polygons to obtain a similar result for the number of zeros and poles with a given $p$-adic absolute value.

We conclude the paper by giving, in Section 7, an algorithm for the simpler problem of computing the zeta function modulo $p$. This algorithm shares several characteristic features with the general $p$-adic method for computing the full zeta function that is presented in [Lauder and Wan 2008] in this volume. Section 7 may thus serve as an introduction to that article.

All algorithms in this paper are deterministic. Probabilistic algorithms will not be discussed. Time is measured in bit operations.

## 2. Generalities on computing zeta functions

Let $X$ be an algebraic set defined over a finite field $\mathbb{F}_q$ of $q$ elements of characteristic $p$. For computational purposes, we may assume that $X$ is affine, i.e., that it is the subset of affine $n$-space $\mathbb{A}^n$ defined by a system of polynomial equations:

$$\begin{cases} f_1(x_1,\ldots,x_n) = 0, \\ \quad \vdots \\ f_m(x_1,\ldots,x_n) = 0, \end{cases}$$

where $f_i \in \mathbb{F}_q[x_1,\ldots,x_n]$. Let

$$X(\mathbb{F}_q) = \{x = (x_1,\ldots,x_n) \in \mathbb{F}_q^n \mid f_1(x) = \ldots = f_m(x) = 0\}$$

be the finite set of $\mathbb{F}_q$-rational points on $X$. It is clear that the cardinality $\#X(\mathbb{F}_q)$ is effectively computable.

For algorithmic purposes, "giving" $\mathbb{F}_q$ means specifying $p$ as well as an irreducible polynomial $h$ in one variable over $\mathbb{F}_p$ that defines $\mathbb{F}_q$, so that $q$ equals $p^{\deg h}$; elements of $\mathbb{F}_q$ are then represented as polynomials of degree less than $\deg h$ in a formal zero of $h$, with coefficients from $\mathbb{F}_p$. Giving $X$ means specifying a system of $m$ defining polynomials $f_i$ in $n$ variables with coefficients in $\mathbb{F}_q$. Let $d$ be the maximum of the total degrees of the polynomials $f_i$. Then the dense input size for $X$ is $O(m\binom{d+n}{n}\log q)$, which is $O(m(d+1)^n \log q)$. Our first fundamental problem is the following.

PROBLEM 2.1. *Given $\mathbb{F}_q$ and $X$, compute the number $\#X(\mathbb{F}_q)$ in time polynomial in the dense input size $O(m(d+1)^n \log q)$.*

This problem is trivial if $q$ is fixed, so we may assume that $q$ is large. In theory, the problem of counting $X(\mathbb{F}_q)$ can be reduced to the zero-dimensional case. Namely, let $Y$ be the zero-dimensional algebraic set defined by

$$\{f_1 = \ldots = f_m = 0, \ x_1^q - x_1 = \ldots = x_n^q - x_n = 0\}.$$

Then it is clear that

$$\#X(\mathbb{F}_q) = \#Y(\mathbb{F}_q).$$

Following a suggestion of Eisenbud and Sturmfels, one may now compute a Gröbner basis for $Y$, and its cardinality equals $\#Y(\mathbb{F}_q)$. However, as $q$ gets large, the cases where the Gröbner basis computation can be done efficiently are likely to become increasingly exceptional. (See [Eisenbud 1995] for Gröbner bases.)

Let $\bar{\mathbb{F}}_q$ denote a fixed algebraic closure of $\mathbb{F}_q$. For each positive integer $k$, let $\mathbb{F}_{q^k}$ denote the unique subfield of $\bar{\mathbb{F}}_q$ with $q^k$ elements. Let $\#X(\mathbb{F}_{q^k})$ denote the number of $\mathbb{F}_{q^k}$-rational points on $X$. The following problem is harder but more interesting than Problem 2.1.

PROBLEM 2.2. *Given $\mathbb{F}_q$ and $X$, compute the sequence of numbers $\#X(\mathbb{F}_{q^k})$ ($k = 1, 2, \ldots$).*

It may not be clear how a finite algorithm can compute an infinite sequence of numbers, but this will be clarified below. As we shall see, one can encode the entire sequence in a suitably defined generating function, which turns out to be a rational function. This so-called *zeta function* of $X$ is finite in nature and can thus be written down in a finite amount of time. Actually doing this for given $X$ is the content of Problem 2.2.

A *geometric point* of $X$ is an $\bar{\mathbb{F}}_q$-rational point of $X$. From the equality

$$\bigcup_{k=1}^{\infty} X(\mathbb{F}_{q^k}) = X(\bar{\mathbb{F}}_q)$$

we see that each geometric point of $X$ will be counted somewhere in the sequence of numbers $\#X(\mathbb{F}_{q^k})$. This may explain why many of the subtle geometric invariants associated with an algebraic variety $X$ can be read off from its zeta function, in addition to a wealth of arithmetic information.

DEFINITION 2.3. The zeta function of $X$ is the generating function

$$Z(X) = Z(X, T) = \exp\left( \sum_{k=1}^{\infty} \frac{T^k}{k} \#X(\mathbb{F}_{q^k}) \right).$$

The $q$-th power Frobenius map $\mathrm{Frob}_q$ is the permutation of the set $X(\bar{\mathbb{F}}_q)$ of geometric points of $X$ defined by

$$\mathrm{Frob}_q: x = (x_1, \ldots, x_n) \mapsto x^q = (x_1^q, \ldots, x_n^q).$$

The *degree* of a geometric point $x$ is defined to be the smallest positive integer $d$ such that

$$\mathrm{Frob}_q^d(x) = x$$

or, equivalently, such that $x \in X(\mathbb{F}_{q^d})$. A *closed point* over $\mathbb{F}_q$ is the orbit of a geometric point under $\mathrm{Frob}_q$. All geometric points belonging to a given closed point have the same degree, and this common degree is called the *degree* of the closed point. We denote by $|X|$ the set of closed points of $X$ over $\mathbb{F}_q$, and, for each positive integer $k$, by $M_k(X)$ the number of closed points of $X$ of degree $k$. Since each closed point of degree $k$ consists of exactly $k$ points in $X(\mathbb{F}_{q^k})$, one deduces

$$\#X(\mathbb{F}_{q^k}) = \sum_{d|k} d M_d(X).$$

Considering the logarithmic derivative of the zeta function, one finds the *Euler product* expansion

$$Z(X) = \prod_{k=1}^{\infty} \frac{1}{(1-T^k)^{M_k(X)}} = \prod_{x \in |X|} \frac{1}{1-T^{\deg(x)}} \in 1 + T\mathbb{Z}[\![T]\!].$$

As the Weil conjectures [1949] predict, the zeta function is a rational function. The first proof, given by Dwork [1960], used $p$-adic analysis. The second proof, given by Grothendieck, used the theory of $\ell$-adic cohomology, where $\ell$ is a prime number different from $p$. These two proofs pioneered the general $p$-adic and $\ell$-adic study of zeta functions over finite fields.

THEOREM 2.4. *The zeta function $Z(X)$ is a rational function, i.e., it belongs to $\mathbb{Q}(T)$. If we write*

$$Z(X,T) = \frac{R_1(X,T)}{R_2(X,T)}, \quad (R_1, R_2) = 1, \quad R_i \in 1 + T\mathbb{Q}[T],$$

*then we have $R_i \in 1 + T\mathbb{Z}[T]$.*

The rationality of $Z(X)$ has an interesting consequence for the numbers $\#X(\mathbb{F}_{q^k})$, as follows. Let $\beta_i$ and $\gamma_j$ denote the reciprocal zeros of $R_1(X)$ and $R_2(X)$, respectively, so that $R_1(X) = \prod_i (1-\beta_i T)$ and $R_2(X) = \prod_j (1-\gamma_j T)$. Then one finds

$$\sum_{k=1}^{\infty} \#X(\mathbb{F}_{q^k}) T^k = T \frac{d \log Z(X,T)}{dT} = \sum_j \frac{\gamma_j T}{1-\gamma_j T} - \sum_i \frac{\beta_i T}{1-\beta_i T}.$$

This implies

$$\#X(\mathbb{F}_{q^k}) = \sum_j \gamma_j^k - \sum_i \beta_i^k \quad \text{for all } k \geq 1.$$

As a corollary, one deduces that for each positive integer $k$ one has

$$Z(X \otimes \mathbb{F}_{q^k}) = \frac{\prod_i (1-\beta_i^k T)}{\prod_j (1-\gamma_j^k T)}.$$

The integrality of the coefficients of $R_i$ can be deduced from the rationality of $Z(X)$ and the following elementary result.

LEMMA 2.5. *Let $f \in 1 + T\mathbb{Z}[\![T]\!]$ be a rational function. Write*

$$f = \frac{f_1}{f_2}, \quad (f_1, f_2) = 1 \quad f_i \in 1 + T\mathbb{Q}[T].$$

*Then $f_i \in 1 + T\mathbb{Z}[T]$.*

PROOF. This is usually derived from the so-called lemma of Fatou; see [Katz 1971]. Here we include two additional proofs.

One proof uses the Newton polygon or the Weierstrass factorization theorem. Suppose some prime number $\ell$ occurs in the common denominator of the coefficients of $f_1$. Then the theory of Newton polygons shows that $f_1$ has an $\ell$-adic zero in the open unit disk $|T|_\ell < 1$. But the power series $f \in 1 + T\mathbb{Z}[\![T]\!]$

is clearly analytic and nonzero in the open unit disk $|T|_\ell < 1$. This gives the desired contradiction.

A second proof, suggested by Hendrik Lenstra, uses Gauss's lemma for power series. The *content* cont$(g)$ of a nonzero power series $g = \sum_i a_i T^i \in \mathbb{Z}[\![T]\!]$ is defined to be the greatest common divisor of its coefficients $a_i$. Call $g$ *primitive* if cont$(g) = 1$ or, equivalently, if $g$ is not in the kernel of the natural map $\mathbb{Z}[\![T]\!] \to \mathbb{F}_\ell[\![T]\!]$ for any prime number $\ell$. Since the rings $\mathbb{F}_\ell[\![T]\!]$ are domains, the product of any two primitive power series is primitive. One deduces

$$\text{cont}(g_1 g_2) = \text{cont}(g_1)\text{cont}(g_2),$$

which is Gauss's lemma for power series.

Lenstra's proof of Lemma 2.5 then proceeds as follow. Write $f = g_1/g_2$, where $g_i \in \mathbb{Z}[T]$ and $(g_1, g_2) = 1$ in $\mathbb{Q}[T]$. It is clear that cont$(f) = 1$. The relation $g_1 = g_2 f$ implies that cont$(g_1) = $ cont$(g_2)$. Cancelling this common factor, we may assume cont$(g_1) = $ cont$(g_2) = 1$. Since $g_1$ and $g_2$ are relatively prime over $\mathbb{Q}$, there is a positive integer $n$ such that

$$n \in g_1 \mathbb{Z}[T] + g_2 \mathbb{Z}[T] \subset g_2 \mathbb{Z}[\![T]\!],$$

the last inclusion because $g_1 = g_2 f$. Write $n = hg_2$ with $h \in \mathbb{Z}[\![T]\!]$. Then

$$n = \text{cont}(n) = \text{cont}(h)\text{cont}(g_2) = \text{cont}(h).$$

Hence $h$ is divisible by $n$. We conclude that $g_2(0) = \pm 1$. This implies 2.5. □

In order to actually compute the zeta function, it is useful to know an upper bound for the total degree deg $R_1 + $ deg $R_2$ of the zeta function. The following explicit bound was proved by Bombieri [1978].

THEOREM 2.6. *The total degree of $Z(X)$ satisfies*

$$\deg R_1 + \deg R_2 < (4d+9)^{n+m},$$

*where*

$$d = \max_{1 \leq j \leq m} \deg(f_j).$$

Bombieri's bound is a general purpose bound. Its proof depends on Dwork's $p$-adic method. It can be improved in various ways, especially when one takes the Newton polytope of the defining polynomials $f_i$ into account, as was done by Adolphson and Sperber [1988]. The bound is reasonably good as a function of $d$, but the dependence on $m$ can probably be significantly improved.

An easy consequence is the following result.

COROLLARY 2.7. *The zeta function $Z(X)$ is effectively computable.*

This corollary is obvious in the special case that the zeta function $Z(X)$ is known to be a polynomial, or known to be the reciprocal of a polynomial, up to some trivial known factors. In the general case, one can deduce Corollary 2.7 from Theorem 2.4 and Theorem 2.6 in several ways, for instance by using the results of Berlekamp and Massey on linear recurring sequences (see [Blahut 1998] for more detail). Here we use a simple linear algebra argument explained to me by Hendrik Lenstra.

Let $D_1$ and $D_2$ be upper bounds for the degree of the numerator and the denominator of $Z(X)$, respectively. For instance, we can take $D_i = D = (4d+9)^{n+m}$, by Theorem 2.6. Compute the first $D_1 + D_2 + 1$ terms of the power series
$$Z(X) = 1 + z_1 T + z_2 T^2 + \cdots + z_{D_1+D_2} T^{D_1+D_2} + \cdots$$
by explicitly counting $X(\mathbb{F}_{q^k})$ for $k \leq D_1 + D_2$. Write $a_i$, $b_i$ for the coefficients of $R_1$ and $R_2$ to be determined:
$$R_1(X) = 1 + a_1 T + \cdots + a_{D_1} T^{D_1},$$
$$R_2(X) = 1 + b_1 T + \cdots + b_{D_2} T^{D_2}.$$

The congruence
$$R_2(X) Z(X) \equiv R_1(X) \pmod{T^{D_1+D_2+1}}$$
gives a system of linear equations in the $a_i$'s and the $b_i$'s. This system has at least one rational solution, and using linear algebra we can find one. Denote it by
$$(a'_1, \ldots, a'_{D_1}; b'_1, \ldots, b'_{D_2}).$$
Let
$$R'_1(X) = 1 + a'_1 T + \cdots + a'_{D_1} T^{D_1},$$
$$R'_2(X) = 1 + b'_1 T + \cdots + b'_{D_2} T^{D_2}.$$
Then, the congruence
$$R'_2(X) Z(X) \equiv R'_1(X) \pmod{T^{D_1+D_2+1}}$$
holds as well, and therefore
$$R_2(X) R'_1(X) \equiv R_2(X) R'_2(X) Z(X) \equiv R'_2(X) R_1(X) \pmod{T^{D_1+D_2+1}}.$$
Since each of $R_2 R'_1$ and $R'_2 R_1$ has degree at most $D_1 + D_2$, we deduce $R_2 R'_1 = R'_2 R_1$, so
$$\frac{R'_1(X)}{R'_2(X)} = \frac{R_1(X)}{R_2(X)} = Z(X).$$
Removing the greatest common factor of $R'_1(X)$ and $R'_2(X)$, one obtains the reduced form of $Z(X)$. The proof is complete.

The above effective algorithm immediately implies the following.

COROLLARY 2.8. *For fixed $n, m, d, q$, the number $\#X(\mathbb{F}_{q^k})$ can be computed in time bounded by a polynomial in $k$.*

We now estimate the output size of any algorithm computing $Z(X)$. Trivially, $\#X(\mathbb{F}_{q^k}) \leq \#\mathbb{A}^n(\mathbb{F}_{q^k}) = q^{nk}$. Hence $T \cdot d \log Z(X, T)/dT$ converges as a complex power series for $|T| < q^{-n}$, so its reciprocal poles $\beta_i$ and $\gamma_j$ are bounded by $q^n$ in absolute value. Let $D$ denote the total degree of $Z(X)$. Then $R_1(X) = \prod_i (1 - \beta_i T)$ and $R_2(X) = \prod_j (1 - \gamma_j T)$ have altogether $O(D)$ coefficients, each of which is $O(2^D q^{nD})$. Then, regardless of the input size for $X$, the output size of the algorithm is $O(nD^2 \log q)$. There is no general formula for the total degree $D$. However, for fixed $m$, Bombieri's degree bound $(4d + 9)^{n+m}$ is reasonably good, and it is comparable to the dense input size $O(m(d + 1)^n \log q)$.

We shall be concerned with the case that $m$ is fixed (or small). If $m$ is large, then the problem of computing $\#X(\mathbb{F}_q)$ is of a totally different, more combinatorial nature. The fundamental question that we consider is the following.

PROBLEM 2.9. *Given $X$ with fixed $m$, compute the zeta function $Z(X)$ in time bounded by a polynomial in $(d + 1)^n \log q$.*

*Remarks.* If $X$ has a sizable automorphism group, then one can often speed up the computation of $Z(X)$ by using a suitable equivariant theory. Examples include diagonal hypersurfaces, certain modular varieties, and certain Calabi–Yau hypersurfaces. In this paper, we do not assume that $X$ is given with any additional structure of this sort.

Currently, the theory of zeta functions over finite fields comprises only two types of methods that are powerful enough to prove the rationality of the zeta function in the general case. These are the $\ell$-adic method and the $p$-adic method, where $\ell$ denotes a prime number different from the characteristic $p$ of the finite field $\mathbb{F}_q$. It is thus natural to try and exploit these general methods for algorithmic purposes.

In the $\ell$-adic method one attempts to compute $Z(X) \bmod \ell^k$ using a suitable $\ell$-adic trace formula. The zeta function $Z(X)$ can be recovered from its reduction modulo a single large prime power $\ell^k$, or from its reductions modulo many small primes $\ell$ via the Chinese remainder theorem. Unfortunately, the available $\ell$-adic trace formula is in the general case not yet effective. Thus the use of the $\ell$-adic method is currently restricted to special varieties, such as curves and abelian varieties. In the cases in which it can be used, the $\ell$-adic method usually results in a polynomial time algorithm if $d, m$, and $n$ are fixed; see [Schoof 1985; Elkies 1998; Poonen 1996] for the first examples. It is an important open problem to make the $\ell$-adic method effective in the general case.

In the $p$-adic method one attempts to compute $Z(X)$ mod $p^k$ using a $p$-adic trace formula, where $p^k$ is chosen so large that one can recover the zeta function $Z(X)$ from its reduction modulo $p^k$. There are many $p$-adic trace formulas. All of them can be made effective, although not all of them result in efficient algorithms. The general feeling is that the $p$-adic method is quite efficient if the characteristic $p$ is suitably small, regardless of the size of the field $\mathbb{F}_q$ of definition and regardless of the degree of $X$. In [Lauder and Wan 2008] in this volume, we present a $p$-adic algorithm that proves the following theorem.

THEOREM 2.10. *There is an algorithm that, given $X$, computes the zeta function $Z(X)$ in time bounded by a polynomial in $2^m d^{n^2} p^n (\log q)^n$.*

For fixed $m$ and $n$, and small $p$—say, $p = O((d \log q)^c)$ for some positive constant $c$—the algorithm of Theorem 2.10 runs in polynomial time, and it thus provides a partial solution to Problem 2.9. In Section 7 below, we illustrate some of the basic ideas of the $p$-adic method by treating an algorithm for the easier problem of computing $Z(X)$ mod $p$. For more details see [Lauder and Wan 2008].

## 3. Reduction to hypersurfaces

For the computation of the zeta function, the general case of an affine algebraic set $X$ defined by a system

$$f_1(x) = \ldots = f_m(x) = 0$$

of $m$ polynomial equations in $n$ variables can be reduced to the case of an affine hypersurface defined by one single equation. In the present section we discuss various ways in which this reduction can be accomplished; not all of them are very efficient.

The quickest theoretical reduction depends on the observation that any algebraic set is birational to an affine hypersurface. One then continues by induction on the dimension. As it stands, this method is not very explicit. It may be of interest to make it both explicit and efficient.

A second method, which is explicit, exploits the inclusion-exclusion principle, as follows. For a subset $I \subset \{1, 2, \ldots, m\}$, let $H(I)$ be the affine hypersurface defined by

$$\prod_{i \in I} f_i = 0,$$

and let $H(I)^c$ be the complement of $H(I)$ in $\mathbb{A}^n$. Thus,

$$H(I) = \bigcup_{i \in I} \{f_i = 0\}, \qquad H(I)^c = \bigcap_{i \in I} \{f_i \neq 0\}.$$

In particular, we have $H(\emptyset) = \emptyset$ and $H(\emptyset)^c = \mathbb{A}^n$. The inclusion-exclusion principle implies that

$$Z(X) = \prod_{I \subset \{1,2,\ldots,m\}} Z(H(I)^c, T)^{(-1)^{\#I}}.$$

Since

$$Z(H(I)^c, T) = \frac{1}{(1 - q^n T) Z(H(I), T)},$$

we conclude that for $m > 0$ we have

$$Z(X) = \prod_{\substack{I \subset \{1,2,\ldots,m\} \\ I \neq \emptyset}} Z(H(I), T)^{(-1)^{\#I-1}},$$

where the factor 1 corresponding to $I = \emptyset$ has been dropped. Each factor of the above product is now the zeta function of an affine hypersurface. Note that this reduction uses $2^m - 1$ hypersurfaces, but they are all in the original affine space $\mathbb{A}^n$.

If we apply Theorem 2.6 to each factor in the above identity, then we find that the total degree of $Z(X)$ is bounded by

$$\sum_{k=1}^{m} \binom{m}{k} (4kd + 9)^{n+1} < 2^m (4md + 9)^{n+1}.$$

In some cases where $m$ is large this is better than what one obtains by applying Theorem 2.6 directly.

If one is willing to work with the slightly more general situation of $L$-functions of exponential sums, then one can use the single polynomial

$$g(x, y) = y_1 f_1(x) + \cdots + y_m f_m(x)$$

in $n + m$ variables. Let $\zeta_p$ denote a fixed primitive $p$-th root of unity in an extension field of $\mathbb{Q}$. For each positive integer $k$, define the exponential sum

$$S_k(g) = \sum_{x_i, y_j \in \mathbb{F}_{q^k}} \zeta_p^{\text{Tr}_k(g(x,y))},$$

where $\text{Tr}_k$ denotes the absolute trace from $\mathbb{F}_{q^k}$ to the prime field $\mathbb{F}_p$. The $L$-function associated to $g$ is defined to be

$$L(g, T) = \exp\left( \sum_{k=1}^{\infty} \frac{S_k(g)}{k} T^k \right).$$

It is straightforward to check that

$$\#X(\mathbb{F}_{q^k}) = \frac{1}{q^{mk}} S_k(g).$$

This gives the desired reduction

$$Z(X) = L\left(g, \frac{1}{q^m}T\right).$$

Replacing $f_i$ by $af_i$, one also deduces that

$$Z(X) = L\left(ag, \frac{1}{q^m}T\right)$$

for each nonzero $a \in \mathbb{F}_q$.

One can avoid the $L$-function in the above reduction by using the zeta function of the following Artin–Schreier hypersurface in $\mathbb{A}^{m+n+1}$:

$$Y: z^p - z = y_1 f_1(x) + \cdots + y_m f_m(x).$$

In fact, a direct calculation gives that

$$\#Y(\mathbb{F}_{q^k}) = \sum_{a \in \mathbb{F}_p} S_k(ag) = q^{(m+n)k} + \sum_{a \in \mathbb{F}_p^*} S_k(ag).$$

It follows that

$$Z(Y, q^{-m}T) = \frac{1}{1-q^n T} \cdot \prod_{a \in \mathbb{F}_p^*} L(ag, q^{-m}T) = \frac{1}{1-q^n T} Z(X)^{p-1}.$$

We obtain the formula

$$Z(X) = \left((1-q^n T) \cdot Z(Y, q^{-m}T)\right)^{1/(p-1)}.$$

For large $p$ this reduction is not likely to be very efficient.

One can also use the affine hypersurface $H \subset \mathbb{A}^{m+n}$ defined by

$$H: g(x, y) = y_1 f_1(x) + \cdots + y_m f_m(x) = 0.$$

As S. Gao observed, one has

$$\#H(\mathbb{F}_q) = q^{m+n-1} + \#X(\mathbb{F}_q)(q^m - q^{m-1}).$$

This shows that for the purpose of counting rational points, we can work with a single hypersurface in the affine space $\mathbb{A}^{m+n}$. In terms of zeta functions, Gao's formula says that

$$Z(H, T) = \frac{Z(X, q^m T)}{(1 - q^{m+n-1}T)Z(X, q^{m-1}T)}.$$

One can inductively solve for $Z(X)$ in terms of $Z(H, T)$. Doing this from the complex point of view, one gets the infinite complex product

$$Z(X) = \prod_{k=0}^{\infty} (1 - q^{n-1-k}T) \cdot \prod_{k=0}^{\infty} Z(H, q^{-m-k}T).$$

Doing it from the $p$-adic point of view, one gets the infinite $p$-adic product

$$\frac{1}{Z(X)} = \prod_{k=0}^{\infty}(1-q^{n+k}T) \cdot \prod_{k=0}^{\infty} Z(H, q^{1-m+k}T).$$

There is a standard manner, as in Hilbert's tenth problem [Matiyasevich 1993], to define $X$ by means of a system of equations of degree two. For example, if one of the equations defining $X$ has a term $ax_1 x_3^3 x_4^2$, with $a \in \mathbb{F}_q$, then one can introduce new variables $x_{1,3}$, $x_{3,3}$, $x_{4,4}$, $x_{1,3,3,3}$, as well as new equations

$$x_{1,3} = x_1 x_3, \qquad x_{3,3} = x_3^2, \qquad x_{4,4} = x_4^2, \qquad x_{1,3,3,3} = x_{1,3} x_{3,3},$$

and replace the term $ax_1 x_3^3 x_4^2$ by $ax_{1,3,3,3} x_{4,4}$; and one can proceed similarly with other terms. If one next applies Gao's reduction, then one obtains a hypersurface $H$ that is defined by a cubic polynomial.

An amusing application of Gao's formula is the reduction of the Hasse–Weil meromorphy conjecture to the case of a cubic hypersurface. Let the polynomials $f_i$ have integer coefficients, and let the affine algebraic sets $X$ and $H$ be defined as above, but now over $\mathbb{Z}$. Let

$$\zeta(X, z) = \prod_{p \text{ prime}} Z(X \otimes \mathbb{F}_p, p^{-z})$$

be the global complex Hasse–Weil zeta function of $X$; it is defined for complex numbers $z$ whose real part is sufficiently large. The Hasse–Weil conjecture asserts that $\zeta(X)$ can be extended to a meromorphic function on all of $\mathbb{C}$. Gao's formula implies that

$$\zeta(H, z) = \frac{\zeta(X, z-m)\zeta(z+1-m-n)}{\zeta(X, z-m+1)},$$

where $\zeta(z)$ is the Riemann zeta function. From this relation, one deduces by induction or iteration that if the Hasse–Weil conjecture is valid for all cubic hypersurfaces $H$, then it is valid for all algebraic sets $X$. (By contrast, for Hilbert's tenth problem on diophantine equations over the integers, one only obtains a reduction to quartic hypersurfaces.) If we write

$$\zeta(X, z) = \sum_{k=1}^{\infty} \frac{a_k(X)}{k^z}$$

as a Dirichlet series, then Corollary 2.7 shows that each coefficient $a_k(X)$ is effectively computable.

## 4. Hypersurface examples

In this section, we focus on the crucial case of hypersurfaces. We discuss them by increasing dimension.

EXAMPLE 1. Let $X$ be the zero-dimensional hypersurface defined by $f(x) = 0$, where $f(x)$ is a nonconstant monic polynomial over $\mathbb{F}_q$ in one variable. Write

$$f(x) = P_1(x)^{k_1} \cdots P_e(x)^{k_e},$$

where the $P_i(x)$ are pairwise distinct monic irreducible polynomials in $\mathbb{F}_q[x]$ and the $k_i$ are positive integers. Then the Euler product reads

$$Z(X) = \prod_{j=1}^{e} \frac{1}{1 - T^{\deg P_j(x)}}.$$

It is not hard to show that one can compute $Z(X)$ in polynomial time using the Frobenius map. Note that if one factors the polynomial $f(x)$ first, one does not get a polynomial time algorithm for computing $Z(X)$. This is because there is currently no known (deterministic) polynomial time algorithm for factoring univariate polynomials over $\mathbb{F}_q$ if $p$ is large, see [Wan 1999] for a further discussion and for the close relation of $Z(X)$ to various algorithms for factoring univariate polynomials over finite fields. This example also occurred as an exercise ascribed to Lenstra in [Cohen 1993, Chap. 6, Exerc. 8].

EXAMPLE 2. Let $f(x_1, x_2) \in \mathbb{F}_q[x_1, x_2]$ be of degree $d$, and suppose that the homogenization of $f(x_1, x_2)$ defines a smooth projective plane curve $C_d$ over $\mathbb{F}_q$. The genus $g$ of the curve $C_d$ is well known to be

$$g = \frac{(d-1)(d-2)}{2}.$$

From the Riemann–Roch theorem one can deduce (see [Monsky 1970])

$$Z(C_d, T) = \frac{P(C_d, T)}{(1-T)(1-qT)},$$

where $P(C_d, T)$ is a polynomial of degree $2g$. Weil proved that we further have

$$P(C_d, T) = \prod_{j=1}^{2g}(1 - \alpha_j T), \quad |\alpha_j| = \sqrt{q}, \quad \alpha_j \alpha_{2g+1-j} = q.$$

The Riemann–Roch theorem also shows that the special value $P(C_d, 1)$ has the following arithmetic meaning:

$$P(C_d, 1) = \#J(C_d)(\mathbb{F}_q) \in \mathbb{Z}_{>0},$$

where $J(C_d)$ is the Jacobian variety of $C_d$, which is a $g$-dimensional abelian variety over $\mathbb{F}_q$. All these results hold for any smooth projective geometrically irreducible curve over $\mathbb{F}_q$, not just for plane curves.

The $\ell$-adic method can be made effective in the case of curves and abelian varieties. One can then use the Chinese remainder theorem as mentioned before. In this way, one obtains an algorithm for computing $Z(C_d, T)$ with running time $O((\log q)^{\Delta_d})$, where $\Delta_d$ is in general an exponential function of $d$; see [Schoof 1985; Pila 1990]. Thus, for fixed $d$, the $\ell$-adic method computes the zeta function $Z(C_d, T)$ in polynomial time, although the algorithm is still doubly exponential in $d$. For hyperelliptic curves, the exponent $\Delta_d$ has been improved to a polynomial in $d$; see [Adleman and Huang 1996]. On the other hand, using the $p$-adic algorithm in [Lauder and Wan 2008], one can compute the zeta function $Z(C_d, T)$ in time $(dp \log q)^{O(1)}$, which is polynomial in $d$ but exponential in $\log p$. In particular, the zeta function $Z(C_d, T)$ can be computed in polynomial time if $p = O((d \log q)^{O(1)})$. This example is important because of its many applications in number theory and cryptography; see [Koblitz 1989; Blake et al. 2000]. For special types of curves in small characteristic, more practical versions of various $p$-adic algorithms have been designed by a number of authors; see [Satoh 2000; Kedlaya 2001; Lauder and Wan 2002; Denef and Vercauteren 2002], and the references listed in those papers. Restricting Problem 2.9 to plane curves, we obtain the following.

PROBLEM 4.1. *Given a smooth projective plane curve $C_d$ over $\mathbb{F}_q$, compute $Z(C_d, T)$ in time $O((d \log q)^c)$, where $c$ is an explicit absolute positive constant.*

EXAMPLE 3. Let $f(x_1, \ldots, x_n) \in \mathbb{F}_q[x_1, \ldots, x_n]$ be of degree $d$, and suppose that the homogenization of $f$ defines a smooth projective hypersurface $H_d$ of dimension $n - 1$ over $\mathbb{F}_q$. Then by the Weil conjectures [Deligne 1974] we can write

$$Z(H_d, T) = \frac{P(H_d, T)^{(-1)^n}}{\prod_{j=0}^{n-1}(1 - q^j T)},$$

where $P(H_d, T) \in 1 + T\mathbb{Z}[T]$ is a polynomial of degree

$$D = \frac{1}{d}\{(d-1)^{n+1} + (-1)^{n+1}(d-1)\}$$

(see [Monsky 1970]) and

$$P(H_d, T) = \prod_{j=1}^{D}(1 - \alpha_j T), \quad |\alpha_j| = \sqrt{q}^{n-1}, \quad \alpha_j \alpha_{D+1-j} = q^{n-1}.$$

This higher dimensional example is undoubtedly more difficult than the previous example of curves, and it has attracted a smaller amount of attention. Both theoretically and computationally, and from an applied point of view, our understanding of this case leaves a great deal to be desired. In particular, the $\ell$-adic method has not been made effective for higher dimensional smooth projective hypersurfaces. We do know that by the $p$-adic algorithm in [Lauder and Wan 2008], the zeta function $Z(H_d, T)$ can be computed in time $(d^n p \log q)^{O(n)}$ for any $(n-1)$-dimensional hypersurface $H_d$, not necessarily smooth or projective. This gives a polynomial time algorithm for small $p$ and fixed $n$. In the smooth projective case, it should be possible to improve the exponent $O(n)$ by finer $p$-adic cohomological methods. In fact, Lauder [2004b; 2004a] has gone further and used the deformation method on the cohomology space to show that the zeta function can be computed in time $(d^n p \log q)^{O(1)}$ for suitable smooth projective $(n-1)$-dimensional hypersurface $H_d$. It would be interesting to explore possible applications of higher dimensional hypersurfaces. As a special case of Problem 2.9, we have the following.

PROBLEM 4.2. *Given a smooth projective $(n-1)$-dimensional hypersurface $H_d$ defined over $\mathbb{F}_q$, compute $Z(H_d, T)$ in time bounded by a polynomial in $(d+1)^n \log q$.*

Lauder's recent results solve this problem if $p$ is small.

## 5. Pure weight decomposition

In this section, we consider the problem of computing the numbers of zeros and poles of $Z(X)$ with a given complex absolute value, for any algebraic set $X$ defined over $\mathbb{F}_q$. Let $R(X, T)$ be the numerator or the denominator of the zeta function $Z(X)$.

Over the complex numbers $\mathbb{C}$, we can write

$$R(X, T) = \prod_i (1 - \alpha_i T),$$

where each $\alpha_i$ is a nonzero algebraic integer. Define the *weight* of a nonzero complex number $\alpha$ by

$$w(\alpha) = \log_q(\alpha \bar{\alpha}),$$

where $\bar{\alpha}$ denotes the complex conjugate of $\alpha$. An elementary archimedean estimate shows that

$$\#X(\mathbb{F}_{q^k}) = O(q^{k \dim X}),$$

where the implied constant depends on the degree of $X$. As in Section 2, one deduces the elementary estimate

$$|\alpha_i| \leq q^{\dim X}, \quad w(\alpha_i) \leq 2 \dim X.$$

The following much deeper result is a consequence of Deligne's main theorem on the Weil conjectures [Deligne 1980].

THEOREM 5.1. *The weights of the reciprocal roots $\alpha_i$ of $R(X,T)$ are integers in the interval $[0, 2\dim X]$. That is, for each $\alpha_i$, there is an integer $w_i$ with $0 \leq w_i \leq 2\dim X$ such that*

$$\alpha_i \bar{\alpha}_i = q^{w_i}.$$

*In particular, each $\alpha_i$ is an $\ell$-adic unit for all prime numbers $\ell \neq p$. Furthermore, each $\alpha_i$ and its Galois conjugates have the same weight.*

The last part of the theorem can be deduced from the first part in an elementary manner, as shown in the following result of Lenstra.

LEMMA 5.2. *Let $f \in \mathbb{Q}[T]$ be an irreducible polynomial. Suppose $\alpha$ and $\beta$ are two complex roots of $f$ with $\alpha\bar{\alpha} \in \mathbb{Q}$ and $\beta\bar{\beta} \in \mathbb{Q}$. Then $\alpha\bar{\alpha} = \beta\bar{\beta}$.*

PROOF. Let $\alpha\bar{\alpha} = a$, $\beta\bar{\beta} = b$, and $\deg(f) = n > 0$. We need to show that $a = b$. Since there is a field isomorphism $\mathbb{Q}(\alpha) \cong \mathbb{Q}(\beta)$ that sends $\alpha$ to $\beta$, we have

$$N_{\mathbb{Q}(\alpha)/\mathbb{Q}}(\alpha) = N_{\mathbb{Q}(\beta)/\mathbb{Q}}(\beta),$$

where $N$ denotes the norm map. From $a \in \mathbb{Q}$ and $\mathbb{Q}(\alpha) = \mathbb{Q}(\bar{\alpha})$ one deduces

$$a^n = N_{\mathbb{Q}(\alpha)/\mathbb{Q}}(a) = N_{\mathbb{Q}(\alpha)/\mathbb{Q}}(\alpha) N_{\mathbb{Q}(\bar{\alpha})/\mathbb{Q}}(\bar{\alpha}) = N_{\mathbb{Q}(\alpha)/\mathbb{Q}}(\alpha)^2.$$

Similarly,

$$b^n = N_{\mathbb{Q}(\beta)/\mathbb{Q}}(\beta)^2.$$

Putting the above together, one deduces that $a^n = b^n$. Since $a \geq 0$ and $b \geq 0$, we conclude that $a = b$. $\square$

For an integer $w$ with $0 \leq w \leq 2\dim X$, let

$$R(w, X, T) = \prod_{w(\alpha_i) = w} (1 - \alpha_i T).$$

This is called the *pure weight $w$ part* of $R(X,T)$. By the above theorem of Deligne, each $R(w, X, T)$ is a polynomial in $1 + T\mathbb{Z}[T]$. The *pure weight decomposition* is

$$R(X,T) = \prod_{w=0}^{2\dim X} R(w, X, T), \quad R(w, X, T) \in 1 + T\mathbb{Z}[T].$$

THEOREM 5.3. *Given the zeta function $Z(X)$, one can compute all pure parts $R(w, X, T)$ in polynomial time.*

PROOF. By the LLL factorization algorithm [Lenstra et al. 1982], the polynomial $R(X, T)$ can be factored as a product of irreducible polynomials in polynomial time:

$$R(X, T) = \prod_i g_i(T),$$

where each factor

$$g_i \in 1 + T\mathbb{Z}[T]$$

is irreducible over $\mathbb{Q}$. Write

$$g_i(T) = 1 + a_{i1}T + \cdots + a_{ie_i}T^{e_i} = \prod_{j=1}^{e_i}(1 - \beta_{ij}T), \quad a_{ie_i} \neq 0.$$

By Deligne's theorem 5.1, each $\beta_{ij}$ is pure of some integer weight $w_i$. Hence $a_{ie_i}^2 = \prod_j \beta_{ij}\bar{\beta}_{ij} = q^{e_i w_i}$, and therefore

$$a_{ie_i} = \pm\sqrt{q}^{w_i e_i} \in \mathbb{Z}.$$

One can recover the integer weight $w_i$ from

$$w_i = 2\frac{\log_q |a_{ie_i}|}{e_i}.$$

The pure weight $w$ part of $Z(X)$ is then

$$R(w, X, T) = \prod_{w_i = w} g_i(T). \qquad \square$$

Clearly, for each irreducible factor $g_i(T)$ of $R(X, T)$, the map $\beta_{ij} \mapsto \bar{\beta}_{ij} = q^{w_i}/\beta_{ij}$ permutes the reciprocal roots of $g_i(T)$. This gives the following functional equation.

COROLLARY 5.4. *For each $w$, the pure weight part $R(w, X, T)$ satisfies*

$$R(w, X, 1/(q^w T)) = \pm q^{-wd(w,R)/2} T^{-d(w,R)} R(w, X, T),$$

*where $d(w, R)$ denotes the degree of $R(w, X, T)$.*

The following is also immediate from Theorem 5.3.

COROLLARY 5.5. *Given the zeta function, the degrees $d(w, R)$ of all pure weight parts $R(w, X, T)$ can be computed in polynomial time.*

The pure degrees $d(w, R)$ contain important geometric information about the variety $X$. For instance, if $R_2$ denotes the denominator of $Z(X)$, then

$$d(2 \dim X, R_2)$$

is the number of top-dimensional components of $X \otimes \bar{\mathbb{F}}_q$. If $X$ is a geometrically irreducible curve and $R_1$ denotes the numerator of $Z(X)$, then $d(1, R_1)$ is twice the genus of the nonsingular model of $X$. If $X$ is smooth and projective, the pure degrees $d(w, R)$ are geometric invariants, that is, they depend only on $X \otimes \bar{\mathbb{F}}_q$. In fact, in this smooth projective case, they are given by the $\ell$-adic Betti numbers, as shown in the proof of the following result.

COROLLARY 5.6. *For any smooth projective variety $X$ defined over $\mathbb{F}_q$, the $\ell$-adic Betti numbers of $X$ can be effectively computed, where $\ell \neq p$.*

PROOF. For a smooth projective variety $X$ over $\mathbb{F}_q$, the $\ell$-adic trace formula states

$$Z(X) = \prod_{i=0}^{2 \dim X} \det(I - (\mathrm{Frob}|H^i(X \otimes \bar{\mathbb{F}}_q, \mathbb{Q}_\ell))T)^{(-1)^{i-1}},$$

where Frob denotes the geometric Frobenius map. By the Weil conjectures as proved in [Deligne 1974], each (complex) eigenvalue of Frob acting on $H^i$ has weight equal to $i$. Thus, the $i$-th Betti number

$$B_i(X, \ell) = \dim_{\mathbb{Q}_\ell} H^i(X \otimes \bar{\mathbb{F}}_q, \mathbb{Q}_\ell)$$

is given by the formula

$$B_i(X, \ell) = \begin{cases} d(i/2, R_1) & \text{if } i \text{ is odd,} \\ d(i/2, R_2) & \text{if } i \text{ is even,} \end{cases}$$

where $R_1$ and $R_2$ are the numerator and denominator of $Z(X)$, respectively. The corollary follows. □

Let $Y$ be a smooth projective scheme over $\mathbb{Z}$ and let $X = Y \otimes \mathbb{F}_p$ be the reduction modulo $p$ of $Y$. The cohomological comparison theorem shows that for all large primes $p$, the pure degrees $d(w, R(X, T))$ depend only on the geometry of $Y \otimes \bar{\mathbb{Q}}$, not on the chosen large prime $p$.

If $X$ is singular or open, the pure degrees $d(w, R)$ may not be geometric invariants, as it may happen that a root and a pole of $Z(X)$ have a quotient that is a root of unity. On the other hand, it is clear that the pure difference degrees $d(w, R_1) - d(w, R_2)$ are geometric invariants, where we recall that $R_1$ and $R_2$ denote the numerator and the denominator of $Z(X)$, respectively.

It would be interesting to know if the pure degrees $d(w, R)$ can be computed in polynomial time without the zeta function being given. Even for a singular plane curve this seems to be unknown. In the case of a smooth projective complete intersection there is a well-known formula showing that the pure degrees can be computed from the total degree of $Z(X)$ and the dimension of $X$; see Example 3 of Section 4 for the case of a smooth projective hypersurface and [Deligne and Katz 1973, pp. 39–61] for the general smooth projective complete intersection case.

For arbitrary $X$ over $\mathbb{F}_q$, not necessarily smooth or projective, Grothendieck [1968] has shown that for $\ell \neq p$ there is a similar $\ell$-adic formula in terms of $\ell$-adic cohomology with compact support:

$$Z(X) = \prod_{i=0}^{2 \dim X} \det(I - (\text{Frob}| H_c^i(X \otimes \bar{\mathbb{F}}_q, \mathbb{Q}_\ell))T)^{(-1)^{i-1}}.$$

But in this generality, even the conjectured independence of the $\ell$-adic Betti numbers on $\ell$ is unknown. The following result of Katz [2001] provides weak evidence in this direction. It gives an explicit upper bound for the $\ell$-adic Betti numbers with compact support that is independent of $\ell$. It is obtained by means of an inductive reduction to the Bombieri–Adolphson–Sperber degree bound for $Z(X)$, which in turn is $p$-adic in nature.

THEOREM 5.7. *Let the polynomials $f_1, \ldots, f_m$ form a system of defining equations for the affine algebraic set $X$, and put $d = \max_i \deg(f_i)$. Then, for every prime number $\ell \neq p$, we have*

$$\sum_{i \geq 0} \dim_{\mathbb{Q}_\ell} H_c^i(X \otimes \bar{\mathbb{F}}_q, \mathbb{Q}_\ell) \leq 2^{m+2}(md+3)^{n+1}.$$

Turning to global zeta functions, it may be of interest to point out that the real parts of the zeros of the classical Riemann zeta function $\zeta(z)$ are not known to be effectively computable in a finite region. To make this precise, denote, for real numbers $w \in (0, 1)$ and $t > 0$, by $d(w, \zeta; t)$ the number of zeros of $\zeta(z)$ that lie on the line segment

$$\text{Re}(z) = w, \quad 0 \leq \text{Im}(z) \leq t.$$

For any given $t$, there are only finitely many $w$ for which $d(w, \zeta; t) > 0$. The following computational problem is now analogous to finding the pure weight decomposition: given $t$, determine the finitely many positive integers among the numbers $d(w, \zeta; t)$, for $0 \leq w \leq 1$, as well as approximations to the corresponding numbers $w$. No effective algorithm for doing this is currently available. The main difficulty is caused by the possibility of a multiple zero or zeros that are very close to each other, see [Odlyzko 1994]. Of course, the Riemann hypothesis says that $d(w, \zeta; t) = 0$ for all $t$ and all $w \neq 1/2$.

## 6. Pure slope decomposition

In the previous section, we considered the problem of purity decomposition of $Z(X)$ from the complex point of view. One can also consider the purity decomposition from a nonarchimedean point of view. If $\ell$ is a prime number different from $p$, then Deligne's Theorem 5.1 shows that $Z(X)$ is already pure from an $\ell$-adic point of view. Thus, we will consider the remaining $p$-adic case in this section. Let $R(X, T)$ again denote either the numerator or the denominator of $Z(X)$.

Let $\mathbb{C}_p$ be the completion of a fixed algebraic closure of $\mathbb{Q}_p$. Over $\mathbb{C}_p$, we can write
$$R(X, T) = \prod_i (1 - \alpha_i T),$$
where each $\alpha_i$ is a nonzero algebraic integer in $\mathbb{C}_p$. Define the slope of a nonzero element $\alpha \in \mathbb{C}_p$ by
$$s(\alpha) = \mathrm{ord}_q(\alpha) = -\log_q |\alpha|_p,$$
where $|\cdot|_p$ denotes the $p$-adic absolute value, normalized such that $|p| = 1/p$. The slopes $s(\alpha_i)$ are nonnegative rational numbers since the $\alpha_i$ are algebraic integers. One immediately derives the bound
$$0 \leq s(\alpha_i) \leq s(\alpha_i \bar{\alpha}_i) = w(\alpha_i) \leq 2 \dim X.$$

This bound can be improved somewhat. Deligne's integrality theorem [1973, pp. 384–400] states that $q^{\dim X}/\alpha_i$ is an algebraic integer. We deduce the following result.

THEOREM 6.1. *The slopes $s(\alpha_i)$ are rational numbers in $[0, \dim X]$.*

Other than in the complex absolute value case, $\alpha_i$ and its Galois conjugates over $\mathbb{Q}$ may have different slopes. Of course, each $\alpha_i$ and its Galois conjugates over $\mathbb{Q}_p$ do have the same slope. The slopes $s(\alpha_i)$ are not integers (or half integers) in general. They are merely rational numbers. It would be interesting to get good bounds for the denominators of the slopes $s(\alpha_i)$.

For a rational number $s$ with $0 \leq s \leq \dim X$, let
$$R(s, X, T) = \prod_{s(\alpha_i) = s} (1 - \alpha_i T).$$
This is called the pure slope $s$ part of $R(X, T)$. We have the $p$-adic purity decomposition
$$R(X, T) = \prod_{s \in \mathbb{Q}} R(s, X, T).$$
The question is then to understand each pure slope part $R(s, X, T)$.

Note that in general, the pure slope parts $R(s, X, T)$ do not have coefficients in $\mathbb{Z}$ any more. The theory of Newton polygons implies that $R(s, X, T)$ is a polynomial in $1 + T(\bar{\mathbb{Q}} \cap \mathbb{Z}_p)[T]$, as we shall see now.

Write

$$R(X, T) = 1 + a_1 T + \cdots + a_e T^e = \prod_{i=1}^{e}(1 - \alpha_i T), \quad a_e \neq 0.$$

DEFINITION 6.2. The Newton polygon $\mathrm{NP}(R)$ of $R(X, T)$ is the lower convex hull in the plane $\mathbb{R}^2$ of the points

$$(k, \mathrm{ord}_q(a_k)), \quad k = 0, 1, \ldots, e.$$

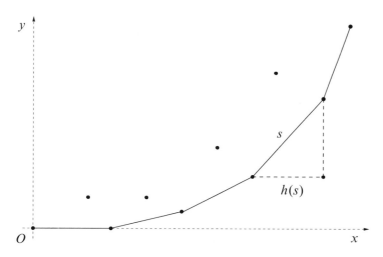

**Figure 1.** Newton polygon

A basic property of the Newton polygon is the following result; see [Koblitz 1984].

THEOREM 6.3. *The polynomial $R(X, T) \in 1 + T\mathbb{Z}[T]$ has exactly $h$ reciprocal zeros $\alpha_i$ with slope $\mathrm{ord}_q(\alpha_i) = s$ (counting multiplicities) if and only if $\mathrm{NP}(R)$ has a side of slope $s$ and horizontal length $h$. Furthermore, the coefficients of $R(s, X, T)$ are in $\mathbb{Z}_p$ for each $s$.*

The following is an immediate consequence.

COROLLARY 6.4. *Let $d(s, R)$ denote the degree of $R(s, X, T)$. Then, $d(s, R)$ is the horizontal length of the slope $s$ side of $\mathrm{NP}(R)$. In particular, when $Z(X)$ is given, the $p$-adic pure degrees $d(s, R)$ can be computed in polynomial time.*

It would be interesting to know if the $p$-adic pure degrees $d(s, R)$ can be computed in polynomial time without the zeta function being given. This amounts to computing the Newton polygons of the numerator and the denominator of the zeta function. Even in a very well-behaved situation such as the smooth projective hypersurface case, we do not have a complete answer in general; see [Wan 2004] for a theoretical introduction to Newton polygons for zeta functions and $L$-functions. In the smooth projective case, the $p$-adic pure degrees $d(s, R)$ are geometric invariants.

If $X$ is the good reduction modulo $p$ of some smooth projective scheme over $\mathbb{Z}$, the $p$-adic pure degrees $d(s, R)$ depend not just on the geometry of the generic fibre $Y \otimes \bar{\mathbb{Q}}$, but also on the chosen prime $p$. Thus, the $p$-adic pure degrees $d(s, R)$ contain arithmetic information on $Y$. They are related to but much deeper than the topological Hodge numbers of $Y \otimes \mathbb{Q}$ defined in terms of the De Rham cohomology of $Y \otimes \mathbb{Q}$; see [Mazur 1972] for an introductory account. Describing the variation of $d(s, R)$ as $p$ varies is a very subtle arithmetic problem, already in the special case that $Y \otimes \mathbb{Q}$ is an elliptic curve.

As a polynomial with coefficients in $\mathbb{Z}_p$, the pure slope part $R(s, X, T)$ cannot be written down in a finite amount of time. However, given $R(X, T)$, one can compute the pure slope parts $R(s, X, T)$ modulo any given power of $p$ in polynomial time using the Newton polygon and Hensel lifting. Note that here we do not factor $R(X, T)$ into a product of irreducible factors over $\mathbb{Q}_p$, which is a harder problem.

## 7. Zeta functions modulo $p$

In [Lauder and Wan 2008] a $p$-adic algorithm for computing the zeta function $Z(X)$ is given, where $p$ is the characteristic of $\mathbb{F}_q$. In this final section, we describe some of the basic ideas behind that algorithm by showing how $Z(X)$ may be computed modulo $p$. The method expands the outline given in [Wan 1999].

Without loss of generality we may restrict to hypersurfaces. Thus, let $X$ be the affine hypersurface defined by a polynomial $f(x_1, \ldots, x_n)$ over $\mathbb{F}_q$ of total degree $d$ in $n$ variables. Let $A_d$ be the $\mathbb{F}_q$-vector space of polynomials in $x_1, \ldots, x_n$ of total degree at most $d$ that are divisible by the product $x_1 \cdots x_n$:

$$A_d = (x_1 \cdots x_n \mathbb{F}_q[x_1, \ldots, x_n])_{\leq d}.$$

It has a row basis $\vec{e}$ consisting of monomials

$$\vec{e} = \{x^u \mid u = (u_1, \ldots, u_n), \ u_i \geq 1, \ |u| \leq d\},$$

where

$$x^u = x_1^{u_1} \cdots x_n^{u_n}, \quad |u| = u_1 + \cdots + u_n.$$

One computes that

$$\dim_{\mathbb{F}_q} A_d = \binom{d}{n}.$$

DEFINITION 7.1. Let $\sigma$ be the $p$-th power Frobenius map acting on $\mathbb{F}_q$:

$$\sigma(a) = a^p, \quad a \in \mathbb{F}_q.$$

Let $\psi_p$ be the $\sigma^{-1}$-linear operator on the $\mathbb{F}_q$-vector space $\mathbb{F}_q[x_1, \ldots, x_n]$ defined by

$$\psi_p\left(\sum_u a_u x^u\right) = \sum_u \sigma^{-1}(a_u) \psi_p(x^u),$$

where

$$\psi_p(x^u) = \begin{cases} x^{u/p}, & \text{if } p|u, \\ 0, & \text{otherwise.} \end{cases}$$

Define

$$\psi_q = \psi_p^r = \psi_p \circ \cdots \circ \psi_p$$

to be the $r$-th iterate of $\psi_p$, where $q = p^r$.

Since $\sigma^r$ is the identity on $\mathbb{F}_q$, the operator $\psi_q$ is actually $\mathbb{F}_q$-linear, although $\psi_p$ is only $\sigma^{-1}$-linear. The operator $\psi_q$ is a left inverse of the $q$-th power Frobenius map on $\mathbb{F}_q[x] = \mathbb{F}_q[x_1, \ldots, x_n]$. For any polynomial $g \in \mathbb{F}_q[x]$, multiplication by $g$ is also an $\mathbb{F}_q$-linear map from $\mathbb{F}_q[x]$ to itself.

LEMMA 7.2. *The $\mathbb{F}_q$-linear composed operator $\psi_q \circ f^{q-1}$ maps the finite dimensional $\mathbb{F}_q$-subspace $A_d$ of $\mathbb{F}_q[x]$ to itself; here $f^{q-1}$ denotes multiplication by $f^{q-1}$. Similarly, the $\sigma^{-1}$-linear composed operator $\psi_p \circ f^{p-1}$ maps $A_d$ to itself, and we have the relation*

$$\psi_q \circ f^{q-1} = (\psi_p \circ f^{p-1})^r.$$

PROOF. Let $h \in A_d$. Then $h$ has degree at most $d$, so $f^{q-1}h$ has degree at most $d(q-1) + d = dq$. Thus, the degree of $\psi_q(f^{q-1}h)$ is at most $d$. Furthermore, if $h$ is divisible by $x_1 \cdots x_n$, then $\psi_q(f^{q-1}h)$ is also divisible by $x_1 \cdots x_n$. This proves that $\psi_q(f^{q-1}h) \in A_d$. The proof of the second part of the lemma is the same. To prove the last part, we write

$$q - 1 = (p-1) + p(p-1) + \cdots + p^{r-1}(p-1).$$

This gives

$$f(x)^{q-1} = \prod_{i=0}^{r-1} f^{\sigma^i}(x^{p^i})^{p-1},$$

where $f^{\sigma^i}(x)$ denotes the polynomial obtained by applying $\sigma^i$ to the coefficients of $f$. Using the easily checked relation

$$\psi_p \circ g^\sigma(x^p) = g(x) \circ \psi_p,$$

one deduces

$$\psi_q \circ f^{q-1} = \psi_p^r \circ \prod_{i=0}^{r-1} f^{\sigma^i}(x^{p^i})^{p-1} = (\psi_p \circ f^{p-1})^r. \qquad \square$$

THEOREM 7.3. *Let $X$ be the affine hypersurface defined by a polynomial $f(x_1, \ldots, x_n)$ over $\mathbb{F}_q$ of total degree $d$ in $n$ variables. Then we have the congruence formula*

$$(Z(X)^{(-1)^n} \bmod p) = \det(I - (\psi_q \circ f^{q-1} | A_d)T).$$

Before we prove this result, we first recall some facts on $L$-functions in characteristic $p$. Let $g \in \mathbb{F}_q[x] = \mathbb{F}_q[x_1, \ldots, x_n]$. For a geometric point $x \in \mathbb{A}^n(\bar{\mathbb{F}}_q)$, the product $g(x)g(x^q) \cdots g(x^{q^{\deg(x)-1}})$ is an element of $\mathbb{F}_q$ that clearly depends only on the orbit (the closed point) of $x$ under the $q$-th power Frobenius map. We define the $L$-function of $g$ by

$$L(g, T) = \prod_x \frac{1}{1 - g(x)g(x^q) \cdots g(x^{q^{\deg(x)-1}})T^{\deg(x)}} \in 1 + T\mathbb{F}_q[\![T]\!],$$

where $x$ runs over the set of closed points of the affine space $\mathbb{A}^n$ over $\mathbb{F}_q$.

For a real number $c$, write $A_c$ for the finite dimensional $\mathbb{F}_q$-subspace of $\mathbb{F}_q[x]$ generated by the monomials of total degree at most $c$ that are divisible by the product $x_1 \cdots x_n$. If the total degree of $g$ is at most $e$, one checks as in the above lemma that the operator $\psi_q \circ g$ maps the subspace $A_{e/(q-1)}$ to itself. Furthermore, the (matrix of the) induced map $\psi_q \circ g$ on the quotient vector space $x_1 \cdots x_n \mathbb{F}_q[x]/A_{e/(q-1)}$ is strictly triangular with respect to the monomial basis $\{x^u | u_i \geq 1, |u| > e/(q-1)\}$. Thus, the composed operator $\psi_q \circ g$ acting on the infinite dimensional $\mathbb{F}_q$-vector space $x_1 \cdots x_n \mathbb{F}_q[x]$ has a well defined characteristic power series $\det(I - (\psi_q \circ g | x_1 \cdots x_n \mathbb{F}_q[x])T)$, which is given by the polynomial

$$\det(I - (\psi_q \circ g | x_1 \cdots x_n \mathbb{F}_q[x])T) = \det(I - (\psi_q \circ g | A_{e/(q-1)})T).$$

The characteristic $p$ version of Dwork's trace formula for the affine space $\mathbb{A}^n$, as given in [Wan 1996], implies that

$$L(g, T)^{(-1)^{n-1}} = \det(I - (\psi_q \circ g | x_1 \cdots x_n \mathbb{F}_q[x])T)$$

and thus

$$L(g, T)^{(-1)^{n-1}} = \det(I - (\psi_q \circ g | A_{e/(q-1)})T).$$

The reader is referred to [Lauder and Wan 2008] to see the Dwork trace formula over the $n$-torus, which is cleaner than the Dwork trace formula over the affine $n$-space $\mathbb{A}^n$.

We now return to the proof of the theorem. Taking $g = f^{q-1}$ and $e = d(q-1)$, we deduce that $e/(q-1) = d$ and

$$L(f^{q-1}, T)^{(-1)^{n-1}} = \det(I - (\psi_q \circ f^{q-1}|A_d)T).$$

For a geometric point $x \in \mathbb{A}^n(\bar{\mathbb{F}}_q)$, one has

$$g(x)g(x^q) \cdots g(x^{q^{\deg(x)-1}}) = f(x)^{q^{\deg(x)}-1},$$

which is 0 or 1 according as $x \in X(\bar{\mathbb{F}}_q)$ or not. It follows that $L(f^{q-1}, T)$ is the reduction modulo $p$ of the zeta function of the complement of $X$ in $\mathbb{A}^n$. Hence for $n > 0$ we obtain

$$L(f^{q-1}, T) = (1/Z(X) \bmod p).$$

Substituting this into the above formula for $L(f^{q-1}, T)^{(-1)^{n-1}}$, we obtain the theorem.

COROLLARY 7.4. *The zeta function $Z(X)$ modulo $p$ can be computed in time bounded by a polynomial in $p\binom{d}{n} \log q$.*

PROOF. Recall that $q = p^r$. Let $M_1$ be the matrix of the $\sigma^{-1}$-linear map $\psi_p \circ f^{p-1}$ with respect to the row monomial basis $\vec{e}$ of $A_d$. That is,

$$(\psi_p \circ f^{p-1})(\vec{e}) = \vec{e} M_1.$$

Then, by the $\sigma^{-1}$-linearity of $\psi_p \circ f^{p-1}$, we deduce

$$(\psi_p \circ f^{p-1})^2(\vec{e}) = (\psi_p \circ f^{p-1})(\vec{e} M_1) = \vec{e} M_1 M_1^{\sigma^{-1}},$$

where $M_1^{\sigma^{-1}}$ is obtained from $M_1$ by applying $\sigma^{-1}$ to each entry (and similarly with $M_1^{\sigma^{-i}}$ below). By iteration, one finds that the matrix of the $\mathbb{F}_q$-linear map

$$\psi_q \circ f^{q-1} = (\psi_p \circ f^{p-1})^r$$

with respect to the row basis $\vec{e}$ is given by

$$M_r = M_1 M_1^{\sigma^{-1}} \cdots M_1^{\sigma^{-(r-1)}}.$$

The matrix $M_1$ can be written down in time $(p\binom{d}{n} \log q)^{O(1)}$. It follows that the matrix $M_r$ can also be computed in time $(p\binom{d}{n} \log q)^{O(1)}$. The zeta function modulo $p$ is essentially just the characteristic polynomial of the matrix $M_r$:

$$(Z(X)^{(-1)^n} \bmod p) = \det(I - M_r T).$$

The corollary is proved. Alternatively, applying $\sigma^{r-1}$ to the above congruence formula, we have

$$(Z(X)^{(-1)^n} \bmod p) = \det(I - M_1^{\sigma^{r-1}} \cdots M_1^{\sigma} M_1 T).$$

This formula may be slightly more efficient from a computational point of view.

If $p$ is small, that is, $p = (\binom{d}{n} \log q)^{O(1)}$, the above corollary gives a polynomial time algorithm for computing $Z(X)$ modulo $p$. If $p$ is large, we do not get a polynomial time algorithm. □

## Acknowledgment

It is a pleasure to thank S. Gao, H.W. Lenstra, Jr., and A.M. Odlyzko for several interesting discussions. Special thanks are due to the referee for many helpful suggestions.

## References

[Adleman and Huang 1996] L. M. Adleman and M.-D. A. Huang, "Counting rational points on curves and abelian varieties over finite fields", pp. 1–16 in *Algorithmic number theory* (ANTS-II) (Talence, 1996), edited by H. Cohen, Lecture Notes in Comput. Sci. **1122**, Springer, Berlin, 1996.

[Adolphson and Sperber 1988] A. Adolphson and S. Sperber, "On the degree of the $L$-function associated with an exponential sum", *Compositio Math.* **68**:2 (1988), 125–159.

[Blahut 1998] R. E. Blahut, "Decoding of cyclic codes and codes on curves", pp. 1569–1633 in *Handbook of coding theory*, vol. II, edited by V. S. Pless and W. C. Huffman, North-Holland, Amsterdam, 1998.

[Blake et al. 2000] I. F. Blake, G. Seroussi, and N. P. Smart, *Elliptic curves in cryptography*, London Mathematical Society Lecture Note Series **265**, Cambridge University Press, Cambridge, 2000.

[Bombieri 1978] E. Bombieri, "On exponential sums in finite fields, II", *Invent. Math.* **47**:1 (1978), 29–39.

[Cohen 1993] H. Cohen, *A course in computational algebraic number theory*, Graduate Texts in Mathematics **138**, Springer, Berlin, 1993.

[Deligne 1974] P. Deligne, "La conjecture de Weil, I", *Inst. Hautes Études Sci. Publ. Math.* no. 43 (1974), 273–307.

[Deligne 1980] P. Deligne, "La conjecture de Weil, II", *Inst. Hautes Études Sci. Publ. Math.* no. 52 (1980), 137–252.

[Deligne and Katz 1973] P. Deligne and N. Katz, *Groupes de monodromie en géométrie algébrique* (SGA 7 II), Lecture Notes in Math. **340**, Springer, Berlin, 1973.

[Denef and Vercauteren 2002] J. Denef and F. Vercauteren, "An extension of Kedlaya's algorithm to Artin–Schreier curves in characteristic 2", pp. 308–323 in *Algorithmic number theory* (ANTS-V) (Sydney, 2002), edited by C. Fieker and D. R. Kohel, Lecture Notes in Comput. Sci. **2369**, Springer, Berlin, 2002.

[Dwork 1960] B. Dwork, "On the rationality of the zeta function of an algebraic variety", *Amer. J. Math.* **82** (1960), 631–648.

[Eisenbud 1995] D. Eisenbud, *Commutative algebra with a view toward algebraic geometry*, Graduate Texts in Math. **150**, Springer, New York, 1995.

[Elkies 1998] N. D. Elkies, "Elliptic and modular curves over finite fields and related computational issues", pp. 21–76 in *Computational perspectives on number theory* (Chicago, IL, 1995), edited by D. A. Buell and J. T. Teitelbaum, AMS/IP Stud. Adv. Math. **7**, Amer. Math. Soc., Providence, RI, 1998.

[Grothendieck 1968] A. Grothendieck, "Formule de Lefschetz et rationalité des fonctions $L$", pp. 279:1–15 in *Séminaire Bourbaki* 1964/65, North Holland, Amsterdam, 1968. Reprinted Soc. Math. France, Paris, 1995 (with 1965/66 volume).

[Katz 1971] N. M. Katz, "On a theorem of Ax", *Amer. J. Math.* **93** (1971), 485–499.

[Katz 2001] N. M. Katz, "Sums of Betti numbers in arbitrary characteristic", *Finite Fields Appl.* **7**:1 (2001), 29–44.

[Kedlaya 2001] K. S. Kedlaya, "Counting points on hyperelliptic curves using Monsky–Washnitzer cohomology", *J. Ramanujan Math. Soc.* **16**:4 (2001), 323–338. Errata in **18**:4 (2003), 417–418.

[Koblitz 1984] N. Koblitz, *p-adic numbers, p-adic analysis, and zeta-functions*, 2nd ed., Graduate Texts in Mathematics **58**, Springer, New York, 1984.

[Koblitz 1989] N. Koblitz, *Hyperelliptic cryptosystems*, vol. 1, 1989.

[Lauder 2004a] A. G. B. Lauder, "Counting solutions to equations in many variables over finite fields", *Found. Comput. Math.* **4**:3 (2004), 221–267.

[Lauder 2004b] A. G. B. Lauder, "Deformation theory and the computation of zeta functions", *Proc. London Math. Soc.* (3) **88**:3 (2004), 565–602.

[Lauder and Wan 2002] A. Lauder and D. Wan, "Computing zeta functions of Artin-Schreier curves over finite fields", pp. 34–55, 2002.

[Lauder and Wan 2008] A. Lauder and D. Wan, "Counting points on varieties over finite fields of small characteristic", pp. 579–612 in *Surveys in algorithmic number theory*, edited by J. P. Buhler and P. Stevenhagen, Math. Sci. Res. Inst. Publ. **44**, Cambridge University Press, New York, 2008.

[Lenstra et al. 1982] A. K. Lenstra, H. W. Lenstra, Jr., and L. Lovász, "Factoring polynomials with rational coefficients", *Math. Ann.* **261**:4 (1982), 515–534.

[Matiyasevich 1993] Y. V. Matiyasevich, "Hilbert's Tenth Problem", (1993).

[Mazur 1972] B. Mazur, "Frobenius and the Hodge filtration", *Bull. Amer. Math. Soc.* **78** (1972), 653–667.

[Monsky 1970] P. Monsky, *p-adic analysis and zeta functions*, Lectures in Mathematics **4**, Kinokuniya, Tokyo, 1970.

[Odlyzko 1994] A. M. Odlyzko, "Analytic computations in number theory", pp. 451–463 in *Mathematics of computation 1943–1993: a half-century of computational mathematics* (Vancouver, 1993), edited by W. Gautschi, Proc. Sympos. Appl. Math. **48**, Amer. Math. Soc., Providence, RI, 1994.

[Pila 1990] J. Pila, "Frobenius maps of abelian varieties and finding roots of unity in finite fields", *Math. Comp.* **55**:192 (1990), 745–763.

[Poonen 1996] B. Poonen, "Computational aspects of curves of genus at least 2", pp. 283–306 in *Algorithmic number theory* (ANTS-II) (Talence, 1996), edited by H. Cohen, Lecture Notes in Comput. Sci. **1122**, Springer, Berlin, 1996.

[Satoh 2000] T. Satoh, "The canonical lift of an ordinary elliptic curve over a finite field and its point counting", *J. Ramanujan Math. Soc.* **15**:4 (2000), 247–270.

[Schoof 1985] R. Schoof, "Elliptic curves over finite fields and the computation of square roots mod $p$", *Math. Comp.* **44**:170 (1985), 483–494.

[Wan 1996] D. Wan, "Meromorphic continuation of $L$-functions of $p$-adic representations", *Ann. of Math.* (2) **143**:3 (1996), 469–498.

[Wan 1999] D. Wan, "Computing zeta functions over finite fields", pp. 131–141 in *Finite fields: theory, applications, and algorithms* (Waterloo, ON, 1997), edited by R. C. Mullin and G. L. Mullen, Contemp. Math. **225**, Amer. Math. Soc., Providence, RI, 1999.

[Wan 2004] D. Wan, "Variation of $p$-adic Newton polygons for $L$-functions of exponential sums", *Asian J. Math.* **8**:3 (2004), 427–471.

[Weil 1949] A. Weil, "Numbers of solutions of equations in finite fields", *Bull. Amer. Math. Soc.* **55** (1949), 497–508.

DAQING WAN
DEPARTMENT OF MATHEMATICS
UNIVERSITY OF CALIFORNIA
IRVINE, CA 92697-3875
UNITED STATES
dwan@math.uci.edu

# Counting points on varieties over finite fields of small characteristic

ALAN G. B. LAUDER AND DAQING WAN

ABSTRACT. We present a deterministic polynomial time algorithm for computing the zeta function of an arbitrary variety of fixed dimension over a finite field of small characteristic. One consequence of this result is an efficient method for computing the order of the group of rational points on the Jacobian of a smooth geometrically connected projective curve over a finite field of small characteristic.

## CONTENTS

| | |
|---|---:|
| 1. Introduction | 579 |
| 2. Additive character sums over finite fields | 583 |
| 3. $p$-adic theory | 584 |
| 4. Analytic representation of characters | 588 |
| 5. Dwork's trace formula | 595 |
| 6. Algorithms | 602 |
| Acknowledgments | 611 |
| References | 611 |

## 1. Introduction

The purpose of this paper is to give an elementary and self-contained proof that one may efficiently compute zeta functions of arbitrary varieties of fixed dimension over finite fields of suitably small characteristic. This is achieved via the $p$-adic methods developed by Dwork in his proof of the rationality of the zeta

---

*Mathematics Subject Classification:* 11Y16, 11T99, 14Q15.

*Keywords:* variety, finite field, zeta function, algorithm.

Alan Lauder gratefully acknowledges the support of the EPSRC (Grant GR/N35366/01) and St John's College, Oxford, and thanks Richard Brent. Daqing Wan is partially supported by the NSF and the NSFC.

function of a variety over a finite field [Dwork 1960; 1962]. Dwork's theorem shows that it is in principle possible to compute the zeta function. Our main contribution is to show how Dwork's trace formula, Bombieri's degree bound [1978] and a semilinear reduction argument yield an efficient algorithm for doing so. That $p$-adic methods may be used to efficiently compute zeta functions for small characteristic was first suggested in [Wan 1999; 2008], where Wan gives a simpler algorithm for counting the number of solutions to an equation over a finite field modulo small powers of the characteristic.

We now give more details of our results. For $q = p^a$, where $p$ is a prime number and $a$ a positive integer, let $\mathbb{F}_q$ denote a finite field with $q$ elements. Let $\bar{\mathbb{F}}_q$ denote an algebraic closure of $\mathbb{F}_q$, and $\mathbb{F}_{q^k}$ the subfield of $\bar{\mathbb{F}}_q$ of order $q^k$. Denote by $\mathbb{F}_q[X_1, \ldots, X_n]$ the ring of all polynomials in $n$ variables over $\mathbb{F}_q$.

For a polynomial $f \in \mathbb{F}_q[X_1, \ldots, X_n]$, we denote by $N_k$ the number of solutions to the equation $f = 0$ with coordinates in $\mathbb{F}_{q^k}$. The zeta function of the variety defined by $f$ is the formal power series in $T$ with nonnegative integer coefficients

$$Z(f/\mathbb{F}_q)(T) = \exp\left(\sum_{k=1}^{\infty} \frac{N_k T^k}{k}\right).$$

Dwork's theorem asserts that $Z(f/\mathbb{F}_q)$ is a rational function $r(T)/s(T)$ with integer coefficients. From this it follows that knowledge of explicit bounds $\deg(r) \leq D_1$ and $\deg(s) \leq D_2$, and of the values $N_k$ for $k = 1, 2, \ldots, D_1 + D_2$ is enough to efficiently determine $Z(f/\mathbb{F}_q)$; see [Wan 2008]. The Bombieri degree bound tells us that $\deg(r) + \deg(s) \leq (4d+9)^{n+1}$; see [Bombieri 1978; Wan 2008], and so in particular we may take $D_1 = D_2 = (4d+9)^{n+1}$. Each number $N_k$ can be computed in a naive fashion by straightforward counting, using $q^{nk}$ evaluations of the polynomial $f$. Thus one may compute the zeta function of a variety, but the naive method described requires a number of steps that is exponential in the parameters $d^n$ and $\log q$, where $d$ is the total degree of $f$. The dense input size of $f$ is $O((d+1)^n \log q)$, and the size of the zeta function is polynomial in $O((d+1)^n \log q)$, by the Bombieri degree bound. We prove the following theorem.

THEOREM 1. *There exist an explicit deterministic algorithm and an explicit polynomial $P$ such that for any $f \in \mathbb{F}_q[X_1, \ldots, X_n]$ of total degree $d$, where $q = p^a$ and $p$ is prime, the algorithm computes the zeta function $Z(f/\mathbb{F}_q)(T)$ of $f$ in a number of bit operations which is bounded by $P(p^n d^{n^2} a^n)$.*

In particular, this computes the zeta function of a polynomial in a fixed number of variables over a finite field of "small characteristic" in deterministic polynomial time. Our result makes no assumption of nonsingularity on the variety defined by the polynomial. Also, we shall explicitly describe all the algorithms in this paper, rather than just prove their existence, and we assume that the finite field $\mathbb{F}_q$ itself is presented as input via an irreducible polynomial of degree $a$ over the prime field $\mathbb{F}_p$, as explained in Section 3.

With regard to the exponents in the algorithm, we state these precisely in Theorem 37. For now we observe that if $p$, $d$ and $n$ are fixed then the time required to compute the number of points in $\mathbb{F}_q$ is $O(a^{3n+7})$, with space complexity $O(a^{2n+4})$. Here we are ignoring logarithmic factors (see also Proposition 36). All our complexity estimates are made using standard methods for multiplication in various rings, and can be modestly reduced with faster methods.

We also present refinements to this result based upon the ideas of Adolphson and Sperber [1987], and indeed from the outset will follow their approach, as it involves little extra complication. This refinement takes into account the terms which actually occur in the polynomial $f$ rather than working solely with the total degree. We shall need more definitions: the support of a polynomial $f$ is the set of exponents $r = (r_1, \ldots, r_n)$ of nonzero terms $a_r X_1^{r_1} \ldots X_n^{r_n}$ which occur in $f$, thought of as points in $\mathbb{R}^n$. The Newton polytope of $f$ is defined to be the convex hull in $\mathbb{R}^n$ of the support of $f$. Our refined version of Theorem 1 essentially replaces the parameter $d^n$ with the normalised volume of the Newton polytope of $f$ (see Proposition 35 and Section 6.3.3).

Zeta functions may also be defined for a finite collection of polynomials, and we next describe how our results may be extended to this case. An affine variety $V$ over $\mathbb{F}_q$ is the set of common zeros in $(\bar{\mathbb{F}}_q)^n$ of a set of polynomials $f_1, f_2, \ldots, f_r$. An analogous zeta function $Z(V/\mathbb{F}_q)$ may then be defined in terms of the number of solutions in each finite extension field of $\mathbb{F}_q$. These numbers may be computed using an inclusion-exclusion argument involving the polynomials $\prod_{i \in S} f_i$, where $S$ is a subset of $\{1, 2, \ldots, r\}$; see [Wan 2008]. Theorem 1 then easily yields the following.

COROLLARY 2. *There exist an explicit deterministic algorithm and an explicit polynomial $Q$ with the following property. Let $f_1, \ldots, f_r \in \mathbb{F}_q[X_1, \ldots, X_n]$ have total degrees $d_1, \ldots, d_r$ respectively, where $q = p^a$ and $p$ is prime. Denote by $V$ the affine variety defined by the common vanishing of these polynomials, and define $d = \sum_{i=1}^{r} d_i$. The algorithm computes the zeta function $Z(V/\mathbb{F}_q)(T)$ in a number of bit operations which is bounded by $Q(p^n d^{n^2} a^n 2^r)$.*

Thus one has an efficient algorithm for computing the zeta function of an arbitrary affine variety over $\mathbb{F}_q$ assuming the characteristic, dimension and the number of defining polynomials are fixed. More generally still, an arbitrary variety is defined through patching together suitable affine varieties. Zeta functions for such general varieties may be defined. These zeta functions may be computed using the above ideas provided explicit data are given on how to construct them from affine patches. As an example, the zeta functions of arbitrary projective varieties or toric varieties may be computed in this way.

Conceptually our algorithm is rather straightforward. The zeta function can be expressed in terms of the "characteristic power series" (Fredholm determinant) of a certain "lifting of Frobenius" which acts on an infinite dimensional $p$-adic Banach space constructed by Dwork. Under modular reduction, we obtain an operator acting on a finite dimensional vector space. Unfortunately, this operator cannot be computed efficiently directly from its definition; however, it can be expressed as a product of certain semilinear operators each of which can be computed efficiently if the characteristic $p$ is small. This last step is thus of crucial importance in deriving an efficient algorithm. The same idea is used in [Wan 1999; 2008] in a simpler situation. In more concrete language, the algorithm requires one to construct a certain "semilinear" finite matrix and compute the "linear" matrix which is the product of the Galois conjugates of the semilinear matrix. The number of points is then read off from the trace of the final linear matrix. The zeta function can be computed from the characteristic polynomial of the final linear matrix. The semilinear matrix itself is defined over a certain finite "$p$-adic lifting" of the original finite field.

In the literature algorithms have already been developed for computing zeta functions of curves and abelian varieties [Adleman and Huang 1996; Elkies 1998; Pila 1990; Schoof 1985; 1995]. They use the theory originally developed by Weil for abelian varieties, whereas we use Dwork's more general and simpler $p$-adic theory. For example, in the case of a smooth geometrically irreducible projective plane curve of degree $d$ over a field of size $q = p^a$ these algorithms have a time complexity which grows as $(\log q)^{C_d}$, where $C_d$ grows exponentially in the degree $d$. Given an absolutely irreducible bivariate polynomial $f$ of degree $d$ over $\mathbb{F}_q$, the zeta function of the unique smooth projective curve birational to the affine curve defined by $f$ may be computed in time polynomial in $d$, $p$ and $a$ using our approach. Thus our more general method is far better in terms of the degree $d$ if the characteristic $p$ is small, since the running time has polynomial growth in $d$ (but much worse if $p$ is large). The zeta function immediately gives the order of the group of rational points on the Jacobian; see [Wan 2008].

COROLLARY 3. *There exist an explicit deterministic algorithm and an explicit polynomial $R$ with the following property. Let $V$ be a geometrically irreducible affine curve defined by the vanishing of polynomials $f_1, \ldots, f_r \in \mathbb{F}_q[X_1, \ldots, X_n]$ of total degrees $d_1, \ldots, d_r$ respectively, where $q = p^a$ and $p$ is prime. Denote by $\tilde{V}$ the unique smooth projective curve birational to the affine curve $V$, and let $d = \sum_{i=1}^{r} d_i$. The algorithm computes the order of the group of rational points on the Jacobian of $\tilde{V}$ in a number of bit operations bounded by $R(p^n d^{n^2} a^n 2^r)$.*

Thus one may compute the order of the group of rational points on the Jacobian of a smooth geometrically irreducible projective curve over a finite field of small characteristic in deterministic polynomial time, provided the number of variables and the number of defining equations are fixed. (The case $r = 1$ and $n = 2$ corresponds to that of being given a possibly singular plane model of the curve.) In particular, this answers a question posed in [Poonen 1996], where it is attributed to Katz and Sarnak. We note that recently a similar result for special classes of plane curves was independently obtained in [Gaudry and Gürel 2001; Kedlaya 2001], using the Monsky–Washnitzer method. Also, a different $p$-adic approach for elliptic curves has been developed in [Satoh 2000].

This paper is written primarily for theoretical computational interest, in obtaining a deterministic polynomial time algorithm for computing the zeta function in full generality if $p$ is small. It can certainly be improved in many ways for practical computations. What we have done in this paper is to work on the easier but more flexible "chain level". A general improvement in the smooth case, is to work on the cohomology level. Then there are several related $p$-adic cohomology theories available, each leading to a somewhat different version of the algorithm. Again, as indicated in [Wan 1999; Wan 2008], these $p$-adic methods are expected to be practical only for small $p$. (In work subsequent to that in the present paper, the first author established deterministic polynomial time computability of zeta functions for smooth projective hypersurfaces in small characteristic and varying dimension, under mild restrictions [Lauder 2004].)

## 2. Additive character sums over finite fields

The most natural objects of study in Dwork's theory are certain additive character sums over finite fields. In this section we introduce these sums, and explain their connection to varieties.

An additive character $\Psi$ is a mapping from $\mathbb{F}_{q^k}$ to the group of units of some commutative ring $S$ with identity 1 such that

$$\Psi(x+y) = \Psi(x)\Psi(y) \quad \text{for } x, y \in \mathbb{F}_{q^k}.$$

We say that it is nontrivial if

$$\Psi(x) \neq 1 \quad \text{for some } x \in \mathbb{F}_{q^k}.$$

Let $\{\Psi_k\}_{k \geq 1}$ be any family of mappings with each $\Psi_k$ a nontrivial additive character from $\mathbb{F}_{q^k}$ to some extension ring of the integers whose image is a group of order $p$ with elements summing to zero. We assume that the family $\{\Psi_k\}$ forms a tower of characters in the sense that for each $k \geq 2$,

$$\Psi_k = \Psi_1 \circ \mathrm{Tr}_{\mathbb{F}_{q^k}/\mathbb{F}_q}.$$

In our application, the ring $S$ will be taken to be a certain $p$-adic ring.

For the remainder of the paper we will use multi-index notation. Specifically, we let $X^u$ represent the monomial $X_0^{u_0} X_1^{u_1} \ldots X_n^{u_n}$ for an integer vector $u = (u_0, u_1, \ldots, u_n)$; let $x$ be an $(n+1)$-tuple $(x_0, x_1, \ldots, x_n)$ of field elements; and $X$ the list of indeterminates $X_0, X_1, \ldots, X_n$. Observe here that we have introduced an extra indeterminate $X_0$. Even though the polynomial $f$ whose zeta function we wish to compute is in the $n$ variables $X_1, \ldots, X_n$, in Dwork's theory the extra indeterminate arises naturally, as we are about to see.

LEMMA 4. *Let $f \in \mathbb{F}_q[X_1, \ldots, X_n]$ and let $N_k^*$ denote the number of solutions to $f = 0$ in the affine torus $(\mathbb{F}_{q^k}^*)^n$. Then*

$$\sum_{x \in (\mathbb{F}_{q^k}^*)^{n+1}} \Psi_k(x_0 f(x_1, \ldots, x_n)) = q^k N_k^* - (q^k - 1)^n,$$

*where $x = (x_0, x_1, \ldots, x_n)$.*

PROOF. For any $u \in \mathbb{F}_{q^k}$,

$$\sum_{x_0 \in \mathbb{F}_{q^k}} \Psi_k(x_0 u) = \begin{cases} 0 & \text{if } u \in \mathbb{F}_{q^k}^*, \\ q^k & \text{if } u = 0. \end{cases}$$

This is a standard result from the theory of additive character sums [Lidl and Niederreiter 1986, p. 168]. Thus $\sum \Psi_k(x_0 f(x_1, \ldots, x_n))$, where the sum is taken over points $x \in \mathbb{F}_{q^k} \times (\mathbb{F}_{q^k}^*)^n$, equals $q^k N_k^*$. Removing the contribution of $(q^k - 1)^n$ from the terms with $x_0 = 0$ in this sum gives the required result. □

Our next step will be to find an alternative formula for the left-hand side of the equation in Lemma 4. This is achieved in Proposition 11, which leads us eventually to Dwork's trace formula (Theorem 26).

## 3. $p$-adic theory

### 3.1. $p$-adic rings.

We first introduce notation for the $p$-adic rings we shall need, before explaining how to construct them and compute in them. Let $\mathbb{Q}_p$ be the field of $p$-adic rationals, and $\mathbb{Z}_p$ the ring of $p$-adic integers (see [Koblitz 1984]). Denote by $\Omega$ the completion of an algebraic closure of $\mathbb{Q}_p$. Select $\pi \in \Omega$ with $\pi^{p-1} = -p$ and define $R_1 = \mathbb{Z}_p[\pi]$, a totally ramified extension of $\mathbb{Z}_p$ of degree $p - 1$. By binomial expansion and Hensel's lifting lemma, one sees that the equation $(1 + \pi t)^p = 1$ has exactly $p$ distinct solutions $t$ in $R_1$. In particular, $R_1$ contains all $p$-th roots of unity. The motivation behind the introduction of $R_1$ is that to define an additive character of order $p$, we need a small $p$-adic ring which contains a primitive $p$-th root of unity.

Let $R_0$ denote the ring of integers of the unique unramified extension of $\mathbb{Q}_p$ in $\Omega$ of degree $a$, where $q = p^a$. Finally, let $R$ be the compositum ring of $R_1$ and $R_0$. We have the diagram of ring extensions

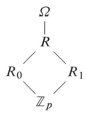

The residue class ring of $R_0$ is $\mathbb{F}_q$, and we shall "lift" the coefficients of the polynomial $f \in \mathbb{F}_q[X_1, \ldots, X_n]$ to this characteristic zero ring, using the Teichmüller lifting. Precisely, the Teichmüller lift $\omega(x)$ of a nonzero element $x \in \bar{\mathbb{F}}_q$ is defined as the unique root of unity in the maximal unramified extension of $\mathbb{Q}_p$ which is congruent to $x$ modulo $p$ and has order coprime to $p$. We define $\omega(0) = 0$. Thus the compositum $R$ contains both the lifting of the coefficients of $f$ and the image of an additive character we shall construct.

## 3.2. Algorithmic aspects

### 3.2.1. Construction and lifting Frobenius.
We assume that $\mathbb{F}_q$ is presented as the quotient $\mathbb{F}_p[y]/(h)$, where $h(y)$ is a monic, irreducible polynomial of degree $a$ over the prime field $\mathbb{F}_p$. For any positive integer $N$ we describe how the quotient ring $R/(p^N)$ may be constructed: Lift the polynomial $h$ to an integer polynomial $\hat{h}$ whose coefficients lie in the open interval $(-p/2, (p+1)/2)$. Take the unramified extension $R_0$ to be $\mathbb{Z}_p[\mu] = \mathbb{Z}_p[y]/(\hat{h})$. Elements in $R_0/(p^N)$ can now be represented as linear combinations over $\mathbb{Z}_p/(p^N)$ of the basis elements $1, \mu, \ldots, \mu^{a-1}$. (Recall that $\mathbb{Z}_p/(p^N)$ can be identified with $\mathbb{Z}/(p^N)$.) The extension $R/(p^N) = R_0[\pi]/(p^N)$ is easily constructed by adjoining an element $\pi$ and specifying the relation $\pi^{p-1} = -p$.

We define a lifting of the Frobenius automorphism on $\mathbb{F}_q$ to an automorphism of $R$ which is the identity on $R_1$. Define the map $\tau : R \to R$ by setting $\tau(\mu)$ to be the unique root of $\hat{h}$ which is congruent to $\mu^p$ modulo $p$. Define $\tau(\pi) = \pi$, and extend to the whole of $R$ by insisting $\tau$ is an automorphism. We extend $\tau$ to act on $R[\![X]\!]$ coefficient-wise, fixing monomials. Here $R[\![X]\!]$ is the ring of all power series in the indeterminates $X = X_0, \ldots, X_n$ with coefficients from $R$.

### 3.2.2. Complexity of arithmetic.
In this section we bound the complexity of the basic arithmetic operations in the ring $R/(p^N)$, along with that of computing the map $\tau$ and Teichmüller lifts. These estimates are all the simplest possible, and can be improved using more advanced methods. The reader may wish to skip the proof of the next lemma, and refer back when required in Section 6.3.2.

LEMMA 5. *Elements in $R/(p^N)$ can be represented using $O(paN \log p)$ bits. Addition and subtraction can be performed in $O(paN \log p)$ bit operations, and multiplication and inversion of units in $O((paN \log p)^2)$ bit operations. The Teichmüller lifting to $R/(p^N)$ of a finite field element can be computed in $O((a \log p)^3 N^2)$ bit operations. For any $1 \leq i \leq a-1$, the map $\tau^i$ on $R/(p^N)$ may be evaluated using $O(p(aN \log p)^2)$ bit operations. (For the powers of $\tau$ we require a total of $O(a^4 N^2 (\log p)^3)$ bits of precomputation.)*

PROOF. Elements in $R/(p^N)$ can be written as

$$\sum_{i=0}^{a-1} \sum_{j=0}^{p-2} c_{ij} \pi^j \mu^i, \tag{3-1}$$

where the coefficients $c_{ij}$ belong to the ring $\mathbb{Z}_p/(p^N)$ of size $p^N$. The bit size of such an expression is $a(p-1)\log(p^N) = O(paN \log p)$. Addition and subtraction are straightforward, just involving the addition of integers and reduction modulo $p^N$. Likewise, multiplication of two expansions of the form (3–1) is straightforward, using the reduction relations $\pi^{p-1} = -p$ and $\hat{h}(\mu) = 0$.

For Teichmüller lifting and inversion we shall use Newton iteration with quadratic convergence. Specifically, we shall use Newton lifting with respect to the prime $\pi$ in $R$, and define $l = (p-1)N$ so that $\pi^l = (-p)^N$. Given a polynomial $\phi(Y) \in R[Y]$ and an element $g_0 \in R$ such that $\phi(g_0) \equiv 0 \mod \pi$ and $\phi'(g_0)$ is invertible (modulo $\pi$) with inverse $s_0$ modulo $\pi$, this algorithm computes an element $g \in R/(\pi^l)$ such that $\phi(g) \equiv 0 \mod \pi^l$ and $g \equiv g_0 \mod \pi$ (compare with [von zur Gathen and Gerhard 1999, Algorithm 9.22]). For $i \geq 1$, assuming that $g_{i-1}$ and $s_{i-1}$ have been found, we define $g_i = g_{i-1} - \phi(g_{i-1})s_{i-1} \mod \pi^{2^i}$ and $s_i = 2s_{i-1} - \phi'(g_i)s_{i-1}^2 \mod \pi^{2^i}$. As in [von zur Gathen and Gerhard 1999, Theorem 9.23] one checks that $g_i \equiv g_0 \mod \pi$, $\phi(g_i) \equiv 0 \mod \pi^{2^i}$ and $s_i \equiv \phi'(g_i)^{-1} \mod \pi^{2^i}$ at the $i$-th step. Thus after $\lceil \log_2(l) \rceil$ steps we shall have found the required approximate root $g = g_{\lceil \log_2(l) \rceil}$. The $i$-th step involves three additions/subtractions and three multiplications in the ring $R/(\pi^{2^i})$, which requires $O((2^i a \log p)^2)$ bit operations, along with evaluation of the polynomials $\phi$ and $\phi'$ modulo $\pi^{2^i}$. Let $c(\phi, i)$ be the complexity of these latter operations. Thus the total complexity is $O((paN \log p)^2 + c(\phi))$ bit operations, where $c(\phi) = \sum_{i=1}^{\lceil \log_2(l) \rceil} c(\phi, i)$. To compute approximate roots in the ring $R_0$ rather than $R$, one can lift using the prime $p$ rather than $\pi$ and obtain a complexity of $O((aN \log p)^2 + c(\phi))$.

Suppose now we are given a unit $x \in R/(p^N)$. We compute a Newton lifting starting from an element $g_0 \in R/(\pi)$ such that $xg_0 \equiv 1 \mod \pi$; that is, we use the equation $\phi(Y) = xY - 1$. Now $R/(\pi)$ is just the finite field $\mathbb{F}_q$ and so an inverse $g_0$ of $x$ modulo $\pi$ can be computed in $O((a \log p)^2)$ bit operations [von zur Gathen and Gerhard 1999, Corollary 4.6]. (Here $\phi'(Y) = x$ and $s_0 = g_0$, and so in fact we only need to iterate the formula for $s_i$.) In this case $c(\phi, i)$ is just one multiplication and a subtraction modulo $\pi^{2^i}$. By the above paragraph the total complexity is then $O((paN \log p)^2)$ for inversion. For Teichmüller lifts we use the same approach, only with the polynomial $\phi(Y) = Y^{q-1} - 1$ and lifting in $R_0$ via the element $p$. Here using a fast exponentiation routine we find that $c(\phi, i)$ involves $O(\log q)$ multiplications and a subtraction in $R_0/(p^{2^i})$. Thus $c(\phi) = O((aN \log p)^2 \log q)$ which gives the Teichmüller lifting estimate.

We precompute $\tau$ on the basis elements $1, \mu, \ldots, \mu^{a-1}$. To do this, recall that $\tau(\mu) \in R_0$ is defined as the unique root of $\hat{h}$ which is congruent modulo $p$ to $\mu^p$. This may be approximated modulo $p^N$ by Newton lifting in $R_0$ with respect to $p$ using the polynomial $\phi(Y) = \hat{h}(Y)$ and the initial value $g_0 = \mu^p$. (Finding $\mu^p$ takes $O(a^2 (\log p)^3)$ bit operations and this is absorbed in the stated precomputation estimate.) Here $c(\phi, i)$ is $O(a)$ additions and multiplications in $R_0/(p^{2^i})$, using Horner's method for polynomial evaluation [von zur Gathen and Gerhard 1999, p. 93]. Thus the Newton lifting estimate gives a complexity of $O(a(aN \log p)^2)$ bit operations. Using $\tau(\mu^i) = (\tau(\mu))^i$ one can now find the image of all basis elements in a further $O(a(aN \log p)^2)$ bit operations. One stores this information as a matrix for $\tau$ acting on $R_0/(p^N)$ as an $\mathbb{Z}_p/(p^N)$-module with basis the powers of $\mu$. The map $\tau$ can now be computed on any element in $R_0/(p^N)$ in $O(a^2(N \log p)^2)$ bit operations using linear algebra. By taking powers of the matrix, matrices for the maps $\tau^i$ for $1 \leq i \leq a-1$ can also be found in $O(a^3 a(N \log p)^2)$ bit operations. Thus the total precomputation is bounded by $O(a^4 N^2 (\log p)^3)$, and each evaluation of $\tau^i$ on $R_0/(p^N)$ takes $O((aN \log p)^2)$ bit operations. Finally, to evaluate $\tau$ on $R/(p^N)$ one writes elements of $R$ on the $R_0$-basis $1, \pi, \ldots, \pi^{p-2}$ and applies $\tau$ component-wise. □

### 3.3. $p$-adic valuations and convergence of power series.
Denote by ord the additive valuation on $\Omega$ normalised so that $\mathrm{ord}(p) = 1$. Thus $\mathrm{ord}(\pi) = 1/(p-1)$. Define a $p$-adic norm $|\cdot|_p$ on $\Omega$ by $|x|_p = p^{-\mathrm{ord}(x)}$. The set of all $x \in \Omega$ with $|x|_p \leq 1$ (equivalently $\mathrm{ord}(x) \geq 0$) is called the closed unit disk. Given any formal power series $\sum_r A_r X^r$, where $A_r \in \Omega$, we say it converges at a point $x = (x_0, \ldots, x_n) \in \Omega^{n+1}$ if the sequence of partial sums $\sum_{r, |r|<e} A_r x^r$ tends to a limit under the $p$-adic norm. (Here $|r| = \sum_{i=0}^n r_i$.) This sequence of partial sums will converge if and only if the summands $A_r x^r$ tend to zero $p$-adically (that is, are divisible in the ring of integers of $\Omega$ by increasingly large powers of $p$), as $|r|$ goes to infinity. In particular, if there is a real number $c > 0$ such that $\mathrm{ord}(A_r) \geq c|r|$ for all $r$, then the series will certainly converge for all points $x$ which are Teichmüller liftings of points over $\bar{\mathbb{F}}_q$. Throughout the paper we shall use the additive valuation ord rather than the $p$-adic norm $|\cdot|_p$ itself.

## 4. Analytic representation of characters

**4.1. Dwork's splitting functions.** We now need to find a suitable $p$-adic expression for a nontrivial additive character from $\mathbb{F}_q$ to $R_1$. In the case of complex characters, this is done via the exponential function. However, the radius of convergence of the exponential function in $\Omega$ is not large enough; in particular, it does not converge on the Teichmüller lifting of all the points in $\mathbb{F}_q$. Instead we use the power series constructed by Dwork using the exponential function. (The reader may find the discussion of the related Artin–Hasse function on [Koblitz 1984, pp. 92–93] helpful.)

Let $\mu$ be the Möbius function. Taking the logarithmic derivative, one checks that the exponential function has the product expansion

$$\exp z = \sum_{k=0}^{\infty} \frac{z^k}{k!} = \prod_{k=1}^{\infty} (1-z^k)^{-\mu(k)/k}.$$

This can be rewritten as

$$\exp z = \prod_{(k,p)=1}^{\infty} (1-z^k)^{-\mu(k)/k}(1-z^{kp})^{\mu(k)/(kp)}.$$

It follows that

$$\exp\left(z + \frac{z^p}{p}\right) = \prod_{(k,p)=1}^{\infty} (1-z^k)^{-\mu(k)/k}(1-z^{kp^2})^{\mu(k)/(kp^2)}. \tag{4-1}$$

Replacing $z$ by $\pi z$ in the above relation and noting that $\pi^p = -p\pi$, we define a power series in $z$ by

$$\theta(z) = \exp(\pi z - \pi z^p).$$

Writing $\theta(z) = \sum_{r=0}^{\infty} \lambda_r z^r$ we see that $\lambda_r = \pi^r/r!$ for $r < p$, and we shall shortly show that all $\lambda_r$ lie in $R_1$. From (4-1) we get the product expansion

$$\theta(z) = \prod_{(k,p)=1}^{\infty} (1-\pi^k z^k)^{-\mu(k)/k}(1-\pi^{kp^2} z^{kp^2})^{\mu(k)/(kp^2)}. \tag{4-2}$$

By the binomial expansion, the first factor

$$(1-\pi^k z^k)^{-\mu(k)/k} = \sum_{j=0}^{\infty} (-1)^j \binom{-\mu(k)/k}{j} \pi^{jk} z^{jk} = \sum_{j=0}^{\infty} b_j(k) z^{kj}$$

is a power series in $z^k$. Now for $(k, p) = 1$, we have, from [Koblitz 1984, p. 82],

$$-\mu(k)/k \in \mathbb{Z}_p, \quad \binom{-\mu(k)/k}{j} \in \mathbb{Z}_p.$$

Thus for $j > 0$ the coefficient of $z^{jk}$ satisfies

$$\operatorname{ord}(b_j(k)) \geq \frac{jk}{p-1} > \frac{p-1}{p^2} jk.$$

Similarly, for the second factor, we write

$$(1 - \pi^{kp^2} z^{kp^2})^{\mu(k)/(kp^2)} = \sum_{j=0}^{\infty} (-1)^j \binom{\mu(k)/kp^2}{j} (\pi z)^{jkp^2}$$

$$= \sum_{j=0}^{\infty} c_j(kp^2) z^{jkp^2}.$$

Now for $j > 0$ we have $\operatorname{ord}(j!) < j/(p-1)$, see [Koblitz 1984, p. 79]. Thus for $j > 0$ and $k$ positive and coprime to $p$, the coefficient of $z^{jkp^2}$ satisfies

$$\operatorname{ord}(c_j(kp^2)) > \frac{jkp^2}{p-1} - 2j - \frac{j}{p-1} \geq \frac{p-1}{p^2} jkp^2.$$

Putting the above two inequalities together, we conclude that for $r > 0$ the coefficients $\lambda_r$ of $\theta(z)$ satisfy

$$\operatorname{ord}(\lambda_r) > \frac{(p-1)r}{p^2}, \quad \lambda_r \in R_1. \tag{4-3}$$

This shows that the power series $\theta(z)$ is convergent in the disk $|z|_p < 1 + \varepsilon$ for some $\varepsilon > 0$. In particular, $\theta(z)$ converges on the closed unit disk, and Definition 6 makes sense.

In the proof of the next lemma we shall use the fact that $\operatorname{ord}(\lambda_r) \geq 2/(p-1)$ for $r \geq 2$, and so

$$\theta(z) \equiv 1 + (\pi z) \bmod (\pi z)^2. \tag{4-4}$$

This can be seen as follows: The Artin–Hasse exponential function [Koblitz 1984, p. 93]

$$E(z) := \prod_{(k,p)=1}^{\infty} (1 - z^k)^{-\mu(k)/k} = \exp\left(z + \frac{z^p}{p} + \frac{z^{p^2}}{p^2} + \cdots\right)$$

has coefficients in $\mathbb{Z}_p$, because each factor in the product expansion does. Since $E(\pi z) \equiv \theta(z) \bmod z^{p^2}$ we see

$$\operatorname{ord}(\lambda_r) \geq \frac{r}{p-1}$$

for $0 \leq r < p^2$. This estimate combined with (4-3) gives (4-4).

DEFINITION 6 (DWORK'S SPLITTING FUNCTION). Let

$$\Phi_k(z) = \prod_{i=0}^{ak-1} \theta(z^{p^i}) \in R_1[\![z]\!],$$

and

$$\Psi_k = \Phi_k \circ \omega : \mathbb{F}_{q^k} \to R_1,$$

where $\omega$ is the Teichmüller map. (Recall that $q = p^a$.)

(That $\Psi_k$ has image in $R_1$ can be seen as follows: Let $R_1[\omega(\mathbb{F}_{q^k})]$ be $R_1$ adjoined the image of $\omega$ on $\mathbb{F}_{q^k}$. Then $R_1[\omega(\mathbb{F}_{q^k})]$ is an unramified extension of $R_1$ of degree $k$. The Galois group of the corresponding quotient field extension is generated by $\tau$. The map $\tau$ acts on a Teichmüller point $\omega(x)$ as $\tau(\omega(x)) = \omega(x)^p$. Hence it fixes the element $\Psi_k(x)$ for $x \in \mathbb{F}_{q^k}$, and so $\Psi_k(x) \in R_1$.)

LEMMA 7. *The maps $\Psi_k$ form a tower of nontrivial additive characters from the fields $\mathbb{F}_{q^k}$ to the ring $R_1$.*

PROOF. (This is the case $s = 1$ on [Dwork 1962, pp. 55–57].) We first show that $\theta(1)$ is a primitive $p$-th root of unity. By (4–4) we see $\theta(1) \neq 1$. As a formal power series in $z$,

$$\theta(z)^p = \exp(p\pi z)\exp(-p\pi z^p).$$

Now, $\theta(z)$, $\exp(p\pi z)$ and $\exp(-p\pi z^p)$ are all convergent in $|z|_p < 1 + \varepsilon$ for some $\varepsilon > 0$. We can thus substitute $z = 1$ and find that

$$\theta(1)^p = \exp(p\pi)\exp(-p\pi) = 1.$$

Thus $\theta(1)$ is a primitive $p$-th root of unity in $R_1$.

Next, for $\gamma \in R_0$ with $\gamma^{p^{ak}} = \gamma$, we claim that

$$\prod_{i=0}^{ak-1} \theta(\gamma^{p^i}) = \theta(1)^{\gamma + \gamma^p + \cdots + \gamma^{p^{ak-1}}}.$$

Using (4–4) it is clear that both sides are congruent to

$$1 + \pi(\gamma + \gamma^p + \cdots + \gamma^{p^{ak-1}})$$

modulo $\pi^2$. To prove the claim, it remains to prove that both sides are $p$-th roots of unity. The right side is a $p$-th root of unity since $\theta(1)$ is a $p$-th root of unity. The $p$-th power of the left side is

$$\prod_{i=0}^{ak-1} \theta(\gamma^{p^i})^p = \exp(p\pi \sum_{i=0}^{ak-1}(\gamma^{p^i} - \gamma^{p^{i+1}})) = \exp(p\pi\gamma)\exp(-p\pi\gamma^{p^{ak}}) = 1.$$

Thus, the left side is also a $p$-th root of unity. The claim is proved. Note that the individual factor $\theta(\gamma)$ is not necessarily a $p$-th root of unity. We conclude that

$$\Psi_k(x) = \theta(1)^{\mathrm{Tr}_k(x)}$$

for any $x \in \mathbb{F}_{q^k}$, where $\mathrm{Tr}_k$ is the trace function from $\mathbb{F}_{q^k}$ to $\mathbb{F}_p$, and the exponent is thought of as an integer. $\square$

NOTE 8. The infinite sum

$$\exp(\pi(z - z^p)) = \sum_{k=0}^{\infty} \frac{(\pi(z - z^p))^k}{k!}$$

is convergent for $|z|_p < 1$, but not necessarily convergent for $|z|_p = 1$. If $|z|_p = 1$, then it is possible that $|z - z^p|_p = 1$ and for such $z$ the above infinite sum does not converge. Since the infinite sum does not converge everywhere on the disk $|z|_p \leq 1$, one cannot simply substitute $z = 1$ into the above infinite sum and get the contradiction that $\theta(1) = 1$. There is no contradiction here!

We now define another power series related to our original polynomial $f$ whose relevance will become apparent in Proposition 11.

DEFINITION 9. Let $f$ be the polynomial whose zeta function we wish to compute, and write

$$X_0 f = \sum_{j \in J} \bar{a}_j X^j,$$

where $J$ is the support of $X_0 f$. Let $a_j$ be the Teichmüller lifting of $\bar{a}_j$. Let $F$ be the formal power series in the indeterminates $X$ with coefficients in $R$ given by

$$F(X) = \prod_{j \in J} \theta(a_j X^j).$$

Let $F^{(a)}(X)$ be the formal power series in the indeterminates $X$ with coefficients in $R$ given by

$$F^{(a)}(X) = \prod_{j \in J} \prod_{s=0}^{a-1} \theta((a_j X^j)^{p^s}).$$

The relation between $F(X)$ and $F^{(a)}(X)$ is clear.

LEMMA 10. *Let the power series $F^{(a)}$ and $F$ be as in Definition 9. Then*

$$F^{(a)}(X) = \prod_{i=0}^{a-1} \tau^i(F(X^{p^i})),$$

*where the map $\tau$ acts coefficient-wise on the power series $F$.*

The power series $F^{(a)}(X)$ relates to rational point counting in the following way.

PROPOSITION 11. *Let* $f \in \mathbb{F}_q[X_1, \ldots, X_n]$ *and let* $F^{(a)} \in R[\![X]\!]$ *be as in Definition* 9. *Then*

$$q^k N_k^* - (q^k - 1)^n = \sum_{x^{q^k-1}=1} F^{(a)}(x) F^{(a)}(x^q) \ldots F^{(a)}(x^{q^{k-1}}),$$

*where* $N_k^*$ *denotes the number of solutions to the equation* $f = 0$ *in the affine torus* $(\mathbb{F}_{q^k}^*)^n$, *and the sum is taken over the Teichmüller lifting of points on the affine torus* $(\mathbb{F}_{q^k}^*)^{n+1}$.

PROOF. For any point $\bar{x}$ in $(\mathbb{F}_{q^k})^{n+1}$ with Teichmüller lifting $x$ we have

$\Psi_k(\bar{x}_0 f(\bar{x}_1, \ldots, \bar{x}_n))$

$$= \Psi_k\left(\sum_{j \in J} \bar{a}_j \bar{x}^j\right) = \prod_{j \in J} \Psi_k(\bar{a}_j \bar{x}^j) = \prod_{j \in J} \Phi_k(a_j x^j)$$

$$= \prod_{j \in J} \prod_{i=0}^{ak-1} \theta((a_j x^j)^{p^i}) = \prod_{j \in J} \prod_{i=0}^{k-1} \prod_{s=0}^{a-1} \theta((a_j x^j)^{q^i p^s})$$

$$= \prod_{i=0}^{k-1} \prod_{j \in J} \prod_{s=0}^{a-1} \theta((a_j^{q^i} x^{jq^i})^{p^s}) = \prod_{i=0}^{k-1} \prod_{j \in J} \prod_{s=0}^{a-1} \theta((a_j (x^{q^i})^j)^{p^s})$$

$$= F^{(a)}(x) F^{(a)}(x^q) \ldots F^{(a)}(x^{q^{k-1}}),$$

where $F^{(a)}(X)$ is given by

$$F^{(a)}(X) = \prod_{j \in J} \prod_{s=0}^{a-1} \theta((a_j X^j)^{p^s}).$$

(We pause to justify the steps above: the first four equalities follow straight from definitions and from the homomorphic property of $\Psi_k$; the fifth and sixth by rearrangement; and the seventh since $a_j$ satisfies $a_j^q = a_j$.)

Thus we have

$$\sum_{\bar{x} \in (\mathbb{F}_{q^k}^*)^{n+1}} \Psi_k(\bar{x}_0 f(\bar{x}_1, \ldots, \bar{x}_n)) = \sum_{x^{q^k-1}=1} F^{(a)}(x) F^{(a)}(x^q) \ldots F^{(a)}(x^{q^{k-1}}),$$

where the latter sum is over the Teichmüller lifting in $\Omega^{n+1}$ of points in $(\mathbb{F}_{q^k}^*)^{n+1}$. Combining this with Lemma 4 gives us the result. □

## 4.2. Decay rates and weight functions.

We now describe the decay rates of the coefficients of the power series $F^{(a)}$ and $F$. Specifically, we obtain lower bounds for the $p$-adic order of the coefficients of the power series $F$ expressed in terms of a certain weight function on integer vectors.

Write $F = \sum_r F_r X^r$, where the sum is over nonnegative integer vectors in $\mathbb{Z}_{\geq 0}^{n+1}$. Let $A$ be the $(n+1) \times |J|$ matrix whose columns are $j = (j_0, \ldots, j_n) \in J$. Then from Definition 9 one sees

$$F_r = \sum_u \prod_{j \in J} \lambda_{u_j} a_j^{u_j}, \qquad (4\text{--}5)$$

where the outer sum is over all $|J|$-tuples $u = (u_j)$ of nonnegative integers such that

$$Au = r, \qquad (4\text{--}6)$$

thinking of $u$ and $r$ as column vectors. Since $j_0 = 1$ for all $(j_0, \ldots, j_n) \in J$, the first row of the matrix $A$ is the vector $(1, 1, \ldots, 1)$. The first equation in the above linear system is then

$$\sum_{j \in J} u_j = r_0. \qquad (4\text{--}7)$$

Now $F_r$ is zero if (4–6) has no solutions. Otherwise, since $\mathrm{ord}(\lambda_{u_j} a_j^{u_j}) = \mathrm{ord}(\lambda_{u_j})$ we get from (4–3), (4–5) and (4–7)

$$\mathrm{ord}(F_r) \geq \inf_u \left\{ \sum_{j \in J} \frac{(p-1)u_j}{p^2} \right\} = \frac{p-1}{p^2} r_0, \qquad (4\text{--}8)$$

where the inf is over all nonnegative integer vector solutions $u$ of (4–6). We now define a weight function $w$ such that (4–8) gives estimates on $\mathrm{ord}(F_r)$ in terms of this weight function.

Let $\delta_1 \subset \mathbb{R}^n$ denote the convex hull of the support of $f$ (the set of exponents of nonzero terms). Let $\delta_2 \subset \mathbb{R}^n$ be the convex hull of the origin and the $n$ points

$$(d, 0, \ldots, 0), \; (0, d, \ldots, 0), \; \ldots, \; (0, \ldots, 0, d),$$

where $d$ is the total degree of $f$. We call $\delta_1$ the Newton polytope of $f$; the polytope $\delta_2$ is just a simplex containing $\delta_1$.

DEFINITION 12. Let $\delta$ be any convex polytope with integer vertices such that $\delta_1 \subseteq \delta \subseteq \delta_2$. Denote by $\Delta$ the convex polytope in $\mathbb{R}^{n+1}$ obtained by embedding $\delta$ in $\mathbb{R}^{n+1}$ via the map $x \mapsto (1, x)$ for $x \in \mathbb{R}^n$, and taking the convex hull with the origin. Denote by $C(\Delta)$ the cone generated in $\mathbb{R}^{n+1}$ as the positive hull of $\Delta$. Thus $C(\Delta)$ is the union of all rays emanating from the origin and passing through $\Delta$.

Ultimately, in Sections 6.3.3 and 6.4, we shall only be interested in the simplest choice of polytope $\delta = \delta_2$ (although the choice $\delta = \delta_1$ leads to the most refined algorithm). Letting $\Delta_1$ denote the polytope in $\mathbb{R}^{n+1}$ obtained by choosing $\delta = \delta_1$ we see that $C(\Delta_1)$ is the cone generated by the exponents of nonzero terms in $X_0 f$. Equation (4–6) has no nonnegative integer (or even real) solutions when $r$ does not lie in $C(\Delta_1)$. Thus for any choice of $\delta (\supseteq \delta_1)$ and corresponding $\Delta (\supseteq \Delta_1)$, all exponents of $F$ lie in the cone $C(\Delta)$.

DEFINITION 13. Define a weight function $w$ from $\mathbb{R}^{n+1}$ to $\mathbb{R} \cup \{\infty\}$ in the following way: For $r = (r_0, r_1, \ldots, r_n) \in \mathbb{R}^{n+1}$ define

$$w(r) = \begin{cases} r_0 & \text{if } r \in C(\Delta), \\ \infty & \text{otherwise.} \end{cases}$$

In particular $w(r)$ is a nonnegative integer for any $r \in C(\Delta) \cap \mathbb{Z}^{n+1}$.

NOTE 14. Choosing $\delta = \delta_1$ corresponds to working with the weight function of Adolphson and Sperber [Adolphson and Sperber 1987], and taking $\delta = \delta_2$ to Dwork's original weight function [Dwork 1960]. Dwork's weight function can equivalently be defined as $w(r) = r_0$ if $r_1 + \cdots + r_n \leq r_0 d$ and $\infty$ otherwise, where $d$ is the total degree of $f$.

The weight function has a simple geometric interpretation: For a real number $c$ define $c\Delta = \{cx \mid x \in \Delta\}$. The next lemma is straightforward.

LEMMA 15. *For any point $r \in C(\Delta)$ we have that $w(r)$ is the smallest nonnegative number $c$ such that $r \in c\Delta$. If $r \notin C(\Delta)$ then $w(r) = \infty$.*

By (4–8) and the sentence preceding Definition 13, we have:

LEMMA 16.
$$\operatorname{ord}(F_r) \geq w(r) \frac{p-1}{p^2}.$$

The proof of the next lemma is straightforward.

LEMMA 17. *Let $r, r' \in \mathbb{R}^{n+1}$ and $k$ a nonnegative integer. Then $w(kr) = kw(r)$ and $w(r + r') \leq w(r) + w(r')$. In particular when $w(r') \neq \infty$,*

$$w(kr - r') \geq kw(r) - w(r').$$

We shall work in certain subrings of $R[\![X]\!]$ defined in terms of the weight function.

DEFINITION 18. Define $L_\Delta$ to be the subring of $R[\![X]\!]$ given by

$$L_\Delta = \left\{ \sum_{r \in C(\Delta) \cap \mathbb{Z}^{n+1}} A_r X^r \,\bigg|\, A_r \in R \right\}.$$

Thus, $L_\Delta$ is just the ring of all power series over $R$ whose terms have exponents in the cone $C(\Delta)$. Certainly $F \in L_\Delta$ and from Lemma 10 we see easily that $F^{(a)} \in L_\Delta$.

LEMMA 19. *The power series $F$ and $F^{(a)}$ belong to $L_\Delta$.*

This concludes all results in this section which shall be essential to the proof of our modular version of Dwork's trace formula. We conclude with a definition and some comments which we will refer to in the analysis of the running time of our algorithm.

DEFINITION 20. For any positive real number $b$, define a set of power series by

$$L_\Delta(b) = \left\{ \sum_r A_r X^r \in L_\Delta \,\middle|\, \mathrm{ord}\, A_r \geq bw(r) \right\}.$$

The set $L_\Delta(b)$ is easily seen to be a subring of $L_\Delta$. Elements in $L_\Delta(b)$ for large $b$ can be thought of as having fast decaying coefficients. Such rings will reduce to rings of small dimension modulo small powers of $p$.

We have

$$F \in L_\Delta\left(\frac{p-1}{p^2}\right), \quad F^{(a)} \in L_\Delta\left(\frac{p-1}{qp}\right). \tag{4-9}$$

The first inequality is immediate from Lemma 16 and the second follows since

$$\tau^i(F(X^{p^i})) \in L_\Delta\left(\frac{p-1}{p^i p^2}\right)$$

for each $0 \leq i \leq a-1$. Thus for $a > 1$ the coefficients of $F^{(a)}$ decay more slowly than those of $F$ itself.

## 5. Dwork's trace formula

### 5.1. Lifting Frobenius.
We now introduce Dwork's "left inverse of Frobenius" mapping $\psi_p$ on the ring $R[\![X]\!]$.

DEFINITION 21. Let $\psi_p$ be defined on the monomials in $R[\![X]\!]$ by

$$\psi_p(X^r) = \begin{cases} X^{r/p} & \text{if } p \mid r, \\ 0 & \text{otherwise}, \end{cases}$$

and extend $\psi_p$ by $\tau^{-1}$-linearity to all of $R[\![X]\!]$. That is,

$$\psi_p\left(\sum_r A_r X^r\right) = \sum_r \tau^{-1}(A_r) \psi_p(X^r) = \sum_r \tau^{-1}(A_{pr}) X^r.$$

Here $p|r$ means that $p$ divides all of the entries in the integer vector $r$. This map is a left inverse of the "Frobenius" map on the ring $R[\![X]\!]$ which takes a power series $\sum_r A_r X^r$ to $\sum_r \tau(A_r) X^{pr}$.

DEFINITION 22. Let $\alpha_a$ be the map from $R[\![X]\!]$ to itself defined as

$$\alpha_a = \psi_p^a \circ F^{(a)}.$$

Precisely, this is the map which is the composition of multiplication by the power series $F^{(a)}$ followed by the mapping $\psi_p^a$ on the ring $R[\![X]\!]$. (Notice that $\psi_p^a$ just acts as

$$\psi_p^a \left( \sum_r A_r X^r \right) = \sum_r A_{qr} X^r$$

since $\tau^{-a}(A_r) = A_r$.) Let the map $\alpha$ from $R[\![X]\!]$ to itself be defined as

$$\alpha = \psi_p \circ F.$$

Thus $\alpha$ is multiplication by $F$ followed by the mapping $\psi_p$.

We have the following result relating these two maps, which shall be of crucial importance in our derivation of an efficient algorithm.

LEMMA 23. *With $\alpha_a$ and $\alpha$ as in Definition 22 we have*

$$\alpha_a = \alpha^a,$$

*where the second exponent is a power under composition.*

PROOF. Firstly let $H \in R[\![X]\!]$ and denote by $\psi_p \circ H(X^p)$ the map composed of multiplication by $H(X^p)$ followed by $\psi_p$. We claim that

$$\psi_p \circ H(X^p) = \tau^{-1}(H(X)) \circ \psi_p. \tag{5-1}$$

To see this write $H = \sum_r H_r X^{pr}$. Then

$$\psi_p \circ H(X^p) = \sum_r \tau^{-1}(H_r)(\psi_p \circ X^{pr}).$$

Here the infinite series is interpreted as a mapping. Now $\psi_p \circ X^{pr} = X^r \circ \psi_p$ as these two maps are $\tau^{-1}$-linear and agree on monomials. Hence we have $\psi_p \circ H(X^p) = \sum_r \tau^{-1}(H_r)(X^r \circ \psi_p) = \tau^{-1}(H(X)) \circ \psi_p$.

Next we claim that for any $b \geq 1$ and power series $H$ we have

$$\psi_p^b \circ \prod_{i=0}^{b-1} \tau^i(H(X^{p^i})) = (\psi_p \circ H(X))^b.$$

We prove this by induction, the result trivially holding if $b = 1$. For $b > 1$, by $b - 1$ applications of (5–1) we get

$$\psi_p^b \circ \prod_{i=0}^{b-1} \tau^i(H(X^{p^i})) = (\psi_p \circ H(X)) \circ \left( \psi_p^{b-1} \circ \prod_{i=0}^{b-2} \tau^i(H(X^{p^i})) \right).$$

The second claim then follows by induction. Putting $H = F$ and $b = a$ we get the required result. □

The map $\alpha_a$ is linear and continuous, in the sense that

$$\alpha_a(\sum_r A_r X^r) = \sum_r A_r \alpha_a(X^r)$$

for any element $\sum_r A_r X^r \in R[\![X]\!]$. The map $\alpha$ is $\tau^{-1}$-linear and continuous, in the sense that

$$\alpha(\sum_r A_r X^r) = \sum_r \tau^{-1}(A_r) \alpha(X^r).$$

From Lemma 19 the next lemma follows easily.

LEMMA 24. *The subring $L_\Delta$ is stable under both $\alpha$ and $\alpha_a$.*

Both maps when restricted to the subring $L_\Delta$ are determined by their action on the monomials $X^r$ (with $w(r) < \infty$), which we now consider.

**5.2. Matrix representations of mappings.** Recall that $C(\Delta)$ is the cone in $\mathbb{R}^{n+1}$ from Definition 12. The set

$$\Gamma_\Delta = \{X^u \mid u \in C(\Delta) \cap \mathbb{Z}_{\geq 0}^{n+1}\}$$

written as a row vector, is a *formal basis* for the space $L_\Delta$. Precisely, this means that any power series in $L_\Delta$ may be written in exactly one way as an infinite sum $\sum_u A_u X^u$ with $X^u$ in the above set. Notice that this is different from the usual notion of a basis in linear algebra, since we allow infinite combinations of basis elements. It is also different from the notion of an "orthonormal basis" in the literature, where one requires that the coefficient $A_u$ goes to zero as $w(u)$ goes to $\infty$ (see [Wan 2000] for a more detailed discussion of these notions).

By Lemma 24, both $\alpha$ and $\alpha_a$ send the ring $L_\Delta$ to itself. We define certain matrices associated to the maps $\alpha$ and $\alpha_a$ restricted to $L_\Delta$ with regard to the formal row basis $\Gamma_\Delta$ of monomials.

DEFINITION 25. Let the infinite matrices $M$ and $M_a$ have columns describing the images of the monomials $X^v \in L_\Delta$ under the maps $\alpha$ and $\alpha_a$ with respect to our formal row basis $\Gamma_\Delta$:

$$\alpha(\Gamma_\Delta) = \Gamma_\Delta M, \quad \alpha_a(\Gamma_\Delta) = \Gamma_\Delta M_a.$$

Specifically, the $(u, v)$-th entries of $M$ and $M_a$ for $u, v \in C(\Delta)$ are $m_{uv}$ and $m_{uv}^{(a)}$, respectively, where

$$m_{uv} = \tau^{-1}(F_{pu-v})$$
$$m_{uv}^{(a)} = F_{qu-v}^{(a)}.$$

Here $F^{(a)} = \sum_r F_r^{(a)} X^r$ and as before $F = \sum_r F_r X^r$, and we take the coefficients of exponents $r$ with negative entries to be zero.

(We have not ordered the basis as yet; however, we shall choose a convenient ordering in the proof of Theorem 28.) We have, by Lemmas 16 and 17

$$\operatorname{ord}(F_{pu-v}) \geq \frac{p-1}{p^2} w(pu - v) \geq \frac{p-1}{p}\left(w(u) - \frac{1}{p} w(v)\right), \tag{5-2}$$

and certainly $\operatorname{ord}(m_{uv}) = \operatorname{ord}(\tau^{-1}(F_{pu-v})) = \operatorname{ord}(F_{pu-v})$.

The matrix powers $M_a^k$ and $M^k$ are defined for every positive integer $k$, since the entries in $M_a^k$, say, are just finite sums of the entries in $M_a$ and $M_a^{k-1}$. This follows for $M_a$, say, since all entries $m_{uv}^{(a)}$ in $M_a$ are zero when the vector $qu - v$ contain negative entries. We define the trace of an infinite matrix to be the sum of its diagonal entries, when this sum converges, and $\infty$ when the sum does not converge. We shall see shortly that the trace of the infinite matrix $M_a^k$ is finite. We write this as $\operatorname{Tr}(M_a^k)$.

THEOREM 26 (DWORK'S TRACE FORMULA). *Let $f \in \mathbb{F}_q[X_1, \ldots, X_n]$ and let $\alpha_a$ be the mapping on the ring $R[\![X]\!]$ given as $\alpha_a = \psi_p^a \circ F^{(a)}$, where $F^{(a)}$ and $\psi_p$ are described in Definitions 9 and 21. Let $M_a$ denote the infinite matrix representing the map $\alpha_a$ restricted to the subring $L_\Delta$ as described in Definition 25. For $k \geq 1$, denote by $N_k^*$ the number of solutions to the equation $f = 0$ in the torus $(\mathbb{F}_{q^k}^*)^n$. Then*

$$(q^k - 1)^{n+1} \operatorname{Tr}(M_a^k) = q^k N_k^* - (q^k - 1)^n.$$

PROOF. (The following is a hybrid of the matrix proof given in [Wan 1996] and the original argument from [Dwork 1960].)

By Proposition 11 we have

$$q^k N_k^* - (q^k - 1)^n = \sum_{x^{q^k-1}=1} F^{(a)}(x) F^{(a)}(x^q) \ldots F^{(a)}(x^{q^{k-1}}),$$

where the sum is over all $(n+1)$-tuples of $(q^k - 1)$st roots of unity in $\Omega$ (namely the Teichmüller lifting in $\Omega^{n+1}$ of points on the torus $(\mathbb{F}_{q^k}^*)^{n+1}$).

We first consider the case $k = 1$. Since $F^{(a)} \in L_\Delta$, we can write $F^{(a)}(X) = \sum_r F_r^{(a)} X^r$, where the sum is over all lattice vectors $r$ which belong to $C(\Delta)$.

Then the latter sum is

$$\sum_{x^{q-1}=1} F^{(a)}(x) = \sum_{x^{q-1}=1} \sum_r F_r^{(a)} x^r = \sum_r F_r^{(a)} \sum_{x^{q-1}=1} x^r = (q-1)^{n+1} \sum_{r,(q-1)|r} F_r^{(a)}$$

$$= (q-1)^{n+1} \sum_s F_{(q-1)s}^{(a)} = (q-1)^{n+1} \mathrm{Tr}(M_a).$$

Here by $(q-1)|r$ we mean that $q-1$ divides every entry in the vector $r$. Also, we use the fact that for any integer $r_i$ [Koblitz 1984, p. 120]

$$\sum_{x_i^{q-1}=1} x_i^{r_i} = \begin{cases} q-1 & \text{if } (q-1)|r_i, \\ 0 & \text{otherwise.} \end{cases}$$

For $k > 1$, define $M_{ak}$ to be the matrix for the map $\psi_p^{ak} \circ \prod_{i=0}^{k-1} F^{(a)}(X^{q^i})$ with respect to the formal basis $\Gamma_\Delta$. Then by an analogous argument to that in the case $k=1$ we see

$$\sum_{x^{q^k-1}=1} \prod_{i=0}^{k-1} F^{(a)}(x^{q^i}) = (q^k-1)^{n+1} \mathrm{Tr}(M_{ak}).$$

Using (5–1) in the proof of Lemma 23 one sees that

$$\psi_p^{ak} \circ \prod_{i=0}^{k-1} F^{(a)}(X^{q^k}) = (\psi_p^a \circ F^{(a)}(X))^k.$$

Since both maps are linear it follows that $M_{ak} = M_a^k$, and the theorem is proved. $\square$

To compute the trace of the matrix in Dwork's formula we shall use the following matrix identity derived from Lemma 23.

LEMMA 27. *Let $M_a$ and $M$ denote the matrices for the maps $\alpha_a$ and $\alpha$ described above. Then*

$$M_a = \prod_{i=0}^{a-1} \tau^{-i}(M) = M M^{\tau^{-1}} \cdots M^{\tau^{-(a-1)}},$$

*where the map $\tau^{-i}$ acts entry wise on the matrix $M$.*

PROOF. First suppose that $N_1$ and $N_2$ are matrices representing maps $\beta_1$ and $\beta_2$ on some subspace of $R[\![X]\!]$. We assume that $\beta_1$ is $\tau^{-1}$-linear and $\beta_2$ is $\tau^{-j}$-linear for some $j \in \mathbb{Z}$. Then it is not difficult to prove that the matrix for the $\tau^{-(j+1)}$-linear map $\beta_1 \circ \beta_2$ is just $N_1 \tau^{-1}(N_2)$.

By Lemma 23 we have $\alpha_a = \alpha^a$. We claim that for any positive integer $b$ the matrix for the map $\alpha^b$ is $\prod_{i=0}^{b-1} \tau^{-i}(M)$. The result is trivially true if $b=1$. For

$b > 1$ we have $\alpha^b = \alpha \circ \alpha^{b-1}$. By induction the matrix for $\alpha^{b-1}$ is $\prod_{i=0}^{b-2} \tau^{-i}(M)$. By $\tau^{-1}$-linearity of $\alpha$ it follows from the observations in the preceding paragraph that the matrix for $\alpha^b$ is $M\tau^{-1}(\prod_{i=0}^{b-2} \tau^{-i}(M)) = \prod_{i=0}^{b-1} \tau^{-i}(M)$. The required result now follows by taking $b = a$. □

We note in passing that $M_a$ in Lemma 27 also equals

$$\tau^{a-1}(S)\tau^{a-2}(S)\ldots S,$$

where $S = \tau(M)$ is the matrix with $(u, v)$-th entry simply $F_{pu-v}$. Although this is slightly more desirable from a practical point of view we shall not use this expression for $M_a$.

### 5.3. Modular reduction of the trace formula.
We now examine the reduction of the Dwork trace formula modulo a power $p^N$ of $p$. Observe that both sides in the trace formula, and all the entries in the matrices $M$ and $M_a$, are elements of $R$, and so this reduction is defined.

We first recall some notation: Let $C(\Delta)$ be the cone in $\mathbb{R}^{n+1}$ from Definition 12, and $L_\Delta$ denote the ring of power series over $R$ whose monomials have exponents lying in $C(\Delta)$ (Definition 18). Let $M$ denote the matrix for the $\tau^{-1}$-linear map $\alpha = \psi_p \circ F$ with respect to the formal row basis $\Gamma_\Delta$ (Definition 25). Here $\psi_p$ is the "left inverse of Frobenius" given in Definition 21 and $F$ is the power series obtained from the polynomial $f$ as in Definition 9.

THEOREM 28. *Let $N$ denote any positive integer and $A_N$ the finite square matrix over the finite ring $R/(p^N)$ obtained by reducing modulo $p^N$ all those entries in $M$ whose rows and columns are indexed by vectors $u \in C(\Delta) \cap \mathbb{Z}_{\geq 0}^{n+1}$ with $w(u) < (p/(p-1))^2 N$. Then*

$$(q^k - 1)^{n+1} \mathrm{Tr}\left(\left(\prod_{i=0}^{a-1} \tau^{-i}(A_N)\right)^k\right) = q^k N_k^* - (q^k - 1)^n \mod p^N,$$

*where $N_k^*$ is the number of solutions to the equation $f = 0$ in the affine torus $(\mathbb{F}_{q^k}^*)^n$. Moreover, the size of $A_N$ is $W = \#(t\Delta)$, where $t = \lceil p^2 N/(p-1)^2 \rceil - 1$ and $\#(t\Delta)$ is number of lattice points in a dilation by a factor $t$ of the polytope $\Delta$.*

PROOF. For any finite or infinite matrix $L$ with coefficients in $R$, we define $\bar{L}$ to be the matrix obtained by reducing all its entries modulo $p^N$. Thus $\bar{L}$ has entries in $R/(p^N)$.

The theorem will follow from the Dwork trace formula (Theorem 26) once we find a suitable expression for the reduction modulo $p^N$ of the trace of the matrix $M_a^k$. This is equal to the trace of the matrix $\overline{M_a^k} = (\overline{M_a})^k$. By Lemma 27 this matrix can be computed as a matrix product from the matrix $\bar{M}$.

By inequality (5–2) every entry $m_{uv} = \tau^{-1}(F_{pu-v})$ in $M$ satisfies

$$\mathrm{ord}(m_{uv}) \geq \frac{p-1}{p}\left(w(u) - \frac{w(v)}{p}\right).$$

Define $t$ to be the greatest integer less than $(p/(p-1))^2 N$. When $w(u) \geq w(v)$ and $w(u) > t$ we have

$$\frac{p-1}{p}\left(w(u) - \frac{w(v)}{p}\right) \geq \frac{(p-1)^2}{p^2} w(u) \geq \frac{(p-1)^2}{p^2}(t+1) \geq N. \quad (5\text{–}3)$$

Recall that we have not yet ordered the basis. Now choose any total ordering on the basis set such that for distinct lattice points $u, v \in C(\Delta)$, the monomial $X^v$ comes before $X^u$ if $w(v) < w(u)$. Thus, by the inequalities in (5–3) and the choice of ordering of the basis, the matrix $\overline{M}$ is of the form

$$\begin{pmatrix} A_N & B_N \\ 0 & C_N \end{pmatrix}, \quad (5\text{–}4)$$

where $C_N$ is a strictly upper triangular infinite matrix and $A_N$ is the finite square matrix indexed by lattice points $u \in C(\Delta)$ such that $w(u) \leq t$. The size of $A_N$ is the number of lattice points $u$ with $w(u) \leq t$. By Lemma 15 this is exactly the number of lattice points in the polytope $t\Delta$.

By Lemma 27 and modular reduction, one has

$$(\overline{M_a})^k = \left(\prod_{i=0}^{a-1} \tau^{-i}(\overline{M})\right)^k.$$

By (5–4) we see that $(\overline{M_a})^k$ is of the form

$$\begin{pmatrix} \left(\prod_{i=0}^{a-1} \tau^{-i}(A_N)\right)^k & B'_N \\ 0 & C'_N \end{pmatrix},$$

where $C'_N = \left(\prod_{i=0}^{a-1} \tau^{-i}(C_N)\right)^k$ is strictly upper triangular.

Hence the trace of $(\overline{M_a})^k$ equals the trace of the finite matrix

$$\left(\prod_{i=0}^{a-1} \tau^{-i}(A_N)\right)^k. \quad (5\text{–}5)$$

The theorem now follows from (5–5) and Theorem 26. □

## 6. Algorithms

In this section we present an algorithm for counting points based upon Theorem 28, and complete the proofs of the results in the introduction. This is a relatively straightforward matter, although the precise complexity estimates require a little care.

### 6.1. Toric point counting algorithm. We first give the algorithm.

ALGORITHM 29 (TORIC POINT COUNTING). Input: Positive integers $a, k, n, d$ and a prime $p$; a polynomial $f$; a polytope $\Delta$. (Here $f$ is a polynomial in $n$ variables of total degree $d$ with coefficients in the field $\mathbb{F}_q$, where $q = p^a$. The polytope $\Delta$ is as in Definition 12. We assume a model of $\mathbb{F}_q$ is given as in Section 3.2.1.)
Output: The number of solutions $N_k^*$ to the equation $f = 0$ in the torus $(\mathbb{F}_{q^k}^*)^n$.

Step 0: Set $N = (n+1)ak$, where $q = p^a$.

Step 1: Compute the polynomial $F$ mod $p^N$ in the ring $(R/(p^N))[X]$, where $F$ is the power series in Definition 9. Specifically, writing $X_0 f = \sum_{j \in J} \bar{a}_j X^j$ we have $F = \prod_{j \in J} \theta(a_j X^j)$ mod $p^N$. Here $a_j$ is the Teichmüller lifting of $\bar{a}_j$ and $\theta(z) = \exp(\pi(z - z^p))$. (See Section 3 for a description of the ring $R/(p^N)$ and the element $\pi$.)

Step 2: Construct the matrix $A_N$ which occurs in the statement of Theorem 28. Specifically, the matrix $A_N$ is indexed by pairs $(u, v)$, where $u$ and $v$ are lattice points in the dilation by a factor $t$ of the polytope $\Delta$, and $t = \lceil p^2 N/(p-1)^2 \rceil - 1$. The $(u, v)$-th entry of $A_N$ is $\tau^{-1}$ of the coefficient of $X^{pu-v}$ in the polynomial $F$ mod $p^N$. The action of $\tau$ is as described in Section 3.

Step 3: Compute the product $\left(\prod_{i=0}^{a-1} \tau^{-i}(A_N)\right)^k$. Let $T$ denote the trace of this product.

Step 4: Output

$$N_k^* = q^{-k}[((q^k - 1)^{n+1} T + (q^k - 1)^n) \bmod p^N],$$

where the square brackets denote the smallest nonnegative residue modulo $p^N$.

In the algorithm we assume that the polynomial is presented as input explicitly via its list of nonzero terms. The manner of presentation of $\Delta$ only affects the time required to find all lattice points in the dilated polytope $t\Delta$. For concreteness, let us say it is presented via its list of vertices, although any other reasonable presentation would suffice.

## 6.2. Proof of correctness of the algorithm.
We know by Theorem 28 that

$$q^k N_k^* - (q^k - 1)^n = (q^k - 1)^{n+1} \operatorname{Tr}\left(\left(\prod_{i=0}^{a-1} \tau^{-i}(A_N)\right)^k\right) \bmod p^N$$

in the ring $R/(p^N)$. Thus,

$$q^k N_k^* = (q^k - 1)^n + (q^k - 1)^{n+1} T \bmod p^N.$$

The left-hand side is a nonnegative integer. The first term on the right-hand side is an integer. The second term on the right-hand side is the reduction modulo $p^N$ of the trace of $M_a^k$, which is known to be an integer by Dwork's trace formula (Theorem 26), and thus it is also an integer. Since $N_k^* \leq (q^k - 1)^n$ it follows that the left-hand side is smaller than $q^{k(n+1)}$. Hence in the case $N \geq ak(n+1)$ we must have

$$q^k N_k^* = [((q^k - 1)^{n+1} T + (q^k - 1)^n) \bmod p^N].$$

The proof is complete.

## 6.3. Complexity analysis.
We shall use big-$O$ and soft-$O$ notation in our analysis of the complexity of the above algorithm. If $C_1$ and $C_2$ are real functions we write $C_1 = O(C_2)$ if $|C_1| \leq c(|C_2| + 1)$ for some positive constant $c$. We write $C_1 = \tilde{O}(C_2)$ if $C_1 = O(C_2 \log(|C_2| + 1)^{c'})$ for some constant $c'$. Thus in the latter notation one ignores logarithmic factors.

### 6.3.1. Ring operations.
We shall first of all count the number of operations in the ring $R/(p^N)$ required in Steps 1, 2, 3, ignoring for the time being any other auxiliary computations. More precisely, because of the complexity bounds in Lemma 5 it is convenient for our analysis to define a "ring operation" to be either arithmetic or the evaluation of the map $\tau^i$, for $1 \leq i \leq a-1$, in the ring $R/(p^N)$ (excluding precomputation). In Section 6.3.2 we shall add back in the small contribution from computing Teichmüller liftings and also the precomputation required for the maps $\tau^i$. Similarly, in Section 6.3.3 we shall account for the remaining operations in Steps 1, 2, 3, arising mainly from computations with the exponents of polynomials (at this stage we will restrict the input polytope $\Delta$ to avoid complications from convex geometry). The contributions from Steps 0 and 4 are easily seen to be absorbed into the other estimates, and we shall not mention them again.

Our running time will be in terms of the parameters $t, \tilde{t}, W, \tilde{W}$. Here

$$t = \left\lceil \left(\frac{p}{p-1}\right)^2 N \right\rceil - 1, \quad W = \#(t\Delta),$$

$$\tilde{t} = \left\lceil \frac{p^2}{p-1} N \right\rceil - 1, \quad \tilde{W} = \#(\tilde{t}\Delta),$$

where $N = ak(n+1)$ and the # operator counts lattice points in convex sets. The sizes of the sets $W$ and $\widetilde{W}$ are the number of lattice points in certain "truncated" cones. Since $\tilde{t}$ is about $p$ times as large as $t$, the integer $\widetilde{W}$ will be around $p^{n+1}$ times as large as $W$, since we are working in $n+1$ dimensional space. The integer $W$ is precisely the size of the matrix which occurs in Step 2. The integer $\widetilde{W}$ will turn out to be the maximum number of terms in the polynomial we compute in Step 1. Define $L_\Delta((p-1)/p^2) \bmod p^N$ to be the ring of polynomials obtained by reducing the coefficients of power series in $L_\Delta((p-1)/p^2) \subseteq R[\![X]\!]$ modulo $p^N$. Then $\widetilde{W}$ is the number of monomials which occur in the finite ring $L_\Delta((p-1)/p^2) \bmod p^N$.

For Step 1 we have the following estimate.

LEMMA 30. *Let $F$ be the power series given in Definition 9 and $N$ a positive integer. The polynomial $F$ mod $p^N$ may be computed in*

$$O(|J|\widetilde{W}^2)$$

*operations in the ring $R/(p^N)$.*

PROOF. By (4–3), $\theta(z) \bmod p^N$ is a polynomial of degree not greater than $p^2 N/(p-1)$. Thus we can obtain $\theta(z)$ via the formula

$$\theta(z) = \exp(\pi z)\exp(-\pi z^p)$$

by computing the first $O(pN)$ terms in the expansion for $\exp(\pi z)$, substituting $z = -z^p$, and one multiplication of polynomials. Note that $\exp(\pi z)$ has $p$-adic integral coefficients. Thus $\theta(z)$ can be found in time $O((pN)^2)$ operations in the ring $R/(p^N)$ using standard polynomial arithmetic.

By the first inequality in (4–9) we have $F \in L_\Delta((p-1)/p^2)$. Thus

$$F \bmod p^N \in L_\Delta((p-1)/p^2) \bmod p^N.$$

One may then compute $F \bmod p^N$ directly from Definition 9 in $|J|-1$ multiplications of polynomials of the form $\theta(a_j X^j) \bmod p^N$. Each such polynomial lies in the ring $L_\Delta((p-1)/p^2) \bmod p^N$, because $\mathrm{ord}(a_j) = 0$, $w(j) = 1$ and the coefficients of $\theta$ decay at a suitable rate. Hence all computations required in computing $F \bmod p^N$ involve polynomials in this ring. Such polynomials have at most $\widetilde{W}$ terms. Exactly $|J|-1$ multiplications are required. Thus the complexity is $O(|J|\widetilde{W}^2)$ ring operations. Noting that $pN = O(\widetilde{W})$ we have the result. □

NOTE 31. It is crucial here that we only need to compute $F \bmod p^N$ and not $F^{(a)} \bmod p^N$, as one might attempt to do using a more naive approach. The latter polynomial has very high degree ($O(q)$) because of the slow decay rate of the coefficients of $F^{(a)}$.

With regard to Step 2, given that the polynomial $F$ mod $p^N$ has already been computed the only task required is to identify those pairs of points $(u, v)$ such that $u, v \in t\Delta$, compute the integer point $pu - v$, and copy $\tau^{-1}$ of the term $F_{pu-v}$ mod $p^N$ from $F$ mod $p^N$ into the correct position in the matrix. Thus no arithmetic operations in the ring are required here, except $W^2$ computations of $\tau^{-1} = \tau^{a-1}$. These arithmetic operations can safely be ignored since in Lemma 30 we have already counted $O(|J|\widetilde{W}^2)$ ring operations. Computation of the appropriate indices $(u, v)$ does require one to find all lattice points in certain polytopes and we return to that in Section 6.3.3.

Finally, for Step 3 we have the following estimate.

LEMMA 32. *With the notation as in the statement of Theorem* 28, *the product*

$$\left(\prod_{i=0}^{a-1} \tau^{-i}(A_N)\right)^k$$

*can be computed given the matrix $A_N$ in*

$$O(W^3 \log(ak))$$

*operations in the ring $R/(p^N)$.*

PROOF. A fast square-and-multiply style algorithm may be used to compute the power $\alpha^a$ in $O(\log a)$ "matrix ring operations". Specifically, working with the matrix representations one may compute $\alpha^{r+s}$ from $\alpha^r$ and $\alpha^s$ using

$$\prod_{i=0}^{r+s-1} \tau^{-i}(A_N) = \prod_{i=0}^{r-1} \tau^{-i}(A_N)\left(\tau^{-r}\left(\prod_{i=0}^{s-1} \tau^{-i}(A_N)\right)\right).$$

Now the case $r = s = 2^c$ for some $c$ gives us the "square" step (computing $\alpha^{2^{c+1}}$ from $\alpha^{2^c}$) and the case $r = 2^c$ and $s$ arbitrary the "multiply" step. These two operations may be combined to give a fast exponentiation method in a straightforward way. The time required to compute a matrix for $\alpha^a$ from one for $\alpha$ is thus $O(\log a)$ "matrix ring operations". By matrix ring operations we mean multiplication of matrices of size $W$ over $R/(p^N)$, and also computing $\tau^{-i}(B)$ for some $1 \leq i \leq a-1$ and matrix $B$ of this form. The former requires $O(W^3)$ operations in the ring $R/(p^N)$ using standard algorithms. Since $\tau^{-i} = \tau^{a-i}$ the latter may be computed in $O(W^2)$ ring operations (that is, applications of a power of $\tau$). Thus the time for computing a matrix for $\alpha^a$ from one for $\alpha$ is $O(W^3 \log a)$.

Having obtained a matrix for $\alpha^a$ one may then compute a matrix for $\alpha^{ak}$ using the standard square-and-multiply algorithm. This requires $O(\log k)$ matrix multiplications, that is $O(W^3 \log k)$ operations in $R/(p^N)$. Thus the total time required is as claimed. □

The exponent 3 for multiplication of matrices can be improved to around 2.4 using faster methods [von zur Gathen and Gerhard 1999, p. 330].

Gathering these results we find the following.

LEMMA 33. *The running time of Algorithm 29 is*

$$O(\widetilde{W}^2|J| + W^3 \log(ak))$$

*ring operations. Here $|J|$ is the number of nonzero terms in $f$, and $\widetilde{W}$ and $W$ are defined as at the start of Section 6.3.1, with $N = ak(n+1)$.*

### 6.3.2. Bit complexity arising from ring operations.

Using Lemma 5 one may now calculate the number of bit operations in the algorithm which arise from operations in the ring $R/(p^N)$. In this section we shall also count the small contribution from computing the Teichmüller lifting of the coefficients of $f$, and also the precomputation required for powers of $\tau$, which it was convenient to ignore in Section 6.3.1,

DEFINITION 34. Let the polytope $\Delta$ from Definition 12 have dimension $\tilde{n} \leq n+1$. Let $V(\Delta)$ be the $\tilde{n}$-dimensional volume of $\Delta$. Denote by $v = \tilde{n}!V(\Delta)$ the "normalised" volume of $\Delta$.

Since $f$ is nonzero we have $\tilde{n} \geq 1$ and certainly $v > 0$.

PROPOSITION 35. *The running time of Algorithm 29 is*

$$\widetilde{O}(a^{3n+7} k^{3n+5} n^{3n+5} v^3 p^{2n+4})$$

*bit operations plus the contribution from operations outside of the ring $R/(p^N)$. The space complexity in bits is*

$$\widetilde{O}(a^{2n+4} k^{2n+3} n^{2n+3} v^2 p)$$

*plus the contribution from operations outside of $R/(p^N)$.*

PROOF. To compute the complexity first observe that $t = \lceil p^2 N/(p-1)^2 \rceil - 1 \leq 4N$. Thus $t, N = O(akn)$ in the algorithm. Also $\tilde{t} = O(pN) = O(pt)$. Now

$$W = \#(t\Delta) \leq \tilde{n}!V(t\Delta) + \tilde{n} = \tilde{n}!t^{\tilde{n}}V(\Delta) + \tilde{n} = O(vt^{n+1}).$$

Here we have used the Blichfeldt bound $\#(P) \leq m!V(P) + m$ for any $m$-dimensional polytope $P$ (see [Goodman and O'Rourke 1997, p. 144]), the fact $V(t\Delta) = t^{\tilde{n}}V(\Delta)$ since $\Delta$ is $\tilde{n}$-dimensional, and also that $\tilde{n} \leq n+1$. Similarly $\widetilde{W} = \#(\tilde{t}\Delta) = O(v\tilde{t}^{n+1}) = O(vt^{n+1}p^{n+1})$.

Thus from Lemma 33 the number of ring operations is

$$O((vt^{n+1}p^{n+1})^2(v+n) + (t^{n+1}v)^3 \log(ak))$$

since $|J| \leq \#(\Delta) - 1 \leq v + n$. Thus the bit complexity which arises from ring operations is by Lemma 5

$$O(((vt^{n+1}p^{n+1})^2(v+n) + (t^{n+1}v)^3 \log(ak))(paN \log p)^2).$$

Tidying up and ignoring logarithmic factors we get

$$\tilde{O}(v^3 t^{2n+2} p^{2n+4} a^2 N^2 + v^3 t^{3n+3} p^2 a^2 N^2).$$

The second term is dominant in all factors except $p$. For simplicity we take the estimate of

$$\tilde{O}(v^3 t^{3n+3} p^{2n+4} a^2 N^2).$$

Now putting $t, N = O(ank)$ we get

$$\tilde{O}(v^3 a^{3n+7} k^{3n+5} n^{3n+5} p^{2n+4}).$$

This is the total bit complexity which arises from "ring operations", as defined at the start of Section 6.3.1. There remains the contribution from computing the Teichmüller liftings of the coefficients of $f$ in Step 1, and also the precomputation required for the map $\tau$. By Lemma 5, this is easily seen to be absorbed in the above estimate.

With regard to the space complexity, this is dominated by the space required to store the matrix $A_N$, which is $O(W^2)$ ring elements. Putting

$$W = O(v(ank)^{n+1})$$

and using Lemma 5 gives us the result. $\square$

One may replace $n$ in the exponents in Proposition 35 by $\tilde{n} - 1$; however, this only gives an improvement when the Newton polytope is not full-dimensional.

### 6.3.3. Bit complexity arising from auxiliary operations.
It remains to bound the complexity which arises from operations outside of the ring $R/(p^N)$ in Steps 1 and 2. In Step 1 manipulation of exponents of polynomials will add an extra term $O(\log(pNd))$ to the running time, which can safely be ignored.

In Step 2 one is required to find all lattice points which lie in $t\Delta$, for $t = O(ank)$. The complexity of this step will depend upon the input polytope $\Delta$. For a "general" $\Delta$ one requires methods from computational convex geometry which are not in the spirit of the present exposition. Thus our total bit complexity estimate for Algorithm 29 will just be as in Proposition 35 "plus the contribution from finding all lattice points in $t\Delta$".

At this stage for simplicity we shall restrict to the choice of $\delta = \delta_2$ in Definition 12. Thus we take $\Delta = \Delta_2$ as the convex hull in $\mathbb{R}^{n+1}$ of the origin and the $n+1$ points

$$(1, 0, \ldots, 0), (1, d, 0, \ldots, 0), \ldots, (1, 0, \ldots, 0, d).$$

For this case the required set of lattice points is

$$\{(r_0, r_1, \ldots, r_n) \mid r_1 + \cdots + r_n \le dr_0 \le dt\}$$

and so no computations are required here. Also, now $v$ equals $d^n$ and directly from Proposition 35 we get the following result.

PROPOSITION 36. *Let Algorithm* 29' *be exactly as Algorithm* 29 *only with input restricted to the choice of polytope* $\Delta = \Delta_2$ *described in the preceding paragraph. The total running time of Algorithm* 29' *is*

$$\tilde{O}(a^{3n+7} k^{3n+5} n^{3n+5} d^{3n} p^{2n+4})$$

*bit operations. The total space complexity in bits is*

$$\tilde{O}(a^{2n+4} k^{2n+3} n^{2n+3} d^{2n} p).$$

We shall use this restricted version of Algorithm 29 in the proofs of our main results in the next section.

**6.4. Proofs of the results in the Introduction.** To compute the number of points on the affine variety defined by a polynomial $f$ one simply uses the torus decomposition of $\mathbb{F}_{q^k}^n$. Specifically, for any subset $S \subseteq \{1, 2, \ldots, n\}$ let $G_k^S$ denote the set of points

$$\{(x_1, \ldots, x_n) \mid x_i \in \mathbb{F}_{q^k}, x_i = 0 \iff x \in S\}.$$

Denote by $f^S$ the polynomial obtained from $f$ by setting to zero all indeterminates $X_i$ which occur in $f$ for $i \in S$. Denote by $N_k^S$ the number of solutions of $f^S = 0$ in the torus $G_k^S$ of dimension $n - |S|$. Then $N_k = \sum_S N_k^S$, where the sum is over all subsets of $\{1, 2, \ldots, n\}$. Each number $N_k^S$ can be computed using Algorithm 29'. (If some $f^S$ is identically zero or has degree 0 then $N_k^S = (q^k - 1)^{n-|S|}$ or 0, respectively, and Algorithm 29' is not required!) Thus by $2^n$ applications of this algorithm we obtain $N_k$ as desired.

Now to obtain the whole zeta function $Z(f/\mathbb{F}_q)$ it suffices to count $N_k$ for all $k = 1, \ldots, \deg(r) + \deg(s)$, where

$$Z(f/\mathbb{F}_q)(T) = \frac{r(T)}{s(T)}$$

with $r$ and $s$ coprime polynomials in $1 + T\mathbb{Z}[T]$. More precisely, it is enough to know upper bounds $\deg(r) \le D_1$ and $\deg(s) \le D_2$, and compute $N_k$ for $k = 1, \ldots, D_1 + D_2$. Then use the linear algebra method described prior to [Wan 2008, Corollary 2.8], which we now supplement with further details.

Let $u(T) = 1 + \sum_{i=1}^{D_1} u_i T^i$ and $v(T) = 1 + \sum_{i=1}^{D_2} v_i T^i$ have indeterminate coefficients. Write

$$Z(f/\mathbb{F}_q)(T) = 1 + z_1 T + z_2 T^2 + \cdots.$$

This power series has nonnegative integer coefficients and it can easily be computed modulo $T^{D_1+D_2+1}$ given $N_k$ for $k = 1, \ldots, D_1 + D_2$. The equation

$$v(T) Z(f/\mathbb{F}_q) \equiv u(T) \bmod T^{D_1+D_2+1}$$

defines a linear system $Ax = y$, where $A$ is a known square $D_1 + D_2$ integer matrix and $y$ a known integer column vector. The entries in $A$ and $y$ are just coefficients from the power series $Z(f/\mathbb{F}_q) \bmod T^{D_1+D_2+1}$. Let $b$ be a bound on their bit length. The unknown entries in $x$ are the coefficients of $u$ and $v$. By [Wan 2008] the set of all solutions to this system consists of precisely those vectors $x$ derived by specialising the coefficients of $u(T)$ and $v(T)$ to equal those of $d(T)r(T)$ and $d(T)s(T)$, respectively, for some $d(T) \in 1 + T\mathbb{Z}[T]$ with degree at most $\min(D_1 - \deg(r), D_2 - \deg(s))$. In particular, the system has a unique solution (i.e. $\det(A) \neq 0$) if and only if either $\deg(r) = D_1$ or $\deg(s) = D_2$ (or both). The determinant $\det(A)$ can be computed using the small primes method in [von zur Gathen and Gerhard 1999, Algorithm 5.10] in a number of bit operations bounded by $\tilde{O}((D_1 + D_2)^4 b^2)$ (see [von zur Gathen and Gerhard 1999, Theorem 5.12]). Now assume that $\det(A) \neq 0$ and so the system has a unique solution, namely the unknown vector containing the integer coefficients of $r$ and $s$. Let $B$ be a bound on the bit length of these coefficients. Find the unique solution to the linear system modulo enough small primes which do not divide $\det(A)$, and recover this integer solution using the Chinese remainder theorem. Precisely, work modulo a collection of such primes whose product has bit length greater than $B$. This second step requires $\tilde{O}((D_1 + D_2)^4 B^2)$ bit operations using Gaussian elimination (this can be improved with a Padé approximation algorithm [von zur Gathen and Gerhard 1999, Section 5.9]). Values for $b$ and $B$ may be deduced from the bound $N_k \leq q^{nk}$. Specifically, one may show from this that the absolute values of the reciprocal zeros of $r$ and $s$ are all $\leq q^n$, and so we can take $B = O((D_1 + D_2)n \log q)$. Also, we can take $b = O((D_1 + D_2)^2 n \log q)$. If in the above we find $\det(A) = 0$ then we must have $\deg(r) < D_1$ and $\deg(s) < D_2$. In this case one must first reduce $D_2$, say, and compute determinants until the correct value $D_2 = \deg(s)$ is found (then $\det(A) \neq 0$ and the above method works).

By the refinement of Bombieri's degree bound [1978] from [Adolphson and Sperber 1987, Equation (1.13)], the "total degree" $\deg(r) + \deg(s)$ is bounded by $2^{n+1} 6^{n+1} (n+1)! V(\Delta_1)$, where $\Delta_1$ is the polytope in $\mathbb{R}^{n+1}$ derived from the Newton polytope of $f$ (see the paragraph following Definition 12). Certainly

$(n+1)!V(\Delta_1) \leq d^n$. Hence we may take $D_1, D_2 = 2^{4n+4}d^n$ and so $D_1 + D_2 = 2^{4n+5}d^n$.

THEOREM 37. *Let $f$ be a polynomial in $n$ variables of total degree $d > 0$ over $\mathbb{F}_q$, where $q = p^a$. The full zeta function $Z(f/\mathbb{F}_q)$ can be computed deterministically in*
$$\tilde{O}(2^{13n^2}a^{3n+7}d^{3n^2+9n}p^{2n+4})$$
*bit operations. (Here we use $\tilde{O}$ notation which ignores logarithmic factors, as defined at the start of Section 6.3.)*

PROOF. From Proposition 36 and the torus decomposition method, the bit complexity of computing $N_k$ for $k = 1, \ldots, 2^{4n+5}d^n$ is
$$\tilde{O}\left(\sum_{k=1}^{2^{4n+5}d^n} (a^{3n+7}k^{3n+5}n^{3n+5}d^{3n}p^{2n+4})2^n\right).$$
(The contribution from recovering $Z(f/\mathbb{F}_q)$ from the $N_k$ is absorbed in this estimate.) Tidying up the factor in $n$ we get the claimed result. □

Since we may assume that $d > 1$ we have $2^{n^2} = O(d^{n^2})$, and Theorem 1 now follows.

The proof of Corollary 2 was explained in the introduction, and we finish with some comments on Corollary 3. By Weil's theorem, the zeta function of the smooth projective curve $\tilde{V}$ from Corollary 3 is of the form
$$Z(\tilde{V})(T) = \frac{P(T)}{(1-T)(1-qT)}$$
for some polynomial $P(T)$ whose reciprocal roots have complex absolute value $q^{1/2}$. Since the (possibly singular) affine curve $V$ and the smooth projective curve $\tilde{V}$ differ in only finitely many closed points, we deduce that the zeta function of $V$ is of the form
$$Z(V)(T) = \frac{P(T)Q(T)}{1-qT},$$
where $Q(T)$ is a rational function whose zeros and poles are roots of unity. This zeta function, and in particular the rational function $P(T)Q(T)$, may be computed within the time bound in Corollary 3 by Corollary 2. In terms of the pure weight decomposition [Wan 2008], the polynomial $P(T)$ (respectively, $Q(T)$) is exactly the pure weight 1 (respectively, weight 0) part of the product $P(T)Q(T)$, and can be recovered quickly from $P(T)Q(T)$ via the LLL polynomial factorization algorithm. In our current special case, one can proceed directly without using the LLL-factorization algorithm. By repeatedly removing the common factor of the numerator of $P(T)Q(T)$ with $T^s - 1$ for $\phi(s)$ (Euler

totient function) not greater than the total degree of $P(T)Q(T)$, the desired polynomial $P(T)$ can be recovered. The order of the group of rational points on $\widetilde{V}$ is simply $P(1)$; see [Wan 2008]. Thus for fixed dimension and finite field one can compute the order of the group of rational points on the Jacobian of a smooth projective curve in time polynomial in the degree $d$.

## Acknowledgments

The authors are pleased to thank Colin McDiarmid and Bernd Sturmfels for answering some questions on convex geometry. Many excellent suggestions for improving the paper were made by the anonymous referee, and were incorporated by the authors. They are especially grateful for this help.

## References

[Adleman and Huang 1996] L. M. Adleman and M.-D. A. Huang, "Counting rational points on curves and abelian varieties over finite fields", pp. 1–16 in *Algorithmic number theory* (ANTS-II) (Talence, 1996), edited by H. Cohen, Lecture Notes in Comput. Sci. **1122**, Springer, Berlin, 1996.

[Adolphson and Sperber 1987] A. Adolphson and S. Sperber, "Newton polyhedra and the degree of the $L$-function associated to an exponential sum", *Invent. Math.* **88**:3 (1987), 555–569.

[Bombieri 1978] E. Bombieri, "On exponential sums in finite fields, II", *Invent. Math.* **47**:1 (1978), 29–39.

[Dwork 1960] B. Dwork, "On the rationality of the zeta function of an algebraic variety", *Amer. J. Math.* **82** (1960), 631–648.

[Dwork 1962] B. Dwork, "On the zeta function of a hypersurface", *Inst. Hautes Études Sci. Publ. Math.* no. 12 (1962), 5–68.

[Elkies 1998] N. D. Elkies, "Elliptic and modular curves over finite fields and related computational issues", pp. 21–76 in *Computational perspectives on number theory* (Chicago, IL, 1995), edited by D. A. Buell and J. T. Teitelbaum, AMS/IP Stud. Adv. Math. **7**, Amer. Math. Soc., Providence, RI, 1998.

[von zur Gathen and Gerhard 1999] J. von zur Gathen and J. Gerhard, *Modern computer algebra*, Cambridge University Press, New York, 1999.

[Gaudry and Gürel 2001] P. Gaudry and N. Gürel, "An extension of Kedlaya's algorithm for counting points on superelliptic curves", pp. 480–494 in *Advances in Cryptology - ASIACRYPT 2001*, edited by C. Boyd, Lecture Notes in Comp. Sci. **2248**, Springer, Berlin, 2001.

[Goodman and O'Rourke 1997] J. E. Goodman and J. O'Rourke (editors), *Handbook of discrete and computational geometry*, CRC Press, Boca Raton, FL, 1997.

[Kedlaya 2001] K. S. Kedlaya, "Counting points on hyperelliptic curves using Monsky–Washnitzer cohomology", *J. Ramanujan Math. Soc.* **16**:4 (2001), 323–338.

[Koblitz 1984] N. Koblitz, *p-adic numbers, p-adic analysis, and zeta-functions*, 2nd ed., Graduate Texts in Mathematics **58**, Springer, New York, 1984.

[Lauder 2004] A. G. B. Lauder, "Counting solutions to equations in many variables over finite fields", *Found. Comput. Math.* **4**:3 (2004), 221–267.

[Lidl and Niederreiter 1986] R. Lidl and H. Niederreiter, *Introduction to finite fields and their applications*, Cambridge University Press, Cambridge, 1986.

[Pila 1990] J. Pila, "Frobenius maps of abelian varieties and finding roots of unity in finite fields", *Math. Comp.* **55**:192 (1990), 745–763.

[Poonen 1996] B. Poonen, "Computational aspects of curves of genus at least 2", pp. 283–306 in *Algorithmic number theory* (ANTS-II) (Talence, 1996), edited by H. Cohen, Lecture Notes in Comput. Sci. **1122**, Springer, Berlin, 1996.

[Satoh 2000] T. Satoh, "The canonical lift of an ordinary elliptic curve over a finite field and its point counting", *J. Ramanujan Math. Soc.* **15**:4 (2000), 247–270.

[Schoof 1985] R. Schoof, "Elliptic curves over finite fields and the computation of square roots mod $p$", *Math. Comp.* **44**:170 (1985), 483–494.

[Schoof 1995] R. Schoof, "Counting points on elliptic curves over finite fields", *J. Théor. Nombres Bordeaux* **7**:1 (1995), 219–254.

[Wan 1996] D. Wan, "Meromorphic continuation of $L$-functions of $p$-adic representations", *Ann. of Math.* (2) **143**:3 (1996), 469–498.

[Wan 1999] D. Wan, "Computing zeta functions over finite fields", pp. 131–141 in *Finite fields: theory, applications, and algorithms (Waterloo, ON, 1997)*, Contemp. Math. **225**, Amer. Math. Soc., Providence, RI, 1999.

[Wan 2000] D. Wan, "Rank one case of Dwork's conjecture", *J. Amer. Math. Soc.* **13**:4 (2000), 853–908.

[Wan 2008] D. Wan, "Algorithmic theory of zeta functions over finite fields", pp. 551–578 in *Surveys in algorithmic number theory*, edited by J. P. Buhler and P. Stevenhagen, Math. Sci. Res. Inst. Publ. **44**, Cambridge University Press, New York, 2008.

ALAN G. B. LAUDER
MATHEMATICAL INSTITUTE
OXFORD UNIVERSITY
OXFORD OX1 3QD
UNITED KINGDOM
lauder@maths.ox.ac.uk

DAQING WAN
DEPARTMENT OF MATHEMATICS
UNIVERSITY OF CALIFORNIA
IRVINE, CA 92697-3875
UNITED STATES
dwan@math.uci.edu

# Congruent number problems and their variants

JAAP TOP AND NORIKO YUI

ABSTRACT. The congruent number problem asks if a natural number $n$ can be realized as the area of a right-angled triangle with rational sides. This problem is related to the existence of rational points on some elliptic curve defined over $\mathbb{Q}$. We present a survey on this problem and several variants, with special emphasis on modularity and other arithmetic questions.

## CONTENTS

| | |
|---|---|
| 1. Introduction | 613 |
| 2. Precursors to the congruent number problem | 614 |
| 3. The classical congruent number problem | 615 |
| 4. A generalized congruent number problem | 621 |
| 5. The $2\pi/3$-congruent number problem | 626 |
| 6. The rational cuboid problems | 629 |
| 7. The semi-perfect rational cuboid problem, I | 630 |
| 8. The semi-perfect rational cuboid problem, II | 633 |
| Acknowledgments | 635 |
| References | 635 |

## 1. Introduction

This survey discusses some innocent-looking longstanding unsolved problems: the *congruent number problem*, and the *perfect rational cuboid problem*,

---

*Mathematics Subject Classification:* Primary 14J27, 14J28, 14J15; Secondary 11E25, 11G05, 11G40.

*Keywords:* Congruent numbers, elliptic curves, rational points, rational cuboids, K3 surfaces, Kummer surfaces, elliptic modular surfaces, theta series, modular (cusp) forms.

Yui was partially supported by a Research Grant from Natural Sciences and Engineering Research Council of Canada (NSERC), and by MSRI Berkeley, CRM Barcelona, MPIM Bonn and FIM ETH Zürich.

as applications of algorithmic number theory. These problems are indeed very old; the congruent number problem dates back to the time of the Greeks; and the perfect rational cuboid problem to the time of Euler or earlier. Accordingly, there are a large number of articles attempting to solve the problems with various different approaches, most of which use elementary number theoretic methods.

We take a geometric approach to the problems, by reformulating them as arithmetic questions (e.g., the existence of rational points) on certain curves and surfaces. We first consider the classical congruent number problem and a generalization to arbitrary rational triangles (not necessarily right-angled), and in particular the $2\pi/3$-congruent number problem. The main results here were obtained by Tunnell [1983], Long [2004, §7], and S.-i. Yoshida [2001; 2002]. All these problems are recapitulated as the problem of finding rational points on elliptic curves and/or on elliptic K3 surfaces defined over $\mathbb{Q}$. Next, the "semi-perfect" rational cuboid problem will be discussed. Some results here may be found in an unpublished paper of Beukers and van Geemen [1995], in Ronald van Luijk's master's thesis [van Luijk 2000], and in Narumiya and Shiga's report [2001] on [Beukers and van Geemen 1995]. Again the problems are recapitulated as the problem of finding rational lines and points on certain K3 surfaces (e.g., Kummer surfaces of product type) defined over $\mathbb{Q}$.

The expositions of the problems discussed here involve properties of elliptic K3 surfaces, and modular forms of integral and half-integral weight for some arithmetic subgroups of $PSL_2(\mathbb{Z})$. An extensive list of literature on these topics is included. We have tried to emphasize material which is not readily available elsewhere.

## 2. Precursors to the congruent number problem

The problem of finding all *Pythagorean triples*, i.e., all triples of integers $(a, b, c)$ with $c \neq 0$ and $a^2 + b^2 = c^2$, is easily seen to be equivalent to the problem of finding all pairs $(r, s)$ of rational numbers satisfying $r^2 + s^2 = 1$. There is a well known geometric construction for all such pairs: the equation $x^2 + y^2 = 1$ defines a circle of radius 1 and center $(0, 0)$ in the $(x, y)$-plane. The rational points $(r, s)$ on it arise as intersection points with a line through $(-1, 0)$ having a rational (or infinite) slope. Explicitly, the equation of such a line with slope $t \neq \infty$ is $y = t(x + 1)$ and this yields as second point of intersection

$$(r, s) = \left(\frac{1-t^2}{1+t^2}, \frac{2t}{1+t^2}\right).$$

It follows that there exist infinitely many Pythagorean triples $(a, b, c)$, even with the additional constraint $\gcd(a, b, c) = 1$.

REMARK 2.1. In contrast, the Diophantine equation $X^n + Y^n - Z^n = 0$ of degree $n \geq 3$ has no nontrivial solutions in integers $(a, b, c)$. Here, by a nontrivial solution we mean a triple of integers $(a, b, c)$ with $abc \neq 0$ satisfying the equation. This is the celebrated proof of Fermat's Last Theorem [Wiles 1995; Taylor and Wiles 1995]. More generally, one considers a generalized Fermat equation $X^p + Y^q - Z^r = 0$ where $p, q, r$ are natural numbers, and asks for solutions $(a, b, c) \in \mathbb{Z}^3$ with $abc \neq 0$, $\gcd(a, b, c) = 1$. Darmon and Granville [1995] showed that when $1/p + 1/q + 1/r < 1$, then a generalized Fermat equation has only finitely many such solutions. The interested reader is referred to [Darmon 1997] and [Kraus 1999] for a survey on generalized Fermat equations. Some quite recent developments may be found in [Beukers 1998; Bruin 1999; 2000].

The result of Darmon and Granville is based on an ingenious application of a theorem of Faltings [1983] on the set of solutions of certain Diophantine equations:

THEOREM 2.2. *Let $C$ be a smooth, geometrically irreducible, projective curve defined over $\mathbb{Q}$ of genus at least 2. Then the set $C(\mathbb{Q})$ of rational points is finite.*

A natural question arising from this theorem is, how to *find* all rational points on a curve of genus at least 2. Recently remarkable progress has been made on this problem, using a method of Coleman [1985] and Chabauty [1941]. Chabauty's theorem asserts that if $C$ is a smooth projective curve of genus $g \geq 2$ defined over a number field $K$, and if the Jacobian of $C$ has Mordell–Weil rank $< g$ over $K$, then $C(K)$ is finite. Coleman [1985] gave an effective bound on the cardinality of the set $C(K)$. For instance, for $K = \mathbb{Q}$, the proof of Coleman's Corollary 4.6 readily gives the bound (compare [Joshi and Tzermias 1999]) $\#C(\mathbb{Q}) \leq \#\tilde{C}(\mathbb{F}_p) + 2g - 2$ provided that $p$ is a rational prime $> 2g$ such that $C$ has good reduction $\tilde{C}$ at $p$ (and of course, the Jacobian of $C$ should have rank $< g$). An explicit example where this is used to prove that all solutions to a certain Diophantine equation have been found, is given by Grant [1994]. Nils Bruin discusses in his thesis [1988] techniques which allow one to apply Chabauty's method in situations where the rank is not smaller than the genus.

We will now focus on congruent number problems. This is done in the sections 3, 4 and 5 below where we discuss, respectively, the congruent number problem, a generalized congruent number problem, and the $2\pi/3$-congruent number problem.

## 3. The classical congruent number problem

DEFINITION 3.1. A square-free natural number $n \in \mathbb{N}$ is called a *congruent number* if it occurs as the area of a right-angled triangle with rational length sides. In other words, $n$ is a congruent number if and only if there is a right-angled triangle with rational sides $X, Y, Z \in \mathbb{Q}$ such that

$$X^2 + Y^2 = Z^2, \quad XY = 2n.$$

EXAMPLES. $n = 5$ is a congruent number as there is a rational right-angled triangle with sides $3/2, 20/3$ and $41/6$. Similarly, 41 is a congruent number as there is a rational right-angled triangle with sides $123/20, 40/3$ and $881/60$. Zagier has shown that 157 is a congruent number, since there is a rational right-angled triangle with sides

$$X = \frac{157841 \cdot 4947203 \cdot 526771095761}{2 \cdot 3^2 \cdot 5 \cdot 13 \cdot 17 \cdot 37 \cdot 101 \cdot 17401 \cdot 46997 \cdot 356441},$$

$$Y = \frac{2^2 \cdot 3^2 \cdot 5 \cdot 13 \cdot 17 \cdot 37 \cdot 101 \cdot 157 \cdot 17401 \cdot 46997 \cdot 356441}{157841 \cdot 4947203 \cdot 526771095761},$$

$$Z = \frac{20085078913 \cdot 1185369214457 \cdot 9425458255024420419074801}{2 \; 3^2 \; 5 \; 13 \; 17 \; 37 \; 101 \; 17401 \; 46997 \; 356441 \; 157841 \; 4947203 \; 526771095761}.$$

Determining whether a given square-free natural number is congruent is the *congruent number problem*. It has not yet been solved in general. A wonderful textbook on this subject was written by Koblitz [1993].

REMARK 3.2. The congruent number problem may be formulated equivalently in terms of "squares in arithmetic progressions": a natural number $n$ is a congruent number if and only if the equation: $\gamma^2 - \beta^2 = \beta^2 - \alpha^2 = n$ is solvable in rational numbers $\alpha, \beta, \gamma$. For instance, Fibonacci found a solution for $n = 5$ ($\alpha = 31/12, \beta = 41/12$ and $\gamma = 49/12$). The transition from this problem to the congruent number problem is easy: $\alpha = (Y - X)/2, \beta = Z/2, \gamma = (Y + X)/2$.

We recall the translation of the congruent number problem into arithmetic questions concerning elliptic curves. For the necessary background on elliptic curves, the reader is referred to the text by Bjorn Poonen [2008] in this volume. Let $C_n$ denote the curve with equation $y^2 = x^3 - n^2 x$.

PROPOSITION 3.3. *Let $n$ be a square-free natural number. The following statements are equivalent*:

(i) *$n$ is a congruent number*;
(ii) *the elliptic curve $C_n$ has a rational point $(x, y)$ with $y \neq 0$*;
(iii) *the elliptic curve $C_n$ has infinitely many rational points*;
(iv) *the Mordell–Weil group $C_n(\mathbb{Q})$ has rank $\geq 1$.*

PROOF. (i) $\Rightarrow$ (ii): Suppose that $n$ is a congruent number. Then there is a right-angled triangle with rational sides $X, Y, Z$ and $XY = 2n$. Put $x := (Z/2)^2$ and

$y := Z(X-Y)(X+Y)/8$. This defines a point $(x, y) \in C_n(\mathbb{Q})$ with $y \neq 0$, as is readily verified.

(ii) $\Rightarrow$ (iii): To prove this, one needs to observe that the only nontrivial torsion points $(x, y)$ in $C_n(\mathbb{Q})$ are the ones with $y = 0$. This follows, e.g., by using that for any prime $p$ not dividing $2n$ reduction modulo $p$ injects the torsion subgroup of $C_n(\mathbb{Q})$ into $\tilde{C}_n(\mathbb{F}_p)$ and the latter group has order $p + 1$ for all such $p \equiv 3 \bmod 4$.

(iii) $\Rightarrow$ (iv): This follows immediately from the Mordell–Weil theorem.

(iv) $\Rightarrow$ (i): If the rank of $C_n(\mathbb{Q})$ is positive, then certainly a rational point $(x, y)$ on $C_n$ exists with $y \neq 0$. Put

$$X = \left| \frac{(x+n)(x-n)}{y} \right|, \quad Y = 2n \left| \frac{x}{y} \right|, \quad Z = \left| \frac{x^2+n^2}{y} \right|.$$

Then $X, Y, Z > 0$ and

$$X^2 + Y^2 = Z^2, \quad XY = 2n,$$

so $n$ is a congruent number. $\square$

REMARK 3.4. (1) There is no known algorithm guaranteed to compute the rank of $C_n(\mathbb{Q})$. Nevertheless, Nemenzo [1998] calculated all $n < 42553$ for which $C_n(\mathbb{Q})$ is infinite, hence all congruent numbers below this bound. Similarly, Elkies [1994; 2002] computed that for all natural numbers $n < 10^6$ which are $\equiv 5, 6,$ or $7 \bmod 8$, the group $C_n(\mathbb{Q})$ has positive rank.

(2) The elliptic curve $C_n$ has complex multiplication by the ring $\mathbb{Z}[\sqrt{-1}]$. This means that the endomorphism ring of $C_n$ is isomorphic to $\mathbb{Z}[\sqrt{-1}]$. The $j$-invariant of $C_n$ is $j = 12^3$ and the discriminant of $C_n$ is $\Delta = (2n)^6$.

(3) The elliptic curve $C_n : y^2 = x^3 - n^2 x$ is the quadratic twist of the elliptic curve $C_1 : y^2 = x^3 - x$ by $\sqrt{n}$. In fact, $(x, y) \mapsto (x/n, y/(n\sqrt{n}))$ yields an isomorphism over $\mathbb{Q}(\sqrt{n})$ from $C_n$ to $C_1$.

(4) There are many quite old results on the rank of $C_n(\mathbb{Q})$ for special classes of integers $n$. For instance, Nagell [1929, pp. 16, 17] has a very short and elementary proof of the fact that this rank is zero in case $n = p$ is a prime number $\equiv 3 \bmod 8$. Hence such primes are noncongruent numbers. Nagell also points out that the same technique shows that $1, 2$ and all $n = 2q$ with $q$ a prime $\equiv 5 \bmod 8$ are noncongruent.

(5) On the positive side, Heegner [1952] used the fact that $C_1$ is isogenous to $X_0(32)$ (which is also an elliptic curve) plus the theory of complex multiplication to construct a non-torsion point in $C_1(\mathbb{Q}\sqrt{-2p})$ for an arbitrary prime number $p \equiv 3 \bmod 4$. This implies that the rank of $C_{2p}(\mathbb{Q})$ is positive for such

primes, hence all $n = 2p$ with $p$ prime $\equiv 3 \bmod 4$ are congruent. Heegner's method was later extended by P. Monsky [1984]. For example, he showed that primes $\equiv 5, 7 \bmod 8$ are congruent. Since primes $\equiv 3 \bmod 8$ are noncongruent by Nagell's result mentioned above, this only leaves the primes $\equiv 1 \bmod 8$. Here the situation is still unknown. For instance, 17 is known to be noncongruent and 41 is congruent.

To be able to say more about the rank $r$ of the Mordell–Weil group of $C_n(\mathbb{Q})$, one invokes the conjecture of Birch and Swinnerton-Dyer. For this, one needs the $L$-series of $C_n/\mathbb{Q}$. Recall that $n$ is assumed to be a square-free integer. This $L$-series is for $\mathrm{Re}(s) > 3/2$ defined by

$$L(C_n, s) = \prod_{p \nmid 2n} (1 - a_p p^{-s} + p^{-2s})^{-1}$$

where $a_p := p + 1 - \#\tilde{C}_n(\mathbb{F}_p)$. In fact, put

$$g(q) := \eta(q^4)^2 \eta(q^8)^2 = \sum_{n=1}^{\infty} b_n q^n,$$

with $\eta(q) = q^{1/24} \prod_{n=1}^{\infty}(1 - q^n)$ the Dedekind eta function. Then $g$ is a cusp form of weight 2 for $\Gamma_0(32)$. Define

$$L(g, \chi, s) = \sum_{\substack{m=1, \\ \gcd(m,N)=1}}^{\infty} \chi(m) b_m m^{-s}$$

for $\chi : (\mathbb{Z}/N\mathbb{Z})^* \to \mathbb{C}^*$ any primitive Dirichlet character modulo $N$. Then

$$L(C_n, s) = L(g, \chi_n, s),$$

in which $\chi_n$ is the nontrivial quadratic character corresponding to the extension $\mathbb{Q}(\sqrt{n})/\mathbb{Q}$. It follows from this, that $L(C_n, s)$ extends to an analytic function on all of $\mathbb{C}$.

REMARK 3.5. The modularity theorem of Wiles [1995], Taylor and Wiles [1995], and Breuil, Conrad, Diamond and Taylor [Breuil et al. 2001] shows that analogous statements hold for an arbitrary elliptic curve $E/\mathbb{Q}$: the $L$-series $L(E, s)$ (defined analogously to that of $C_n$, see [Silverman 1986, Appendix C, §16]) equals $L(h, s)$ for some cusp form $h$ of weight 2 for a group $\Gamma_0(N)$. In particular, this implies that $L(E, s)$ has an analytic continuation to the entire complex plane. The conjecture of Birch and Swinnerton-Dyer for elliptic curves over $\mathbb{Q}$ (which was already formulated a long time before this analytic continuation was known to exist) predicts how $L(E, s)$ behaves near $s = 1$.

CONJECTURE [Birch and Swinnerton-Dyer 1965]. *The expansion of $L(E,s)$ at $s = 1$ has the form $L(E,s) = c(s-1)^r +$ higher order terms, with $c \neq 0$ and $r$ the rank of $E(\mathbb{Q})$. In particular, $L(E,1) \neq 0$ if and only if the Mordell–Weil group $E(\mathbb{Q})$ is finite.*

This is in fact a *weak* form of the Birch and Swinnerton-Dyer conjecture, and we will call it the BSD Conjecture. From results in [Breuil et al. 2001; Bump et al. 1990; Coates and Wiles 1977; Gross and Zagier 1986; Kan 2000; Kolyvagin 1988; Murty and Murty 1991; Taylor and Wiles 1995; Wiles 1995], the BSD Conjecture is known to be true if $L(E,s)$ vanishes to order $\leq 1$ at $s=1$. However, the general case is still wide open, and this is in fact one of the seven Millennium Prize Problems announced by the Clay Mathematics Institute with $1 million prizes. A more thorough treatment on modular forms can be found in the article of Stein [2008] in this volume. As a first application to congruent numbers, one can observe (using the sign in the functional equation which relates $L(C_n, s)$ to $L(C_n, 2-s)$; see [Koblitz 1993, p. 84]) that for square-free $n > 0$ the order of vanishing of $L(C_n, s)$ at $s = 1$ is *odd* precisely when $n \equiv 5, 6, 7 \mod 8$. Hence for these $n$ we certainly have that $L(C_n, 1) = 0$, which by the BSD conjecture should imply that all such $n$ are congruent numbers. At present, no proof of this is known, however.

Here is a characterization of congruent numbers due to Tunnell [1983], assuming the validity of the BSD conjecture. We denote by $\mathfrak{S}_k(N)$ the space of cusp forms of weight $k$ with respect to the congruence subgroup $\Gamma_0(N)$. Moreover define

$$f(q) := \sum_{n=1}^{\infty} a_f(n) q^n = q \prod_{n=1}^{\infty} (1-q^{8n})(1-q^{16n}) \sum_{n \in \mathbb{Z}} q^{2n^2}$$

and

$$f'(q) := \sum_{n=1}^{\infty} a_{f'}(n) q^n = q \prod_{n=1}^{\infty} (1-q^{8n})(1-q^{16n}) \sum_{n \in \mathbb{Z}} q^{4n^2}.$$

These are in fact elements of $\mathfrak{S}_{3/2}(128)$ and $\mathfrak{S}_{3/2}(128, \chi_2)$, respectively (compare [Koblitz 1993, Ch. IV] for precise definitions). We owe the formulation used in (v) below to Noam Elkies.

THEOREM 3.6. *Let $n$ be a square-free natural number. Assuming the validity of the BSD Conjecture for $C_n$, the following statements are equivalent:*

(i) *$n$ is a congruent number;*
(ii) *$C_n(\mathbb{Q})$ is infinite;*
(iii) *$L(C_n, 1) = 0$;*
(iv) *For $n$ odd, $a_f(n) = 0$ and for $n$ even, $a_{f'}(n/2) = 0$;*

(v) *If n is odd, then there are as many integer solutions to $2x^2 + y^2 + 8z^2 = n$ with z even as there are with z odd.*
*If n is even, the analogous statement holds for $x^2 + y^2 + 8z^2 = n/2$.*

PROOF. (i) $\iff$ (ii): This is proved in Proposition 3.3.

(ii) $\iff$ (iii): This is the BSD Conjecture for $C_n$. As remarked earlier, the implication (ii) $\Rightarrow$ (iii) is in fact known without assuming any conjectures [Coates and Wiles 1977].

(iii) $\iff$ (iv): This was proved in [Tunnell 1983], based on results by Shimura [1973] and Waldspurger [1981]. Recall that $L(C_n, s)$ equals $L(g, \chi_n, s)$ where $g$ is a cusp form of weight 2 for $\Gamma_0(32)$. Tunnell shows that the modular forms $f$ and $f'$ of weight $3/2$ for $\Gamma_0(128)$ are both related under a correspondence described by Shimura to the modular form $g$. A formula due to Waldspurger [1981] allows him to conclude that $L(C_n, 1)$ vanishes precisely when $a_f(n)$ (resp. $a_{f'}(n/2)$) vanishes. For further details, see [Tunnell 1983] and [Koblitz 1993, IV, §4].

(iv) $\iff$ (v): This follows by expressing $f$ and $f'$ in terms of theta functions (compare [Tunnell 1983]):

$$f(q) = \sum_{x,y,z \in \mathbb{Z}} q^{2x^2+y^2+32z^2} - \frac{1}{2} \sum_{x,y,z \in \mathbb{Z}} q^{2x^2+y^2+8z^2}$$

and

$$f'(q) = \sum_{x,y,z \in \mathbb{Z}} q^{4x^2+y^2+32z^2} - \frac{1}{2} \sum_{x,y,z \in \mathbb{Z}} q^{4x^2+y^2+8z^2}.$$

The translation from the ternary forms used here to the criterion given in (v) is an amusing elementary exercise. □

EXAMPLE. (1) Tunnell showed that for $p$ prime $\equiv 3 \mod 8$, the Fourier coefficient $a_f(p)$ is $\equiv 2 \pmod 4$, hence it is nonzero. This implies, using the full force of Theorem 3.6 (no BSD conjecture is needed here) that such primes are noncongruent, providing a new proof of Nagell's half a century older result.

(2) Since it is relatively easy to calculate the number of representations of a not too large $n$ by the ternary forms mentioned in Theorem 3.6, assuming BSD one can decide whether $n$ is congruent at least for all $n < 10^9$.

REMARK 3.7. In general, one may find forms of weight $3/2$ such as the ones in Theorem 3.6 (iv) as follows. Starting from an arbitrary eigenform of weight 2, check whether it is in the image of the Shimura map using a criterion due to Flicker [1980]. Then finding a form of weight $3/2$ that maps to the given weight 2 form is in principle reduced to a finite amount of computation. This is because the spaces of modular (cusp) forms of fixed weight and level are finite-dimensional. (For instance, the space $\mathfrak{S}_{3/2}(128)$ has dimension 3, and testing

## 4. A generalized congruent number problem

In this section, we will consider a generalized congruent number problem which asks if a natural number $n$ can occur as the area of any rational triangle with some given angle $\theta$. The exposition is partly based on the article by Ling Long [2004, §7]. Classically, a triangle with rational sides and rational area is called a *Heron triangle*. Heron of Alexandria proved almost 2000 years ago that the area $n$ of a triangle with sides $a, b$ and $c$ satisfy $n^2 = s(s-a)(s-b)(s-c)$, where $s = (a+b+c)/2$. Moreover, he provided the example $a = 13, b = 14, c = 15$ which shows that 84 is the area of a Heron triangle. The subject was much studied in the first half of the 17th century, with contributions by famous mathematicians such as François Viète, C.G. Bachet and Frans van Schooten, jr. Basically, they constructed examples of Heron triangles by gluing right-angled triangles along a common side. In the 19th century, many problems concerning Heron triangles were discussed in the British journal *Ladies' Diary*. Dickson [1934, pp. 191–201] mentions numerous results on Heron triangles. A natural number $n$ occurs as the area of a Heron triangle if and only if positive rational numbers $a, b, c$ and a real number $\theta$ with $0 < \theta < \pi$ exist such that

$$a^2 = b^2 + c^2 - 2bc\cos\theta \quad \text{and} \quad 2n = bc\sin\theta.$$

The equations imply that $(\cos\theta, \sin\theta)$ must be a rational point $\neq (\pm 1, 0)$ on the upper half of the unit circle, hence a rational number $t > 0$ exists such that

$$\sin\theta = \frac{2t}{1+t^2} \quad \text{and} \quad \cos\theta = \frac{t^2-1}{t^2+1}.$$

Now fix a rational number $t > 0$.

DEFINITION 4.1. An integer $n$ is called $t$-congruent if positive rational numbers $a, b, c$ exist such that

$$a^2 = b^2 + c^2 - 2bc\frac{t^2-1}{t^2+1} \quad \text{and} \quad 2n = bc\frac{2t}{1+t^2}.$$

The case $t = 1$ corresponds to the classical congruent number problem. The $t$-congruent number problem, which asks whether a given integer is $t$-congruent, can be reformulated as an arithmetic question of certain elliptic curves. Basically this was done using elementary methods by D. N. Lehmer [1899/1900] a century ago. Recently Ling Long constructed a family of elliptic curves corresponding to $t$-congruent numbers.

PROPOSITION 4.2. *Let $t$ be a positive rational number and $n \in \mathbb{N}$. The following statements are equivalent*:

(i) *$n$ is a $t$-congruent number.*
(ii) *Either both $n/t$ and $t^2+1$ are nonzero rational squares, or the elliptic curve $C_{n,t} : y^2 = x(x - n/t)(x + nt)$ has a rational point $(x, y)$ with $y \neq 0$.*

PROOF. (i) $\Rightarrow$ (ii): Suppose that $n$ is a $t$-congruent number. Then there exist positive rational numbers $a, b, c$ satisfying the two equations given in Definition 4.1. The second of these equations can be written as $n/t = bc/(1+t^2)$. Using this, it easily follows that $(x, y) := (a^2/4, (ab^2 - ac^2)/8)$ is a point in $C_{n,t}(\mathbb{Q})$. It satisfies $y \neq 0$, unless we have $b = c$. In the latter case one verifies that $t^2 + 1 = (2b/a)^2$ and $n/t = (a/2)^2$.

(ii) $\Rightarrow$ (i): Suppose first that $n/t$ and $t^2 + 1$ are nonzero rational squares. Then $a = 2\sqrt{n/t}$ and $b = c = \sqrt{n(t^2+1)/t}$ show that $n$ is $t$-congruent. For the other case, if $P = (x, y) \in C_{n,t}(\mathbb{Q})$ with $y \neq 0$ is a rational point on $C_{n,t}$, then

$$a = \left|\frac{x^2 + n^2}{y}\right|, \quad b = \left|\frac{(x+nt)(x-n/t)}{y}\right|, \quad c = n\left|\frac{x(1/t+t)}{y}\right|$$

show that $n$ is $t$-congruent. $\square$

EXAMPLE. Consider $n = 12$, $t = 4/3$. The torsion subgroup of $C_{12,4/3}(\mathbb{Q})$ is given by $C_{12,4/3}(\mathbb{Q})_{tors} \simeq \mathbb{Z}/2\mathbb{Z} \times \mathbb{Z}/4\mathbb{Z}$. Two generators are $(0, 0)$ and $(-6, 30)$. The latter point is not a 2-torsion point, and corresponds to a rational triangle with sides $5, 5, 6$ with area 12. Note that, contrary to the situation for classical congruent numbers, here a torsion point in $C_{n,t}(\mathbb{Q})$ leads to $n$ being $t$-congruent.

Several people have given proofs of the fact that any natural number $n$ occurs as the area of some Heron triangle. A quite elementary proof was obtained by Fine [1976], who in his paper also mentions an even simpler proof by S. and P. Chowla. Also H. Cohen showed a proof of this to Ling Long and Noriko Yui in the summer of 2000. All proofs are in the spirit of F. van Schooten's 17th century work on Heron triangles: consider two positive rational numbers $r, s$, both $\neq 1$. Gluing the two Pythagorean triangles with sides $(2, |r - r^{-1}|, r + r^{-1})$ and $(2, |s - s^{-1}|, s + s^{-1})$ along their common side with length 2 yields a Heron triangle with area $|r - r^{-1}| + |s - s^{-1}|$. It remains to find suitable $r, s$. After first multiplying $n$ by a square (which amounts to scaling a triangle), we may assume $n > 6$. Then $r = 2n/(n-2)$, $s = (n-2)/4$ give area $(n+2)^2/(4n)$.

Scaling the corresponding triangle by a factor $2n/(n+2)$ results in one with area $n$. We have proved:

THEOREM 4.3. *Any square-free natural number $n$ can be realized as a $t$-congruent number for some $t \in \mathbb{Q}_{>0}$.* □

In terms of Proposition 4.2 (considering $n$ as a variable, using $r, s$ as above, and setting $a = 2n(r + r^{-1})/(n+2)$ and $b = 2n(s + s^{-1})/(n+2)$ and $c = 2n(r - r^{-1} + s - s^{-1})/(n+2)$), this can be interpreted as the fact that the curve $C_{n,(n-2)/4}$, given by $y^2 = x(x - 4n/(n-2))(x + n(n-2)/4)$ contains a $\mathbb{Q}(n)$-rational point $(x, y)$ with $y \neq 0$. In fact, $((-n+2)/2, (n^2-4)/4)$ is such a point. It has infinite order in the group $C_{n,(n-2)/4}(\mathbb{Q}(n))$, as follows from the fact that it specializes for $n = 0$ to $(1, -1)$ on the curve given by $y^2 = x^3$. The latter point has infinite order, as follows from [Silverman 1986, III Prop. 2.5].

Long considered $C_{n,t} : y^2 = x(x - n/t)(x + nt)$ as a surface over the $t$-line. Note that geometrically, this defines the same surface for every nonzero $n$: the map $(x, y) \mapsto (x/n, y/(n\sqrt{n}))$ defines an isomorphism from $C_{n,t}$ to $C_{1,t}$. For this reason we will only consider $n = 1$. A general introduction to elliptic surfaces as considered here, may be found in [Shioda 1990].

PROPOSITION 4.4. *Let $C_{1,t} : y^2 = x(x - 1/t)(x + t)$ and let $\Phi_1 : \mathcal{E}_1 \to \mathbb{P}_t^1$ be the smooth minimal model coresponding to $C_{1,t} \to \mathbb{P}_t^1 : (x, y, t) \mapsto t$.*

(a) *$\mathcal{E}_1$ is a singular elliptic K3 surface, and $\Phi_1$ has exactly four singular fibres, which are of Kodaira type $I_2, I_2, I_4^*, I_4^*$, respectively.*
(b) *The Mordell–Weil group of $C_{1,t}$ over $\mathbb{C}(t)$ is isomorphic to $\mathbb{Z}/2\mathbb{Z} \oplus \mathbb{Z}/2\mathbb{Z}$.*
(c) *In the Shioda–Inose classification of singular K3 surfaces, $\mathcal{E}_1$ corresponds to the even positive definite binary quadratic form $2x^2 + 2y^2$.*
(d) *The L-series of $\mathcal{E}_1$ is given by*

$$L(\mathcal{E}_1, s) = \zeta(s-1)^{18} L(\chi_4, s-1)^2 L(f, s)$$

*where $f \in \mathcal{S}_3(\Gamma_0(1024), (2/p))$ is the twist of $\eta(q^4)^6 = q \prod_{n=1}^{\infty}(1 - q^{4n})^6$ by the quadratic character $(\frac{2}{p})$ and $\chi_4$ is the nontrivial Dirichlet character modulo 4.*

PROOF. Since most of the notions used in the statement have not been defined here, we only sketch the proof and meanwhile, explain these notions a bit more.

(a,b) The elliptic surface $C_{1,t}$ is birationally equivalent to the elliptic surface $y^2 = x(x-t)(x+t^3)$ by the birational map $(x, y) \mapsto (x/t^2, y/t^3)$. Using Tate's algorithm [Birch and Kuyk 1975], one can read off the singular fibres of $\Phi_1$ from the latter equation. They occur at $t = \pm\sqrt{-1}, 0, \infty$ and are of type $I_2, I_2, I_4^*$ and $I_4^*$, respectively. From this, one concludes that $\mathcal{E}_1$ has Euler characteristic 24 and hence is an elliptic K3 surface.

The Shioda–Tate formula (see [Shioda 1990]) asserts that the rank of the Néron–Severi group (which is by definition the group of 1-cycles modulo algebraic equivalence) of $\mathscr{E}_1$ equals $2 + \sum_{v \in \Sigma}(m_v - 1) + r$ where $\Sigma$ is the finite set of points $v \in \mathbb{P}^1_t$ such that the fiber of $\Phi_1$ over $v$ is a reducible curve; $m_v$ denotes its number of irreducible components. Moreover, $r$ denotes the rank of the group of sections of $\Phi_1$ (which equals the rank of $C_{1,t}(\mathbb{C}(t))$).

For an elliptic K3 surface in characteristic 0, the rank of the Néron–Severi group is at most 20. By definition, a singular K3 surface is one for which this rank is maximal. In the present case we have

$$2 + \sum_{v \in \Sigma}(m_v - 1) + r = 2 + (2-1) + (2-1) + (9-1) + (9-1) + r = 20 + r \leq 20$$

hence $r = 0$ and $\mathscr{E}_1$ is a singular K3 surface.

The torsion subgroup of $C_{1,t}(\mathbb{C}(t))$ certainly contains a group $\mathbb{Z}/2\mathbb{Z} \oplus \mathbb{Z}/2\mathbb{Z}$. Moreover, the torsion subgroup injects in the torsion subgroup of any fibre of $\Phi_1$. Since $\mathscr{E}_1$ contains a fiber of type $I_4^*$, whose torsion subgroup we read off from the tables in [Birch and Kuyk 1975] to be isomorphic to $\mathbb{Z}/2\mathbb{Z} \oplus \mathbb{Z}/2\mathbb{Z}$, we conclude that the Mordell–Weil group is isomorphic to $\mathbb{Z}/2\mathbb{Z} \oplus \mathbb{Z}/2\mathbb{Z}$.

(c) Shioda and Inose [1977] (see also [Inose 1978] for a very explicit description) have shown that there is a one-to-one correspondence between singular K3 surfaces and $SL_2(\mathbb{Z})$-equivalence classes of positive definite even integral binary quadratic forms. Moreover, under this correspondence the discriminant of a quadratic form (by which we mean the determinant of the associated symmetric matrix) coincides up to sign with the determinant of the corresponding Néron–Severi lattice. In case $r = 0$, this determinant equals the product over the singular fibres of the number of irreducible components with multiplicity one, divided by the square of the order of the torsion subgroup of the Mordell–Weil group. In our case, this yields $2 \cdot 2 \cdot 4 \cdot 4/16 = 4$. The unique equivalence class of forms with this discriminant is that of $2x^2 + 2y^2$.

(d) One way to interpret this assertion is that for every odd prime $p$, the number $\#\widetilde{\mathscr{E}}_1(\mathbb{F}_p)$ of points equals $p^2 + 1 + 18p + \chi_4(p) + (\frac{2}{p})a_p$ where $a_p$ is the coefficient of $q^p$ in $\eta(q^4)^6$. The factor $\zeta(s-1)^{18} L(\chi_4, s-1)^2$ in the assertion refers to the fact that the Néron–Severi group has rank 20 and is generated by 18 rational 1-cycles and two conjugate ones over $\mathbb{Q}(i)$. The fact that the remaining factor (in case of a singular K3 surface) is the $L$-series of some modular form of weight 3 follows from a general result of Livné [1995]. Arguing as in [Stienstra and Beukers 1985] (or alternatively, as in [Livné 1987] and [Peters et al. 1992, §4]), one can verify that this modular form is as given in the proposition. □

The (unique) singular K3 surface corresponding to the form $2x^2+2y^2$ has been studied by many authors, including Vinberg [1983] and Inose [1976].

Next, we consider the family of elliptic curves

$$C_{t,(t-2)/4} : y^2 = x(x - 4t/(t-2))(x + t(t-2)/4),$$

which appeared in the proof of Theorem 4.3.

PROPOSITION 4.5. $C_{t,(t-2)/4}$ *defines a (smooth, relatively minimal) elliptic surface* $\Phi_2 : \mathcal{E}_2 \to \mathbb{P}_t^1$.

(a) $\mathcal{E}_2$ *is a singular elliptic K3 surface, and* $\Phi_2$ *has exactly five singular fibres, which are of Kodaira type* $I_2, I_2, I_4, I_0^*, I_4^*$, *respectively*.

(b) *The Mordell–Weil group of* $C_{t,(t-2)/4}$ *over* $\mathbb{C}(t)$ *is isomorphic to*

$$\mathbb{Z} \oplus \mathbb{Z}/2\mathbb{Z} \oplus \mathbb{Z}/2\mathbb{Z}.$$

(c) *The L-series of* $\mathcal{E}_2$ *is given by* $L(\mathcal{E}_2, s) = \zeta(s-1)^{18} L(\chi_4, s-1)^2 L(f, s)$, *where* $f \in \mathfrak{S}_3(\Gamma_0(20), (-5/p))$.

PROOF. This is quite analogous to the proof of Proposition 4.4. Tate's algorithm shows that there are two $I_2$-fibres at the roots of $t^2 - 4t + 20 = 0$, an $I_4$-fibre at $t = \infty$, an $I_0^*$-fibre at $t = 0$ and an $I_4^*$-fibre at $t = 2$. It follows that $\mathcal{E}_2$ is a K3 surface.

We already saw that $C_{t,\frac{t-2}{4}}(\mathbb{Q}(t))$ contains a point of infinite order. So the Mordell–Weil rank $r$ is $\geq 1$. The Shioda–Tate formula in this case yields

$$19 + 1 \leq 19 + r \leq 20,$$

hence $r = 1$ and the surface is a singular K3. For the torsion part of the Mordell–Weil group of $\mathcal{E}_2$, apply the same argument as in Proposition 4.4. Hence the Mordell–Weil group of $\mathcal{E}_2$ is isomorphic to $\mathbb{Z} \oplus \mathbb{Z}/2\mathbb{Z} \oplus \mathbb{Z}/2\mathbb{Z}$.

The determinant of the Néron–Severi lattice in this case equals the product over all bad fibers of the number of components with multiplicity one, multiplied by the height of a generator of the Mordell–Weil group modulo torsion, and divided by the square of the order of the torsion subgroup of the Mordell–Weil group. This yields $16h(P)$ where $h$ is the height and $P$ a generator. For a given point, this height can be calculated using an algorithm of Shioda [1990, Thm. 8.6]; in the present case it yields the answer $5/4$ which both implies that the point we have is indeed a generator, and that the determinant is 20. Note that there exist precisely two inequivalent forms here: $2x^2 + 10y^2$ and $4x^2 + 4xy + 6y^2$. Which of these corresponds to $\mathcal{E}_2$ can possibly be settled by determining a finite morphism to some Kummer surface.

The statement concerning the $L$-series can be proven analogously to the previous case. Here we find

$$f(q) = q+2q^2-4q^3-4q^4-5q^5-8q^6+4q^7-24q^8-11q^9-10q^{10}+16q^{12}$$
$$+8q^{14}+20q^{15}-16q^{16}-22q^{18}+20q^{20}-16q^{21}-44q^{23}+96q^{24}-100q^{25}$$
$$+152q^{27}-16q^{28}-22q^{29}+40q^{30}+160q^{32}-20q^{35}+O(q^{36}).$$

The form $f(q)$ cannot be written as a product of $\eta$-functions, as follows from [Dummit et al. 1985] where a complete list of weight 3 newforms that can be written in such form is given. □

## 5. The $2\pi/3$-congruent number problem

Fujiwara [1998] and Kan [2000] considered a variant of the congruent number problem, called the $\theta$-congruent number problem. Suppose that there is a triangle with rational sides containing an angle $\theta$. Then $\cos\theta$ is a rational number, so write $\cos\theta = s/r$ with $r, s \in \mathbb{Z}$, $|s| \leq r$, $\gcd(r, s) = 1$. Note that $\sin\theta = \frac{1}{r}\sqrt{r^2-s^2}$, hence the following makes sense.

DEFINITION 5.1. Suppose $\theta$ is a real number with $0 < \theta < \pi$, such that $\cos\theta = s/r$ with $r, s \in \mathbb{Z}$, $|s| \leq r$, $\gcd(r, s) = 1$. A natural number $n$ is called $\theta$-*congruent* if $n\sqrt{r^2-s^2}$ occurs as the area of a triangle with rational sides and an angle $\theta$.

In terms of the cosine rule and a formula for the area of a triangle, using the same notations, $n$ is $\theta$-congruent precisely when positive rational numbers $a, b, c$ exist such that $c^2 = a^2 + b^2 - 2abs/r$ and $2nr = ab$. The $\theta$-congruent number problem is the problem of describing all $\theta$-congruent integers $n$. Our exposition is based on [Yoshida 2001; 2002; Fujiwara 1998; Kan 2000].

REMARK 5.2. The classical congruent number problem is a special case of the $t$-congruent number problem, obtained by taking $t = 1$. The $t$-congruent number problem is a special case of the $\theta$-congruent number problem in the following sense. Write $t = k/\ell$ for integers $\ell \geq k \geq 1$ with $\gcd(k, \ell) = 1$. Take the unique real number $\theta$ such that $\cos\theta = (t^2-1)/(t^2+1)$, $0 < \theta < \pi$. Then $\cos\theta$ is written in lowest terms as $(\ell^2 - k^2)/(\ell^2 + k^2)$ in case one of $k, \ell$ is even, respectively $((\ell^2 - k^2)/2) / ((\ell^2 + k^2)/2)$ in case both $k, \ell$ are odd. So it follows that an integer $n$ is $\theta$-congruent precisely when $k\ell n$ is $t$-congruent (in case both $k, \ell$ are odd), respectively $2k\ell n$ is $\theta$-congruent (in case one of $k, \ell$ is even). In particular, the classical congruent number problem also equals the $\pi/2$-congruent number problem.

PROPOSITION 5.3. *Let $n > 0$ be an integer. The following statements are equivalent.*

(i) $n$ is a $2\pi/3$-congruent number;
(ii) the elliptic curve $C_n : y^2 = x^3 - 2nx^2 - 3n^2 x = x(x+n)(x-3n)$ contains a point $(x, y)$ with $y \neq 0$;
(iii) the Mordell–Weil group $C_n(\mathbb{Q})$ has rank $\geq 1$.

PROOF. (i) $\iff$ (ii): Suppose that $n$ is a $2\pi/3$-congruent number. Then there exist positive rational numbers $a, b, c$ such that $c^2 = a^2 + b^2 + ab$ and $ab = 4n$. Substituting $b = 4n/a$ in the first equality and multiplying by $a^2$ yields $(ca)^2 = a^4 + 4na^2 + 16n^2$. Now put $x = (ca + a^2 + 2n)/2$ and $y = a(ca + a^2 + 2n)/2$. Then $(x, y)$ is a rational point on $C_n$ with $y \neq 0$.

Conversely, if $(x, y)$ is a rational point with $y \neq 0$ on $C_n$, then after possibly changing the sign of $y$ we have that

$$a = \frac{(x+n)(x-3n)}{y} = \frac{y}{x} > 0 \quad \text{and} \quad b = 4n\frac{x}{y} > 0 \quad \text{and} \quad c = \frac{x^2 + 3n^2}{|y|}$$

show that $n$ is a $2\pi/3$-congruent number.

(ii) $\iff$ (iii): Since all 2-torsion on $C_n$ is rational, the torsion subgroup $C_n(\mathbb{Q})_{\text{tors}}$ is isomorphic to a group of the form $(\mathbb{Z}/2\mathbb{Z}) \times (\mathbb{Z}/2M\mathbb{Z})$ for some $M \in \{1, 2, 3, 4\}$ by a theorem of Mazur [1978]. Hence if the torsion subgroup of $C_n(\mathbb{Q})$ is unequal to the 2-torsion subgroup, then a rational torsion point of order 3 or 4 exists. The 3-division polynomial of $C_n$ is $n^4(3(\frac{x}{n})^4 - 8(\frac{x}{n})^3 - 18(\frac{x}{n})^2 - 9)$, which is irreducible over $\mathbb{Q}$. The 4-division polynomial is $4y(x-n)(x+3n)(x^2+3n^2)(x^2-6nx-3n^2)$. All its rational zeroes correspond to points of order 2. Therefore the Mordell–Weil group $C_n(\mathbb{Q})$ has rank $\geq 1$ if and only if $C_n(\mathbb{Q})$ has a rational point $(x, y)$ with $y \neq 0$. $\square$

Similarly, all of the $\pi/3$-congruent numbers can be characterized. The proof is left to the interested reader as an exercise.

PROPOSITION 5.4. *Let $n > 0$ be an integer. Then the following statements are equivalent.*

(i) $n$ is a $\pi/3$-congruent number;
(ii) the elliptic curve $C_{-n} : y^2 = x(x-n)(x+3n)$ has a rational point $(x, y)$ with $y \neq 0$;
(iii) the Mordell–Weil group $C_{-n}(\mathbb{Q})$ has rank $\geq 1$.

REMARK 5.5. (1) The $j$-invariant of the elliptic curves $C_n$ and $C_{-n}$ is $\frac{2^4 13^3}{3^2} \notin \mathbb{Z}$. This implies that they have no complex multiplication. In this respect, these variants are different from the classical congruent number problem; compare Remark 3.4(2).

(2) The elliptic curves $C_n$ and $C_{-n}$ are quadratic twists of $C_1$.

The weak form of the conjecture of Birch and Swinnerton-Dyer for the elliptic curves $C_n$ and $C_{-n}$ plus modularity of $C_1$ allow one to obtain results analogous to the result of Tunnell [1983] for the $\pi/2$-congruent number problem. There is a modular form of weight $3/2$ on some congruence subgroup of $\mathrm{PSL}(2,\mathbb{Z})$ such that the vanishing of the $n$-th coefficient in its Fourier expansion gives a criterion for $2\pi/3$-congruent numbers. A prototypical result is given in the following theorem.

THEOREM 5.6. *Let $n$ be a positive square-free integer such that $n \equiv 1, 7$ or $13$ (mod 24). Assume the validity of the conjecture of Birch and Swinnerton-Dyer for $C_n : y^2 = x(x+n)(x-3n)$. The following statements are equivalent.*

(i) *$n$ is a $2\pi/3$-congruent number;*
(ii) *$C_n(\mathbb{Q})$ has infinitely many rational points;*
(iii) *$L(C_n, 1) = 0$;*
(iv) *$a_f(n) = 0$, where $a_f(n)$ is the $n$-th Fourier coefficient of the modular form $f$ of weight $3/2$ for $\Gamma_0(576)$ defined by*

$$f(q) = \sum_{n=1}^{\infty} b(n) q^n = \sum_{x,y,z \in \mathbb{Z}} q^{Q_1(x,y,z)} - \sum_{x,y,z \in \mathbb{Z}} q^{Q_2(x,y,z)} - G_2,$$

*where*

$$Q_1(x,y,z) = x^2 + 3y^2 + 144z^2, \quad Q_2(x,y,z) = 3x^2 + 9y^2 + 16z^2$$

*and*

$$G_2 = \frac{1}{2} \sum_{n \in \mathbb{Z}} \chi_{-3}(n) n \, q^{n^2} + 4 \sum_{n \in \mathbb{Z}} \chi_{-3}(n) n \, q^{4n^2} + 8 \sum_{n \in \mathbb{Z}} \chi_{-3}(n) n \, q^{16n^2}.$$

*Here $\chi_{-3}$ is the nontrivial Dirichlet character modulo 3.*

PROOF. This is due to Yoshida [2002]. Note in particular that the BSD conjecture is only used in (iii) $\Rightarrow$ (ii).

Also, note that for square-free $n > 1$, the form $G_2$ does not contribute to $a_f(n)$. Hence, assuming BSD, the theorem claims that $n$ is $2\pi/3$-congruent if and only if the number of representations of $n$ by $x^2 + 3y^2 + 144z^2$ equals the number of representations by $3x^2 + 9y^2 + 16z^2$. □

Considering the sign in the functional equation for $L(C_{\pm n}, s)$, the BSD conjecture predicts the following.

PROPOSITION 5.7 [Yoshida 2001]. *Let $n$ be a square-free natural number. Assume the BSD conjecture for $C_n$ and $C_{-n}$.*

(a) *If $n \equiv 5, 9, 10, 15, 17, 19, 21, 22, 23$ (mod 24), then $n$ is $2\pi/3$-congruent.*
(b) *If $-n \equiv 3, 6, 11, 17, 18, 21, 22, 23$ (mod 24), then $n$ is $\pi/3$-congruent.*

As for the classical congruent number problem, so-called Heegner point constructions (compare [Kan 2000]) can be used to show that certain types of numbers are indeed $2\pi/3$-congruent (or $\pi/3$-congruent).

THEOREM 5.8. *If $p$ be a prime such that $p \equiv -1$ (mod 24), then $p, 2p$ and $3p$ are $2\pi/3$-congruent and $p, 2p$ and $6p$ are $\pi/3$-congruent.*

REMARK 5.9. Some negative results may be obtained by showing directly that the rank of $C_{\pm 1}(\mathbb{Q})$ is zero for certain sets of numbers $n$, or alternatively, by checking that $L(C_n, 1)$ or $L(C_{-n}, 1)$ is nonzero.

(a) Let $p$ be a prime such that $p \equiv 7, 13$ (mod 24). Then a direct computation in the spirit of [Silverman 1986, X, Prop. 6.2] reveals that $p$ is not $2\pi/3$-congruent. Alternatively, the $p$-th Fourier coefficient $a_p(f)$ of the appropriate modular form $f$ is given by

$$\#\{(x, y, z) \in \mathbb{Z}^3 \mid x^2 + 3y^2 + 144z^2 = p\}$$
$$- \#\{(x, y, z) \in \mathbb{Z}^3 \mid 3x^2 + 9y^2 + 16z^2 = p\},$$

which is congruent to 4 (mod 8). Hence $a_p(f) \neq 0$, and consequently $p$ is not a $2\pi/3$-congruent number.

(b) A similar argument shows that a prime $p \equiv 5, 7, 19$ (mod 24) is not $\pi/3$-congruent. (See [Goto 2001; 2002; Kan 2000; Yoshida 2001; 2002] for more general $n$ like $2p, 3p, 6p$ or $pq, 2pq$, etc.)

## 6. The rational cuboid problems

DEFINITION 6.1. We say that a cuboid $\mathcal{H}$ is a *perfect* rational cuboid if the sides and the three face diagonals and the body diagonal are all integers.

The existence of a perfect rational cuboid is easily seen to be equivalent to the existence of a rational point with nonzero coordinates on the surface $\mathcal{S}_{\mathcal{H}}$ in $\mathbb{P}^6$ defined by

$$X^2 + Y^2 = P^2, \ Y^2 + Z^2 = Q^2, \ Z^2 + X^2 = R^2, \ X^2 + Y^2 + Z^2 = W^2.$$

REMARK 6.2. The problem of whether a perfect rational cuboid $\mathcal{H}$ exists was known to Euler and is still unsolved. It is shown by Korec [1984] that no perfect rational cuboid $\mathcal{H}$ exists with shortest side $\leq 10^6$. In December 2004, B. Butler improved the search up to smallest side $\leq 2.1 \cdot 10^{10}$, not finding any examples. Ronald van Luijk [van Luijk 2000] showed that the surface $\mathcal{S}_{\mathcal{H}}$ *is of general type*.

If one relaxes the rationality requirement for one of the seven coordinates, then the resulting problem of finding *semi-perfect* rational cuboids turns out to be more tractable. We will take this direction in this survey. From a geometric point of view, this means passing from the surface $\mathcal{S}_{\mathcal{H}}$ to a quotient by some automorphism. We first formulate the problems that we will consider.

PROBLEM 1. Find semi-perfect rational cuboids $\mathcal{H}$ with rational face diagonals $P, Q$ and a rational body diagonal $W$ (dropping the integrality condition for $R$). In other words, find rational points with nonzero coordinates on the surface in $\mathbb{P}^5$ defined by
$$X^2 + Y^2 = P^2, \quad Y^2 + Z^2 = Q^2, \quad X^2 + Y^2 + Z^2 = W^2.$$

EXAMPLE. $(X, Y, Z, P, Q, W) = (104, 153, 672, 185, 680, 697)$ is a solution to Problem 1. In Section 7 below we show that infinitely many solutions exist.

PROBLEM 2. Find semi-perfect rational cuboids $\mathcal{H}$ with rational face diagonals $P, Q, R$ (relaxing the integrality condition for $W$). In other words, find rational points with nonzero coordinates on the surface in $\mathbb{P}^5$ defined by
$$X^2 + Y^2 = P^2, \quad Y^2 + Z^2 = Q^2, \quad Z^2 + X^2 = R^2.$$

EXAMPLE. Some small solutions are $(X, Y, Z, P, Q, R) = (44, 117, 240, 125, 267, 244)$, $(231, 160, 792, 281, 808, 825)$, and $(748, 195, 6336, 773, 6339, 6380)$. We show in Section 8 below that infinitely many such solutions exist.

The "semi-perfect" rational cuboid problems have attracted considerable attention. There are many papers on the problem, such as [Colman 1988]; see the references in [van Luijk 2000].

## 7. The semi-perfect rational cuboid problem, I

In this section we find solutions to Problem 1 following [Narumiya and Shiga 2001; Beukers and van Geemen 1995; van Luijk 2000] and we discuss arithmetic properties (e.g., $L$-series, modularity) of the associated varieties.

Using affine coordinates, Problem 1 becomes:

PROBLEM 1A. Find nonzero rational numbers $x, y, z, q, w$ satisfying
$$x^2 + y^2 = 1, \quad y^2 + z^2 = q^2, \quad 1 + z^2 = w^2. \tag{$*$}$$

The substitutions
$$x = \frac{1-t^2}{1+t^2}, \quad y = \frac{2t}{1+t^2}, \quad z = \frac{2s}{1-s^2}, \quad w = \frac{1+s^2}{1-s^2}, \quad q = \frac{2u}{(1+t^2)(1-s^2)}$$
give a birational map over $\mathbb{Q}$ from the surface defined by $(*)$ to the surface defined by
$$u^2 = (s^2 t^2 + 1)(s^2 + t^2).$$

We now apply a series of birational maps (changes of variables) over $\mathbb{Q}$ in order to bring this equation into a familiar form.

(i) The change of variables $(v_1, t_1, u_1) := (st, t, tu)$ transforms our equation into $u_1^2 = (v_1^2 + 1)(v_1^2 + t_1^4)$.

(ii) The change of variables $(v_2, t_2, u_2) := (v_1, 1/(t_1 - 1), u_1/(t_1 - 1)^2)$ transforms the resulting equation into $u_2^2 = (v_2^2 + 1)(t_2^4 v_2^2 + (t_2 + 1)^4)$.

(iii) In the new variables $(v_3, t_3, u_3) := (v_2, (v_2^2 + 1)t_2, (v_2^2 + 1)u_2)$ this becomes $u_3^2 = t_3^4 + 4t_3^3 + 6(v_3^2 + 1)t_3^2 + 4(v_3^2 + 1)^2 t_3 + (v_3 + 1)^3$.

(iv) Using $(v_4, t_4, u_4) := (v_3, t_3 + 1, u_3)$ one obtains

$$u_4^2 = t_4^4 + 6v_4^2 t_4^2 + 4v_4^2(v_4^2 - 1)t_4 + v_4^2(v_4^4 - v_4^2 + 1).$$

(v) One transforms this quartic into a cubic as explained in [Cassels 1991, p. 35]. Explicitly, with $(x_1, y_1, v_5) := (-2u_4 + 2t_4^2 + 6v_4^2, 4t_4 u_4 - 4t_4^3 - 12 t_4 v_4^2, v_4)$, the equation becomes

$$y_1^2 - 8v_5^2(v_5^2 - 1)y_1 = x_1^3 - 12v_5^2 x_1^2 - 4v_5^2(v_5^4 - 10v_5^2 + 1)x_1.$$

(vi) Next, put $(x_2, y_2, v_6) := (x_1 + 4v_5^2, y_1 - 4v_5^2(v_5^2 - 1), v_5)$. This transforms the equation into $y_2^2 = x_2^3 - 4v_6^2(v_6^2 + 1)^2 x_2$.

(vii) Finally, the change of variables

$$(x_3, y_3, z_3) := \left( \frac{x_2}{2v_6(v_6^2 + 1)}, \frac{y_2}{v_6(v_6^2 + 1)}, 2v_6 \right)$$

gives the equation

$$y_3^2 = z_3(z_3^2 + 4)x_3(x_3^2 - 1).$$

PROPOSITION 7.1. (a) *Take* $C_1 : w_1^2 = x(x^2 - 1)$ *and* $E_2 : w_2^2 = z(z^2 + 4)$, *and let* $\iota : ((x, w_1), (z, w_2)) \mapsto ((x, -w_1), (z, -w_2))$ *be the* $[-1]$*-map on the abelian surface* $C_1 \times E_2$ *and* $\iota'$ *the* $[-1]$*-map on* $C_1 \times C_1$. *Then the algebraic surface* $S$ *given by* $y^2 = x(x^2 - 1)z(z^2 + 4)$ *is birational to the Kummer surface* $\mathrm{Kum}(C_1 \times E_2) = (C_1 \times E_2)/\iota$.

(b) *The elliptic curves* $C_1$ *and* $E_2$ *are 2-isogeneous over* $\mathbb{Q}$. *The Kummer surface* $X := (C_1 \times C_1)/\iota'$, *defined by* $\eta^2 = \xi(\xi^2 - 1)\zeta(\zeta^2 - 1)$, *is a double cover of* $S$.

PROOF. (a) This is clear from the definitions.

(b) The 2-isogeny $C_1 \to E_2$ and its dual isogeny $E_2 \to C$ are well known; see, e.g., [Silverman and Tate 1992, p. 79]. Explicitly, the isogeny from $C_1$ to $E_2$ is given as

$$(x, w_1) \mapsto \left( \frac{w_1^2}{x^2}, \frac{w_1(x^2 + 1)}{x^2} \right).$$

The $2:1$ map from $X = \mathrm{Kum}(C_1 \times C_1)$ to $S$ is then given by

$$x = \xi, \quad z = \frac{\eta^2}{\zeta^2 \xi(\xi^2 - 1)}, \quad y = \frac{\eta(1 + \zeta^2)}{\zeta^2}. \qquad \square$$

REMARK 7.2. B. van Geemen pointed out to us that the surface $S$ is birational to the quartic Fermat surface. Hence it corresponds in the Shioda–Inose classification to the form $8x^2 + 8y^2$. A nice summary of the arithmetic of this surface, including a description of its Néron–Severi group and its $L$-series, is provided in [Pinch and Swinnerton-Dyer 1991]. An explicit birational map from the quartic Fermat surface to $S$ is given in the (hand-written, Japanese) doctoral thesis of Masumi Mizukami, written around 1980.

The above considerations show that to find solutions to Problem 1A, one may construct rational points on the Kummer surface $X = \mathrm{Kum}(C_1 \times C_1)$. To describe such points, first note that a rational point on $X$ lifts to a pair $(P, Q)$ of points in $C_1 \times C_1$, defined over some quadratic extension $K/\mathbb{Q}$. If $\sigma$ denotes conjugation in $K/\mathbb{Q}$, then the image of $(P, Q)$ being rational precisely means that $(\sigma(P), \sigma(Q)) = \pm(P, Q)$. Hence either $P$ and $Q$ are both in $C_1(\mathbb{Q})$ (which means they are points of order 2 on $C_1$), or they are both rational points of infinite order on a quadratic twist $C_n$ of $C_1$. This discussion is summarized as follows.

THEOREM 7.3. *Suppose that $n$ is a nonzero integer, and $(a, b)$ and $(c, d)$ rational points on $C_n$ : $y^2 = x^3 - n^2 x$. Then*

$$\left( \frac{a}{n}, \frac{c}{n}, \frac{bd}{n^3} \right)$$

*is a rational point on the Kummer surface $X$ : $w^2 = x(x^2 - 1)z(z^2 - 1)$. Conversely, every non-trivial rational point on $X$ is obtained like this.* $\square$

Note that the above result links congruent numbers to semi-perfect rational cuboids: from a pair of rational right-angled triangles with area $n$, one can construct a semi-perfect rational cuboid. In general, the problem of describing the set of rational points on a Kummer surface $\mathrm{Kum}(E_1 \times E_2)$ of a product of two elliptic curves has been studied by Kuwata and Wang [1993].

The Kummer surface $X$ of $C_1 \times C_1$ has been studied extensively, e.g., in [Keum and Kondō 2001; Shioda and Inose 1977; Vinberg 1983; Ahlgren et al. 2002]. Here are some of its properties:

THEOREM 7.4. *Let $X$ be the Kummer surface given by $w^2 = x(x^2-1)z(z^2-1)$.*

(a) *$X$ is a singular K3 surface. Its Néron–Severi lattice has discriminant $-16$.*
(b) *$X$ corresponds in the Shioda–Inose classification to the positive definite even binary quadratic form $4x^2 + 4y^2$.*

(c) *The L-series of X is*

$$L(X, s) = \zeta(s-1)^{19} L(\chi_4, s-1) L(f, s),$$

with

$$f(q) = \eta(q^4)^6 \in \mathfrak{S}_3(\Gamma_0(8), (2/\cdot)).$$

## 8. The semi-perfect rational cuboid problem, II

We now consider Problem 2 of finding nonzero integers satisfying

$$X^2 + Y^2 = P^2, \quad Y^2 + Z^2 = Q^2, \quad Z^2 + X^2 = R^2.$$

Euler recorded in 1772 the following parametric solution to this system:

$$X = 8\lambda(\lambda^2 - 1)(\lambda^2 + 1),$$
$$Y = (\lambda^2 - 1)(\lambda^2 - 4\lambda + 1)(\lambda^2 + 4\lambda + 1),$$
$$Z = 2\lambda(\lambda^2 - 3)(3\lambda^2 - 1),$$
$$P = (\lambda^2 - 1)(\lambda^4 + 18\lambda + 1),$$
$$Q = (\lambda^2 + 1)^6,$$
$$R = 2\lambda(5\lambda^4 - 6\lambda^2 + 5).,$$

The system of equations defines a surface $\mathcal{W} \subset \mathbb{P}^5$. Bremner [1988] has shown that $\mathcal{W}$ is birational to the quartic surface given by

$$(X^2 - Y^2)(Z^2 - R^2) = 2YZ(X^2 - R^2).$$

He has produced pencils of elliptic curves on this quartic surface, and some rational curves on such pencils yield new parametric solutions over $\mathbb{Q}$ of degree 8 (different from that of Euler's) to Problem 2. A somewhat similar approach to that of Bremner was taken by Narumiya and Shiga [2001]. We briefly sketch their approach. Using affine coordinates, the surface is given by

$$x^2 + y^2 = 1, \quad y^2 + z^2 = q^2, \quad z^2 + x^2 = r^2.$$

Put $t = y/(x+1)$, so that

$$x = \frac{1-t^2}{1+t^2} \quad \text{and} \quad y = \frac{2t}{1+t^2}.$$

Next, change coordinates to $(t, x_0, x_2, x_3) := (t, z(1+t^2), q(1+t^2), r(1+t^2))$. The surface is now given by

$$x_0^2 + 4t^2 = x_2^2, \quad x_0^2 + (1-t^2)^2 = x_3^2.$$

We will denote this surface, regarded as a family of curves over the $t$-line, by $\mathcal{R}_t$.

LEMMA 8.1. *The family $E_t/\mathbb{Q}$ defined by $y^2 = x(x+4t^2)(x+(1-t^2)^2)$ is birational over $\mathbb{Q}$ to $\mathcal{R}_t$.*

PROOF. Put $M = 4t^2$ and $N = (1-t^2)^2$. Then the map $\mathcal{R}_t \to E_t$ is given by $(x_0, x_2, x_3) \mapsto (x, y)$ with

$$x = MN(x_3 - x_2)/\{(M-N)x_0 + Nx_2 - Mx_3\},$$
$$y = MN(N-M)/\{(M-N)x_0 + Nx_2 - Mx_3\}.$$

The inverse map $E_t \to \mathcal{R}_t$ is given by

$$x_0 = \{y^2 - M(x+N)^2\}/(2(x+N)y),$$
$$x_2 = \{y^2 + M(x+N)^2\}/(2(x+N)y),$$
$$x_3 = \{y^2 + N(x+M)^2\}/(2(x+M)y).$$

The maps are clearly defined over $\mathbb{Q}$. □

Using a similar analysis as given in Section 7, Narumiya and Shiga [2001] showed that over $\mathbb{Q}(\sqrt{2})$, the surface $E_t$ is birational to the Kummer surface associated with a product of two isogenous elliptic curves with CM by $\mathbb{Z}[\sqrt{-2}]$. An explicit rational curve in $E_t$ is then used to show the following.

PROPOSITION 8.2 [Narumiya and Shiga 2001]. *One parametric solution to Problem 2 is given by*

$$X = -2(\lambda^2 - 4\lambda + 5)^2(\lambda^2 - 5\lambda + 5)(\lambda^2 - 5),$$
$$Y = -4\lambda(\lambda - 2)(2\lambda - 5)(\lambda^2 - 4\lambda + 5)(\lambda^2 - 5\lambda + 5),$$
$$Z = \lambda(\lambda - 1)(\lambda - 2)(\lambda - 3)(\lambda - 5)(2\lambda - 5)(3\lambda - 5),$$
$$P = -2(\lambda^2 - 4\lambda + 5)(\lambda^2 - 5\lambda + 5)(\lambda^4 - 4\lambda^3 + 8\lambda^2 - 20\lambda + 25),$$
$$Q = \lambda(\lambda - 2)(2\lambda - 5)(-5\lambda^4 + 48\lambda^3 - 166\lambda^2 + 240\lambda - 125),$$
$$R = 2\lambda^8 - 26\lambda^7 + 14\lambda^6 - 446\lambda^5 + 1066\lambda^4 - 2230\lambda^3 + 3525\lambda^2 - 3250\lambda + 1250.$$

We remark that Narumiya and Shiga's method implies that $E_t$ defines a singular K3 surface and has Mordell–Weil rank 2. One section of infinite order is given by $x = 4t^2$ (defined over $\mathbb{Q}(\sqrt{2})$). This is found using that $E_t$ is a double cover of the rational elliptic surface defined by $y^2 = x(x+4s)(x+(1-s)^2)$. Using the table in [Oguiso and Shioda 1991], the Mordell–Weil group of the latter surface is seen to be isomorphic to $\mathbb{Z} \oplus (\mathbb{Z}/2\mathbb{Z})^2$, with a generator modulo torsion of height $1/4$. It is then easily verified using [Shioda 1990] that $x = 4s$ defines such a generator. A second point on $E_t$ is found by using the complex multiplication acting on the associated Kummer surface.

## Acknowledgments

This text is based on a talk by Yui at the Clay Mathematics Institute Introductory Workshop (August 14–23, 2000) of the MSRI program on algorithmic number theory. Sincere thanks go to CMI for its generous support, and to MSRI for the hospitality. Parts of the article were written at CRM Barcelona, at Max-Planck-Institut für Mathematik Bonn, and at FIM ETH Zürich, during visiting professorships of the second author from January to March, April to May, and June 2001, respectively. She is especially indebted to H. Shiga, N. Narumiya, S.-i. Yoshida, and Ling Long for providing her with their respective papers (see bibliography), Ling Long also checked many calculations and suggested improvements. William Stein carried out calculations on some of the modular forms involved. Several mathematicians have given suggestions, and some have read earlier versions of this paper pointing out some inaccuracies and improvements. These include Joe Buhler, Imin Chen, Henri Cohen, Chuck Doran, Noam Elkies, Bert van Geemen, Yasuhiro Goto, Fernando Gouvêa, Ian Kiming, Kenichiro Kimura, Shigeyuki Kondo, Jyoti Sengupta, Jean-Pierre Serre, Peter Stevenhagen and Don Zagier. Last but not least, we thank Masanobu Kaneko and Takeshi Goto for reading the galley proof and pointing out typos and drawing attention to [Goto 2001; 2002].

## References

[Ahlgren et al. 2002] S. Ahlgren, K. Ono, and D. Penniston, "Zeta functions of an infinite family of $K3$ surfaces", *Amer. J. Math.* **124**:2 (2002), 353–368.

[Basmaji 1996] J. Basmaji, *Ein Algorithmus zur Berechnung von Hecke-Operatoren und Anwendungen auf modulare Kurven*, Thesis, GHS-Essen (now Univ. Duisburg-Essen), 1996.

[Beukers 1998] F. Beukers, "The Diophantine equation $Ax^p + By^q = Cz^r$", *Duke Math. J.* **91**:1 (1998), 61–88.

[Beukers and van Geemen 1995] F. Beukers and B. van Geemen, "Rational cuboids", preprint, Universiteit Utrecht, 1995.

[Birch and Kuyk 1975] B. Birch and W. Kuyk (editors), *Modular functions of one variable, IV*, Lecture notes in mathematics **476**, Springer, Berlin, 1975.

[Birch and Swinnerton-Dyer 1965] B. J. Birch and H. P. F. Swinnerton-Dyer, "Notes on elliptic curves. II", *J. Reine Angew. Math.* **218** (1965), 79–108.

[Bremner 1988] A. Bremner, "The rational cuboid and a quartic surface", *Rocky Mountain J. Math.* **18**:1 (1988), 105–121.

[Breuil et al. 2001] C. Breuil, B. Conrad, F. Diamond, and R. Taylor, "On the modularity of elliptic curves over **Q**: wild 3-adic exercises", *J. Amer. Math. Soc.* **14**:4 (2001), 843–939.

[Bruin 1999] N. Bruin, "The Diophantine equations $x^2 \pm y^4 = \pm z^6$ and $x^2 + y^8 = z^3$", *Compositio Math.* **118**:3 (1999), 305–321.

[Bruin 2000] N. Bruin, "On powers as sums of two cubes", pp. 169–184 in *Algorithmic number theory* (Leiden, 2000), Lecture Notes in Comput. Sci. **1838**, Springer, Berlin, 2000.

[Buhler 2006] J. Buhler, "$L$-series in algorithmic number theory", in *Surveys in algorithmic number theory*, edited by J. P. Buhler and P. Stevenhagen, Math. Sci. Res. Inst. Publ. **44**, Cambridge University Press, New York, 2006.

[Bump et al. 1990] D. Bump, S. Friedberg, and J. Hoffstein, "Nonvanishing theorems for $L$-functions of modular forms and their derivatives", *Invent. Math.* **102**:3 (1990), 543–618.

[Cassels 1991] J. W. S. Cassels, *Lectures on elliptic curves*, London Mathematical Society Student Texts **24**, Cambridge University Press, Cambridge, 1991.

[Chabauty 1941] C. Chabauty, "Sur les points rationnels des courbes algébriques de genre supérieur à l'unité", *C. R. Acad. Sci. Paris* **212** (1941), 882–885.

[Coates and Wiles 1977] J. Coates and A. Wiles, "On the conjecture of Birch and Swinnerton-Dyer", *Invent. Math.* **39**:3 (1977), 223–251.

[Coleman 1985] R. F. Coleman, "Effective Chabauty", *Duke Math. J.* **52**:3 (1985), 765–770.

[Colman 1988] W. J. A. Colman, "On certain semiperfect cuboids", *Fibonacci Quart.* **26**:1 (1988), 54–57.

[Darmon 1997] H. Darmon, "Faltings plus epsilon, Wiles plus epsilon, and the generalized Fermat equation", *C. R. Math. Rep. Acad. Sci. Canada* **19**:1 (1997), 3–14.

[Darmon and Granville 1995] H. Darmon and A. Granville, "On the equations $z^m = F(x, y)$ and $Ax^p + By^q = Cz^r$", *Bull. London Math. Soc.* **27**:6 (1995), 513–543.

[Dickson 1934] L. E. Dickson, *History of the theory of numbers*, vol. II, G. E. Stechert, New York, 1934.

[Dummit et al. 1985] D. Dummit, H. Kisilevsky, and J. McKay, "Multiplicative products of $\eta$-functions", pp. 89–98 in *Finite groups—coming of age* (Montreal, 1982), edited by J. McKay, Contemp. Math. **45**, Amer. Math. Soc., Providence, RI, 1985.

[Elkies 1994] N. D. Elkies, "Heegner point computations", pp. 122–133 in *Algorithmic number theory (ANTS-I)* (Ithaca, NY, 1994), edited by L. Adleman and M.-D. Huang, Lecture Notes in Comput. Sci. **877**, Springer, Berlin, 1994.

[Elkies 2002] N. D. Elkies, "Curves $Dy^2 = x^3 - x$ of odd analytic rank", pp. 244–251 in *Algorithmic number theory (ANTS-V)* (Sydney, 2002), edited by C. Fieker and D. R. Kohel, Lecture Notes in Comput. Sci. **2369**, Springer, Berlin, 2002.

[Faltings 1983] G. Faltings, "Endlichkeitssätze für abelsche Varietäten über Zahlkörpern", *Invent. Math.* **73**:3 (1983), 349–366.

[Fine 1976] N. J. Fine, "On rational triangles", *Amer. Math. Monthly* **83**:7 (1976), 517–521.

[Flicker 1980] Y. Z. Flicker, "Automorphic forms on covering groups of GL(2)", *Invent. Math.* **57**:2 (1980), 119–182.

[Fujiwara 1998] M. Fujiwara, "$\theta$-congruent numbers", pp. 235–241 in *Number theory* (Eger, 1996), edited by K. Győry et al., de Gruyter, Berlin, 1998.

[Goto 2001] T. Goto, "Calculation of Selmer groups of elliptic curves with rational 2-torsions and $\theta$-congruent number problem", *Comment. Math. Univ. St. Paul.* **50**:2 (2001), 147–172.

[Goto 2002] T. Goto, *A study on the Selmer groups of the elliptic curves with a rational 2-torsion*, doctoral thesis, Kyushu Univ., 2002. See http://www.ma.noda.tus.ac.jp/u/tg/files/thesis.pdf.

[Grant 1994] D. Grant, "A curve for which Coleman's effective Chabauty bound is sharp", *Proc. Amer. Math. Soc.* **122**:1 (1994), 317–319.

[Gross and Zagier 1986] B. H. Gross and D. B. Zagier, "Heegner points and derivatives of $L$-series", *Invent. Math.* **84**:2 (1986), 225–320.

[Heegner 1952] K. Heegner, "Diophantische Analysis und Modulfunktionen", *Math. Z.* **56** (1952), 227–253.

[Inose 1976] H. Inose, "On certain Kummer surfaces which can be realized as nonsingular quartic surfaces in $P^3$", *J. Fac. Sci. Univ. Tokyo Sect. IA Math.* **23**:3 (1976), 545–560.

[Inose 1978] H. Inose, "Defining equations of singular $K3$ surfaces and a notion of isogeny", pp. 495–502 in *Proceedings of the International Symposium on Algebraic Geometry* (Kyoto, 1977), edited by M. Nagata, Kinokuniya, Tokyo, 1978.

[Joshi and Tzermias 1999] K. Joshi and P. Tzermias, "On the Coleman–Chabauty bound", *C. R. Acad. Sci. Paris Sér. I Math.* **329**:6 (1999), 459–463.

[Kan 2000] M. Kan, "$\theta$-congruent numbers and elliptic curves", *Acta Arith.* **94**:2 (2000), 153–160.

[Keum and Kondō 2001] J. Keum and S. Kondō, "The automorphism groups of Kummer surfaces associated with the product of two elliptic curves", *Trans. Amer. Math. Soc.* **353**:4 (2001), 1469–1487.

[Koblitz 1993] N. Koblitz, *Introduction to elliptic curves and modular forms*, vol. 97, 2nd ed., Graduate Texts in Mathematics, Springer, New York, 1993.

[Kolyvagin 1988] V. A. Kolyvagin, "Finiteness of $E(\mathbf{Q})$ and $\mathrm{SH}(E, \mathbf{Q})$ for a subclass of Weil curves", *Izv. Akad. Nauk SSSR Ser. Mat.* **52**:3 (1988), 522–540. In Russian; translated in *Math. USSR. Izv.* **32** (1989), 523–542.

[Korec 1984] I. Korec, "Nonexistence of a small perfect rational cuboid, II", *Acta Math. Univ. Comenian.* **44/45** (1984), 39–48.

[Kraus 1999] A. Kraus, "On the equation $x^p + y^q = z^r$: a survey", *Ramanujan J.* **3**:3 (1999), 315–333.

[Kuwata and Wang 1993] M. Kuwata and L. Wang, "Topology of rational points on isotrivial elliptic surfaces", *Internat. Math. Res. Notices* **1993**:4 (1993), 113–123.

[Lehmer 1899/1900] D. N. Lehmer, "Rational triangles", *Ann. of Math.* (2) **1**:1-4 (1899/1900), 97–102.

[Livné 1987] R. Livné, "Cubic exponential sums and Galois representations", pp. 247–261 in *Current trends in arithmetical algebraic geometry* (Arcata, CA, 1985), edited by K. Ribet, Contemp. Math. **67**, Amer. Math. Soc., Providence, RI, 1987.

[Livné 1995] R. Livné, "Motivic orthogonal two-dimensional representations of Gal($\overline{\mathbf{Q}}/\mathbf{Q}$)", *Israel J. Math.* **92**:1-3 (1995), 149–156.

[Long 2004] L. Long, "On Shioda-Inose structures of one-parameter families of K3 surfaces", *J. Number Theory* **109**:2 (2004), 299–318.

[van Luijk 2000] R. van Luijk, *On perfect cuboids*, Doctoraalscriptie, Universiteit Utrecht, 2000. See http://www.math.leidenuniv.nl/reports/2001-12.shtml.

[Mazur 1978] B. Mazur, "Rational isogenies of prime degree", *Invent. Math.* **44**:2 (1978), 129–162.

[Morrison 1984] D. R. Morrison, "On $K3$ surfaces with large Picard number", *Invent. Math.* **75**:1 (1984), 105–121.

[Murty and Murty 1991] M. R. Murty and V. K. Murty, "Mean values of derivatives of modular $L$-series", *Ann. of Math.* (2) **133**:3 (1991), 447–475.

[Nagell 1929] T. Nagell, *L' Analyse indéterminée de degré supérieur*, vol. 39, Gauthier-Villars, Paris, 1929.

[Narumiya and Shiga 2001] N. Narumiya and H. Shiga, "On certain rational cuboid problems", *Nihonkai Math. J.* **12**:1 (2001), 75–88.

[Nemenzo 1998] F. R. Nemenzo, "All congruent numbers less than 40000", *Proc. Japan Acad. Ser. A Math. Sci.* **74**:1 (1998), 29–31.

[Oguiso and Shioda 1991] K. Oguiso and T. Shioda, "The Mordell-Weil lattice of a rational elliptic surface", *Comment. Math. Univ. St. Paul.* **40**:1 (1991), 83–99.

[Peters et al. 1992] C. Peters, J. Top, , and M. van der Vlugt, "The Hasse zeta function of a $K3$ surface related to the number of words of weight 5 in the Melas codes", *J. Reine Angew. Math.* **432** (1992), 151–176.

[Pinch and Swinnerton-Dyer 1991] R. G. E. Pinch and H. P. F. Swinnerton-Dyer, "Arithmetic of diagonal quartic surfaces, I", pp. 317–338 in *L-functions and arithmetic* (Durham, 1989), London Math. Soc. Lecture Note Ser. **153**, Cambridge Univ. Press, Cambridge, 1991.

[Poonen 2008] B. Poonen, "Elliptic curves", pp. 183–207 in *Surveys in algorithmic number theory*, edited by J. P. Buhler and P. Stevenhagen, Math. Sci. Res. Inst. Publ. **44**, Cambridge University Press, New York, 2008.

[Shimura 1973] G. Shimura, "On modular forms of half integral weight", *Ann. of Math.* (2) **97** (1973), 440–481.

[Shioda 1990] T. Shioda, "On the Mordell-Weil lattices", *Comment. Math. Univ. St. Paul.* **39**:2 (1990), 211–240.

[Shioda and Inose 1977] T. Shioda and H. Inose, "On singular $K3$ surfaces", pp. 119–136 in *Complex analysis and algebraic geometry*, edited by W. Baily and T. Shioda, Iwanami Shoten, Tokyo, and Cambridge University Press, New York, 1977.

[Silverman 1986] J. H. Silverman, *The arithmetic of elliptic curves*, Graduate Texts in Mathematics **106**, Springer, New York, 1986.

[Silverman and Tate 1992] J. H. Silverman and J. Tate, *Rational points on elliptic curves*, Undergraduate Texts in Mathematics, Springer, New York, 1992.

[Stein 2008] W. Stein, "An introduction to computing modular forms using modular symbols", pp. 641–652 in *Surveys in algorithmic number theory*, edited by J. P. Buhler and P. Stevenhagen, Math. Sci. Res. Inst. Publ. **44**, Cambridge University Press, New York, 2008.

[Stienstra and Beukers 1985] J. Stienstra and F. Beukers, "On the Picard-Fuchs equation and the formal Brauer group of certain elliptic $K3$-surfaces", *Math. Ann.* **271**:2 (1985), 269–304.

[Taylor and Wiles 1995] R. Taylor and A. Wiles, "Ring-theoretic properties of certain Hecke algebras", *Ann. of Math.* (2) **141**:3 (1995), 553–572.

[Tunnell 1983] J. B. Tunnell, "A classical Diophantine problem and modular forms of weight $3/2$", *Invent. Math.* **72**:2 (1983), 323–334.

[Vinberg 1983] È. B. Vinberg, "The two most algebraic $K3$ surfaces", *Math. Ann.* **265**:1 (1983), 1–21.

[Waldspurger 1981] J.-L. Waldspurger, "Sur les coefficients de Fourier des formes modulaires de poids demi-entier", *J. Math. Pures Appl.* (9) **60**:4 (1981), 375–484.

[Wiles 1995] A. Wiles, "Modular elliptic curves and Fermat's last theorem", *Ann. of Math.* (2) **141**:3 (1995), 443–551.

[Yoshida 2001] S.-i. Yoshida, "Some variants of the congruent number problem, I", *Kyushu J. Math.* **55**:2 (2001), 387–404.

[Yoshida 2002] S.-i. Yoshida, "Some variants of the congruent number problem, II", *Kyushu J. Math.* **56**:1 (2002), 147–165.

JAAP TOP
INSTITUUT VOOR WISKUNDE EN INFORMATICA
P.O.BOX 800
9700 AV GRONINGEN
THE NETHERLANDS
top@math.rug.nl

NORIKO YUI
DEPARTMENT OF MATHEMATICS AND STATISTICS
QUEEN'S UNIVERSITY
KINGSTON, ONTARIO
CANADA K7L 3N6
yui@mast.queensu.ca

# An introduction to computing modular forms using modular symbols

WILLIAM A. STEIN

ABSTRACT. We explain how weight-two modular forms on $\Gamma_0(N)$ are related to modular symbols, and how to use this to explicitly compute spaces of modular forms.

## Introduction

The definition of the spaces of modular forms as functions on the upper half plane satisfying a certain equation is very abstract. The definition of the Hecke operators even more so. Nevertheless, one wishes to carry out explicit investigations involving these objects.

We are fortunate that we now have methods available that allow us to transform the vector space of cusp forms of given weight and level into a concrete object, which can be explicitly computed. We have the work of Atkin–Lehner, Birch, Swinnerton-Dyer, Manin, Mazur, Merel, and many others to thank for this (see, e.g., [Birch and Kuyk 1975; Cremona 1997; Mazur 1973; Merel 1994]). For example, we can use the Eichler–Selberg trace formula, as extended in [Hijikata 1974], to compute characteristic polynomials of Hecke operators. Then the method described in [Wada 1971] gives a basis for certain spaces of modular forms. Alternatively, we can compute $\Theta$-series using Brandt matrices and quaternion algebras as in [Kohel 2001; Pizer 1980], or we can use a closely related geometric method that involves the module of enhanced supersingular elliptic curves [Mestre 1986]. Another related method of Birch [1991] is very fast, but gives only a piece of the full space of modular forms. The power of the modular symbols approach was demonstrated by Cremona in his book [1997], where he systematically computes a large table of invariants of all elliptic curves of conductor up to 1000 (his online tables [Cremona undated] go well beyond 100,000).

Though the above methods are each beautiful and well suited to certain applications, we will only discuss the modular symbols method further, as it has many advantages. We will primarily discuss the theory in this summary paper, leaving an explicit description of the objects involved for other papers. Nonetheless, there is a definite gap between the theory on the one hand, and an efficient running machine implementation on the other. To implement the algorithms hinted at below requires making absolutely everything completely explicit, then finding intelligent and efficient ways of performing the necessary manipulations. This is a nontrivial and tedious task, with room for error at every step. Fortunately, Sage [2008] has extensive capabilities for computing with modular forms and includes Cremona's programs; we will give a few examples below. See also the author's MAGMA [Bosma et al. 1997] package for computing with modular forms and modular symbols.

In this paper we will focus exclusively on the case of weight-2 modular forms for $\Gamma_0(N)$. The methods explained here extend to modular forms of integer weight greater than 2; for more details see [Stein 2007; Merel 1994].

Section 1 contains a brief summary of basic facts about modular forms, Hecke operators, and integral homology. Section 2 introduces modular symbols, and describes how to compute with them. Section 3 outlines an algorithm for constructing cusp forms using modular symbols in conjunction with Atkin–Lehner theory.

This paper assumes some familiarity with algebraic curves, Riemann surfaces, and homology groups of compact surfaces. A few basic facts about modular forms are recalled, but only briefly. In particular, only a roundabout attempt is made to motivate why one might be interested in modular forms; for this, see many of the references in the bibliography. No prior exposure to modular symbols is assumed.

## 1. Modular forms and Hecke operators

All of the objects we will consider arise from the modular group $\mathrm{SL}_2(\mathbf{Z})$ of two-by-two integer matrices with determinant equal to one. This group acts via linear fractional transformations on the complex upper half plane $\mathfrak{h}$, and also on the extended upper half plane

$$\mathfrak{h}^* = \mathfrak{h} \cup \mathbf{P}^1(\mathbf{Q}) = \mathfrak{h} \cup \mathbf{Q} \cup \{\infty\}.$$

See [Shimura 1971, § 1.3–1.5] for a careful description of the topology on $\mathfrak{h}^*$. A basis of neighborhoods for $\alpha \in \mathbf{Q}$ is given by the sets $\{\alpha\} \cup D$, where $D$ is a disc in $\mathfrak{h}$ that is tangent to the real line at $\alpha$. Let $N$ be a positive integer and consider the group $\Gamma_0(N)$ of matrices $\begin{pmatrix} a & b \\ c & d \end{pmatrix} \in \mathrm{SL}_2(\mathbf{Z})$ such that $N \mid c$. This group acts on $\mathfrak{h}^*$ by linear fractional transformations, and the quotient $\Gamma_0(N) \backslash \mathfrak{h}^*$ is a Riemann surface, which we denote by $X_0(N)$. Shimura showed [1971, §6.7] that $X_0(N)$ has a canonical structure of algebraic curve over $\mathbf{Q}$.

A *cusp form* is a function $f$ on $\mathfrak{h}$ such that $f(z)dz$ is a holomorphic differential on $X_0(N)$. Equivalently, a cusp form is a holomorphic function $f$ on $\mathfrak{h}$ such that

(a) the expression $f(z)dz$ is invariant under replacing $z$ by $\gamma(z)$ for each $\gamma \in \Gamma_0(N)$, and
(b) $f(z)$ is holomorphic at each element of $\mathbf{P}^1(\mathbf{Q})$, and moreover $f(z)$ tends to 0 as $z$ tends to any element of $\mathbf{P}^1(\mathbf{Q})$.

The space of cusp forms on $\Gamma_0(N)$ is a finite dimensional complex vector space, of dimension equal to the genus $g$ of $X_0(N)$. Viewed topologically, as a 2-dimensional real manifold, $X_0(N)(\mathbf{C})$ is a $g$-holed torus.

Condition (b) in the definition of $f(z)$ means that $f(z)$ has a Fourier expansion about each element of $\mathbf{P}^1(\mathbf{Q})$. Thus, at $\infty$ we have

$$f(z) = a_1 e^{2\pi i z} + a_2 e^{2\pi i 2z} + a_3 e^{2\pi i 3z} + \cdots$$
$$= a_1 q + a_2 q^2 + a_3 q^3 + \cdots,$$

where, for brevity, we write $q = q(z) = e^{2\pi i z}$.

EXAMPLE 1.1. Let $E$ be the elliptic curve defined by the equation $y^2 + xy = x^3 + x^2 - 4x - 5$. For $p \neq 3, 13$, let $a_p = p + 1 - \#\tilde{E}(\mathbf{F}_p)$, where $\tilde{E}$ is the reduction of $E$ mod $p$, and let $a_3 = -11$, $a_{13} = 1$. For $n$ composite, define $a_n$ using the relations at the end of Section 3. Then

$$f = q + a_2 q^2 + a_3 q^3 + a_4 q^4 + a_5 q^5 + \cdots = q + q^2 - 11q^3 + 2q^5 + \cdots$$

is the $q$-expansion of a modular form on $\Gamma_0(39)$. The Shimura–Taniyama conjecture, which is now a theorem (see [Breuil et al. 2001]) asserts that any $q$-expansion constructed as above from an elliptic curve over $\mathbf{Q}$ is a modular form. We define the above elliptic curve and compute the associated modular form $f$ using Sage as follows:

```
sage: E = EllipticCurve([1,1,0,-4,-5]); E
Elliptic Curve defined by y^2 + x*y  = x^3 + x^2 - 4*x - 5
over Rational Field
sage: E.q_eigenform(10)
q + q^2 - q^3 - q^4 + 2*q^5 - q^6 - 4*q^7 - 3*q^8 + q^9 + O(q^10)
```

The Hecke operators are a family of *commuting* endomorphisms of $S_2(N)$, which are defined as follows. The complex points of the open subcurve $Y_0(N) = \Gamma_0(N)\backslash\mathfrak{h}$ are in bijection with pairs $(E, C)$, where $E$ is an elliptic curve over $\mathbf{C}$ and $C$ is a cyclic subgroup of $E(\mathbf{C})$ of order $N$. If $p \nmid N$ then there are two natural maps $\pi_1$ and $\pi_2$ from $Y_0(pN)$ to $Y_0(N)$; the first, $\pi_1$, sends $(E, C)$

to $(E, C')$, where $C'$ is the unique cyclic subgroup of $C$ of order $N$, and the second, $\pi_2$, sends a point $(E, C) \in Y_0(N)(\mathbf{C})$ to $(E/D, C/D)$, where $D$ is the unique cyclic subgroup of $C$ of order $p$. These maps extend in a unique way to maps from $X_0(pN)$ to $X_0(N)$:

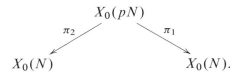

The $p$-th *Hecke operator* $T_p$ is $(\pi_1)_* \circ (\pi_2)^*$; it acts on most objects attached to $X_0(N)$, such as divisors and cusp forms. There is a Hecke operator $T_n$ for every positive integer $n$, but we will not need to consider those with $n$ composite.

EXAMPLE 1.2. There is a basis of $S_2(39)$ so that

$$T_2 = \begin{pmatrix} 0 & 2 & -1 \\ 1 & -2 & 1 \\ 0 & -1 & 1 \end{pmatrix} \text{ and } T_5 = \begin{pmatrix} 1 & -1 & -1 \\ -2 & 2 & -2 \\ -3 & -1 & -1 \end{pmatrix}.$$

Notice that these matrices commute, and that 1 is an eigenvalue of $T_2$, and 2 is an eigenvalue of $T_5$. We compute each of the above matrices and verify that they commute using Sage as follows:

```
sage: S = CuspForms(39)
sage: T2 = S.hecke_matrix(2); T2
[ 0  2 -1]
[ 1 -2  1]
[ 0 -1  1]
sage: T5 = S.hecke_matrix(5); T5
[ 1 -1 -1]
[-2  2 -2]
[-3 -1 -1]
sage: T2*T5 == T5*T2
True
```

The first homology group $H_1(X_0(N), \mathbf{Z})$ is the group of singular 1-cycles modulo homology relations. Recall that topologically $X_0(N)$ is a $g$-holed torus, where $g$ is the genus of $X_0(N)$. The group $H_1(X_0(N), \mathbf{Z})$ is thus a free abelian group of rank $2g$ (see, e.g., [Greenberg and Harper 1981, Ex. 19.30]), with two generators corresponding to each hole, as illustrated in the case $N = 39$ in Diagram 1.

# AN INTRODUCTION TO COMPUTING MODULAR FORMS USING MODULAR SYMBOLS

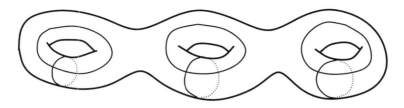

$$H_1(X_0(39), \mathbf{Z}) \cong \mathbf{Z} \times \mathbf{Z} \times \mathbf{Z} \times \mathbf{Z} \times \mathbf{Z} \times \mathbf{Z}$$

**Figure 1.** The homology of $X_0(39)$.

The Hecke operators $T_p$ act on $H_1(X_0(N), \mathbf{Z})$, and integration defines a nondegenerate Hecke-equivariant pairing

$$\langle\, ,\, \rangle : S_2(N) \times H_1(X_0(N), \mathbf{Z}) \to \mathbf{C}.$$

Explicitly, for a path $x$,

$$\langle f, x \rangle = 2\pi i \int_x f(z) dz,$$

where the integral may be viewed as a complex line integral along an appropriate piece of the preimage of $x$ in the upper half plane. The pairing is Hecke equivariant in the sense that for every prime $p$, we have $\langle fT_p, x \rangle = \langle f, T_p x \rangle$. As we will see, modular symbols allow us to make explicit the action of the Hecke operators on $H_1(X_0(N), \mathbf{Z})$; the above pairing then translates this into a wealth of information about cusp forms.

For a more detailed survey of the basic facts about modular curves and modular forms, we urge the reader to consult the book [Diamond and Shurman 2005] along with the survey paper [Diamond and Im 1995]. For a discussion of how to draw a picture of the ring generated by the Hecke operators, see [Ribet and Stein 2001, § 3.8].

## 2. Modular symbols

The modular symbols formalism provides a presentation of $H_1(X_0(N), \mathbf{Z})$ in terms of paths between elements of $\mathbf{P}^1(\mathbf{Q})$. Furthermore, a trick due to Manin gives an explicit finite list of generators and relations for the space of modular symbols.

The *modular symbol* defined by a pair $\alpha, \beta \in \mathbf{P}^1(\mathbf{Q})$ is denoted $\{\alpha, \beta\}$. As illustrated in Figure 2, this modular symbol should be viewed as the homology class, relative to the cusps, of a geodesic path from $\alpha$ to $\beta$ in $\mathfrak{h}^*$. The homology group relative to the cusps is a slight enlargement of the usual homology group,

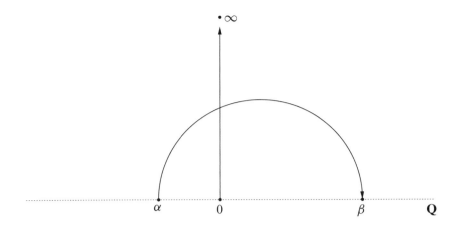

**Figure 2.** The modular symbols $\{\alpha, \beta\}$ and $\{0, \infty\}$.

in that we allow paths with endpoints in $\mathbf{P}^1(\mathbf{Q})$ instead of restricting to closed loops.

Motivated by this picture, we declare that modular symbols satisfy the following homology relations: if $\alpha, \beta, \gamma \in \mathbf{Q} \cup \{\infty\}$, then

$$\{\alpha, \beta\} + \{\beta, \gamma\} + \{\gamma, \alpha\} = 0.$$

Furthermore, we quotient out by any torsion, so, e.g., $\{\alpha, \alpha\} = 0$ and $\{\alpha, \beta\} = -\{\beta, \alpha\}$.

Denote by $\mathcal{M}_2$ the free abelian group with basis the set of symbols $\{\alpha, \beta\}$ modulo the three-term homology relations above and modulo any torsion. There is a left action of $\mathrm{GL}_2(\mathbf{Q})$ on $\mathcal{M}_2$, whereby a matrix $g$ acts by

$$g\{\alpha, \beta\} = \{g(\alpha), g(\beta)\},$$

and $g$ acts on $\alpha$ and $\beta$ by a linear fractional transformation. The space $\mathcal{M}_2(N)$ of *modular symbols for* $\Gamma_0(N)$ is the quotient of $\mathcal{M}_2$ by the submodule generated by the infinitely many elements of the form $x - g(x)$, for $x$ in $\mathcal{M}_2$ and $g$ in $\Gamma_0(N)$, and modulo any torsion. A *modular symbol for* $\Gamma_0(N)$ is an element of this space. We frequently denote the equivalence class that defines a modular symbol by giving a representative element.

Manin [1972] proved there is a natural injection $H_1(X_0(N), \mathbf{Z}) \to \mathcal{M}_2(N)$. The image of $H_1(X_0(N), \mathbf{Z})$ in $\mathcal{M}_2(N)$ can be identified as follows. Let $\mathcal{B}_2(N)$ denote the free abelian group whose basis is the finite set $\Gamma_0(N)\backslash\mathbf{P}^1(\mathbf{Q})$. The *boundary map* $\delta: \mathcal{M}_2(N) \to \mathcal{B}_2(N)$ sends $\{\alpha, \beta\}$ to $[\beta] - [\alpha]$, where $[\beta]$ denotes the basis element of $\mathcal{B}_2(N)$ corresponding to $\beta \in \mathbf{P}^1(\mathbf{Q})$. The kernel $\mathcal{S}_2(N)$ of $\delta$ is the subspace of *cuspidal* modular symbols. An element of $\mathcal{S}_2(N)$ can

AN INTRODUCTION TO COMPUTING MODULAR FORMS USING MODULAR SYMBOLS 647

be thought of as a linear combination of paths in $\mathfrak{h}^*$ whose endpoints are cusps, and whose images in $X_0(N)$ are a linear combination of loops. We thus obtain a map $\varphi : \mathcal{S}_2(N) \to H_1(X_0(N), \mathbf{Z})$.

THEOREM 2.1. *The map $\varphi$ given above defines a canonical isomorphism*

$$\mathcal{S}_2(N) \cong H_1(X_0(N), \mathbf{Z}).$$

### 2.1. Manin's trick.
In this section, we describe a trick of Manin that shows that the space of modular symbols can be computed.

By reducing modulo $N$, one sees that the group $\Gamma_0(N)$ has finite index in $SL_2(\mathbf{Z})$. Let $r_0, r_1, \ldots, r_m$ be distinct right coset representatives for $\Gamma_0(N)$ in $SL_2(\mathbf{Z})$, so that

$$SL_2(\mathbf{Z}) = \Gamma_0(N) r_0 \cup \Gamma_0(N) r_1 \cup \cdots \cup \Gamma_0(N) r_m,$$

where the union is disjoint. For example, when $N$ is prime, a list of coset representatives is

$$\begin{pmatrix} 1 & 0 \\ 0 & 1 \end{pmatrix}, \begin{pmatrix} 1 & 0 \\ 1 & 1 \end{pmatrix}, \begin{pmatrix} 1 & 0 \\ 2 & 1 \end{pmatrix}, \begin{pmatrix} 1 & 0 \\ 3 & 1 \end{pmatrix}, \ldots, \begin{pmatrix} 1 & 0 \\ N-1 & 1 \end{pmatrix}, \begin{pmatrix} 0 & -1 \\ 1 & 0 \end{pmatrix}.$$

In general, the right cosets of $\Gamma_0(N)$ in $SL_2(\mathbf{Z})$ are in bijection with the elements of $\mathbf{P}^1(\mathbf{Z}/N\mathbf{Z})$ (see [Cremona 1997, §2.2] for complete details).

The following trick of Manin [1972, §1.5] (see also [Cremona 1997, §2.1.6]) allows us to write every modular symbol as a $\mathbf{Z}$-linear combination of symbols of the form $r_i\{0, \infty\}$. In particular, the finitely many symbols $r_i\{0, \infty\}$ generate $\mathcal{M}_2(N)$.

Because of the relation $\{\alpha, \beta\} = \{0, \beta\} - \{0, \alpha\}$, it suffices to consider modular symbols of the form $\{0, b/a\}$, where the rational number $b/a$ is in lowest terms. Expand $b/a$ as a continued fraction and consider the successive convergents in lowest terms:

$$\frac{b_{-2}}{a_{-2}} = \frac{0}{1}, \quad \frac{b_{-1}}{a_{-1}} = \frac{1}{0}, \quad \frac{b_0}{a_0} = \frac{b_0}{1}, \ldots, \frac{b_{n-1}}{a_{n-1}}, \frac{b_n}{a_n} = \frac{b}{a}$$

where the first two are added formally. Then

$$b_k a_{k-1} - b_{k-1} a_k = (-1)^{k-1},$$

so that

$$g_k = \begin{pmatrix} b_k & (-1)^{k-1} b_{k-1} \\ a_k & (-1)^{k-1} a_{k-1} \end{pmatrix} \in SL_2(\mathbf{Z}).$$

Hence

$$\left\{ \frac{b_{k-1}}{a_{k-1}}, \frac{b_k}{a_k} \right\} = g_k\{0, \infty\} = r_i\{0, \infty\},$$

for some $i$, is of the required special form.

EXAMPLE 2.2. Let $N = 11$, and consider the modular symbol $\{0, 4/7\}$. We have
$$\frac{4}{7} = 0 + \cfrac{1}{1 + \cfrac{1}{1 + \cfrac{1}{3}}},$$
so the partial convergents are
$$\frac{b_{-2}}{a_{-2}} = \frac{0}{1}, \quad \frac{b_{-1}}{a_{-1}} = \frac{1}{0}, \quad \frac{b_0}{a_0} = \frac{0}{1}, \quad \frac{b_1}{a_1} = \frac{1}{1}, \quad \frac{b_2}{a_2} = \frac{1}{2}, \quad \frac{b_3}{a_3} = \frac{4}{7}.$$
Thus
$$\{0, 4/7\} = \{0, \infty\} + \{\infty, 0\} + \{0, 1\} + \{1, 1/2\} + \{1/2, 4/7\}$$
$$= \begin{pmatrix} 1 & -1 \\ 2 & -1 \end{pmatrix} \{0, \infty\} + \begin{pmatrix} 4 & 1 \\ 7 & 2 \end{pmatrix} \{0, \infty\}$$
$$= 2 \cdot \left[ \begin{pmatrix} 1 & 4 \\ 1 & 5 \end{pmatrix} \{0, \infty\} \right].$$

**2.2. Manin symbols.** As above, fix coset representatives $r_0, \ldots, r_m$ for $\Gamma_0(N)$ in $SL_2(\mathbf{Z})$. Denote the modular symbol $r_i\{0, \infty\}$ by $[r_i]$. The symbols $[r_0]$, $\ldots, [r_m]$ are called *Manin symbols*, and they are equipped with a right action of $SL_2(\mathbf{Z})$, which is given by $[r_i]g = [r_j]$, where $\Gamma_0(N)r_j = \Gamma_0(N)r_i g$. Recall that $SL_2(\mathbf{Z})$ is generated by the two matrices $\sigma = \begin{pmatrix} 0 & -1 \\ 1 & 0 \end{pmatrix}$ and $\tau = \begin{pmatrix} 1 & -1 \\ 1 & 0 \end{pmatrix}$ (see Theorem 2 of [Serre 1973, VII.1.2]).

THEOREM 2.3 (MANIN). *The Manin symbols $[r_0], \ldots, [r_m]$ satisfy the relations*
$$[r_i] + [r_i]\sigma = 0,$$
$$[r_i] + [r_i]\tau + [r_i]\tau^2 = 0.$$
*Furthermore, these relations generate all relations (modulo torsion relations).*

This theorem, proved in [Manin 1972, §1.7], provides a finite presentation for the space of modular symbols.

**2.3. Hecke operators on modular symbols.** When $p$ is a prime not dividing $N$, define
$$T_p\{\alpha, \beta\} = \begin{pmatrix} p & 0 \\ 0 & 1 \end{pmatrix} \{\alpha, \beta\} + \sum_{r \bmod p} \begin{pmatrix} 1 & r \\ 0 & p \end{pmatrix} \{\alpha, \beta\}.$$
As mentioned before, this definition is compatible with the integration pairing $\langle, \rangle$ of Section 1, in the sense that $\langle fT_p, x \rangle = \langle f, T_p x \rangle$. When $p \mid N$, the definition is the same, except that the matrix $\begin{pmatrix} p & 0 \\ 0 & 1 \end{pmatrix}$ is dropped.

For example, when $N = 11$ we have
$$T_2\{0, 1/5\} = \{0, 2/5\} + \{0, 1/10\} + \{1/2, 3/5\} = -2\{0, 1/5\}.$$

L. Merel [1994] gave a description of the action of $T_p$ directly on Manin symbols $[r_i]$ (see also [Cremona 1997, §2.4]). For example, when $p = 2$ and $N$ is odd, we have

$$T_2([r_i]) = [r_i]\begin{pmatrix} 1 & 0 \\ 0 & 2 \end{pmatrix} + [r_i]\begin{pmatrix} 2 & 0 \\ 0 & 1 \end{pmatrix} + [r_i]\begin{pmatrix} 2 & 1 \\ 0 & 1 \end{pmatrix} + [r_i]\begin{pmatrix} 1 & 0 \\ 1 & 2 \end{pmatrix}.$$

## 3. Computing the space of modular forms

In this section we describe how to use modular symbols to construct a basis of $S_2(N)$ consisting of modular forms that are eigenvectors for every element of the ring $\mathbf{T}'$ generated by the Hecke operator $T_p$, with $p \nmid N$. Such eigenvectors are called *eigenforms*.

Suppose $M$ is a positive integer that divides $N$. As explained in [Lang 1995, VIII.1–2], for each divisor $d$ of $N/M$ there is a natural *degeneracy map* $\beta_{M,d}$: $S_2(M) \to S_2(N)$ given by $\beta_{M,d}(f(q)) = f(q^d)$. The *new subspace* of $S_2(N)$, denoted $S_2(N)^{\text{new}}$, is the orthogonal complement with respect to the Petersson inner product of the images of all maps $\beta_{M,d}$, with $M$ and $d$ as above.

The theory of Atkin and Lehner [1970] asserts that, as a $\mathbf{T}'$-module, $S_2(N)$ is built up as follows:

$$S_2(N) = \bigoplus_{M|N,\, d|N/M} \beta_{M,d}(S_2(M)^{\text{new}}).$$

To compute $S_2(N)$ it thus suffices to compute $S_2(M)^{\text{new}}$ for each positive divisor $M$ of $N$.

We now turn to the problem of computing $S_2(N)^{\text{new}}$. Atkin and Lehner [1970] also proved that $S_2(N)^{\text{new}}$ is spanned by eigenforms, each of which occurs with multiplicity one in $S_2(N)^{\text{new}}$. Moreover, if $f \in S_2(N)^{\text{new}}$ is an eigenform then the coefficient of $q$ in the $q$-expansion of $f$ is nonzero, so it is possible to normalize $f$ so that coefficient of $q$ is 1. With $f$ so normalized, if $T_p(f) = a_p f$, then the $p$-th Fourier coefficient of $f$ is $a_p$. If $f = \sum_{n=1}^{\infty} a_n q^n$ is a normalized eigenvector for all $T_p$, then the $a_n$, with $n$ composite, are determined by the $a_p$, with $p$ prime, by the following formulas: $a_{nm} = a_n a_m$ when $n$ and $m$ are relatively prime, and $a_{p^r} = a_{p^{r-1}} a_p - p a_{p^{r-2}}$ for $p \nmid N$ prime. When $p \mid N$, $a_{p^r} = a_p^r$. We conclude that in order to compute $S_2(N)^{\text{new}}$, it suffices to compute all systems of eigenvalues $\{a_2, a_3, a_5, \ldots\}$ of the Hecke operators $T_2, T_3, T_5, \ldots$ acting on $S_2(N)^{\text{new}}$. Given a system of eigenvalues, the corresponding eigenform is $f = \sum_{n=1}^{\infty} a_n q^n$, where the $a_n$, for $n$ composite, are determined by the recurrence given above.

In light of the pairing $\langle\,,\,\rangle$ introduced in Section 1, computing the above systems of eigenvalues $\{a_2, a_3, a_5, \ldots\}$ amounts to computing the systems of eigenvalues of the Hecke operators $T_p$ on the subspace $V$ of $\mathcal{S}_2(N)$ that corresponds to the new subspace of $S_2(N)$. For each proper divisor $M$ of $N$ and each divisors $d$ of $N/M$, let $\phi_{M,d} : \mathcal{S}_2(N) \to \mathcal{S}_2(M)$ be the map sending $x$ to $\begin{pmatrix} t & 0 \\ 0 & 1 \end{pmatrix} x$. Then $V$ is the intersection of the kernels of all maps $\phi_{M,d}$.

The computation of the systems of eigenvalues of a collection of commuting diagonalizable endomorphisms involves standard linear algebra techniques, such as the computation of characteristic polynomials and kernels of matrices. There are, however, several tricks that greatly speed up this process, some of which are described in [Stein 2000, § 3.5.4].

EXAMPLE 3.1. All forms in $S_2(39)$ are new. Up to Galois conjugacy, the eigenvalues of the Hecke operators $T_2$, $T_3$, $T_5$, and $T_7$ on $\mathcal{S}_2(39)$ are $\{1, -1, 2, -4\}$ and $\{a, 1, -2a-2, 2a+2\}$, where $a^2 + 2a - 1 = 0$. (Note that these eigenvalues occur with multiplicity two.) Thus $S_2(39)$ has dimension 3, and is spanned by

$$f_1 = q + q^2 - q^3 - q^4 + 2q^5 - q^6 - 4q^7 + \cdots,$$
$$f_2 = q + aq^2 + q^3 + (-2a-1)q^4 + (-2a-2)q^5 + aq^6 + (2a+2)q^7 + \cdots,$$

and the Galois conjugate of $f_2$. We compute $f_1$ and $f_2$ using Sage as follows:

```
sage: CuspForms(39).newforms('a')
[q + q^2 - q^3 - q^4 + 2*q^5 + O(q^6),
 q + a1*q^2 + q^3 + (-2*a1 - 1)*q^4 + (-2*a1 - 2)*q^5 + O(q^6)]
```

## 3.1. Summary. 
To compute the $q$-expansion, to some precision, of each eigenforms in $S_2(N)$, we use the degeneracy maps so that we only have to solve the problem for $S_2(N)^{\text{new}}$. Here, using modular symbols we compute the systems of eigenvalues $\{a_2, a_3, a_5, \ldots\}$, then write down each of the corresponding eigenforms $q + a_2 q^2 + a_3 q^3 + \cdots$.

# References

[Atkin and Lehner 1970] A. Atkin and J. Lehner, "Hecke operators on $\Gamma_0(m)$", *Math. Ann.* **185** (1970), 134–160.

[Birch 1991] B. J. Birch, "Hecke actions on classes of ternary quadratic forms", pp. 191–212 in *Computational number theory* (Debrecen, 1989), de Gruyter, Berlin, 1991.

[Birch and Kuyk 1975] B. Birch and W. Kuyk (editors), *Modular functions of one variable. IV*, Lecture Notes in Mathematics **476**, Springer, Berlin, 1975.

[Bosma et al. 1997] W. Bosma, J. Cannon, and C. Playoust, "The Magma algebra system. I. The user language", *J. Symbolic Comput.* **24**:3–4 (1997), 235–265.

[Breuil et al. 2001] C. Breuil, B. Conrad, F. Diamond, and R. Taylor, "On the modularity of elliptic curves over **Q**: wild 3-adic exercises", *J. Amer. Math. Soc.* **14**:4 (2001), 843–939.

[Cremona 1997] J. Cremona, *Algorithms for modular elliptic curves*, Second ed., Cambridge University Press, Cambridge, 1997. Available at http://www.maths.nott.ac.uk/personal/jec/book/.

[Cremona undated] J. Cremona, Elliptic curves data, undated. Available at http://www.warwick.ac.uk/~masgaj/ftp/data/.

[Diamond and Im 1995] F. Diamond and J. Im, "Modular forms and modular curves", pp. 39–133 in *Seminar on Fermat's Last Theorem*, Providence, RI, 1995.

[Diamond and Shurman 2005] F. Diamond and J. Shurman, *A first course in modular forms*, Graduate Texts in Mathematics **228**, Springer, New York, 2005.

[Greenberg and Harper 1981] M. Greenberg and J. Harper, *Algebraic topology: A first course*, Benjamin/Cummings, Reading, MA, 1981.

[Hijikata 1974] H. Hijikata, "Explicit formula of the traces of Hecke operators for $\Gamma_0(N)$", *J. Math. Soc. Japan* **26**:1 (1974), 56–82.

[Kohel 2001] D. R. Kohel, "Hecke module structure of quaternions", pp. 177–195 in *Class field theory—its centenary and prospect* (Tokyo, 1998), edited by K. Miyake, Adv. Stud. Pure Math. **30**, Math. Soc. Japan, Tokyo, 2001.

[Lang 1995] S. Lang, *Introduction to modular forms*, Springer, Berlin, 1995. With appendixes by D. Zagier and W. Feit, Corrected reprint of the 1976 original.

[Manin 1972] J. Manin, "Parabolic points and zeta functions of modular curves", *Izv. Akad. Nauk SSSR Ser. Mat.* **36** (1972), 19–66.

[Mazur 1973] B. Mazur, "Courbes elliptiques et symboles modulaires", pp. 277–294, Exp. No. 414 in *Séminaire Bourbaki, 24ème année* (1971/1972), Lecture Notes in Math. **317**, Springer, Berlin, 1973.

[Merel 1994] L. Merel, "Universal Fourier expansions of modular forms", pp. 59–94 in *On Artin's conjecture for odd 2-dimensional representations*, Springer, 1994.

[Mestre 1986] J.-F. Mestre, "La méthode des graphes: exemples et applications", pp. 217–242 in *Proceedings of the international conference on class numbers and fundamental units of algebraic number fields (Katata)*, Nagoya Univ., Nagoya, 1986.

[Pizer 1980] A. Pizer, "An algorithm for computing modular forms on $\Gamma_0(N)$", *J. Algebra* **64**:2 (1980), 340–390.

[Ribet and Stein 2001] K. Ribet and W. Stein, "Lectures on Serre's conjectures", pp. 143–232 in *Arithmetic algebraic geometry* (Park City, UT, 1999), IAS/Park City Math. Ser. **9**, Amer. Math. Soc., Providence, RI, 2001.

[Sage 2008] W. Stein and the Sage Group, *Sage:* open source mathematical software (Version 3.0), 2008. Available at http://www.sagemath.org.

[Serre 1973] J.-P. Serre, *A course in arithmetic*, Graduate Texts in Mathematics **7**, Springer, New York, 1973.

[Shimura 1971] G. Shimura, *Introduction to the arithmetic theory of automorphic functions*, Iwanami Shoten, Tokyo, and Princeton University Press, Princeton, NJ, 1971.

[Stein 2000] W. Stein, *Explicit approaches to modular abelian varieties*, Ph.D. thesis, University of California, Berkeley, 2000.

[Stein 2007] W. Stein, *Modular forms, a computational approach*, Graduate Studies in Mathematics **79**, American Mathematical Society, Providence, RI, 2007. With an appendix by Paul E. Gunnells.

[Wada 1971] H. Wada, "Tables of Hecke operations, I", pp. 10 in *Seminar on modern methods in number theory*, Inst. Statist. Math., Tokyo, 1971.

WILLIAM A. STEIN
ASSOCIATE PROFESSOR OF MATHEMATICS
UNIVERSITY OF WASHINGTON
wstein@gmail.com